								2 HELIUM He 4.00
			5 BORON B 10.81	6 CARBON C 12.01	7 NITROGEN N 14.01	8 OXYGEN O 15.999	9 FLUORINE F 18.998	10 NEON Ne 20.18
			13 ALUMINUM Al 26.98	14 SILICON Si 28.08	15 PHOSPHORUS P 30.97	16 SULFUR S 32.06	17 CHLORINE Cl 35.45	18 ARGON Ar 39.95
28 NICKEL Ni 58.69	29 COPPER Cu 63.55	30 ZINC Zn 65.38	31 GALLIUM Ga 69.72	32 GERMANIUM Ge 72.59	33 ARSENIC As 74.92	34 SELENIUM Se 78.96	35 BROMINE Br 79.90	36 KRYPTON Kr 83.80
46 PALLADIUM Pd 106.42	47 SILVER Ag 107.87	48 CADMIUM Cd 112.41	49 INDIUM In 114.82	50 TIN Sn 118.69	51 ANTIMONY Sb 121.75	52 TELLURIUM Te 127.60	53 IODINE I 126.90	54 XENON Xe 131.29
78 PLATINUM Pt 195.08	79 GOLD Au 196.97	80 MERCURY Hg 200.59	81 THALLIUM Tl 204.38	82 LEAD Pb 207.2	83 BISMUTH Bi 208.98	84 POLONIUM Po (209)	85 ASTATINE At (210)	86 RADON Rn (222)

63 EUROPIUM Eu 151.96	64 GADOLINIUM Gd 157.25	65 TERBIUM Tb 158.92	66 DYSPROSIUM Dy 162.50	67 HOLMIUM Ho 164.93	68 ERBIUM Er 167.26	69 THULIUM Tm 168.93	70 YTTERBIUM Yb 173.04	71 LUTETIUM Lu 174.97
95 AMERICIUM Am (243)	96 CURIUM Cm (247)	97 BERKELIUM Bk (247)	98 CALIFORNIUM Cf (251)	99 EINSTEINIUM Es (252)	100 FERMIUM Fm (257)	101 MENDELEVIUM Md (258)	102 NOBELIUM No (259)	103 LAWRENCIUM Lr (260)

1

A SEARCH TO UNDERSTAND

Chapter 1

CHEMISTRY
A Search to Understand

CHEMISTRY
A Search to Understand

ANNA J. HARRISON | EDWIN S. WEAVER

Mount Holyoke College

Harcourt Brace Jovanovich, Publishers
and its subsidiary, Academic Press

San Diego New York Chicago Austin Washington, D.C.
London Sydney Tokyo Toronto

ISBN: 0-15-506476-2

Library of Congress Catalog Card Number: 87-81155

Printed in the United States of America

The authors wish to thank CRC Press, Inc., for excerpts of tables taken from *CRC Handbook of Chemistry and Physics*, ed. Robert C. Weast, Melvin J. Astle, and William H. Beyer, 64th ed. (Boca Raton: CRC Press, 1974). Reprinted by permission of Robert C. Weast.

Cover: *Tomoe* (comma) motif used in Japanese textile design.
Adapted from a seventeenth- or eighteenth-century *Noh* costume.

Illustration Credits

Text illustrations by Mel Erikson Art Services.

P. 427 (*bottom*), Courtesy Harvard Apparatus; p. 429 (*right*), Courtesy Dr. Karl A. Piez; p. 431 (*bottom left*), Photo by Edward Leigh, courtesy John Kendrew; p. 431 (*bottom right*), Courtesy Medical Research Council, Laboratory of Molecular Biology, Cambridge; p. 434, Courtesy Medical Research Council, Laboratory of Molecular Biology, Cambridge; p. 463, Courtesy Harvard Apparatus; p. 490, Adapted from a drawing by Bunji Tagawa; p. 491 (*top*), Courtesy Dr. Vincent T. Marchesi; p. 491 (*bottom*), Adapted from Singer, S. J., "Architecture and Topography of Biologic Membranes," *Hospital Practice*, 8, 81–90 (May 1973). Adapted from drawings by Bunji Tagawa.

DEDICATION

To all who are intellectually curious about chemistry but not professionally driven to develop technical competence in chemistry

FOREWORD

There are two searches to understand:

- the search to understand chemical phenomena and
- the search to understand the relation of chemicals and chemical technology to social, economic, and political issues.

This book is the pursuit of the first in order that individuals may pursue the second throughout their life spans.

A Letter to Those Who Teach

It has been the good fortune of both authors to teach first courses in chemistry for college students who are scientifically oriented and highly motivated to pursue careers in science, and also to teach a one-semester first course for college students who are intellectually curious but not professionally driven to develop technical competence in chemistry. We find both groups to be challenging and rewarding to teach, even though their interests and needs may be quite different. This book is based on handouts developed over a number of years for the one-semester course designed to serve the general student.

We use an informal narrative style and an orchestrated approach. It is our intent to introduce various phenomena rather briefly and then return again and again to these same phenomena in relation to various concepts and models—also to introduce concepts and models rather briefly and then return to those concepts and models again and again in relation to various phenomena. This approach evolved over time as we endeavored to discover how far these students could pursue a given topic at one time without overloading their intellectual circuits. It is an approach that is highly acceptable and rewarding to these general students. We suspect it is an approach that science-oriented students might also find more acceptable and more rewarding than the more conventional encyclopedic approach. It is our intent to use scientific terms in context at least once before defining or discussing the terms in detail. You will not find a chapter on the Periodic Table, but you will find that discussions of the properties of elements are closely tied to the Periodic Table. You will not find a chapter on equilibrium, but you will find that properties of systems are frequently explored in terms of equilibria. As you might guess, all of this leads to a table of contents that is a bit unusual. The selection and sequence of topics were guided in part by what these students seem to know and what they seem to want to know next. The selection was also guided by what we think these students need to know in order to continue to extend their knowledge of things chemical throughout their life span, at least at the level of the mass media. It is this last consideration that leads us to be particularly alert to those aspects of chemistry essential to an understanding of the nature and properties of large molecules—both naturally occurring and synthetic.

We recognize that many students whom we would like to serve may have limited backgrounds in mathematics and science and that they may also be diffident about their capacity to do science. Numerous background sections are dispersed throughout the chapters—particularly in the earlier chapters. These sections are set off from the text so that they do not interfere with the flow of the narrative. These sections are there for students to use in whatever manner is appropriate to their needs. In a similar manner, there are numerous gratuitous information sections and a few editorials. The gratuitous information sections are exactly that. Some are there for fun, and some are there to provide more technical information for the students who may want it. The editorials are used to raise value-laden questions having to do with the relation of science, engineering, and technology to societal issues. They are there to encourage students to think about these issues and to become, in time, participants in the resolution of societal issues.

The immediate goals pursued in the text are to enable students

- to discover chemical phenomena,
- to have some experience with the investigation of chemical phenomena,
- to have some experience with assessing the validity of experimental studies,
- to have some experience with the use of models to correlate and rationalize experimental results,
- to have some experience with the use of models to predict chemical properties that have yet to be experimentally investigated,
- to discover that they can understand things chemical and that they enjoy the process of extending their own knowledge, and
- to discover that they can extend their knowledge independent of the classroom.

The long-term goal is, of course, to enable students to continue to extend their knowledge of things chemical throughout their lives.

The strategy is to develop those aspects of chemistry essential to the development of some understanding of (1) the nature of forces that hold atoms together in molecules and ions, (2) the nature of forces that hold molecules and ions together in liquids and solids of either pure compounds or mixtures of compounds, and (3) the nature of forces between atoms and groups of atoms in a single large molecule that impose the three-dimensional conformation on that large molecule. Throughout, there is a very strong emphasis on the concept of the mole. Chemistry is presented as a quantitative science, although the text in general does not focus on numerical operations.

The initial introduction to chemistry is through the use of the ball–stick–spring model to predict compounds of carbon, hydrogen, and oxygen and through a critique of the powers and limitations of the model. This approach assumes that students have some concept of atoms and molecules. Whether those concepts are correct or not is quite immaterial. Those problems can be resolved later. Through this approach, students (1) encounter a few compounds that are already known to them by name, (2) encounter structural isomers and discover that they have distinctly different properties, and (3) encounter *cis–trans* isomers and discover that these, too, have distinctly different properties. In later chapters, these compounds are used in topics such as the exploration of combustion reactions and the writing of chemical equations for complete combustion reactions, the exploration of the nature of the chemical bond, and the exploration of phase relations. The most attractive characteristics of the approach are that the students have fun and that all students can be successful if they are willing to spend a little time putting models together.

We use exercises as a means of encouraging students to think and to focus that thought process. To that end, we use a rather small number of exercises and encourage every student to work on all of them. We recognize that others may want to use more problems, and to that end additional exercises for each chapter are given in the appendix.

In designing a one-semester course, hard decisions must be made about what topics are to be deleted and to what depth other topics will be explored. It is highly improbable that many could agree on what should be omitted and what should be included. This book is "overwritten." Twenty-two chapters, even short chapters, are too many for most one-semester courses. We do not use all of them in any one semester. We consider those dealing with large molecules and nuclear reactions to be gratuitous information chapters. They are there to be used in whatever way you and your students wish to use them. To many students, these are the "icing on the cake." Students have developed adequate backgrounds to puzzle their way through them. If they are interested, they will. Many topics relate to significant societal issues.

An unusually detailed index is provided to assist the reader in locating topics and in tracing the orchestration of investigations, concepts, principles, and models to higher and higher levels of integration and comprehension.

A small companion publication, *Notes to Teachers*, can be obtained from the publisher upon request. It consists of two sections. The first delineates for each chapter what we are endeavoring to accomplish in the chapter and how we are endeavoring to do it. This section also suggests some lecture demonstrations. The second section deals with the laboratory experience that is an integral part of our course. In this, we delineate experiments that we have used from time to time. They are, again, there to indicate what we endeavor to accomplish in the laboratory. The section on the laboratory experience is in no sense a laboratory manual. It is an example of what is done in one institution, and it is there to be used in any manner that is helpful to you. In using it, please take care to ensure your own safety and the safety of your students. If you do not have a laboratory with your course, you will have to make do with lecture demonstrations. Very simple demonstrations can be very effective and rewarding to the students.

What of the students who do well in a course based on this book and then decide to "go on" in science? It is our experience that they can complete successfully in a chemistry course for students who have a substantial background in high school chemistry and also compete successfully in the first semester of organic chemistry. Even though it is not the intent of this course to recruit students into the sciences, students who discover an interest in chemistry do manage to bridge the gap between this course and the majors program, and we have had students go the full doctoral route in chemistry, in related sciences, and in related professional fields.

ACKNOWLEDGMENTS

We gratefully acknowledge the contributions made by

- family and friends, for support and encouragement throughout the enterprise;
- all of those students who have taught us so much about teaching;
- the Alfred P. Sloan Foundation, for a grant to Mount Holyoke College that released time for the early exploratory years in developing a first course in chemistry for the general student;
- Phyllis Brauner and David Henderson, who used earlier drafts of chapters of this book in their own courses at Simmons College and Trinity College;
- members of the Mount Holyoke College Chemistry Department, who have used the materials on which this book is based;
- Seyhan Eğe, professor of chemistry at the University of Michigan, Ann Arbor; David Henderson, associate professor of chemistry at Trinity College, Hartford, Connecticut; and Lucy T. Pryde, professor of chemistry at Southwestern College, Chula Vista, California, who read and critiqued the entire manuscript;
- HBJ staff members Tom Thompson, Richard Morel, Cate DaPron, Kay Kaylor, Diane Pella, James Chadwick, Susan Holtz, and Lynne Bush;
- and, in particular, Doris D'Antonio, who supported us throughout by editing and typing not only the manuscript of the book but also all of those drafts of chapters used as handouts in the preceding years.

Anna J. Harrison
Edwin S. Weaver

CONTENTS

CHEMISTRY
A Search to Understand

Plants and animals are complex arrays of chemical substances in an environment made up of a complex system of chemical substances on the crust of a planet of chemical substances. The term "chemical substance" is used for emphasis. All substances are chemicals. Many familiar things, such as orange juice and carbonated beverages, are mixtures of substances—mixtures of chemicals. Ordinary sugar, sucrose, is a single substance—a single chemical. The term "chemical substance" is redundant. Either the term "chemical" or the term "substance" is adequate and appropriate. If anything can be smelled, tasted, touched, weighed, or confined to a container, it is made up of at least one chemical, possibly a number of chemicals. If the color or the odor or the taste changes, it is very probable that one or more chemical changes have taken place. Even the physiological perceptions of color, odor, taste, touch, and sound involve chemical changes. All of the natural sciences investigate systems made up of chemicals.

Chemistry a Process of Investigation

Chemistry is the investigation of the structure of chemicals, the manner in which chemicals react to form other chemicals, and the energy changes that accompany the changes. For example, when methane gas reacts with oxygen gas, water, and carbon dioxide are formed and energy is released as heat and light. Another way to make the same statement: methane gas burns in oxygen gas with a hot blue flame. The products of the reaction are water and carbon dioxide. The heat and light produced are both evidence of the energy change.

Chemical changes have been and are extensively investigated and our understanding of how to control chemical changes continues to grow. Chemical industries and many other industries control chemical changes to produce industrial products. Included among these products are the synthetic polymers, which are widely used in fabricating clothing, containers, floor covering, and now the bodies of cars; the detergents, which have largely replaced the soaps; the fertilizers, which are the bases of our agricultural productivity; and the pharmaceuticals, which are so important in modern medicine. Through the use of these pharmaceuticals, much can be done to control or at least modify the balance of chemicals within the body.

Search to Understand

There are two "searches to understand." One search is to understand the structure of chemicals, the manner in which chemicals convert to other chemicals, and the energy changes that accompany these changes. This is the search to understand chemistry. This search is the expertise of chemists. The other search is to understand the relations of chemicals and chemical reactions to the quality of life of this and succeeding generations. To understand these relations requires not only an understanding of chemical change and biological change but also an understanding of our social, economic, and political systems and processes. Questions related to the quality of life are value questions. Consequently, the control, the regulation, of the use and distribution of chemicals, both naturally occurring and synthetic chemicals, must be achieved through political processes. All citizens are directly or indirectly involved—directly involved in voting on initiatives and referenda, indirectly involved in electing candidates to represent us and make our value judgments for us. This search to understand the relation of chemical change to the quality of life is diligently being pursued on many fronts, and you cannot escape being a part of the search.

The primary intent of this book is to assist you in your search to understand chemical structures, the conversion of chemicals into other chemicals, and the energy changes that are a part of those conversions in order that each of you may participate in the search to understand the relation of chemicals and chemical reactions to the quality of life.

In spite of a superabundance of chemical transformations (chemical conversions), there is no uniquely logical point at which to begin a study of chemistry. An arbitrary starting point must be chosen. The point chosen here is a model. It is an old model and in many ways a limited model. It is also a useful model, used by practicing chemists and other scientists with full awareness of its limitation. In Chapter 2, we look at this model and its uses. Then we look at the limitations of the model and proceed in Chapter 3 to modify and extend the model to make it more useful.

Model

A model, as the term is used here, is a picture or a concept or a mathematical formulation or a three-dimensional construct that serves to correlate known phenomena and predict other phenomena. Social scientists and natural scientists use a great variety of models. Models are useful: all models fall short of reality. The deviations of models from reality frequently raise interesting and productive questions. Much research is involved with proposing, testing, modifying, and extending models. In this sense, modeling is a continuous process that encompasses both hypotheses and theories, a continuous process that does not make the distinction between hypothesis and theory.

Language of Chemistry

The approach to the language of chemistry in this book is to use terms in context at least once, in some cases many times, before the term is discussed. This is the natural way to encounter new terms and to extend your vocabulary. Our experience is that students acquire the language of chemistry effectively and rather painlessly through this approach. Through the use of terms in context, you will acquire increasingly specific connotations of the terms used.

In the next chapter you will encounter numerous concepts and terms. Muddle through, doing the best you can. All of these concepts and terms will occur again and again. Confusion is a normal step in the learning process. The next steps are to identify specific points of confusion, raise questions, and seek to resolve confusion bit by bit. Learning is an active process. The approach to the presentation of materials in the text is frequently to raise rhetorical questions and then proceed to explore these questions in the following sentences and paragraphs—and even chapters. The term "rationalize," meaning to relate to rational principles, is used frequently; the words "explain" and "why" are used infrequently.

It is suggested that you approach each chapter by reading it rather quickly, using the graphs and the diagrams and putting together the models, if three-dimensional models are involved. This is equivalent to a lecture with slides and demonstrations. There is no time to stop, to contemplate, to resolve confusion. Having some idea of where the chapter leads, go back and work through the chapter sentence by sentence and paragraph by paragraph. Few individuals can comprehend scientific materials in a single reading.

Structure of the Book

Scramble Exercises

A Look Ahead

At the end of most chapters there are two sections designed to help you identify and resolve confusion. One is the Scramble Exercises. These are not ordinary exercises but scramble exercises. When you read an ordinary exercise, you frequently know what to do and all that you have to do is go do it. When you read a scramble exercise, one that is really appropriate for you, you will have no idea how to do it and you will have to scramble, to struggle unceremoniously, and to make do with what is available to you. This is learning at its best. The other section, A Look Ahead, takes a brief look at how the material in the chapter relates to the next chapter or to future topics and delineates the concepts and skills you should have at your command in order to proceed.

Background Information

One potential source of confusion is the extent of the background material that you need in order to deal with the topics in the text. This may be material that you have well in hand, it may be material that you have had in hand at some time but are now uncertain of, and in a few cases it may be material that you have never encountered. To minimize confusion related to background information, a number of short sections dealing with these background materials have been inserted in the text. The position of each background section in the text precedes the place in the text where the information will be needed. This presentation of background material as a series of bits and pieces is based on the assumption that you will read and work through background sections if they are brief, focused, and readily available. In this chapter there are two background sections—one on large and small numbers and one on length. The first is essential to the second. In Chapter 2, the unit of length "nanometer" is used. With this background material, the use of nanometers should not be a barrier when you encounter it. All background information is set in two-column format and titled Background Information to make it easy to distinguish from the primary text. Read each background section before proceeding to the next chapter. Study these background sections and refer to them in whatever way is appropriate to your needs. Early chapters contain a number of background sections, later chapters very few.

Gratuitous Information

Editorial Comment

Also distinctly set off from the primary text in two-column format are other sections designated Gratuitous Information and Editorial Comment. The primary text is in no way dependent upon these sections. The gratuitous information is just for fun. The editorial comments are our endeavor to enable you to discover and consider some of the questions involved in the search to understand the relations of chemicals and chemical transformation to the quality of life. It is not important whether you agree or disagree with these editorial comments. It is important that you discover some of the issues, endeavor to understand these issues, and explore your role as an individual in the political resolution of these issues.

LARGE AND SMALL NUMBERS

Atoms are very small in size and have very small masses but there are great numbers of atoms. The magnitudes of these very small dimensions and these very large numbers are outside of our day-to-day experience with numbers. Concepts of the magnitudes of these numbers can only be developed by repeated experience with them, particularly with the notation used to express them. This section deals with the conventional exponential notation used to express large and small quantities.

An example of a large number: A cup of water contains approximately 7×10^{24} molecules of water. The quantity 7×10^{24} is read seven times ten to the twenty-fourth power and can be written

7 000 000 000 000 000 000 000 000.

The spaces between the sets of three zeros make it easier to keep track of the decimal point. Such a count-off always starts with the decimal point. The number could also be written

7000000000000000000000000.

Should you live to be 100 and spend all your days counting, you could not count to 7×10^{24}.

The notation 7×10^{24} indicates 7 multiplied by 10 twenty-four times. To multiply by 10 once (by 10^1) moves the decimal point one place to the right: $7. \times 10^1 = 70$. To multiply by 10 twice moves the decimal point two places to the right: $7. \times 10^2 = 700$. To multiply by 10 twenty-four times moves the decimal point twenty-four places to the right. Check the position of the decimal place in the expressions given previously for 7×10^{24}.

An example of a very small quantity: On the average, a molecule of water has a mass of 2.99×10^{-23} grams. (Not all molecules of water have the same mass, also incorrectly called weight. More about this later—much later.) The quantity 2.99×10^{-23} is read two point nine nine times ten to the negative twenty-third power and can be written as the decimal

0.000 000 000 000 000 000 000 0299

or

0.0000000000000000000000299

The notation 2.99×10^{-23} indicates 2.99 divided by 10 twenty-three times. To divide 2.99 by 10 once moves the decimal point one place to the left: $2.99 \times 10^{-1} = 0.299$. To divide 2.99 by 10 three times moves the decimal point three places to the left: $2.99 \times 10^{-3} = 0.00299$. To divide 2.99 by 10 twenty-three times moves the decimal point twenty-three places to the left. Check the position of the decimal point in the decimal quantity previously written for 2.99×10^{-23}.

In using this exponential notation to express a quantity, there are two conventions (common practices) in expressing the quantity: (1) The number must be 1 or more than 1 but less than 10 and (2) the power of 10 must be a whole number (an integer). Examples:

312 would be written 3.12×10^2

18 200 would be written 1.8200×10^4

0.000978 would be written 9.78×10^{-4}

0.0123 would be written 1.23×10^{-2}

2.42 could be written 2.42×10^0

(To multiply by 10 zero times is not to multiply by 10 at all.)

Note that 1.82×10^4 is the largest of the five quantities given above and 9.78×10^{-4} is the smallest. Using the above conventions, it is the power of ten that is the key to the magnitude of the number. Note also that 18 200 is numerically equal to each of the following six expressions:

1820.0×10	0.18200×10^5
182.00×10^2	0.018200×10^6
18.200×10^3	0.0018200×10^7

These forms are not used. Only in 1.8200×10^4 does the number lie between 1 and 10.

Calculations using numbers expressed in this exponential notation will be addressed in another background section in Chapter 2.

SCIENCE, ENGINEERING, TECHNOLOGY, AND THE EXPECTATIONS OF SOCIETY

The developments of modern technological societies such as ours have been based upon the benefits derived from the use of science, engineering, and technology to fulfill basic human needs for food, shelter, and clothing and to enhance the quality of life in a great multiplicity of ways. Benefits derived from the use of chemistry, chemical engineering, and chemical technology are a very significant part of our heritage. Today, we understand a great deal about the nature of chemical processes in biological organisms, including ourselves, and about chemical processes in the inanimate universe in which we live. We use chemical engineering to bring about chemical processes that convert natural resources and waste materials into more useful materials and we use chemical technology to mass-produce goods and services that can be profitably marketed.

The expansion of knowledge of biochemical processes enhances the capacity of biochemists, pharmacologists, and physicians to design pharmaceuticals to modify abnormal chemical balances characteristic of specific illnesses. This same expansion of knowledge enhances the capacity of biochemists, toxicologists, and physicians to identify and assess the biochemical nature of toxic responses to chemicals that are a part of the natural environment and to chemicals that have been introduced into our environment during the production, distribution, use, and disposal of products of technology.

The benefits derived by society from technology have been and are tremendous. The burdens (frequently called risks) placed upon society by technology have been and are very significant. The two may be closely coupled. No one would deny the benefits inherent in adequate food and improved health care. Even so, many of our current societal issues are related to the unprecedented population of the world today. It is probably true that every technological innovation, regardless of how great its positive impact upon society, also has a negative impact on society. This is a statement for which there can be no proof. It is, however, a statement for which it is very difficult to find an exception—perhaps impossible to find an exception.

There are at least two fundamental questions. What burdens is society willing to accept to have the benefits of specific technological innovation? To what degree can scientists, engineers, and technological institutions create technological options that enhance the benefits and minimize the burdens? When we vote in national, state, and local elections, we take positions on the first question when we vote on initiatives and referenda and when we vote for candidates who will become our surrogates and make value decisions for us. The primary goal of this book is to introduce you to chemical phenomena in such a way that you can and will continue to expand your knowledge and understanding of chemistry and of societal issues involving chemical phenomena throughout your life span, at least at the level of the mass media. It is essential that each of us understands these issues in order that the positions we take individually are, in fact, consistent with our personal values. This is not to say that all of those who understand the nature of a societal issue will agree on the course of action to be taken. Far from it. In matters having to do with the quality of life, personal values are a determining factor in the decisions made. The nature and magnitude of our collective investment of talent, energy, and other resources will to a large degree determine the answer to the second question.

LENGTH OR DISTANCE

The units of length used in science are the meter and multiples of the meter. The system of measurements based upon the meter, the metric system, is also the accepted system of measurements in commerce throughout the industralized world—with the exception of the United States.

The word "meter" is derived from the Greek *metron*, meaning measure. Our use of the word meter to designate a unit of length is not to be confused with the use of the word in the analysis of the rhythm of verse and music.

A meter stick is used in much the same way as a yardstick. The smaller divisions on the meter stick are usually centimeters and millimeters—100 centimeters and 1000 millimeters. The centimeter is shorter than the meter.

1 centimeter is 10^{-2} meter, also written

1 centimeter is 1×10^{-2} meter

1 cm is 10^{-2} m

1 cm is 0.01 m

The millimeter is even smaller:

1 millimeter is 10^{-3} meter, also written

1 millimeter is 1×10^{-3} meter

1 mm is 10^{-3} m

1 mm is 0.001 m

The prefixes *centi-* and *milli-* are multiplicative prefixes. Some of the multiplicative prefixes are given in the table at right along with the factors they represent and the symbols that represent them.

Only the five prefixes printed in boldface type will be extensively used in this text. Used with the meter, these units of length are

1 decimeter is 10^{-1} meter, also written

1 decimeter is 0.1 meter

1 dm is 10^{-1} m

1 centimeter is 10^{-2} meter, also written

1 centimeter is 0.01 meter

1 cm is 10^{-2} m

1 millimeter is 10^{-3} meter, also written

1 millimeter is 0.001 meter

1 mm is 10^{-3} m

1 nanometer is 10^{-9} meter, also written

1 nanometer is 0.000 000 001 meter

1 nm is 10^{-9} m

1 kilometer is 10^{3} meters, also written

1 kilometer is 1000 meters

1 km is 10^{3} m

TABLE OF MULTIPLICATIVE PREFIXES

Prefix	Factor	Symbol
deci	10^{-1}	d
centi	10^{-2}	c
milli	10^{-3}	m
micro	10^{-6}	μ
nano	10^{-9}	n
pico	10^{-12}	p
deka	10^{1}	da
hecto	10^{2}	h
kilo	10^{3}	k
mega	10^{6}	M
giga	10^{9}	G
tera	10^{12}	T

The first four units of length given above are very small. The nanometer is particularly useful in discussing the sizes of atoms and the separation between atoms in molecules.

Another unit of length, Angstrom, Å, has historically been used in discussing the sizes of atoms.

1 Angstrom is 10^{-10} meter

1 Å is 10^{-10} m

The Angstrom is not a part of the multiplicative system and its use is now discouraged.

The meter is a bit longer than a yard (39.4 inches) and the kilometer is a bit longer than six tenths of a mile (0.62 miles).

Measurements of length made using the meter and units derived from the meter cannot be more precise than the definition of the meter. To define the meter is a technical matter. To use the meter does not require an understanding of the definition. You should understand that great care has been taken in defining the meter and that the definitions of units such as the meter are changed from time to time to utilize the increased precision of measurements made possible by the development of new methodologies. The current definition of the meter utilizes the very high precision with which the velocity of light and the frequency of light can be measured.

Today, the meter is defined in terms of a specific number of wavelengths of light in the spectrum of a particular element under specified conditions. To be more specific, the meter is the length equal to 1.65076373×10^6 wavelengths in vacuum of the radiation corresponding to the transition between the levels $2p_{10}$ and $5d_5$ of the krypton-86 atom. This is a very sophisticated definition involving measurements of exceptional precision. It is not expected that you understand this definition at this time or that you memorize this definition ever.

You should understand that the manner in which the number of wavelengths of light is expressed, the 1.65076373×10^6 indicates that all of those nine figures are experimentally significant. In the laboratory you will discover how challenging it is to make measurements in which four figures are experimentally significant. If only four figures had been experimentally significant, the number of wavelengths would have been expressed $1.65\underset{\cdot}{1} \times 10^6$, a value much less precise than 1.65076373×10^6.

It is frequently necessary to convert lengths measured in one unit to the same lengths expressed in another unit, for example, 32.2 centimeters to meters. Many students can make the conversion to 0.322 meters very readily but there are others who have great difficulty making this transition. If you have difficulty, work through the next paragraphs. You may find them helpful.

Length expressed in meters is directly proportional to the same length expressed in centimeters:

$$[\text{length in meter}] \propto [\text{length in centimeters}]$$

This statement can be converted into an equality, an equation, by the introduction of the 1×10^{-2} factor given in the Table of Multiplicative Prefixes. The factor 1×10^{-2} has the units meters per centimeters.

$$\begin{bmatrix}\text{length in}\\ \text{meters}\end{bmatrix} = \begin{bmatrix}1 \times 10^{-2}\ \dfrac{\text{meters}}{\text{centimeters}}\end{bmatrix}\begin{bmatrix}\text{length in}\\ \text{centimeters}\end{bmatrix}$$

This equation is of the type $y = ax$ where y is the length in meters, x is the same length in centimeters, and a is the proportionality constant 1×10^{-2} meters/centimeters. Note that both sides of the equation have the same unit, the meter.

An example: A length of 250 centimeters is a length of 2.50 meters.

$$\text{length in meters} = 1 \times 10^{-2}\ \frac{\text{meters}}{\cancel{\text{centimeters}}} \times 250\ \cancel{\text{centimeters}}$$

Another example: A length of 2.50 centimeters is 0.0250 meters.

$$\text{length in meters} = 1 \times 10^{-2}\ \frac{\text{meters}}{\cancel{\text{centimeters}}} \times 2.50\ \cancel{\text{centimeters}}$$

Similarly, length in meters is directly proportional to length in millimeters and the two are related by the equation

$$\begin{bmatrix}\text{length in}\\ \text{meters}\end{bmatrix} = \begin{bmatrix}1 \times 10^{-3}\ \dfrac{\text{meters}}{\text{millimeters}}\end{bmatrix}\begin{bmatrix}\text{length in}\\ \text{millimeters}\end{bmatrix}$$

Other useful equations relating units of length:

$$\begin{bmatrix}\text{length in}\\ \text{meters}\end{bmatrix} = \begin{bmatrix}1 \times 10^{-6}\ \dfrac{\text{meters}}{\text{micrometers}}\end{bmatrix}\begin{bmatrix}\text{length in}\\ \text{micrometers}\end{bmatrix}$$

Micrometers are also called microns.

$$\begin{bmatrix}\text{length in}\\ \text{meters}\end{bmatrix} = \begin{bmatrix}1 \times 10^{-9}\ \dfrac{\text{meters}}{\text{nanometers}}\end{bmatrix}\begin{bmatrix}\text{length in}\\ \text{nanometers}\end{bmatrix}$$

$$\begin{bmatrix}\text{length in}\\ \text{meters}\end{bmatrix} = \begin{bmatrix}1 \times 10^{3}\ \dfrac{\text{meters}}{\text{kilometers}}\end{bmatrix}\begin{bmatrix}\text{length in}\\ \text{kilometers}\end{bmatrix}$$

Check these equations using the definitions for the various multiplicative prefixes.

Direct proportions are discussed in more detail in Background Information 11-2 in Chapter 11.

Scramble Exercises

There are few scramble exercises for this chapter. There will be ample opportunity later to sort out concepts introduced in this chapter.

1. Show that the equation

$$\left[\text{length in millimeters}\right] = \left[1 \times 10^{3}\ \frac{\text{millimeters}}{\text{meters}}\right]\left[\text{length in meters}\right]$$

is equivalent to the equation

$$\left[\text{length in meters}\right] = \left[1 \times 10^{-3}\ \frac{\text{meters}}{\text{millimeters}}\right]\left[\text{length in millimeters}\right].$$

Both are equations for a direct proportion.

2. The background materials are your responsibility. Work with them to the degree you consider appropriate. Make up exercises for yourself. For example, (a) convert 123.4 to $1.234 \times 10^{?}$ and check to see that the power of ten you choose maintains the equality $123.4 = 1.234 \times 10^{?}$, and (b) express 2.3 nanometers as __________ meters. Learn to test yourself.

To the degree possible, always check the manipulation of numbers in the chapters. In doing so, you may discover errors in the text, you may discover errors in the way you manipulate numbers, and you will develop confidence in your ability to handle numbers.

Additional exercises are given in the Appendix.

A Look Ahead

In Chapter 2, a specific model is presented. This model is used to explore the structure of molecules containing atoms of three elements: carbon, hydrogen and oxygen. Two concepts are carried forward from Chapter 1.

(1) Chemistry is the investigation of
- the structures of chemicals,
- the manner in which chemicals react to produce other chemicals, and
- the energy changes associated with chemical reaction.

(2) A model is a picture or a concept or a mathematical formulation or a three-dimensional construct or some combination of the above that correlates known phenomena and can be used to predict other phenomena.

The model presented in Chapter 2 correlates information about thousands of compounds of carbon, hydrogen, and oxygen.

On to Chapter 2. This is where the fun begins.

2
ONE MODEL OF ATOMS AND MOLECULES

Chapter 2

An Introduction to Compounds of Carbon, Hydrogen, and Oxygen

The Model

The model elected here as the starting point deals with the grouping of **atoms of elements** into **molecules of compounds** in terms of satisfying characteristic bonding sites of the atoms. Initially, the use of the model will be limited to three elements: hydrogen, carbon, and oxygen. In this model,

Bonding Sites

- the **hydrogen atom** is considered to have a single bonding site,
- the **carbon atom** to have four bonding sites uniformly distributed about the carbon atom, and
- the **oxygen atom** to have two bonding sites so oriented as to establish an angle of a little more than 90 degrees in bonding to two other atoms.

The one ground rule is that each bonding site of an atom must be used to bond to a bonding site of another atom.

Ball and Stick Model

It is important to understand the spatial distribution of atoms in molecules and it is common practice, even an essential practice, for experienced scientists to build molecular models. Frequently, balls drilled with the appropriate number and orientation of holes and short sticks of the appropriate diameter to fit the holes are used to represent the atoms and the bonds. The balls are frequently color coded: black for carbon, red for oxygen, and white or some other color for hydrogen. Modeling clay or gum drops with bits of toothpicks serve equally well to explore the three-dimensional nature of molecules, but these structures tend to sag and they can be sticky. To understand the spatial relations of atoms in molecules, it is absolutely essential to work with three-dimensional models. If you work with three-dimensional models, you will be amazed at how much you will understand. If you do not work with three-dimensional models, you will become hopelessly confused. To watch someone put models together is helpful but not nearly as helpful as building the models yourself. College bookstores usually carry small model sets made of plastic. The benefits derived from their use justify the investment.

Water

In terms of this model, the simplest compound of oxygen and hydrogen would require two atoms of hydrogen to satisfy the two bonding sites of an oxygen atom. This is the molecule of water, H_2O. (Read H-two-O.) (Figure 2-1)

Methane

The simplest compound of carbon and hydrogen would require four atoms of hydrogen to satisfy the four bonding sites of a carbon atom. This is the molecule of methane, CH_4. (Read C-H-four.) (Figure 2-1)

Figure 2-1

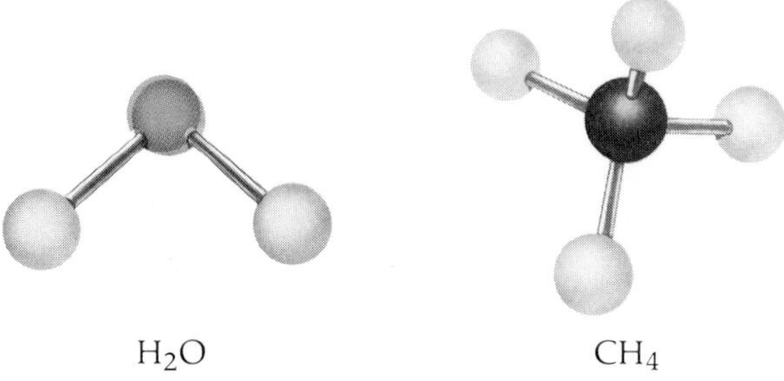

BALL AND STICK MODELS OF WATER, H_2O, AND METHANE, CH_4.

The models show the water molecule to be flat and bent. The three atoms cannot be forced into a straight line. The centers of the three atoms, the three balls, lie in a plane and the molecule of water is said to be **planar.** The water molecule, of course, does not really lie in a plane: the atoms take up space. It is the centers of the three atoms that lie in a plane.

In marked contrast to this, the molecule of methane, CH_4, is bulky. The model of the methane molecule can be placed on a plane surface with any three of the hydrogen atoms constituting the base but the five atoms cannot be forced into a plane. The model predicts the molecule of methane to be nonplanar.

2-1 *Background Information*

TEMPERATURE

In science, temperatures are measured in degrees Celsius and in kelvin. The relation between these two temperature scales is given by

$$[\text{temperature kelvin}] = \left[\begin{matrix}\text{temperature in} \\ \text{degrees Celsius}\end{matrix}\right] + 273.16$$

$$K = {}^\circ C + 273.16$$

The sizes of units are the same; the two scales begin at quite different temperatures. Zero degrees Celsius is 273.16 kelvin. Zero kelvin is -273.16 degrees Celsius.

The above statements are in keeping with current editorial practices. Celsius is capitalized; kelvin is not. The abbreviation for Celsius is capital C; the abbreviation for kelvin is capital K. The word degree and the degree sign are used in connection with the Celsius scale; the word degree and the degree sign are not used in connection with the kelvin scale. There is no obvious logic to these editorial practices. This is just the confusion that results when two sets of conventions meet.

More information about temperature and the measurement of temperature is given in Gratuitous Information 2-1, Temperature and the Measurement of Temperature.

The most common representations of these molecules on paper are diagrams with circles representing the atoms and straight lines representing the bonds:

molecule of water   molecule of methane

or structural formulas that omit the circles:

```
H—O            H
  |            |
  H        H—C—H
               |
               H
```

Both are somewhat inadequate. The structural formula of methane is particularly misleading. It looks flat and gives the impression that the five atoms lie in a plane. The ball and stick model predicts methane to be nonplanar.

The next consideration is to explore the above predictions in terms of what is known experimentally about water and methane. Both compounds are well known and have been extensively studied by a great variety of instrumental methods. Water, H_2O, is the most widely distributed compound on the surface of the earth. Water is essential for life as we know it. Under a pressure of one atmosphere, water freezes at a temperature of exactly zero degrees Celsius and boils at exactly one hundred degrees Celsius. In fact, the freezing point and the boiling point of water are used to define these two temperatures and the magnitude of the degree Celsius, the unit used to measure temperature. (Gratuitous Information 2-1, Temperature and the Measurement of Temperature.)

Physical Properties of Water

Room temperature varies between 15 and 30 °C and water is normally considered to be a liquid. To the chemist, the ice that exists below 0 °C, the freezing point of water, is simply the solid phase of water. At all temperatures, some molecules of water also exist in the vapor phase: the higher the temperature the greater the water molecule population (concentration) that can exist in the vapor phase. Water strongly absorbs light in certain regions of the infrared spectrum and also in the high-energy region of the ultraviolet spectrum. Water does not absorb in the visible region of the spectrum and is consequently said to be colorless. Two other properties of water that will become significant to later discussions are the dipole moment of the molecule (1.85 debye) and the dielectric constant of the liquid phase (78.5 at 25 °C). It is not necessary to understand now about the absorption of light and about dipole moments and dielectric constants. The point, for the moment, is that compounds have a number of **physical properties** and that the magnitude of these properties can be experimentally determined. The dipole moment of water is closely related to the attraction between molecules of water, H_2O, that leads to the molecules of water coming together to give the liquid phase and also the solid phase. The dielectric constant of water is closely related to the effectiveness of water as a solvent for sodium chloride (table salt) and many other compounds.

The dimensions of the water molecule are well known. On the average, the distance between the center of the oxygen atom and the center of a hydrogen atom is 0.096 nanometers (nm). On the average, the hydrogen–oxygen–hydrogen angle is 104.5°. "On the average" is used in the above statements since there is clear-cut evidence from the interpretation of the absorption of infrared light and other physical measurements that the bond lengths undergo very small but rapid periodic variations and the hydrogen–oxygen–hydrogen bond angle also undergoes very small but rapid periodic variations.

Chemical Properties of Water

Water is extremely stable insofar as decomposition into hydrogen and oxygen is concerned. The decomposition of water can, however, be brought about by very high temperatures or at room temperature by supplying electrical energy. Water also reacts with a great number of other substances quite readily at room temperature.

Methane, CH_4, is a gas at room temperature and room pressure. Under a pressure of one atmosphere, the boiling point of liquid methane is −164 °C and the freezing point is −182 °C. Consequently, methane is a gas under normal room conditions.

2-1 Gratuitous Information

TEMPERATURE AND THE MEASUREMENT OF TEMPERATURE

Two objects are said to be at different temperatures if heat flows from one to the other when the two objects are placed in contact. In measuring temperature, there are two problems to be addressed. One is the measurement of differences in temperature and the other is where to begin a scale of measurement. Every scale of measurement must have a zero. What is the significance of zero on a temperature scale?

Anders Celsius, an early astronomer, dealt with the problem of measuring differences in temperature. The Celsius scale measures differences in temperature from the temperature at which water freezes under a pressure of one atmosphere. (Both the boiling point and the freezing point of a liquid depend upon the pressure of the atmosphere. Consequently, the atmospheric pressure must be controlled.) The zero on the Celsius scale is set at the freezing point of water under a pressure of one atmosphere. To do this, the thermometer is placed in a mixture of ice and water under a pressure of one atmosphere and the response of the thermometer is marked. The magnitude of the division on the scale is set by using another reference temperature. This second reference temperature is the boiling point of water under a pressure of one atmosphere. The response of the thermometer to the temperature of boiling water is assigned the value 100 degrees Celsius. One one-hundredth of the difference between the response at the freezing point of water and the response at the boiling point of water then becomes the measure of one degree Celsius. Using this measure for one degree, the Celsius scale is extended above 100° and below 0°. Temperatures on the Celsius scale may be positive, higher than the freezing point of water, or negative, lower than the freezing point of water. Our normal body temperature is 37.0 °C above the freezing point of water.

The kelvin scale, named in honor of William Thomson, Baron Kelvin, a physicist, uses the same unit as the Celsius scale but sets zero for the scale at a much lower temperature, −273.16 °C. Studies of the responses of gases to changes in temperature and studies of the conversion of heat into work indicate that there really is a zero temperature—a temperature below which there are no other temperatures. This is the zero kelvin. There are no negative temperatures on the kelvin scale. Temperatures below 0.001 kelvin have been reached experimentally.

The Fahrenheit temperature scale, which has been commonly used in English-speaking areas of the world, sets the freezing point of water under a pressure of one atmosphere at exactly 32° F and the boiling point of water under a pressure of one atmosphere at exactly 212° F and divides the interval between those two temperatures into

Methane is one of the compounds in the mixture of compounds that make up natural gas and is a common fuel. Methane absorbs light in a very limited region of the infrared spectrum and in the very high-energy region of the ultraviolet spectrum but is transparent in the intervening regions of the spectrum, including the visible spectrum. Methane is colorless. The dimensions of the molecule are well known: the average carbon–hydrogen bond length is 0.109 nm and the average hydrogen–carbon–hydrogen angle is 109.5°. Here again, the frequency and the magnitude of the periodic variations in bond length and in bond angle have been studied and are known to have characteristic values. The dipole moment of the gas molecule is zero debye; the dielectric constant of the gas phase at one atmosphere pressure is 1.0009 at 0 °C and of the liquid phase is 1.70 at −173 °C. Note the marked differences between these values and the corresponding values for water.

Physical Properties of Methane

The methane molecule, CH_4, is rather stable with respect to decomposition into smaller units and is also relatively unreactive at room temperature with other substances. It can, however, be decomposed at high temperatures and it reacts readily with a number of substances, such as oxygen, at high temperatures. Methane is a fuel.

Chemical Properties of Methane

---

180 equal units—the one degree Fahrenheit unit. One degree Fahrenheit is smaller than one degree Celsius. The relation between Fahrenheit temperature and Celsius temperature is given by

$$\left[\begin{matrix}\text{temperature}\\ \text{Fahrenheit}\end{matrix}\right] - 32 = \left[\begin{matrix}\text{temperature}\\ \text{Celsius}\end{matrix}\right] \times \frac{180}{100}$$

$$°\text{F} = \left(°\text{C} \times \frac{9}{5}\right) + 32$$

The manner in which the zero on the Fahrenheit scale was originally defined need not concern us here, but it may be of interest to note that the intent had been to define zero on the Fahrenheit scale in terms of the temperature obtained with a mixture of snow and salt.

Any measurable property that changes reproducibly with temperature can be made the basis of a thermometer. The property most commonly used in commercial thermometers is the volume of a liquid such as mercury or ethanol. The visibility of colorless liquids such as ethanol is increased by the addition of a soluble dye, frequently a red dye. As the temperature of the liquid changes, the volume of the liquid changes. Most of the liquid is in the bulb of the thermometer. The total change in volume with change in temperature is observed by the movement of the surface of the liquid in the capillary stem of the thermometer. The larger the bulb and the smaller the diameter of the capillary the greater the sensitivity of the thermometer.

The resistance of a metal to the flow of an electric current increases as the temperature increases and "resistance thermometers" based on this property are frequently used, particularly at the low temperatures at which many of the usual thermometer fluids become solids. Mercury, for example, freezes at −38.9 °C. More scientists than would care to admit it have been embarrassed by attempting to measure temperatures lower than this with a mercury thermometer. Fortunately, mercury contracts on freezing (most substances do), so those individuals have been spared the further embarrassment of coming out with a broken thermometer.

The precise measurement of temperature, particularly temperatures well removed from the temperatures used to define the scale, presents problems due to the dependency of the property of the substance being used as the basis for the temperature measurement on the temperature range. For example, the change in volume for a sample of liquid mercury is not the same for a change in temperature from 0 °C to 100 °C as it is from 100 °C to 200 °C. Such problems are not insurmountable. They just require careful attention to attain the high accuracy required for some scientific work.

The mass (weight) of the hydrogen atom is much less than the mass of either the carbon atom or the oxygen atom; the oxygen atom is heavier than the carbon atom. The water molecule, H_2O, has a greater mass than the methane molecule, CH_4. The difference, however, is not great and on the scale of measurement used, water is 18 as compared to 16 for methane. The contrasts in the physical properties other than mass are much more striking.

Hydrogen Peroxide

A slightly more complicated molecule, H_2O_2 (read H-two-O-two), is also predicted by the model. In this molecule, each of the two oxygen atoms uses one bonding site in bonding with the other and the second bonding site of each oxygen atom is satisfied by an atom of hydrogen. This is the hydrogen peroxide molecule. According to the model, the centers of the four atoms do not necessarily lie in a plane—although they could. This follows from the freedom of the HO groups to rotate about the oxygen–oxygen bond (the stick) between the two oxygen atoms and give an infinite number of orientations. Two are shown below.

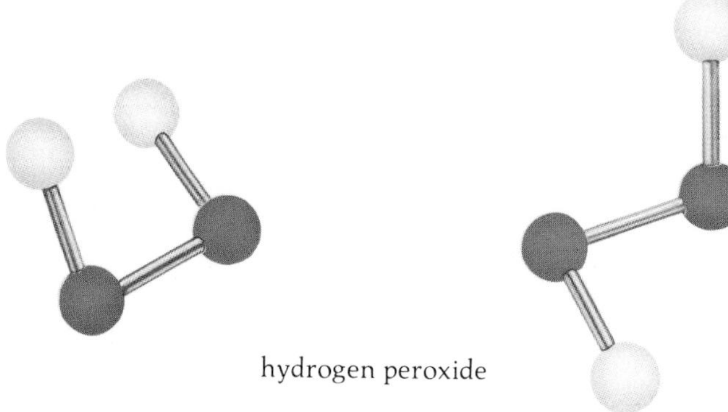

hydrogen peroxide

2-2 *Background Information*

AREA

Units of area are derived from units of length. For example, the area of a rectangle is given by the product of the length times the width:

$$\text{area} = \text{length} \times \text{width}$$

The area of a circle is given by the product of pi times the radius of the circle squared:

$$\text{area} = \pi \times \text{radius}^2 = 3.1416 \times \text{radius}^2$$

If the length and the width of the rectangle are measured in meters, the area of the rectangle computed using those units will be given in meters times meters or meters^2 (read either meters squared or square meters).

If the radius of the circle is used in centimeters, then the area of the circle is given in centimeters^2. Note that the area of a square one meter on a side can be expressed as 1 meter^2 and also as 10 000 centimeters^2 and also as $1. \times 10^4$ centimeters^2. If the above statement is not obvious to you, pull it apart and get help as a last resort.

If the United States ever manages to go along with the use of the metric system in commerce, land areas will be measured in hectares. The hectare is the area equivalent to the area in a square 100 meters on a side. The word hectare is derived from the Greek *hekaton*, hundred. The hectare is approximately 2.5 acres.

Area can be expressed in units with special names such as acre and hectare; area in scientific work is usually expressed in $(\text{unit of length})^2$.

The corresponding compound of carbon and hydrogen with the two atoms of carbon bonded to each other is C_2H_6, ethane. Both compounds, hydrogen peroxide and ethane, are well known. Hydrogen peroxide (H_2O_2) is a colorless liquid that is frequently marketed and used as a dilute aqueous (water) solution. Hydrogen peroxide molecules, H_2O_2, decompose into simpler molecules even at room temperature. In this reaction, two molecules of hydrogen peroxide, H_2O_2, produce two molecules of water, H_2O, and one molecule of oxygen gas, O_2. This reaction is concisely stated by the expression: Hydrogen peroxide in aqueous solution decomposes to give liquid water and oxygen gas.

$$2\,H_2O_2(aq) \longrightarrow 2\,H_2O(l) + 1\,O_2(g)$$

The rate at which this reaction occurs can be extremely slow, but it can also be greatly accelerated by a number of compounds (catalysts). The foaming observed when 3% aqueous solution of hydrogen peroxide is placed on a wound is caused by the escape of bubbles containing oxygen gas molecules, O_2. One of these very small bubbles contains more than 1×10^{17} molecules of oxygen (100 000 000 000 000 000 molecules of oxygen), O_2, in the gas phase.

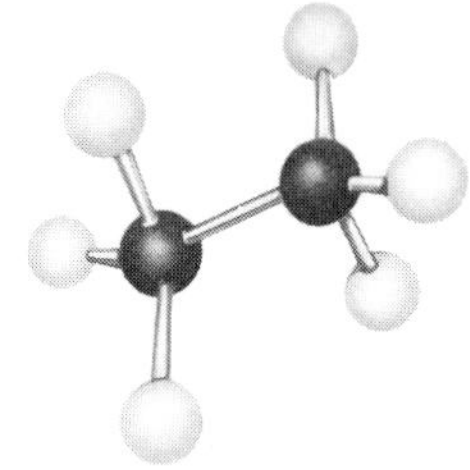

ethane

The model predicts the two-carbon compound, C_2H_6. This is ethane, a compound found in natural gas along with methane and other compounds of carbon and hydrogen. The formula for ethane is frequently written as

$$\begin{array}{ccccccc} & & \text{H} & & \text{H} & & \\ & & | & & | & & \\ \text{H} & — & \text{C} & — & \text{C} & — & \text{H} \\ & & | & & | & & \\ & & \text{H} & & \text{H} & & \end{array}$$

Ethane

and also as CH_3CH_3. The first representation can be misleading since it could be interpreted to mean that the molecule of ethane is planar (flat). A ball and stick model gives a better representation of the three-dimensional nature of the compound. Each carbon carries a pinwheel of three hydrogen atoms. By rotating the pinwheels around the carbon–carbon bond, it is possible to bring four atoms into a single plane, but not more than four atoms: either two atoms of hydrogen and the two atoms of carbon, or four atoms of hydrogen. The carbon–carbon bond distance is 0.154 nm on the average. The carbon–hydrogen average bond distance is essentially that found in methane, 0.109 nm. Pure ethane is quite stable with respect to decomposition into smaller molecules at room temperature, but it can be decomposed at high temperatures and it reacts rapidly with oxygen gas at high temperatures. (Ethane burns in air but the reaction has to be initiated by a spark or another flame.)

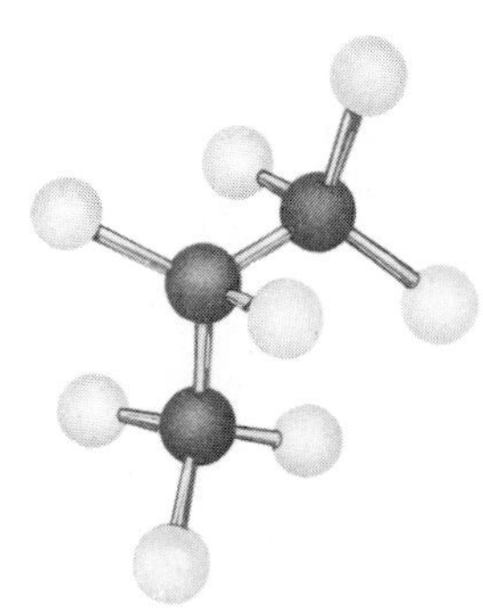

propane

Although the model in no sense rules out compounds such as H_2O_3 and H_2O_4, with a series of oxygen atoms bonded to each other, the experimental fact is that these compounds are so extremely unstable that they are not prepared in the laboratory. On the other hand, molecules such as

$$\begin{array}{ccccccccc} & & \text{H} & & \text{H} & & \text{H} & & \\ & & | & & | & & | & & \\ \text{H} & — & \text{C} & — & \text{C} & — & \text{C} & — & \text{H} \\ & & | & & | & & | & & \\ & & \text{H} & & \text{H} & & \text{H} & & \end{array} \qquad \underset{\text{propane}}{CH_3CH_2CH_3} \qquad \text{and} \qquad \begin{array}{ccccccccccc} & & \text{H} & & \text{H} & & \text{H} & & \text{H} & & \\ & & | & & | & & | & & | & & \\ \text{H} & — & \text{C} & — & \text{C} & — & \text{C} & — & \text{C} & — & \text{H} \\ & & | & & | & & | & & | & & \\ & & \text{H} & & \text{H} & & \text{H} & & \text{H} & & \end{array} \qquad \underset{\text{butane}}{CH_3CH_2CH_2CH_3}$$

are well-known components of natural gas and are the principal compounds in the small tanks of "liquid gas" used to supply limited fuel requirements in regions removed from gas lines. The three-carbon compound is propane; the four-carbon compound is butane. Note that the names of these compounds of hydrogen and carbon (hydrocarbons) all end in the suffix *-ane*. The series of hydrocarbons called alkanes have the general formula $CH_3(CH_2)_nCH_3$ where n takes on integer values: $n = 0$ for ethane, $n = 1$ for propane, and $n = 2$ for butane. For $n = 3$ (the five-carbon molecule) or

Alkanes

2-3 *Background Information*

VOLUME

Units of volume are also derived from units of length using the same multiplicative prefixes. For example, the volume of a rectangular solid is the product of the area of the base times the height:

$$\text{volume} = (\text{length} \times \text{width}) \times \text{height}$$

The volume of a cylinder is the area of the base times the height:

$$\text{volume} = \pi\ \text{radius}^2 \times \text{height}$$
$$= 3.1416\ \text{radius}^2 \times \text{height}$$

The volume of a sphere is the product of 4/3 pi times the radius cubed:

$$\text{volume} = 4/3\ \pi\ \text{radius}^3 = 4/3 \times 3.1416 \times \text{radius}^3$$

In any of the above, if all of the distances are measured in meters, the volume is given in meters^3, or cubic meters. If all of the distances are measured in centimeters, the volume computed in these units will be given in centimeters^3, or cubic centimeters. You should be able to show that 1 meter^3 is 10^6 centimeters^3.

In chemistry, the cubic decimeter is extensively used as a unit of volume. It is, however, used under another name. This name is the liter, abbreviated L. Using the multiplicative prefix *milli-*,

1 milliliter is 10^{-3} liter, also written

1 milliliter is .001 liter

1 mL is 10^{-3} L

and

1000 milliliters are 1 liter

1000 mL are 1 L

The capital *L* is used in the abbreviation for liter (L) and milliliter (mL) to avoid the confusion between the number one and the letter l in typing.

The liter, the cubic decimeter, can also be expressed in terms of cubic centimeters. The decimeter is 10 centimeters and the volume of a cube 1 decimeter on each edge can be expressed in terms of decimeters^3 or centimeters^3:

$$\begin{aligned}\text{volume} &= 1\ \text{decimeter} \times 1\ \text{decimeter} \times 1\ \text{decimeter}\\ &= 1\ \text{decimeter}^3\\ &= 10\ \text{centimeters} \times 10\ \text{centimeters} \times 10\ \text{centimeters}\\ &= 1000\ \text{centimeters}^3\end{aligned}$$

In summary,

1000 milliliters is 1 liter (L), also 1 decimeter^3,

also 1000 centimeters^3

and

1 milliliter (mL) is 1 centimeter^3 (cm^3)

One milliliter is the exact equivalent of one cubic centimeter and the volumes milliliter and cubic centimeter are used interchangeably.

The liter is a little larger than the U.S. liquid quart (1.06 quarts). Using equations for the direct proportion, the relation between liters and milliliters can be expressed as either

$$\begin{bmatrix}\text{volume}\\ \text{in liters}\end{bmatrix} = \begin{bmatrix}1 \times 10^{-3}\ \dfrac{\text{liters}}{\text{milliliters}}\end{bmatrix}\begin{bmatrix}\text{volume in}\\ \text{milliliters}\end{bmatrix}$$

or

$$\begin{bmatrix}\text{volume in}\\ \text{milliliters}\end{bmatrix} = \begin{bmatrix}1 \times 10^{3}\ \dfrac{\text{milliliters}}{\text{liters}}\end{bmatrix}\begin{bmatrix}\text{volume}\\ \text{in liters}\end{bmatrix}$$

greater, these hydrocarbons are named in terms of the total number of carbon atoms: the five-carbon hydrocarbon is pentane; the six-carbon hydrocarbon is hexane.

Structural Isomers

Figure 2-2 gives ball and stick models for the two structural isomers of butane. Focus on their three-dimensional structures. In structures such as these, first determine the relation of the carbon atoms to each other.

Note that

```
    H   H   H   H
    |   |   |   |
H — C — C — C — C — H        CH3CH2CH2CH3
    |   |   |   |
    H   H   H   H
```

is not the only arrangement for four atoms of carbon and ten atoms of hydrogen that satisfies the model. An equally satisfactory arrangement is

```
    H   H   H
    |   |   |
H — C — C — C — H
    |   |   |                          CH3CHCH3
    H   |   H          also written        |
        |                                 CH3
    H — C — H
        |
        H
```

Both molecules are said to have the same **molecular formula** (C_4H_{10}) but to have different **structural formulas.** The first structural formula for butane is referred to as a "straight-chain" hydrocarbon, in spite of the fact that there is nothing straight about

Figure 2-2

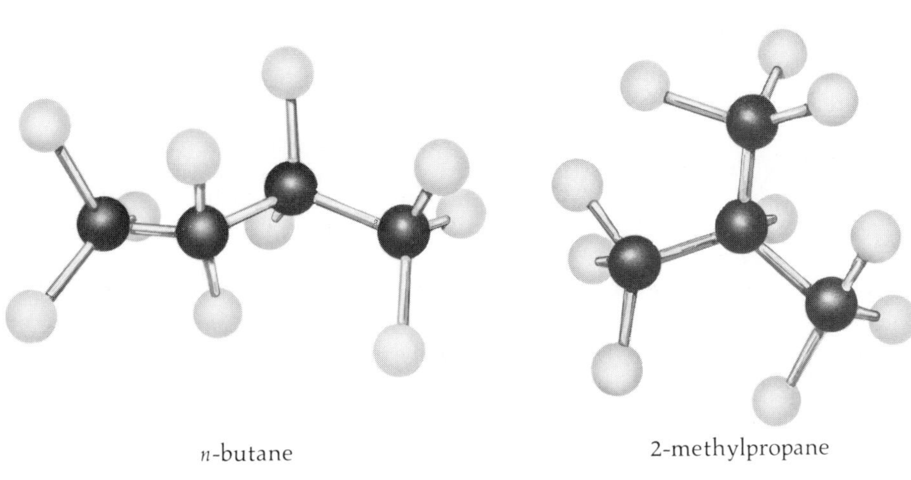

n-butane

2-methylpropane

BALL AND STICK MODELS OF THE TWO STRUCTURAL ISOMERS OF BUTANE: *n*-BUTANE AND 2-METHYLPROPANE.

a chain of four carbon atoms in which each carbon–carbon–carbon bond angle is approximately 109°. The straight-chain compound is called normal butane, *n*-butane. The term straight-chain comes from the appearance of the structural representation

```
    H  H  H  H
    |  |  |  |
 H—C—C—C—C—H
    |  |  |  |
    H  H  H  H
```

in which the four carbon atoms seem to be a straight chain:

```
   |  |  |  |
 —C—C—C—C—
   |  |  |  |
```

See Figure 2-2 for the three-dimensional model—or better still, build your own model.

The second representation, the branched chain butane, is called either *iso*butane or 2-methylpropane (a methyl group, $-CH_3$, on the second carbon of propane). 2-Methylpropane is now the preferred name. Both compounds, *n*-butane and 2-methylpropane, are known and have different physical properties and different chemical reactivities. Their boiling points and freezing points under a pressure of one atmosphere are

	***n*-butane**	**2-methylpropane**
boiling point (°C)	−0.5	−12
freezing point (°C)	−139	−160

Compounds that have the same molecular formula but different structural formulas are known as **structural isomers.** 2-Methylpropane and *n*-butane are structural isomers. They are different compounds. In considering structural isomers, focus on the skeleton of the molecule. It is the differences in the sequence of bonding of the atoms that are characteristic of structural isomers.

```
C—C—C—C          C—C—C
                    |
                    C
```

n-butane skeleton   2-methylpropane skeleton

The additional hydrogen atoms can be added very easily to fill out the remaining bonding sites.

Iso-octane

One of the isomers of octane, C_8H_{18}, one of the eight-carbon hydrocarbons, is of considerable interest as a standard for the rating of fuels used in internal combustion engines. The compound is known as "iso-octane." This is an ambiguous name since there are 18 octane structural isomers, all with the molecular formula C_8H_{18}. The particular structural isomer used as a standard of performance for internal combustion

engines has the skeleton

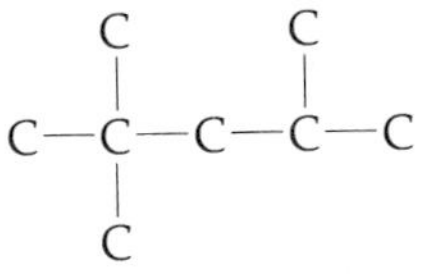

$$\begin{array}{ccccccccc} & & C & & & & C & & \\ & & | & & & & | & & \\ C & - & C & - & C & - & C & - & C \\ & & | & & & & & & \\ & & C & & & & & & \end{array}$$

and the formula

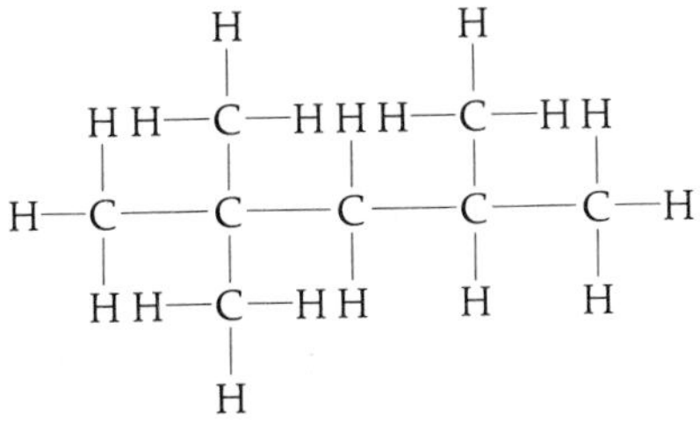

$$\begin{array}{ccccccccccccccc} & & & & & H & & & & & H & & & & \\ & & & & & | & & & & & | & & & & \\ & & H & H & - & C & - & H & H & - & C & - & H & H & \\ & & | & & & | & & | & & & | & & & | & \\ H & - & C & - & - & C & - & C & - & - & C & - & - & C & - H \\ & & | & & & | & & | & & & | & & & | & \\ & & H & H & - & C & - & H & H & & H & & & H & \\ & & & & & | & & & & & & & & & \\ & & & & & H & & & & & & & & & \end{array}$$

also written

$$\begin{array}{l} \quad\;\; CH_3 \;\; CH_3 \\ \qquad | \qquad\; | \\ CH_3CCH_2CHCH_3 \\ \qquad | \\ \quad\;\; CH_3 \end{array}$$

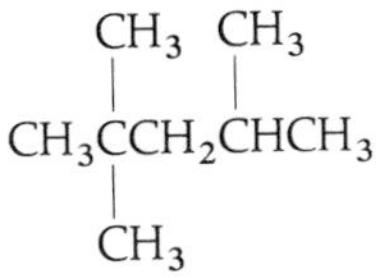

The full name of the compound is 2,2,4-trimethylpentane. The name is based upon the five-carbon chain, or *n*-pentane structure, to which the three methyl groups, three $-CH_3$ groups, are attached. The carbon atoms in the chain are numbered 1 through 5:

$$C^1-C^2-C^3-C^4-C^5$$

The 2,2,4-trimethyl- portion of the name designates the position of the three methyl groups, $-CH_3$. Two methyl groups are bonded to carbon-2 and one methyl group to carbon-4 of the pentane chain:

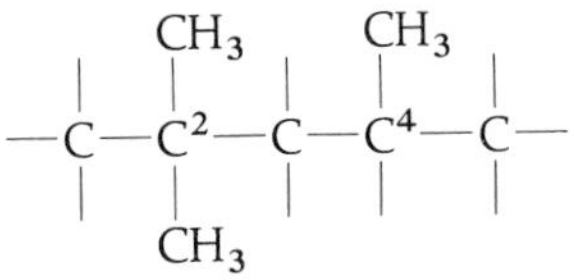

$$\begin{array}{ccccccccccc} & & & CH_3 & & & & CH_3 & & & \\ & | & & | & & | & & | & & | & \\ - & C & - & C^2 & - & C & - & C^4 & - & C & - \\ & | & & | & & | & & | & & | & \\ & & & CH_3 & & & & & & & \end{array}$$

The remaining nine bonding sites of the carbon atoms are taken up by nine hydrogen atoms. (The methyl group, $-CH_3$, derives its name from methane, CH_4.) The branched structure endows this compound with the combustion characteristics that yield a smooth operation of piston engines. The details of the combustion reactions involved are indeed complex.

Cyclic Hydrocarbons

The model also predicts that five or more carbon atoms can be fitted together to give a closed ring. Add hydrogen atoms, two per carbon atom, and the four bonding sites of each carbon atom are satisfied. For the five-carbon cyclic compound, the skeleton is

```
     C
   /   \
  C     C
  |     |
  C-----C
```

and the structural formula is

```
      H     H
       \   /
   H    C      H
    \  / \    /
     C     C
   / |     | \
  H  |     |  H
     C-----C
   /  \   /  \
  H    H H    H
```

2-4 *Background Information*

KEEPING TRACK OF POWERS OF TEN

The expression $2 \times 10^3 \times 10^5$ indicates that 2 is multiplied by 10 three times and also by 10 five more times:

$$2 \times 10^3 \times 10^5 = 2 \times 10^8$$

The rule in the multiplication of 10^a by 10^b is to add the exponents *a* and *b*. (The expression 10^a is read 10 to the *a*th power.)

$$10^a \times 10^b = 10^{a+b}$$

Either exponent *a* or exponent *b* or both exponent *a* and exponent *b* could, of course, be negative:

$$10^2 \times 10^{-8} = 10^{2+(-8)} = 10^{-6}$$

The quotient of 2×10^3 divided by 10^5 is 2×10^{-2}:

$$2 \times 10^3 \div 10^5 = 2 \times 10^{-2}$$

The rule in division is to subtract the exponent of the divisor:

$$10^a \div 10^b = 10^{a-b}$$

$$10^3 \div 10^5 = 10^{3-5} = 10^{-2}$$

Here again, *a* and/or *b* could be negative:

$$10^3 \div 10^{-5} = 10^{3-(-5)} = 10^8$$

The solution of the complex expression

$$\frac{2.4 \times 10^{-21} \times 1.2 \times 10^7}{8.0 \times 10^{-3} \times 5.0 \times 10^9}$$

is 7.2×10^{-22}. Test yourself by endeavoring to show that the 7.2×10^{-22} is correct. If you need help, follow through the following series of steps. A number of other series of steps could be equally valid.

Regroup the expression into the product of two fractions:

$$\frac{2.4 \times 1.2}{8.0 \times 5.0} \times \frac{10^{-21} \times 10^7}{10^{-3} \times 10^9}$$

Combine terms in the numerators and the denominators of the two fractions:

$$\frac{2.88}{40.0} \times \frac{10^{-14}}{10^6}$$

Evaluate each fraction:

$$0.072 \times 10^{-20}$$

Convert the 0.072 into 7.2 times 10 to a power:

$$7.2 \times 10^{-2} \times 10^{-20}$$

Combine the exponential terms:

$$7.2 \times 10^{-22}$$

If you need practice, make up expressions and solve them. In each case, find a second path—a second series of steps—and solve again. If the two paths do not give the same answer, you know that there is at least one error. If you cannot find the error or errors, you may have to ask someone to help you find what went wrong. Stick with it. You are going to need this skill.

This cyclic compound has the molecular formula C_5H_{10} and is known as cyclopentane. It is a colorless liquid (transparent in the visible region of the spectrum) that under a pressure of one atmosphere boils at 49.2 °C and freezes at −93.9 °C. Cyclopentane, C_5H_{10}, is not an isomer of the noncyclic (acyclic) pentane, C_5H_{12}. They not only have different structures, they also have different molecular formulas: C_5H_{10} versus C_5H_{12}. This difference in number of hydrogen atoms is, of course, a consequence of the bonding sites of carbon used in closing the chain to give the cyclic structure.

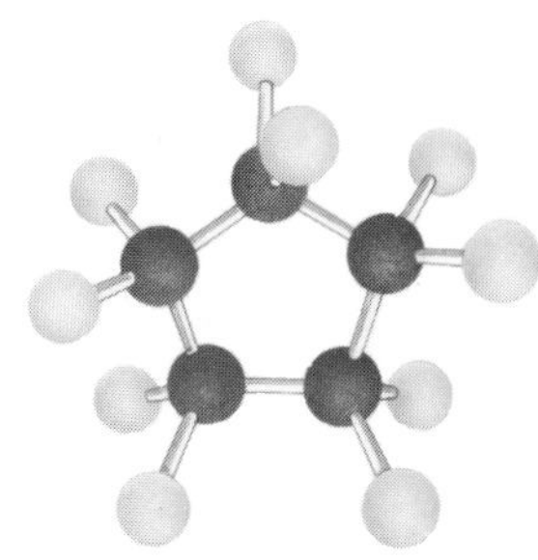

cyclopentane

All hydrocarbons have very limited solubilities in water at room temperature and are said to be hydrophobic (water-hating). All hydrocarbons absorb in the infrared and the high energy ultraviolet regions of their spectra but not in the visible portion of their spectra. All hydrocarbons are colorless.

The number of compounds predicted by the model becomes interesting when molecules containing atoms of the three elements carbon, hydrogen, and oxygen are considered. The simplest of these is a molecule that contains only one atom of carbon, one atom of oxygen, and an appropriate number of hydrogen atoms to satisfy the other bonding sites. This compound has the structural formula

```
     H
     |
  H—C—O—H      or      CH3OH
     |
     H
```

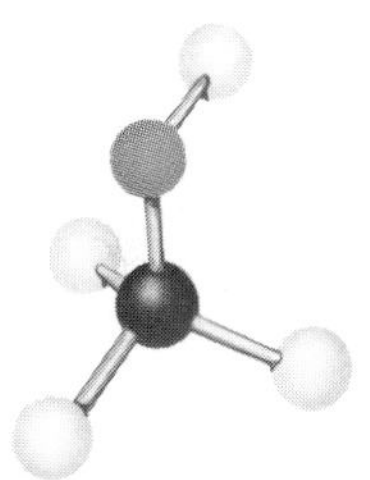

methanol

and the molecular formula CH_4O.

Alcohols

The preferred name of the compound is methanol but it is also known as methyl alcohol and wood alcohol. The first name indicates its close relation to methane, CH_4, with the suffix *-ol* indicating an alcohol with the hydroxyl group, —OH, bonded to a carbon atom. The second name combines the name of the methyl group, $—CH_3$, and the word alcohol. The last name reflects the fact that this alcohol is formed in the destruction of wood by heating in the absence of air.

The model predicts two structural isomers for the molecular formula, C_2H_6O (Figure 2-3). The two isomers are

```
    H  H                      H     H
    |  |                      |     |
 H—C—C—O—H        and      H—C—O—C—H
    |  |                      |     |
    H  H                      H     H
  CH3CH2OH                  CH3OCH3
   ethanol                dimethyl ether
```

The first is another alcohol, the second an ether. The preferred name for the alcohol is ethanol. Note the relation to ethane, CH_3CH_3. Another name, ethyl alcohol, uses the name of the ethyl group, CH_3CH_2, and the word alcohol. The third name, grain alcohol, arises from the formation of CH_3CH_2OH in the fermentation of grain. Ethanol is the alcohol associated with beverages. It is a drug—the most widely used drug throughout the word, having been discovered and used by every major civilization.

The great sensitivity of biological systems to small differences in the structure of molecules is demonstrated by the difference in response to ethanol, CH_3CH_2OH,

Figure 2-3

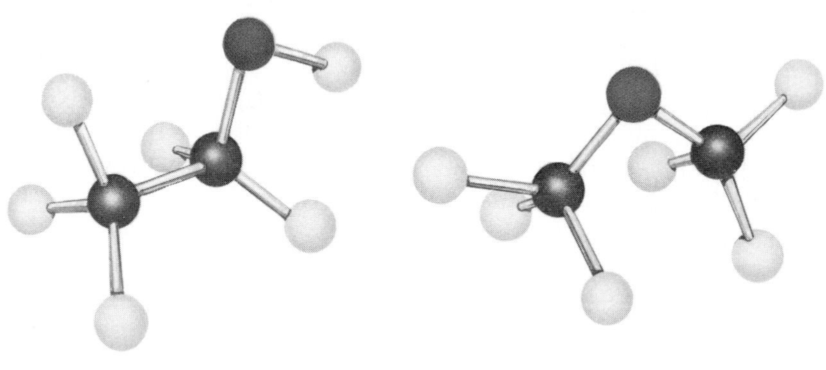

BALL AND STICK MOLECULAR MODELS OF ETHANOL AND DIMETHYL ETHER.

and methanol, CH_3OH. The consumption of small quantities of the ethanol leads to temporarily modified perception and motor control. The consumption of even smaller quantities of methanol leads to permanent blindness and, in some cases, death within a few hours.

Ethers

The other structural isomer of C_2H_6O, the ether, has the name dimethyl ether, CH_3OCH_3. The oxygen bonding the two carbon atoms, C—O—C, makes it an ether. The two methyl groups make it dimethyl ether. This is not the compound that is commonly known as ether and used as an anesthetic. That ether is diethyl ether:

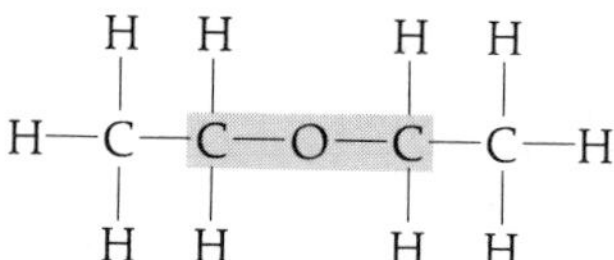

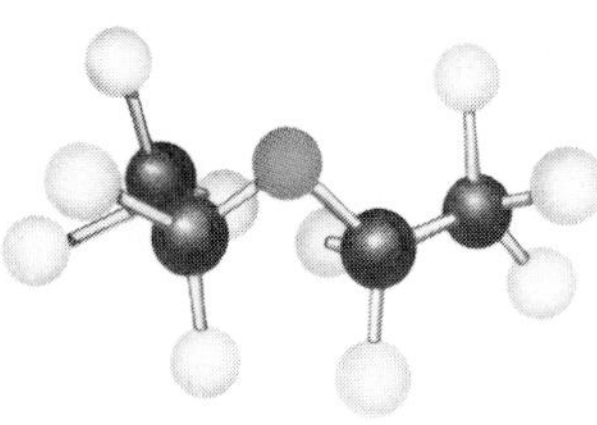

diethyl ether

also written $CH_3CH_2OCH_2CH_3$.

Some of the physical properties of the two structural isomers, ethanol and dimethyl ether, are given in Table 2-1.

Compare the boiling point and the freezing point of ethanol with the boiling point and the freezing point of dimethyl ether. Both the boiling point and the freezing point of ethanol are higher than those of dimethyl ether. The boiling point of ethanol (78.5 °C) is higher than room temperature (approximately 20 °Celsius). Therefore, ethanol is not a gas at room temperature. The freezing point of ethanol (−117.3 °C) is below room temperature. Therefore, ethanol is not a solid at room temperature. Ethanol must be a liquid at room temperature. What about dimethyl ether? The boiling point of dimethyl ether (−23 °C) is below room temperature. At room temperature, dimethyl ether is a gas. Since you do not really know what dipole moments and dielectric constants are, all that you can say about the dipole moments and the dielectric constants is that these quantities are greater for ethanol than for dimethyl

Table 2-1

PHYSICAL PROPERTIES OF THE STRUCTURAL ISOMERS OF C_2H_6O

property	ethanol CH_3CH_2OH	dimethyl ether CH_3OCH_3
boiling point (°C at one atmosphere pressure)	78.5	−23
freezing point (°C at one atmosphere pressure)	−117.3	−139
dipole moment (debye) for the gas molecule	1.69	1.30
dielectric constant of liquid at 25°C	24.30	5.02
color	colorless	colorless

ether. The molecules of the two structural isomers contain the same number of atoms of carbon, hydrogen, and oxygen. The molecule of ethanol has the same mass (weight) as the molecule of dimethyl ether. The only difference between the molecules of the two compounds is the sequence of bonding of the atoms. In a later chapter, we will look to that difference in arrangement to understand the differences in properties of the two compounds.

An **alcohol** contains the hydroxyl group —OH attached to a carbon atom. An **ether** contains an oxygen atom bonded to two carbon atoms:

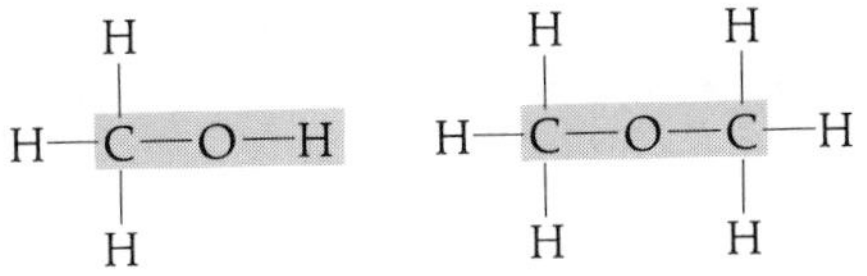

```
     H              H     H
     |              |     |
  H—C—O—H        H—C—O—C—H
     |              |     |
     H              H     H
```

Bond Angles

Bond angles and bond lengths have been given for a number of compounds. Initially, these were introduced just to indicate that chemists have been able to determine experimentally a great deal about molecules. This bond angle and bond length information is compiled in Table 2-2 and Table 2-3 for quick inspection. First, consider Table 2-2, Approximate Bond Angles. Glance quickly at the values in the body of the table. What is most obvious about those angles? The most frequently occurring value is 109° and all bond angles are within the range 100° to 110°. Look to the headings across the top of the table to identify the angle and to the list at the left of the table to identify the compound. The angle HCH, which could also be written ∠HCH, identifies the angle between one C—H bond and another C—H bond involving the same carbon atom.

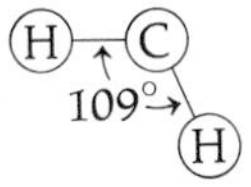

(H)—(C)
109°
(H)

For the compounds included in the table, this angle of approximately 109° is characteristic of all bond angles with a carbon atom at the apex. It is the bond angles

Table 2-2

APPROXIMATE BOND ANGLES

compounds	angles (in degrees)							
	<HOH	<HOC	<HOO	<COC	<HCH	<HCO	<HCC	<CCC
H_2O	104							
H_2O_2			100					
CH_3OH		109			109	109		
CH_3CH_2OH		109				109	109	
CH_3OCH_3				110		109		
CH_4					109			
CH_3CH_3					109		109	
$CH_3CH_2CH_3$					109		109	109

with an oxygen atom at the apex that exhibit the variation from 100° to 110°. But even these values do not depart very far from the 109° value and some are in fact 109°. The significance of 109 degrees is of great interest to chemists and we shall pursue the rationalization of this angle in later chapters.

A quick inspection of Table 2-3, Approximate Bond Lengths, reveals that the range in bond lengths, 0.096 to 0.154 nanometers, is greater than the range of bond

Table 2-3

APPROXIMATE BOND LENGTHS

compounds	lengths (in nanometers)				
	O—H	C—H	O—C	O—O	C—C
H_2O	0.096				
H_2O_2	0.096			0.148	
CH_3OH	0.096	0.110	0.143		
CH_3CH_2OH	0.096	0.110	0.143		0.154
CH_3OCH_3		0.110	0.143		
CH_4		0.109			
CH_3CH_3		0.111			0.154
$CH_3CH_2CH_3$		0.111			0.154

angles given in Table 2-2. There is also some indication that bond lengths, such as the O—H bond, have the same value irrespective of the compound within which the bond occurs. These bond lengths are

Bond Lengths

O—H bond	0.096 nm
C—H bond	0.109 to 0.111 nm
O—C bond	0.143 nm
C—C bond	0.154 nm

To have confidence that the O—C bond length is 0.143 nm irrespective of the compound, the measured O—C bond length in more than three compounds should be known. In the same sense, the information in Table 2-3 is inadequate to establish that the C—C bond length is 0.154 nm irrespective of the compounds.

What can be said of the model? Does it serve a useful purpose? What are its limitations? Is it misleading? Is it magic or does it simply regenerate the information that was the basis of the design of the model in the first place?

Critique of the Ball and Stick Model

The model makes it possible to write out the structural formulas for literally thousands of compounds of carbon, hydrogen, and oxygen and predicts the three-dimensional distribution of the atoms in these molecules. With the exception of long chains of oxygen atoms, so many of these compounds are known that it is highly probable that the others could be prepared. The hydrocarbons, the alcohols, and the ethers are important classes of compounds and the model is extremely useful.

At the same time; the model as we have used it is a very limited model in that it gives no information about the nature of the bonds that hold the atoms together in molecules; no rationalization for the characteristic number of bonding sites, no rationalization for the characteristic orientation in space of these bonding sites; no rationalization for the great differences in stability of compounds with long chains of oxygen atoms and of compounds with long chains of carbon atoms; and no rationalization for the great differences in physical properties of compounds, including the differences in physical properties of structural isomers.

The model as presented here in terms of balls and sticks of the same length is misleading in that it represents the molecules as being rigid and all bond lengths as being the same. There are also many well-known compounds, such as carbon monoxide, CO; carbon dioxide, CO_2; formaldehyde, CH_2O; acetone, C_3H_6O; acetylene, C_2H_2; and benzene, C_6H_6, not predicted by this ball and stick model.

There is absolutely no magic in the model. It is simply a generalized statement of that which experiments had shown to be true for a great number of compounds: hydrogen atoms form one bond, oxygen atoms form two bonds with a more or less characteristic orientation in space, and carbon atoms form four bonds with a characteristic orientation in space. The limitations of the model in no way detract from its usefulness as a means of approaching and classifying vast numbers of hydrocarbons, alcohols, and ethers without the intolerable burden of memorizing the formula and the structure of each compound.

The model can be modified to bring it closer to reality by using sticks of lengths proportional to the known lengths of bonds such as the O—H bond, the H—C bond, and the C—C bond. The model must be extended if it is to be useful in representing other types of compounds.

Scramble Exercises

To cope with several of the following exercises, it will be necessary to build three-dimensional models of the molecular structures being proposed. Two structures written on paper may seem to be structural isomers. This can be tested by building the two structures. If by rotation about the bonds (sticks) the structures can be manipulated into forms that can be superimposed, then the two structures are identities and not isomers. If the two structures cannot be twisted into superimposable forms, the two structures are structural isomers.

1. Write out structural formulas for at least three isomers of pentane, C_5H_{12}. (Check for identities by building the models.)

2. Write out structural formulas for at least three isomers of octane, C_8H_{18}. (Check for identities by building models, unless you are quite confident that your three proposed structures are indeed structural isomers.)

3. Point out the relation of the names ethanol and ethyl alcohol for CH_3CH_2OH to the corresponding hydrocarbon, CH_3CH_3.

4. There are two structural isomers of the three-carbon alcohol. The molecular formula is C_3H_8O. Propose structures for these two isomers. One isomer is called either normal propyl alcohol or 1-propanol. The other isomer is called isopropyl alcohol or 2-propanol. Match these names with your proposed structures. Commercial rubbing alcohol is frequently an aqueous solution of isopropyl alcohol.

5. A number of compounds have the molecular formula $C_4H_{10}O_2$. Write the structural formula for

   (a) an isomer that has two alcohol groups;
   (b) an isomer that has two ether linkages;
   (c) an isomer that has both an alcohol group and an ether linkage;
   (d) an isomer that has a peroxide linkage. (Peroxides are a class of compounds that includes hydrogen peroxide, H_2O_2. A peroxide must contain the peroxide linkage, —O—O—, with other elements satisfying the two remaining sites.)

6. Dioxane is a particular structural isomer of $C_4H_8O_2$. There is no way for you to predict the specific isomer, but do the best you can to propose the structure on the basis of the following information: dioxane is a cyclic compound (a ring compound), it contains two ether linkages, and the molecule has a high degree of symmetry.

7. Propose several other structural isomers for the molecular formula $C_4H_8O_2$ and identify alcohol groups, ether linkages, and peroxide linkages wherever they appear in the proposed structural isomers.

Additional exercises are given in the Appendix.

A Look Ahead

In the preceding pages a great number of terms and concepts have been introduced. All of these will reappear repeatedly. At this time, there is no reason to attempt to define these terms or rigorously formulate these concepts. As these terms are repeatedly used in context and as the concepts repeatedly recur in the discussion of other systems, they will become a part of your vocabulary and part of a method of thinking. Much of the language of science and many of the concepts of science can be acquired through use in much the same way other languages and other concepts are acquired.

In order to continue, it would be helpful

(1) to be able to use the terms or give examples of molecular formulas, structural isomers, and structural formulas;

(2) to be able to write the structural formulas of methane, ethane, methanol, ethanol, and dimethyl ether;

(3) to be able to recognize the methyl group, $-CH_3$; the ethyl group, $-CH_2CH_3$; the alcohol group,

$$-\overset{|}{\underset{|}{C}}-O-H$$

the ether linkage

$$-\overset{|}{\underset{|}{C}}-O-\overset{|}{\underset{|}{C}}-$$

and the peroxide linkage, $-O-O-$; and

(4) to be able to put together ball and stick models to represent a variety of molecules and to translate these structures onto paper.

It is not at all necessary to become concerned with the naming of complicated structures.

In the next chapter, a very simple extension will be made to the model. This extension is to add springs of the same diameter and length as the sticks. (In small plastic models, bent plastic links are added instead of springs.) In all other ways, the model and the ground rule remain the same. This very simple extension makes the model applicable to thousands and thousands of additional compounds. Many of these compounds are biologically significant.

3

EXTENSION OF THE MODEL OF ATOMS AND MOLECULES

Chapter 3

Ball-Stick-Spring Model

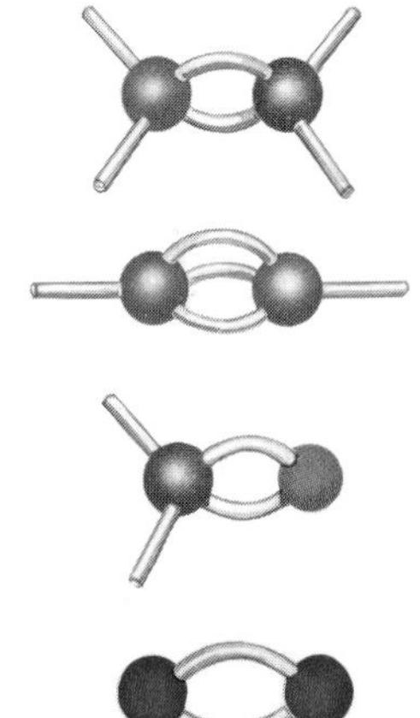

The ball and stick model explored in the previous chapter can be extended to encompass many more compounds of the elements carbon, hydrogen, and oxygen by the replacement of some of the sticks with rather stiff springs of the same diameter and the same length as the sticks. Springs are added to the model to make the model more comprehensive. These springs permit the formation of more than one bond (more than one connecting link) between any two atoms that have at least two bonding sites. These multiple bonds include

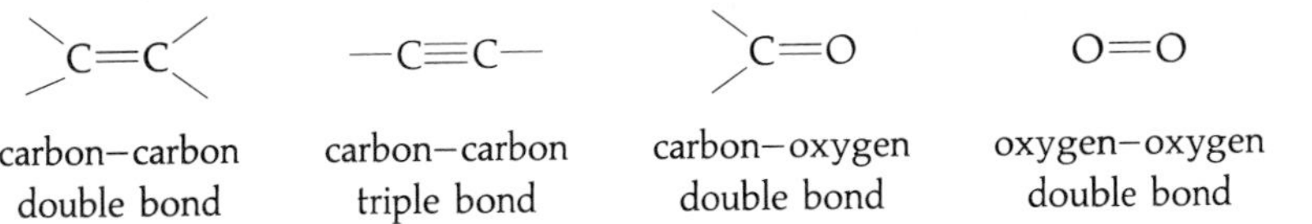

This chapter explores some of the compounds of the elements carbon, hydrogen, and oxygen that contain multiple bonds: double bonds and triple bonds.

More Compounds of Carbon and Hydrogen

UNSATURATED HYDROCARBONS

Ethene

To begin to explore the added capabilities of the model introduced by the springs, use one pair of springs to build a structure for the molecule ethene, C_2H_4. For comparison, again build the structure for ethane, C_2H_6, using only sticks. (In the small plastic sets, curved plastic linkages are supplied instead of springs.) The models of C_2H_4 and C_2H_6 are given in Figure 3-1.

The contrast between the flat C_2H_4 structure and the bulky C_2H_6 structure is striking, and a detailed inspection of the double bond structure for C_2H_4 is in order. The double bond, the two bent springs or the two bent plastic linkages, does not allow rotation about the carbon–carbon double bond and the centers of the six atoms in the molecule are constrained to a plane. The molecule C_2H_4 is said to be planar. (No molecule is really planar. The atoms of the elements take up space and it is the centers of the atoms that are constrained to the plane.)

Bond Angles in Ethene

The positions of the holes in the balls representing carbon atoms dictate that all of the hydrogen–carbon–hydrogen bond angles, ∠HCH, in ethane and the hydrogen–carbon–hydrogen bond angles, ∠HCH, in ethene be the same and have the value 109.5°. The model, however, predicts that the four hydrogen–carbon–carbon bond angles, ∠HCC, in ethene are much larger. Since the molecule is predicted to be planar, the hydrogen–carbon–carbon bond angles, ∠HCC, predicted are readily

Figure 3-1

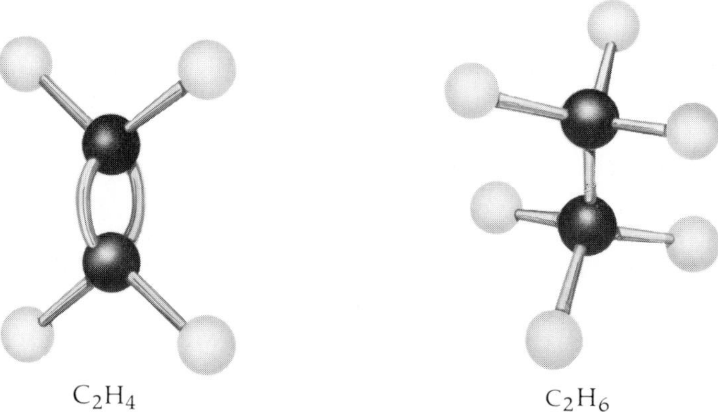

THE BALL, STICK, AND SPRING MODEL OF ETHENE, C_2H_4, AND THE BALL-STICK MODEL OF ETHANE, C_2H_6.

calculated. See Figure 3-2 for two schematic representations of the ethene molecule. The sum of the angles about a point in a plane is 360°. If one angle is 109.5° and the two other angles are equal to each other, each of these angles must be 125.2°.

The model predicts that all carbon–hydrogen bond lengths in ethane and ethene are the same and that the carbon–carbon double bond distance, C═C, in ethene is shorter than the carbon–carbon single bond distance, C—C, in ethane.

Bond Lengths in Ethene

Experimental values for the bond angles and bond lengths in the ethene molecule have been determined through the analysis of the infrared light absorbed by ethene. The molecule is planar and the carbon–carbon double bond distance, C═C, is 0.134 nm, as compared to 0.154 nm for the carbon–carbon single bond distance, C—C, in ethane. The experimental values for the bond angles in ethene are $\angle HCH = 117.3°$ and $\angle HCC = 121.4°$.

Figure 3-2

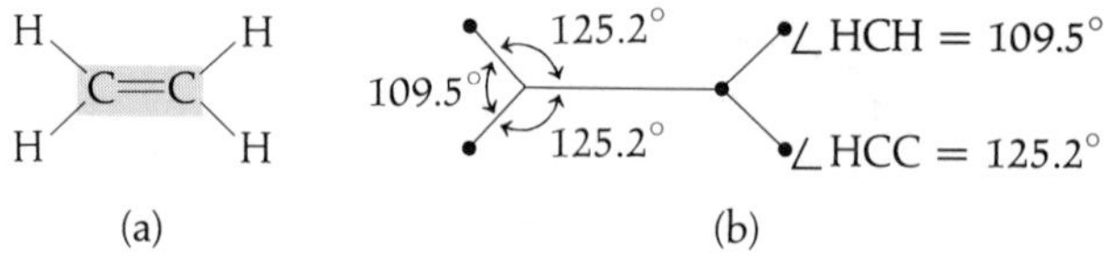

H H C═C H H

(a)

109.5° 125.2° 125.2° $\angle HCH = 109.5°$ $\angle HCC = 125.2°$

(b)

SCHEMATIC REPRESENTATION OF THE STRUCTURE OF ETHENE, C_2H_4, AS PREDICTED BY THE MODEL. (a) Showing atoms and bonds; (b) showing bond angles determined by lines connecting the centers of the atoms.

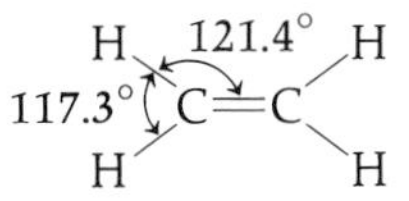

$$\begin{array}{c} H \quad 121.4^\circ \quad H \\ 117.3^\circ \; C{=}C \\ H \qquad\quad H \end{array}$$

These angles are significantly different from the 109.5° and 125.2° values predicted by the model. The experimental values for the bond angles in ethene are more nearly equal and are approximately 120°.

Saturated and Unsaturated Hydrocarbons

Ethene, also called ethylene, and ethane are, of course, hydrocarbons—compounds of the elements carbon and hydrogen. Ethane, C_2H_6, is classified as a saturated hydrocarbon and ethene, C_2H_4, with its double bond is classified as an unsaturated hydrocarbon. Ethene is not as fully saturated with hydrogen atoms as

Background Information 3-1

PHASES AND PHASE CHANGES

Gases are believed to be made up of comparatively small units called molecules that have very little attraction for each other and consequently roam comparatively free within the confines of the container. There is a great deal of space between the molecules of a gas.

At room temperature, **liquids** are believed to be made up of small units called molecules that are attracted to each other to a sufficient degree to come together in a compact fluid body that settles to the bottom of the container due to the earth's gravitational field. At high temperatures—above 300 °C—some liquids may be made up of ions. More about ions later.

Solids, on the other hand, are believed to be made up of units (either molecules or ions) whose attractive forces have brought them together in comparatively rigid bodies. Solids that are highly ordered and exhibit plane surfaces that intersect at characteristic angles are called crystals. Powders are finely divided solids and exhibit individual particles of solids when viewed with an ordinary microscope. Gases and liquids viewed in the same manner do not show individual particles and appear to be continuous. Molecules we have been discussing are too small to be seen as individual molecules by an ordinary optical microscope.

Just to know the phase of a compound at room temperature and one atmosphere pressure provides significant information about the nature of the compound. The most obvious property of any sample is its **phase.** Is the sample a solid, a liquid, a gas, or some combination of these phases? Many gases can be converted into liquids by simply compressing them into smaller volumes:

$$\text{gas} \longrightarrow \text{liquid}$$

Other gases may require a decrease in temperature as well as an increase in pressure to convert them into liquids. Under suitable conditions of pressure (high) and temperature (low), all gases can be converted into liquids. In turn, at sufficiently low temperatures, almost all liquids can be converted into solids:

$$\text{liquid} \longrightarrow \text{solid}$$

The solids so produced can be reconverted into the liquid phase and also the vapor phase (gas)

$$\text{solid} \longrightarrow \text{liquid} \longrightarrow \text{gas}$$

by reversing the conditions (increasing the temperature and/or decreasing the pressure). These phase changes are reversible. Unless otherwise specified, the classification of substances as gases, liquids, and solids is made at room conditions, about 20 °C and one atmosphere pressure.

Although all gases at room temperature and pressure may be converted to liquids and solids by suitable changes in the temperature and pressure, not all solids at room conditions can be readily converted to liquids and gases by increasing the temperature and/or reducing the pressure. Extremely high temperatures may be required and, in some cases, a compound simply decomposes into other substances under the high-energy condition of the elevated temperatures. Example: the starches decompose, char, before they melt.

ethane, and one of the chemical properties of ethene directly related to its double bond is that, under appropriate experimental conditions, ethene reacts with hydrogen gas to give ethane gas:

$$\underset{\text{ethene gas}}{\begin{matrix} H & & H \\ | & & | \\ C & = & C \\ | & & | \\ H & & H \end{matrix}} \quad + \quad \underset{\text{hydrogen gas}}{H\text{—}H} \quad \longrightarrow \quad \underset{\text{ethane gas}}{\begin{matrix} & H & & H & \\ & | & & | & \\ H\text{—} & C & \text{—} & C & \text{—}H \\ & | & & | & \\ & H & & H & \end{matrix}}$$

$$CH_2CH_2(g) + H_2(g) \longrightarrow CH_3CH_3(g)$$

The appropriate experimental conditions for this reaction to proceed are high temperature and high pressure without a catalyst. Somewhat lower temperatures and pressures are required with a catalyst. The names of **saturated hydrocarbons** have the suffix *-ane*. The names of **unsaturated hydrocarbons** with a double bond have the suffix *-ene*.

Structural Isomers of Butene

There are several structural isomers of butene, C_4H_8, the four-carbon hydrocarbon with one double bond. One of these, the four-carbon chain with the double bond between carbon-2 and carbon-3 (called 2-butene),

$$\text{—}\overset{|}{\underset{|}{C^1}}\text{—}\overset{|}{C^2}=\overset{|}{C^3}\text{—}\overset{|}{\underset{|}{C^4}}\text{—}$$

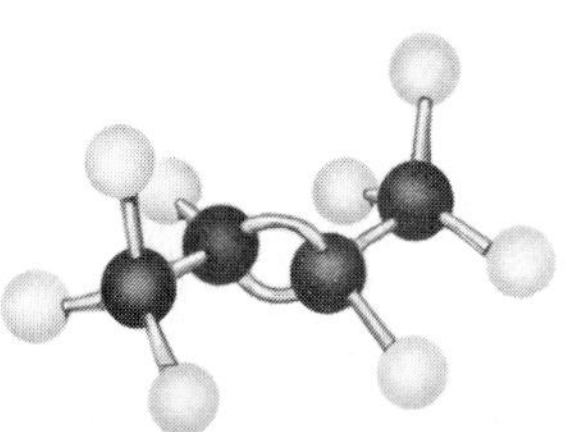

trans-2-butene

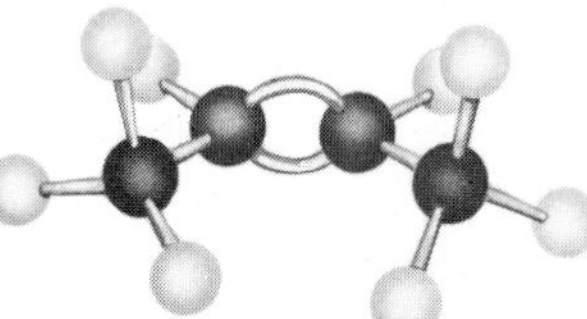

cis-2-butene

is of particular interest. To understand 2-butene, get out your model set and build a structure. For comparison, also build the structure of *n*-butane, the four-carbon, straight-chain, saturated hydrocarbon, $CH_3CH_2CH_2CH_3$. Perhaps the most obvious characteristics of the structure of 2-butene are that the four carbon atoms are not in a straight line and that the centers of the four carbon atoms and two hydrogen atoms are constrained by the double bond to a plane. A total of six atoms, the six atoms most closely related to the double bond, lie in a plane. There are two configurations of 2-butene. Endeavor to put together another configuration of 2-butene that cannot be superimposed on your first configuration. The two configurations of 2-butene are

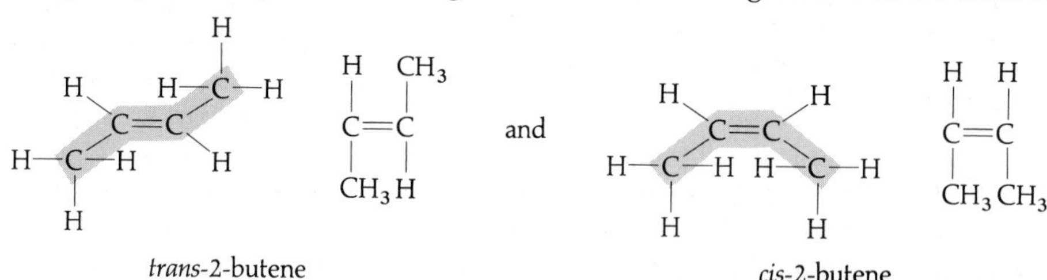

trans-2-butene and *cis*-2-butene

Note that in the first, the *trans*-2-butene, the two methyl groups are diagonally across the double bond from each other and that in the second, the *cis*-2-butene, the two methyl groups are on the same side of the double bond. These prefixes originally came from the analogy to transalpine, across the alps, and cisalpine, on the same side of the alps.

Exploration with the models of 1-butene, the four-carbon chain with the double bond between the first and second carbon atoms,

$$C^1=C^2\text{—}C^3\text{—}C^4$$

demonstrates that there is only one structure. There are no *trans* and *cis* isomers of 1-butene.

What of 3-butene, the four-carbon chain with a double bond between the third and fourth carbon atoms?

$$C^1—C^2—C^3═C^4$$

Exploration through the building of structures establishes that 3-butene is identical with 1-butene. Try it.

A comparison of physical properties of 1-butene, *cis*-2-butene, and *trans*-2-butene demonstrates that these three structures are indeed different compounds. Their melting points and boiling points at one atmosphere pressure are given below.

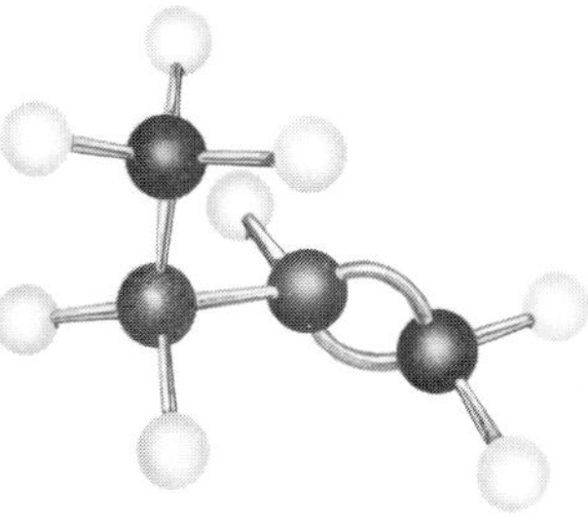

1-butene

	1-butene	***cis*-2-butene**	***trans*-2-butene**
melting point (°C)	−185	−139	−106
boiling point (°C)	−6	−4	1

Background Information 3-2

SPEED AND VELOCITY

Some of the properties of gases, liquids, and solids are believed to be closely related to motion of the molecules and the ions of which gases, liquids, and solids are composed. In discussing these phenomena, we rely heavily on the concepts of speed, velocity, momentum, and kinetic energy as developed in physics.

For the moment, we shall restrict our consideration to uniform straight-line motion: the direction of motion does not change and the speed is constant. Speed is simply the rate of change of distance:

$$\text{speed} = \frac{\text{distance traversed}}{\text{time interval}} \text{ or } \frac{\text{change in distance}}{\text{change in time}}$$

If the speed is constant, the value of the quotient

$$\frac{\text{distance traversed}}{\text{time interval}}$$

remains unchanged; for a different time interval, the distance traversed is also different but the quotient remains the same. Speed can be expressed in any combination of units of length and units of time, such as the following:

$$\frac{\text{meters}}{\text{seconds}}$$

also written meters/seconds or meters seconds^{-1}. The last two are read meters per second and meters seconds to the negative one and abbreviated m/sec and m sec^{-1}.

The average speed of molecules of oxygen in air at room temperature is of the order of 500 meters per second, also written 500 m/sec or 500 m sec^{-1}. The average speed of those same molecules of oxygen could also be expressed as 5×10^4 cm sec^{-1}. This is a bit faster than 1000 miles per hour, 1000 miles hour^{-1}. The dimensions of speed are said to be units of length divided by units of time or length divided by time (length/time or length time^{-1}).

Speed is a scalar quantity (magnitude only) and gives no information about the direction of motion. Velocity, on the other hand, is a vector quantity and specifies the direction of motion as well as the speed. A car traveling at 50 miles per hour has speed of 50 miles per hour. If the direction of motion is also specified—such as "going north"—the velocity of the car is 50 miles per hour moving to the north. There is, however, a great deal of confusion in the use of the terms speed and velocity.

The variation in these values establishes that these three structural isomers are different compounds. Note that they are all gases. (All boiling points are below room temperature.) In these compounds, the bond lengths and the bond angles are essentially those found in ethene for the region of the double bond and essentially those found in ethane for the region of the carbon–carbon single bonds.

Another unsaturated four-carbon hydrocarbon, 1,3-butadiene, is of structural interest and commercial value. The molecules of this compound contain two double bonds, hence the *-diene* suffix. The numerals 1 and 3 designate the positions of the double bonds.

1,3-Butadiene

To understand the structure of the molecules of this compound, bring out the models again. Several representations of 1,3-butadiene are given below.

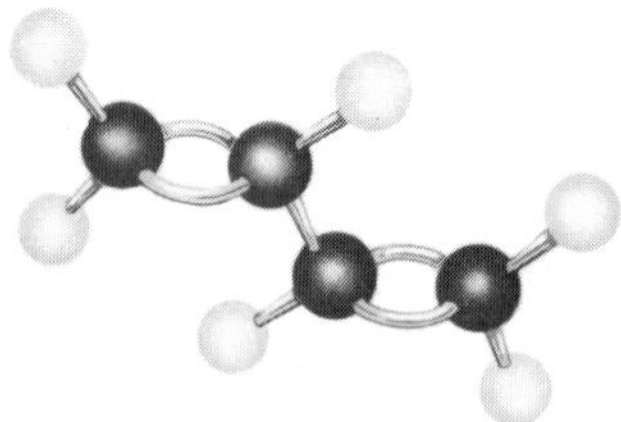

1,3-butadiene

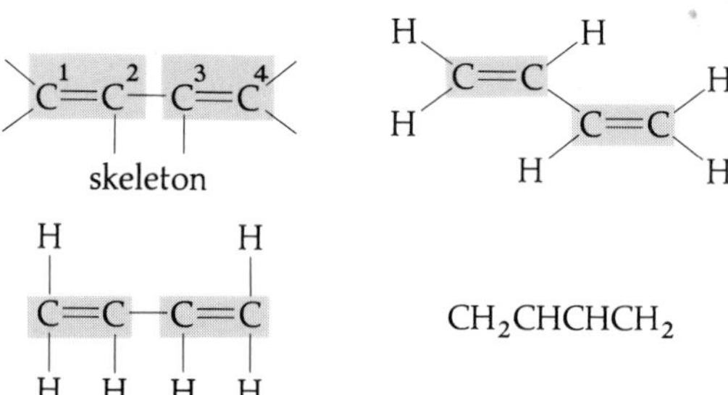

Note that the presence of double bonds in $CH_2CHCHCH_2$ is signaled by the number of hydrogen atoms on the various carbon atoms. Compare with the $CH_3CH_2CH_2CH_3$ representation for the corresponding saturated butane. 1,3-Butadiene is explored further in Scramble Exercise 9.

Triple Bonds

Ethyne

The simplest hydrocarbon with a carbon–carbon triple bond, C≡C, is ethyne, C_2H_2. Note the suffix *-yne* indicating the triple bond. This compound is also commonly called acetylene. The structure of ethyne is explored in Scramble Exercise 10.

In summary, hydrocarbons of the types explored in Chapter 2 and 3 are known as alkanes, alkenes, and alkynes.

class	characteristic	naming
alkanes, saturated	only single bonds, C—C	suffix *-ane* as in ethane, CH_3CH_3
alkenes, unsaturated	at least one carbon–carbon double bond, C═C	suffix *-ene* as in ethene, CH_2CH_2
alkynes, unsaturated	at least one carbon–carbon triple bond, C≡C	suffix *-yne* as in ethyne, CHCH

Some unsaturated hydrocarbons are extensively used as starting compounds (called feed stocks) in the commercial production of synthetic fibers, plastics, and rubbers. Examples: polyethylene and butadiene rubber. (More about this in Chapter 21, Synthetic Polymers.) The commercial source of hydrocarbons, both saturated and unsaturated, is petroleum.

Background Information 3-3

ACCELERATION

Acceleration has to do with the rate at which the velocity of motion changes with time. For this discussion, we will restrict our consideration to motion along a straight line—no change in direction. An object at rest may acquire an increasing velocity of motion if the object is continuously subjected to a net force along the line of motion.

Acceleration is the rate of change of velocity:

$$\text{acceleration} = \frac{\text{change in velocity}}{\text{change in time}}$$

If the acceleration is continuously changing, the above expression can give only an average acceleration for the period of time used in the denominator.

Since velocity is itself a rate of change of distance with time,

$$\text{velocity} = \frac{\text{change in distance}}{\text{change in time}}$$

acceleration is a rate of change of a rate of change.

Most students find acceleration a very difficult concept. Mathematically, acceleration is best approached through calculus. In the next paragraph, the paragraph that is bracketed, the language of calculus is used. If you do not have the background to understand this language, read the paragraph anyway. Read the paragraph for fun. It is always interesting to meet a new language. In this book we do not need to calculate acceleration, but some understanding of the concept of acceleration will be useful.

[Velocity, v, is the first derivative of distance, x, with time, t, as shown here:

$$v = \frac{dx}{dt}$$

Acceleration, a, is the first derivative of velocity, v, with time, t:

$$a = \frac{dv}{dt}$$

Consequently, acceleration, a, is the second derivative of distance, x, with time, t:

$$a = \frac{d^2x}{dt^2}$$]

Our concern at the moment is not with the numerical evaluation of acceleration but with the units used to express acceleration.

Velocity is measured in units of length divided by units of time:

$$\frac{\text{units of length}}{\text{units of time}}$$

Acceleration is measured in units of velocity divided by units of time:

$$\frac{\text{units of velocity}}{\text{units of time}}$$

Consequently, acceleration is measured in units of length divided by units of time twice:

$$\frac{\text{units of length}}{(\text{units of time})^2}$$

For example, acceleration can be measured in meters per second per second,

$$\frac{\text{meters}}{\text{seconds}^2}$$

also written meters/seconds^2 (m/sec^2) or meters seconds^{-2} (m sec^{-2}). Since velocity is a vector quantity, acceleration also has direction as well as magnitude.

More Compounds of Carbon, Hydrogen, and Oxygen

The extended model encompasses the carbon–oxygen double bond, C=O, and predicts great numbers of compounds, many of which are of biological significance. The configuration of atoms

$$\gt C{=}O$$

Carbonyl Group

is known as the **carbonyl** group. Compounds containing this group are known as aldehydes, ketones, acids or esters, depending upon the atoms or groups of atoms attached at the two other bonding sites of carbon in the carbonyl group. All four of these classes of compounds are unsaturated compounds. All four of these classes of compounds are important biologically and commercially.

ALDEHYDES

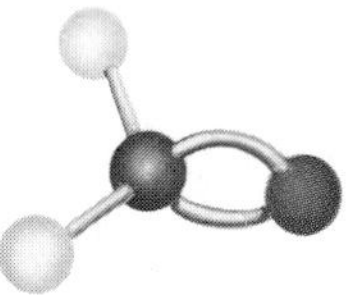

formaldehyde

The two simplest compounds of carbon, hydrogen, and oxygen involving two bonds between a carbon atom and an oxygen atom are

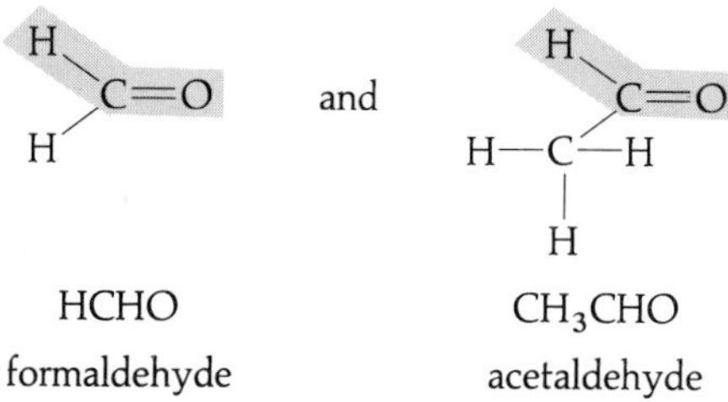

$HCHO$ formaldehyde and CH_3CHO acetaldehyde

Note the manner in which the characteristic aldehyde group

$$H{-}C{=}O$$

is written in the structural formulas $HCHO$ and CH_3CHO. To understand the configuration of these molecules, build the structures.

Formaldehyde

Formaldehyde is a planar molecule: the centers of the four atoms lie in a plane. The experimental values for the bond angles are ∠HCH 118° and ∠HCO 121°. Note that these are not the 109.5° for ∠HCH and 125.2 for ∠HCO predicted by the model (see Scramble Exercise 2) but they are very similar to the bond angles found experimentally in ethene. In the formaldehyde molecule, the bond angles around the carbon of the carbonyl group are approximately equal, approximately 120°. The average carbon–hydrogen bond length is again 0.107 nm. The carbon–oxygen bond distance is 0.123 nm—short in comparison to the 0.143-nm bond length found for carbon–oxygen single bond lengths in alcohols and ethers.

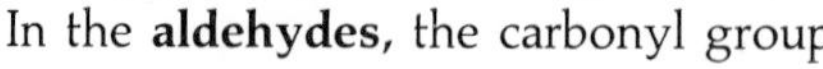

In the **aldehydes**, the carbonyl group

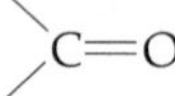

is always bonded to a hydrogen atom and either to another hydrogen atom, as in formaldehyde, or to a carbon atom, as in acetaldehyde.

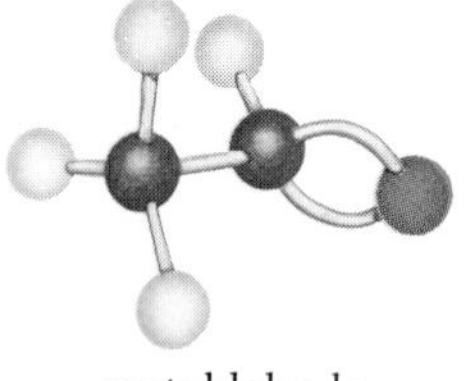

acetaldehyde

KETONES

The two simplest ketones are

CH_3COCH_3 or $(CH_3)_2CO$

acetone
or
dimethyl ketone
or
2-propanone

and

$CH_3CH_2COCH_3$

ethyl methyl ketone
or
2-butanone

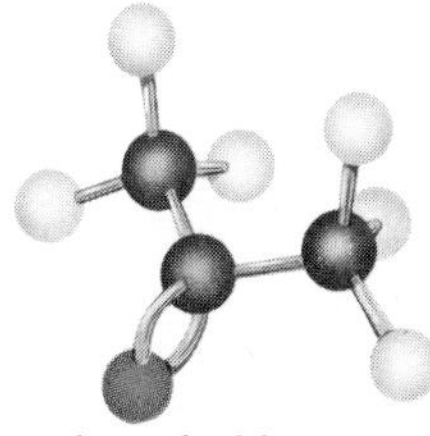

dimethyl ketone

The models predict that the four atoms

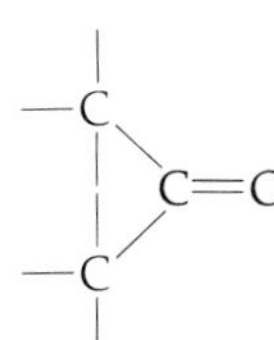

are restricted to a plane. Experimentally, the bond angles around the carbon atom of the carbonyl group are approximately 120°. The carbon–oxygen double bond distance is approximately that found in formaldehyde.

In the **ketones,** the carbonyl group

$>C{=}O$

is always bonded to two carbon atoms, and the technical name of the compound ends in the suffix *-one.*

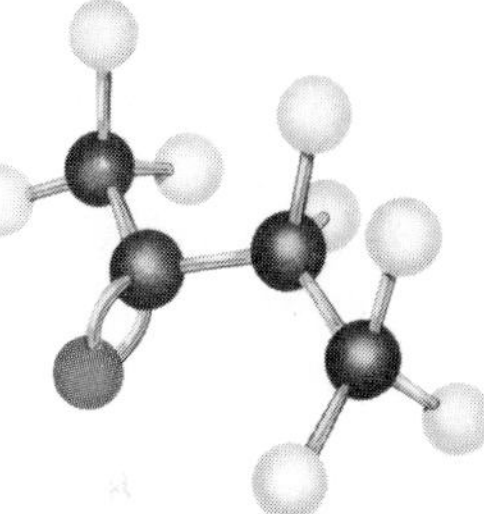

ethyl methyl ketone

CARBOXYLIC ACIDS

The three simplest carboxylic acids are

$HCOOH$
formic acid

CH_3COOH
acetic acid

CH_3CH_2COOH
propionic acid

formic acid

acetic acid

propionic acid

The characteristic group in these acids is the carboxylic acid group:

Carboxylic Acid Group

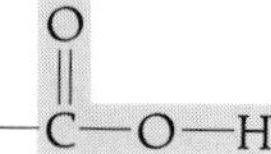

Learn to look for this combination of atoms and take care not to confuse this group with some combination of the characteristic groups of aldehydes, ketones, alcohols, and peroxides. As shown in the two examples, either a hydrogen atom or a carbon atom may be bonded to the fourth bonding site of the carbon atom in the carboxyl group. Acids of this type are known as **carboxylic acids.** The carboxylic acid group

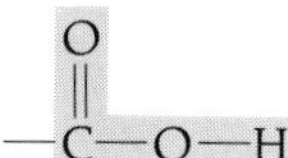

is also written —COOH. Do not confuse this symbolism with the symbolism for the peroxides (Chapter 2). Count up the bonding sites for each atom and it is quite clear that —COOH cannot be a peroxide. If it were a peroxide, two of the four bonding sites of carbon would be unaccounted for.

3-1 *Editorial Comment*

OBSERVATIONS, MEASUREMENT, AND SCIENTIFIC INFORMATION

Observations are made through the use of the senses. Measurements are observations made through the use of instruments. Instruments such as rulers and thermometers can be very simple. Instruments such as spectrophotometers, mass spectrometers, and gas chromatographs can be quite complex.

In scientific work, it is much more informative to know the melting point of a compound in degrees Celsius than to know that some individual considers the melting point to be high. It is much more informative to know the absorption spectra of a solution than to know that some individual considers the solution to be blue.

The experimental bond lengths and bond angles given in Chapters 2 and 3 were not measured with rulers and compasses. Molecules are much too small for those instruments. The experimental bond lengths and bond angles in many cases have been calculated from measurements made with spectrophotometers of the infrared light absorbed by the various compounds. Most chemists understand in principle how bond lengths and bond angles are determined experimentally, but most chemists do not make these types of measurements. Measured quantities, such as bond lengths and bond angles, are determined by specialists and the validity of the measurements is checked by specialists. Most chemists accept reported values and use them in their work until they have reason to question the validity of the reported values. If evidence is obtained that indicates a reported value is in error, steps must be taken to redetermine the quantity. The value is challenged and reevaluated by either the investigator who challenged the value or by another investigator who has access to the appropriate equipment and is a specialist in the interpretation of the data obtained.

All physical, chemical, and biological properties of all elements and compounds must be experimentally determined. Such measurements go on continuously in laboratories throughout the world as a means of acquiring the capability to control chemical processes and produce products with desired physical, chemical, and biological properties or as a means of testing evolving models in the search to extend our understanding of the nature of matter. Although these goals for research may arise from quite different motives, the results of the investigations are mutually supportive and the results are published in scientific journals,

ESTERS

Three of the simplest esters are

```
    O     H             H   O       H             H   O       H   H
    ‖     |             |   ‖       |             |   ‖       |   |
H — C — O — C — H   H — C — C — O — C — H   H — C — C — O — C — C — H
          |             |           |             |           |   |
          H             H           H             H           H   H
```

$HCOOCH_3$ — methyl ester of formic acid, or methyl formate

CH_3COOCH_3 — methyl ester of acetic acid, or methyl acetate

$CH_3COOCH_2CH_3$ — ethyl ester of acetic acid, or ethyl acetate

Note the relation of each of the above esters to each of the corresponding acids given below.

```
    O                H   O                H   O
    ‖                |   ‖                |   ‖
H — C — O — H    H — C — C — O — H    H — C — C — O — H
                     |                    |
                     H                    H
```

formic acid — acetic acid — acetic acid

---

unless there are believed to be national security advantages or economic advantages or personal advantages in maintaining secrecy. An ethic of scientists is that the productivity of scientists is enhanced by the open exchange of scientific knowledge.

The American Chemical Society monitors approximately 12 thousand scientific journals and scientific proceedings published annually and abstracts approximately 400 thousand chemically related articles. Most of these articles are, of course, published in chemical journals, but many are published in journals of physics, biology, medicine, agriculture, astronomy, geology, psychology, art, engineering, and technology. The abstracts are made accessible to the public worldwide through the American Chemical Society publication *Chemical Abstracts* and also through the American Chemical Society computer bank. The *Chemical Abstracts* registry now lists more than seven million chemical substances—including mixtures.

Offsetting the challenge of keeping pace with the great volume and diversity of chemical knowledge that is generated each year is a growing number of annual review publications and comprehensive symposia presented at scientific meetings.

Most scientists think in terms of generalizations (models) and depend heavily on handbooks for the numerical values of specific properties of elements and compounds. In handbooks, compounds are listed alphabetically and it is convenient to be at home with the nomenclature established by the International Union of Pure and Applied Chemistry, IUPAC. However, an adequate background in vocabulary is gradually acquired through use and there is no virtue in rushing off to memorize rules. *Chemical Abstracts* and many handbooks also include formula indices listed in order of increasing number of carbon atoms. For example, under C_2H_6O, all isomers having this molecular formula are cited and cross referenced by name to the various isomers. The number of known compounds now exceeds four million.

Essentially all—perhaps all—numerical values given in this book can be found in handbooks. A chemist working in a specific area comes to know a number of the numerical values frequently used. In most courses, students are not expected to memorize very many numerical values. Some you will learn through repeated use. Example: the number of bonding sites for carbon, hydrogen, and oxygen and a few frequently reoccurring bond angles—possibly some bond lengths.

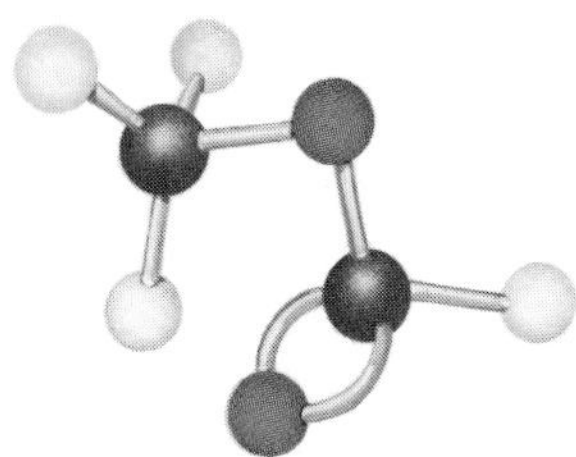

methyl formate

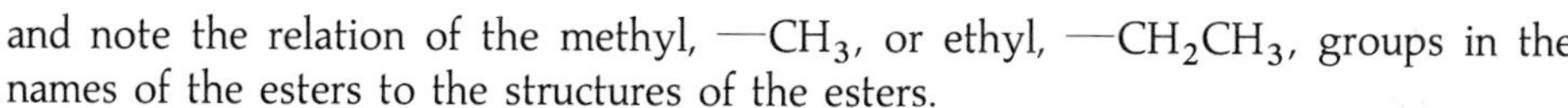

and note the relation of the methyl, $-CH_3$, or ethyl, $-CH_2CH_3$, groups in the names of the esters to the structures of the esters.

The esters are very similar to the acids with the exception that the hydrogen atom of the carboxylic acid group

$$-\overset{\overset{\displaystyle O}{\|}}{C}-O-H$$

is replaced by the carbon atom of a methyl group or some other similar group. In **esters,** the characteristic group of atoms is

$$-\overset{\overset{\displaystyle O}{\|}}{C}-O-\overset{|}{\underset{|}{C}}-$$

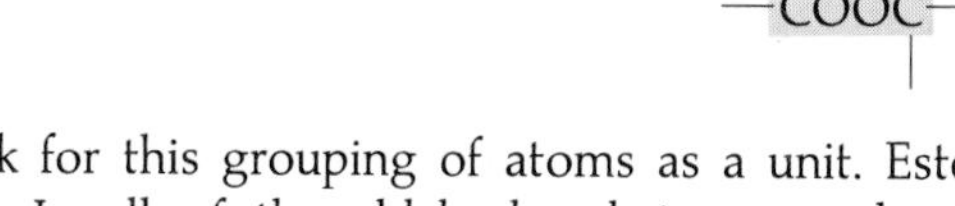

methyl acetate

also written

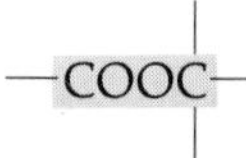

$$-COO\overset{|}{\underset{|}{C}}-$$

Look for this grouping of atoms as a unit. Esters are neither ethers nor peroxides.

In all of the aldehydes, ketones, carboxylic acids, and esters, experimental measurements show that four atoms are constrained to a plane. These are the oxygen atom and the carbon atom of the carbonyl group, and the two atoms connected to the carbon atom of the carbonyl group. Experimental measurements also establish that the three bond angles at the carbonyl carbon are approximately 120°.

$$120^\circ \; C{=}O \; 120^\circ \; 120^\circ$$

The carbon–oxygen double bond length is approximately 0.123 nm (as compared to the carbon–oxygen single bond distance in alcohols of 0.143 nm).

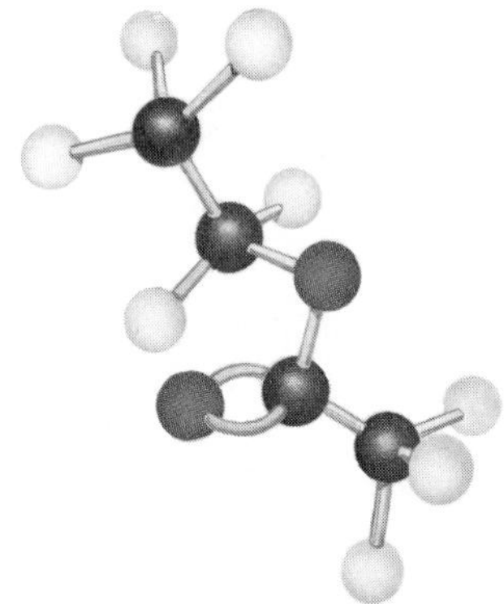

ethyl acetate

Aldehydes, ketones, acids, and esters are frequently encountered: aqueous solutions of formaldehyde have been extensively used to preserve biological samples; acetone is frequently used as a solvent to remove paints and varnishes, including nail polish; acetic acid is the principal ingredient other than water in vinegar; white vinegar is frequently an aqueous (water) solution of acetic acid; the ethyl ester of acetic acid and many other esters are responsible for the "fruity" odors of natural products.

Assessment of the Model

The model predicts a vast array of compounds made up of just the three elements carbon, hydrogen, and oxygen. The number of compounds seems to be without limit, ranging in size from molecules of just a few atoms to molecules containing thousands of atoms made up of various combinations of long chains of carbon atoms, branched chains of carbon atoms and closed rings of carbon atoms, sprinkled with double bonds and triple bonds along with oxygen-containing groups and linkages. These compounds encompass many of the vast array of compounds synthesized by pharmaceutical houses and other chemical industries and include many of the plastics and synthetic fibers, such as the polyethylenes and the polyesters, that are produced in such tremendous quantities. This production of compounds of carbon, hydrogen, and oxygen by industry is dwarfed by the production of compounds such as fats, sugars, starches, and cellulose that goes on continuously in the biological world. All of these are compounds of the three elements carbon, hydrogen, and oxygen.

The ball-stick-spring model is a very convenient guide to the structural formulas of a huge number of compounds, although it is incapable of predicting the 120° bond angle that repeatedly occurs with carbon–carbon double bonds (unsaturated hydrocarbons) and with carbon–oxygen double bonds (carbonyl groups, carboxylic acids, and esters). The model is also misleading in that the model shows the two bonds (the two springs) of a double bond to be equivalent. The interpretation of the ultraviolet absorption spectrum of unsaturated hydrocarbons and carbonyl compounds indicates that the two bonds of a double bond are in fact quite different. More about this later.

A few well-known molecules are also a contradiction to the predictions of the model. There is no way to rationalize (to understand) carbon monoxide, CO, in terms of the ball, stick, and spring model. (Carbon dioxide, CO_2, is no problem.) Another troublesome molecule is the oxygen molecule, O_2, itself. Oxygen seems to fit very nicely with a double bond between the two oxygen atoms, but, as we shall see later, the magnetic properties of liquid oxygen indicate that the bonding between the atoms is not the double bond predicted by this model.

The model does not address the nature of the bonds themselves. Certainly they are not wooden sticks, wire springs, or plastic links several centimeters long. Presumably, an understanding of the nature of the bonds must be sought in the structure of the atoms themselves. The ball, stick, and spring model is an extremely useful model—also a limited model.

Scramble Exercises

1. The interpretation of the infrared absorption spectra of carbon dioxide, CO_2, indicates that the carbon dioxide molecule is a linear molecule ($\angle OCO = 180°$). Is this in keeping with or contradictory to the shape of the molecule that is predicted by the construction of a ball and spring model? Build the model of this molecule.

2. Show that the model predicts 125° for the hydrogen–carbon–oxygen angle, $\angle HCO$, in the formaldehyde molecule, HCHO.

3. Propose structural formulas for two acyclic (noncyclic) isomers each of (a) C_2H_6O, (b) C_2H_4O, and (c) $C_2H_4O_2$. Classify each as to type of compound (alcohol, ether, peroxide, aldehyde, ketone, acid, or ester). Some of the structures that can be proposed are extremely unstable compounds, but that is another question: What determines the stability of compounds? This question is not addressed by the model.

4. Propose structural formulas for methyl amine, CH_3NH_2, and for ethyl amine, $CH_3CH_2NH_2$. In these compounds, nitrogen atoms use three bonding sites. Point out the similarities between alcohols and amines and relate differences in these compounds of oxygen and nitrogen to the position of the two elements in the Periodic Table. See A Look Ahead (page 50) for help with this one. If you do not have a ball with three holes for the nitrogen atom, substitute a ball normally used for the carbon atom and use only three of the four holes.

5. Predict the formula of dimethylamine and write out its structure. It is an isomer of ethylamine. Point out the similarity of this compound to dimethyl ether; also point out the difference or differences between the two compounds.

6. The formulas of glucose and fructose are frequently written

```
                                   H
                                   |
   H—C=O                        H—C—OH
     |                             |
   H—C—OH                          C=O
     |                             |
  HO—C—H                       HO—C—H
     |            and              |
   H—C—OH                       H—C—OH
     |                             |
   H—C—OH                       H—C—OH
     |                             |
   H—C—OH                       H—C—OH
     |                             |
     H                             H
  glucose                       fructose
```

Both of these have the molecular formula $C_6H_{12}O_6$ and are structural isomers. Classify each of these isomeric hexoses (six-carbon sugars) in terms of the oxygen-containing groups. (Are the isomers acids, alcohols, or aldehydes, for example, or some combination of these?) One of the isomers is frequently classified as an aldose sugar (an aldehyde sugar) and the other as a ketose sugar (a ketone sugar).

7. Propose the structural formula for a carboxylic acid that has the molecular formula $C_3H_6O_2$. Write out the structural formula for an ester that has the molecular formula $C_3H_6O_2$. There are two esters. One is used as a nail polish remover.

8. Propose structural formulas for two other isomers predicted by the model for $C_3H_6O_2$. There are a great number of isomers, particularly if carbon–carbon double bonds and ring compounds, with oxygen either in or outside the ring, are included. Build the models to check out the structures proposed.

9. In 1,3-butadiene there are two carbon–carbon double bonds. Each of these double bonds requires that the centers of six atoms lie in a plane. Identify the two sets of six atoms that are planar. The angle between two planes is called the **dihedral**

angle. Does the model impose any restrictions upon this angle in 1,3-butadiene? Check your conclusion by building a ball, stick, and spring model.

$$\begin{array}{c} \text{H} \quad\quad\quad \text{H} \\ \text{C}=\text{C} \quad\quad \text{H} \\ \text{H} \quad\quad \text{C}=\text{C} \\ \text{H} \quad\quad\quad \text{H} \end{array}$$

10. The simplest of the carbon–carbon triple bond compounds is acetylene, or ethyne, C_2H_2. Experimental measurements indicate that the centers of the four atoms lie in a straight line, and the molecule is said to be linear. What does the ball and stick model predict? The experimental carbon–carbon triple bond length is 0.120 nm.

11. One of the most interesting compounds of carbon and hydrogen is benzene, C_6H_6. This molecule is frequently treated as a six-carbon cyclic compound with alternating double and single carbon–carbon bonds around the ring. Write a structural formula that is consistent with the model and predict how many of the twelve atoms lie in a single plane. Check your prediction by building the model. Benzene has been extensively used in industrial processes and has been a component of many consumer products, such as shoe polish, paint thinner and unleaded gasoline. Concern over toxic responses to exposure to this compound has led to severe regulation of the level of exposure to this compound in the work place and of its use in consumer products. One exception to this regulation: the percent of benzene in gasoline is not restricted and the greatest risk of exposure to benzene now occurs at filling stations. Some benzene is a natural component of gasoline prepared from petroleum. The benzene content in gasoline was increased when regulations called for the level of tetraethyl lead, $Pb(CH_2CH_3)_4$, in gasoline to be reduced—before the toxicity of benzene was recognized. Both benzene and tetraethyl lead improve the combustion properties of hydrocarbons in piston engines. Both pose biological risks.

    The properties of benzene, C_6H_6, are so unusual, it is not considered to be an alkene—a carbon–hydrogen compound containing at least one double bond. Benzene and related compounds are known as aromatic hydrocarbons.

Additional exercises for Chapter 3 are given in the Appendix.

A Look Ahead

You should now have in hand the concepts and skills to enable you

- to propose isomeric structures for a given molecular formula containing carbon, hydrogen, and oxygen;
- to identify various types of compounds, given the structural formulas as acids, alcohols, aldehydes, esters, ethers, saturated and unsaturated hydrocarbons, or combinations of these;
- to identify *cis* and *trans* isomers of alkenes;

- to point out structures that require more than three atoms lie in the same plane and the bond angles characteristic of these planar configurations and;
- to compare the bond lengths of single bonds and multiple bonds involving the same atoms.

In the next chapter, we shall explore rapid reactions of oxygen gas with compounds of carbon, hydrogen, and oxygen. These rapid reactions are also known as combustion and as fire.

In succeeding chapters we shall explore the structures of atoms in order to explore the nature of chemical bonds within molecules; the nature of the forces between molecules that determine whether a compound is a gas, a liquid, or a solid; and the nature of the forces that determine the solubility of a compound in solvents—including water but in no sense limited to water.

The compounds that have large molecules can be depended upon to occur in the solid phase and in many cases to decompose into smaller molecules as the temperature is raised, before the temperature reaches a melting point. The smaller the molecule the greater the probability the compound will be a gas. The situation is not that simple, however. Structural isomers may have quite different boiling points and freezing points. These facts raise questions about the nature of the forces between molecules that hold molecules together in the fixed orientations of the solid phase, or hold molecules together without this fixed orientation in the liquid phase, while the molecules of a gas or vapor phase roam free.

Periodic Table

Perhaps the most useful correlation in chemistry is a table of the elements known as the *Periodic Table*. For easy and frequent accessibility, this table has been placed inside the front cover of this book. It contains all the known elements arranged in a pattern such that elements of similar properties fall in vertical columns while the order of the elements in the rows maintains an approximate sequence of increasing masses of the individual atoms. This grouping of elements results in a table that is bottom heavy, with the first row (period) containing only 2 elements, the second and third periods (rows) containing only 8 elements, and the fourth and fifth periods containing 18 elements. The sixth period contains 32 elements and the seventh period builds toward 32 elements. Such very long periods are difficult to represent on paper from the standpoint of type size, and it is conventional to cut out a block of 14 elements in the sixth and seventh periods and set these at the bottom of the table. The number of elements in the seventh period (row) depends entirely on how far nuclear scientists have proceeded in making and proving the identity of new elements. Prior to the rapid advances in the 1940's, the known elements ended with uranium, $Z = 92$ (symbol U). The numbers of elements (2, 8, 18, and 32) in the various periods of the Periodic Table are anchor points for any model that deals with the structure of atoms.

The three elements that have been so much a part of this discussion are among the lighter elements: hydrogen in the first period; carbon and oxygen in the second period. Three other elements of this region of the Periodic Table are also extremely important to biological systems. Nitrogen, which falls between carbon and oxygen in the second period, frequently exhibits three bonding sites and forms compounds such as ammonia, NH_3; methylamine, CH_3NH_2, dimethylamine, $(CH_3)_2NH$; and trimethylamine, $(CH_3)_3N$. Sulfur falls below oxygen in the third period and forms compounds such as hydrogen sulfide, H_2S (H_2O chemically speaking is hydrogen oxide); thioalcohols, CH_3CH_2SH; and thioethers, CH_3SCH_3. The prefix *thio-* is derived from the Greek word for sulfur. Nitrogen atoms are a part of all amino acids and are consequently present in proteins. Sulfur atoms also play an important role in many proteins and in the vulcanization of rubber. Phosphorus, third period and below nitrogen, is an essential part of the nucleic acids, RNA and DNA. Although phosphorus does form compounds using three bonding sites—for example phosphine, PH_3—it does not appear in this form in biological materials. Five bonding sites are much more common for phosphorus. Nitrogen also frequently forms compounds through the use of five bondings sites, as well as the three bonding sites used in ammonia, NH_3, and the amines, such as ethylamine, $CH_3CH_2NH_2$.

In general, the elements appearing in vertical columns have similar properties while the elements appearing in horizontal rows have properties that change continuously from one end of the period to the other end of the period. Watch for similarities and differences. Generalizations are interesting; they are also very useful in understanding the nature of the chemical world and reducing the burden of memorizing facts.

As you work with various atoms in class and in the laboratory, mark their positions in the Periodic Table. You can use what you learn about these elements to understand other elements.

4
FIRE

Chapter 4

The use of fire to prepare food, to provide warmth, and to clear the undergrowth of forests has been an integral part of early civilizations. The combustion of fuels, primarily the fossil fuels (natural gas, petroleum, and coal), continues to be the chief source of energy to propel vehicles and to carry on industrial processes. In some countries, the combustion of wood and other biomass materials is still the primary source of heat energy. It is estimated that it was not until 1972 that the energy produced by nuclear power stations exceeded that obtained from the combustion of wood.

The flame has a strong emotional impact and has taken on a great variety of symbolic meanings throughout time. It is the symbol of youth and growth, the symbol of leadership and the pursuit of excellence, the symbol of knowledge and the pursuit of truth, the symbol of life, the symbol of hope and eternal life—as well as the symbol of clean heat. Multiple flames, particularly large flames, against a dark and chaotic background have quite a different emotional impact: fire is also the symbol of destruction, disaster, war, and the fate of the damned.

Reaction of Hydrogen Gas, H_2, with Oxygen Gas, O_2

Combustion of Hydrogen Gas in Air

Hydrogen gas, H_2, flowing slowly through a small-diameter metal or glass tip into the air can be ignited with a match. The chemical reaction continues at the orifice with an almost colorless but a very hot flame. The product is water in the vapor phase. This reaction of hydrogen gas, H_2, with oxygen gas, O_2, under appropriate conditions to give water, heat, and light can be expressed more briefly using symbols

$$\text{hydrogen gas} + \text{oxygen gas} \longrightarrow \text{water vapor} + \text{energy}$$

$$H_2(g) + O_2(g) \longrightarrow H_2O(g) + \text{energy}$$

where "energy" is used as a collective term for the two forms of energy: heat and light.

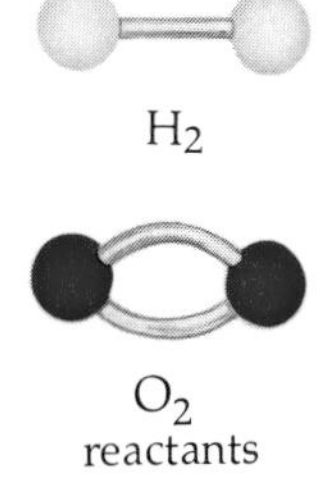

reactants

Clearly this chemical reaction involves breaking the bonding of the two atoms of oxygen in the oxygen molecule, O_2: the bonds between these two atoms must be broken if only one atom of oxygen ends up in each water molecule. The two atoms of oxygen from a molecule of oxygen gas, O_2, must find their homes in different molecules of water. Each molecule of oxygen, O_2, leads to the formation of two molecules of water, H_2O. If the earlier model of the water molecule

$$\begin{array}{l} H-O \\ \quad\;\; | \\ \quad\;\; H \end{array}$$

is accepted, the bond between the two hydrogen atoms in the hydrogen molecule, H_2, must also be broken to form the new bonds between a hydrogen atom and an oxygen atom. The two hydrogen atoms in a molecule of water do not necessarily come from the same hydrogen molecule. Whatever the nature of bonds in molecules, bonds are broken and bonds are formed in the chemical reaction of hydrogen gas, H_2, with oxygen gas, O_2, to form water.

H_2O
product

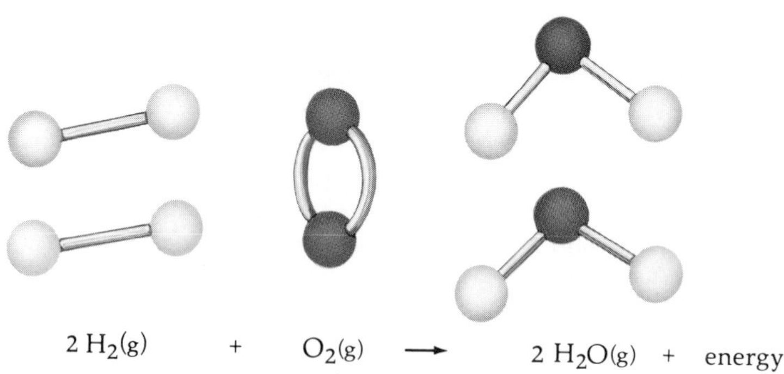

$2\,H_2(g)$ + $O_2(g)$ ⟶ $2\,H_2O(g)$ + energy

Including the relative number of molecules involved, the symbolic expression for the reaction becomes

$$2\,H_2(g) + O_2(g) \longrightarrow 2\,H_2O(g) + \text{energy}$$

Read: Two molecules of hydrogen gas react with one molecule of oxygen gas to give two molecules of water vapor and energy. This is a bookkeeping statement. All atoms in the reactants are accounted for in the products. The statement represents a chemical equality and is called a **chemical equation.** The expression is said to be balanced. It records an event but says absolutely nothing about the manner in which the event proceeds: what bonds are broken first or indeed why bonds break at all. The expression is more complete when the amount of energy released is numerically expressed for the quantities of hydrogen and oxygen reacted.

Hydrogen Gas as a Fuel

Hydrogen gas has been and is an important fuel for rockets and space ships. If we can develop technologically feasible and economically feasible sources of hydrogen gas and delivery systems for tanks of hydrogen gas, it may become an important fuel in our economy as the need to find substitutes for fossil fuels and to protect the quality of the environment accelerates. If the energy is available, water can be converted to hydrogen gas and oxygen gas:

$$\text{energy} + 2\,H_2O(l) \longrightarrow 2\,H_2(g) + O_2(g)$$

Electrolysis of Water

This reversal of the combustion reaction can readily be carried out by an electrical process using any source of water provided that there is also a source of electrical energy. One proposal is to use sea water and the energy from either sunlight or the winds to generate the electrical energy. The seriousness with which this possibility has been considered is indicated by the fact that the term "hydrogen economy" has been introduced by the mass media into the public vocabulary. It should be noted that the use of hydrogen gas does not increase our energy resources. More energy will be required to produce the hydrogen gas and to transport the hydrogen gas to the locations where it will be marketed than will be obtained from the combustion reaction. The concept of the "hydrogen economy" is attractive simply because the hydrogen gas can be stored, transported, and used on demand and its combination to form water in no way pollutes the atmosphere. Sunlight and winds cannot be stored. To produce and store hydrogen gas is simply one means of storing energy as chemical energy that can be converted to heat energy on combustion. To charge batteries is another means of storing energy as chemical energy that can be withdrawn upon demand as electrical energy. At present, it is not technically and economically feasible to store large quantities of electrical energy using batteries. Small quantities can be stored—but not large quantities. In Tokyo, all taxi cabs are required to use hydrogen gas as the fuel.

Explosive Mixtures of Hydrogen and Oxygen Gases

If hydrogen gas is allowed to mix with air before the reaction is initiated by a spark or flame, the reaction spreads very rapidly, seemingly instantaneously, through the mixture; the released energy raises the temperature rapidly throughout the mixture; there is a consequent rapid increase in pressure; and the gases expand rapidly, taking the walls of the container with them unless the container walls can sustain the pressure. Such a rapid, uncontrolled reaction is called an explosion and can be extremely violent. How violent is determined in part by the relative number of molecules of hydrogen and molecules of oxygen and by the number of nonreactive nitrogen molecules from the air in the mixture. The maximum temperature and the maximum explosion are attained with a mixture made up of two molecules of hydrogen for every molecule of oxygen, without any other molecules present, as would be predicted from the chemical equation. If either reactant is present in great excess, the reaction on ignition may be quite mild. The potential hazard of explosive mixtures being formed due to hydrogen leaks into the air is reduced by the rapid upward diffusion of hydrogen gas in air. Hydrogen molecules are the lightest of all known molecules. Consequently hydrogen molecules tend to diffuse faster than the molecules of any other fuel. This dissipation of explosive mixtures is an important factor in the evaluation of the feasibility of a hydrogen economy. The total energy produced by a reaction of hydrogen gas with the oxygen molecules of the air is, of course, dependent upon the quantity (the number of molecules) of hydrogen involved in the reaction.

Mechanism of the Hydrogen Gas–Oxygen Gas Reaction

What is the significance of the necessity to use a spark or flame to initiate the combustion reaction which then sustains itself? The initial flame or spark brings at least a small portion of the reaction mixture to a high temperature. It is believed that this high-energy situation "loosens up" (increases the vibration of the atoms in molecules) and breaks some chemical bonds in at least a few molecules of one or more of the reactants. This enables the reaction to proceed, frequently through a number of steps, to the final product. The sequence of reactions proposed for the combustion of hydrogen involves reactions such as the following:

$$
\begin{array}{lcll}
\text{energy} + O_2 & \longrightarrow & O + O & \text{initiating reaction (formation of free radicals)} \\
O + H_2 & \longrightarrow & OH + H & \\
OH + H_2 & \longrightarrow & H_2O + H & \text{continuing reactions} \\
H + O_2 & \longrightarrow & HO + O & \\
H + H & \longrightarrow & H_2 & \text{terminating reactions} \\
H + OH & \longrightarrow & H_2O &
\end{array}
$$

4-1 *Background Information*

MASS AND WEIGHT

The kilogram and the gram may soon replace the pound and the ounce as units of mass in commerce in the United States. The kilogram is approximately equivalent to 2.2 pounds (2.205 lb) and the gram is approximately equivalent to 0.04 ounces (0.0353 oz).

One liter of water at 4 °C has a mass of one kilogram.

The names of units of mass less than the gram and units of mass greater than the gram use the word gram with the appropriate multiplicative prefix. (Background Information 1-2 on length in Chapter 1.)

The units of mass most commonly used by chemists are

1 kilogram is 1000 grams	1 kg is 1×10^3 g
1 milligram is 0.001 gram	1 mg is 1×10^{-3} g
1 microgram is 0.000 001 gram	1 μg is 1×10^{-6} g
1 nanogram is 0.000 000 001 gram	1 ng is 1×10^{-9} g

The concepts of mass and weight are sources of great confusion compounded by the common use of an inexact vocabulary. Two masses attract each other. This is the basis of gravity. Release your grasp on your pencil and you will not be surprised that it moves toward the center of the earth. On this planet, the **weight** is the **force** that tends to make the pencil fall—move toward the center of the planet. The magnitude of this force (the weight) depends upon the distance from the center of the earth and the mass of the pencil. Climb to the top of Mount Everest and the pencil weighs less; send it to a skylab and its weight approaches zero. But the mass of the pencil has remained unchanged, unless, of course, you have been using the pencil to take notes along the way. The mass of an object is a fundamental property of the object. Actually, it is the rest mass that is the fundamental property, but let us not get involved at this point with the mass change associated with very high-speed motion.

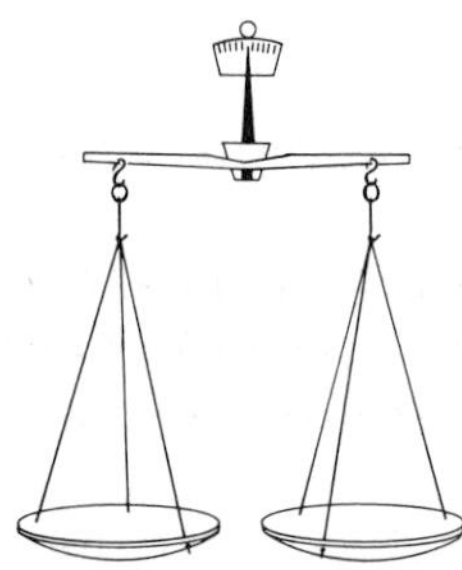

The process of weighing can be carried out by balancing two forces. With a balance of the goddess of justice type, the pencil to be weighed is placed on one pan and masses of known value are added to the other pan until the total gravitational force acting on the added masses equals the gravitational force acting on the pencil. Under this balanced condition, the sum of the added masses equals the mass of the pencil provided (1) the two balance arms are the same length, (2) the balance was in balance before either the pencil or the masses were added, and (3) no other forces are acting on either the pencil or the added masses. Air currents, magnetic fields, electrostatic charges, and mechanical vibration can invalidate the comparison.

Note that according to the model previously discussed, oxygen atoms, hydrogen atoms, and hydroxyl groups, OH, have unsatisfied bonding sites and would therefore be expected to be quite reactive. These atoms and groups of atoms with unsatisfied bonding sites are called free radicals. In the end, all of these free radicals combine; water, H_2O, and an excess of either hydrogen gas, H_2, or oxygen gas, O_2, remain.

All combustion reactions must produce energy to sustain the high temperature and consequently sustain the production of free radicals.

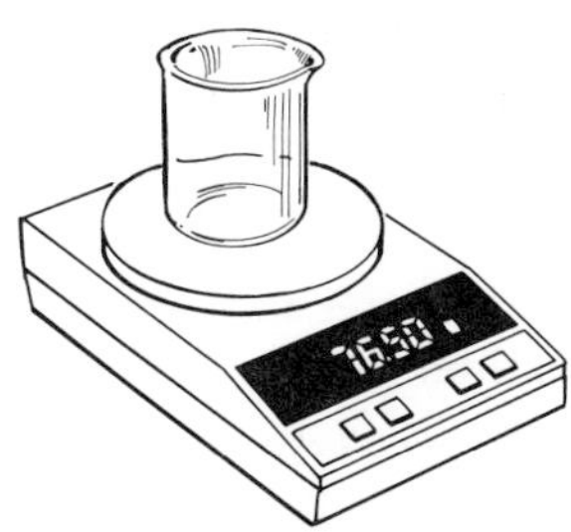

A balance of the goddess of justice type is conceptually simple but laborious to use. Modern balances frequently have only one pan and give a digital readout of the mass of the object placed on that pan. The only operations required of the operator are to position the pencil on the pan and, in some cases, turn a few knobs. In these digital readout balances, the response of the balance to the gravitational force acting on the pencil is determined indirectly by a physical response within the balance to the gravitational force acting upon the pencil. Each balance has been calibrated in terms of its responses to gravitational forces using known masses. The balance electronically translates the gravitational force acting upon the pencil into the mass of the pencil and displays that numerical value. The design, calibration, and maintenance of single-pan digital readout balances are the realm of experts and competent technicians. The use of the single-pan digital readout balance places very minimal demands upon the user and facilitates investigations of phenomena.

The confusion of the terms mass and weight is compounded by the use of the word pound to designate a unit of mass and also to designate a unit of force. After all of this, the term weight will undoubtedly be incorrectly used occasionally in this book as it is in most other books. If we were purists, we would refer to boxes of masses, not to boxes of weights. You just can't put forces in boxes.

Frequently used equations based upon the multiplicative prefixes:

$$[\text{mass in grams (g)}] = \left[1 \times 10^{3} \frac{\text{grams}}{\text{kilograms}}\right][\text{mass in kilograms (kg)}]$$

$$[\text{mass in grams (g)}] = \left[1 \times 10^{-3} \frac{\text{grams}}{\text{milligrams}}\right][\text{mass in milligrams (mg)}]$$

$$[\text{mass in grams (g)}] = \left[1 \times 10^{-6} \frac{\text{grams}}{\text{micrograms}}\right][\text{mass in micrograms } (\mu\text{g})]$$

$$[\text{mass in grams (g)}] = \left[1 \times 10^{-9} \frac{\text{grams}}{\text{nanograms}}\right][\text{mass in nanograms (ng)}]$$

Reactions of Compounds of Carbon, Hydrogen, and Oxygen with Oxygen Gas

Combustion of Methane

If methane gas, CH_4, emerges through a small orifice into the atmosphere, a reaction of methane with the oxygen of the air can be initiated with a flame or a spark. If there is an ample supply of oxygen gas, the methane gas burns with a light blue, almost colorless, flame; heat is evolved and the products are exclusively water and carbon dioxide, CO_2:

$$CH_4(g) + O_2(g) \longrightarrow H_2O(g) + CO_2(g) + \text{energy}$$

Complete Combustion

To balance the books, focus on the one molecule of methane. This one molecule of methane would generate two molecules of water (to use up the four atoms of hydrogen) and one molecule of carbon dioxide. This makes a requirement of a total of four atoms of oxygen (two molecules of oxygen, O_2) for the process:

$$CH_4(g) + 2\ O_2(g) \longrightarrow 2\ H_2O(g) + CO_2(g) + \text{energy}$$

It is the above expression that is the equality—the chemical equation. The equation is read: One molecule of methane gas reacts with two molecules of oxygen gas under appropriate conditions to give two molecules of water vapor, one molecule of carbon dioxide gas, and energy. This reaction leading to water and carbon dioxide as the only products is referred to as **complete combustion.**

Incomplete Combustion

If the relative abundance of oxygen available to react with the methane is limited, there may not be adequate oxygen for complete combustion to give the products carbon dioxide and water exclusively. The situation can become complex. An unpredictable mixture of substances results: carbon, carbon monoxide, hydrogen gas, unreacted methane, and possibly some intermediates such as methyl alcohol, formaldehyde, and formic acid, as well as water and carbon dioxide.

C   CO   H—H $H{-}\underset{H}{\overset{H}{C}}{-}H$ $H{-}\underset{H}{\overset{H}{C}}{-}OH$ $(H)(H)C{=}O$ $(H)(HO)C{=}O$ $H{-}O{-}H$ $O{=}C{=}O$

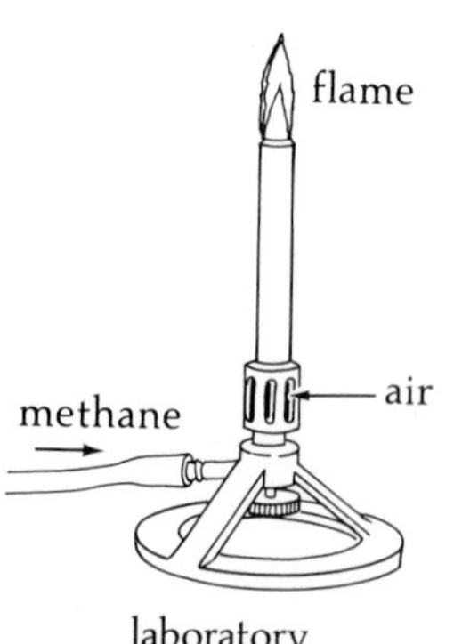

laboratory gas burner

It is even possible that fragments of molecules (free radicals) unite to give much more complex molecules. All of the products of incomplete combustion could, under appropriate experimental conditions, react with more oxygen to give carbon dioxide and water. In an incomplete combustion reaction, it is impossible to predict the relative quantities of the products or to write a single chemical equation. Quite different reactions may be going on in different parts of the flame. Atoms of carbon tend to collect to give clusters of carbon atoms and these hot clusters of carbon glow giving a yellow luminous flame. The flame is sooty unless the carbon clusters can react at the edge of the flame with oxygen of the air to give the colorless gases carbon monoxide or carbon dioxide.

For the purpose of considering the quantity of oxygen involved in a reaction giving products other than water and carbon dioxide, assume that experimental conditions for an incomplete combustion reaction of methane can be controlled to give the products water, carbon monoxide, and hydrogen (no carbon dioxide):

$$CH_4(g) + O_2(g) \longrightarrow H_2O(g) + CO(g) + H_2(g) + \text{energy}$$

The bookkeeping is correct as the expression stands and the above expression is the equation for this specific incomplete combustion reaction. In this reaction, one molecule of methane requires one molecule of oxygen. In the complete combustion of methane to give carbon dioxide and water, discussed earlier, one molecule of methane required two molecules of oxygen. The complete combustion of one molecule of methane requires twice as much oxygen as this specific incomplete reaction. Complete combustion reactions always utilize more oxygen than incomplete combustion reactions: an inadequate supply of oxygen in close proximity to the fuel molecules leads to incomplete combustion. All other gaseous hydrocarbons and gaseous compounds of carbon, hydrogen, and oxygen flowing through an orifice into air can also be ignited. They all burn in a very similar manner. Under conditions of complete combustion, the only products are water and carbon dioxide and the flame is essentially odorless. Under conditions of incomplete combustion, a number of intermediate compounds, such as aldehydes, ketones, and carbon monoxide, may be formed. Hydrogen gas and large quantities of carbon black (clusters of carbon atoms) may also be formed. It is the intermediate compounds that are associated with the odor of incomplete burning. Water, carbon dioxide, and carbon monoxide are odorless.

Writing Equations for Complete Combustion

To write the equation for the complete combustion of *n*-butane, the four-carbon, straight-chain, saturated hydrocarbon, write down the formulas of the reactants and the products

$$CH_3CH_2CH_2CH_3(g) + O_2(g) \longrightarrow H_2O(g) + CO_2(g) + \text{energy}$$

and then get on with the bookkeeping. The four atoms of carbon in the butane molecule generate four molecules of carbon dioxide and the ten atoms of hydrogen in the butane molecule generate five molecules of water (two atoms of hydrogen per molecule of water):

$$CH_3CH_2CH_2CH_3(g) + O_2(g) \longrightarrow 5\ H_2O(g) + 4\ CO_2(g) + \text{energy}$$

How many atoms of oxygen are required for this complete combustion of the one molecule of butane? Five atoms of oxygen for the five molecules of water and eight atoms of oxygen for the four molecules of carbon dioxide: the total number of atoms of oxygen required is thirteen. This is a bit of bad luck. Atoms of oxygen come two at a time in molecules of oxygen, O_2. To end run this problem, write the equation using twice as much butane, two molecules of butane, to give twice as much water, ten molecules of water, and twice as much carbon dioxide, eight molecules of carbon dioxide:

$$2\ CH_3CH_2CH_2CH_3(g) + O_2(g) \longrightarrow 10\ H_2O(g) + 8\ CO_2(g) + \text{energy}$$

Now the bookkeeping requires twenty-six atoms of oxygen or thirteen molecules of oxygen, O_2, and the equation, the equality, is

$$2\ CH_3CH_2CH_2CH_3(g) + 13\ O_2(g) \longrightarrow 10\ H_2O(g) + 8\ CO_2(g) + \text{energy}$$

In arriving at this equation, the expression has been written out four times to delineate the process. In practice, it is not necessary to write out the intermediate steps in the thought process. Write down the formulas for the reactants and the products and insert the bookkeeping numbers in that one statement. In the end, mentally check the atom count on the two sides of the equation.

	lefthand side				righthand side	
	$2\ CH_3CH_2CH_2CH_3$	$+\ 13\ O_2$	$\longrightarrow$	$10\ H_2O$	$+\ 8\ CO_2$	+ energy
atoms of carbon	8	+ 0	=	0	+ 8	
atoms of hydrogen	20	+ 0	=	20	+ 0	
atoms of oxygen	0	+ 26	=	10	+ 16	

All of the atoms in the reactants have been accounted for in the products. The equality exists and the expression is the chemical equation.

The larger the number of molecules of oxygen required per molecule of the fuel for complete combustion, the more likely it is that incomplete combustion will occur. Sufficient numbers of molecules of oxygen may not be available in the region of the flame where the temperature is high enough for the chemical reactions to occur.

4-2 *Background Information*

FORCE

A net force acting on an object leads to acceleration: an object at rest begins to move and continues to move at an increasingly rapid rate; an object in motion changes rate of motion or direction of motion. The larger the force, the greater the acceleration: the greater the mass of an object, the greater the force required to achieve a given acceleration. The fundamental relation from physics is that force is the product of mass times acceleration:

$$\text{force} = \text{mass} \times \text{acceleration}$$

$$f = ma$$

If the mass is expressed in kilograms, kg, and the acceleration is expressed in meters per second per second, m $\sec^{-2}$, the units of force are kg m $\sec^{-2}$. This rather complex relation of units is designated as a unit of force, the newton. The name of the unit recognizes the Newton of falling apple fame. As a unit of force, it is spelled newton—without a capital letter.

$$[\text{force in newtons}] = \text{kilograms} \times \frac{\text{meters}}{\text{seconds}^2}$$

A **force of one newton** gives a mass of one kilogram an acceleration of one meter per second per second.

If the mass is expressed in grams, g, and the acceleration is expressed in centimeters per second per second, cm $\sec^{-2}$, the units of force are g cm $\sec^{-2}$. This combination of units is designated as a unit of force, the dyne.

$$[\text{force in dynes}] = \text{grams} \times \frac{\text{centimeters}}{\text{seconds}^2}$$

A **force of one dyne** gives a mass of one gram an acceleration of one centimeter per second per second.

Note that 1 newton equals 1×10^5 dynes and that, in general, force is calculated from units of mass × units of lengths divided by units of time squared:

$$\frac{\text{mass} \times \text{length}}{\text{time}^2}$$

The relation between force in dynes and force in newtons is given by the equation

$$[\text{force in dynes}] = \left[1 \times 10^5\ \frac{\text{dynes}}{\text{newtons}}\right][\text{force in newtons}]$$

If acetylene gas (ethyne gas), H—C≡C—H, flowing through a small orifice into the atmosphere is lighted, it might be expected to burn with a clean, almost colorless flame to give carbon dioxide and water:

$$2\,H{-}C{\equiv}C{-}H(g) + 5\,O_2(g) \longrightarrow 4\,CO_2(g) + 2\,H_2O(g) + \text{energy}$$

Combustion of Acetylene

Instead, the combustion of acetylene in this uncontrolled fashion is exceptionally untidy. Copious quantities of soot are formed and the odors of some of the products of incomplete combustion are unpleasant. The uncontrolled reaction is very incomplete unless air or oxygen gas is premixed with the acetylene before it emerges from the orifice of a burner or a torch.

Under conditions of complete combustion, a much higher temperature is produced using an oxygen–acetylene mixture than using an air–acetylene mixture. Air contains approximately 20% oxygen with the remainder being primarily nitrogen. For every 20 molecules of oxygen, there are approximately 80 molecules of nitrogen. In the air–acetylene mixture, part of the heat generated by the combustion reaction goes into raising the temperature of the nitrogen molecules. Consequently, the temperature of an air–acetylene flame never attains the high temperature of an oxygen–acetylene flame.

Combustion of Structural Isomers

A study of the combustion of small samples of liquid hydrocarbons or liquid compounds of carbon, hydrogen, and oxygen in open, heat-resistant dishes reveals diverse behavior—even of structural isomers such as diethyl ether, $CH_3CH_2OCH_2CH_3$, and 1-butanol, $CH_3CH_2CH_2CH_2OH$. Both are liquids at room temperature and one atmosphere pressure. When a burning wooden splint or a burning long match is brought near the surface of a small volume (1 milliliter) of diethyl ether, the compound ignites immediately and burns rapidly with an almost colorless blue flame until all the liquid is consumed. The flame is so colorless and the reaction so rapid that the flame itself may not be evident unless observed closely. Treated in the same manner, 1-butanol remains unchanged: no flame is initiated by the burning splint. However, if the dish is first warmed until small bubbles begin to form in the liquid and then the burning splint is brought up, 1-butanol burns with a large sooty flame and a black coating appears inside the rim of the dish. With both the diethyl ether and the 1-butanol, the flame, the region of the chemical reaction, is above the surface of the liquid. The chemical reaction occurs in the gas phase, the vapor phase, above the liquid, not in the liquid or on the surface of the liquid. It is in the gas phase, the vapor phase, that the molecules of the compounds get together with oxygen molecules of the air at the temperature required to initiate combustion.

incomplete combustion

The vaporization of diethyl ether and 1-butanol is a physical change and can be represented

$$CH_3CH_2OCH_2CH_3(l) \longrightarrow CH_3CH_2OCH_2CH_3(g)$$
$$CH_3CH_2CH_2CH_2OH(l) \longrightarrow CH_3CH_2CH_2CH_2OH(g)$$

Diethyl ether vaporizes very readily in an open dish at room temperature and diethyl ether is said to be a very volatile liquid. The 1-butanol isomer, on the other hand, vaporizes much more slowly (more slowly than water at the same temperature) and 1-butanol is described as a nonvolatile liquid. The rate of vaporization of all liquids is increased by raising the temperature of the liquid. The boiling point of diethyl ether is 35 °C and the boiling point of 1-butanol is 117 °C, both at one atmosphere pressure. The more volatile the liquid, the lower its boiling point. The difference in volatility of the two structural isomers is consistent with the necessity to preheat the alcohol and not the ether to obtain combustion of the liquids in the open dish. The marked difference in the completeness of the two reactions cannot be rationalized by differences in volatility of the two isomers or by differences in the number of molecules of oxygen that would be required for the complete combustion of one molecule of diethyl ether and one molecule of 1-butanol. See Scramble Exercise 4. Presumably, an understanding of the difference in the products of the reaction must be sought in the differences in bonding within the structures of the two isomers.

Lamps

Lamps use liquid fuels that do not burn in an open dish. Lamps depend upon wicks to conduct the fuel into the upper region of the wick where it is converted into the vapor phase by the heat of the flame; it is the combustion of the vapor that then becomes the flame. This movement of a liquid in the wick against gravity depends upon the attractive forces between molecules of the liquid and the molecules that constitute the woven or twisted fibers of the wick. The net result is that the fluid moves between the closely spaced fibers. This same capillary action phenomenon is observed when an edge of a towel, ink blotter, or filter paper touches a liquid that wets the fibers. The liquid rises between the closely spaced fibers above the surface

4-1 *Gratuitous Information*

OXY–ACETYLENE TORCHES

An operating oxy–acetylene torch is an awesome thing. As acetylene gas under pressure is supplied to the torch from one tank, oxygen gas under pressure is supplied to the torch from a second tank; it is the mixture of acetylene and oxygen that emerges from the orifice of the torch. The gases hiss as they escape the orifice and the flame roars as the mixture reacts just beyond the orifice. The appearance of the flame depends upon the composition of the acetylene–oxygen mixture. When no oxygen is introduced into the torch, the flame is voluminous and sooty as combustion

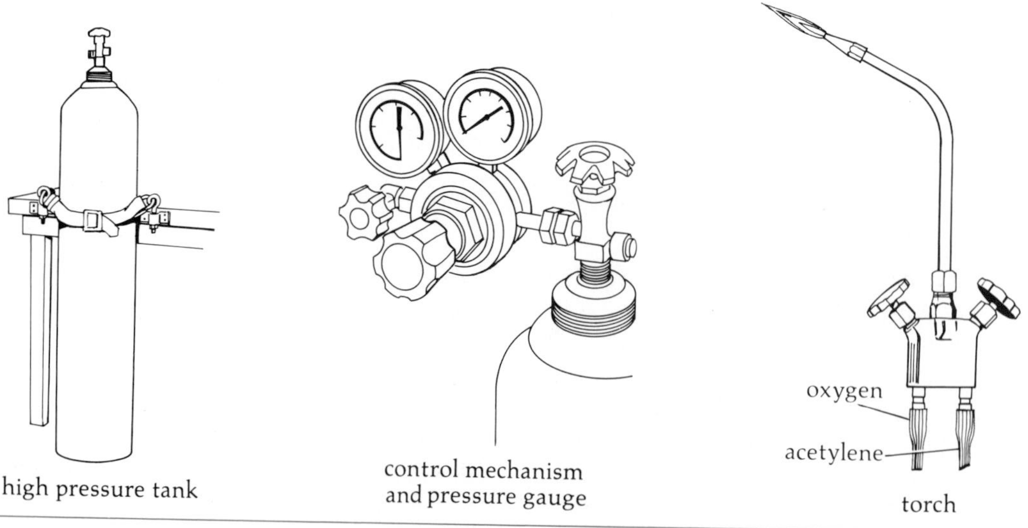

high pressure tank

control mechanism
and pressure gauge

torch

of the liquid. For the flame to be luminous, the combustion process must produce solid carbon within a region of the flame by incomplete combustion. These hot carbon particles radiate light in the same manner as any other hot solid, such as a white-hot poker or the filament in an incandescent light bulb. If the lamp is not to smoke, these particles of carbon, clusters of carbon atoms, must be oxidized to carbon monoxide or carbon dioxide—perferably the latter—in the outer region of the flame. Carbon monoxide is very toxic. The CO molecule, by reactions with the hemoglobin in red blood cells, usurps positions on the hemoglobin molecule essential to the transport of oxygen by hemoglobin.

Candles

Candles operate much as lamps do except that candles are made of paraffin or wax and the heat of the burning candle must convert the wax to liquid before that liquid can feed through the wick to the high-temperature region of the flame where combustion of the paraffin vapor occurs. Air currents, as we all know, may lead to an untidy meltdown of the sidewalls of what is normally a very tidy pool of liquid paraffin or wax at the base of the wick. Paraffin is derived from the fractionation of petroleum and is a mixture of hydrocarbon molecules containing large numbers of carbon atoms. Tallows and waxes derived from plants and animals are esters. In all cases, the molecules contain long carbon–carbon chains and are very large in comparison with the molecules of the gas and liquid compounds we have been exploring.

---

proceeds with the oxygen of the surrounding air. As oxygen is added through the torch, the flame becomes less voluminous and less sooty. With sufficient oxygen, the flame becomes a much smaller, well-defined conical flame with at least two identifiable conical regions within the flame. The inner cone exhibits no color and is cold. The inner cone is simply the outflowing gas mixture that has not reached a temperature sufficiently high to initiate the combustion reaction and so is really not a part of the flame at all. The chemical reactions occur in one or more regions external to this inner cone. If the flow of oxygen is increased, the inner dark cone enlarges and it can become so large the flame is blown out. The flow of gases is simply too great for the temperature to be maintained at a sufficiently high level to sustain combustion.

The high temperature of the well-defined flame (complete or essentially complete combustion) is capable of melting and vaporizing iron and steel. Used as a cutting torch, the oxy-acetylene torch is even more dramatic. The cutting action is initiated by directing the well-defined oxygen–acetylene flame at the spot on the edge of the plate where the cutting is to begin. As the iron melts and sparks fly, a second reaction can be initiated by directing a stream of oxygen gas towards the molten iron. The cutting torch has a separate orifice in the center of the torch head for this purpose. The reaction is the combustion of iron to give an oxide of iron:

$$\underset{\text{iron}}{3\,\mathrm{Fe(l)}} + 2\,\mathrm{O_2(g)} \longrightarrow \underset{\text{iron oxide}}{\mathrm{Fe_3O_4(s)}} + \text{energy}$$

After the reaction of iron with oxygen is well underway, the supply of acetylene to the torch can be turned off without diminishing the fireworks display arising from the blast of oxygen gas against the molten metal. The heat generated by the reaction of iron with oxygen is adequate to sustain the high temperature required for the iron–oxygen reaction to continue and a space in the iron plate opens up as the iron is dispersed in a shower of sparks of molten iron and solid iron oxide.

Protective clothing and a face shield must be worn as protection from the heat, the shower of flying particles, and the ultraviolet light emitted by the white-hot metal.

The symbol Fe is derived from the Latin *ferrum*, meaning iron. There are three common oxides of iron, all solids: FeO black, Fe_2O_3 red, and Fe_3O_4 black.

The reaction of nonvolatile solids such as charcoal brickettes with oxygen molecules does occur at the surface of the brickette. This accounts for the absence of a flame and for the very hot surface of the brickette. It also accounts for the high temperature the solid must attain for the reaction to become self-sustaining. Quick-starting charcoal has been treated with combustible compounds that are more volatile than the carbon of the charcoal.

Uncontrolled Combustion Reactions

We live in a world of fires waiting to happen. Our atmosphere is 20% oxygen and much of our buildings, furnishings, and clothing is combustible. Fire departments endeavor to prevent the initiation of uncontrolled fires and to interrupt the process of combustion when uncontrolled fires do occur. There are several obvious lines of approach: provide a way for vaporized material to escape from a burning building before an explosion occurs, cut off the source of the combustible material, cut off the supply of oxygen, reduce the temperature to diminish the quantities of the fuel getting into the vapor phase, and reduce the temperature to reduce the rate of the chemical reaction. Even now, a great deal is not known about the reactions of many building materials under the extremely high-energy conditions of destructive fires. It is known that many fatalities are the consequence of inhaling the compounds formed by incomplete combustion, and federally supported research is currently underway

4-3 *Background Information*

ELECTRICAL CHARGES

Electrical charges come in two varieties, designated as positive charges and negative charges. In chemistry, the most useful unit of charge is the magnitude of the charge on the electron. The charge on the electron is designated as negative. In this system, the charge on the electron is 1−, and an equal but opposite charge is 1+.

The electrostatic force between two electrons, each with a charge of 1−, is a force of repulsion and the two electrons are said to repel each other. The force tends to make the two electrons move away from each other.

The electrostatic force between an electron with a charge of 1− and a positive particle with a charge of 1+ is a force of attraction; the electron and the positive particle attract each other. This force of attraction makes the electron and the positive particle tend to move toward each other.

The electrostatic force between two positive particles is a force of repulsion. The two positive particles repel each other and the two positive particles tend to move away from each other. This is summarized: Like charges repel; unlike charges attract.

The diagram in the next column represents two particles separated by distance.

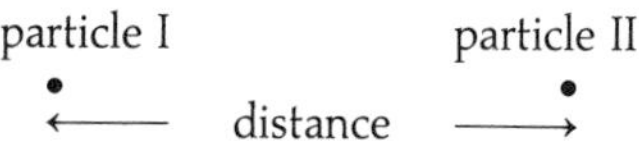

The magnitude of the force acting on particle I and also on particle II is dependent upon

(1) the charge on particle I, designated as charge I;
(2) the charge on particle II, designated as charge II;
(3) the distance between the two particles, designated as distance; and
(4) a property of the material between the two particles, designated as the dielectric constant.

The relation among these five quantities is given by the equation

$$\text{force} = \frac{\text{charge I times charge II}}{\text{dielectric constant times distance}^2}$$

Force is the product of charge I multiplied by charge II divided by the product of the dielectric constant multiplied by the square of the distance separating the two particles:

to assess the behavior of both the conventional and the newer synthetic materials during large-scale fires.

Controlled Combustion Reactions

The design of burners and internal combustion engines, including automobile engines and jet engines, is a series of exercises in the design of devices to control combustion reactions to produce, at economically feasible costs, maximum temperatures and minimum quantities of products other than carbon dioxide and water. It is desirable for the fuel to be free of elements other than carbon, hydrogen, and oxygen, for the fuel to be dispersed through the oxygen as a gas or as very small droplets or as a very fine powder, for the proportion of oxygen to fuel to favor complete combustion, and for the rate of delivery to the combustion chamber to be controlled. Among the devices that have appeared are precombustion purification systems for the fuels, compressors to compress the gases and valves to control their flow, preheaters to vaporize fuels, mills to pulverize solids, atomizers to disperse liquids into droplets, carburetors to provide appropriate oxygen to fuel ratios, burner heads with multiple orifices to direct the rapid flow of gases, afterburners and catalytic converters to attain more complete combustion to CO_2 and H_2O, precipitators to remove ash from stack gases, and scrubbers (water showers) to dissolve water-soluble stack gases.

In 1984, the Environmental Protection Agency estimated that, nationwide, wood-burning stoves in homes accounted for 44% of air pollution, fireplaces 3%, commercial

---

$$\text{force} = \frac{\text{charge I} \times \text{charge II}}{\text{dielectric constant} \times \text{distance}^2}$$

The force acting on each particle is directly dependent upon the magnitude of charge I and also directly dependent upon charge II provided the distance separating the two particles remains unchanged. If the charge on particle I is doubled, the force is doubled. If, in addition, the charge on particle II is tripled, the force acting on each particle is now six times the original force acting on each particle.

The force acting on each particle is inversely dependent upon the square of the distance separating the two charges (distance^2) provided the charges remain unchanged. If the distance separating the two particles is doubled, distance^2 is now four times as great and the force acting on each particle is now one-fourth the original force. (Distance^2 is in the denominator; distance^2 is a divisor.) The force acting upon charged particles is diminished by increasing the separation between the particles. (Inverse proportions are discussed in some detail in Background Information 11-3, Chapter 11.)

The dielectric constant for a vacuum is exactly 1, and for a special case of two charged particles in a vacuum, the equation becomes

$$\text{force} = \frac{\text{charge I} \times \text{charge II}}{\text{distance}^2}$$

This is the situation encountered in the next chapter in the exploration of the internal structure of atoms. The dielectric constants for dry gases at 25 °C and one atmosphere pressure are slightly greater than 1. For example, the dielectric constant for hydrogen gas, H_2, is 1.0003. The dielectric constant for liquid water is exceptionally high, 78.5, and an adequate model for the structure of the H_2O molecule and the structure of the liquid must rationalize this high value. Many of the properties of solvents and solutions are closely related to the dielectric property of the solvent.

In this discussion, we have not faced into the development of a consistent set of units for the force equation. This can and must be done to explore many topics. For the explorations undertaken in this book, it is only necessary to understand the manner in which electrostatic force is dependent upon relative magnitudes of the charges involved, the distance separating the charges, and the dielectric constant of the medium within which the charges are distributed. It can be pointed out, however, that the force equation defines the dielectric constant.

incinerators 3%, open fires 7%, cars and trucks 25%, ships, planes, farm machines and other equipment 15%, and coke ovens 3%. Air quality control regulations focus upon the sources of pollution most significant in regions of high pollution—usually high-density population areas and high-density industrial areas.

Scramble Exercises

1. Methanol, CH_3OH, was at one time used in laboratories as the fuel for lamps to produce small quantities of heat. Covered small electrical heaters are safer and an alcohol lamp in a laboratory would be difficult to find today. Methanol is formed in the fermentation of plant materials and the feasibility of its production from agricultural waste materials makes methanol–gasoline mixtures an attractive liquid fuel.

   Try your hand at writing the equation for the complete combustion of methanol with the oxygen molecules, O_2, in the air. The only products are carbon dioxide and water. In doing the bookkeeping, don't forget the atom of oxygen in each molecule of methanol, CH_3OH.

4-1 *Editorial Comment*

VOCABULARY

We continuously expand and refine the vocabularies of the sciences in order to describe and analyze scientific phenomena and to explore scientific concepts. To the degree the richer and more specific language enhances our capacity to communicate with others and with ourselves, we enhance our abilities to extend both factual knowledge and cognitive knowledge, particularly the latter. The growth of the sciences and the growth of scientific terminology are simultaneous evolutionary processes. For any individual, the level of comprehension of a science and the level of awareness of the full connotation of the vocabulary of the science grow simultaneously.

As a science develops, many concepts develop for a time but are clearly shown to be inadequate later. Many of the terms that developed with these concepts gradually fade from our vocabulary. Other terms take on additional meanings to encompass more comprehensive concepts. At one time, the term oxidation was used to designate reactions with oxygen gas, such as those explored in this chapter. Today, the term oxidation is used in a much more comprehensive sense to encompass not only reactions with oxygen gas but also many reactions that do not involve oxygen. There is a commonality about all of these reactions not recognized at the time the term came into use. This broader concept of oxidation is addressed in Chapter 16.

Many terms are precisely defined and consistently used. These include dielectric constant, dipole moment, and pH. Patience: we shall get to the definition of these terms later.

It is the intent of this book to define a minimum number of terms and depend upon the repeated use of terms in context to develop workable connotations. There are risks in this approach but the authors believe the benefits greatly exceed the risks. The mind must be able to run free, to enjoy interrelations, and to dream of new concepts. The scientist seeks the satisfaction of developing and testing the concepts suggested by those dreams. The capacity to work with precise detail develops in due time.

Be alert to the use of terms and enjoy the language itself. Skill in its use develops more rapidly than you might think. It may be helpful to make a list of new words and scratch them off when you feel secure in their use.

There are inconsistencies in the use of the terms "observation" and "conclusion." You may encounter these and you are entitled to an explanation. If an aqueous solution of hydrochloric acid is added to a piece of metal at room temperature and colorless bubbles rise from the surface of the metal to break through the surface of the water solution, a student is encouraged to make the observation that color-

2. In the combustion of *thio-* compounds such as thiomethanol, CH_3SH, the sulfur becomes sulfur dioxide gas, SO_2. For complete combustion, the products are CO_2, H_2O, and SO_2. Try writing the equation for the complete combustion of thiomethanol. Biological materials contain a number of *thio-* compounds and this sulfur has been carried over in varying degrees in fossil fuels. The sulfur dioxide formed on combustion is a serious source of atmospheric pollution.

3. Hydrogen gas, H_2, and carbon monoxide, CO, are both useful compounds for industrial processes. Under controlled conditions of temperature, pressure, and relative numbers of molecules, it is possible to obtain the partial combustion of a hydrocarbon to give primarily these compounds. Assuming that these are the only products, complete the following equation for the partial combustion of ethane, C_2H_6.

$$\_\_\_\_ C_2H_6(g) + \_\_\_\_ O_2(g) \longrightarrow \_\_\_\_ CO(g) + \_\_\_\_ H_2(g)$$

Also write the equation for the complete combustion of ethane to give carbon dioxide and water and compare the relative number of oxygen molecules, O_2, used for the incomplete and the complete combustion of one molecule of ethane.

---

less bubbles are formed and draw the conclusion that (1) a gas has been formed, (2) the gas is colorless, and (3) a chemical reaction has occurred. On the other hand, it would be completely acceptable to the community of chemists for an experienced chemist (1) to make "the observation" that hydrogen gas has been produced and (2) to proceed to draw conclusions about the nature of the metal. This is a double standard based upon differences in creditability acquired through previous experience and accumulated knowledge.

Terms such as high and low, large and small, fast and slow, concentrated and dilute, reactive and nonreactive, and soluble and insoluble have no place in scientific discussions unless it is quite clear what the points of reference are. Fast with respect to what? Such terms are in fact very useful in communication with other scientists who understand the context within which the terms are used. To the uninitiated, these terms are much more troublesome since that which is fast at one moment may be slow the next moment in quite a different context. To say that a reaction is fast means something quite different to a geologist interested in the geological processes that produce petroleum than to a combustion engineer interested in the combustion reactions of petroleum in the cylinder of an automobile. The time interval of interest to the geologist is of the order of thousands of years; the time interval of interest to the engineer is of the order of fractions of seconds, such as nanoseconds.

An exact statement of the quantity of a compound reacting per century or per millisecond under specified conditions of temperature and pressure is also troublesome. The terminology may have to be complex and the magnitude of the numerical value may have very little meaning without values for related phenomena for comparison.

The maximum in communication occurs when numerical values are precisely stated and the recipient of the information has previous knowledge of the order of magnitude of values encountered with other substances under the same conditions or with the same substances under different conditions. The carbon–carbon double bond, C═C, length of 0.134 nanometers has added meaning if you also know that the carbon–carbon single bond, C—C, length is 0.154 nanometers. The 117 °C boiling point at one atmosphere pressure for 1-butanol has added meaning if you also know that, under a pressure of one atmosphere, diethyl ether, a structural isomer, boils at 35 °C and water boils at 100 °C. The parallel statement can be made for sporting events. Greg Lougainis' score of 754 in diving at the 1984 Olympics has added meaning if you also know that no other diver had ever scored in the 700's for that event.

4. (a) Write the equation for the complete combustion of diethyl ether, $CH_3CH_2OCH_2CH_3$, with the oxygen molecules, O_2, of the air.
(b) Write the equation for the complete combustion of 1-butanol, $CH_3CH_2CH_2CH_2OH$, with the oxygen molecules, O_2, of the air.
(c) Compare the equations for the complete combustion of the two structural isomers diethyl ether and 1-butanol.

5. Learn to be your own taskmaster. In the complete combustion of compounds of carbon, hydrogen, and oxygen with oxygen molecules, O_2, the only products are water and carbon dioxide. Given the formula of any compound of carbon, hydrogen, and oxygen, you should be able to write the equation for the complete combustion of that compound. All you have to do is to supply the formula for the compound and fill in the four blanks with the appropriate numbers.

$$\_\_\_\text{(formula of compound)} + \_\_\_O_2 \longrightarrow \_\_\_H_2O(g) + \_\_\_CO_2(g)$$

4-2 *Gratuitous Information*

COMBUSTION REACTIONS AND ACID RAIN

In addition to the products of incomplete combustion of compounds of carbon, hydrogen, and oxygen, two other types of air pollutants are commonly produced during the combustion of fuels and other materials. These are the oxides of nitrogen and the oxides of sulfur.

Nitrogen monoxide gas, NO, is produced by the reaction of nitrogen molecules of the air with oxygen molecules of the air at the high temperatures produced during combustion reactions of compounds of carbon, hydrogen, and oxygen:

$$\text{energy} + N_2(g) + O_2(g) \longrightarrow \underset{\substack{\text{nitrogen monoxide} \\ \text{(also called nitric oxide)}}}{2\,NO(g)}$$

The reaction of nitrogen gas with oxygen gas does not release energy—neither light nor heat. Instead, energy must be supplied in order for the nitrogen molecules and oxygen molecules to react. The reaction of nitrogen molecules and oxygen molecules does not generate the energy necessary to sustain the reaction of nitrogen molecules with oxygen molecules. As we all know, air—essentially a mixture of 80 molecules of nitrogen for every 20 molecules of oxygen—does not explode or burn.

The reaction of nitrogen molecules with oxygen molecules to produce nitrogen monoxide occurs as the molecules of nitrogen and oxygen move through the high-temperature region of a flame. Even then, only a very small fraction of the molecules of nitrogen and oxygen react and we might ignore the reaction if it were not for the impact of the nitrogen monoxide on air quality. Nitrogen monoxide is a colorless gas. It reacts at room temperature with oxygen molecules to give nitrogen dioxide, a brown gas with an irritating odor.

$$\underset{\substack{\text{nitrogen} \\ \text{monoxide}}}{2\,NO(g)} + O_2(g) \longrightarrow \underset{\substack{\text{nitrogen} \\ \text{dioxide}}}{2\,NO_2(g)}$$

In discussing air pollutants, oxides of nitrogen are usually designated as a group by NO_x. The expression NO_x is not a chemical formula. It is instead a notation that includes NO, NO_2, and other oxides of nitrogen. Oxides of nitrogen react with water to give acidic solutions.

Nitrogen monoxide is also produced by the reaction of the nitrogen and oxygen molecules of the air in the high-energy conditions produced by lightning. Throughout the ages, this conversion of atmospheric nitrogen, N_2, into water-soluble compounds has been the natural mechanism of making nitrogen compounds, which are essential to plant growth, accessible to plants. The conversion of nitrogen molecules, N_2, into compounds of nitrogen is known as nitrogen fixation. There is an ample supply of the element nitrogen throughout the world, but compounds of nitrogen are in limited supply.

Practice writing equations of this type until you have no doubt you can write equations for complete combustion of compounds of carbon, hydrogen, and oxygen and also no doubt that you can write them reasonably rapidly.

Additional exercises for Chapter 4 are given in the Appendix.

A Look Ahead

In Chapters 2 and 3 we explored the structures and formulas of compounds of the elements carbon, hydrogen, and oxygen in terms of the ball, stick, and spring model for compounds. These same compounds will be further explored later in terms of other models based upon modern concepts of the structures of atoms. All of these models are useful to chemists. Each model also has its limitations and chemists use the model most helpful in addressing a specific question. In Chapters 5 and 6 we shall explore the structures of atoms in order to seek an understanding of the forces that

---

Sulfur dioxide is produced by the combustion of compounds containing sulfur, such as the thioethers. All plants and animals utilize some sulfur-containing compounds. Consequently, such compounds are widely distributed in the biosphere and in the fossil fuels. The actual numbers of atoms of sulfur present in these materials are quite small in comparison to the numbers of atoms of carbon or hydrogen or oxygen, but the sulfur compounds are widely distributed.

The element sulfur is a yellow solid. Molten sulfur vaporizes readily and on ignition burns with a blue flame. You may have observed this flame if you have watched coal burn in a fireplace grate. Small localized blue flames shoot out from the solid coal. These flames are the combustion of sulfur vapor. The product is sulfur dioxide, a colorless gas:

$$S(g) + O_2(g) \longrightarrow \underset{\text{sulfur dioxide}}{SO_2(g)}$$

It is irritating to breathe and has a characteristic taste. In the formation of the coal, small pockets of solid sulfur are deposited in the coal and it is these small reservoirs of sulfur that melt and vaporize to give the rather attractive blue flames observed when coal is burned. Sulfur is the brimstone in the biblical description of hell.

By the further reaction of sulfur dioxide with oxygen, sulfur trioxide, SO_3, is formed. Here again, the general notation SO_x is used in the discussion of air pollution to include both SO_2 and SO_3. The notation SO_x is not a chemical formula. Both sulfur dioxide, SO_2, and sulfur trioxide, SO_3, react with water to give acidic solutions. In most localities where acid rain occurs, it is the oxides of sulfur that are considered to be the primary source of that acidity.

The United States has almost unlimited supplies of coal and large investments of federal money (taxpayers' money) and private enterprise capital have gone into the technology to "gasify" coal and to "liquify" coal—to convert solid coal to compounds that are gases at room temperature and one atmosphere pressure or to convert solid coal to compounds that are liquids at room temperature and one atmosphere pressure. Fuels that are gases or liquids are attractive since we have in place a well-developed distribution system of pipe lines to deliver gases and liquids. Automobiles are presently designed to utilize gases and liquids but not solids. Gases and liquids are also easier to purify than solid coal, even powdered coal. The future of coal gasification and coal liquification in this country is largely dependent upon how economically competitive energy produced using gasified coal and liquefied coal is with energy produced through the use of other fuels or other processes. Included in the cost of producing energy is the cost of meeting air quality standards.

hold atoms together in characteristic spatial orientations in molecules and also to explore the forces between molecules that account for the differences in physical properties, such as the boiling points of structural isomers, for example, diethyl ether (35 °C) and 1-butanol (117 °C).

In this chapter we explored one type of chemical reactions: the reactions of compounds of carbon, hydrogen, and oxygen with oxygen molecules, O_2, at elevated temperature. All of these combustion reactions release energy as heat and light and are capable of sustaining the elevated temperature once the chemical reaction has been initiated. In this exploration, no effort was made to relate the quantity of energy released to the quantity of the compound that reacted. This quantitative relation cannot be addressed until we have a system of expressing the quantity of the compound that reacted. To talk in terms of the numbers of molecules of the compound that reacted with oxygen is removed from the reality of the experimental measurements we can make. Molecules are too small to count. The easiest quantity to measure is mass. We cannot proceed further until we have a relation between the mass of a sample of an element and the number of atoms of that element in the sample, and also a relation between the mass of a sample of a compound and the number of molecules of the compound in the sample.

In Chapter 5 we shall explore the system of atomic masses used in general by scientists and also a number used by chemists to express quantity. This unit, this number, is called the mole and is used by chemists in much the same way that grocers use the unit dozen. The mole is a number.

We have explored a number of reactions in which reactants have become products. These have been chemical reactions. We shall continue to explore reactions in which reactants become products. Most of these will be chemical reactions but there will also be a very significant group of reactions to give products that are classified as nuclear reactions. In both chemical reactions and nuclear reactions, reactants react to give products. In both chemical reactions and nuclear reactions, energy changes are involved. For chemical reactions, we write chemical equations. For nuclear reactions, we write nuclear equations.

To write equations of any type, we must know

(1) what substances react (reactants),
(2) what substances are formed (products), and
(3) the formulas to represent all reactants and products.

From there on, it is simply a matter of writing down the formulas and then adjusting the numbers of units of reactants and products to preserve the equality. In combustion reactions, all of the reactants and products other than energy have been molecules. Many chemical reactions involve reactants and products that are not molecules. In writing equations for these reactions, additional precautions must be taken to preserve the equality. The fundamental approach to equation writing remains the same.

Nuclear reactions are involved with atoms, not molecules. Actually, nuclear reactions are involved with only a portion of the atoms, the nuclei of the atoms. The fundamental approach to equation writing for nuclear reactions remains the same as the approach to writing equations for chemical reactions: know the reactants, know the products, preserve the equality. Nuclear reactions are explored in Chapter 22.

5

A MODEL FOR THE STRUCTURE OF ATOMS I: THE NUCLEUS

Chapter 5

Atomic Numbers, Atomic Masses, Isotopes, and the Mole

Atoms are extremely small: the numbers of molecules and atoms in even small samples are extremely large. A drop of water contains more than 1×10^{21} molecules of water, H_2O (made up of more than 2×10^{21} atoms of hydrogen and more than 1×10^{21} atoms of oxygen). With each breath, we inhale about 1×10^{21} molecules of oxygen, O_2; about 4×10^{21} molecules of nitrogen, N_2; about 5×10^{19} atoms of argon, Ar; about 5×10^{17} atoms of neon, Ne; about 5×10^{15} atoms of krypton, Kr, about 2×10^{15} atoms of helium, He; about 5×10^{14} atoms of xenon, Xe; about 2×10^{14} molecules of hydrogen, H_2; and unpredictable numbers of molecules of carbon dioxide, CO_2; carbon monoxide, CO; various oxides of nitrogen, NO_x; sulfur dioxide, SO_2; formaldehyde, H_2CO; and ammonia, NH_3; plus an unpredictable number of particles of dust pollen, bacteria, viruses, and goodness knows what else. The points here are that (1) there are lots of molecules, (2) molecules must be small, and (3) even if the concentration of a substance in the atmosphere is only of the order of one part per billion (abbreviated 1 ppb and numerically expressed 1 in 1 000 000 000 or 1 in 10^9), this very low concentration corresponds to about 5×10^{12} molecules inhaled with each breath. All numbers, of course, depend upon how deeply you inhale, but if you breathe at all, the numbers are impressive.

Size

The extremely small masses and the extremely small dimensions of atoms have been and are a real challenge to investigators. It is just not possible to select an atom and measure it or to count out one by one a number of atoms and go weigh them. These problems have demanded more imaginative approaches, supported by the best instruments that can be designed and built. Over the past two centuries, investigations have ranged from rather crude measurements on gases using the simplest of equipment to the current, very precise indirect measurements using a host of increasingly sophisticated instruments and methodologies.

Mass

Experimental results are rationalized in terms of a model that treats an atom as a diffuse **atmosphere of electrons** surrounding a very small and very dense **nucleus.** An atom is electrically neutral: the electrons are negative and the nucleus is positive. This chapter focuses upon charge and mass. The next chapter focuses upon the distribution of electrons in the relatively large diffuse atmosphere of electrons.

Electron Atmosphere

Nucleus

Hydrogen, the simplest atom, has only one electron, with a charge of 1− and a nucleus, with a charge of 1+. This nucleus of the hydrogen atom is also called the **proton.** The mass of the proton is almost two thousand times the mass of the electron. The atoms of each element have a characteristic number of electrons and the same characteristic positive charge on the nucleus. This characteristic number is known as the **atomic number** of that element and is frequently designated by the letter Z. These atomic numbers also correspond to the sequence of the elements in the Periodic Table. An atom of carbon (Z = 6), the sixth element in the Periodic Table, has an atmosphere of six electrons and a nuclear charge of 6+. The mass of the carbon atom is about four thousand times the total mass of the six electrons. Clearly, the mass of an atom is concentrated in the nucleus.

Atomic Number

This model is now so completely accepted, it is customary to define an element in terms of the particular charge on the nuclei of its atoms; that is, to define the element in terms of its atomic number. Atomic numbers are frequently given with the symbol for the element: hydrogen $_1H$, carbon $_6C$, and nitrogen $_7N$.

Current developments of models concern the distribution of the electrons in the electronic atmospheres about the nuclei and the structure of the nuclei themselves. The volume occupied by the nucleus of an atom is very small in comparison to the volume of the total atom. The diameters of nuclei are of the order of 1×10^{-14} meters (1×10^{-5} nanometers) and the diameters, of atoms are of the order of 1×10^{-10} meters (1×10^{-1} nanometers). The diameter of the atom is of the order of 10 000 times the diameter of the nucleus. The hydrogen atom is the smallest atom. The uranium atom, Z = 92, is the largest known naturally occurring atom on this planet.

Dimensions

5-1 *Editorial Comment*

HOW PRECISE ARE EXPERIMENTAL MEASUREMENTS?

How exact are experimental measurements? This simple question has no simple answer. In fact, it can only be answered for each specific situation. The important point is to report the experimental value in such a way that the best estimate of the experimental uncertainty is made clear. To take a very simple case, the two experimental values 2.0 grams and 2 grams are not equivalent. The zero following the decimal indicates that the value of the mass is closer to 2.0 than it is to either 1.9 or 2.1 grams. To report 2 grams simply means that the mass is closer to 2 grams than it is to either 1 gram or 3 grams. For the purpose of calculation, the experimental value of 2. grams cannot be converted to 2.0 grams. To do so would be a lack of integrity in the use of the data.

For many purposes, a measurement of 2 grams may be entirely adequate. The experienced scientist assesses the precision required and proceeds accordingly. Wisdom along these lines seems to be acquired through misadventure and there is probably not a scientist who has not been caught out on not measuring some quantity with sufficient precision. The masses 132.1 and 132.4 grams (four significant figures) may look impressive, but if the value of interest to the experiment is the difference between these two values, the 0.3 gram difference (one significant figure) may not hold up very well with the other measurements. In this case, it might have been well to use a more sensitive balance to obtain values such as 132.0876 and 132.4352 grams (seven significant figures). The experimental difference is then 0.3476 grams (four significant figures)—a value of much greater significance than the 0.3 gram.

One indication of the reliability of an experimental result is the reproducibility of the results. Reproducibility becomes increasingly impressive when the experiment is carried out by more than one experimenter, the sample sizes are varied, more than one instrument is used, and the experimental method is varied. The skill and care of the experimenter are, of course, important, but instrumental limitations and simplifying assumptions in the design of the experiment may place severe limits on the validity of the results even when dial readings on repeated runs are reassuring. No experimental measurement can be absolutely exact. That which makes the work scientific is the assessment and the report

All of the atoms of an element do not have the same mass. However, the atoms of any one element have only a limited number of values of mass. The best known variations in carbon atoms are designated as carbon-twelve (carbon-12); carbon-thirteen (carbon-13); and carbon-fourteen (carbon-14). These various atoms of carbon are **isotopes** of carbon. Carbon-12 is by far the most common isotope of carbon on the earth. In a natural sample, almost 99% of the atoms of carbon are the carbon-12 isotope, a bit more than 1% is the carbon-13 isotope (1.11%), and there is a mere trace of the carbon-14 isotope. Other isotopes of carbon may be formed by nuclear reactions but these isotopes decompose very rapidly after they are formed. The exact proportion of carbon isotopes in a sample depends to some degree upon the source and previous history of that sample. The word isotope means same place: the same place in the Periodic Table; the same atomic number.

Isotopes

The development of a system of atomic masses followed a fascinating but tortuous course. Fundamentally, the approach has been to assign a value of mass to one type of atom and then to evaluate the masses of all other atoms relative to this arbitrarily chosen reference. Over the years hydrogen atoms, oxygen atoms, and carbon atoms have been used in turn as the reference. The early comparisons were achieved by measurements on gases using very primitive equipment and a considerable degree of ingenuity in both the use of the equipment and the interpretation of the data.

---

of the significance of the experimental value. Good scientific work may be limited by the state of the art (our scientific concepts, our methodologies, and our instruments) and, in many cases, is less exact than the public may expect. Many experiments can be carried out with a precision of 0.1% but a great number may have a uncertainty of 1% or even 10%. There are other experiments for which the power of 10 may be just about as far as it is possible to go at this time.

Contrasted with the above, there are some quantities that are assigned exact values. For example, the atomic mass of the carbon-12 isotope is assigned the value of exactly twelve and may be written 12.000 . . . with any number of zeros. The dielectric constant for a vacuum has been assigned the value of exactly 1. The freezing point of water and the boiling point of water under specified conditions have also been assigned exact temperatures on the Celsius scale. There are many other exactly defined quantities. By definition, there are exactly 1000 grams in 1 kilogram. All of these are assigned values, not experimental values.

The numbers in the formula for a molecule such as C_2H_4 are integers (whole numbers). The coefficients used in balancing equations are integers. These integers follow from our concept of molecules being specific groupings of atoms.

The broad uncertainty associated with the relative atomic mass of naturally occurring isotopic mixtures of the element sulfur, 32.06 ± 0.03, is a consequence of variations in isotopic distribution in naturally occurring sulfur samples. The average relative mass of sulfur atoms is known to be as low as 32.03 and as high as 32.09 depending upon the source of the sample.

As you work in the laboratory, give attention to how precise your measurements are. The precision of the measurements you make depends in part upon your skills and the care you take, but that precision also depends upon the method of measurement used and the instruments available to you. It is quite possible for you to work with great precision using methodologies that are flawed and instruments that are inaccurate. Measurements can be precise in that the results are reproducible but inaccurate in the sense that all values are wrong. Always suspect the calibration of an instrument. Investigators should always seek ways to check the reliability of instruments and methodologies.

Currently, the carbon-12 isotope is used as the standard reference for all atomic masses. Today, instruments known as mass spectrometers are used to determine the relation of the masses of other atoms to this reference. The carbon-12 isotope is arbitrarily assigned the value of exactly twelve (12.000 . . . with as many zeros as anyone cares to write). Using a mass spectrometer, it is a comparatively simple matter to show, for example, that an isotope of magnesium ($Z = 12$) has a mass 1.9988 times (almost twice) the mass of a carbon-12 atom. This isotope of magnesium, therefore, takes on the value 23.985 (1.9988×12) for the relative mass of this isotope of magnesium, as compared to the exact value of 12 assigned to carbon-12. In this manner, the relative masses of the various isotopes of the various elements are obtained with impressive precision. The above isotope of magnesium is known as magnesium-24, where 24 is the nearest whole number to the relative atomic mass of 23.985.

Carbon-12 as a Standard of Mass

Some scientists prefer to attach units of mass to the numbers presented in the previous paragraph as a purely relative scale. There is no reason this cannot be done, just so long as the unit of mass is tailored to serve this particular end. Following this scheme, one atom of the carbon-12 isotope is assigned a mass of exactly 12 "atomic mass units," abbreviated amu. In making this assignment, another unit of mass, the atomic mass unit has been defined. There is no obvious relation of this atomic mass unit to the gram. In this system of atomic mass units, if carbon-12 has a mass of exactly 12 atomic mass units, then the magnesium-24 isotope has a mass of 23.985 atomic mass units. In this book, we will not use atomic mass units; we will instead use atomic masses as relative numbers based upon the assignment of exactly 12 to the carbon-12 isotope.

The unit of measurement of quantity used by chemists is the number of atoms of carbon-12 in exactly 12 grams of carbon-12. If the 23.985 correctly compares the mass of one atom of magnesium-24 to the mass of one atom of carbon-12, then the number of atoms in 23.985 grams of magnesium-24 must be the same as the number of atoms in exactly 12 grams of carbon-12. This number, which must be determined experimentally, would be expected to be very large: atoms have small masses and neither the 12 grams of carbon-12 nor the almost 24 grams of magnesium is an infinitesimal sample by any means. The currently accepted value of this number is 6.0225×10^{23}. In terms of magnitude, the 6.0225 part of the number is of very little consequence in comparison to the exponential quantity 10^{23}. Just write out the number obtained by multiplying 6 by 10 twenty-three times and admire it. The value 6.02×10^{23} (three significant figures) is adequate for most discussions and will be used in the following sections.

Avogadro's Number

Mole

This number of atoms of carbon-12 in exactly 12 grams of carbon-12 is known as **Avogadro's number** in honor of the Italian scientist who first recognized that such a number must play a significant role in chemical relations. Although Amedeo Avogadro recognized that there must be such a number and that it was a very important number, the numerical value was first determined long after his death. This number of atoms or of any other item (6.02×10^{23} atoms or 6.02×10^{23} of any other item) is given the name **mole,** abbreviated mol. A mole of atoms is just as real as a dozen of eggs, a ream of paper, a gross of paper clips, and a case of Orange Crush. The mole is a specific number of items. A mole of crickets is 6.02×10^{23} crickets, never mind whether there really are that many crickets in existence. A mole of carbon-12 is 6.02×10^{23} atoms of carbon-12 and has a mass of exactly 12 grams. A mole of magnesium-24 is 6.02×10^{23} atoms of magnesium-24 and has a mass of 23.985 grams.

The most important skill in the chemist's game is to be able to think in terms of moles. Moles are the units with which the game is played and this unit of quantity will always be with us throughout this text.

Natural Mixtures of Isotopes

So far, the discussion has been in terms of particular isotopes. Most chemical reactions involve the natural mixtures of isotopes and it is important to be able to deal with these in a simple, straightforward manner. For the natural mixture of isotopes of carbon, only carbon-12 (98.89%) and carbon-13 (1.11%) need to be considered. The quantities of the other isotopes are negligible insofar as the determination of average masses is concerned. On the carbon-12 scale, the relative value for carbon-13 is shown by mass spectrometer measurements to be 13.003, in comparison to the exact value 12 assigned to carbon-12. One mole of carbon-13, 6.02×10^{23} atoms of carbon-13, has a mass of 13.003 g. If the distribution of the isotopes in the natural sample is taken into consideration, the average value of the relative mass would be between 12 and 13. One mole of the isotopic mixture, a total of 6.02×10^{23} atoms of the mixture of the two isotopes, would have a mass between 12 and 13 grams. More precisely, the average relative mass is 12.011 and the mass of 6.02×10^{23} atoms of this isotopic mixture is 12.011 grams. The value 12.011 follows directly from the distribution of isotopes in the natural sample, as shown in the following calculations.

For 10 000 atoms of the natural sample, 9 889 atoms are carbon-12 and 111 atoms are carbon-13. A much larger sample would be required to assure the presence of even one atom of carbon-14 or any other isotope of carbon.

isotope	**natural abundance**	**relative mass**		**number of atoms in a 10 000-atom sample**		**total mass, on relative mass scale**
carbon-12	98.89%	12.000	×	9889	=	118,668
carbon-13	1.11%	13.003	×	111	=	1,443
carbon-14	trace			negligible		______
				total for 10 000 atoms		120,111
				average		12.0111

It is these average values of relative masses for the natural mixtures of isotopes that are most directly useful to those interested in chemical reactions. These are the values listed in the table of Atomic Numbers and Relative Atomic Masses (inside back cover) and also given in the Periodic Table (inside front cover).

The accumulation of evidence concerning the composition of isotopic mixtures has revealed that the variations in compositions of isotopic mixtures of some elements are greater than the variations that had been encountered previously. A few years ago, the accepted value for carbon was 12.01115 ± 0.00005. A recent publication of the American Chemical Society gives the value 12.01. When a more precise value is required, the isotopic composition of the particular sample would be determined by mass spectrometric measurements.

The point has been made that the mass of an atom is concentrated in the nucleus of the atom, that the size of the nucleus is small in comparison to the size of the atom, and that the nucleus of the atom is always positive with a charge numerically equal to the atomic number of the element. Much is known about the characteristics of the nuclei of atoms and research programs that are both extensive and expensive are in progress to develop a comprehensive model for the structure of nuclei. At the moment, the picture of the nucleus is quite complicated. (Gratuitous Information 22-1). Not the least of the problems is to rationalize how a total positive charge as great as 100 times the magnitude of the charge of the electron can be confined to a volume as small as the volume of the nucleus of an atom. In chemical reactions, the nuclei of atoms remain unchanged, and chemists can get along with a very limited model of the nucleus. In this limited model, the nucleus is considered to be made up of two types of particles called **nucleons.** These two types of nucleons are **protons**

Chemical Reactions

5-1 *Background Information*

ENERGY

A moving automobile has **kinetic energy,** energy of motion. If the moving automobile collides with an object, the kinetic energy of the automobile will, in part, become kinetic energy of the object or of fragments of the object it strikes.

A bucket of rocks supported by a rope has **potential energy.** Release the rope and the bucket moves toward the center of the earth with the bucket moving faster and faster as more and more of the potential energy becomes kinetic energy. A stretched rubber band has potential energy. This potential energy can, in part, become the kinetic energy of a toy plane by snapping the plane into flight.

Two electrical charges in the vicinity of each other have potential energy. Given the freedom to do so, these charges will move either toward each other or away from each other, depending upon whether the two charges have different signs or the same sign. In the process, the potential energy, due to the force between the two particles, becomes kinetic energy.

Energy takes many forms. The sun radiates energy as units of electromagnetic radiation. Units of energy of electromagnetic radiation are called **photons.** The absorption of photons by the earth warms the earth. The emission of photons by the earth cools the earth. The absorption of some photons by plants supplies the energy required for the chemical reactions known as photosynthesis to take place.

The kinetic energy of a bullet is provided by the chemical energy of the mixture of compounds with which the cartridge was charged.

Falling water has both potential and kinetic energy. With the appropriate gadgetry, part of this energy can be used to turn wheels and generate electrical energy.

The list of transformations is endless. Muscles convert chemical energy to kinetic energy even to the extent that an athlete can jump a bar at a height of more than seven feet.

All of the forms of energy are carefully defined quantities that are related by well-known equations of physics.

Three of the most commonly used units to measure energy are the **joule** (pronounced jewel, also jowl by some), the **calorie,** and the **erg.** If a force of one newton acts through a distance of one meter, the energy expended is one joule.

$$\text{force} \times \text{distance} = \text{energy} \quad \text{(the general relation)}$$

$$\text{newtons} \times \text{meters} = \text{joules} \quad \text{(with specific units)}$$

If a force of one dyne acts through a distance of one centimeter, the energy expended is one erg:

$$\text{dynes} \times \text{centimeters} = \text{ergs} \quad \text{(with specific units)}$$

and **neutrons.** In nuclear reactions, the nuclei of atoms do undergo change, and detailed models of nuclear structure are needed to rationalize these reactions.

Nuclear Reactions

The masses, on the carbon-12 relative scale, for the nucleons (the proton and the neutron) and also for the electron and the hydrogen atom are given below. The charges, in units equivalent to the magnitude of the charge of the electron, are also included.

Nucleons

	relative mass	**charge**
proton	1.007277	1+
neutron	1.008665	0
electron	0.000549	1−
hydrogen atom	1.007825	0

One calorie, 1 cal, is now defined to be 4.184 joules. At one time, the calorie was defined as the energy required to raise the temperature of one gram of water one degree Celsius. This definition turned out to be difficult to use since the energy required to raise one gram of water from 15° to 16° Celsius is not the same as the energy required to raise one gram of water from 75° to 76° Celsius. The calorie, Cal, used in the discussion of diets is really the kilocalorie, kcal (1000 calories). Note the distinction in the abbreviation cal for a calorie and Cal or kcal for a kilocalorie. In scientific work, the kilocalorie is abbreviated kcal, not Cal. Useful quantitative relations between energy units expressed as equations are

$$[\text{energy in ergs}] = \left[1 \times 10^7 \frac{\text{ergs}}{\text{joule}}\right][\text{energy in joules}]$$

$$[\text{energy in joules}] = \left[4.184 \frac{\text{joules}}{\text{cal}}\right][\text{energy in calories}]$$

$$[\text{energy in calories}] = \left[1 \times 10^3 \frac{\text{cal}}{\text{kcal}}\right][\text{energy in kilocalories}]$$

Reprinted with special permission of King Features Syndicate, Inc.

POTENTIAL ENERGY ACCORDING TO HUGO

Note that, on the relative scale of exactly 12 for carbon-12, the nucleons (the protons and the neutrons) have relative masses of slightly more than 1.00 and that the electrons have much smaller masses (0.000549). Also note that the proton has a charge of 1+ and that the neutron has no charge.

How do the relative masses and charges of these three particles relate to the structures of the atoms of elements and the relative masses of the isotopes of an element? The atomic number of the element carbon is 6, $Z = 6$. All atoms of carbon, $_{6}C$, are neutral and have electron atmospheres of 6 electrons and nuclei that contain 6 protons. The nuclei of the atoms also contain neutrons. For carbon-12, the number of neutrons is 6; for carbon-13, the number of neutrons is 7; and for carbon-14, the number of neutrons is 8. Carbon-12 contains 12 nucleons, carbon-13 contains 13 nucleons, and carbon-14 contains 14 nucleons. The symbols used to represent these isotopes are $^{12}_{6}C$, $^{13}_{6}C$, and $^{14}_{6}C$. The subscript is the atomic number (the number of electrons in the electron atmosphere and also the number of protons in the nucleus), and the superscript is the number of nucleons (the total number of protons and neutrons) in the nucleus. The symbolism for the magnesium-24 isotope is $^{24}_{12}Mg$ (12 electrons, 12 protons, and 12 neutrons). The symbolism for the hydrogen atom is $^{1}_{1}H$ (1 electron and 1 proton). [When less was known about the nuclei of atoms, the superscript (24 in $^{24}_{12}Mg$, for example) was called the mass number—the integer, the whole number, that approximates the relative atomic mass of the isotope. It follows, from the fact that the relative masses of both neutrons and protons are approximately 1, that the mass number of an isotope and the number of nucleons in an atom of that isotope must be the same. Throughout this book, we shall use the term number of nucleons rather than the older term mass number.]

One further question: How does the sum of the relative masses for the electrons, the protons, and the neutrons in an atom relate to the experimental value of the relative atomic mass of the atom? The answer may be quite a surprise to you. For carbon-12, $^{12}_{6}C$, the sum of the relative masses is

		relative masses
6 electrons	6 × 0.000549	0.003294
6 protons	6 × 1.007277	6.043662
6 neutrons	6 × 1.008665	6.051990
	total relative mass $^{12}_{6}C$	12.098946

The sum of the relative masses of the constituent parts of carbon-12, $^{12}_{6}C$, is 12.098946. The experimental relative mass for the carbon-12 atom is, by definition of the scale of atomic masses, exactly 12. This is a very significant difference. The experimental value of the relative mass of the atom, the 12.00000 in this case, is always less than the value calculated from the sum of the relative mass of all the particles. This **mass difference** is related to the stability of the nucleus. Mass changes do occur in nuclear reactions, and these mass changes are discussed in some detail in Chapter 22, Nuclear Reactions.

Mass Difference

In chemical reactions where the nuclei of atoms remain unchanged, we are only concerned with the experimental values of relative mass determined through mass spectrometer measurements for the isotopes. The availability of values of atomic masses for the various elements, the concept of the mole, and the numerical value of Avogadro's number are powerful tools in the investigation of the nature of chemical change.

Mass of 1 Mole of Carbon-12

The actual mass in grams of one atom of carbon-12 can readily be evaluated: one mole of carbon-12 has a mass of exactly 12 grams and contains 6.02×10^{23} atoms. Simply divide the total mass, the twelve grams, by the number of atoms (6.02×10^{23} atoms) that make up the twelve grams of carbon-12.

$$12.00 \text{ grams} \div 6.02 \times 10^{23} \text{ atoms}$$

$$\frac{12.00 \text{ grams}}{6.02 \times 10^{23} \text{ atoms}} = \frac{12.00 \times 10^{-23} \text{ grams}}{6.02 \text{ atoms}} = 1.99 \times 10^{-23} \frac{\text{grams}}{\text{atom}}$$

Mass of 1 Atom of Carbon-12

One atom of carbon-12 has a mass of 1.99×10^{-23} grams. As expected, this is an extremely small mass. (The notation 1.99×10^{-23} is 1.99 divided by 10 twenty-three times.)

Mass Relations Derived from Chemical Equations

With a system of atomic masses, the chemical equation now takes on additional meaning: the quantities of reactants and products can be expressed in terms of moles and in terms of grams of each reactant and each product. The equation for the reaction of hydrogen gas with oxygen gas to give water has been discussed in terms of two molecules of hydrogen reacting with one molecule of oxygen to give two molecules of water. This equation is a bookkeeping equality

$$2\,H_2(g) + O_2(g) \qquad 2\,H_2O(g) + \text{energy}$$

and there is no reason the equation cannot be "multiplied through" by any number. If Avogadro's number, 6.02×10^{23}, is used, the equation becomes

$$2 \times 6.02 \times 10^{23}\,H_2(g) + 6.0 \times 10^{23}\,O_2(g) \longrightarrow 2 \times 6.02 \times 10^{23}\,H_2O(g) + \text{energy}$$

Since 6.02×10^{23} molecules of H_2 is one mole of H_2, etc., the above expression can now be read: Two moles of hydrogen molecules react with one mole of oxygen molecules to give two moles of water molecules and energy.

$$2 \text{ moles } H_2(g) + 1 \text{ mole } O_2(g) \longrightarrow 2 \text{ moles } H_2O(g) + \text{energy}$$

Heats of many chemical reactions have been experimentally determined and these values are readily available in handbooks. The value reported for the formation of one mole of water by the above reaction is 57.80 kcal. This information can be incorporated in the expression involved in forming two moles of water in place of the descriptive term "energy":

$$2 \text{ moles } H_2(g) + 1 \text{ mole } O_2(g) \longrightarrow 2 \text{ moles } H_2O(g) + 115.60 \text{ kcal}$$

The expression stated above in terms of moles of reactants and products can also be translated into an expression in terms of grams of reactants and products:

$$4.04 \text{ grams hydrogen} + 32.00 \text{ grams oxygen} \longrightarrow 36.04 \text{ grams water} + 115.60 \text{ kcal}$$

These numerical values of mass follow directly from the atomic masses given in the Periodic Table for the normal isotope distribution of the elements hydrogen and oxygen. One mole of hydrogen gas molecules, H_2, contains two moles of hydrogen atoms, H. One mole of hydrogen atoms has the mass 1.01 grams. Therefore, each mole of hydrogen gas molecules, H_2, has a mass of $2 \times 1.01 = 2.02$ grams and the two moles of hydrogen gas, H_2, required by the equation have a mass of 4.04 grams.

For 1 mole oxygen gas molecules, O_2:

$$2 \text{ moles oxygen atoms} \times 16.00 \frac{\text{grams}}{\text{mole oxygen atoms}} = 32.00 \text{ grams}$$

For 2 moles water, H_2O; consider first one mole of water:

$$1 \text{ mole water} \begin{cases} 2 \times 1.01 = 2.02 \text{ grams hydrogen} \\ 1 \times 16.00 = \underline{16.00} \text{ grams oxygen} \end{cases}$$
$$18.02 \text{ grams water}$$

The mass of 2 moles water is therefore 36.04 grams.

This mass relation, sometimes incorrectly called "weight relation," is experimentally very useful since masses can be so easily determined in the laboratory. Given a modern balance, it is as easy to determine the mass of a sample as it is to step on a bathroom scale to determine your own mass.

The mass relation

$$4.04 \text{ grams hydrogen} + 32.00 \text{ grams oxygen} \longrightarrow 36.04 \text{ grams water} + 115.80 \text{ kcal}$$

is again (or still) a bookkeeping equality and, consequently, the entire expression can be multiplied or divided by an arbitrarily chosen number. For example,

1. dividing the equality by 4.04,

$$1.00 \text{ gram hydrogen} + 7.92 \text{ grams oxygen} \longrightarrow 8.92 \text{ grams water} + 28.66 \text{ kcal}$$

2. dividing the equality by 32.00,

$$0.13 \text{ gram hydrogen} + 1.00 \text{ gram oxygen} \longrightarrow 1.13 \text{ grams water} + 3.62 \text{ kcal}$$

3. dividing the equality by 36.04,

$$0.11 \text{ gram hydrogen} + 0.89 \text{ gram oxygen} \longrightarrow 1.00 \text{ gram water} + 3.21 \text{ kcal}$$

The above operations may seem a bit peculiar, but they serve very useful ends. In the first, the relation is expressed in terms of 1.00 gram of the reactant hydrogen: it gives the mass of oxygen needed to react with the 1.00 gram of hydrogen and also the mass of water formed and the energy released. In the second, the relation is expressed in terms of 1.00 gram of the reactant oxygen: it gives the mass of hydrogen needed to react with 1.00 gram of oxygen and the mass of water produced and the energy released. In the third, the relation is expressed in terms of 1.00 gram of the product water: it gives the grams of hydrogen and the grams of oxygen required and the energy released in the production of 1.00 gram of water.

Mass Change in Chemical Reactions

In chemical reactions, the total experimentally detectable mass of the reactants equals the total detectable mass of the products. This, you will find, is in marked contrast to the type of reactions known as nuclear reactions.

SPECTRA

Visible light is only a small portion of the entire range of electromagnetic radiation. Even within the visible portion there are great variations in the electromagnetic radiation, as is shown by the fact that a glass prism can disperse light from the sun into a spectrum of colors ranging from red on one end through orange, yellow, green, and blue to violet on the other end. The full spectrum of the electromagnetic radiation of the sun extends beyond the ends of the visible spectrum that are set by the limits of our ability to respond physiologically. Beyond the red "end" lies the infrared region and beyond the violet "end" lies the ultraviolet region.

The total range of electromagnetic radiation goes far beyond this to include many other well-known radiation regions:

Radio Waves
Microwaves
Infrared Light
Visible Light
Ultraviolet Light
X-rays
γ-rays (Gamma rays)

The above are convenient names to refer to various regions of the total electromagnetic spectrum. There are, however, no breaks between the various regions. As with the visible region of the spectrum, the total spectrum changes continuously from one end to the other.

With the exception of a very hot object, such as a star, a given source of electromagnetic radiation does not emit the entire electromagnetic spectrum from radio waves to gamma rays, γ-rays. The spectrum that a source emits is the **emission spectrum** of that source.

See the color plate on the back cover of this book for the visible portion of the emission spectra of an incandescent lamp, a neon gas lamp, a mercury vapor lamp, and a sodium vapor lamp. The whitehot metal filament of the incandescent light bulb produces a continuum of wavelengths that extends beyond the longest wavelengths (red) and also extends beyond the shortest wavelengths (violet) that our eyes respond to as colors. The gases emit only specific wavelengths, which appear in their spectra as bright lines. The wavelengths emitted are uniquely characteristic of each element.

Each line in a bright-line spectrum is the image of the narrow slit through which light is admitted into the spectrograph. If the slit is wide, the lines are wide; if the slit is narrow, the lines are narrow. If the entrance of light to the instrument had been through a round hole, the emission spectrum obtained would be a bright spot spectrum. Visible light entering the instrument is dispersed according to wavelength either by a glass prism or by a diffraction grating (a system of closely spaced, equidistant, and parallel lines ruled on a polished surface).

All substances absorb electromagnetic radiation. No substance, however, absorbs the entire electromagnetic spectrum from radio waves to gamma rays. The electromagnetic radiation that a substance does absorb is the **absorption spectrum** of that substance. A substance that absorbs the entire visible region of the spectrum is black. A substance that does not absorb any region of the visible spectrum is colorless and is called white if there are a multitude of surfaces from which light is reflected. All of the compounds of carbon, hydrogen, and oxygen discussed in Chapters 2 and 3 are colorless. A substance that absorbs only a part of the visible spectrum is a colored substance, that color being the consequence of the regions of the visible spectrum not absorbed.

Electromagnetic radiation is emitted and absorbed in units of energy called photons. A radio wave photon has very little energy compared to a gamma ray photon. The absorption of a photon by a molecule may provide the energy needed to initiate a chemical reaction. This is the role of light in photosynthesis. The absorption of a photon in the ultraviolet region provides much more energy than the absorption of a photon in the visible region.

The absorption of a photon in the infrared region provides even less energy than a photon in the visible region. The infrared photons seldom provide the energy needed for chemical reactions to take place. Their absorption manifests itself by an increase in temperature, and infrared lamps are frequently referred to as heat lamps.

Mass Change in Nuclear Reactions

In nuclear reactions, the nuclei of atoms undergo change, with transformations into other isotopes of the same element in some cases and with transformations into other elements in other cases. In nuclear reactions, there is a detectable mass change when the masses of the reactants and the products are compared. This mass difference in nuclear reactions is associated with the energy of the nuclear reaction.

In chemical reactions, there are also energy changes. Why is this energy change not associated with a detectable mass difference? The answer lies in the magnitude of the energy change. The energy changes associated with chemical reactions are of the order of 1 000 000 joules (approximately 240 000 cal) or less for reactions involving a few moles of product. The energy changes associated with nuclear reactions can be of the order of one million times as much—or even more. The mass changes in nuclear reactions are explored in Chapter 22.

In discussing energy changes, chemists use both the calorie and the joule. One calorie is equivalent to 4.184 joules. (We shall use the value 4.18.) In the United States, the use of the calorie and kilocalorie has been more common. In the rest of the world, the joule and kilojoule are used almost exclusively, and scientists in this country are gradually shifting to the use of those units.

Energy Equivalent of Mass

The relationship between mass changes and energy changes is expressed as follows: energy change measured in joules is equal to the product of the mass change expressed in kilograms multiplied by the square of the velocity of light expressed in meters per second.

$$\begin{pmatrix}\text{energy change}\\ \text{in joules}\end{pmatrix} = \begin{pmatrix}\text{mass change}\\ \text{in kilograms}\end{pmatrix} \times \begin{pmatrix}\text{velocity of light in}\\ \text{meters per second}\end{pmatrix}^2$$

This equation, $E = mc^2$, is known as the Einstein equation. Units other than those shown above can be used. The one requirement is that the units on the lefthand side are the same as the units on the righthand side of the equation.

The velocity of light, c, is a well-known experimental quantity, 3.00×10^8 meters per second, and it is quite straightforward to show that a 1 000 000-joule (1.0×10^6 joules) energy change corresponds to a mass change of 1.1×10^{-11} kilograms. This mass change can also be expressed as 1.1×10^{-8} grams or 11 nanograms. To detect a mass change associated with a chemical reaction as small as 11 nanograms exceeds the capability of our instruments. It is important to recognize that there is no experimental evidence to indicate that mass changes are or are not associated with chemical reactions. The expected mass changes associated with the energy changes characteristic of chemical reactions are just too small to be experimentally measured. It is generally accepted that the Einstein relation is applicable to all energy changes and that the corresponding mass change may or may not be experimentally detectable.

Scramble Exercises

Use atomic numbers and the atomic masses of the normal isotopic mixtures from the Periodic Table, inside front cover, as needed.

1. Always check the calculations indicated in the text as you work through a section. For example:

Show that the diameter of nuclei of 1×10^{-14} meters is equivalent to 1×10^{-5} nanometers.

Follow through the numerical calculations to find the mass of one atom of carbon-12. (The value of 10^0 is 1, and 10^0 can be introduced as a multiplier of any number anytime without changing the value of the quantity multiplied. Note the sign of the power of 10 in the answer.)

Show that a change in energy of 1 000 000 joules, according to the Einstein equation, is equivalent to a change in mass of 11 nanograms.

2. Two isotopes of uranium that have played and continue to play a significant role in international politics and our domestic economy are uranium-235, ${}^{235}_{92}U$, and uranium-238, ${}^{238}_{92}U$. In terms of neutrons and protons, what are the structures of the nuclei of the atoms of these isotopes? What is the number of electrons in the electron atmosphere of the atoms of these isotopes?

3. Another isotope that has been much in the news is strontium-90, ${}^{90}_{38}Sr$. How does the nuclear structure of this isotope compare with ${}^{88}_{38}Sr$, the most common isotope of strontium? How does the electron atmosphere of ${}^{90}_{38}Sr$ compare with the electron atmosphere of ${}^{88}_{38}Sr$?

4. The ingestion of any isotope of strontium leads to its deposition in bone structures. This is a chemical property of the electron structure of the element. What correlation can be made between the position of strontium in the Periodic Table and the experimental fact that ingested compounds of strontium lead to the deposit of strontium in the bone structure? The ingestion of compounds of calcium is essential to the maintenance of healthy bone structure.

5. Several of the following exercises are a rather detailed exploration of the mass relations (the stoichiometric relations) for the complete combustion of methane, CH_4, in oxygen, O_2.

(a) Start by writing the equation for the complete combustion of methane, CH_4. Get this correct! Everything else depends upon this equation. The only products are carbon dioxide and water.

(b) How many moles of methane, how many moles of oxygen, how many moles of carbon dioxide, and how many moles of water are represented by the equation you have just written?

_____ moles methane + _____ moles oxygen ⟶ _____ moles carbon dioxide + _____ moles water

(c) Convert b into a statement of mass relations (a stoichiometric relation).

_____ grams methane (CH_4) + _____ grams oxygen (O_2) ⟶ _____ grams carbon dioxide (CO_2) + _____ grams water (H_2O)

(For all chemical reactions, the equality requires that the total mass of reactants equal the total mass or products.)

6. The following problem depends upon the relations written out in Exercise 5. This problem, like all other problems of this type, can be solved by a number of routes. Two routes are suggested here: the first route pivots directly from the mole relation in 5(b); the second route pivots directly from the mass relations stated in 5(c). A number of other routes are possible and equally proper.

   Problem: How many grams of water are formed by the complete combustion of 4.8 grams of methane?

   Route 1: Convert 4.8 grams of methane to moles of methane. (Will this answer be more than 1 mole or less than 1 mole?) Use the relation in 5(b) to find the number of moles of water that corresponds to this number of moles of methane. Convert the moles of water to grams of water.

   Route 2: Use the relation in 5(c) to find the grams of water that are formed from the complete combustion of 1 gram of methane and then the number of grams of water formed by the complete combustion of 4.8 grams of methane.

   Cross-check: The moment of truth comes when the results of route 1 and route 2, or any other valid route, are compared. If they do not check, then it is back to the beginning to check the equation in 5(a) and all the following steps. If the values obtained by the two routes do not check, you know you have at least one error. If values obtained by the two routes do check, there is still a bit of doubt. You may have made two compensating errors—for example, you may have used an incorrect value for an atomic mass in both routes.

7. In the complete combustion of one mole of methane (approximately 16 grams), 211 kilocalories of energy are released. Evaluate how much energy in kilocalories would be released in (a) the complete combustion of 1.00 gram of methane and (b) the complete combustion of 4.8 grams of methane.

8. In the complete combustion of one mole of methane, the energy released can also be expressed as 882 kilojoules. Evaluate the energy in kilojoules released in (a) the complete combustion of 1.00 gram of methane and (b) the complete combustion of 4.8 grams of methane.

9. By definition, 1 kilocalorie is 4.18 kilojoules. Cross-check the answers you obtained in 7 and 8.

10. Find the mass of methane that would be required to produce 40 grams of carbon monoxide, CO, according to the reaction of incomplete combustion indicated below. The first step is to complete the chemical equation by conserving the number of atoms of each element in the following statement of reactants and products.

$$\_\_\_\_ CH_4 + \_\_\_\_ O_2 \longrightarrow \_\_\_\_ CO + \_\_\_\_ H_2$$

   Having found one route to solve this problem, now find a second route and cross-check the values obtained by the two routes.

11. Calculate the number of molecules of methane, CH_4, in 4.8 grams of methane. (Should the power of 10 in the answer be positive or negative?)

Additional exercises for Chapter 5 are given in the Appendix.

A Look Ahead

The concepts of atomic masses and moles are essential to the exploration of chemical phenomena. The scramble exercises present types of situations in which these concepts are used. Stay with each of those scramble exercises until you understand thoroughly the problem, how to approach the problem, and how to obtain the numerical value of the answer.

The material in Chapter 6 dealing with the modern model for the distribution of electrons in the electronic atmosphere of atoms is both complex and abstract. If this is your first experience with chemistry, you have no way of knowing where the development leads or how it gets there. There is, also, a very good chance that you have limited familiarity with the background on which it is built.

A rigorous development of the topic is far beyond the scope of this book and only an indication of the development of the concepts can be given here. The use of the results of the development is within the scope of this book and within your capacity to understand at this time. The results are absolutely essential to the pursuit of many of the more interesting aspects of modern chemistry.

It is recommended

(1) that you read the chapter to discover where the development leads,
(2) that you learn to play the game of writing out the notation for the ground state distribution of electrons in atoms of elements with atomic numbers up to at least 20,
(3) that you learn the correlation between these distributions and the positions of the elements in the Periodic Table, and
(4) that you then go back and work through the chapter several times.

Comprehension develops gradually. When you can manage items 2 and 3, there will be no difficulty continuing with later chapters. The concepts introduced in Chapter 6 reappear many times. This is not a time to cut classes. Hearing the material discussed is a great help.

Work on this model. You may find the approach both interesting and fascinating. This model is one of the many great triumphs of the human mind.

6

A MODEL FOR THE STRUCTURE OF ATOMS II: THE ELECTRON ATMOSPHERE

Chapter 6

The most significant single step in the development of a modern model of the atom is credited to Niels Bohr (1913). The Bohr model of the atom was remarkably successful in accounting for the emission spectrum of the hydrogen atom and the formation of a number of compounds. The Bohr model was, however, soon superseded by a more comprehensive wave mechanical approach, first proposed in 1927 by Erwin Schrödinger.

The Bohr model pictures the atom as a dense positive nucleus surrounded by planetary electrons in specific circular and elliptical orbits. Diagrams of the larger atoms are attractive designs in themselves and have become a well-known symbol—almost a symbol of the science and technology of the twentieth century. The Bohr model of the atom was designed to rationalize the emission spectrum of hydrogen gas (color plate on the back cover).

The Bohr Atom

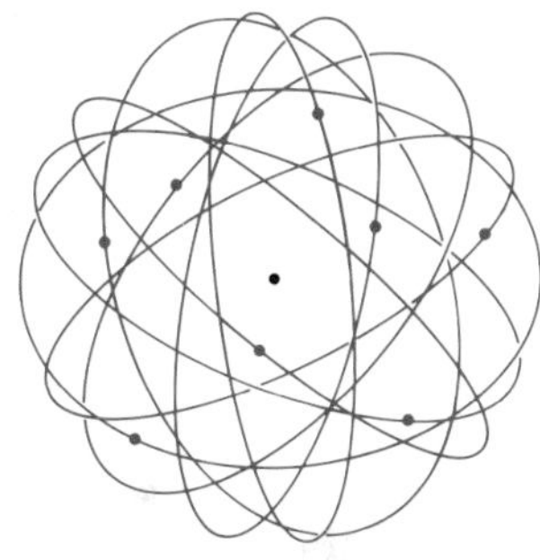

The very significant breakthrough in the concept of atoms resulting from Bohr's studies of the emission spectrum of hydrogen is the treatment of atoms as having a number of discrete energy states. Each atom has a specific energy for each energy state. To change in energy, the atom gains just exactly the energy that separates its energy state from a higher energy state or releases just exactly the energy that separates it from a lower energy state. These energy changes are usually brought about by the absorption or the release of a photon of light. To repeat, an atom has a number of discrete energy states.

An analogy to this system of quantized states is a climber on a rock face that has a limited number of secure foot positions. The climber can move from one position

Gratuitous Information 6-1

BOHR AND HIS NOBEL MEDAL

When Bohr was preparing to flee Denmark prior to the invasion by Germany during World War II, he dissolved the Nobel Medal, which he had received in recognition of his work on the structure of the atom, in a mixture of hydrochloric acid and nitric acid and placed the bottle containing the solution on a lower shelf in the laboratory. After the war, the bottle was still there, the gold was recovered from the acid solution by chemical means, and the medal recast. Gold (classified as a noble metal and also as a coin metal) is one of the most unreactive metals and does not dissolve in either hydrochloric acid or nitric acid, but it does dissolve in a mixture of the two acids known as *aqua regia* (royal water).

to another but there are only a limited number of positions that can be assumed. The positions the climber can assume are inherent in the structure of the rock face. The energy states, the orbits the electron can assume, are inherent in the structure of the atom.

The Schrödinger Approach

The Schrödinger approach also considers the atom to be made up of a dense positive nucleus surrounded by an atmosphere of moving electrons. The electrons are, however, not considered as particles moving in specific orbits. The great contribution of the Schrödinger approach is to describe the properties of the electron, a particle, in terms of a wave phenomenon. This seems a strange approach to those more accustomed to classical mechanics than to wave mechanics.

The properties of waves are really quite well understood and their characteristics are described by well-known mathematical expressions. Even the properties of a moving basketball could be treated as wave phenomena and students in more advanced courses are frequently assigned problems dealing with a particle confined to a box as a wave phenomenon.

Atomic Orbital

The wave mechanical approach generates the concept of regions about the nucleus that are likely to be populated with electrons and regions that are unlikely to be populated with electrons. There is a high probability for the electron to be in the populated regions and a low probability for the electron to be in the other regions. An analogy: the regions of high probability for an individual tend to be the places where the individual sleeps, works, plays, and eats. These are the places of high probability for the body at least. The region of high probability for the electron is called an **orbital.** One very significant difference between the behavior of an electron and that of a person is that electrons are moving continuously and also moving at extremely high speeds. Where an electron is and what it is doing are in no way dependent upon the time of day. The mathematics of the wave mechanical approach predicts the probability of finding an electron in a specific region. To most chemists, this region of high probability for the electron is called an orbital; to some chemists, the orbital is the mathematical equation from which these probabilities are calculated.

The end product of the wave mechanical approach is a set of mathematical expressions known as wave functions. These functions are also known as *psi* (Greek letter ψ) functions simply because they are frequently written as *psi* equal to a rather complex expression involving quantities such as the charge of the electron, the atomic number, and the coordinates of the electrons with respect to the nucleus of the atom. It is these wave functions that define the orbital. Or, as it is more frequently stated by those who carry out these calculations, these wave functions are the orbitals. It is the square of the wave functions, ψ^2, from which the probabilities for an electron, at a specific position in reference to the nucleus, are derived. Note again, these regions of high probability for electrons are commonly called orbitals by many chemists. Polar coordinates are used in these wave functions rather than Cartesian coordinates. By placing the nucleus of the atom at the origin and expressing the position of each electron in terms of a distance from the origin along a radius identified by two angles, the mathematical operations are greatly simplified.

Types of Atomic Orbitals

Many types of atomic orbitals emerge from the wave mechanical approach. Only four of these are of importance to most chemical problems. These four are designated as *s*-orbitals, *p*-orbitals, *d*-orbitals, and *f*-orbitals. The formation of compounds of hydrogen and carbon and the reactions of these compounds are rationalized in terms of an s-orbital of the hydrogen atom and in terms of an s-orbital and three p-orbitals of the carbon atom. The success of ammonia in removing blue compounds that frequently build up on bathtubs where the faucet drips is rationalized in terms of d-orbitals of the copper atoms in the deposit and a p-orbital of the nitrogen atom in the ammonia molecule, NH_3.

Background Information 6-1

WAVES AND PARTICLES: PARTICLES AND WAVES

Shock waves, including sound waves, are transmitted through solids, liquids, and gases. Ocean waves and other waves evident on the surfaces of liquids are transmitted by those liquids. None of these types of waves are transmitted through a vacuum and it is not these types of waves that are being considered here.

Visible light and all other spectral regions of electromagnetic radiation, including microwaves and radiowaves, are transmitted through a vacuum and also through solids, liquids, and gases unless the photons of electromagnetic radiation are absorbed by energy state transitions within the solids, liquids, and gases. In the above sentence, electromagnetic radiation has been referred to as waves (microwaves and radiowaves) and also as particles (photons of electromagnetic radiation).

Whatever photons are, some of their properties can best be described as waves and some of their properties can best be described as particles. Whatever electrons are, some of their properties can best be described as particles and some of the properties of moving electrons can best be described as waves.

Today, these apparent contradictions in the properties of photons and electrons are not considered to be a contradiction in the properties of photons and electrons but to be instead inherent in the carry-over of the language of the nineteenth-century concepts of waves and particles into the analysis of the results of twentieth-century investigations.

The properties of all moving particles, including basketballs, can be treated mathematically as waves. Such treatments do not, however, lead to very useful results for large particles. The wavelength for a basketball, moving at any perceptible rate, turns out to be much shorter than cosmic rays and even shorter for rapidly moving basketballs. These wavelengths are much shorter than the diameters of atoms; consequently, basketball waves cannot be experimentally detected. For the very small moving particles such as electrons, their wave properties are experimental realities.

In the discussion of light and other regions of the electromagnetic spectrum, the terms velocity, wavelength, frequency, and energy are often used. These are all mathematically related.

The velocity of light is an experimentally determined quantity. For a vacuum, the value is 3.00×10^8 meters/sec. (Light travels 3.00×10^8 meters in one second in a vacuum.) For air, the velocity is slightly lower but not significantly lower insofar as we will be concerned in this book. The velocity of light in liquids and solids is lower than the velocity of light in air.

Wavelength is the length of the wave measured along a straight line from any point on one wave to the corresponding point on the next wave. The frequency is the number of waves per second. Wavelengths and frequency are related:

$$\text{wavelength} \times \text{frequency} = \text{velocity of light}$$

$$\left[\begin{matrix}\text{wavelength}\\ \text{in meters}\end{matrix}\right] \times \left[\begin{matrix}\text{number of waves}\\ \text{per second}\end{matrix}\right] = 3.0 \times 10^8\ \frac{\text{meters}}{\text{second}}$$

Either the wavelength or the frequency must be experimentally determined.

Energy, the energy of the photon, is directly proportional to the frequency of the radiation and given by the relation

$$\left[\begin{matrix}\text{energy of the}\\ \text{photon in ergs}\end{matrix}\right] = [6.6 \times 10^{-27}\ \text{erg seconds}] \times [\text{frequency in seconds}^{-1}]$$

$$\left[\begin{matrix}\text{energy of the}\\ \text{photon in ergs}\end{matrix}\right] = [6.6 \times 10^{-27} \times \text{frequency}]\ \text{ergs}$$

Figure 6-1

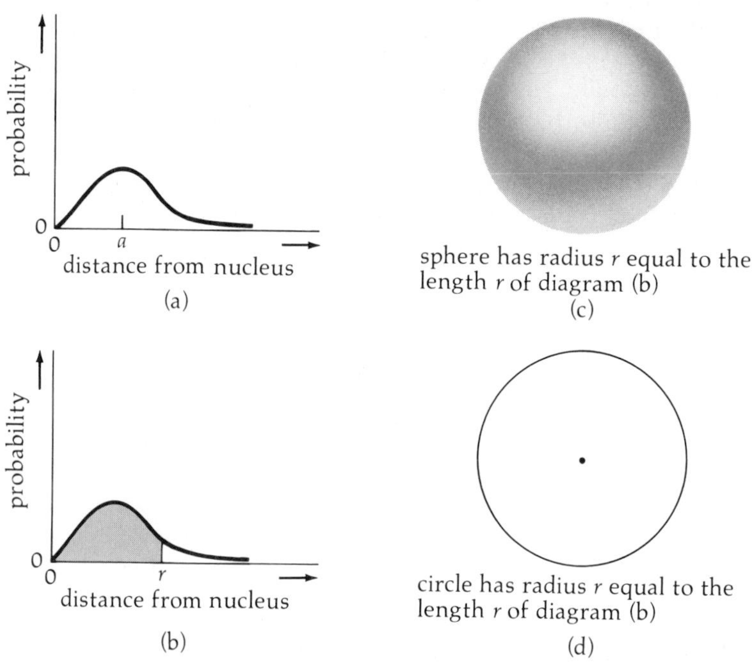

PROBABILITY DISTRIBUTION OF 1*s* ELECTRONS WITH RESPECT TO THE NUCLEUS BASED UPON THE 1*s* *psi* FUNCTION FOR HYDROGEN ATOMS. Representations (a) and (b) are in terms of a graph, representation (c) is in terms of a sketch of a sphere, and representation (d) is in terms of a line drawing of a circle that has the same radius as the sphere.

1s-Orbitals

The *s*-orbital of the hydrogen atom is spherically symmetrical with respect to the nucleus. The probability of the electron being in some small volume changes with distance from the nucleus, but it makes absolutely no difference what direction the distance is measured from the nucleus. Representations of an *s*-orbital are given in Figure 6-1.

Representation (a) is a graph of the probability (vertical axis) of the 1*s* electron being in a shell of some set thickness, such as the skin of an orange or the rubber of a hollow rubber ball, at different distances from the nucleus (horizontal axis). This graph shows both the probability of the electron being close to the nucleus and the probability of the electron being far removed from the nucleus to be small. The probability of the electron being far removed from the nucleus does not become zero, however. It just continues to approach zero. The highest probability is that the electron will be at the distance *a*, diagram (a), from the nucleus. Representations (b), (c), and (d) are based upon this graph.

It is difficult to think about and talk about an atomic orbital that has no finite boundary, an orbital for which there is a probability, even though a very small probability, that the electron is far removed from the nucleus. The practical resolution of this problem is to ignore the very diffuse outer edge of the 1*s*-orbital and to focus upon a spherical volume that encompasses 90% probability for the electron and excludes 10% probability for the electron. In representation (b), Figure 6-1, 90% of the area under the probability curve has been used to determine the distance *r*, measured from the nucleus, to be used as the radius of the sphere, representation (c). The probability of the electron being within the sphere is 90%. The probability of the electron being outside the sphere is 10%. The circle, representation (d), is a line drawing to

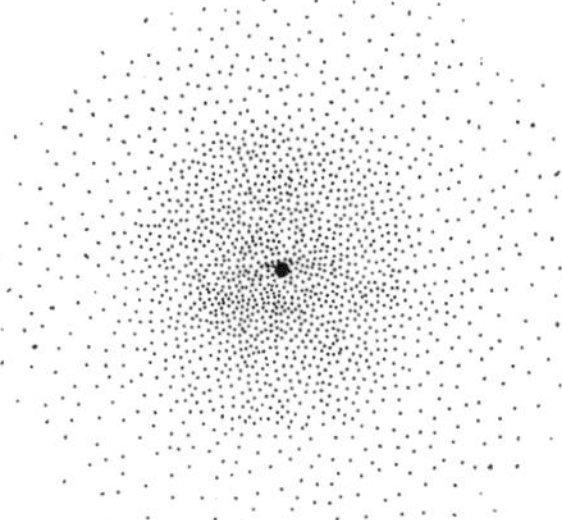

distribution over time of 1*s* electrons about the nucleus

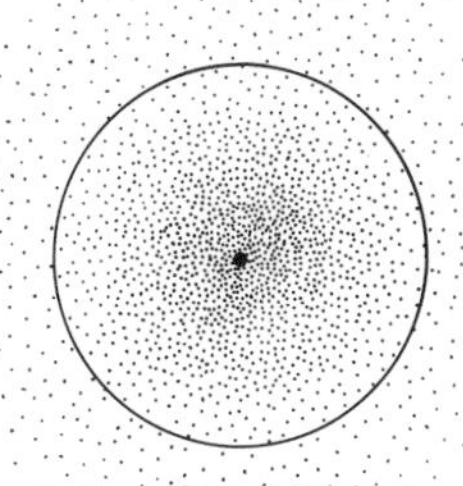

90% probability region for 1*s* electrons enclosed by circle

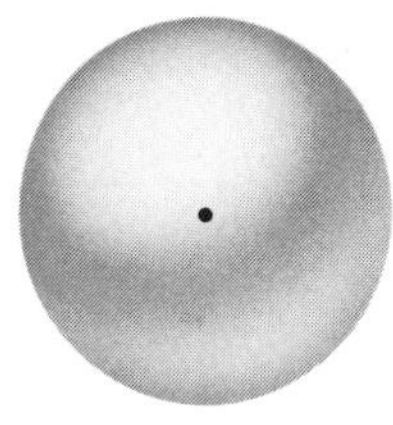

sphere of same diameter as circle representing the 1*s*-orbital

Background Information 6-2

AXIS OF SYMMETRY

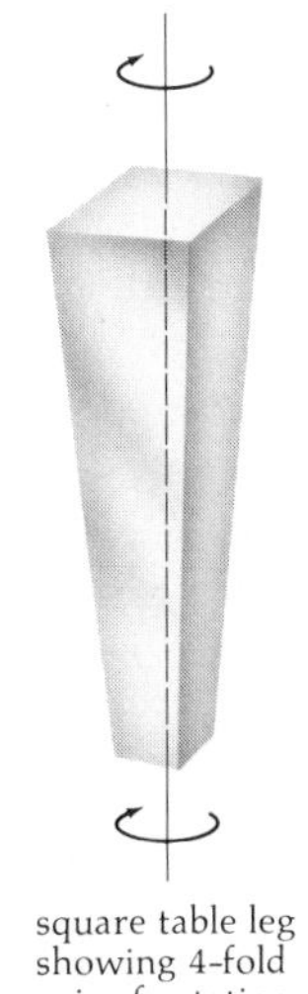

square table leg showing 4-fold axis of rotation

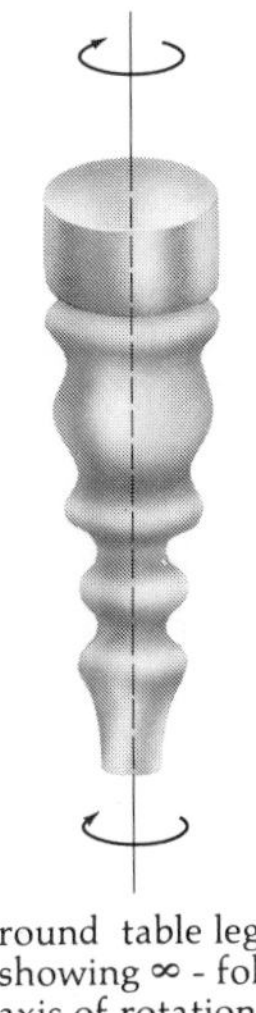

round table leg showing ∞ - fold axis of rotation

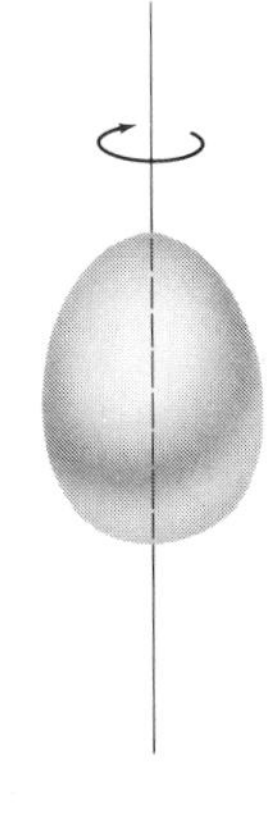

egg showing ∞ - fold axis of rotation

A square table leg has an axis of 4-fold symmetry. An egg, a bowling pin, and a round table leg all have an axis of ∞-fold symmetry. For each of these objects, there is an imaginary line about which the object can be rotated (less than 360°) to give a new position that is indistinguishable from the initial position. For the square table leg, there are four indistinguishable positions in one complete revolution (360°). Hence the term 4-fold axis. For the egg, the bowling pin, and the round table leg, there is an axis of rotation with an infinite number of indistinguishable positions, an axis of ∞-fold symmetry. This assumes that you ignore imperfection in the surface of the egg, the bowling pin, and the table leg.

A *p*-orbital has an axis of ∞-fold symmetry; an *s*-orbital has an infinite number of ∞-fold axes of symmetry.

represent the sphere, and the circle has become the symbol, the pictograph, for the 1*s*-orbital.

p-Orbitals

The *p*-orbital is again a region of high probability for the electron but the general shape of that volume is similar to that of a very fat hourglass (almost round bulbs) with the nucleus at the constriction between the two bulbs. In the same sense that it is customary to represent an *s*-orbital by a line drawing of a circle, it is customary to represent the *p*-orbital by a figure-eight line drawing with the understanding that this drawing represents a three-dimensional figure of the fat hourglass type and that the volume again encompasses a 90% probability of enclosing the electron: 45% on one side of the nucleus and 45% on the other side with a 10% probability that the electron is outside this volume.

6-3 *Background Information*

MOVING CHARGES AND MAGNETIC FIELDS

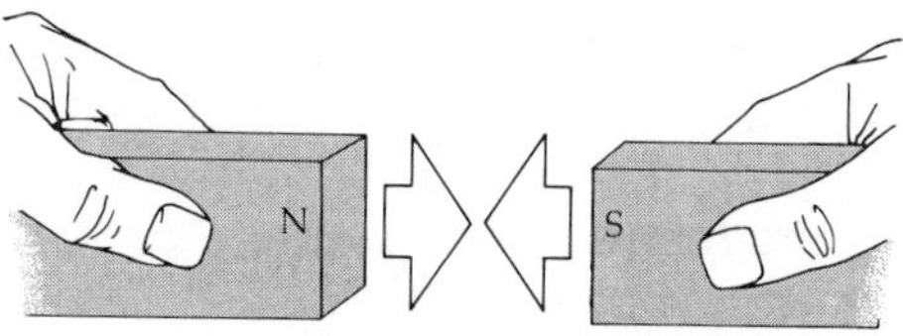

attraction between north pole of one magnet and south pole of another

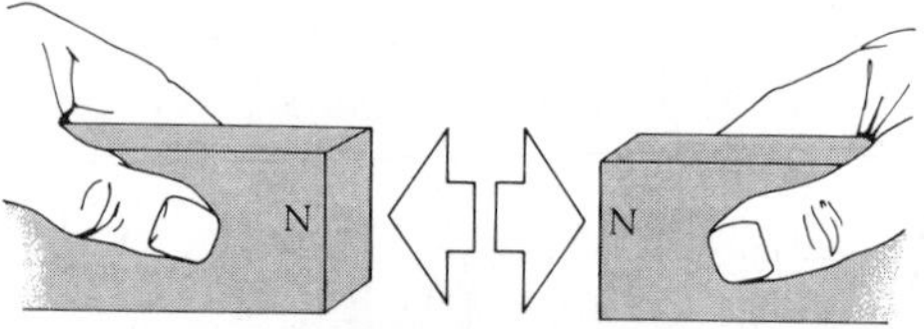

repulsion between north pole of one magnet and north pole of another

Bar magnets and horseshoe magnets are intriguing items. The force of attraction for an iron object is quite apparent when a handheld magnet is brought near an iron object and, by comparison, the absence of this force of attraction when the magnet is brought near a silver or an aluminum object is also apparent. The region around a magnet in which the magnet is effective as a magnet is referred to as the magnetic field of the magnet. The strength of the magnetic field is most pronounced in the region of the two poles of the magnet—one designated as the north pole and the other designated as the south pole. With two handheld bar magnets, the force of attraction between the north pole of the one magnet and the south pole of the other magnet is quite apparent; the force of repulsion between the north pole of one magnet and the north pole of the other magnet is equally apparent—also the force of repulsion between the south pole of one magnet and the south pole of the other magnet.

A charged particle in motion also creates a magnetic field. A wire carrying an electric current is surrounded by a magnetic field. If the direction of the current is reversed, the orientation of the magnetic field is reversed. Electromagnets are the consequence of passing a direct electric current through a coil of wire. The coil produces a more intense magnetic field concentrated in a smaller region than the magnetic field obtained along a straight wire. These intense magnetic fields of electromagnets are very useful in sorting scrap materials and in loading ferromagnetic materials.

The flow of the many electrons along the wire in electromagnets produces the intense magnetic field. The magnetic field produced by the linear motion of a single electron is small indeed. Many electric motors, including electric

This fading off of the probability is difficult to convey in a physical model. A useful model can be made from a cork (the nucleus), a double-ended knitting needle (for mechanical support only), and two large plugs of absorbent cotton. With the cork at the center of the needle, thrust a large ball of cotton over each end of the needle, shape the cotton to create the hourglass figure, and fluff out the edges to give the impression of decreasing probability near the outer edges of the cotton balls. A similar cotton model of an *s*-orbital would simply be a cork (the nucleus) imbedded in a ball of cotton that is fluffed out at the edge. A *p*-orbital has one ∞-fold axis of rotation. In the "knitting needle representation," the knitting needle coincides with that ∞-fold axis of rotation. An *s*-orbital has an infinite number of ∞-fold axes of rotation. In your own work, you must become adept in using the line drawing of a circle to represent an *s*-orbital and the line drawing of a figure eight to represent a *p*-orbital. These are pictographs with which you can communicate with yourself and with others.

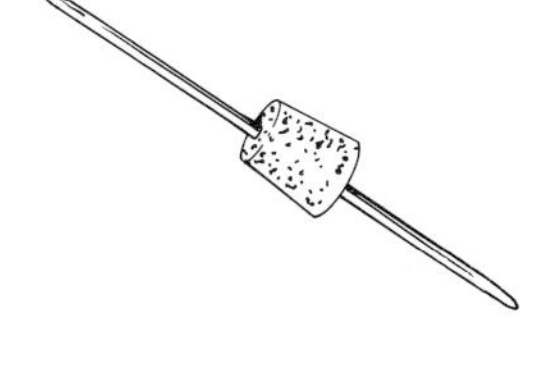

There is an additional point: *p*-orbitals come in sets of three. If the ∞-fold axis of rotation of one orbital is lined up with the *x* axis of an arbitrarily chosen set of Cartesian coordinates with the nucleus at the origin, the ∞-fold axis of rotation of a

---

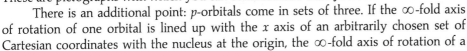

clocks, depend upon passing an alternating current through several coils of wire, some coils mounted on the frame of the clock and some coils mounted on the rotary shaft of the motor. By a clever orientation of these coils and the directions of flow of current, the magnetic fields produced interact to impart motion to the shaft of the motor. The average number of such electrical motors in American homes is impressive. Don't forget to count the record players and the electric toothbrushes as well as clocks, fans, and refrigerator compressors.

A magnetic field is also associated with a spinning charge, the orientation of the magnetic field being determined by the axis of rotation of the charge and the direction of rotation. Reverse the direction of spin and the magnetic field is also reversed. A spinning electron has the properties of a very small bar magnet with a north pole and a south pole.

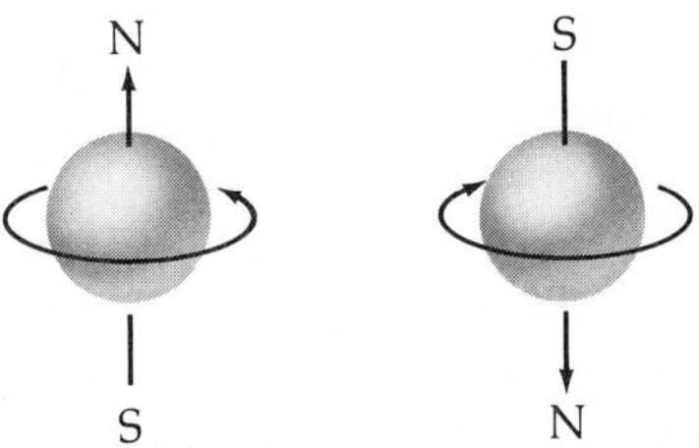

If an electron moving at a constant velocity comes into a magnetic field produced by some external magnet, the motion of the electron will be changed. This change in motion may be a change in rate of motion, a change in direction of motion, or a change in both the rate and the direction of motion, depending upon the orientation of the direction of motion of the electron to the magnetic field. If another moving charged particle has the same charge as an electron and a mass greater than that of the electron, the particle's response to the magnetic field is much less. This dependency on mass is the basis of the mass spectrometers used to determine relative atomic masses.

If a conductor, such as a piece of copper wire, is moved through a magnetic field, movement of electrons is induced in the conductor. This generation of an electric current by the movement of a wire across a magnetic field is the basis of an electric generator. Electric generators are the reverse of electric motors. From the standpoint of construction, they are very similar. In electric motors, an electric current is used to bring about the rotation of the shaft of the motor. In an electric generator, the shaft is rotated mechanically to bring about the flow of an electric current in a metallic conductor.

Electricity delivered by power lines is almost entirely produced by using steam turbines and water turbines to rotate the shafts of electric generators. Steam is generated using heat from combustion furnaces utilizing conventional fuels, particularly fossil fuels, or from nuclear reactors utilizing controlled fission reactions.

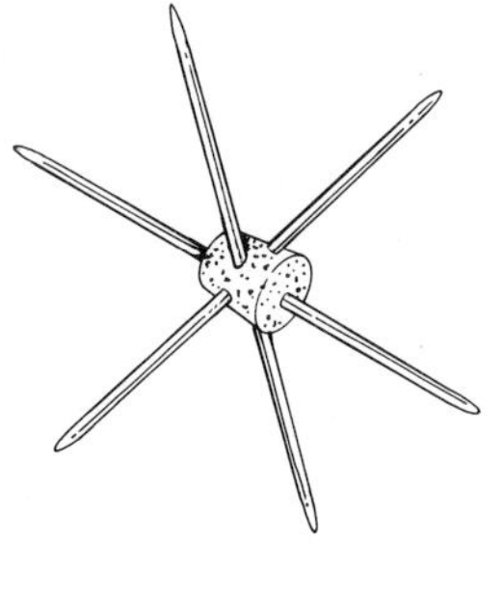

second p-orbital can be lined up with the y axis. In the process, the ∞-fold axis of rotation of the third p-orbital automatically coincides with the z axis. Consequently, the three p-orbitals are frequently designated as p_x- (read p-x), p_y-, and p_z-orbitals.

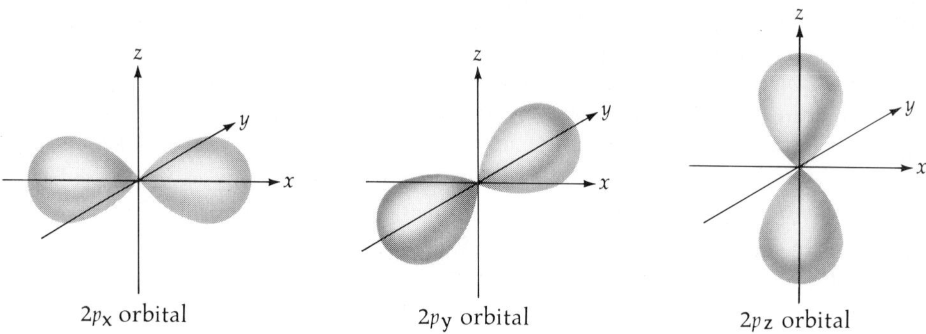

d-Orbitals

The d-orbitals come in sets of five and are more complicated than p-orbitals. Each orbital has a characteristic distribution in space (shape), four being identical in character and the fifth quite different. They also have specific orientations in space with respect to each other and to the p-orbitals.

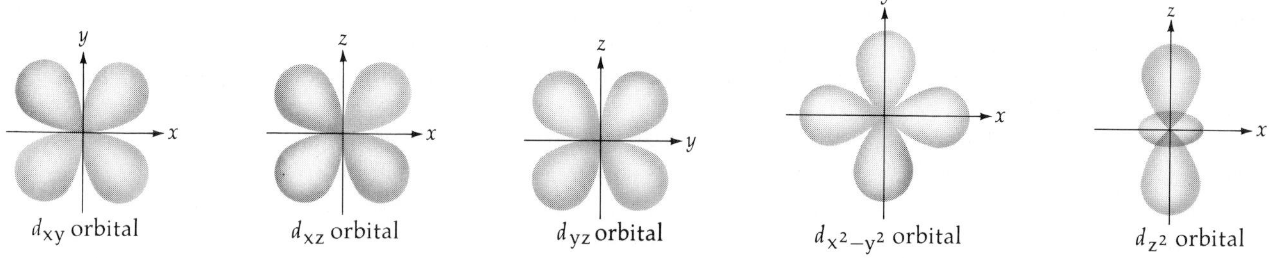

f-Orbitals

The f-orbitals are still more complicated. They come in sets of seven and have specific orientations with respect to each other, to the p-orbitals, and to the d-orbitals. The s-orbital is the simplest. It comes in sets of one, and has the maximum degree of symmetry. In this book, we shall be almost entirely concerned with chemical reactions that are rationalized in terms of s- and p-orbitals.

All of the concepts of s, p, d, and f atomic orbitals and also the concept of quantized energy levels arise directly from the wave mechanical approach to the atom. This is a mathematical approach that is presented here in very general terms. The detail cannot be sufficient to do more than indicate the manner in which the concepts arise. The usefulness of the resulting model of the atom becomes increasingly apparent as the concepts are repeatedly used to rationalize phenomena involving atoms.

The Schrödinger equation assumes that kinetic energy plus potential energy equals the total energy for the atom and expresses this relation in terms of a wave function. As stated before, this wave function is the equation usually written as the Greek *psi*, ψ, equal to an expression involving a number of terms made up of quantities such as the charge on the electron, the mass of the electron, and the displacement of the electron from the nucleus. A wave function, ψ, is said to be a useful solution to the Schrödinger equation if the total energies calculated from the Schrödinger equation, using that expression for ψ in the Schrödinger equation, have some element of consistency with experimental values.

To compare the total energies calculated for energy states with experimental values becomes a bit complicated since the experimental phenomena measured, emission and absorption spectra, give differences in energy as an atom changes from one energy state to another state but do not give the total energy of the atom in any one state. One energy state of the atom is accepted as a reference point and all energies are measured in reference to that particular state. This method of measurement, of course, gives relative energies of the states, not absolute energies. In an emission spectrum, the photon emitted is the energy equivalent to the difference between the energy of the atom before it radiates the photon and the energy of the atom after it radiates the photon. The energy of the atom decreases in the emission of a photon as the atom moves from one energy state to a lower energy state. An atom may also increase in energy by the absorption of a photon. This is the process that produces its absorption spectrum. In order for the photon to be absorbed, the photon must have exactly the energy needed to move the atom to a higher energy state. Under appropriate conditions a sample made up of a large number of atoms of an element radiates photons of a number of specific energies. This is simply the consequence of some atoms making one transition and other atoms making another transition. A sample of that element made up of a large number of atoms also absorbs photons of a number of specific energies—very frequently photons of the same energies as those of the photons emitted in the emission spectrum. The photons absorbed or emitted are so characteristic of an element that they are the "fingerprints" used to identify the presence or absence of an element in the atmosphere of stars and in interstellar space as well as in gaseous samples in the laboratory.

Emission Spectra

Absorption Spectra

For an artist's rendering of the visible emission spectrum of hydrogen atoms, see the spectrum color plate on the back cover. Hydrogen atoms have an exceptionally simple emission spectrum with only four lines in the visible region of the spectrum: a red line, a green line, a blue line, and a violet line. In viewing the emission spectrum of hydrogen with a spectroscope, you may be aware of only the red line and possibly the green line. The sensitivity of our eyes to wavelengths less than 450 nm is very limited and the energy supplied to the hydrogen atoms by the electric discharge tube may also favor the emission of the red photons. Using black and white film in a spectrograph, the photographic image of the hydrogen emission spectrum very clearly shows the four lines in the visible region of the spectrum and additional lines beyond both the red end and the violet end of the visible spectrum.

Emission Spectrum of Hydrogen Atoms

The energy of the red photons in the hydrogen emission spectrum can be evaluated from the experimental value for their wavelength, 656 nanometers. The general relationship, known as the **Planck relation,** is that the energy of a photon is directly proportional to the frequency of the photon. The proportionality constant, known as Planck's constant, is 6.63×10^{-27} erg seconds.

$$\begin{bmatrix}\text{energy photon} \\ \text{in ergs}\end{bmatrix} = 6.63 \times 10^{-27} \text{ erg seconds} \begin{bmatrix}\text{frequency in number of} \\ \text{waves per second}\end{bmatrix}$$

Light travels 3.00×10^8 meters per second and the frequency (the number of waves per second) of the red photon is given by

$$3.00 \times 10^8 \frac{\text{meters}}{\text{second}} \div 656 \times 10^{-9} \text{ meters}$$

Consequently, the frequency of the red photon is 4.57×10^{14} second^{-1} and the energy of the photon is given by

$$6.63 \times 10^{-27} \text{ erg } \cancel{\text{seconds}} \times 4.57 \times 10^{14} \cancel{\text{seconds}^{-1}}$$

and has the value 3.03×10^{-12} ergs or 3.03×10^{-19} joules per photon. The red line observed in the emission spectrum is the effect produced by a great number of photons, not just one photon: each photon with the wavelength 656 nm has the frequency 4.57×10^{14} waves per second and an energy of 3.03×10^{-19} joules. A mole of these red photons would be produced by a mole of hydrogen atoms, each atom undergoing the same energy state transition. The total energy for these 6.02×10^{23} transitions is 1.82×10^5 joules or 182 kilojoules.

The experimental value of the wavelength for photons of the green line of the hydrogen emission spectrum is 486 nm. These photons have a frequency of 6.17×10^{14} seconds^{-1} and each proton has 4.09×10^{-19} joules of energy. For one mole of hydrogen atoms making the energy state transition responsible for these green photons, the total decrease in energy of the mole of hydrogen atoms is 246 kilojoules. Photons corresponding to the blue line have a shorter wavelength, a higher frequency, and a higher energy. Photons corresponding to the violet line have an even shorter wavelength, an even higher frequency, and even higher energy.

Absorption Spectrum of Hydrogen Atoms

The color plate on the back cover includes an artist's rendering of the spectrum of the sun as viewed from the earth. Note the difference between this emission spectrum and the emission spectrum of a very hot solid or liquid on earth. The dark lines in the spectrum of the sun are attributed to the absorption of photons in the continuous emission of the very hot sun by atoms in the cooler outer atmosphere of the sun. Note that four of these dark lines correspond to the four emission lines of hydrogen atoms. This is taken to be evidence of a high concentration of hydrogen atoms in the outer atmosphere of the sun. Whatever energy state transitions occur in atoms to give the emission of photons in a hydrogen electric discharge tube, the reverse energy state transitions occur in atoms of hydrogen to give the absorption of photons in the outer atmosphere of the sun.

Quantum Numbers

In the wave mechanical approach, a multiplicity of energy states arises from the wave functions themselves. The wave functions contain terms that can take on integral values, such as 1, 2, 3, . . . Each value corresponds to an energy state for the atom. These are the quantum levels of the atom and the numbers are known as **principal quantum numbers.** The wave functions generate not one series of quantum numbers but three series of quantum numbers. The number of possible combinations is obviously large. There are, however, restrictions on the values that can be assumed by the second series and by the third series of quantum numbers. The pattern in which these three series of quantum numbers fit together is a delight to behold. It is, however, not necessary to admire its full majesty to pursue the nature of chemical change. This is particularly true since chemists use the orbital notation involving the letters *s*, *p*, *d*, etc. rather than the l and m_l quantum numbers generated by the Schrödinger model.

First Three Series of Quantum Numbers

The first series, called principal quantum numbers, n, take on values 1, 2, 3, . . . These numbers are closely related to the total energy of the atom and correlate with the large difference in energy of some quantum states. The red line in the hydrogen emission spectrum is the transition of hydrogen atoms from the $n = 3$ energy state to the $n = 2$ energy state.

The second series (the l series) of quantum numbers may take on values that introduce additional energy states of the atom. The separations in energy levels introduced by this second series of quantum numbers are in general, but not always, small compared to those introduced by the principal quantum numbers. It is this second series of quantum numbers that is responsible for the difference in character of the wave functions that lead to the concepts of *s*-, *p*-, *d*-, and *f*-orbitals.

The third series (the m_l series) of quantum numbers may take on values that give rise to additional energy states of the atom. This third set of quantum numbers is responsible for the concepts of orientation in space of the orbitals: the p_x-, the p_y-, and p_z-orbitals, for example. The separations in energy levels introduced by this third set of quantum numbers are dependent upon the environment of the atom and are in general small—frequently zero.

The *s*-, *p*-, *d*-, and *f* orbitals that occur with the principal quantum numbers 1, 2, 3, and 4 are given below. The four types of orbitals *s*, *p*, *d*, and *f* are adequate to cope with the chemical phenomena we shall address in this book.

principal quantum number, *n*	**orbitals**	**orientations of orbitals**
1	*s* only	none
2	*s*	none
	p	three: $2p_x$, $2p_y$, and $2p_z$
3	*s*	none
	p	three: $3p_x$, $3p_y$, and $3p_z$
	d	five (known but not specified here)
4	*s*	none
	p	three: $4p_x$, $4p_y$, and $4p_z$
	d	five (known but not specified here)
	f	seven (known but not specified here)

Also, see Gratuitous Information 6-2, Rules from the Schrödinger Model.

Initially, we began the exploration of the nature of atoms in Chapter 5 as a basis to understand the nature of the forces that not only hold atoms together in molecules but also hold them together with specific orientations in space to give isomeric compounds such as *cis*- and *trans*-2-butene. The ball-stick-spring model is very useful, but it is clearly not sticks and springs that hold atoms together in molecules. We turn now to an exploration of the role of electrons in the formation of chemical bonds. Our first task is to relate the distribution of electrons in atomic orbitals to the chemical properties of elements. This we shall approach through the relation of the Periodic Table to the distribution of electrons in atomic orbitals of ground state atoms. This correlation is displayed in Figure 6-6, Schematic Energy Sequence of Atomic Orbitals Appropriate to All Elements Except Hydrogen, and Figure 6-7, General Pattern of the Periodic Table Correlated with the Highest Energy Atomic Orbitals Occupied in Ground State Atoms. The goal is to so thoroughly understand these relations that you will discover that you can recreate both Figures 6-6 and 6-7 from a very few relations that you will learn and remember. These relations are the basis of our understanding of modern chemistry. Figures 6-2, 6-3, 6-4, 6-5, and 6-8 are essential to the development of our understanding of Figures 6-6 and 6-7. Take a look at these two figures now and return to them from time to time as we proceed to work through the background material.

Atomic Energy Levels for Hydrogen

The hydrogen atom is a very unique atom with its atomic number, Z, equal to 1: a nucleus composed of 1 proton and an electron atmosphere of 1 electron. The potential energy relations within the atom are the simplest imaginable: the single electron is attracted by the nucleus. In all other atoms, each electron is attracted by the nucleus and each electron repels all other electrons. Figure 6-2 presents the schematic energy sequence of atomic orbitals of the hydrogen atom. Compare this figure with Figure 6-6 (page 115), the schematic energy sequence of atomic orbitals for atoms other than hydrogen atoms. The great simplicity of energy level diagrams for the hydrogen atom is very striking. In these diagrams, a short horizontal line represents an atomic orbital. As we shall soon see, each of these atomic orbitals can accommodate a maximum of one pair of electrons with antiparallel spins. The identities of the orbitals are given by notations such as $2s$, $2p_x$, $2p_y$, and $2p_z$. In the hydrogen atom, these four atomic orbitals have the same energy. In the atoms of all other elements, Figure 6-6, the three *p*-orbitals p_x, p_y, and p_z have the same energy (placed in the diagram on the same horizontal line corresponding to a particular energy value) and the energy associated with each of these three $2p$-orbitals is greater than the energy associated with the $2s$-orbital. In atoms of elements other than hydrogen, the three $2p$-orbitals are said to be degenerate (to have the same energy). In the hydrogen atom, the $2s$-orbital and the three $2p$-orbitals are degenerate. Note that Figure 6-2 shows all of the atomic orbitals of hydrogen atoms of principal quantum number $n = 3$ to be degenerate. It is this degeneracy of atomic orbitals that is believed to be the basis of the unique simplicity of the emission and absorption spectra of hydrogen atoms and that led to the association of principal quantum numbers with energy states in the early investigations of emission and absorption spectra of hydrogen atoms.

Degenerate Atomic Orbitals

Figure 6-3 presents the actual energies of hydrogen atoms in joules per atom for the energy states of a hydrogen atom with its electron in the various principal quantum levels. The energy values given in this diagram are predicted by both the

Bohr approach and the Schrödinger approach and are supported by the interpretation of the experimentally determined emission and absorption spectra of hydrogen atoms. The energy values for several principal quantum numbers of the hydrogen atoms are

n	**energy in joules per atom**
10	-0.0218×10^{-18}
9	-0.0269×10^{-18}
8	-0.0341×10^{-18}
7	-0.0445×10^{-18}
6	-0.0606×10^{-18}
5	-0.0872×10^{-18}
4	-0.136×10^{-18}
3	-0.242×10^{-18}
2	-0.545×10^{-18}
1	-2.18×10^{-18}

In order to display the change in energy with changing principal quantum numbers, all values of energy are expressed in the above listing as a quantity times 10^{-18} rather than in the more conventional form—for example, 0.0218×10^{-18} rather than 2.18×10^{-20}.

Note that there is a much larger energy separation between the energy states for $n = 1$ and $n = 2$ than there is between the states for $n = 2$ and $n = 3$. As the quantum numbers increase, the states are closer and closer together in energy and the representation of these states on a diagram (the horizontal lines) becomes a troublesome problem. The energy states come so close together that the lines used to represent them begin to overlap in a diagram as n, the principal quantum number, approaches infinity. As the principal quantum number, n, approaches infinity, the energies for the various states approach zero, and for $n = \infty$ the energy is zero.

2s-Orbitals

The wave function that generates a 2*s*-orbital for the hydrogen atom is very similar to the wave function that generates a 1*s*-orbital. Both orbitals have spherical symmetry. The volume of the sphere that encompasses 90% probability for electrons in a 2*s*-orbital is greater than the volume of the sphere that encompasses 90% probability for the 1*s*-orbital. According to this model, an electron is further removed, on the average, from the nucleus in the 2*s* state than in the 1*s* state. The 2*s*-orbital also has more structure. The 1*s*-orbital is a single region of high probability for electrons with a center of symmetry at the nucleus. The 2*s*-orbital is two concentric spherical regions of high probability with a center of symmetry at the nucleus. The outer region is the region of greater probability and is the more useful region in the discussion of many chemical problems. The much smaller spherical region of somewhat lower probability nearer the nucleus is frequently ignored.

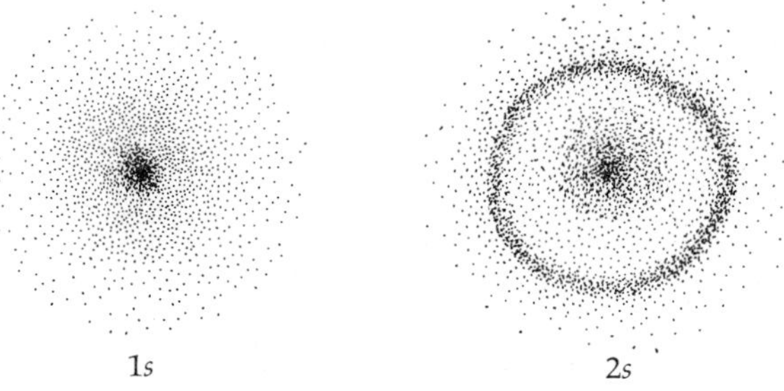

A 3s-orbital has three concentric regions, with the outer region again being the most significant. With larger and larger principal quantum numbers, the *s* electron is further and further removed, on the average, from the nucleus. At $n = \infty$, the divorce is complete and the electron is no longer specifically related to this particular nucleus. The process of divorce is called ionization and the hydrogen nucleus, the proton, is called a hydrogen ion, written $H^{(1+)}$.

Definition of Zero on the Energy Scale

It is this ionic state of hydrogen, with the electron completely separated from the proton, that is taken as the reference state to measure energies for the various quantum states of the hydrogen atom. The state $n = \infty$ for hydrogen is arbitrarily assigned an energy of zero. All other quantum states of the hydrogen atom have less than zero energy on this scale. Consequently, all quantum states of the atom have

Figure 6-2

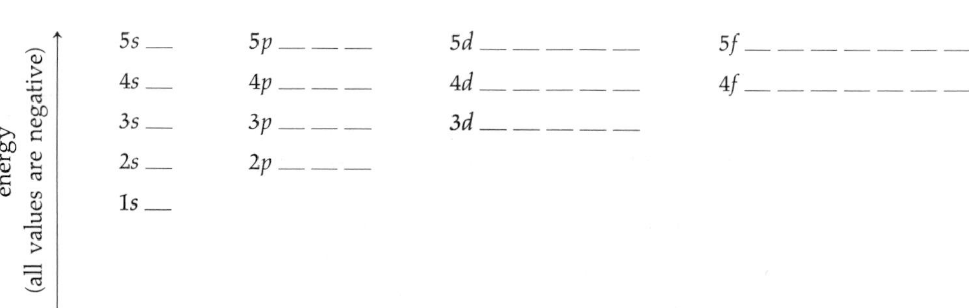

SCHEMATIC ENERGY SEQUENCE OF ATOMIC ORBITALS OF THE HYDROGEN ATOM. Each horizontal line represents an atomic orbital. The energy level sequence for the hydrogen atom is unique in that all of the *s*-, *p*-, *d*-, and *f*-orbitals of a given principal quantum number are degenerate (have the same energy). This high degree of degeneracy is consistent with the uniquely simple emission spectrum of hydrogen. Without this degeneracy, Bohr's task would have been much more demanding—perhaps too demanding. See Figure 6-3 for the energy level diagram of hydrogen atoms in joules per atom.

negative values of energy on this arbitrary scale: the smaller the principal quantum number, the lower the energies of the hydrogen atom. For $n = 1$, the electron occupies the 1*s*-orbital and the atom is said to be in the ground state, the lowest energy state for an isolated hydrogen atom.

Under the energy conditions of the earth's atmosphere, the hydrogen atoms pair off to form diatomic hydrogen molecules, H_2. This is another story. One hydrogen molecule, H_2, has less energy (is more stable) than two hydrogen atoms, as in the equation on the following page:

Figure 6-3

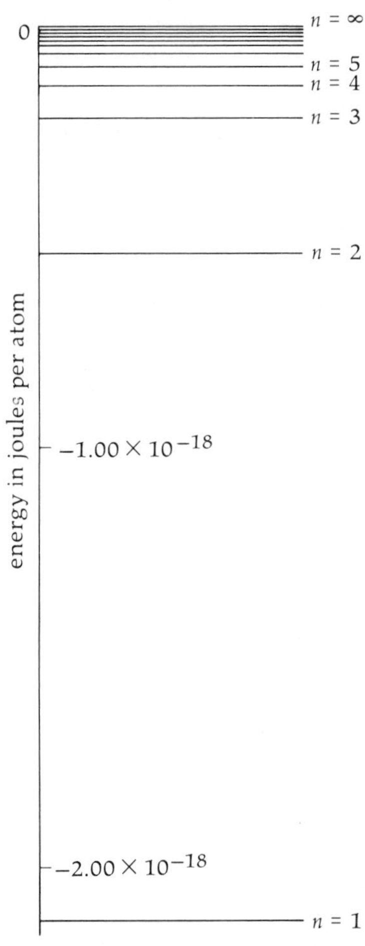

ENERGY LEVEL DIAGRAM OF HYDROGEN ATOMS IN JOULES PER ATOM. This energy level diagram was predicted by both the Bohr model and the Schrödinger approach and is supported by both the emission spectrum and the absorption spectrum of hydrogen atoms.

$$H + H \longrightarrow H_2 + \text{energy}$$

To dissociate hydrogen molecules, H_2, into atoms, H, requires the investment of energy:

$$H_2 + \text{energy} \longrightarrow H + H$$

All of the discussion of hydrogen in this chapter has to do with individual hydrogen atoms—not with diatomic hydrogen molecules.

Emission Spectrum Revisited

In the discussion of the visible emission spectrum of hydrogen atoms, we evaluated the energy of the 656-nm photon (the red photon) to be 3.03×10^{-19} joules per photon and the energy of the 486-nm photon (the green photon) to be 4.09×10^{-19} joules per photon. Note that the energy of the red photon corresponds to the transition of a hydrogen atom from the $n = 3$ state to the $n = 2$ state with a change in energy of

$$-0.545 \times 10^{-18} - (-0.242 \times 10^{-18})$$
$$= -0.303 \times 10^{-18} \text{ or } -3.03 \times 10^{-19} \text{ joules per atom}$$

and that the energy of the green photon corresponds to the transition of a hydrogen atom from the $n = 4$ state to the $n = 2$ state with a change in energy of

$$-0.545 \times 10^{-18} - (-0.136 \times 10^{-18})$$
$$= -0.409 \times 10^{-18} \text{ or } -4.09 \times 10^{-19} \text{ joules per atom}$$

As you may have already guessed, the blue line of the hydrogen emission spectrum is made up of photons derived from the transition of hydrogen atoms from the $n = 5$ energy state to the $n = 2$ energy state and the violet line is made up of photons derived from the transition of hydrogen atoms from the $n = 6$ energy state to the $n = 2$ energy state.

The energy states of the hydrogen atom that we have been discussing—with the electron in principal quantum levels 2, 3, 4, 5, and 6—are **excited states of the hydrogen atom.** The energy state of the hydrogen atom in which the electron occupies the principal quantum level $n = 1$ is the lowest of the quantized energy states of hydrogen atoms. This energy state is at the bottom of the energy level diagram and this energy state is said to be the **ground state hydrogen atom.** In the ground state hydrogen atom, the electron is said to occupy the 1*s* atomic orbital.

Atomic Energy Levels for Other Elements

We shall now turn our attention to the distribution of electrons in atomic orbitals of ground state atoms other than hydrogen atoms. Initially, we shall focus on elements of atomic number $Z = 18$ or less. This encompasses the elements in the first three periods of the Periodic Table: the very short period of two elements and the two short periods of eight elements each. Figure 6-4 is the segment of Figure 6-6 that we shall need in order to discuss the ground state atoms of the elements in the first three periods of the Periodic Table. Although Figures 6-4 and 6-6 can be used to discuss either ground state atoms or excited state atoms, we shall only be concerned with ground state atoms in this discussion.

Fourth Quantum Number

In addition to the three series of quantum numbers that arose from the Schrödinger approach, a fourth quantum number, known as the **spin quantum number,** was introduced separately from the Schrödinger approach. Each electron in the atom has the properties of a spinning charge. The spin quantum number has only two values: $+1/2$ and $-1/2$. That there are only two values is related to the orientation of the spin of the electron. Reverse the direction of the spin of the electron and the sign of the spin quantum number is reversed from $+1/2$ to $-1/2$ or from $-1/2$ to $+1/2$. A spinning electron, a moving charge, produces a magnetic field and therefore has the properties of a small magnet. A reverse in the direction of spin reverses the north and south poles of the magnet.

Pauli Exclusion Principle

In order to correlate the distribution of electrons in an atom with the properties of the atom, a postulate known as the **Pauli exclusion principle,** has been imposed upon the use of the four quantum numbers. Its use makes the model for the distribution of electrons in the atom consistent with experimental evidence. This exclusion principle states that two electrons of the same atom cannot have the same set of quantum numbers. In other words, an orbital such as the $3p_x$-orbital can accommodate only two electrons: one with spin $+1/2$ and the other with spin $-1/2$. These two electrons in a single orbital are said to have antiparallel spins and to be paired. Note that all four quantum numbers are involved in the $3p_x$ example. The 3 is the principal quantum number that is primarily related to energy. The p arises from the second series of quantum numbers and specifies the shape of the orbital. The x arises from the third series and specifies the spatial orientation of the p-orbital by designating one of the three possible p-orbitals. The spin quantum number, the fourth type, with its two possible values, places the limit on the number of electrons in the orbital. The first three quantum numbers designate the orbital. The fourth designates the spin of the electron in the orbital.

The pattern of the Periodic Table began to emerge on the basis of the known properties of elements long before the discovery of the electron. The wave mechan-

Figure 6-4

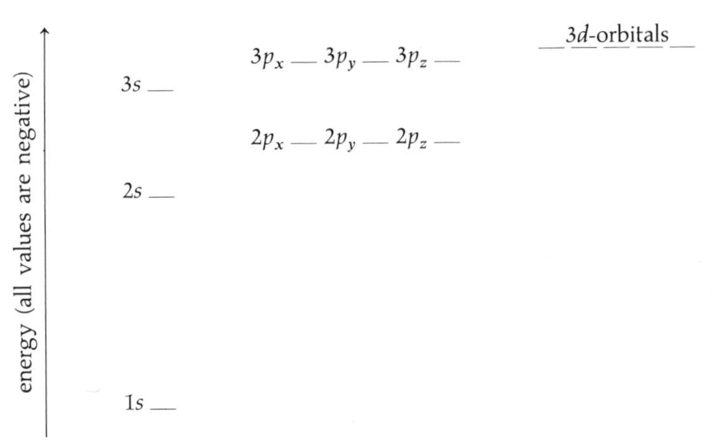

SCHEMATIC ENERGY SEQUENCE OF ATOMIC ORBITALS FOR PRINCIPAL QUANTUM NUMBERS 1, 2, AND 3 FOR ALL ELEMENTS OTHER THAN HYDROGEN.

Figure 6-5

1	1 H																	2 He
2	3 Li	4 Be											5 B	6 C	7 N	8 O	9 F	10 Ne
3	11 Na	12 Mg											13 Al	14 Si	15 P	16 S	17 Cl	18 Ar
4	19 K	20 Ca	21 Sc	22 Ti	23 V	24 Cr	25 Mn	26 Fe	27 Co	28 Ni	29 Cu	30 Zn	31 Ga	32 Ge	33 As	34 Se	35 Br	36 Kr
5	37 Rb	38 Sr	39 Y	40 Zr	41 Nb	42 Mo	43 Tc	44 Ru	45 Rh	46 Pd	47 Ag	48 Cd	49 In	50 Sn	51 Sb	52 Te	53 I	54 Xe
6	55 Cs	56 Ba	57 La *	72 Hf	73 Ta	74 W	75 Re	76 Os	77 Ir	78 Pt	79 Au	80 Hg	81 Tl	82 Pb	83 Bi	84 Po	85 At	86 Rn
7	87 Fr	88 Ra	89 Ac #	104 Rf	105 Ha	106	107	108	109									

* Lanthanide Series →	58 Ce	59 Pr	60 Nd	61 Pm	62 Sm	63 Eu	64 Gd	65 Tb	66 Dy	67 Ho	68 Er	69 Tm	70 Yb	71 Lu
# Actinide Series →	90 Th	91 Pa	92 U	93 Np	94 Pu	95 Am	96 Cm	97 Bk	98 Cf	99 Es	100 Fm	101 Md	102 No	103 Lr

PERIODIC TABLE

ical model of the distribution of electrons in the atom now provides a basis of rationalization of the pattern of the Periodic Table and the value of the model is judged in terms of its contribution to our understanding of the properties of atoms. Our first concern is the correlation between the pattern of the Periodic Table and the distribution of electrons in ground state atoms of the elements. The aspects of the Periodic Table essential to this discussion are repeated in Figure 6-5.

The distribution of electrons in the ground state according to the energy state diagram, Figure 6-6, is straightforward and will become rather obvious. One additional problem does arise, however, in relation to the distrbution of electrons in degenerate orbitals such as $2p_x$, $2p_y$, and $2p_z$. How are two electrons distributed? One in the $2p_x$-orbital and one in the $2p_y$-orbital or two in the $2p_x$-orbital? One in the $2p_x$-orbital and one in the $2p_y$-orbital corresponds to two partially occupied orbitals and an unoccupied $2p_z$-orbital. Two paired electrons in the $2p_x$-orbital corresponds to a filled $2p_x$-orbital and two unoccupied orbitals, $2p_y$ and $2p_z$. The experimental evidence from the magnetic properties of the atoms supports the first alternative. The general rule, known as **Hund's rule,** is that electrons distribute themselves to give a maximum of partially filled degenerate orbitals before pairing to give filled degenerate orbitals. The physical evidence for this is based upon magnetic properties that can be determined experimentally. This distribution of electrons seems reasonable enough. Partially filled degenerate orbitals provide the maximum space for the electrons and thus reduce to a minimum the repulsion due to the like charges of the electrons. The antiparallel spins of paired electrons correspond to reversed magnetic fields and give rise to an attractive interaction between the two electrons. This magnetic interaction is, however, small in comparison to the electrostatic interaction of repulsion arising from the negative charge of the electrons.

Hund's Rule

Distribution of Electrons in Ground State Atoms

Based upon the schematic distribution of energy states shown in Figure 6-4, the Pauli exclusion principle, and Hund's rule, we use the **construction principle,** also called the *Aufbau Prinzip,* to build the ground state structures of atoms starting with helium, Z = 2, and proceeding to more complex atoms. As you work through the following pages, refer to the energy state diagram, Figure 6-4, and the Periodic Table, Figure 6-5—also Figures 6-6 and 6-7 if you like.

The one electron of the ground state hydrogen atom, atomic number Z = 1, would, according to either Figure 6-2 or Figure 6-4, be in the *s*-orbital associated with quantum level $n = 1$. The notation for this ground state of the hydrogen atom is $1s^1$ (read one-s-one). Note that this refers to an isolated hydrogen atom, not to the hydrogen atom in the diatomic molecule, H_2, or a molecule of a compound.

Helium

According to Figure 6-4, the two electrons of the helium atom, Z = 2, would be paired in the *s*-orbital associated with the principal quantum number, $n = 1$. The notation for this ground state of the helium atom is $1s^2$ (read one-s-two). This is the monatomic helium gas molecule, He.

The notation for the ground state of the lithium atom, Z = 3, is $1s^2\ 2s^1$: a pair of electrons in the *s*-orbital associated with principal quantum level 1 and one electron in the *s*-orbital associated with principal quantum number 2.

The 1*s*-orbitals for hydrogen, for helium, and for lithium fall at quite different positions on their respective energy diagrams. The energy of the 1*s*-orbital of lithium is furthest removed from the arbitrary zero energy for lithium at $n = \infty$. The "volume" of the 1*s*-orbital for lithium is also the smallest. These differences are related to the charge of +3 of the lithium nucleus providing a stronger force of attraction for the electrons than the charge of +1 of the hydrogen nucleus and the charge of +2 of the helium nucleus.

Lithium

Lithium, Z = 3, is a metal. It has the bright metallic sheen, the high electrical conductivity, and the high heat conductivity characteristic of a metal. It is also an unusually reactive metal and must be protected from oxygen, water, and a great variety of other substances. At room temperature, lithium is a solid. In the solid, the atoms do not correspond to the ground state of the atom discussed above. In order for the lithium to be a solid, the atoms must be interacting with each other in some fashion. The ground state $1s^2\ 2s^1$ for the lithium atom refers to an isolated atom of lithium in the gas phase, not the metallic state.

Beryllium

The notation for the ground state of the beryllium atom, Z = 4, is $1s^2\ 2s^2$ (read one-s-two, two-s-two). Beryllium is a metal but not as chemically reactive as lithium. At room temperature, beryllium is a solid and the atoms are not the free ground state atoms described by the above notation.

Boron

The notation for the ground state of the boron atom, Z = 5, is $1s^2\ 2s^2\ 2p^1$. There are now three orbitals: the small spherical 1*s*-orbital, the larger spherical 2*s*-orbital, and the hourglass-like 2*p*-orbital. All three overlap in the region near the nucleus, but the *p*-orbital extends its region of high probability in the direction of its ∞-fold axis of rotation beyond the region of high probability of the *s*-orbitals. At room temperature, boron is a chemically unreactive solid that has low electrical conductivity and low heat conductivity. Boron forms the types of compounds characteristic of nonmetals.

Carbon

The notation for the ground state of the carbon atom, Z = 6, is $1s^2\ 2s^2\ 2p_x^{\ 1}\ 2p_y^{\ 1}$. There is an electron in each of two *p*-orbitals. The choice of axes to designate these is immaterial. The notation is also frequently expressed as $1s^2\ 2s^2\ 2p^2$. Carbon has the low electrical conductivity and forms the types of compounds typical of a nonmetal. At room temperature, carbon exists in two highly ordered solid forms: graphite and diamond. In both cases, the atoms are bonded to each other and are not the ground state atoms specified by the above notation.

Nitrogen

The notation for the ground state atom of nitrogen, Z = 7, is $1s^2\ 2s^2\ 2p_x^{\ 1}\ 2p_y^{\ 1}\ 2p_z^{\ 1}$, also written $1s^2\ 2s^2\ 2p^3$. Nitrogen forms the types of compounds characteristic of nonmetals. Nitrogen exists as a diatomic gas, N_2, at room temperature. The N_2 molecule is more stable than two ground state nitrogen atoms.

Oxygen

Fluorine

The notations for the ground state atoms of oxygen, Z = 8, and fluorine, Z = 9, are

oxygen $1s^2\ 2s^2\ 2p_x^{\ 2}\ 2p_y^{\ 1}\ 2p_z^{\ 1}$ or $1s^2\ 2s^2\ 2p^4$

fluorine $1s^2\ 2s^2\ 2p_x^{\ 2}\ 2p_y^{\ 2}\ 2p_z^{\ 1}$ or $1s^2\ 2s^2\ 2p^5$

Both elements form large numbers of compounds of the types characteristic of nonmetals. At room conditions, both are diatomic gases, O_2 and F_2. Both gases, O_2 and F_2, are much more chemically reactive than nitrogen gas, N_2.

The notation for the ground state of the neon atom, Z = 10, is $1s^2\ 2s^2\ 2p_x^{\ 2}\ 2p_y^{\ 2}\ 2p_z^{\ 2}$ or $1s^2\ 2s^2\ 2p^6$. At room temperature, neon is a monatomic gas, Ne, and the above notation correctly represents the neon atoms present in neon gas at room conditions. The monatomic gas molecules of neon and the very marked reluctance of neon to form compounds with other elements suggest that the distribution of electrons in ground state atoms of neon is an exceptionally stable electron configuration.

Neon

Another method that could have been used to display the electron distribution in an atom is to simply fill in the schematic energy level diagram, Figure 6-4, using a small vertical arrow to represent each electron with the direction of the arrow representing the relative orientation of the spins. For the ground state of the oxygen atom, its eight electrons could be represented in the following fashion:

$2p_x$ ⇅ $2p_y$ ↑ $2p_z$ ↑

$2s$ ⇅

$1s$ ⇅

This representation emphasizes the filling of the lower energy levels and the pairing of the electrons. The direction of all of the arrows could have equally well been drawn in the reverse directions.

Clearly, this game can be continued for other elements. The ground state of an atom of sodium, Z = 11, would be represented as $1s^2\ 2s^2\ 2p_x^{\ 2}\ 2p_y^{\ 2}\ 2p_z^{\ 2}\ 3s^1$. This has the five filled orbitals of neon and an additional electron in the *s*-orbital associated with principal quantum number 3. The notation is also frequently written $1s^2\ 2s^2\ 2p^6\ 3s^1$ and also (Ne) $3s^1$. The symbol (Ne) emphasizes the close relation of ten electrons to the neon electron distribution but does not imply that the orbitals represented by (Ne) are identical to those in the neon atom. The charge of +11 on the sodium nucleus, as compared to the +10 on the neon nucleus, is believed to draw the electrons closer to the nucleus and to reduce the energies relative to the corresponding levels in neon. At room temperature, sodium is a solid with a bright metallic luster. It is a good conductor of electricity and of heat. It forms compounds characteristic of metals. The solid is quite reactive and must be protected from oxygen, water, and a number of other subtances. In short, sodium is very similar to lithium, $1s^2\ 2s^1$ or (He) $2s^1$, in its electron structure and also its chemical and physical properties.

Sodium

Magnesium, Z = 12, is a metal and has many properties similar to those of beryllium, Z = 4.

Magnesium

beryllium	$1s^2\ 2s^2$	or	(He) $2s^2$
magnesium	$1s^2\ 2s^2\ 2p^6\ 3s^2$	or	(Ne) $3s^2$

The pattern continues:

boron (Z = 5)	(He) $2s^2\ 2p^1$
aluminum (Z = 13)	(Ne) $3s^2\ 3p^1$
carbon (Z = 6)	(He) $2s^2\ 2p_x^{\ 1}\ 2p_y^{\ 1}$
silicon (Z = 14)	(Ne) $3s^2\ 3p_x^{\ 1}\ 3p_y^{\ 1}$

nitrogen (Z = 7)	(He) $2s^2\ 2p_x^{\ 1}\ 2p_y^{\ 1}\ 2p_z^{\ 1}$
phosphorus (Z = 15)	(Ne) $3s^2\ 3p_x^{\ 1}\ 3p_y^{\ 1}\ 3p_z^{\ 1}$
oxygen (Z = 8)	(He) $2s^2\ 2p_x^{\ 2}\ 2p_y^{\ 1}\ 2p_z^{\ 1}$
sulfur (Z = 16)	(Ne) $3s^2\ 3p_x^{\ 2}\ 3p_y^{\ 1}\ 3p_z^{\ 1}$
fluorine (Z = 9)	(He) $2s^2\ 2p_x^{\ 2}\ 2p_y^{\ 2}\ 2p_z^{\ 1}$
chlorine (Z = 17)	(Ne) $3s^2\ 3p_x^{\ 2}\ 3p_y^{\ 2}\ 3p_z^{\ 1}$
neon (Z = 10)	(He) $2s^2\ 2p_x^{\ 2}\ 2p_y^{\ 2}\ 2p_z^{\ 2}$
argon (Z = 18)	(Ne) $3s^2\ 3p_x^{\ 2}\ 3p_y^{\ 2}\ 3p_z^{\ 2}$

Correlation of Third Period Elements with Second Period Elements

It is indeed a highly ordered system. Three of these first eighteen elements are monatomic gases at room conditions: helium, neon, and argon. There are three other elements that are also monatomic gases at room conditions: krypton, xenon, and radon. The latter three gases fall below helium, neon, and argon in the Periodic Table and have atomic numbers 36, 54, and 86, respectively. Presumably, the atomic structures involving 36, 54, and 86 electrons are also highly stable.

Potassium, Z = 19, and calcium, Z = 20, have properties similar to sodium, Z = 11, and magnesium, Z = 12, respectively, and fall directly below them in the Periodic Table. Continuing the above pattern of electron distribution, the electronic structure of potassium and calcium could be predicted in terms of argon, the nearest monatomic gas.

lithium	(He) $2s^1$	beryllium	(He) $2s^2$
sodium	(Ne) $3s^1$	magnesium	(Ne) $3s^2$
potassium	(Ar) $4s^1$	calcium	(Ar) $4s^2$

Written out in more detail, these would be $1s^2\ 2s^2\ 2p^6\ 3s^2\ 3p^6\ 4s^1$ for potassium and $1s^2\ 2s^2\ 2p^6\ 3s^2\ 3p^6\ 4s^2$ for calcium. The partial energy sequence given in Figure 6-4 does not include the 4*s*-orbital and the diagram is consequently too incomplete to support or to deny the electron distributions written above for potassium and calcium. Spectral measurements do support this ordering of energy levels. Figure 6-6 is an extension of Figure 6-4 to include more orbitals. In the energy sequence in Figure 6-8, the sequence of atomic orbitals is placed on a linear energy scale. In Figure 6-6, the vertical spacings have been distorted to display more openly the interweaving of *d*- and *f*-orbitals of principal quantum numbers 3, 4, and 5 with the *s*- and *p*-orbitals of higher principal quantum numbers. The energy associated with the 4*s*-orbital does lie between the energy associated with the 3*p*-orbitals and the energy associated with the 3*d*-orbitals. This seems to contradict the earlier assumption of the principal quantum numbers being the primary factor in determining relative energies of energy states. It is in keeping with the energy levels of the *s*-orbitals coming closer and closer together at higher principal quantum numbers. Eventually, this energy separation between *s*-orbitals of successive principal quantum numbers is less than the separation

Figure 6-6

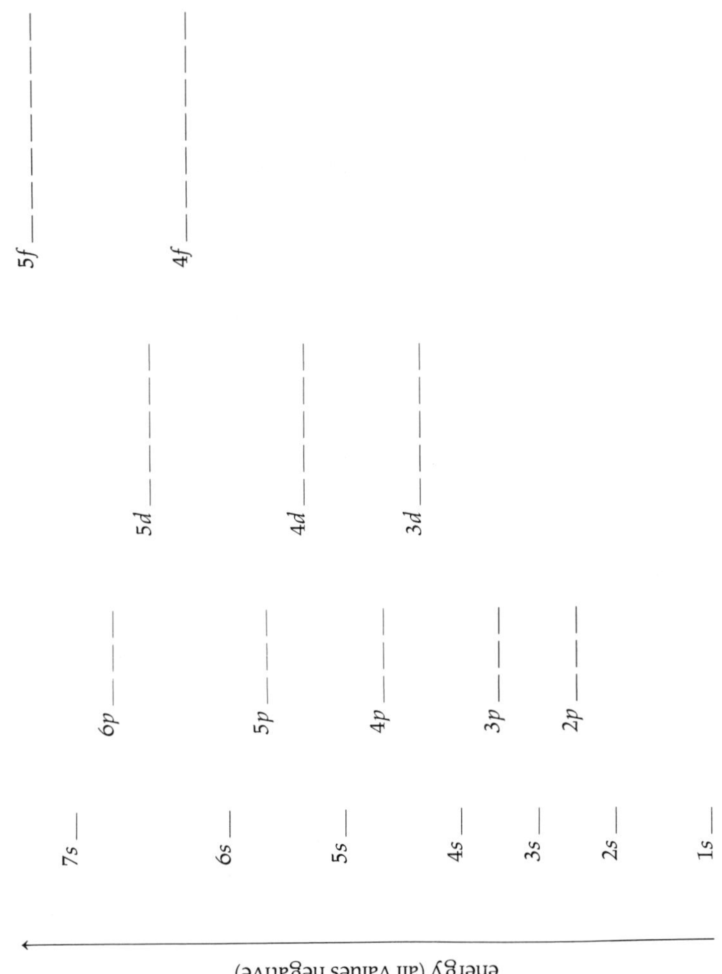

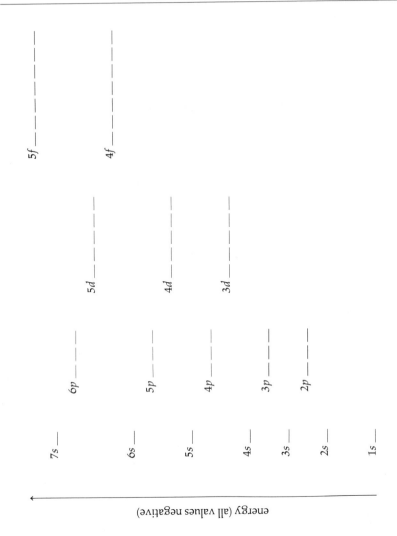

Schematic Energy Sequence of Atomic Orbitals Appropriate to All Elements Other Than Hydrogen. Spacing between *s*-orbitals is distorted to display the interweaving of *d*- and *f*-orbitals of principal quantum numbers 3, 4, and 5 with the *s*- and *p*-orbitals of higher principal quantum numbers. See Figure 6-8 for a more correct energy distribution of the same sequence. Each short horizontal line represents an atomic orbital that can accommodate a maximum of two electrons with antiparallel spins.

of the *s*-orbital and the *d*-orbital of the same principal quantum number: the 4*s*-orbital is closer in energy to the 3*s*-orbital than the 3*d*-orbital is to the 3*s*-orbital.

At the higher principal quantum numbers, there is a great deal of interweaving of the above type and the energy level diagrams become quite complex. This complexity correlates with the pattern of the Periodic Table. These energy level diagrams have been determined experimentally from spectral measurements and can be used mechanically to reconstruct the pattern of the Periodic Table. The following presentation will instead approach from the pattern of the Periodic Table to evolve the pattern of energy levels. This is not the best scientific approach but it is a very convenient way to reconstruct a lot of details by learning the pattern of the Periodic Table, which is comparatively simple:

Larger Atomic Number Elements

one very short period of 2 elements,

two short periods of 8 elements each,

two long periods of 18 elements each,

one very long period of 32 elements, and

one unfinished very long period.

Each period ends with a monatomic gas usually called an inert gas or a noble gas. Figure 6-7 presents the general pattern of the Periodic Table and the correlation of this pattern with the highest occupied atomic orbitals in ground state atoms.

Correlation of Electron Structures of Ground State Atoms with Positions of Elements in the Periodic Table

In the Periodic Table, the eight elements of the short periods are divided 2 and 6, the eighteen elements of the long periods are divided 2, 10, and 6, and the thirty-two elements of the very long periods are divided 2, 14, 10, and 6. The two elements at the beginning of a period are clearly related to *s*-orbitals. The six elements at the end of a period are clearly related to *p*-orbitals. Note that in both cases the highest principal quantum number is the same as the number of the period. The ten elements in the center of the table correlate nicely with the five *d*-orbitals, which are associated with principal quantum number 3 and greater. The fourteen elements set off below the table correlate with the seven *f*-orbitals, which are associated with principal quantum number 4 and greater.

Placing hydrogen and helium, the two elements in the very short period, in the table creates a bit of a problem. On the basis of their 1*s*-orbital electrons, they would be expected to be at the extreme left of the table with the metals. On the basis of its monatomic nature, helium should be with the other monatomic gases at the extreme right of the table. On the basis of having one electron less than a monatomic gas, hydrogen should be next to helium at the right of the table above fluorine, chlorine, and bromine. Helium is always placed at the extreme right of the Periodic Table. Hydrogen is shuttled to and fro across the table in the rationalization of its properties. We are free to choose the position that meets our specific needs.

To write out the notation for the distribution of the electrons in a ground state atom of platinum, $Z = 78$, we could work from the energy level diagrams, Figure 6-6 or Figure 6-8, or from the pattern of the Periodic Table diagram, Figure 6-7. All approaches should yield the same results. We shall now use the pattern of the Periodic Table. The preceding inert gas is xenon, $Z = 54$. Consequently, the electron structure for platinum can be built by starting with the xenon core structure and filling in additional electrons by moving across period 6—a very long period—until the 24

Figure 6-7

	s-block	d-block	p-block	atomic number and name of the last element in each period
Period 1 (very short)	1s		1s	2 helium He
Period 2 (short)	2s		2p	10 neon Ne
Period 3 (short)	3s		3p	18 argon Ar
Period 4 (long)	4s	3d	4p	36 krypton Kr
Period 5 (long)	5s	4d	5p	54 xenon Xe
Period 6 (very long)	6s	* 5d	6p	86 radon Rn
Period 7 (very long)	7s	# 6d		
	two elements	ten elements	six elements	

* 4f-orbitals

\# 5f-orbitals

fourteen elements

General Pattern of the Periodic Table Correlated with the Highest Energy Atomic Orbitals Occupied in Ground State Atoms. Each short horizontal line represents an element.

additional electrons are placed: 2 electrons in the 6s-orbital, 14 electrons in the 4f-orbitals, and 8 electrons in the 5d-orbitals:

$$\text{Pt } (Z = 78) \ (\text{Xe}) \ 6s^2 \ 4f^{14} \ 5d^8$$

If this seems like cheating, start at the beginning of Figure 6-7 and walk through to element 78:

$$\text{Pt } (Z = 78) \ 1s^2 \ 2s^2 \ 2p^6 \ 3s^2 \ 3p^6 \ 4s^2 \ 3d^{10} \ 4p^6 \ 5s^2 \ 4d^{10} \ 5p^6 \ 6s^2 \ 4f^{14} \ 5d^8$$

Figure 6-8

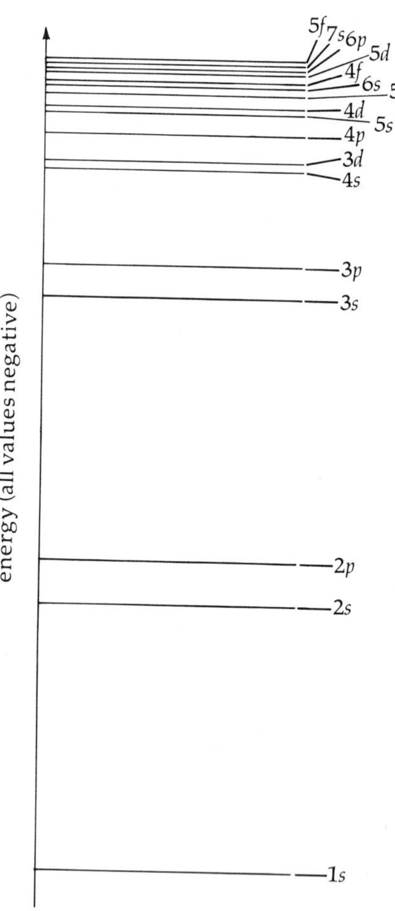

TYPICAL ENERGY LEVEL DIAGRAM OF ATOMIC ORBITALS FOR ALL ATOMS OTHER THAN HYDROGEN. Relative energies are correctly represented. Each horizontal line represents an atomic orbital or a set of degenerate atomic orbitals. The numerical values of energy are a characteristic of the element and are not the same for any two elements.

RULES FROM THE SCHRÖDINGER MODEL

The rules of the Schrödinger model place limits on the values possible for the three quantum numbers. While the principal quantum number, usually symbolized by n, can have any integral value, $n = 1, 2, 3 \ldots$, the possible values of the other two quantum numbers depend on the value of n. The rules are (1) that the second (orbital) quantum number, usually called l, can have the value zero and any other integral value up to $n - 1$ but no higher and (2) that the values of the third (orientation) quantum number (usually called m_l) can have integral values from $-l$ through zero up to $+l$).

The system for values of n up to four is given here.

n	possible l values and the corresponding orbital symbols		possible m_l values
1	0	*s*	0
2	0	*s*	0
	1	*p*	−1, 0, 1
3	0	*s*	0
	1	*p*	−1, 0, 1
	2	*d*	−2, −1, 0, 1, 2
4	0	*s*	0
	1	*p*	−1, 0, 1
	2	*d*	−2, −1, 0, 1, 2
	3	*f*	−3, −2, −1, 0, 1, 2, 3

Note that these simple rules predict, as observed,

1 kind of *s*-orbital,

3 kinds of *p*-orbitals,

5 kinds of *d*-orbitals,

7 kinds of *f*-orbitals.

If you would care to guess what happens next, the next orbital available for $n = 5$ is a *g*-orbital. How many *g*-orbitals would there be?

If you have trouble remembering the ordering of the energy levels (or the shape of the Periodic Table), there is a third mnenomic device that can be helpful. Just write down the orbitals in the following order:

1*s*
2*s* 2*p*
3*s* 3*p* 3*d*
4*s* 4*p* 4*d* 4*f*
5*s* 5*p* 5*d* 5*f*
6*s* 6*p* 6*d* . . .
7*s* 7*p*

The normal or expected order of filling the orbitals to give ground state atoms is found by drawing arrows, as shown below. The order of filling is the sequence in which the arrows cross the orbital notations starting from the top.

1*s*
2*s* 2*p*
3*s* 3*p* 3*d*
4*s* 4*p* 4*d* 4*f*
5*s* 5*p* 5*d* 5*f*
6*s* 6*p* 6*d*
7*s* 7*p*

Let's try element number 55, cesium. The predicted ground state electron configuration, following the simple method above, is

$$1s^2\ 2s^2\ 2p^6\ 3s^2\ 3p^6\ 4s^2\ 3d^{10}\ 4p^6\ 5s^2\ 4d^{10}\ 5p^6\ 6s^1$$

While there are often slight differences from these predictions, it is a convenient memory device.

Note the prediction for chromium, element 24:

$$1s^2\ 2s^2\ 2p^6\ 3s^2\ 3p^6\ 4s^2\ 3d^4$$

The actual ground state configuration for chromium determined experimentally is

$$1s^2\ 2s^2\ 2p^6\ 3s^2\ 3p^6\ 4s^1\ 3d^5$$

This difference is rationalized in terms of the energy difference between the paired *s* electrons ($4s^2\ 3d^4$) and the unpaired and thus spatially dispersed $4s^1\ 3d^5$ configuration.

With any luck at all, all of the superscripts add up to 78. Many of the chemical properties of platinum are rationalized in terms of its eight 5*d* electrons. In using Figure 6-6 or Figure 6-8, simply start at the bottom and fill in 78 electrons, two per orbital with antiparallel spins.

In Gratuitous Information 6-2, there is a third mnemonic device to help you remember the ordering of atomic orbitals according to energy.

The fundamental problem of chemistry is to unravel the relationship between the structure of atoms and the properties of elements and compounds. Much has been accomplished and a great deal of research is in progress. Much remains to be resolved.

In this section, a qualitative look has been taken at a model of the electron distribution in atoms. The model, at least in principle, is a quantitative mathematical model for all atoms of all elements. In practice, the mathematics becomes so complicated that only the simplest atoms can be approached rigorously with even the best computers. The problem arises due to the numbers of electrons involved. Every electron is attracted by the nucleus. Every electron repels every other electron. This leads to extremely complex potential energy relations. In order to survive mathematically, we make simplifying assumptions and, to a remarkable degree, our concepts of complex atoms are built by modifying the rigorous solutions for the hydrogen atom to accommodate more electrons.

Scramble Exercises

1. The element calcium, Ca, is an essential part of compounds that make up the structure of bones. Its atomic number is 20. Propose the complete notation for the electron structure of an atom of calcium in the ground state. Also give the more abbreviated notation in terms of the argon structure.

   On the basis of the Periodic Table, select the two elements that would be expected to have properties most closely related to those of calcium and propose the notation for the electron distribution in the ground state atoms of these two metals.

2. Radium, Ra, element 88, can also be incorporated with calcium in bone structure. This would probably be of only passing interest if it were not for the fact that naturally occurring isotopes of radium are radioactive and their nuclei undergo spontaneous transformations emitting energy. This localized energy is disruptive to the mechanism of producing the components of human blood in bone marrow. Propose the detailed notation and the abbreviated notation for the electron distribution in the ground state radium atom.

3. Sulfur, S, element 16, and selenium, Se, element 34, form a number of compounds of much the same type as oxygen, element 8. All three are incorporated by biological systems. Compounds of selenium are not very abundant in most geographical areas. Where they are more abundant, in some localized geographic areas, they create problems in the development of teeth. Propose the notation for the distribution of electrons in the ground state atoms of these three nonmetals.

4. The metals extensively used in industry as construction materials appear in that central block of ten elements in the periodic table. Many of the most commonly used of these **transition heavy metals** appear in the top row: chromium, Cr (24); manganese, Mn (25); iron, Fe (26); cobalt, Co (27); nickel, Ni (28); and copper, Cu (29). Propose the notation for the electron distribution in ground state atoms of chromium, manganese, iron and cobalt.

5. Tin, Sn, element 50, and lead, Pb, element 82, look like metals and have many of the properties of metals. They form compounds characteristic of metals. They also form compounds characteristic of nonmetals. These elements are known as **metalloids.** Propose the abbreviated notation for the ground state atoms of these elements. Compare these structures with the abbreviated structures for carbon, C, element 6, and silicon, Si, element 14.

6. Check back on the scramble exercises for the previous chapter. To become very secure with that limited number of exercises will serve you well.

7. There are two oxides of copper: CuO and Cu_2O. Both are solids and both are reduced to the pure metal by reaction of hydrogen gas, H_2, at about 400 °C. The only products are copper, Cu, and water, H_2O:

$$\text{oxide of copper (s)} + H_2\text{ (g)} \longrightarrow \text{Cu (s)} + H_2O\text{ (g)}$$

This reaction can be used to determine whether a sample of an oxide of copper is CuO or Cu_2O. For example, a 1.43-gram sample of one of these oxides of copper yields 1.27 grams of copper. See if you can find and use at least one route to determine whether the sample was CuO or Cu_2O.

8. Use the copy of the Periodic Table (inside the front cover) and Figure 6-7 to write out the abbreviated distribution of electrons in atomic orbitals of ground state atoms of lanthanum, $Z = 57$, and hafnium, $Z = 72$. Note that, for both elements, partially filled 5*d*-orbitals and partially filled 4*f*-orbitals are involved. The properties of the lanthanide series of elements and the actinide series of elements (the block of 28 elements set off at the bottom of the table) are rationalized in terms of both *d*-orbitals and *f*-orbitals. This is one of those finer points of chemistry with which we shall not concern ourselves in this book.

Additional exercises for Chapter 6 are given in the Appendix.

A Look Ahead

Scientists rationalize the chemical properties of elements and compounds in terms of the distribution of electrons in the various energy states of atoms and the distribution of electrons in various energy states of molecules. This chapter has addressed the distribution of electrons in atoms. The next step will be to address diatomic molecules—molecules made up of two atoms. The simplest of these is the hydrogen molecule, H_2. The ball and stick model represents the bonding in H_2 as a single bond, H—H, a stick between two balls representing the two hydrogen atoms. The question to be addressed is "In what way do the one proton (the nucleus) and the

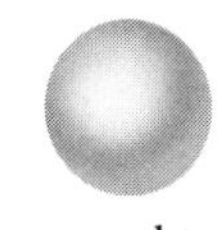

an *s*-orbital

a *p*-orbital

a *d*-orbital

one electron of one atom of hydrogen interact with the proton and the electron of another atom of hydrogen in the molecule H_2?" The two protons repel each other, the two electrons repel each other, each electron attracts both protons, and each proton attracts both electrons. In what way does the combination of these forces hold the atoms together with some bond length (separation between the two nuclei) characteristic of molecules of hydrogen? The only thing going for stability of the H_2 unit is the attractive forces between the electrons and the nuclei and possibly the much smaller spin attraction between the two electrons.

When we can rationalize the stability of the hydrogen molecule, we are well on our way to understanding other diatomic homonuclear molecules, such as chlorine gas, Cl_2, and nitrogen gas, N_2—also heteronuclear molecules, such as hydrogen chloride gas, HCl.

Diatomic molecules are linear. Two nuclei determine a straight line. The need to address bond angles arises with molecules of three or more atoms such as water, H_2O; carbon dioxide, CO_2; ammonia, NH_3; and methane, CH_4.

In order to proceed, you should have confidence that you have the skills, knowledge, and concepts sufficiently well in hand

- to write out the schematic pattern of the Periodic Table (Figure 6-7) giving the notation for the atomic orbitals of highest energy occupied by electrons in ground state atoms and correctly placing the symbols for at least ten elements—such as the inert gases, hydrogen, carbon, nitrogen, sulfur, fluorine, sodium, magnesium, copper, and zinc—on the appropriate blanks in the schematic pattern of the Periodic Table;
- to write out the notation of all electrons in the ground state atom of an element given the atomic number of that element;
- to present the concept of the 1*s*-orbital; and
- to present the concept of the three 2*p*-orbitals.

To develop these skills and acquire confidence, work with a pencil and paper.

Much of modern chemistry is rationalized in terms of the interactions of *s*, *p*, and *d* atomic orbitals.

7

HYDROGEN AND OTHER DIATOMIC MOLECULES: VALENCE BOND APPROACH

Chapter 7

In this chapter, we explore the bonding in diatomic molecules in terms of the model of electron distribution in atoms. The approach used is known as the valence bond model, VB. The molecules discussed are hydrogen, the four diatomic halogens, the four hydrogen halides, and a group of four other diatomic molecules that have multiple bonds.

In the investigation of models for bonding, the adequacy of each model should be assessed in terms of the capacity of the model to provide a basis for understanding the experimentally determined values for the properties of bonds, such as bond lengths and bond strengths; the properties of the molecules, such as polarity; and the properties of macro samples of substances, such as boiling points. In this chapter, we look at some of the many correlations that might be made.

The Diatomic Hydrogen Molecule

Dissociation of Diatomic Hydrogen Molecules

At room conditions, hydrogen gas exists as diatomic molecules, H_2. In the atmosphere of the sun, hydrogen gas exists as atoms, H. The strength of the chemical bond in the diatomic molecules can be experimentally determined. If the bonding involves only a single bond, the bond strength is essentially the energy required to *dissociate,* to separate, one mole of diatomic molecules into two moles of atoms. The reaction for the dissociation of one mole of diatomic hydrogen molecules into two moles of hydrogen atoms is

$$H_2 \longrightarrow 2\,H$$

one mole of diatomic hydrogen molecules ⟶ two moles of hydrogen atoms

6.02×10^{23} diatomic hydrogen molecules ⟶ 12.04×10^{23} hydrogen atoms

At room temperature and one atmosphere pressure, the experimental value is 436 kilojoules per mole or 104 kilocalories per mole of diatomic hydrogen molecules. Information of this type is expressed

$$H_2(g) \longrightarrow 2\,H(g) \qquad \Delta H = 436 \text{ kilojoules (104 kilocalories)}$$

with it being understood that the numerical values relate to the number of moles expressed in the chemical equation. In the symbol ΔH, H does not refer to hydrogen in any way. The symbol ΔH was originally used to represent change in heat content.

Heat is absorbed during the reaction and the heat content of the products (the hydrogen atoms) is greater than the heat content of the reactants (the diatomic hydrogen molecules). Today, the preferred terminology is to read the symbol ΔH as **change in enthalpy.** Energy is absorbed during the reaction, the enthalpy of the products is greater than the enthalpy of the reactants, and the value of ΔH is positive.

Expressed in terms of one molecule of diatomic hydrogen, the change in enthalpy is

$$\frac{436 \text{ kilojoules per mole } H_2}{6.02 \times 10^{23} \text{ molecules } H_2 \text{ per mole } H_2} = 72.4 \times 10^{-23}$$

$$= 7.24 \times 10^{-22} \text{ kilojoules per molecule } H_2$$

The question to be addressed: What is the nature of the chemical bond in the diatomic hydrogen molecule?

Ground State Hydrogen Atom

The hydrogen atom, ${}^1_1H$, with its atomic number of 1 and its one nucleon is the simplest atom. The nucleus is a single proton and the electron atmosphere consists of a single electron. In the ground state atom, the 1*s*-orbital is partially filled and the region of high probability for the electron is a spherical volume surrounding the nucleus. This spherical volume is also said to be the region of the atom with the highest negative charge density. The nucleus is the region of the atom with the highest positive charge density.

The diatomic hydrogen molecule, H_2, has an experimentally determined bond length of 0.075 nanometers, meaning that the centers of the two nuclei are separated, on the average, by the distance 0.075 nm as the two nuclei oscillate (vibrate) between a separation greater than 0.075 nm and a separation less than 0.075 nm. The fundamental question: What is the balance of forces that holds the nuclei at some average bond length? This balance of forces is the chemical bond in the diatomic hydrogen molecule. This balance of forces is, in this model, the equivalent of the "stick" used in the ball and stick model to represent the bond in the diatomic hydrogen molecule, H—H.

Within the diatomic molecule of hydrogen, attractive forces arise through the electrostatic attraction of unlike charges (electrons with nuclei) and the much weaker magnetic attraction of paired electron spins. The repulsive forces arise through the electrostatic repulsion of like charges (electron with electron and nucleus with nucleus). The bond length is the separation between the two nuclei when these forces are balanced.

When a model is used for the atom in which the electrons are in continuous motion and are continuously changing position with respect to the nucleus, it is the distribution of negative charge that is significant in considering the electrostatic attraction between the negative electrons and the positive nuclei. This distribution of negative charge is expressed as the negative charge density. In individual atoms of hydrogen (one electron per atom), the region of high negative charge density (the region of high electron probability) is a volume, spherically symmetrical with respect to the nucleus of the atom, Figure 7-1.

Interaction of One Electron with Two Protons

It is difficult to think about forces between regions of negative charge density and positive nuclei. To elucidate the nature of these forces, it is helpful to explore the force interactions involving three static particles: one electron and two protons.

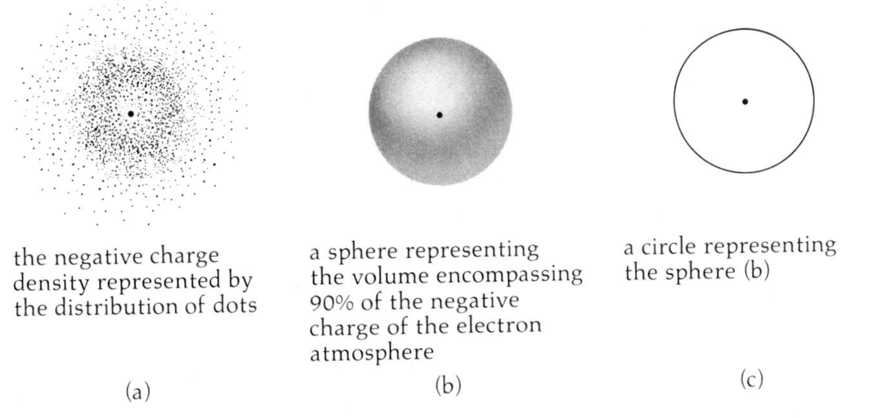

the negative charge density represented by the distribution of dots
(a)
a sphere representing the volume encompassing 90% of the negative charge of the electron atmosphere
(b)
a circle representing the sphere (b)
(c)

THREE SCHEMATIC REPRESENTATIONS OF THE NEGATIVE CHARGE DISTRIBUTION ABOUT THE NUCLEUS OF A GROUND STATE ATOM OF HYDROGEN. The position of the nucleus is indicated by the dot in the center of each diagram. Both (a) and (c) are two-dimensional representations of three-dimensional phenomena. (c) is the pictograph conventionally used to represent the *s* atomic orbital.

In this exploration, the distance separating the two protons will remain fixed at some arbitrary distance.

p^+	p^+
proton I	proton II

The force between these two positive particles is a force of repulsion, indicated below as two vectors (two arrows pointing away from each other).

$$\longleftarrow p^+ \qquad\qquad\qquad p^+ \longrightarrow$$

The magnitude of the force, represented by the length of the arrows, is given by the expression

$$\text{force} = \frac{(\text{charge of proton I}) \times (\text{charge of proton II})}{(\text{distance separating the protons})^2}$$

(The dielectric constant for space, for a vacuum, is 1; consequently the dielectric constant does not appear in the above force equation.)

To explore the interaction of these protons with the electron, insert the electron halfway between the two protons and consider the force between the electron and each proton.

p^+	e^-	p^+
proton I	electron	proton II

The force is a force of attraction. The forces of attraction acting on the two protons are indicated below by two arrows (vectors):

$$p^+ \longrightarrow e^- \longleftarrow p^+$$

The magnitude of each force is given by

$$\text{force} = \frac{(\text{charge on the proton}) \times (\text{charge on the electron})}{(\text{distance separating the proton and the electron})^2}$$

Since the charge of the electron has the same magnitude as the charge of the proton and the distance in this arbitrary configuration separating the two protons is twice as great as the distance separating either proton and the electron, the force tending to move the protons toward the electron, and consequently toward each other, is four times greater than the force of repulsion tending to move the protons further apart. (The factor of 4 in the force arises from the square of the distance of separation in the denominator of the force equation.) An electron between the two nuclei is highly effective in stabilizing the diatomic hydrogen molecule.

Consider a second special static case, starting again with the two protons in the same fixed positions:

p^+	p^+
proton I	proton II

This time, place the electron on the line of centers of the two protons and to the right of both protons:

p^+	p^+	e^-
proton I	proton II	electron

Since the electron is closer to proton II than it is to proton I, the force of attraction between proton II and the electron is greater than the force of attraction between proton I and the electron as indicated by the vectors shown below:

$$p^+ \rightarrow \qquad\qquad p^+ \longrightarrow e^-$$

The presence of the electron in this outside end position tends to increase the separation between the two protons, and an electron in this position does not contribute to the stability of the diatomic molecule. In fact, an electron in this position would destabilize the bonding. These two static arrays represent two extremes. An electron between the two nuclei is effective in forming the diatomic hydrogen molecule; an electron on the line of the two protons and not between the protons is countereffective in forming the diatomic hydrogen molecule.

Consider a third special static case, starting again with the two protons in the same fixed positions:

p^+	p^+
proton I	proton II

This time, place an electron off of the line of centers, such as above the line of centers, but otherwise half-way between the two protons:

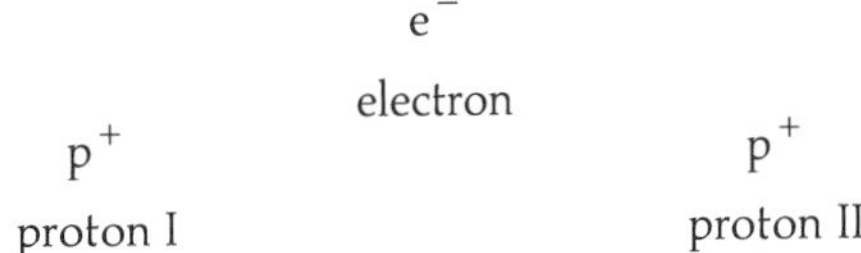

The electron is equidistant from the two protons but further removed from each proton than it would be if it were on the line of centers.

Consequently, the force acting on each proton is less than the force acting on each proton in the previous equidistance case. These forces tend to displace both protons toward the electron. In responding to these forces, the protons would be moved closer to each other but away from the original line of centers. (Another electron on the other side of the line of centers, below the line of centers, would counterbalance this sideways displacement.) An electron off of the line of centers is less effective than an electron on the line of centers in pulling the two protons towards each other.

Negative charge density is the consequence of where electrons can be and the fraction of the time they are there. The forces of attraction for both nuclei arising from a negative charge density between the two nuclei are highly effective in stabilizing the bond. The forces of attraction for both nuclei by negative charge densities in other positions relative to the two nuclei are less effective, and negative charge densities in some positions are countereffective in stabilizing the bond.

Molecular Orbital

In the valence bond approach to the nature of a chemical bond between two atoms, the wave functions of two partially filled atomic orbitals, one from each atom, are mathematically combined to give a wave function for a new orbital that encompasses the two nuclei and is occupied (filled) by the two paired electrons. This new orbital that encompasses two nuclei is known as a **molecular orbital.** The distribution of negative charge density in that orbital is, in the valence bond approach, the bond.

The molecular orbital for the diatomic hydrogen molecule is obtained by combining the wave function of the 1*s*-orbital of one hydrogen atom with the wave function of the 1*s*-orbital of the other hydrogen atom. Figure 7-2 is a schematic representation for the charge distribution in the diatomic molecule of hydrogen. It is to the diatomic hydrogen molecule what Figure 7-1 is to the ground state atom of hydrogen. Two of these figures are two-dimensional representations of three-dimensional phenomena.

sigma-Orbital for the Diatomic Hydrogen Molecule

The molecular orbital shown in Figure 7-2 has an ∞-fold axis of rotation through the two nuclei. A molecular orbital that has an ∞-fold axis of rotation through the two nuclei is designated by the Greek letter equivalent to *s*. The molecular orbital for the diatomic hydrogen molecule is a *sigma*-orbital, also written σ-orbital. In this book, we shall frequently use red in diagrams of molecular orbitals and black in diagrams of atomic orbitals. Other books often use the same color for both. Types of atomic orbitals are designated by letters of the Roman alphabet; types of molecular orbitals are designated by letters of the Greek alphabet.

Figure 7-2

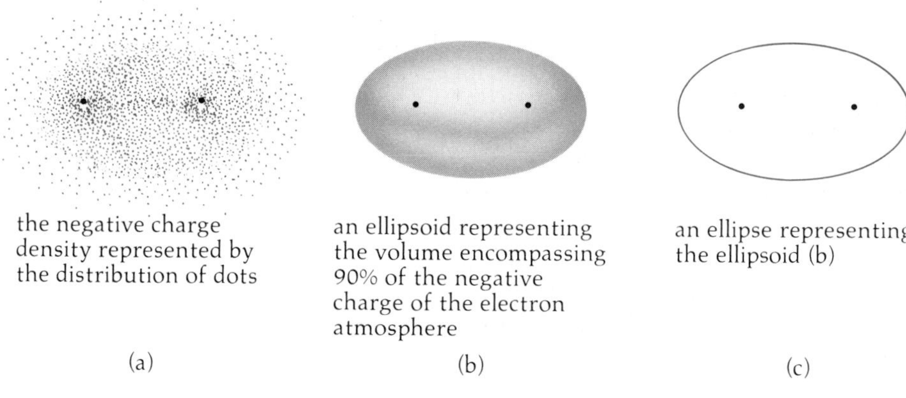

THREE SCHEMATIC REPRESENTATIONS OF THE NEGATIVE CHARGE DISTRIBUTION OF TWO ELECTRONS ABOUT THE TWO ATOMIC NUCLEI IN A DIATOMIC HYDROGEN MOLECULE. The locations of the two nuclei are indicated by the two large dots in each diagram. (a) uses stippling to represent the electron density; (b) is the ellipsoid that encompasses 90% of the negative charge; (c) is the ellipse that has become the conventional pictograph used to represent the *sigma* molecular orbital formed by the overlap of two *s* atomic orbitals. (a) and (c) are two-dimensional representations of the three-dimensional distribution of negative charge.

sigma-Bond in Diatomic Hydrogen

Wave functions, the mathematical expressions that give the distribution of electrons about two nuclei, are called molecular orbitals by those who do these types of calculation. Chemists who use the concepts of molecular orbitals, but usually do not make the calculations, use a much more pictorial language to discuss the formation of molecular orbitals from atomic orbitals. For the diatomic molecule of hydrogen, this language is: The *sigma*-bond of the diatomic hydrogen molecule is formed by the overlap of the partially filled 1*s*-orbitals of the two hydrogen atoms.

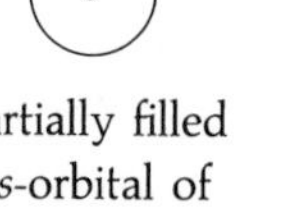

partially filled 1*s*-orbital of one hydrogen atom

partially filled 1*s*-orbital of another hydrogen atom

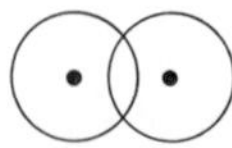

the two partially filled 1*s*-orbitals in the overlap position

the filled σ-orbital of the diatomic hydrogen molecule

This emphasis on the overlap of atomic orbitals focuses upon the buildup of negative charge between the nuclei in the formation of the bond. Don't forget that line drawings such as the above are two-dimensional representations of three-dimensional phenomena. The three-dimensional representation is given below.

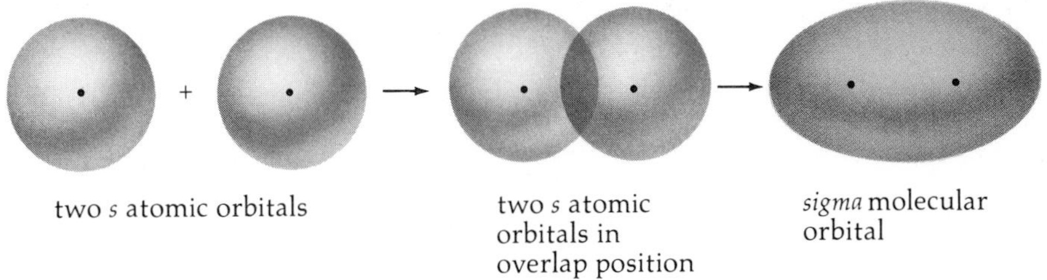

two *s* atomic orbitals — two *s* atomic orbitals in overlap position — *sigma* molecular orbital

In the diatomic hydrogen molecule, the distribution of the two nuclei and the two electrons in space is such that the forces of attraction equal the forces of repulsion. The separation between the two nuclei is the bond length. The distribution of electrons is given by the *sigma*-orbital. The bond is called a *sigma*-bond.

The Diatomic Halogen Molecules

Diatomic Halogens

The family of elements immediately preceding the inert gases in the Periodic Table is the halogen family. The four most plentiful elements in that family are fluorine, chlorine, bromine, and iodine. Each of these elements exists as diatomic molecules at room conditions and each of these elements forms a diatomic molecule with hydrogen. See Table 7-1 for atomic numbers, electron distributions in ground state atoms, formulas of molecules, and also the colors and phases at room conditions for these elements and their hydrogen compounds.

All of these diatomic halogens are destructive to biological tissues. Diatomic chlorine and diatomic iodine are used with care in low concentrations as germicidal agents—for example, chlorinated water, chlorine bleaches, and tincture of iodine. The first three—F_2, Cl_2, and Br_2—must be handled with knowledge and with care, particularly fluorine.

Hydrogen Halides

All of the hydrogen halides dissolve in water to give acidic solutions, and the stomach of each of us generates liters of a dilute aqueous solution of hydrogen chloride, called hydrochloric acid, every day to facilitate the digestion of the food we eat. An aqueous solution of hydrogen fluoride is highly destructive to biological tissues. The production, distribution, use, and disposal of both F_2 and HF are to be carefully supervised. The fluoridation agents added to water and to toothpastes are compounds containing fluorine—not the element itself and not hydrogen fluoride. All of the

Table 7-1

Halogens and Hydrogen Compounds of the Halogens

element	atomic number	electron distribution in the ground state atom	diatomic molecule, color, and phase*	hydrogen compound, name, color, and phase*
fluorine	9	(He) $2s^2\ 2p^5$	F_2 colorless gas	HF hydrogen fluoride colorless gas
chlorine	17	(Ne) $3s^2\ 3p^5$	Cl_2 yellow green gas	HCl hydrogen chloride colorless gas
bromine	35	(Ar) $4s^2\ 3d^{10}\ 4p^5$	Br_2 yellow brown liquid†	HBr hydrogen bromide colorless gas
iodine	53	(Kr) $5s^2\ 4d^{10}\ 5p^5$	I_2 violet solid‡	HI hydrogen iodide colorless gas

* All phases given for room conditions.
† Liquid bromine, Br_2, vaporizes very readily.
‡ Solid iodine, I_2, vaporizes (sublimes) very readily.

halogens and many of the compounds of all the halogens are extensively used in industry.

The immediate question: How do atoms of the halogens bond with each other in the diatomic molecule such as Cl_2?

The valence bond approach focuses upon the utilization of partially filled atomic orbitals to construct and to fill molecular orbitals, orbitals encompassing the two nuclei. All filled atomic orbitals and all of the electrons that fill them are ignored in the valence bond (VB) approach. This is an approximation equivalent to the assumption that whatever the electrons in filled atomic orbitals do in an isolated atom, these electrons continue to do when the atom is a part of a molecule. Electrons in filled atomic orbitals in no way contribute to or inhibit the formation of the bond. In Chapter 16 we shall explore another approach to the chemical bonds that utilizes all electrons of both atoms.

Ground State Halogen Atoms

In the ground state, the atoms of all the halogens have five electrons in three *p*-orbitals of the highest principle quantum level: $2p_x^{\ 2}\ 2p_y^{\ 2}\ 2p_z^{\ 1}$ for fluorine, $3p_x^{\ 2}\ 3p_y^{\ 2}\ 3p_z^{\ 1}$ for chlorine, etc. In the VB approach, it is the partially filled *p*-orbital that is considered to be involved in molecular bonding. In the notation given, this partially filled *p*-orbital has been arbitrarily designated as the p_z-orbital. The partially filled orbital could be equally well designated as the p_x-orbital or the p_y-orbital.

By combining the wave function for the partially filled *p*-orbital of one atom with the wave function for the partially filled *p*-orbital of a second atom, a wave function can be generated for a molecular orbital involving the two nuclei. The two atomic wave functions are combined for the spatial orientation that allows a maximum interaction (a maximum overlap). The pictograph for this orientation is

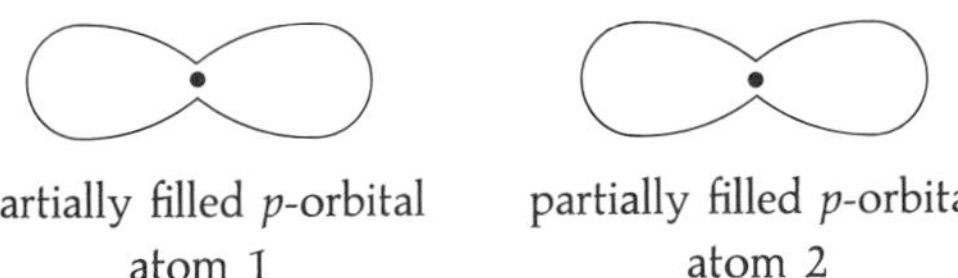

partially filled *p*-orbital atom 1   partially filled *p*-orbital atom 2

The two partially filled atomic orbitals are brought together in the overlap position

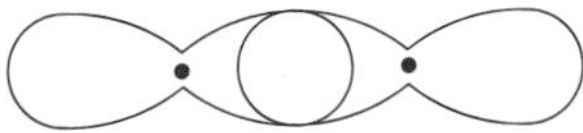

and the fusion of the two atomic orbitals forms the molecular orbital:

filled *sigma*-orbital for a diatomic halogen such as Cl_2

sigma-Orbitals of Diatomic Halogen Molecules

The shape of a molecular orbital formed from two partially filled *p*-orbitals does not have the ellipsoid character of the *sigma*-orbital formed by the overlap of two 1*s*-orbitals in the diatomic hydrogen molecule. The three-dimensional shape is instead more comparable to the figure generated by filling one long balloon with enough air to maintain its inflated shape and then twisting the balloon to give three inflated segments—a small inflated segment at each end and a larger inflated segment in the center. In this analogy, the three segments represent concentrations of negative charge with the two nuclei at the two constrictions formed by the twisting that isolates the segments. It is the distribution of negative charge in the central segment that stabilizes the bond. The distribution of negative charge in the smaller segments at the ends destabilizes the bond.

The molecular orbital formed by the end-to-end overlap of two *p*-orbitals has an ∞-fold axis of rotation through the two nuclei. This molecular orbital is also called a *sigma*-orbital and the distribution of charge in this *sigma*-orbital is the *sigma*-bond. The distance separating the two nuclei and the distribution of electrons is such that the forces of attraction, due to the electrostatic attractive forces between unlike charges (electrons and the nuclei) and the much smaller magnetic attraction between spin paired electrons, equal the forces of repulsion, due to the electrostatic repulsion forces between like charges (nucleus with nucleus and electron with electron).

The bond lengths and the bond energies of the four diatomic halogen molecules and the four hydrogen halide molecules are given in Tables 7-2 and 7-3. Note that, in all cases, the bond lengths are greater than the 0.075-nm bond length for diatomic

Table 7-2

Selected Physical Properties of the Diatomic Halogens

halogen	boiling point* (°C)	bond energy (kilojoules per mole)	bond energy (kilocalories per mole)	bond length (nm)	dipole moment (debye)
fluorine F_2	−188	155	37	0.14	0
chlorine Cl_2	−35	243	58	0.20	0
bromine Br_2	59	192	46	0.23	0
iodine I_2	184	151	36	0.26	0

* At one atmosphere pressure.

Table 7-3

Selected Physical Properties of the Hydrogen Halides

hydrogen halide	boiling point* (°C)	bond energy (kilojoules per mole)	bond energy (kilocalories per mole)	bond length (nm)	dipole moment (debye)
hydrogen fluoride HF	20	565	135	0.09	1.82
hydrogen chloride HCl	−115	431	103	0.13	1.08
hydrogen bromide HBr	−67	364	87	0.14	0.82
hydrogen iodide HI	−35	297	71	0.16	0.44

* At one atmosphere pressure.

hydrogen and that the bond lengths increase with the increasing complexity of the halogen atoms.

In order to better visualize the relative sizes of atoms and molecules, scientists use space-filling models. In these model sets, the dimensions of the pieces representing the atoms are scaled according to the best available experimental values and the pieces are held together by connector links that are not visible when the atoms are in position in the molecule. Figure 7-3 gives these space-filling models for the diatomic hydrogen molecule, the four diatomic halogen molecules, and the four hydrogen halides. Also note that all of the bond energies are less for the halogens than the H—H bond energy of 436 kilojoules per mole and that the F—F bond energy of 155 kilojoules per mole is an anomaly in comparison to the pattern of bond energies for the other diatomic halogens.

Homonuclear Diatomic Halogen Molecules

The diatomic halogen molecules that have been discussed—F_2, Cl_2, Br_2 and I_2—are homonuclear (nuclei of the same element) diatomic halogen molecules. What about heteronuclear (nuclei of different elements) diatomic halogen molecules? Compounds such as bromine chloride, BrCl, iodine chloride, ICl, and iodine bromide, IBr, are known. The formation of a *sigma*-bond using a partially filled *p*-orbital of a bromine atom and a partially filled *p*-orbital of a chlorine atom to form the *sigma*-orbital for the molecule BrCl is exactly parallel to the formation of the *sigma*-orbitals for Br_2 or Cl_2 molecules using partially filled *p*-orbitals of two bromine atoms or partially filled *p*-orbitals of two chlorine atoms.

Heteronuclear Diatomic Halogen Molecules

Figure 7-3

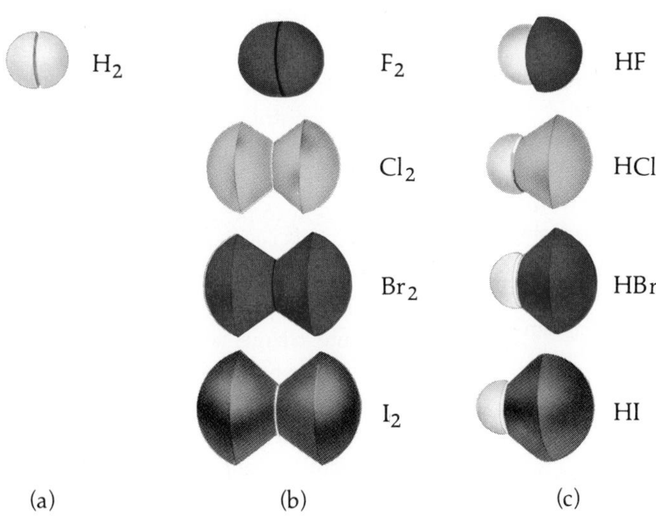

SPACE-FILLING MODELS FOR THE DIATOMIC HYDROGEN MOLECULE, THE FOUR DIATOMIC HALOGEN MOLECULES, AND THE FOUR HYDROGEN HALIDE MOLECULES.

The heteronuclear diatomic halogen molecules differ in one very important respect from the homonuclear diatomic halogen molecules. The homonuclear molecules F_2, Cl_2, Br_2, and I_2 have zero dipole moments. The heteronuclear molecules have very definite dipole moments. For example, chlorine fluoride, ClF, has a dipole

7-1 *Background Information*

DIPOLE MOMENTS OF MOLECULES

The term dipole implies two poles. It is used here in the sense of treating a molecule as having a localized positive charge and a localized negative charge separated by a distance, *d*. The molecule as a whole has a net charge of zero. These localized charges are much smaller than the magnitude of the charge of the electron and are written δ^- and δ^+ in which the small Greek delta, δ, is used to indicate a small increment of change.

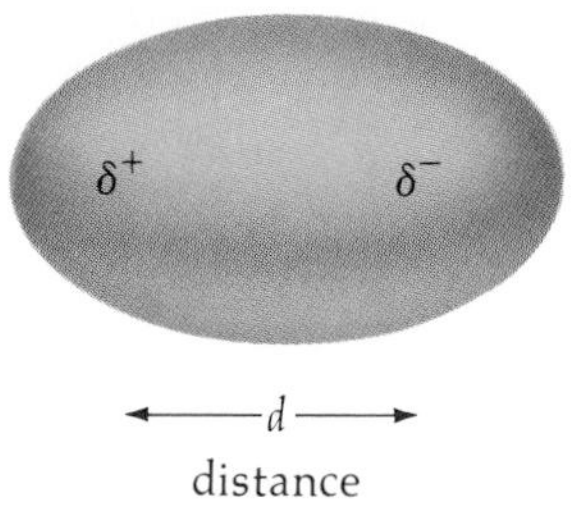

The term moment implies a tendency to turn about a point. This tendency of the molecule to turn arises when the dipole is placed between two charged plates. Any two metallic plates electrically connected to a high-voltge battery will do. The negative plate attracts the positive end of the molecule and repels the negative end, as indicated by the two arrows (vectors) on the righthand side of the diagram.

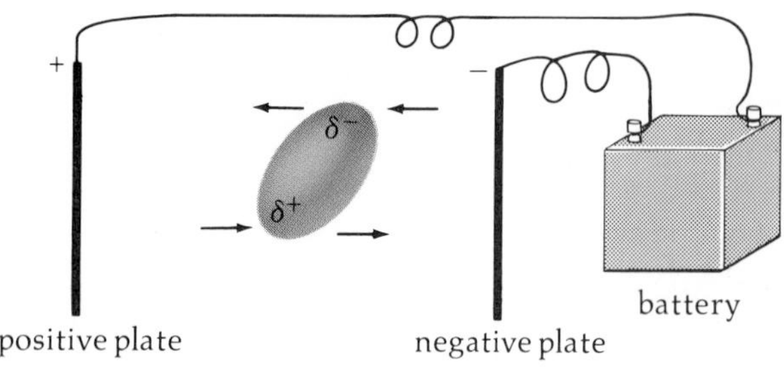

The positive plate attracts the negative end of the dipole and repels the positive end, as indicated by the arrows on the lefthand side. As a consequence of these forces, the molecule tends to turn about a point at its center. Since the two charges are equal in magnitude but opposite in sign, the center of rotation is halfway between the two charges.

Situations in which there is a diffuse distribution of charge can be treated in terms of a center of negative charge and a center of positive charge. In an isolated atom, the center of positive charge is the center of the nucleus and the center of negative charge from the electron atmosphere is the center of the atom. This center of the atom, of course, corresponds to the center of the nucleus, and all isolated atoms are nonpolar. In molecules, the electron charge distribution is a consequence of the distribution of electrons in both the atomic orbitals and the molecular orbitals. In a molecule, the center of the distribution of positive charges (the nuclei) and the center of the distribution of the negative charges (the electrons) are not necessarily the same. When the centers of positive and negative charges are separated by a distance, the molecule is said to possess a dipole moment.

The dipole moment is defined as the product of the charge of one pole times the distance separating the two poles. The units of dipole moments depend upon the units used to measure charge and the units used to measure distance. These need not concern us here since it is the relative values of dipole moments that are essential to the concept of polar and nonpolar bonds in compounds. The unit most frequently used to express dipole moments is the debye, D, named in recognition of Peter Debye, a Dutch physical chemist who investigated dipole moments and the properties of polar compounds. The dipole moment of a molecule cannot be measured directly but can be calculated from the measurements of other physical properties.

moment of 0.88 debyes. We shall return to a discussion of dipole moments, polar bonds, and polar molecules after discussing the hydrogen halides.

The Hydrogen Halide Molecules

According to the valence bond model, *sigma*-bond formation for the hydrogen halides (HF, HCl, HBr, and HI) involves the combination (the overlap) of the partially filled 1*s*-orbital of the hydrogen atom and a partially filled *p*-orbital of the halogen atom. In this *sigma*-bond, as in the previous *sigma*-bonds, the distribution of the two electrons and the two nuclei is such that the forces of attraction equal the forces of repulsion. The orbital again has an ∞-fold axis of rotation through the two nuclei.

The pictographs for the formation of the *sigma*-bond in hydrogen halides are

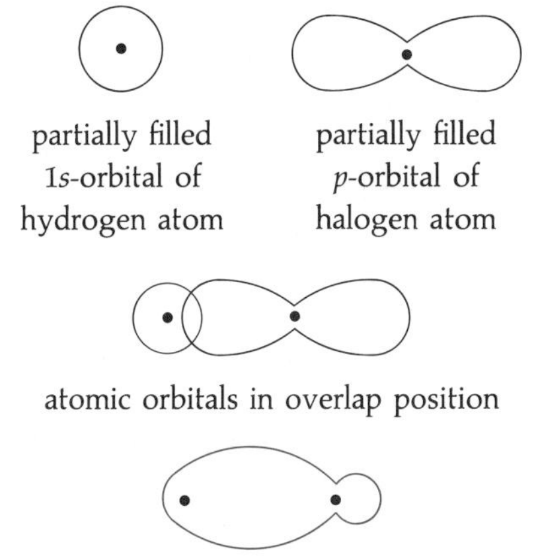

filled *sigma*-orbital
(the *sigma*-bond of the hydrogen halide molecule)

The experimentally determined bond lengths and bond energies are given in Table 7-3. Note that the bond lengths again increase as the halogen atoms become more complex and that the bond lengths for the hydrogen halides are shorter than the bond lengths for the corresponding diatomic halogen molecules, for example, 0.09 nm for HF compared with 0.14 nm for F_2. Note also that the bond energies of the hydrogen halides are much greater than the bond energies of the corresponding diatomic halogen molecules and that the bond energy of HCl (431 kilojoules per mole) is approximately the same as the bond energy of H_2 (436 kilojoules per mole).

Polar and Nonpolar Molecules

The experimental values for the dipole moments of the hydrogen halides range from 1.82 debye for HF to 0.44 debye for HI. Table 7-3. As stated previously, the dipole moment for chlorine fluoride, ClF, is 0.88 debye. All homonuclear diatomic molecules are nonpolar and have experimental dipole moments of zero. All heteronuclear diatomic molecules are polar and have dipole moments greater than zero. A dipole moment of zero is interpreted to mean that charge distribution in the molecule is symmetrical. Phrased in terms of the VB model, the electrons in the molecular orbital are shared equally by the two nuclei. Equal distribution, equal sharing of

Table 7-4

SUMMARY OF ATOMIC ORBITALS USED IN THE FORMATION OF MOLECULAR ORBITALS OF SELECTED DIATOMIC MOLECULES

Molecule	Pictographs for the atomic orbital used	Pictograph of the *sigma*-orbital formed
H_2	1s 1s	
F_2, Cl_2, Br_2, I_2	p p	
HF, HCl, HBr, HI	s p	

electrons, is to be expected in homonuclear diatomic molecules since the atoms are identical. A dipole moment greater than zero is interpreted to mean that the distribution of charge in the molecule is not symmetrical. Phrased in terms of the VB model, the electron pair of the bond is unequally shared by the two nuclei.

The preceding discussion of the valence bond approach to the formation of molecular orbitals from atomic orbitals is summarized in Table 7-4 for the diatomic molecules of hydrogen, the four homonuclear halides, and the four hydrogen halides. In all cases, the molecular orbital formed has an ∞-fold axis of rotation and is designated as a *sigma*-orbital. Note that in the representations (the pictographs) for the atomic orbitals, the pictograph used for all *p*-orbitals is the same regardless of whether the *p*-orbital is a 2*p*-orbital, a 3*p*-orbital, or another *p*-orbital, even though the sizes of these *p*-orbitals are quite different with the 2*p*-orbital being the smallest and the 5*p*-orbital being the largest. In the same sense, the same pictograph is used to represent the *sigma*-orbital in all four homonuclear diatomic molecules of the halogens even though there are great differences in the sizes of the *sigma*-orbitals for the four elements. Likewise, a single pictograph is used to represent the *sigma*-orbitals in the four hydrogen halides. You should be able to distinguish readily among the three pictographs used to represent the diatomic hydrogen molecular orbital, the diatomic halogen molecular orbital, and the hydrogen halide molecular orbital.

Electronegativity Values of Elements

The property of atoms that leads to unequal sharing of electrons between atoms is treated in a semiquantitative manner on an arbitrary scale of **electronegativity**

values. On the basis of the experimental values of a number of physical properties of compounds and of chemical properties of compounds, each element is assigned a numerical value of electronegativity: the larger the numerical value, the greater the tendency of the atom in a molecule to attract electrons. The electronegativities of a few of the elements are given in Table 7-5. The electronegativities follow a pattern of values that is easy to correlate with the positions of the elements in the Periodic Table. In the first short period, the electronegativities start at 1.0 for lithium and increase 0.5 units per element in moving across the table to fluorine, the most electronegative element, with a value of 4.0.

The electronegativity values for all of the inert gas elements are zero. Within other families, the electronegativity values of the elements decrease as the sizes of the atoms increase. Within a period, the element immediately preceding the inert gas element has the highest electronegativity value.

Chlorine Fluoride

Fluorine, electronegativity value of 4.0, is more electronegative than chlorine, electronegativity value of 3.0. Chlorine, with the lower electronegativity value, is said to be electropositive with respect to fluorine. The molecule ClF is polar. The compound of fluorine and chlorine is named chlorine fluoride. The general rule for naming compounds of two elements is to name the more electropositive element first and to change the ending of the more electronegative element to *-ide*. The chlorine end of the chlorine fluoride molecule is slightly positive and the fluorine end of the molecule is slightly negative. The magnitude of these localized charges is much less than the charge of the electron. The molecule as a whole is neutral.

Hydrogen has an electronegativity value of 2.1 and is slightly electronegative with respect to boron and slightly electropositive with respect to carbon. The hydrogen end of a hydrogen halide molecule is slightly positive and the halide end is slightly negative. All molecules, including the molecule of a hydrogen halide, are neutral. A molecule has no net charge.

Table 7-5

ELECTRONEGATIVITY VALUES*

lithium 1.0	beryllium 1.5	boron 2.0	carbon 2.5	nitrogen 3.0	oxygen 3.5	fluorine 4.0
sodium 0.9	magnesium 1.2	aluminum 1.5	silicon 1.8	phosphorus 2.1	sulfur 2.5	chlorine 3.0
potassium 0.8	calcium 1.0			arsenic 2.0	selenium 2.4	bromine 2.8
rubidium 0.8	strontium 1.0				tellurium 2.1	iodine 2.5
cesium 0.7	barium 0.9					

* The electronegativity value for hydrogen is 2.1.

Other Diatomic Molecules

Four other diatomic molecules are considered in this section: N_2, O_2, NO, and CO. All have high energies of dissociation and short bond lengths. All presumably involve multiple bonds. Three—O_2, NO, and CO—severely challenge the adequacy of the valence bond approach. Selected physical properties are given in Table 7-6.

The dissociation energies are exceptionally high:

$$N_2 \longrightarrow N + N \qquad \Delta H = 950 \text{ kilojoules}$$
$$O_2 \longrightarrow O + O \qquad \Delta H = 498 \text{ kilojoules}$$
$$NO \longrightarrow N + O \qquad \Delta H = 626 \text{ kilojoules}$$
$$CO \longrightarrow C + O \qquad \Delta H = 1073 \text{ kilojoules}$$

Of the nine molecules discussed earlier in this chapter, only one, hydrogen fluoride, comes into this range of high energies of dissociation:

$$HF \longrightarrow H + F \qquad \Delta H = 565 \text{ kilojoules}$$

Diatomic Nitrogen

Approximately 80% of the molecules in the atmosphere are diatomic molecules of nitrogen, N_2. The molecules are relatively inactive chemically, and pure nitrogen gas is frequently used when an inert atmosphere is needed to carry out experimental work with compounds that react with oxygen gas. One of the challenges to chemists

Table 7-6

SELECTED PHYSICAL PROPERTIES OF FOUR MULTIPLE BOND DIATOMIC MOLECULES

		energies of dissociation			
molecule	**boiling point* (°C)**	**(kilojoules per mole)**	**(kilocalories per mole)**	**bond length (nm)**	**dipole moment (debye)**
nitrogen N_2	−196	950	227	0.11	0
oxygen O_2	−183	498	119	0.12	0
nitrogen oxide NO	−152	626	150	0.12	0.15
carbon monoxide CO	−191	1073	257	0.11	0.11

* At one atmosphere pressure.

has been and is to convert part of the very plentiful nitrogen of the atmosphere into nitrogen compounds essential to agriculture and industry.

The ground state nitrogen atom

$$\text{nitrogen } (Z = 7)\ \ 1s^2\ 2s^2\ 2p_x^{\ 1}\ 2p_y^{\ 1}\ 2p_z^{\ 1}$$

Orientation of Atomic p-Orbitals

has three partially filled p-orbitals. In any consideration of the interaction, the overlap, of these partially filled orbitals of one atom with partially filled orbitals of another atom, the spatial orientation of these three "hourglass" atomic orbitals with respect to each other is of prime importance. Each of the p-orbitals has an ∞-fold axis of rotation through the nucleus of the atom, and the orientation of the three p-orbitals can be simply stated in terms of the relative orientation of these three axes. The axis of the p_x-orbital and the axis of the p_y-orbital are both perpendicular to the axis of the p_z-orbital. The axis of the p_x-orbital is perpendicular to the axis of the p_y-orbital. The three axes are mutually perpendicular. The p_x axis, the p_y axis, and the p_z axis have the orientation of the x, y, and z axes of Cartesian coordinates. The three knitting needles in the marginal sketch represent these ∞-fold axes of rotation of the three atomic orbitals.

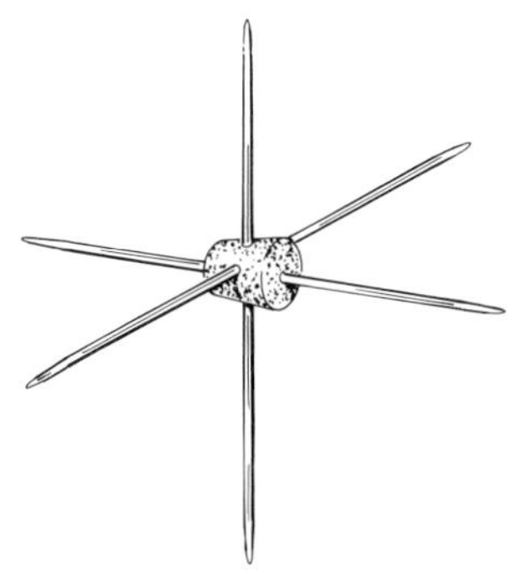

Any one of the partially filled p-orbitals of one nitrogen atom can interact with any one of the partially filled p-orbitals of another nitrogen atom to form a *sigma*-orbital by end-to-end overlap in exactly the same sense that the partially filled p-orbital of one halogen atom can interact with the partially filled p-orbital of another halogen atom to form the *sigma*-orbital in the diatomic halogen molecule.

Formation of Molecular σ-Orbital

pictograph of *sigma*-orbital formed from the p_x atomic orbitals of the two atoms

If the partially filled p-orbitals involved in the formation of a *sigma*-orbital are designated as the atomic p_x-orbitals of each atom, then the axes of rotation of the four remaining partially filled p-orbitals are all perpendicular to the axis of rotation of the *sigma*-orbital. By rotation about the ∞-fold axis of rotation of the σ-orbital, the ∞-fold

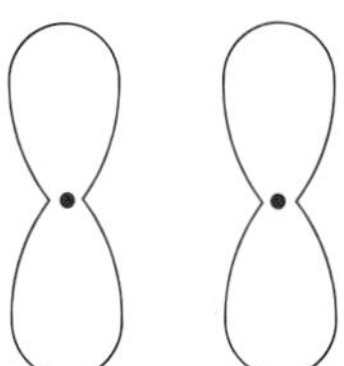

pictograph of two partially filled p_y-orbitals of two separate atoms appropriately oriented to interact

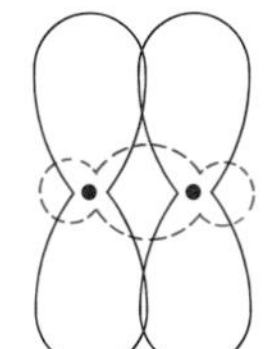

broken line is the pictograph of the σ-orbital

pictograph of two partially filled p_y-orbitals of two atoms being brought closer by the formation of a *sigma*-bond using the p_x-orbitals of the two atoms

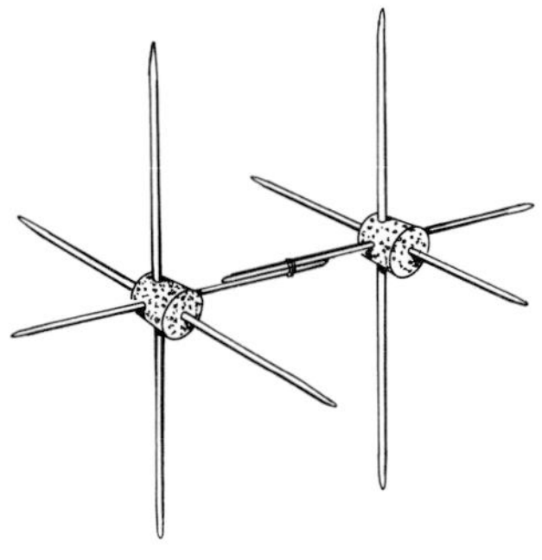

axes of rotation of the two p_y-orbitals can be brought into line with each other so that these two axes are parallel to each other. In the formation of the σ-bond, by the end-to-end overlap of the p_x-orbitals, these partially filled p_y-orbitals are brought close enough to overlap, to interact, in this side-to-side position as shown in the pictograph at the bottom of page 141.

On fusion, this side-to-side overlap of the two hourglass-type orbitals leads to two regions, two volumes, of high electron probability on opposite sides of the ∞-fold axis of rotation of the σ-orbital.

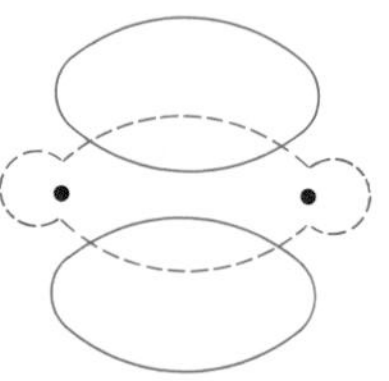

pictograph of the filled σ-orbital (dashed line)
and the filled π-orbital (solid lines)

Formation of One π-Orbital

These two regions of high negative charge density are one molecular orbital. This type of filled orbital is frequently described as the combination of an electron cloud on one side of the filled σ-orbital and an electron cloud on the opposite side of the filled σ-orbital. This type of molecular orbital is designated as a *pi*-orbital (note that *pi* is the Greek equivalent of *p*) and accommodates two electrons with paired spins. Each ellipse in the above pictograph of the π-orbital indicates the volume within which there is 45% probability of finding the electrons that fill this π-orbital. A three-dimensional representation of this π-orbital is given below.

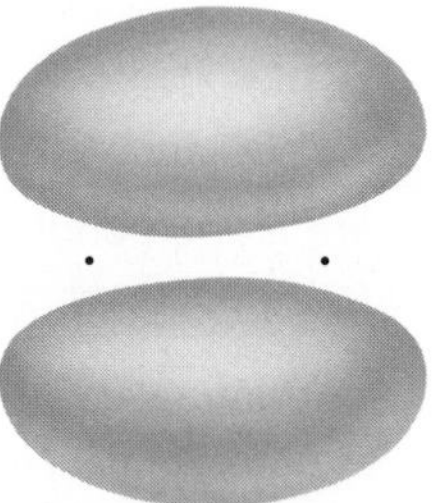

This approach to the formation of a π-bond requires that a σ-bond is formed and that it is the σ-bond that brings the two nuclei close enough together for the side-to-side interaction of two partially filled p-orbitals to take place and form the filled π-orbital. The force arising from the negative charge distribution in the π-orbital contributes to the stability of the molecule and brings the two nuclei still closer together. However, the concentration of negative charge in the π-orbital is not directly between the two nuclei and the π-bond contributes less to the stability of the molecule than the σ-bond does.

At the same time the two nitrogen atoms were rotated to bring the two p_y-orbitals into position for side-to-side overlap, the two p_z-orbitals were also brought into position for side-to-side overlap and the formation of a second π-orbital. In terms of the previous sketch of the σ-orbital and π-orbital, one of the electron clouds for this second filled π-orbital is above the plane of the paper and the other electron cloud is below the plane of the paper. It is presumed that the two π-bonds are formed simultaneously.

Formation of Second π-Orbital

The valence bond approach predicts a triple bond between the two atoms of diatomic nitrogen: one σ-bond and two π-bonds. The ball and spring model also predicts a triple bond between the two atoms of diatomic nitrogen. There is, however, a very significant difference. The valence bond approach predicts that one bond, the σ-bond, is quite different from the other two bonds, the π-bonds. The ball and spring model predicts the three bonds to be equivalent bonds.

The concept of triple bonds for N_2 is consistent with the high experimental value for the energy of dissociation of N_2 and with the short experimental value of the nitrogen–nitrogen bond length.

$$N_2 \longrightarrow N + N \qquad \Delta H = 950 \text{ kilojoules}$$

The bond length of N_2 is 0.11 nm, compared to the nitrogen–nitrogen single bond length of 0.15 nm in many compounds. The high energy of dissociation is consistent with the experimental observation that diatomic nitrogen is chemically inert. For hydrogen, the halogens, and the hydrogen halides, the energies of dissociation of the

Background Information 7-2

PARAMAGNETISM AND DIAMAGNETISM

In spite of what may have been implied in an earlier section, all substances do respond to a magnetic field. Some substances are strongly drawn into a magnetic field; other substances are very weakly rejected by a magnetic field. Substances that are attracted by a magnetic field are classified as paramagnetic and those that are rejected as diamagnetic. The forces involved in paramagnetism are so much greater than the forces involved in diamagnetism that the latter are frequently ignored. Both types of response are very easy to demonstrate experimentally provided a sensitive balance and a strong electromagnet are available. The sample is suspended from the balance arm or from the bottom of the balance pan by a long thread that is passed through a hole in the bottom of the balance case so that the sample can be suspended in the magnetic field of an electromagnet. In this position, the usual weighing procedures can be carried out with the magnet turned off or with the magnet turned on. When the magnet is turned on, the sample is subjected to three force fields: the usual gravitational field of the earth, the usual magnetic field of the earth, and the imposed magnetic field of the magnet. If the sample is pulled into the imposed magnetic field, this added force, in terms of the response of the balance, is equivalent to an increase in mass of the sample. If the sample is rejected, the response is equivalent to a decrease in mass.

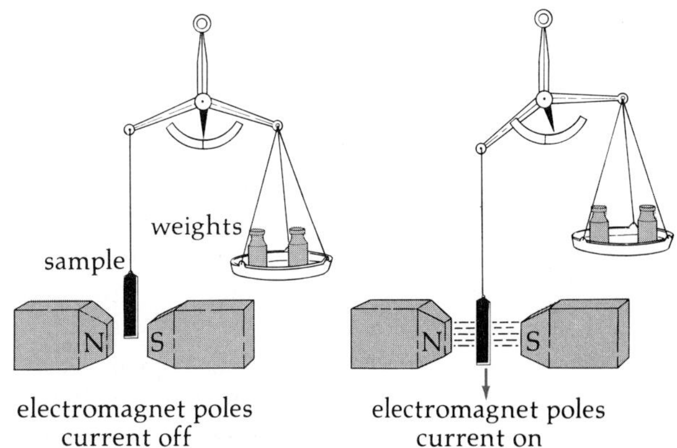

diatomic molecules are essentially the bond energies. These molecules contain only single bonds. (There can be a slight difference between the bond energy for diatomic molecules involving single bonds and the energy of dissociation for these same molecules. This point is discussed further in Chapter 8.) The dissociation energy for diatomic nitrogen involves the rupture of three bonds.

Diatomic Oxygen

Approximately 20% of the molecules of the atmosphere are diatomic oxygen molecules, O_2. Diatomic oxygen is much more chemically reactive than diatomic nitrogen molecules and essential to all biological systems. In red-blooded animals, diatomic oxygen is adsorbed by the hemoglobin of the blood and transported by the hemoglobin throughout the biological system.

The ground state atom of oxygen

$$\text{oxygen } (Z = 8) \;\; 1s^2\, 2s^2\, 2p_x{}^2\, 2p_y{}^1\, 2p_z{}^1$$

has two partially filled p-orbitals. The valence bond model predicts that one partially filled p-orbital of each atom, by end-to-end overlap, forms a filled σ-orbital and that the second partially filled p-orbitals, by side-to-side overlap, form a filled π-orbital. The valence bond approach thus predicts a double bond between the two atoms of oxygen, a σ-bond and a π-bond. The ball and spring model predicted two equal bonds between the two atoms of oxygen.

There is, however, a physical measurement that contradicts the valence bond structure of paired electrons in molecular orbitals. Diatomic oxygen is paramagnetic. It is attracted by a magnetic field. All of the other diatomic substances that have been discussed in this chapter are diamagnetic. They are slightly repelled by a magnetic field. The overwhelming evidence from the measurement of the magnetic properties of many substances is that paramagnetic substances contain unpaired electrons. Consequently, the scientific community rejects the valence bond structure and looks further for a model of bonding to rationalize the properties of diatomic oxygen. Such a model is presented in Chapter 16, Oxygen Gas, the Molecular Orbital Approach to Diatomic Molecules, and Oxidation–Reduction Reactions. This model, called the molecular orbital approach, the MO approach, places all sixteen electrons of the O_2 molecule in molecular orbitals—two of which are only partially filled.

Nitrogen Oxide

Nitrogen monoxide, also called nitric oxide, NO, is formed in air during high-energy conditions—in electrical storms and in high-temperature combustion chambers of automobiles and furnaces. Nitrogen oxide is an odd electron molecule, drawing seven electrons from the nitrogen atom and eight electrons from the oxygen atom. The valence bond approach (VB) is based upon the formation of filled molecular orbitals using two partially filled atomic orbitals, one from each atom, and the VB model is inapplicable for odd electron molecules. The molecular orbital approach (MO) is applicable to both odd electron molecules and even electron molecules. Nitrogen oxide will be discussed in connection with the MO model in Chapter 16.

Carbon Monoxide

Carbon monoxide, CO, is formed in the incomplete combustion of carbon compounds. Its inhalation is very toxic to red-blooded animals. The nature of this toxicity is well understood. The carbon monoxide molecule is adsorbed by hemoglobin in preference to oxygen molecules; thus inadequate quantities of oxygen are transported by the blood to tissues throughout the body. Carbon monoxide has an exceptionally high energy of dissociation, over a thousand kilojoules per mole.

The valence bond approach is based upon the two partially filled *p*-orbitals of each atom: one partially filled *p*-orbital from each atom being used to form a filled σ-orbital and one partially filled *p*-orbital of each atom being used to form a filled π-orbital.

$$\text{carbon } (Z = 6) \quad 1s^2\ 2s^2\ 2p_x^{\ 1}\ 2p_y^{\ 1}$$
$$\text{oxygen } (Z = 8) \quad 1s^2\ 2s^2\ 2p_x^{\ 2}\ 2p_y^{\ 1}\ 2p_z^{\ 1}$$

This is all that is required by the valence bond model to rationalize a double bond between the two atoms. The high energy of dissociation for the model raises the question of whether a third bond could be involved. The filled atomic *p*-orbital of the oxygen atom, in keeping with the VB approach, has remained a filled atomic orbital. In the alignment of the two atomic *p*-orbitals to form the molecular π-orbital, this filled *p*-orbital of the oxygen atom is also aligned with an "unoccupied atomic *p*-orbital" of the carbon atom. The phrase "unoccupied orbital" is commonly used to refer to the space where an orbital might be but no electrons are available to occupy that space. By overlapping the filled *p*-orbital of the oxygen atom with the unoccupied *p*-orbital of the carbon atom, a second π-orbital with two electrons could be formed. With the formation of this second π-orbital, the atoms of the carbon monoxide molecule become triple bonded: one *sigma*-bond and two *pi*-bonds. This is consistent with the high dissociation energy of the molecule. The ball-stick-spring model does not predict the existence of a CO molecule.

All four of these multiple bond diatomic molecules will be considered again in terms of the MO model in the chapter on oxygen.

Lewis Electron Structures

Another method of representing electron structures of molecules is widely used. It is an older approach and it is much less specific as to the nature of the bonds between atoms than the valence bond approach. It was proposed by G. N. Lewis in 1916 and preceded the concepts of the atomic orbitals that arose from wave mechanics after 1923. It is largely a correlation that relates many known formulas of small-atomic-number elements with inert gas structures. In this system of notation, helium is represented He:, neon :N̤̈e:, and argon :Är̤:. Each dot represents an electron. The helium atom (Z = 2) has 2 electrons. The neon atom (Z = 10) has 10 electrons: only 8 electrons are shown in the Lewis structure. These are the electrons in the principle quantum level 2. The argon atom (Z = 18) has 18 electrons; only 8 electrons are shown in the Lewis structure. These are the electrons of principle quantum level 3. Lewis was fascinated by the stability of the inert gas element structures and recognized that a consistent system could be evolved if the chemical bond is considered to be a pair of electrons shared by two atoms. The Lewis representations for the molecules we have just discussed are given on page 146.

atoms	molecules	
H·	H:H	
:F·	:F:F:	H:F:
:Cl·	:Cl:Cl:	H:Cl:
:Br·	:Br:Br:	H:Br:
:I·	:I:I:	H:I:
:N·	:N:::N:	
:O·	:O::O:*	:O:O:†
·C·	:C:::O:	
	:N::O:†	:N:O:†

* Disallowed on the basis of magnetic properties.
† Pseudo Lewis structures.

First, be confident that you understand the Lewis representation for the atoms of the elements. Not all of the electrons are shown—just the *s* and *p* electrons in the highest principle quantum number.

In the Lewis structure for the hydrogen molecule, H:H, with two shared electrons, each hydrogen atom is said to have a share in the two electrons and each atom is said to have acquired the helium atom structure, the structure of an inert gas. Counting shared electrons is a bit like counting posts in a fence. Each neighbor has the benefit of all the posts in the fence.

In the Lewis structure for the diatomic fluorine molecule, :F:F:, with the two shared electrons, each atom is said to have eight electrons (six of its own and two shared) and to have acquired the octet of electrons characteristic of all inert gas molecules other than helium.

In the Lewis structure for hydrogen fluoride, H:F:, by sharing the two electrons, hydrogen is said to have acquired an inert gas structure (helium) and the fluorine to have also assumed an inert gas structure (neon).

The parallel statements can be made for diatomic chlorine and hydrogen chloride—also for bromine, hydrogen bromide, iodine, and hydrogen iodide—using the *s*-electrons and the *p*-electrons of the highest occupied principle quantum number.

The Lewis structures are more complicated for the multiple bond compounds. Each bond involves the sharing of two electrons. The sharing of the six electrons in the nitrogen molecule, :N:::N:, constitutes a triple bond, N≡N. Each atom of nitrogen has five electrons to build into the molecule:

:N· ·N:

Sharing one pair of electrons,

:N:N:

each atom of nitrogen has a share in only six electrons. Sharing a second pair,

:N::N:

each nitrogen has a share in seven electrons. Sharing a third pair of electrons,

:N:::N:

each atom of nitrogen has assumed the octet of the inert gases.

In the Lewis structure of carbon monoxide, there are ten electrons, four from the carbon atom and six from the oxygen atom:

·C·   ·O:

Sharing one pair of electrons,

·C:O:

gives the carbon a share in five electrons and the oxygen a share in seven electrons. Sharing two pairs of electrons,

:C::O:

gives the carbon a share in six electrons, the oxygen eight electrons. For the Lewis structure, each atom should have eight. This can be achieved by sharing a pair of electrons of oxygen with the carbon

:C:::O:

to give a triple bond and an octet of electrons to the carbon as well as the oxygen. In Lewis structures, the two bonds of a double bond are equivalent and the three bonds of a triple bond are equivalent. In the valence bond approach, σ-bonds are quite different in character from π-bonds.

To Lewis, the diatomic molecule of oxygen was quite straightforward:

:O::O:

This is now disallowed by the magnetic properties of oxygen and a pseudo Lewis structure is sometimes written :O:O: to show unpaired electrons. In this structure, the octets are not attained, nor is there a double bond.

Lewis structures, the valence bond model, and the ball–stick–spring model are closely related. For example, the ball–stick–spring structure

```
     H   H   H
     |   |   |
 H—C—C=C—H
     |
     H
```

becomes the Lewis structure

```
  H H  H
  ¨ ¨  ¨
H:C:C::C:H
  ¨
  H
```

The valence bond approach identifies one pair of dots as a σ-bond and two pairs of dots as a σ-bond and a π-bond. Presently, we will attribute the lack of freedom of rotation about the double bond to the two-electron-cloud nature of π-bonds. Rotation of the atoms cannot take place at a double bond without breaking the π-bond and this requires the investment of energy.

To understand the nature of chemical bonds is indeed a challenge. It is our practice to use the various models for bonding to complement each other. The individual most associated with the development and popularization among chemists of the valence bond approach and the concept of electronegativities is Linus Pauling.

Scramble Exercises

1. Starting with the atomic number of chlorine, work through the bonding in the diatomic molecule of chlorine, Cl_2,
   (a) in terms of the valence bond approach and
   (b) in terms of the Lewis structure.
   (c) Also relate the valence bond structure to the Lewis structure.

2. Starting with the atomic numbers for hydrogen and chlorine, work through the bonding in hydrogen chloride, HCl,
   (a) in terms of the valence bond model and
   (b) in terms of the Lewis structure.
   (c) Also relate the valence bond structure to the Lewis structure.

3. Starting with atomic numbers, show that nitrogen monoxide, NO, is an odd electron molecule.

4. Give the name of the diatomic halogen compound with molecules consisting of one atom of bromine and one atom of iodine and predict the distribution of charge in the molecule (predict which end of the molecule is slightly positive, etc.).

5. The energy of dissociation of chlorine fluoride, ClF, is 251 kilojoules per mole:

   $$\text{ClF} \longrightarrow \text{Cl} + \text{F} \qquad \Delta H = 251 \text{ kilojoules}$$

   Show that this relation could also be expressed

   $$\text{ClF} \longrightarrow \text{Cl} + \text{F} \qquad \Delta H = 60. \text{ kilocalories}$$

   This exercise depends entirely upon the definition of the calorie in terms of joules and is an exercise in converting from one set of units to another set of units.

6. Hydrogen gas burns in chlorine gas to give hydrogen chloride gas, heat, and light:

$$H_2(g) + Cl_2(g) \longrightarrow 2\,HCl(g) + \text{energy}$$

The experimental value for the total energy released when one mole of hydrogen reacts with one mole of chlorine is 184 kilojoules at 25 °C. This change in enthalpy, change in heat content, per mole of hydrogen reacted can be expressed

$$H_2(g) + Cl_2(g) \longrightarrow 2\,HCl(g) \qquad \Delta H = -184 \text{ kilojoules}$$

The negative sign indicates that two moles of hydrogen chloride have less enthalpy than one mole of diatomic hydrogen and one mole of diatomic chlorine. This is as it should be if hydrogen gas burns in chlorine gas with the evolution of heat and light.

Show that this change in enthalpy can be calculated within the uncertainty of these numbers from the three energies of dissociation at 25 °C given in the chapter.

$$H_2 \longrightarrow H + H \qquad \Delta H = 436 \text{ kilojoules}$$

$$Cl_2 \longrightarrow Cl + Cl \qquad \Delta H = 243 \text{ kilojoules}$$

$$HCl \longrightarrow H + Cl \qquad \Delta H = 431 \text{ kilojoules}$$

Keep track of the number of moles of bonds in H_2 that are broken, the number of moles of bonds in Cl_2 that are broken, and the number of moles of bonds in HCl that are formed in the reaction

$$H_2 + Cl_2 \longrightarrow 2\,HCl$$

and (a) depend on common sense to determine how to use the energies of dissociation <u>or</u> (b) treat the relations as simultaneous equations. The chemical change

$$HCl \longrightarrow H + Cl \qquad \Delta H = 431 \text{ kilojoules}$$

can be reversed if the sign of ΔH is also reversed:

$$H + Cl \longrightarrow HCl \qquad \Delta H = -431 \text{ kilojoules.}$$

The chemical quantities can be doubled if the value of ΔH is also doubled:

$$2\,H + 2\,Cl \longrightarrow 2\,HCl \qquad \Delta H = -862 \text{ kilojoules.}$$

Better yet, carry out this exercise (a) depending on common sense <u>and</u> also (b) manipulating the quantitative relations as simultaneous equations. This is a fine way to develop understanding and confidence. (This is not an easy exercise to find your way through. Don't be dismayed if you stick, but don't let yourself stay stuck.)

7. Show that the molecules N_2 and CO are isoelectronic—the molecule of N_2 has the same total number of electrons as the molecule of CO. How has this isoelectronic characteristic been reflected in the proposed bonding in the two molecules

(a) according to the VB model, and
(b) according to the Lewis structures?

8. In Tables 7-2 and 7-3, numerical values for several physical properties have been given for the diatomic halogens and the hydrogen halides.

   (a) Identify a physical property for which the trend of the numerical values is an increase from fluorine to iodine (either the diatomic halogen or the hydrogen halide).

   (b) Identify a physical property for which the trend in numerical value is a decrease from fluorine to iodine (either the diatomic halogen or the hydrogen halide).

   (c) Identify a property for which the numerical value for one diatomic halogen or hydrogen halide seems to be an exception to a trend exhibited by the other three diatomic halogens or hydrogen halides.

9. The boiling points given in Tables 7-2 and 7-3 have not been discussed in the chapter. Boiling points are related to the capacity of the molecules to escape from the liquid phase: the more easily they escape, the lower the boiling point. In general, the boiling points are higher for substances made up of high molecular weight molecules and/or for substances made up of molecules that have strong attractive forces for each other. What trends or correlations can you identify or propose concerning the relative values of the boiling points of

   (a) the four diatomic halogens and

   (b) the four hydrogen halides?

10. Speculate on the nature of the chemical bonds in water, H_2O, according to the valence bond approach and the Lewis structure approach. What is your prediction for the

H
  O
H

bond angle? What approach did you use in making the prediction?

Additional exercises for Chapter 7 are given in the Appendix.

A Look Ahead

IONS AND IONIC COMPOUNDS

How great can the degree of unequal sharing of a pair of electrons by two atoms be?

For two identical atoms, the sharing is equal, the bond is nonpolar, and the diatomic molecule is nonpolar.

For two different atoms, the degree to which the pair of electrons is shared unequally depends upon the identities of the two atoms, a maximum degree of unequal sharing occurring with atoms of elements of markedly different electronegativities. One compound of this type is sodium chloride, NaCl, ordinary salt of the type used to season eggs for breakfast and melt ice on the front walk. On the electronegativity value scale, sodium has a value of 0.9 and chlorine a value of 3.0.

The structures of the two ground state atoms

sodium (Z = 11) $1s^2\ 2s^2\ 2p^6\ 3s^1$ or (Ne) $3s^1$

chlorine (Z = 17) $1s^2\ 2s^2\ 2p^6\ 3s^2\ 3p^5$ or (Ne) $3s^2\ 3p^5$

indicate that, according to the valence bond approach, the partially filled 3*s*-orbital of the sodium atom and the partially filled *p*-orbital of the chlorine could form a σ-orbital for the molecule NaCl. The electronegativity values, 0.9 versus 3.0, indicate that the two electrons in this orbital are very unequally shared and that the negative charge builds up in the vicinity of the chlorine atom. The σ-bond is extremely polar. The extreme of this unequal sharing is the capture of the two electrons by the chlorine atom. The resultant chlorine unit has a negative charge equal to the charge of the one electron and the unit is called the chloride ion, Cl^{1-}. This charge is the consequence of the 17 protons in the nucleus and the 18-electron atmosphere. The resultant sodium unit has a positive charge equal in magnitude to the charge of an electron and is a consequence of the 11 protons in the nucleus and the 10-electron atmosphere. It is called the sodium ion, Na^{1+}.

Na^{1+} $1s^2\ 2s^2\ 2p^6$

Cl^{1-} $1s^2\ 2s^2\ 2p^6\ 3s^2\ 3p^6$

The sodium ion is said to have the neon structure and the chloride ion the argon structure.

An endeavor to write a Lewis structure for an NaCl molecule leads to two octet structures, one for the sodium ion and one for the chloride. The Lewis representations for the two atoms are

Na· ·C̤̈l:

In the approach to the rationalization of the bond structure of sodium chloride there are two choices. One is a quadruple bond, Na::::Cl, which is not supported by the crystal structure, the bond energy, and the reactivity of sodium chloride. The other is the two ions

Na^{1+} or $(:\ddot{\underset{..}{Na}}:)^{1+}$	$(:\ddot{\underset{..}{Cl}}:)^{1-}$
sodium ion	chloride ion

The two representations for the sodium ion depend upon the choice of using Na^{1+} to represent the neon structure or showing all eight of the electrons in principal quantum number 2:

$(:\ddot{\underset{..}{Na}}:)^{1+}$

Compounds made up of ions are called ionic compounds. All other compounds are called covalent compounds. All ionic compounds have one physical property in common: they are all solids at room temperature (melting points of at least several hundred degrees Celsius). The aqueous solutions of those ionic compounds that do dissolve in water are also good conductors of electricity. (Not all ionic compounds dissolve in water to any significant degree.)

It is now customary to classify all chemical compounds as nonpolar covalent compounds, polar covalent compounds, and ionic compounds. There is, however, no clear-cut dividing line between very polar covalent compounds and ionic compounds. Note that the names of the classes of chemical compounds—covalent and ionic—are derived from the model used to rationalize their properties. The compounds of carbon, hydrogen, and oxygen that were considered in the earlier chapters are, almost without exception, covalent compounds. Compounds involving the metals are very frequently ionic compounds.

Ions have the same reality as molecules and ions will become increasingly featured characters in the following chapters.

8

VIBRATIONAL MOTION AND THE DISSOCIATION OF DIATOMIC MOLECULES

Chapter 8

In this chapter, we explore the nature of vibrational motion in diatomic molecules and the relation of this vibrational motion to the dissociation of the molecules. An exploration of vibrational motion extends our understanding of the nature of the chemical bond. Many of the properties of molecules that determine our climate and general well-being are related to the vibrational motion of atoms in molecules. These properties of molecules include the capacity of molecules to absorb heat and transfer heat, the capacity of molecules to absorb infrared light (heat lamps) and to emit infrared light (part of the greenhouse effect), and the increased chemical reactivity of molecules at elevated temperatures.

The concept of vibrational motion is a rather simple one but is difficult to express in words. Further, the representation in a diagram of the positions of nuclei in a molecule and the energy of vibration of the molecule is equally challenging. Three figures will be used. The commonly used general representation, Figure 8-3, is difficult to fully comprehand, and slightly different representations, Figures 8-1 and 8-2, are used initially. These two figures at first glance may seem to be more complicated, but trust us. These two representations are straightforward and comprehensible.

Diatomic Chlorine a Homonuclear Molecule

As a start, we shall assume that there is an isolated molecule of chlorine, Cl_2, and that the separation between the nuclei of the two atoms of the chlorine molecule is the normal chlorine–chlorine bond distance of 0.199 nanometers. We shall also assume that we have a pair of very small hands so that we can at will either squeeze the two atoms closer together or pull the two atoms further apart without moving the position of the molecule as a whole. The position of the point halfway between the nuclei will remain unchanged during these manipulations. This position is the center of mass of this homonuclear molecule.

The initial positions of the two nuclei are indicated below by the two dots:

•   •

In this position, the forces within the molecule tending to pull the two nuclei closer together and the forces tending to push the two nuclei further apart are just balanced. The two nuclei repel each other (like charges). All electrons attract both nuclei (unlike charges). Through these attractive forces, electrons can be effective in pulling the two nuclei toward each other—and, as discussed in the introduction to Chapter 7, can also be ineffective, even counterproductive, in pulling the nuclei toward each other. How effective the electron charge is in pulling nuclei toward each other depends upon the position of the electron charge with respect to the line of centers between the two nuclei. In this discussion, we shall use the phrase "effective force of attraction" for the net force along the line of centers arising from the interaction of all of the electrons in the molecule with the two nuclei.

If we use our "hands" to push the two nuclei closer together, the net force within the molecule tends to move the nuclei apart: within the molecule, the force of repulsion between the two nuclei is greater than the effective force of attraction between the electrons and the nuclei. In this compressed position, the forces tending to increase the distance of separation can be indicated by two vectors (two arrows):

$$\longleftarrow\!\bullet \qquad \bullet\!\longrightarrow$$

These are the forces that the "hands" must counterbalance in constraining the molecule to this compressed position.

As the "hands" compressed the molecule in the first place, the forces that the "hands" worked against started at zero and gradually increased throughout the compression process. A reduction in the distance between the two nuclei is accompanied by an increase of the force of repulsion between the two nuclei and may also decrease the effective forces of attraction between the electrons and the nuclei. (Part of the electron charge between the two nuclei is squeezed out.) In working against these forces, the "hands" increased the potential energy of the molecule. As soon as the "hands" are removed, the nuclei begin to move apart and the acquired potential energy begins to become kinetic energy of the nuclei. When the potential energy has been reduced to zero, the net force of repulsion between the two nuclei and the effective force of attraction between the electrons and the nuclei has again become zero. But the kinetic motion of the nuclei has reached a maximum and the motion of the nuclei continues due to the momentum of the nuclei. As the nuclei move further apart, the forces of repulsion between the two nuclei decrease and the effective force of attraction between the electrons and the nuclei becomes the greater force acting on the nuclei. The motions of the nuclei are slowed and eventually the kinetic energy of the nuclei is reduced to zero. All motion stops momentarily with the molecule in an extended position. All of the kinetic energy is again potential energy. The nuclei respond to the force pulling them together,

$$\bullet\!\longrightarrow \qquad \longleftarrow\!\bullet$$

and off the nuclei go on a return trip. This is an isolated molecule: there is no friction and there are no collisions. The process continues without end as the molecule continues to vibrate.

The more energy used in the compression, the greater the amplitude of the vibration: the atoms move from a shorter compressed position to a longer extended position.

Up to this point, we have assumed that we can compress an isolated molecule to any degree that we choose. This is not the situation in nature. There are no "hands" small enough to bring about this compression. There are instead lots of molecules bumping into each other. Some collisions transfer energy from one molecule to another. The vibrational motion of a molecule is determined by the vibrational energy the molecule possesses as a consequence of these collisions and as a consequence of either the absorption or the emission of photons of electromagnetic radiation.

Each molecule has a characteristic frequency of vibration. This characteristic frequency for diatomic chlorine is 1.7×10^{13} vibrations per second. The characteristic frequency of vibration of the bromine molecule, Br_2, is 1.0×10^{13} vibrations per second. The difference is the consequence of the differences in the bond strengths of the respective molecules and the differences in the masses of the chlorine and bromine nuclei.

Characteristic Frequency of Vibration

	bond energies (joules)*	atomic mass (grams)†	frequency vibrations per second
chlorine, Cl_2	2.4×10^5	35.5	1.7×10^{13}
bromine, Br_2	1.9×10^5	79.9	1.0×10^{13}

* Per 6.02×10^{23} molecules.
† Per 6.02×10^{23} atoms.

The natural frequency of vibration of diatomic molecules is of the order of 10^{13} vibrations per second (range of roughly 1×10^{13} to 10×10^{13} sec^{-1}).

All of this and more is represented in a diagram, Figure 8-1, for homonuclear diatomic molecules such as chlorine. To orient your thinking to this diagram, focus

Background Information 8-1

MEASUREMENTS FROM AN ARBITRARILY CHOSEN ZERO

By choosing arbitrary points from which to measure, we come out with numerical values that must be given a positive or negative sign to indicate whether the measured quantity has a value greater than the arbitrarily chosen zero or a value less than the arbitrarily chosen zero. Negative temperatures are the consequence of an arbitrary choice of a reference temperature assigned zero. In the absolute sense, the concept of a negative temperature would not be readily accepted by the scientific community.

The arbitrary choice of a zero may be imposed by an inability to determine experimentally the absolute zero. This problem was true for temperatures for many years but is now solved and it is now possible to evaluate temperatures in terms of this absolute zero, called zero kelvin.

The use of an arbitrarily chosen zero may be one of convenience. It may be easier to measure from some point that is near to the values that are to be determined. It is for this reason that the Celsius temperatures are used, not temperatures kelvin, in dealing with ordinary affairs of life.

The choice of an arbitrary zero for energy measurements is forced upon us for many types of energy situations. Energies relative to those arbitrarily chosen zeros may be expressed as negative values. In the absolute sense, there are no negative energies. Part of the difficulty in the adjustment to "negative energies" in the relative sense is that there is not one arbitrarily chosen zero energy but a great number of arbitrarily chosen zero energies. The zero energy used in discussing the energy states of atoms is not at all the zero energy used in discussing the vibrational energy states of molecules.

Figure 8-1

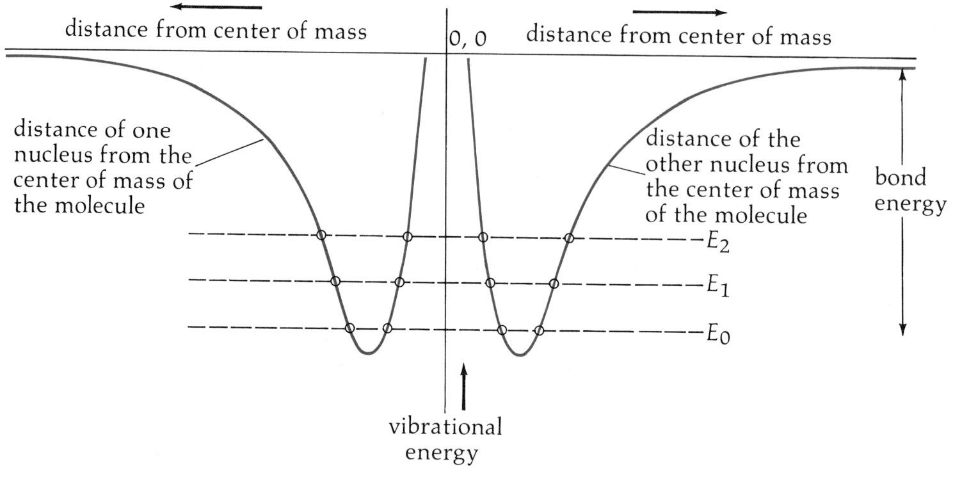

SCHEMATIC DIAGRAM FOR THE VIBRATIONAL ENERGY STATES OF HOMONUCLEAR DIATOMIC MOLECULES. Note that the arbitrarily chosen zero for vibrational energy is the solid horizontal line at the top of this diagram. All vibrational energies are expressed as negative values in reference to this zero.

first on the axes and then upon four circles on a horizontal line, such as the four circles on the broken line E_1. Horizontal displacement to the right of the vertical axis expresses the distance of one nucleus from the center of mass of the molecule. Horizontal displacement to the left expresses the distance of the other nucleus from the center of mass of the molecule. The center of mass of the molecule, and also the center of the bond for homonuclear molecules, is always on the vertical axis. The vertical displacements below the horizontal solid line represent vibrational energies of the molecule relative to the arbitrarily chosen zero for vibrational energy of the molecule under consideration. To repeat, this arbitrary zero is the horizontal solid line (the axis) at the top of the diagram. Consequently, all vibrational energies are expressed as negative values. The negative value in no way indicates that the vibrational energy is less than zero in an absolute sense.

Molecules also have other types of energy, such as rotational energy, transitional kinetic energy, and energy associated with the electronic states. Only vibrational energy of the molecule is represented in Figure 8-1.

The four circles on the line E_1 specify limiting positions of the two nuclei for a molecule with the vibrational energy E_1. For the diatomic chlorine molecule, the two inside circles represent the positions of the two nuclei of the chlorine molecule in the compressed molecule and the two outside circles represent the positions of the two nuclei of the molecule in the extended position.

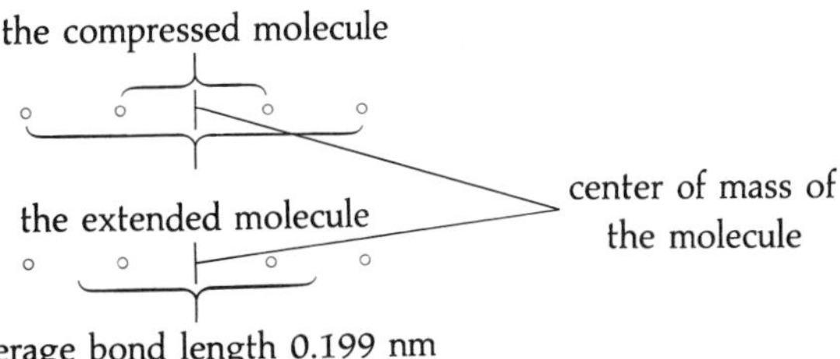

The average of the separation between the nuclei in the compressed and in the extended positions is the 0.199-nm bond length for the chlorine molecule.

For the four circles on the horizontal line E_2, the story is the same. The only difference is the amplitude of the vibration: the compressed separation is shorter and the extended separation is longer. The average of the compressed separation and the extended separation for E_2 remains unchanged from the average of the compressed and extended separations for E_1. The bond length is unchanged. At E_0, the story is the same except that the amplitude of the vibration is even further reduced.

Heteronuclear Molecules such as Hydrogen Chloride

Figure 8-2 is the companion representation for heteronuclear diatomic molecules, such as HCl. The coordinates are the same. The center of the mass of the

Figure 8-2

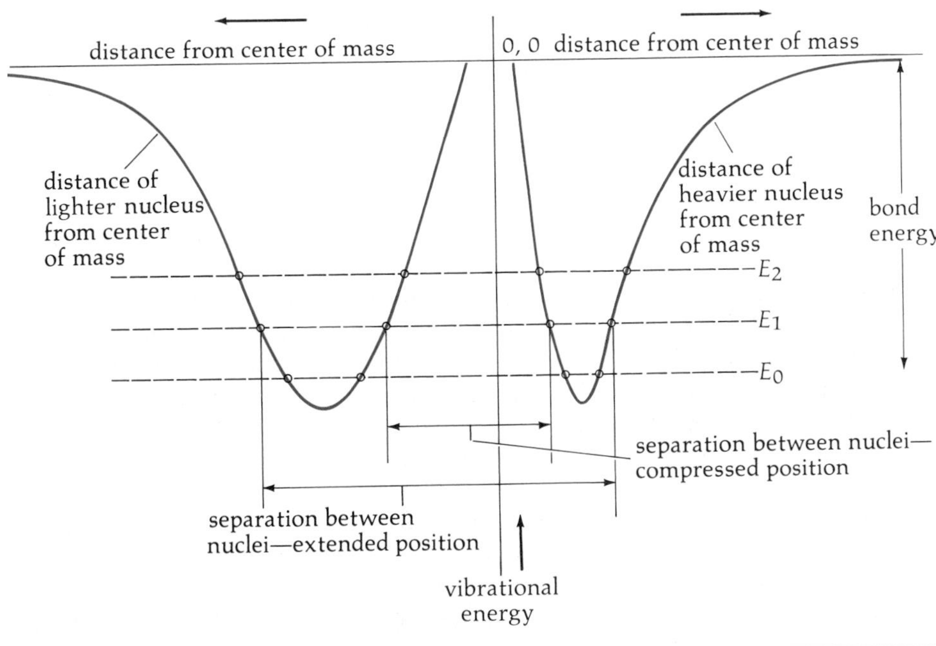

SCHEMATIC DIAGRAM FOR THE VIBRATIONAL ENERGY STATES OF HETERONUCLEAR DIATOMIC MOLECULES.

molecule is always on the vertical axis. But due to the differences in mass of the two nuclei, the center of the bond is never on the vertical axis. In the vibration of a heteronuclear diatomic molecule such as HCl, the center of mass of the molecule is maintained fixed (on the vertical axis) with the lighter nucleus, the hydrogen nucleus, undergoing a greater displacement than the heavier nucleus, the chlorine nucleus. The displacement of the hydrogen nucleus is roughly 35 times the displacement of the chlorine nucleus. For the NO molecule, the displacement of the nitrogen nucleus is greater than the displacement of the oxygen nucleus but only 16/14 times as great. In both cases, the center of mass remains at the position of the vertical axis.

The analysis of infrared spectra—both absorption and emission—led to the concept of quantized vibrational energy states for diatomic molecules. In Figure 8-1 and also in Figure 8-2, E_0, E_1, and E_2 represent three vibrational energy states where the energy of state E_2 is greater than the energy of state E_1 and the energy of state E_1 is greater than the energy of state E_0. The numerical values of energy of the vibrational energy states are a property of the molecule and must be experimentally determined for each diatomic element and each diatomic compound, such as Cl_2 and HCl, respectively. For the transition of a molecule from vibrational energy state E_1 to vibrational energy state E_2, the frequency of vibration remains essentially unchanged but the amplitude of the vibration increases. The transition from E_1 to E_2 requires the specific quantity of energy represented by the vertical separation of E_1 and E_2 on the diagrams.

Changes in Vibrational Energy States

In moving from vibrational energy state E_1 to vibrational energy state E_0, the vibrational frequency remains essentially unchanged but the amplitude of vibration is decreased. The transition from E_1 to E_0 is accompanied by a loss of energy corresponding to the vertical separation between E_1 and E_0 on the diagrams.

In a given sample of molecules, some molecules have energies corresponding to the lowest energy level, E_0, some have energies corresponding to E_1, others to E_2, etc. The distribution of the molecules among the various energy levels depends upon the magnitude of the energy difference between adjacent energy levels and upon the temperature. At low temperatures, the lower energy levels are most highly populated. For most diatomic molecules the lowest energy level, E_0, is the most populated level at room temperature. At higher temperatures, these low energy levels are forsaken for the higher energy levels. All diatomic molecules vibrate. Those in the ground state level, E_0, have the smallest amplitude of vibration. A molecule may acquire the energy to make the transition from energy state E_1 to energy state E_2 or a higher energy state by collision with another molecule and the transfer of energy from that molecule or by the absorption of a photon of electromagnetic radiation. In either case, the energy transferred must match the energy difference between the lower energy state and the higher energy state. The photons that have the energies that match these energy differences are in the infrared region of the spectrum.

General Approach to Homonuclear and Heteronuclear Molecules

Throughout the above discussion, homonuclear molecules have been discussed in terms of Figure 8-1 and heteronuclear molecules have been discussed in terms of Figure 8-2. In these figures, distances are measured for each nuclei from the center of mass of the diatomic molecule. Figure 8-3 is a more general diagram equally applicable to homonuclear and heteronuclear diatomic molecules. In this diagram, distances are measured between the two nuclei. This gives a single quadrant diagram that seems to be a more simple diagram than the two preceding diagrams. It is, however, a more complex diagram to read. Circles represent the distances of separation between the two nuclei at maximum compression and at maximum extension. Crosses represent the bond length—the average between these two extremes. In using Figure 8-3, think in terms of distances between nuclei not in terms of positions of nuclei. In all other respects, the three figures—8-1, 8-2, and 8-3—present the same information. For the vibrational energy state E_1, in Figure 8-3, S_c represents the separation between the nuclei in the compressed position (the minimum separation). S_e represents the separation between the nuclei in the extended position (the maximum separation). The difference between S_e and S_c is the amplitude of the vibration. S_b is the average of S_e and S_c and is the bond length.

The representations given in Figures 8-1, 8-2, and 8-3 are greatly simplified. There are a great number of vibrational energy states. In these figures, only a few are shown. Molecular rotation is also quantized and rotational energy states could be included in these figures. All of the curves in the three figures are nearly symmetrical at the lower energy levels. For simplicity, only a few energy levels—E_0, E_1, etc.—have been marked in Figures 8-1 and 8-2. These are within the approximately symmetrical portion of the curves. The energy difference between E_0 and E_1 is essentially the same as the energy difference between E_1 and E_2. All of the curves become asymmetrical (nonsymmetrical) at higher energies and, as indicated in Figure 8-3, the

Figure 8-3

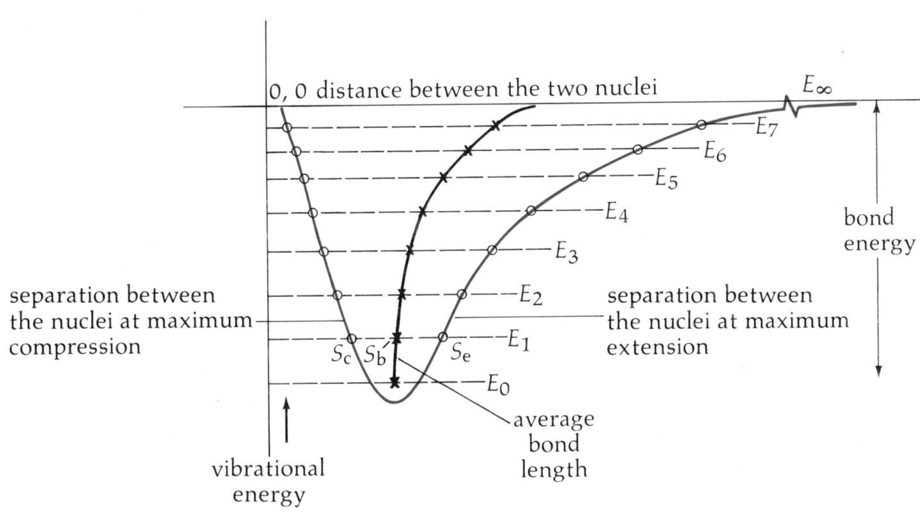

General Diagram for Vibrational Energy States of Diatomic Molecules.

separations between energy levels become smaller, the amplitudes of the vibrations become greater, the bond lengths become greater, and the frequency of vibration decreases. From the standpoint of chemical reactions, this asymmetrical portion of the curve is the most significant portion of the diagrams. As the amplitude of the vibration increases at higher and higher energy levels, the amplitude eventually becomes

8-1 *Gratuitous Information*

TRANSFER OF HEAT

In a room at 20 °C (68° F), most people feel reasonably comfortable. Body temperature is about 37.0 °C (98.6° F) and there is a net transfer of heat from the body to the air of the room. One evidence of this is that the temperature of the room will rise noticeably when a group of people remain in the room.

The skin is continuously bombarded by the molecules in the air—largely molecules of nitrogen, N_2, and molecules of oxygen, O_2. On the average, the molecules of nitrogen and oxygen rebound from the surface of the skin with more energy than they possessed at the instant they collided with the skin. The skin is cooled and the air is warmed.

How do the molecules accommodate this energy (heat) that is transferred from the skin to the molecules? It is believed that this energy is converted into the kinetic energy of motion of either (1) the total molecule or (2) one part of the molecule with respect to another part of the molecule. It is the second type of motion, the vibrational motion, that is the concern of this chapter. For a diatomic molecule, this is simply the motion of one nucleus with respect to the other nucleus resulting in a change in the separation between the two nuclei. (In molecules containing more than two atoms, the motions of nuclei with respect to other nuclei may also lead to changes in bond angles.)

Motion of the total molecule (the first type of motion) can involve motion of the center of mass. The whole molecule moves from one point in space to another point in space; the molecule is translated from one position to another. Motion of the total molecule can also involve rotation (spinning) about the center of mass of the molecule as well as translational motion, which changes the position of the molecule.

In both rotational motion and vibrational motion, the position of the center of mass of the molecule remains unchanged. Translational motion is concerned only with the movement of the center of mass of the molecule. Any one molecule in a gas or liquid can be, and usually is, involved in all three types of motion all the time.

Under normal room conditions, when molecules in the air rebound from our skin, they rebound on the average with greater kinetic energy than they possessed before the collision. This energy is manifest in some combination of increased kinetic energy of translation, increased kinetic energy of rotation, and increased kinetic energy of vibration.

It is interesting to contemplate the fundamental basis of hot air in heating systems. At the furnance, the molecules in the air are jazzed up (increased rotational, vibrational, and translational motion). Later, part of this energy is transferred in collisions with molecules that have less kinetic energy. These "colder molecules" may simply be other molecules in the gas phase, but they can also be molecules in the walls, floors, and furnishings of the room.

Our comfort in a room depends upon a number of factors in addition to the temperature of the air. These factors include air currents, the net evaporation of moisture from the skin, the radiation of electromagnetic radiation from the skin as infrared light, and the absorption of electromagnetic radiation by the skin primarily as infrared light. Absorption of visible light and absorption of ultraviolet light are only minor factors.

You might find it amusing to consider the nature of the wind in terms of the individual molecules. As a child, one of the authors spent a very frustrating afternoon trying to catch a breeze by closing one end flap of a rectangular tent.

so great that at E_∞ there is no return and the atoms fly apart. The bond ruptures at E_∞ and there is no longer a molecule of HCl:

$$HCl \longrightarrow H + Cl$$

$$H{:}\ddot{\underset{\cdot\cdot}{Cl}}{:} \longrightarrow H\cdot + \cdot\ddot{\underset{\cdot\cdot}{Cl}}{:}$$

Dissociation at E_∞

The molecule that had been HCl on dissociation becomes an atom of hydrogen with a partially filled atomic orbital and an atom of chlorine with a partially filled atomic orbital. Since there is no molecule at energy state E_∞ for hydrogen chloride or any other diatomic molecule, the vibrational energy for hydrogen chloride in the energy state E_∞ must be zero and the horizontal solid axis in Figures 8-2 and 8-3 becomes the zero vibrational energy line for hydrogen chloride, from which all vibrational energies of hydrogen chloride molecules are measured.

At sufficiently low temperatures, all diatomic molecules occupy their E_0 vibrational states, the ground vibrational energy states. But as pointed out before, at higher temperatures the molecules in a sample of any diatomic substance are distributed in a number of vibrational energy states. Just exactly what that distribution is depends upon the magnitude of the energy separation between successive energy states and the temperature. At room temperature, about 20 °C, most molecules occupy the E_0 state with fewer molecules in the E_1 state, even fewer in the E_2 state, etc. As the temperature is increased, the population of the E_0 state is diminished and the populations of higher energy states are increased. Above 6000 K, few, if any, chemical bonds exist.

Bond Energies

The bond energies for diatomic molecules, such as Cl_2 and HCl, are designated in Figures 8-1, 8-2, and 8-3 by the vertical line between the solid horizontal axis E_∞ and E_0. The dissociation energies for single bond diatomic molecules discussed in

Gratuitous Information 8-2

HEAT LAMPS AND THE GREENHOUSE EFFECT

Heat lamps emit photons in the infrared region of the spectrum. Part of these photons have the appropriate energies to be absorbed in vibrational energy state transitions and those that are absorbed raise the temperature of the absorbing molecules.

In recent years, there has been public concern as to the effect of the increasing concentration of carbon dioxide in the atmosphere (due to the combustion of fossil fuels) on the temperature of the earth through a process known as the greenhouse effect. One of the processes through which the earth is cooled is the emission of photons of infrared light. This emission is a characteristic of all warm molecules and solids. These photons continue into space unless absorbed. Carbon dioxide, a triatomic molecule, has several modes of vibration, not just the single mode of diatomic molecules, and is effective in absorbing some of the photons by transitions to higher vibrational energy states. These higher energy state molecules emit photons as the higher energy vibrational states return to lower vibrational energy states. The emission of photons by carbon dioxide is randomly distributed: photons are emitted in all directions. Part reach the earth and are absorbed. Through this process, the earth is warmer than it would be if carbon dioxide was not there to absorb and reemit photons. The unresolved questions are the rate at which the concentration of carbon dioxide in the atmosphere is changing and the rate of warming of the earth attributable to the presence of that carbon dioxide.

Chapter 7 are closely related to but not quite identical with these bond energies. Heats of dissociation reported in Chapter 7 were for the diatomic gas at 25 °C. At that temperature, the 6.0×10^{23} molecules of a mole populate primarily the lowest vibrational energy state, E_0. For chlorine gas, Cl_2, at 25 °C, the heat of dissociation is 242 kilojoules (58 kilocalories) per mole:

$$Cl_2 \longrightarrow Cl + Cl \qquad \Delta H = 242 \text{ kilojoules at } 25\ ^\circ\text{C}$$
$$= 58 \text{ kilocalories}$$

Expressed in terms of one molecule of Cl_2, the values are 4.0×10^{-22} kilojoules (5.6×10^{-23} kilocalories). Any of these four values is used to express the average energy of dissociation of the chlorine molecule at 25 °C. But at higher temperatures, the molecules populate higher vibrational energy states and, consequently, the dissociation energies are smaller at these higher temperatures. The bond energy is independent of temperature.

Since at 25 °C the molecules of chlorine are primarily in the ground vibrational state, E_0, the energy of dissociation discussed above is essentially the bond energy of the chlorine–chlorine bond. This would not have been true if the heat of dissociation had been experimentally evaluated at 300 °C. Bond energies, E_∞–E_0, are larger than this heat of dissociation. At 25 °C, bond energies are only slightly larger than heats of dissociation and the terms bond energies and heats of dissociation are frequently used interchangeably for diatomic molecules.

Scramble Exercises

1. Check all calculations given in the chapter—if, in fact, they can be checked. There is always the possibility of errors.

2. The heat of dissociation of hydrogen chloride gas, HCl, at 25 °C is 431 kilojoules (103 kilocalories) per mole. Convert these energies per mole to energies per molecule.

   Relate the energies obtained as closely as you can to the diagrams given in Figures 8-2 and 8-3.

3. Relate the bond length given in Chapter 7 for hydrogen chloride, HCl, to Figures 8-2 and 8-3.

4. The natural frequencies of vibration of the hydrogen halides are

hydrogen fluoride	11.9×10^{13} sec^{-1}
hydrogen chloride	8.7×10^{13}
hydrogen bromide	7.7×10^{13}
hydrogen iodide	6.9×10^{13}

   What correlations, if any, are there between these values and (a) the bond energies, (b) the molecular weights, and (c) the bond lengths of the molecules? (These are complicated relations, but part of the fun in a science is to look for correlations or lack of correlations and to speculate on the nature of the phenomena.)

Additional exercises for Chapter 8 are given in the Appendix.

A Look Ahead

The intent of this chapter has been to enable you to extend your concept of the covalent bond and the process of dissociation. Although the presentation has been in terms of diatomic molecules, the concepts are generally applicable to bonds between pairs of atoms in more complex molecules and it is in the exploration of more complex molecules that the concepts developed here have their greatest value.

9

WATER, AMMONIA, METHANE, AND OTHER COMPOUNDS OF CARBON

Chapter 9

Valence Bond Approach

To understand the properties of molecules, we must endeavor to understand the forces that impose the characteristic spatial arrangements of atoms in molecules. In the ball-stick-spring model, spatial orientation was imposed by the orientation of the holes in the balls. The Lewis electron dot approach addresses the number of bonds between atoms but not the angles made by those bonds. In this chapter, we explore the contribution of the valence bond approach to our understanding of the structure of small molecules more complex than the diatomic molecules discussed in Chapter 7. It takes three atoms to establish a bond angle and at least four atoms to form a nonplanar molecule. The focus in this chapter is upon the rationalization of the number of bonds and the magnitude of bond angles in terms of atomic orbitals of the atoms and molecular orbitals of the molecules.

For each molecule, build the ball-stick-spring model and write the Lewis electron dot structure at the same time we explore the valence bond approach. In working through this chapter, remember that types of atomic orbitals are designated by Roman letters and types of molecular orbitals are designated by Greek letters. In using the valence bond approach, keep in mind that if there is one bond between two atoms, it is a *sigma*-bond—two paired electrons in a σ-orbital. If there is more than one bond between two atoms, one bond is a *sigma*-bond and the other bond or bonds are *pi*-bonds—two paired electrons per π-orbital.

The elements other than hydrogen that will be involved in this presentation are fluorine and the three elements that immediately precede fluorine in the second period of the Periodic Table.

element	**ground state electron distribution**	**bonding sites**	**Lewis representation**
carbon (Z = 6)	(He) $2s^2\ 2p_x^{\ 1}\ 2p_y^{\ 1}$	4	·Ċ̣·
nitrogen (Z = 7)	(He) $2s^2\ 2p_x^{\ 1}\ 2p_y^{\ 1}\ 2p_z^{\ 1}$	3	:Ṇ̇·
oxygen (Z = 8)	(He) $2s^2\ 2p_x^{\ 2}\ 2p_y^{\ 1}\ 2p_z^{\ 1}$	2	:Ȯ̤·
fluorine (Z = 9)	(He) $2s^2\ 2p_x^{\ 2}\ 2p_y^{\ 2}\ 2p_z^1$	1	:F̤̈·

Water, Ammonia, and Methane

water, H_2O

The ball–stick–spring structures and the Lewis structures for the three molecules water, H_2O, ammonia, NH_3, and methane, CH_4, are given below:

```
                          H
                          |
O—H        H—N—H      H—C—H
|             |           |
H             H           H

                          H
 ..            ..         ..
:O:H        H:N:H      H:C:H
 ..            ..         ..
 H             H          H
```

In the Lewis structures, the hydrogen atoms assume the helium structure by sharing a pair of electrons and the central atom assumes the neon structure by having an octet of electrons, some of which are shared with atoms of hydrogen.

Water

Starting with the ground state of the oxygen atom, (He) $2s^2\ 2p_x{}^2\ 2p_y{}^1\ 2p_z{}^1$, the valence bond approach to water is to consider each of the hydrogen atoms to be bonded to the oxygen atom through a *sigma*-bond arising from the overlap of the partially filled *s*-orbital of the hydrogen atom with a partially filled *p*-orbital of the oxygen atom. This is exactly the same approach as that presented in Chapter 7 for the formation of the σ-bond in hydrogen fluoride, HF. The only difference is that the ground state oxygen atom has two partially filled *p*-orbitals, as compared to the one partially filled *p*-orbital of the ground state atom of fluorine. The oxygen forms two σ-orbitals, one with one hydrogen atom and one with the other atom of hydrogen. (Figure 9-1.)

90° Bond Angle

Since the ∞-fold axes of rotation of the p_y- and the p_z-orbitals in oxygen atoms are at right angles, this valence bond approach predicts 90° for the hydrogen–oxygen–hydrogen bond angle. The valence bond model provides a means of approach to the rationalization of those bond angles for which there is so much experimental evidence. This is a great triumph even though the angle predicted here is 90° for water and the experimental value is 104.5°. Clearly the model as we have presented it must be modified and we shall return to the rationalization of the 104.5° value in the water molecule later.

Ammonia

In the ammonia molecule, NH_3, nitrogen utilizes three bonding sites. The experimental value for all hydrogen–nitrogen–hydrogen bond angles is 106.8°. If your bag of balls, sticks, and springs does not have a special ball for nitrogen, substitute a ball normally used for carbon and use only three of the four holes.

Starting with ground state distribution of electrons in the nitrogen atom, (He) $2s^2\ 2p_x{}^1\ 2p_y{}^1\ 2p_z{}^1$, the valence bond approach generates three σ-bonds by the overlap of these partially filled *p*-orbitals of the nitrogen atom with the partially filled *s*-orbitals of three hydrogen atoms. Figure 9-2. Here again, the hydrogen–nitrogen–hydrogen bond angles predicted by the valence bond approach are 90°, not the experimental value 106.8°.

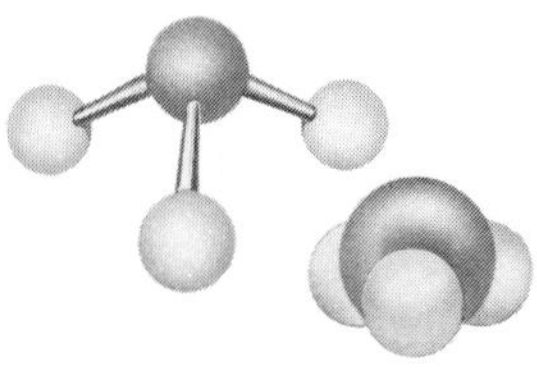

ammonia, NH_3

The ground state electron distribution of electrons in the carbon atom, (He) $2s^2$ $2p_x{}^1\ 2p_y{}^1$, with one filled *s*-orbital and two partially filled *p*-orbitals does not trans-

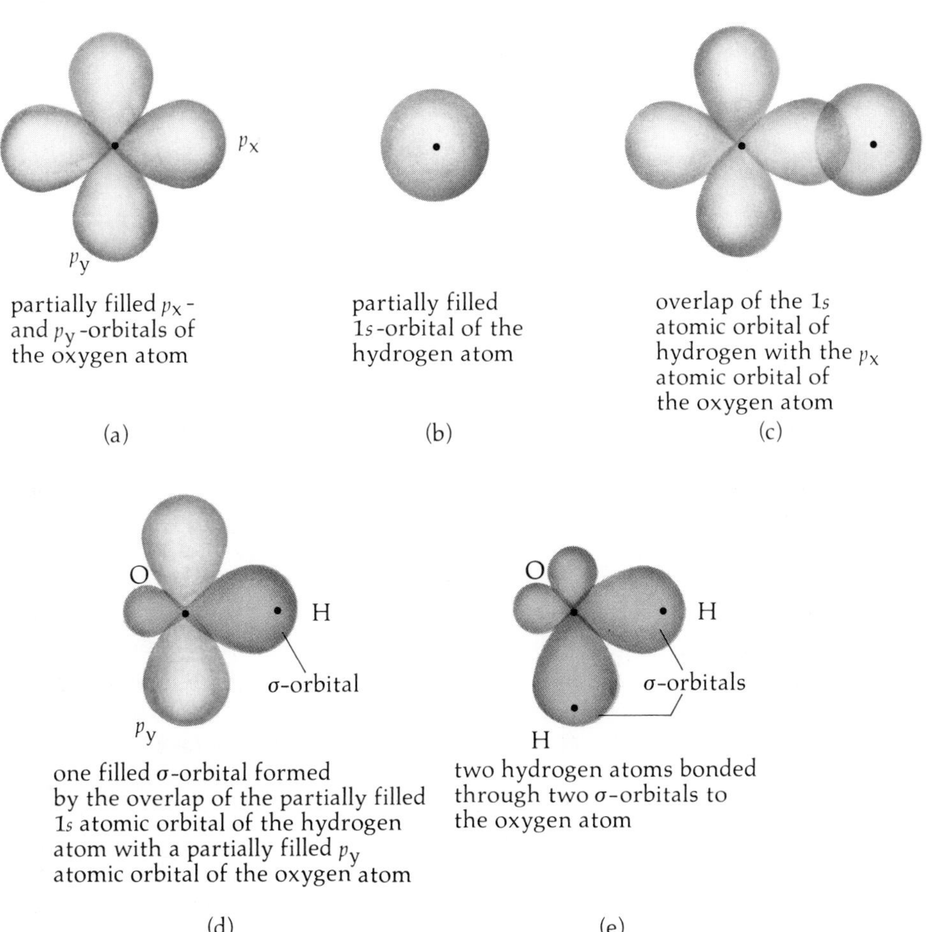

VALENCE BOND APPROACH TO BONDING IN THE WATER MOLECULE. The nuclei of the oxygen atom and the hydrogen atoms are represented by dots. The ∞-fold axis of rotation of the filled $2p_z$ atomic orbital, not shown in the above figures, is perpendicular to the plane of the paper. These drawings indicate volumes within which there is 90% probability for the one electron in a partially filled orbital or for the two electrons in a filled orbital. The *s*-orbital has spherical symmetry; each *p*-orbital and each σ-orbital has an ∞-fold axis of rotation. This approach predicts a bond angle. However, the predicted hydrogen–oxygen bond angle is 90°—much smaller than the experimental value 104.5°.

Figure 9-2

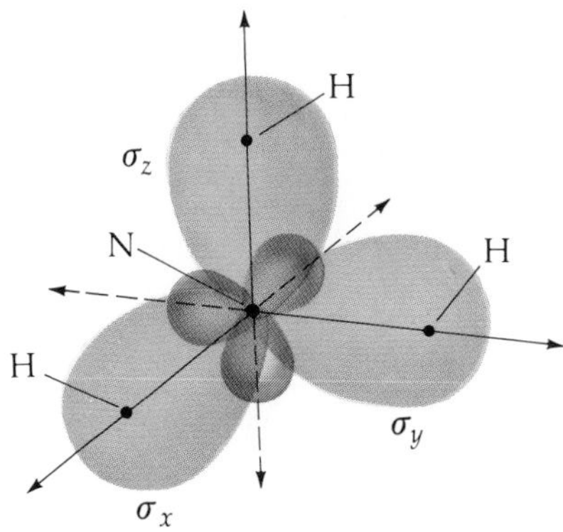

THE THREE FILLED σ-ORBITALS OF THE AMMONIA MOLECULE. The ∞-fold axis of rotation of the orbital designated σ_z is perpendicular to the plane of the paper. The ∞-fold axes of rotation of the other two σ-orbitals lie in the plane of the paper. Nuclei are represented by dots.

form directly to the formation (by overlap of four partially filled atomic orbitals of carbon with partially filled *s*-orbitals of four hydrogen atoms) of four molecular σ-orbitals with the uniform spatial orientation known to be characteristic of the hydrogen atoms in the methane molecule. The experimental value of all bond angles in methane is 109.5°.

Methane

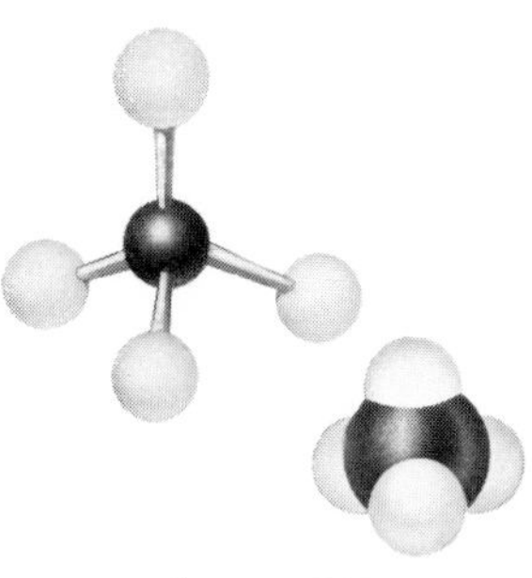

methane, CH_4

To get around these difficulties, we switch from the ground state carbon atom to (He) $2s^1\ 2p_x{}^1\ 2p_y{}^1\ 2p_z{}^1$, which does have four partially filled atomic orbitals—an *s*-orbital and three *p*-orbitals. In this excited state, four partially filled atomic orbitals are available for the formation of four molecular orbitals, but these four partially filled atomic orbitals are not uniformly distributed in space and the four molecular orbitals formed with them would not be uniformly distributed in space. The ∞-fold axes of rotation of the three *p*-orbitals are still mutually perpendicular and the *s*-orbital has no specific orientation in space. In the formation of σ-orbitals by overlap with partially filled *s*-orbitals of hydrogen atoms, the orientation of the three σ-orbitals derived from the *p*-orbitals would have the orientation of the *p*-orbitals and the σ-orbital formed by the overlap of the *s*-orbital of carbon and the *s*-orbital of the fourth hydrogen atom would have to squeeze in somewhere.

sp^3 Hybridization of Atomic Orbitals

To rectify this problem, the wave functions for the 2*s*-orbital, the $2p_x$-orbital, the $2p_y$-orbital, and the $2p_z$-orbital of the excited atom of carbon are mathematically combined to give four new wave functions. The four new wave functions are identical with the exception of the terms that give their orientation about the nucleus. The squares of these four wave functions, ψ^2, express the four probabilities of electron distributions. These four new atomic orbitals are called **hybrid atomic orbitals,** crosses between one s-orbital and three p-orbitals, and are designated as **sp^3-orbitals** (read s-p-3). Figure 9-3 contrasts the shapes of an s-orbital, a p-orbital, and an sp^3-orbital. The shape of each of the hybrid atomic sp^3-orbitals is a disproportionate hourglass with a large volume of high electron probability on one side of the nucleus and a much smaller volume of high electron probability on the other side of the nucleus. The probability of the electrons being on one side of the nucleus in an sp^3 atomic orbital is much greater than the probability of the electrons being on the other side of the nucleus. Frequently, this smaller region of sp^3 hybrid atomic orbitals is ignored in the discussion of the formation of *sigma* molecular orbitals. Each of the four sp^3 atomic orbitals has an ∞-fold axis of rotation. These four axes pass through

***Figure* 9-3**

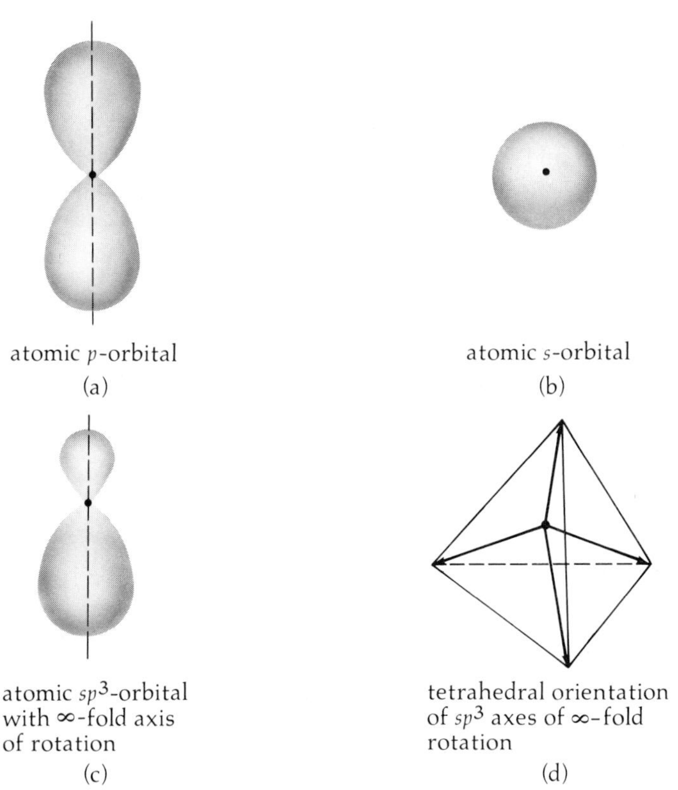

SKETCHES OF ATOMIC ORBITALS ENCOMPASSING THE 90% PROBABILITY DISTRIBUTION FOR THE ELECTRON. (a) Atomic p-orbital; (b) atomic s-orbital; (c) atomic sp^3-orbital with ∞-fold axis of rotation; and (d) tetrahedral orientation of the ∞-fold axes of rotation of the four sp^3-orbitals. The dot in all four figures represents the nucleus of the atom.

the nucleus and are uniformly oriented in space. This distribution is frequently described in terms of a regular tetrahedron with the nucleus of the carbon atom at the center and the four ∞-fold axes of rotation of the atomic orbitals directed to the four corners of the tetrahedron. (Figure 9-3d.) Since the three *p*-orbitals were directed along the three Cartesian coordinates involving the full sweep of space, it is reasonable that the four sp^3 hybrid orbitals are also distributed in the full sweep of space. The corresponding bond angle formed through the use of sp^3 atomic orbitals is 109.5°. Table 2-2 indicates how common this bond angle is. It was, of course, this tetrahedral angle that was built into the spacing of the holes in the black ball for the carbon atom on the basis of empirical knowledge about carbon compounds. A sketch of the three-dimensional representation of four sp^3-orbitals is given in Figure 9-4.

Figure 9-5 illustrates the formation of the molecular orbitals in the methane molecule, CH_4, utilizing the 1*s*-orbitals of hydrogen atoms and the four sp^3-orbitals of the carbon atom. Each of the resultant four molecular orbitals encompasses the carbon nucleus and a hydrogen nucleus. Each of these molecular orbitals is a *sigma*-orbital and has a contour very similar to the contour of the sp^3 atomic orbital from which it is formed, but the σ-orbital is somewhat larger than the sp^3-orbital of the atom. Each σ-orbital has an ∞-fold axis of rotation, and the spatial orientation of this set of four ∞-fold axes of rotation of molecular orbitals is the same as the orientation of the set of four ∞-fold axes of rotation for the four sp^3 atomic orbitals from which they were formed. Each σ-orbital accommodates two electrons with paired spins.

It is indeed doubtful that theoretical chemists and physicists would have developed the concept of four hybrid sp^3-orbitals if the concept of the tetrahedral carbon had not been forced upon them as a means of rationalizing a vast array of experimental facts concerning carbon compounds. The approach to four equivalent tetrahedral orbitals through the mixing of four more familiar atomic orbitals was one of mathematical convenience in that it simplified very complex mathematical problems in the

Figure 9-4

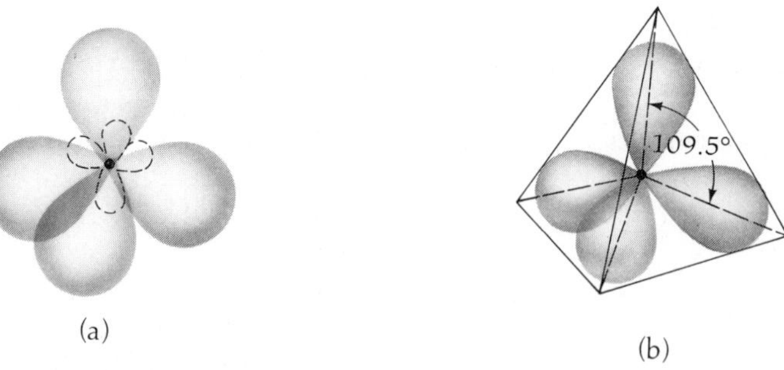

Orientation of the Four sp^3 Hybrid Atomic Orbitals in a Regular Tetrahedron. (a) sketch of the orbitals; (b) sketch showing the 109.5° angle between any pair of the ∞-fold axes of rotation. Note that only the larger lobe of each sp^3-orbital is displayed in the second sketch.

Figure 9-5

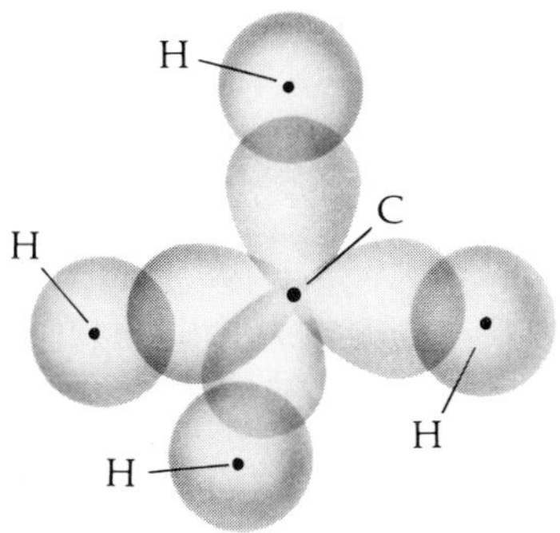

an overlap position of 1*s*-orbitals of hydrogen atoms with sp^3-orbitals of the carbon atom

(a)

sigma-orbitals of the methane molecule with the four ∞-fold axes of rotation indicated

(b)

FORMATION OF THE σ-ORBITALS OF THE CH_4 MOLECULE.

development of new wave functions. From a theoretical point of view, nothing is radically new in the formation of these hybrid atomic orbitals. Atomic wave functions are mathematical expressions. Wave functions for atomic orbitals, one from each atom, are combined to generate the wave functions that are the molecular orbitals. In the formation of the four hybrid atomic sp^3-orbitals, four wave functions of the same atom are combined. The orientation of the ∞-fold axes of rotation of the new hybrid orbitals is the tetrahedral orientation. These four sp^3-orbitals of the carbon atom become the basis of our understanding of single bonds with carbon in compounds such as saturated hydrocarbons, the alcohols, and the ethers.

sp^2 Hybridization

To rationalize the structure, including the double bonds, of compounds of carbon such as unsaturated hydrocarbons, aldehydes, ketones, acids, and esters, another type of hybridization of atomic orbitals of carbon is useful. One of the goals in this case is to achieve orientations of atomic orbitals that are consistent with the 120° bond angles determined experimentally for many of these compounds. In this type of hybridization, the wave functions of the *s*-orbital and the wave functions of two *p*-orbitals (for example, the p_x- and p_y-orbitals) of the excited carbon atom are hybridized to give three new wave functions that are equivalent to each other in shape but not orientation. Various representations of sp^2 hybrid atomic orbitals are given in Figure 9-6. These three hybrid atomic orbitals are designated **sp^2-orbitals.** Each has the same shape as an sp^3-orbital. Each of the sp^2-orbitals has an ∞-fold axis of rotation and these three axes lie in a plane. Each ∞-fold axis of rotation passes through the nucleus and each ∞-fold axis of rotation is oriented toward a corner of an equilateral triangle with the nucleus of the carbon atom at the center of the triangle. Since the plane of the three ∞-fold axes of rotation of the three sp^2-orbitals was determined by the plane of the ∞-fold axes of rotation of the p_x- and p_y-orbitals from which they were formed, the axis of rotation of the unhybridized p_z-orbital is perpendicular to that plane.

sp Hybridization

To rationalize the structure, including the triple bonds of carbon, in compounds such as acetylene, H—C≡C—H, and hydrogen cyanide, H—C≡N, a third type of hybridization is used. In this type, known as *sp*-hybridization, the wave function of the *s*-orbital and the wave function of one *p*-orbital, such as the p_x-orbital, of the excited state carbon atom are combined to give two hybrid atomic orbitals designated as ***sp*-orbitals.** Various representations of *sp* hybrid orbitals are given in Figure 9-7. The shape of an *sp*-orbital is the same as the shape of an sp^3-orbital. Each *sp*-orbital has an ∞-fold axis of symmetry; these two axes pass through the nucleus of the

Figure 9-6

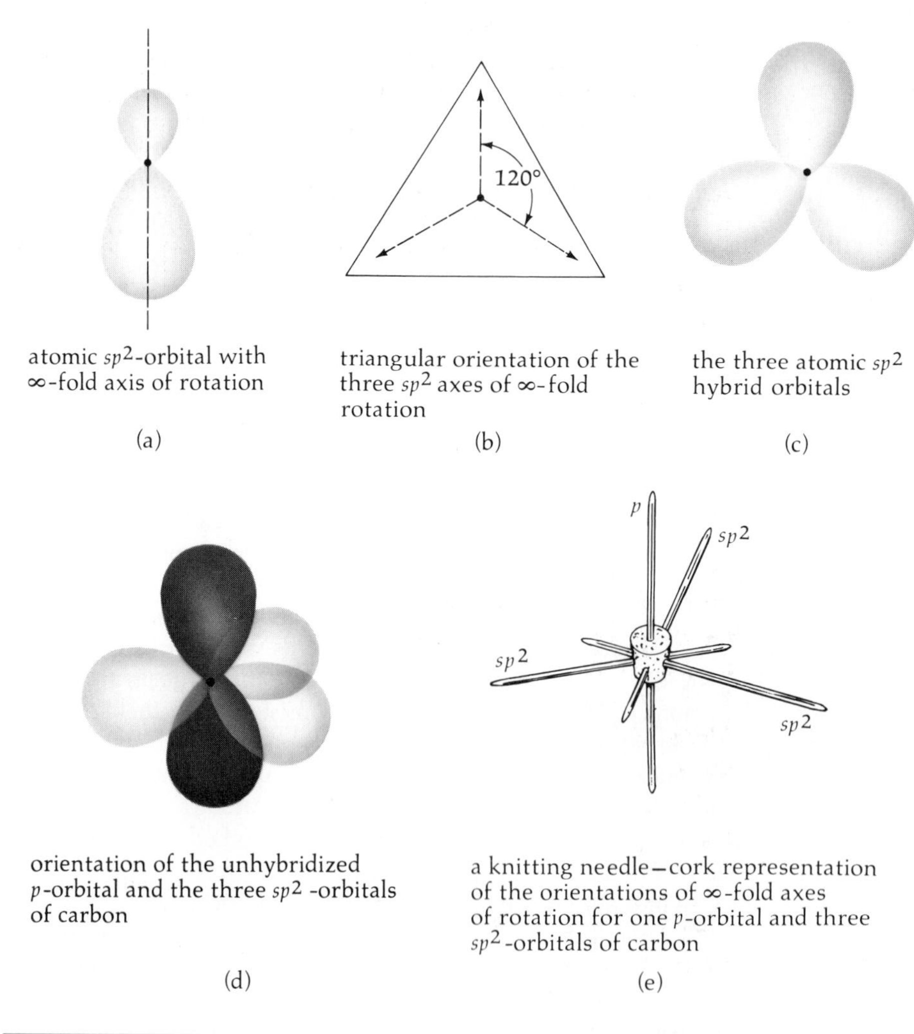

atomic sp^2-orbital with ∞-fold axis of rotation

(a)

triangular orientation of the three sp^2 axes of ∞-fold rotation

(b)

the three atomic sp^2 hybrid orbitals

(c)

orientation of the unhybridized *p*-orbital and the three sp^2-orbitals of carbon

(d)

a knitting needle–cork representation of the orientations of ∞-fold axes of rotation for one *p*-orbital and three sp^2-orbitals of carbon

(e)

REPRESENTATIONS OF sp^2 HYBRID ORBITALS FOR THE CARBON ATOM.

carbon atom and the two axes coincide, but the orbitals are oriented in opposite directions. The orientation of the two axes is said to be linear. Since only one *p*-orbital is involved in the formation of the two *sp*-orbitals, it is not surprising that the ∞-fold axes of rotation of the two *sp*-orbitals are confined to the orientation of the ∞-fold axes of rotation of that *p*-orbital. The ∞-fold axes of rotation of the unhybridized p_y-orbital and the unhybridized p_z-orbital are perpendicular to the axes of the rotation of the *sp*-orbitals. All of the hybrid atomic orbitals have the same shape: a large volume of high electron probability on one side of the nucleus and a much smaller volume of high electron probability on the opposite side of the nucleus.

Figure 9-7

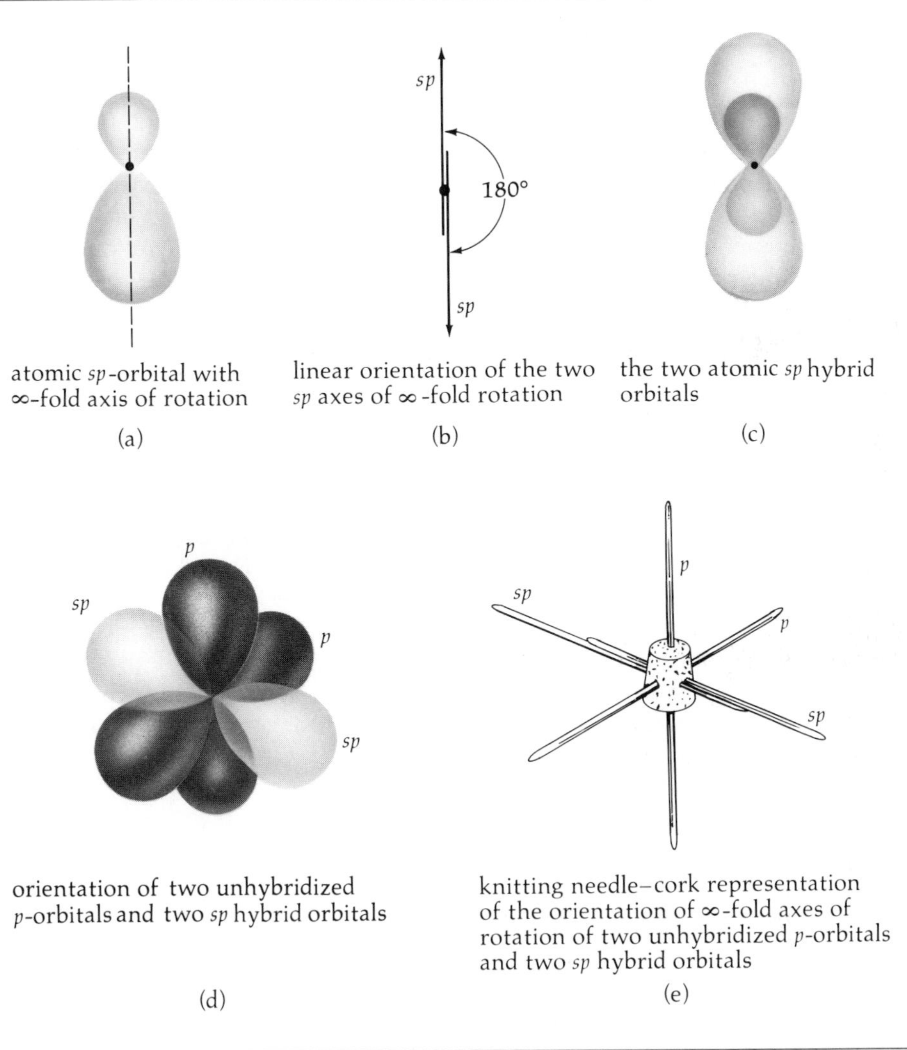

atomic *sp*-orbital with ∞-fold axis of rotation

(a)

linear orientation of the two *sp* axes of ∞-fold rotation

(b)

the two atomic *sp* hybrid orbitals

(c)

orientation of two unhybridized *p*-orbitals and two *sp* hybrid orbitals

(d)

knitting needle–cork representation of the orientation of ∞-fold axes of rotation of two unhybridized *p*-orbitals and two *sp* hybrid orbitals

(e)

REPRESENTATION OF *sp* HYBRID ORBITALS FOR THE CARBON ATOM.

Table 9-1

FORMATION OF HYBRID ATOMIC ORBITALS OF CARBONS ATOMS

atomic orbitals utilized	hybrid atomic orbitals formed	orientations of ∞-fold axes of rotation of hybrid orbitals	*p*-orbitals not involved in hybridization
one *s*-orbital, three *p*-orbitals	four sp^3-orbitals	tetrahedral	none
one *s*-orbital, two *p*-orbitals	three sp^2-orbitals	triangular	one
one *s*-orbital, one *p*-orbital	two *sp*-orbitals	linear	two

All of this seems quite complicated. Once you have accepted the concept that the wave functions for *s*- and *p*-orbitals of the same atom can be mathematically mixed to give hybrid atomic orbitals, the pattern in which they are combined and the pattern of hybrid atomic orbitals generated is quite straightforward. Once you catch on, the pattern can be readily reconstructed. (Table 9-1.) Note that there are always a total of four atomic orbitals for carbon. All can be unhybridized, all can be hybridized, or part can be unhybridized and part hybridized. As indicated in the discussion of methane, partially filled hybrid atomic orbitals of the carbon atom are involved in the formation of σ-orbitals in molecules. As you will soon see, partially filled *p*-orbitals of the carbon atom are involved in the formation of π-orbitals in molecules as the second bond of a double bond and as the second and third bonds of a triple bond.

Methane

As we have already seen, the formation of each of the four σ-bonds in methane, CH_4, is pictured in terms of an interaction of a partially filled 1*s*-orbital of a hydrogen atom with a partially filled sp^3-orbital of the carbon atom. (Figure 9-8.)

It is essential that you become accustomed to using line drawings to represent molecular orbitals and their orientation in space. Most people cannot manage three-dimensional sketches and even your best line drawings may more nearly resemble a daisy. Don't be dismayed. Use them anyway, label as necessary, and write notes to explain what they represent. For example, the molecular orbital picture for methane might look something like this:

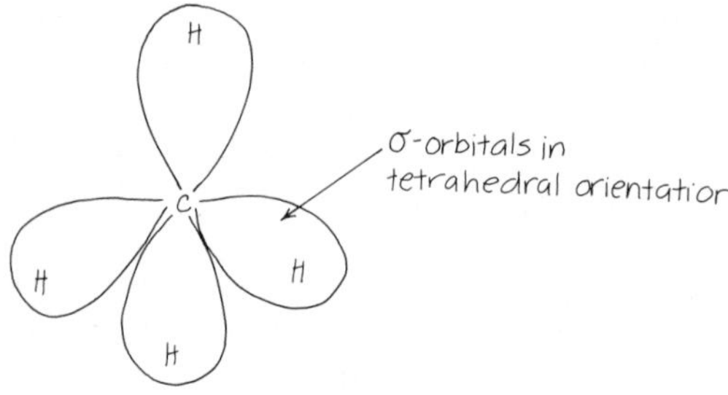

Figure 9-8

(a) one partially filled s-orbital hydrogen atom

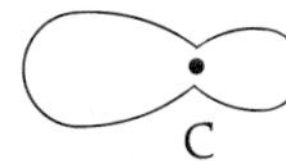

(b) one partially filled sp^3-orbital carbon atom

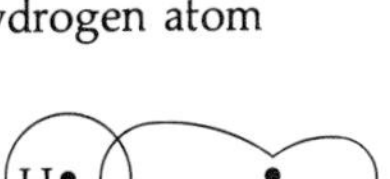

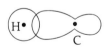

(c) overlap of partially filled atomic orbitals

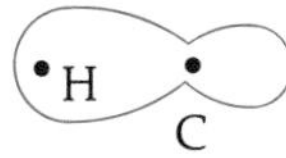

(d) filled σ-orbital of a hydrogen–carbon bond

FORMATION OF ONE *sigma*-BOND IN METHANE. All of the above are two-dimensional representations of three-dimensional phenomena. Atomic nuclei are indicated by dots.

with perhaps the note "In the VB approach, methane has four *sigma*-bonds. Each is formed by the overlap of a partially filled 1*s*-orbital of the hydrogen atom and a partially filled sp^3 hybrid orbital of carbon. The four σ-orbitals are tetrahedrally oriented about the carbon nucleus."

We shall now return to the problem of bond angles in molecules of ammonia and water discussed earlier. The 109.5° angle predicted using sp^3-orbitals for carbon atoms makes the use of sp^3 hybridization for nitrogen atoms attractive in the endeavor to rationalize the experimental hydrogen–nitrogen–hydrogen bond angle of 106.5° in ammonia. Figure 9-9 provides representation of the use of sp^3-orbitals in the rationalization of the experimental bond angles in ammonia. In using four sp^3-orbitals for the nitrogen atom, one sp^3-orbital is filled with two electrons and three sp^3-orbitals are partially filled. Each of the three σ-bonds of the nitrogen atom with a hydrogen atom is formed in the usual way by maximum overlap of a partially filled sp^3-orbital and the partially filled s-orbital of a hydrogen atom. In this approach, the predicted bond angles are 109.5°. The slight compression of these bond angles to 106.5° can be attributed to the electrostatic repulsion between the pair of electrons—the nonbonding electrons—in the filled atomic sp^3-orbital and the electrons in the molecular σ-orbitals.

In the same manner, the formation of σ-bonds in water can be approached in terms of four hybrid atomic sp^3-orbitals for oxygen. (Figure 9-10.) In this case, two of the sp^3-orbitals of the oxygen are filled and two are partially filled. The overlap

Figure 9-9

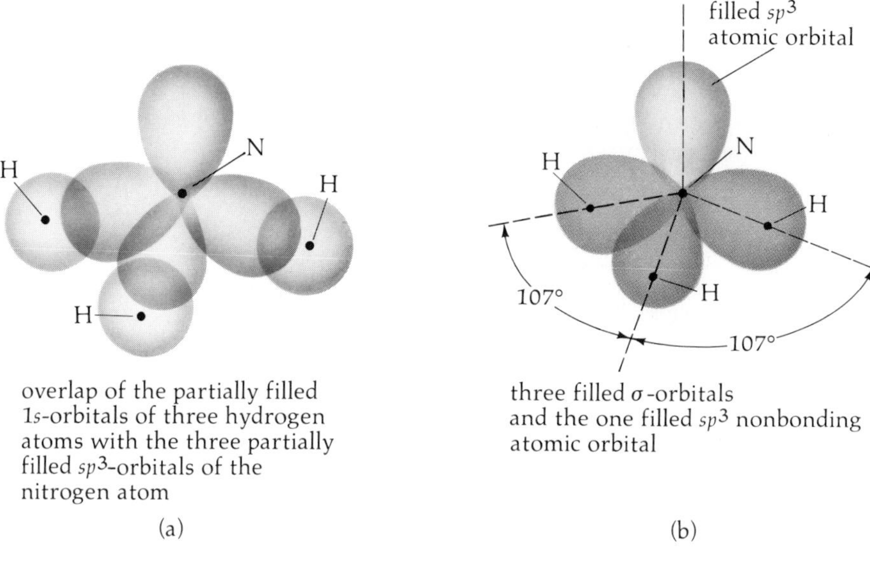

USE OF sp^3-ORBITALS OF THE EXCITED NITROGEN ATOM IN THE FORMATION OF THE AMMONIA MOLECULE. In terms of sp^3 hybrid atomic orbitals, the nitrogen atom has three partially filled sp^3-orbitals and one filled sp^3-orbital.

of the two partially filled sp^3-orbitals with the two partially filled *s*-orbitals of the hydrogen atoms forms the two σ-bonds in water. The predicted hydrogen–oxygen–hydrogen bond angle in this approach is again 109.5°. The experimental value is 104.5° and the compression of the bond angle is attributed to the electrostatic repulsion of nonbonding electrons in the two filled atomic sp^3-orbitals of the oxygen atom and the electrons in the two molecular σ-orbitals.

Bond lengths and bond strengths are given in Table 9-2 for the hydrogen compounds of carbon, nitrogen, oxygen, and fluorine. The electronegativity values for the five elements are hydrogen 2.1, carbon 2.5, nitrogen 3.0, oxygen 3.5, and fluorine 4.0. All of the bonds are polar with the electrons in the σ-orbitals attracted more strongly by carbon, nitrogen, oxygen, and fluorine than by hydrogen and most strongly by fluorine. The least polar bond is the carbon–hydrogen bond. Note the correlation of the electronegativity values of the central atoms with bond energies and the reverse correlation with bond lengths. Correlations are just exactly that. They are not necessarily statements of cause and effect.

Figure 9-10

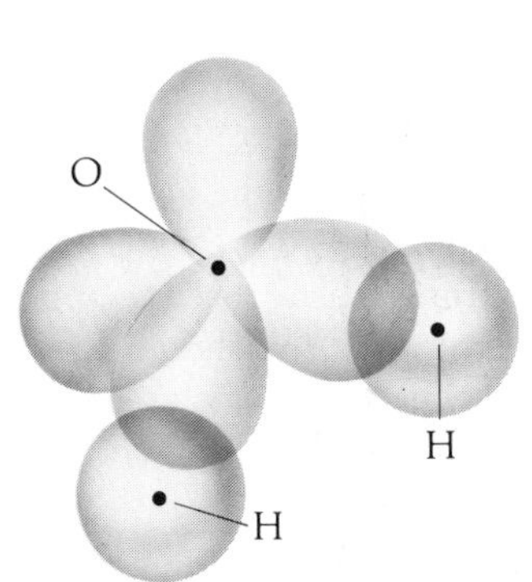

overlap of the partially filled 1*s*-orbitals of two hydrogen atoms with the two partially filled sp^3-orbitals of the oxygen atom

(a)

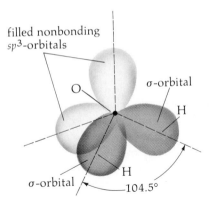

filled nonbonding sp^3-orbitals

O

σ-orbital

H

σ-orbital

H

104.5°

two filled σ-orbitals and two filled sp^3 nonbonding atomic orbitals

(b)

USE OF sp^3-ORBITALS OF THE OXYGEN ATOM IN THE FORMATION OF THE WATER MOLECULE. In terms of sp^3 hybrid atomic orbitals, the oxygen atom has two partially filled sp^3-orbitals and two filled sp^3-orbitals.

Table 9-2

BOND LENGTHS AND BOND STRENGTHS FOR HYDROGEN COMPOUNDS OF CARBON, NITROGEN, OXYGEN, AND FLUORINE

bond	compound	bond length in nanometers	bond strength in kilojoules/mole (kilocalories/mole)
H—C	CH_4	0.109	435 (104)
H—N	NH_3	0.101	431 (103)
H—O	H_2O	0.096	497 (119)
H—F	HF	0.092	564 (135)

Other Compounds of Carbon

Essentially all of the concepts of the valence bond approach used in this chapter have been introduced in Chapter 7 or in the previous section of this chapter. In the remainder of the chapter, a number of specific compounds will be explored. In exploring these, there will be repetition. This repetition provides an opportunity to develop facility and confidence in using the valence bond model. By the time you have finished this chapter, you should be able to translate any of the ball–stick–spring structures you have been building in previous chapters into Lewis electron dot structures and, starting with the atomic numbers of the atoms, check that the appropriate number of electrons have been used in each Lewis structure. You should also be able to translate ball–stick–spring structures into valence bond model analyses of the molecular orbitals in the molecules, including the identification of the origin of each molecular orbital in terms of the atomic orbitals utilized in the formation of the molecular orbital, and correlate the atomic orbitals used with the experimental values of bond angles. The space-filling model sketches in the margin of the page are useful in the development of a sense of the contours of the molecules as a whole.

chloromethane

Compounds of carbon and halogens are numerous. In many of these compounds, halogen atoms such as the chlorine atom, $:\ddot{\underset{..}{Cl}}\cdot$, replace hydrogen atoms in hydrocarbons, alcohols, ethers, aldehydes, ketones, carboxylic acids, and esters. The methane series of chlorine-containing compounds are chloromethane, CH_3Cl; dichloromethane, CH_2Cl_2; trichloromethane, $CHCl_3$; and tetrachloromethane, CCl_4. The last two of these are better known by the public as chloroform, $CHCl_3$, and carbon tetrachloride, CCl_4. Carbon tetrachloride was extensively used as a solvent until concerns about its toxicity led to the regulation of its use. The representations of these compounds by the various models are exactly parallel to the representation for methane. The halogen has a single bonding site and can be represented in the ball–stick–spring model by any ball with a single hole or by using only one hole of a ball with more holes. In the Lewis structures, $:\ddot{\underset{..}{Cl}}\cdot$ is used in place of $H\cdot$.

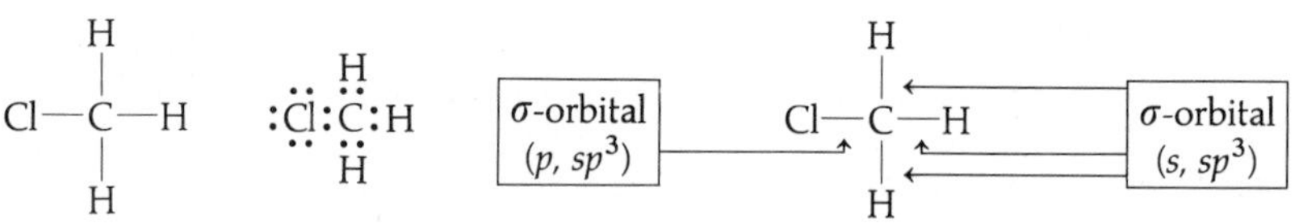

Chloromethane

The Lewis structure for chloromethane uses a total of fourteen electrons: the seven electrons of the chlorine atom, the four electrons of the carbon atom, and a total of three electrons from the three hydrogen atoms. In the molecule, the chlorine atom has an octet of electrons (the argon structure): six that are not shared with another element and two that are shared with the carbon atom. Each hydrogen atom has a pair of electrons (the helium structure) shared with the carbon atom. The carbon atom has an octet of electrons (the neon structure), two being shared with the chlorine atom and six being shared with the three hydrogen atoms.

one partially filled p-orbital of the halogen atom

(a)

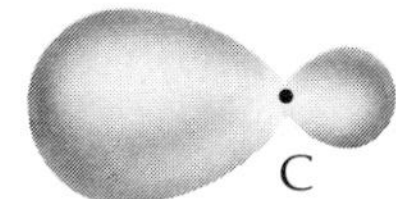

one partially filled sp^3-orbital of the carbon atom

(b)

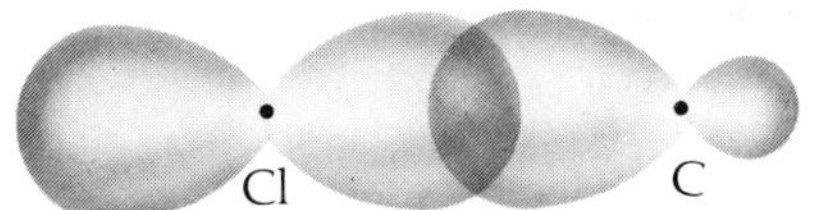

overlap of the partially filled p-orbital of the halogen atom and the partially filled sp^3-orbital of the carbon atom

(c)

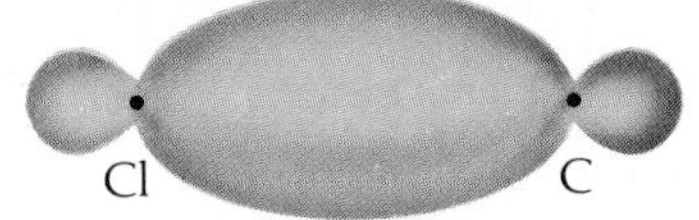

one filled σ-orbital encompassing the chlorine nucleus and the carbon nucleus

(d)

FORMATION OF A *sigma*-BOND BETWEEN A CHLORINE ATOM AND A CARBON ATOM. Dots represent the nuclei of atoms. Involved in the formation of the σ-bond is a partially filled p-orbital of the halogen atom and a partially filled sp^3 orbital of the carbon atom.

The valence bond approach to all of these chloromethanes is to form four σ-orbitals using the overlap of the four partially filled sp^3-orbitals of the carbon atom with either a partially filled p-orbital of the halogen atoms or the partially filled s-orbitals of hydrogen atoms. Figure 9-11 gives the orbitals involved in the formation of a σ-orbital between a chlorine atom and a carbon atom. All molecular orbitals of chloromethane are filled by two electrons of antiparallel spins and all bonds are *sigma*-bonds, each with an axis of ∞-fold symmetry through the line of centers of the two nuclei.

The ball–stick–spring structures of saturated hydrocarbons such as ethane, CH_3CH_3, and propane, $CH_3CH_2CH_3$, are

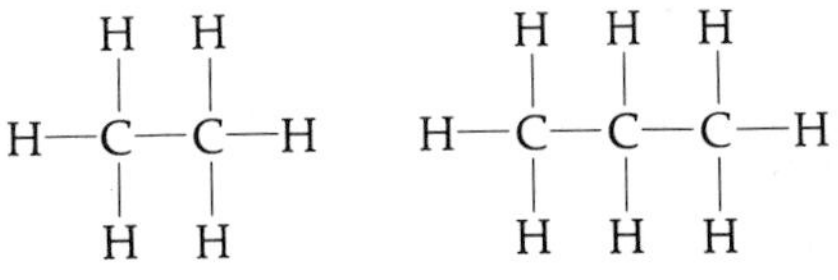

```
   H  H          H  H  H
   |  |          |  |  |
H—C—C—H      H—C—C—C—H
   |  |          |  |  |
   H  H          H  H  H
```

Ethane and Propane

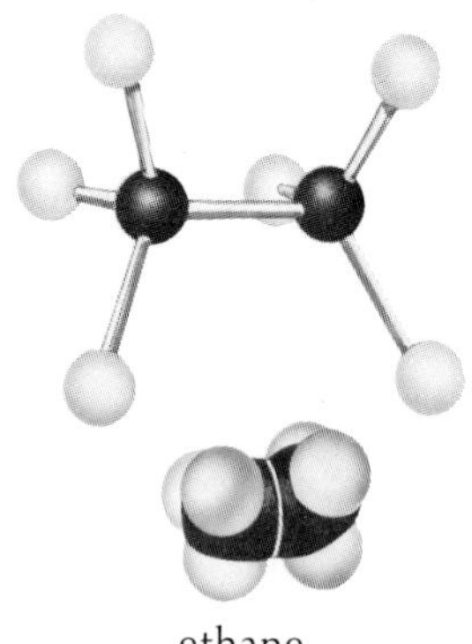

ethane

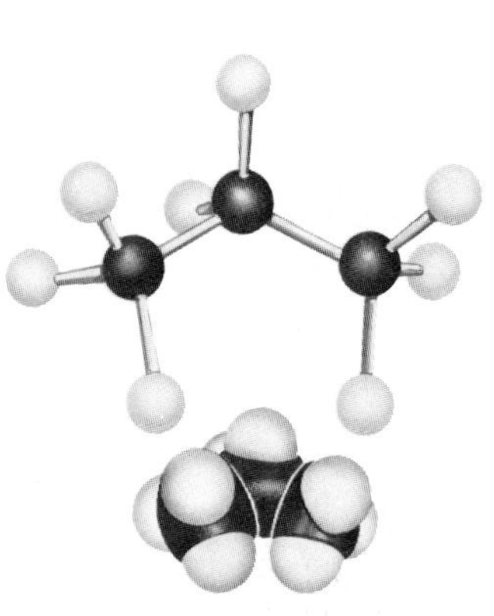

propane

The Lewis structures are

```
  H H           H H H
H:C:C:H       H:C:C:C:H
  H H           H H H
```

The electron counts are

from hydrogen atoms:	$6 \times 1 = 6$	$8 \times 1 = 8$
from carbon atoms:	$2 \times 4 = 8$	$3 \times 4 = 12$
total	14	20
Electrons shown in Lewis structures:	$7 \times 2 = 14$	$10 \times 2 = 20$

Bond angles are all approximately 109.5° in both compounds.

In the valence bond model, each pair of electrons corresponds to a σ-orbital containing two electrons with antiparallel spins. A *sigma*-orbital involving hydrogen and carbon is formed by maximum overlap of the partially filled *s*-orbital of the hydrogen atom and a partially filled sp^3-orbital of the carbon atom. The *sigma*-orbital involving carbon and carbon is formed by maximum end-to-end overlap of one partially filled sp^3-orbital from each of the carbon atoms. (Figure 9-12.) The tetrahedral orientation of the sp^3 atomic orbitals is, of course, consistent with the experimental values of the bond angles. All carbon–hydrogen bonds are polar with the hydrogen atom being very slightly positive. The carbon–carbon bond in ethane is nonpolar. Carbon–carbon bonds in propane are essentially nonpolar. However, the negative charge distributions about an end carbon atom and about the central carbon atom of propane are slightly different.

Clearly this game can continue indefinitely with longer and longer chains of carbon atoms and also with the formation of branched chains of carbon atoms. In the saturated hydrocarbons, all bond angles (<HCH, <HCC, and <CCC) are essentially

Figure 9-12

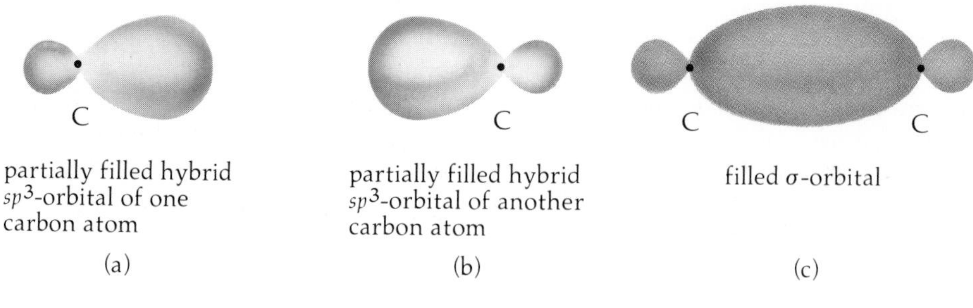

partially filled hybrid sp^3-orbital of one carbon atom
(a)

partially filled hybrid sp^3-orbital of another carbon atom
(b)

filled σ-orbital
(c)

FORMATION OF CARBON–CARBON *sigma*-ORBITAL USING sp^3 ORBITALS.

the tetrahedral angle, 109.5°. The identities of the atomic orbitals used in the formation of the molecular orbitals in ethane are

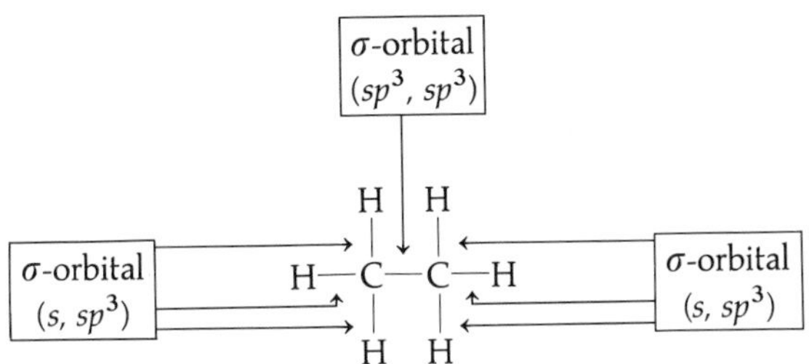

Unsaturated Hydrocarbons

Unsaturated hydrocarbons with double and triple bonds require special consideration. The two molecules considered here are

	ethylene (ethene)	acetylene (ethyne)
	$H_2C{=}CH_2$	H—C≡C—H
	H:C::C:H (with H's)	H:C:::C:H

Electron count:

	ethylene	acetylene
from hydrogen atoms:	4 × 1 = 4	2 × 1 = 2
from carbon atoms:	2 × 4 = 8	2 × 4 = 8
total	12	10
Shown in Lewis structures	12	10

Bond angles:

ethylene	acetylene
all six approximately 120° planar molecule	two, each 180° linear molecule

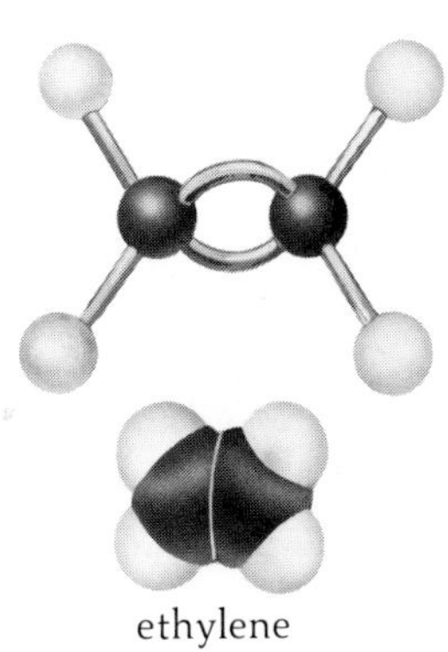
ethylene

Ethylene

The ethylene molecule is planar, has bond angles of approximately 120°, and has a double bond. (Chapter 3.) To achieve the first two characteristics, the valence bond approach requires that each carbon atom form three σ-bonds utilizing three equivalent atomic orbitals with ∞-fold axes of rotation that lie in the same plane and have the triangular orientation. To achieve the double bond between the two carbon atoms, the valence bond approach requires the formation of a *pi*-bond in addition to the *sigma*-bond. The π-bond is formed by the side-to-side overlap of appropriately oriented partially filled *p*-orbitals of the two carbon atoms.

Earlier in this chapter, it was suggested that sp^2-hybridization of one *s*- and two *p*-orbitals of carbon generates exactly the required set of atomic orbitals for carbon—three sp^2-orbitals and one *p*-orbital. The ∞-fold axis of rotation of the unhybridized

Figure 9-13

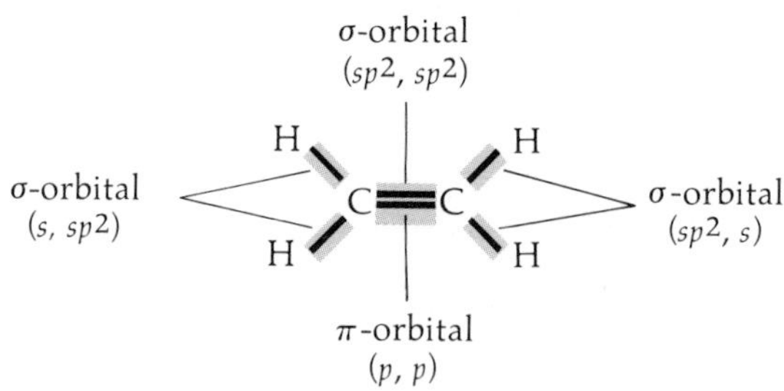

identification of the atomic orbitals used in the formation of molecular orbitals

(a)

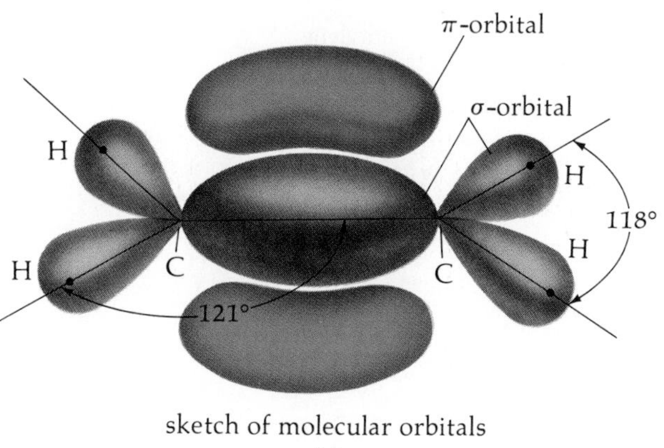

sketch of molecular orbitals

(b)

MOLECULAR ORBITALS IN THE ETHYLENE MOLECULE. The scale of Figure 9-13(b) is larger than that of Figure 9-12 so that more detail may be shown.

p-orbital is at right angles to the plane of the three ∞-fold axes of rotation of the three sp^2-orbitals. (Figure 9-6.) Each of these four atomic orbitals for carbon is partially filled. Figure 9-13 gives the valence bond interpretation of the structural formula of ethylene.

Diagrams for the formation of the σ-orbital and π-orbital for the two atoms of carbon are given in Figure 9-14. The overlap of the *s*-orbital of hydrogen with an sp^2-orbital of carbon in the formation of the hydrogen–carbon σ-orbital in no way differs from the overlap of the *s*-orbital of hydrogen with an sp^3-orbital of carbon in saturated hydrocarbons.

In the double bond, it is the carbon–carbon π-orbital that prevents the free rotation about the carbon–carbon σ-bond. To rotate the two carbon atoms with respect to each other, the overlap of the atomic *p*-orbitals that constitutes the π-bond would have to be broken. This would be equivalent to breaking the π-bond. According to

Figure 9-14

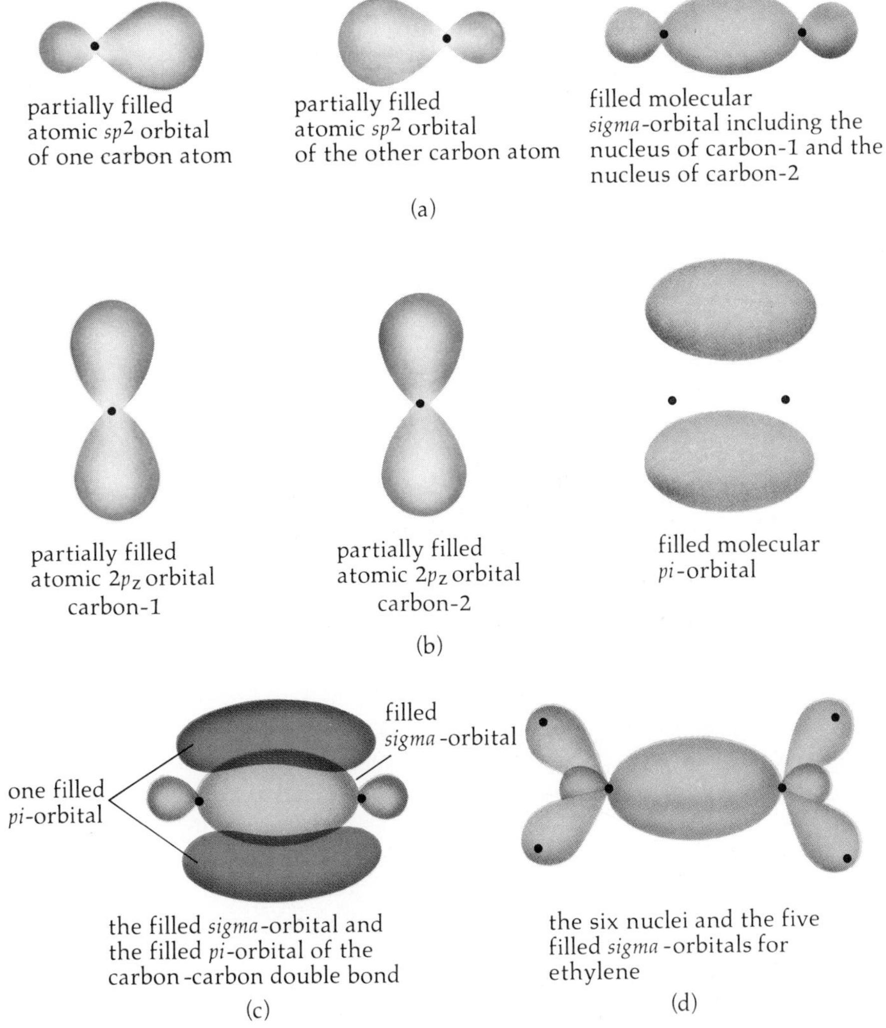

ATOMIC ORBITALS OF THE CARBON ATOMS AND THE MOLECULAR ORBITALS OF ETHYLENE. (a) The atomic sp^2-orbitals of the two carbon atoms and the σ-orbital of the two carbon atoms in ethylene; (b) the atomic p-orbitals of the two carbon atoms and the molecular π-orbital in ethylene. (c) The relative orientation of the σ-orbital and the two halves of the π-orbital; and (d) the six nuclei and the five filled σ-orbitals of ethylene. The π-orbital of the double bond is not shown. One half of that orbital is above the plane of the pepre and the other half is below the plane of the paper.

the valence bond approach, it is this restriction to free rotation that is the basis of *cis*- and *trans*-isomerism. (Scramble Exercise 6.)

Acetylene

The acetylene molecule is linear (bond angles of 180°) and has a triple bond. To achieve linearity, the valence bond approach requires that each carbon atom have two atomic orbitals with the same ∞-fold axes of rotation. This condition is satisfied by two *sp* hybrid orbitals for each carbon atom.

The identification of the atomic orbitals used in the formation of the molecular orbitals in terms of the valence bond approach is marked on the conventional structural formula.

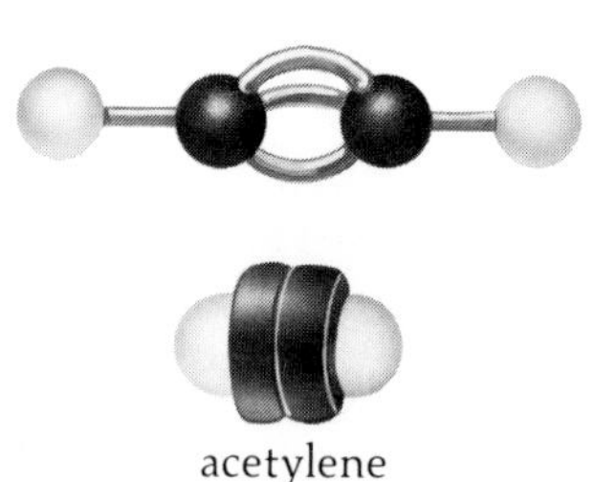
acetylene

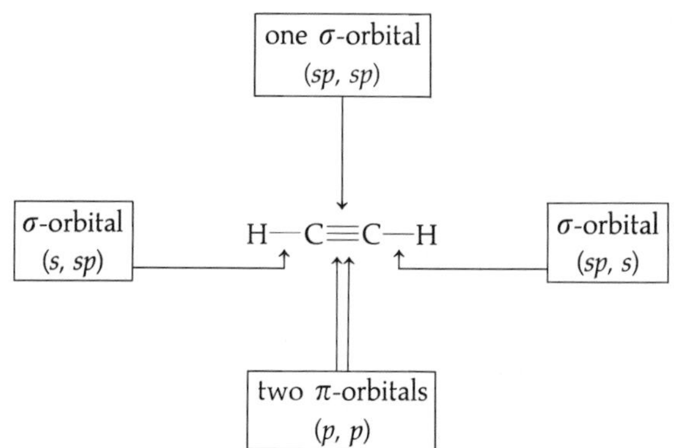

To achieve the triple bond, the valence bond approach requires that each carbon also have two *p*-orbitals with ∞-fold axes of rotation perpendicular to the ∞-fold axes of rotation of the two atomic *sp*-orbitals that have the same ∞-fold axis of rotation. As suggested earlier, the hybridization of the *s*-orbital and one *p*-orbital (for example, the p_x-orbital) of the excited state carbon atom generates the set of four atomic orbitals—two *sp*-orbitals, the p_y-orbital, and the p_z-orbital—required by the valence bond approach.

Each of the *sp*-orbitals has the general shape of sp^3- and sp^2-orbitals. It is the orientation of the two *sp*-orbitals with respect to each other that is unique. The line drawing sketch of two *sp*-orbitals for one carbon atom is

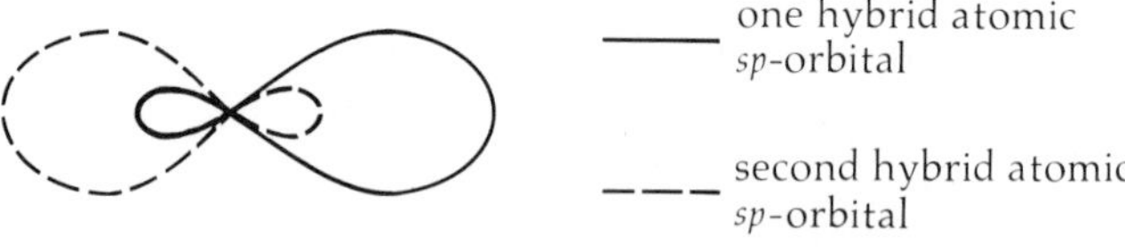

These are frequently sketched without showing the smaller regions,

and care must be taken not to confuse this sketch for these <u>two</u> atomic *sp*-orbitals with the two parts of one atomic *p*-orbital.

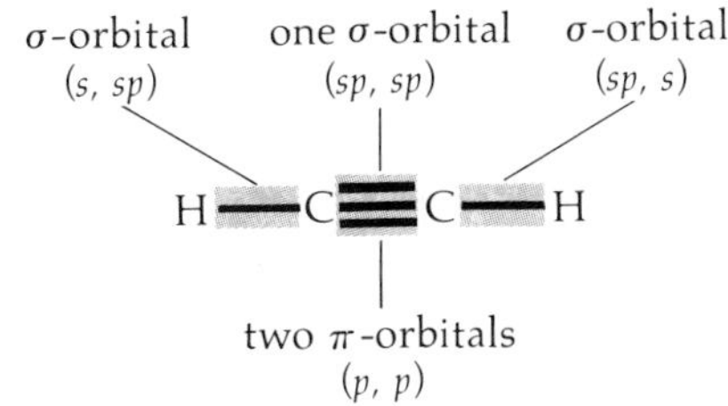

identification of the atomic orbitals used in the formation of molecular orbitals

(a)

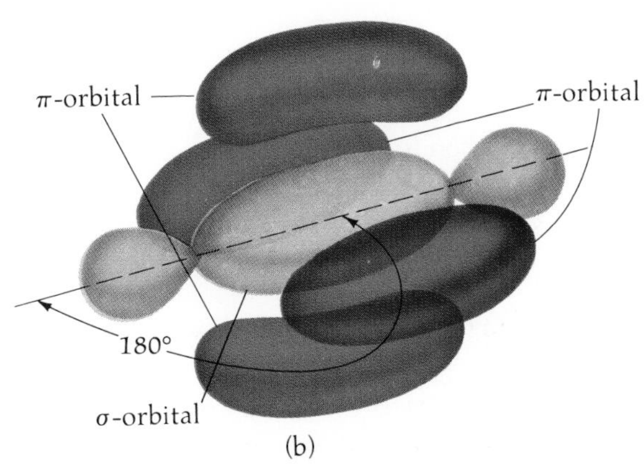

(b)

MOLECULAR ORBITALS IN THE ACETYLENE MOLECULE. The scale of Figure 9-15(b) is larger than those of Figures 9-14(c) and (d) so that more detail may be shown.

In acetylene (Figure 9-15), one of the partially filled *sp*-orbitals is used to form a *sigma*-bond with a hydrogen atom in the usual fashion and the other partially filled *sp*-orbital is used to form a *sigma*-bond with the other carbon atom. Note that this rationalizes the linear configuration of the acetylene molecule. What of the two *pi*-bonds? They are formed in the usual manner by side-to-side overlap of the two partially filled atomic p_y-orbitals (one from each carbon atom) and side-to-side overlap of the two partially filled atomic p_z-orbitals (one from each carbon atom).

Energy of Dissociation of Carbon–Carbon Bonds

The series of compounds ethane, ethylene, and acetylene provide an opportunity to compare the properties of single, double, and triple bonds between two carbon atoms. In Table 9-3, the names of the compounds appear in the center of the table with the experimentally determined carbon–carbon bond lengths and molar energies of dissociation to the left and the valence bond identification of carbon–carbon molecular orbitals and the involved atomic orbitals of carbon to the right. The energies

Table 9-3

BOND LENGTHS AND BOND ENERGIES FOR CARBON–CARBON SINGLE, DOUBLE, AND TRIPLE BONDS

carbon–carbon bond length (nanometers)	energy of dissociation in kilojoules/mole (kilocalories/mole)	compound	carbon–carbon molecular orbitals broken	identity of hybrid atomic orbitals of the carbon atoms involved	number of unhybridized *p*-orbitals per carbon atom
0.154	368 (88)	ethane	1 *sigma*	sp^3	0
0.134	698 (167)	ethylene	1 *sigma*, 1 *pi*	sp^2	1
0.120	961 (230)	acetylene	1 *sigma*, 2 *pi*	sp	2

of dissociation referred to in these polyatomic molecules are the energy required to break the carbon–carbon bonds of 6.02×10^{23} molecules (one mole) of the compound.

```
                 H H            H     H
                 ·· ··          ··    ··
for ethane     H:C:C:H  ⟶   H:C· + ·C:H
                 ·· ··          ··    ··
                 H H            H     H

               H.    .H       H.        .H
for ethylene     :C::C:   ⟶     :C: + :C:
               H˙    ˙H       H˙        ˙H

for acetylene  H:C:::C:H  ⟶  H:C: + :C:H
```

Each π-bond contributes to the energy required for dissociation, but the increment between ethylene and ethane of 330 kilojoules/mole is not as great as the energy of dissociation of ethane, 368 kilojoules/mole, and the increment between acetylene and ethylene of 263 kilojoules/mole is still smaller.

Lengths of Carbon–Carbon Bonds

The addition of a π-bond decreases the bond length from 0.154 nm for ethane to 0.134 nm for ethylene. The addition of the second π-bond in acetylene reduces the bond length to 0.120 nm. This is a smaller decrease than that produced by the first π-bond.

Ethanol and Dimethyl Ether

In alcohols and ethers, such as ethanol, CH_3CH_2OH, and dimethyl ether, CH_3OCH_3, all bond angles with carbon at the apex are approximately 109.5° and with oxygen at the apex approximately 109°.

```
    H  H                H       H
    |  |                |       |
 H—C—C—O—H          H—C—O—C—H
    |  |                |       |
    H  H                H       H

    H H                 H   H
    ·· ·· ··            ··  ··  ··
  H:C:C:O:H           H:C:O:C:H
    ·· ·· ··            ··  ··  ··
    H H                 H   H
```

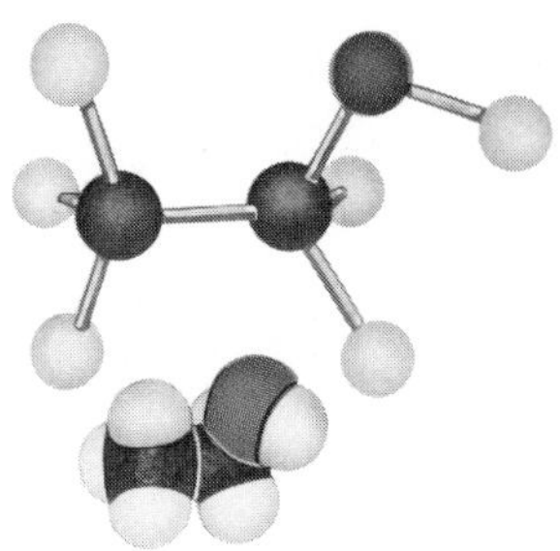
ethanol

The electron count for each isomer:

from six hydrogen atoms:	$6 \times 1 =$	6
from two carbon atoms:	$2 \times 4 =$	8
from one oxygen atom:	$1 \times 6 =$	6
total		20

Shown in the Lewis structures: $10 \times 2 = 20$

In the valence bond approach to both compounds, there are eight filled σ-orbitals. The bond angles suggest sp^3 hybridization of atomic orbitals in both carbon atoms and the oxygen atom.

The carbon–hydrogen σ-orbitals and the carbon–carbon σ-orbital are identical with those of the saturated hydrocarbons. There is a choice in forming the carbon–oxygen and the oxygen–hydrogen σ-orbitals as to the atomic orbitals to use for oxygen. One choice is to use two partially filled p-orbitals of a ground state oxygen atom (He) $2s^2\ 2p_x{}^2\ p_y{}^1\ p_z{}^1$. This predicts a carbon–oxygen–hydrogen bond angle and a carbon–oxygen–carbon bond angle of 90° each. The other choice is to use two partially filled sp^3-orbitals for the oxygen atom. The two other sp^3-orbitals of the oxygen atom are filled by pairs of electrons. The use of sp^3-orbitals predicts the tetrahedral bond angle of 109.5°. The experimental values for the alcohol and the ether are approximately 109°. This experimental value indicates that the atomic orbitals of choice are sp^3-orbitals.

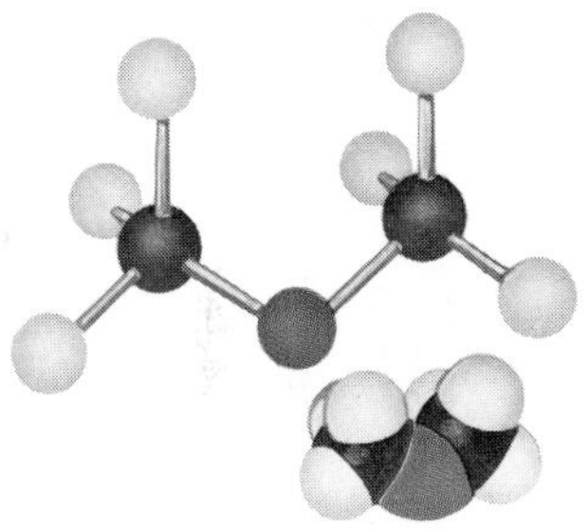
dimethyl ether

The identification of the atomic orbitals used in the formation of the molecular orbitals for ethanol is

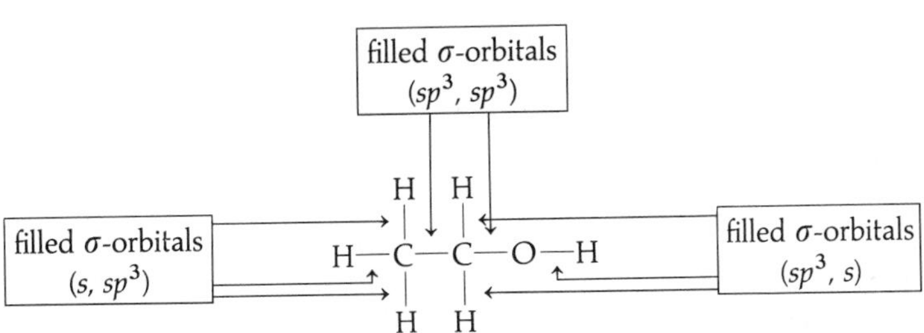

The identification of the atomic orbitals used in the formation of the molecular orbitals for dimethyl ether is

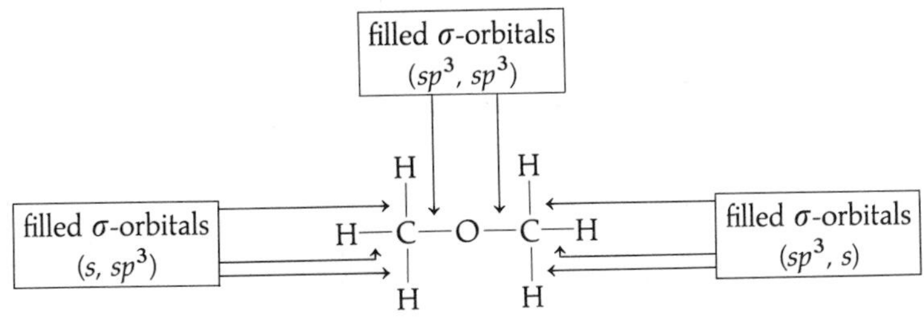

Carbonyl Group

Aldehydes, ketones, carboxylic acids, and esters have structures that seem complicated from the standpoint of the bonding involved. Examples:

acetaldehyde CH_3CHO

```
     H
     |
H—C—C=O
     |   |
     H  H
```

acetic acid CH_3COOH

```
     H
     |
H—C—C=O
     |   |
     H  O
          \
           H
```

acetone (dimethyl ketone) $(CH_3)_2CO$

```
     H
     |
H—C———C=O
     |        |
     H  H—C—H
              |
              H
```

methyl acetate CH_3COOCH_3

```
     H
     |
H—C—C=O
     |   |
     H  O    H
          \  /
           C
          /  \
         H    H
```

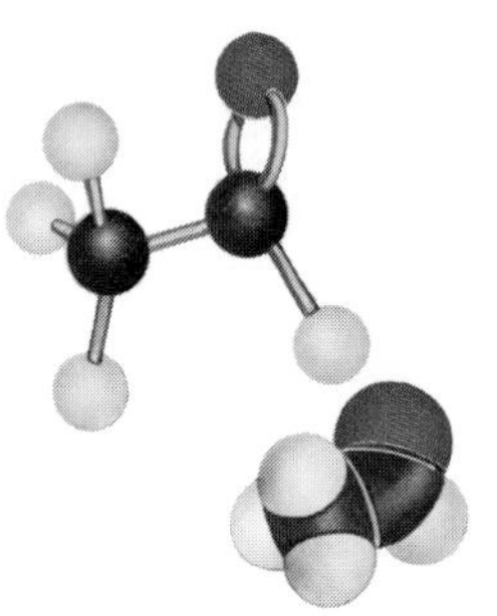
acetaldehyde

To draw out the molecular orbitals involved would be complicated but to identify the types of molecular orbitals and the atomic orbitals from which the molecular orbitals are formed is comparatively easy. All of these molecules contain the carbonyl group

$$>C=O$$

With the exception of that group, all of the molecular orbitals involved in these molecules have been encountered before.

To select the atomic orbitals utilized by the carbon atom in the carbonyl group, we need to know the bond angles at that carbon atom. If the atoms to which the carbonyl group is attached are designated by *X*, we know that these four atoms lie in a plane

```
X
  \
   C=O
  /
X
```

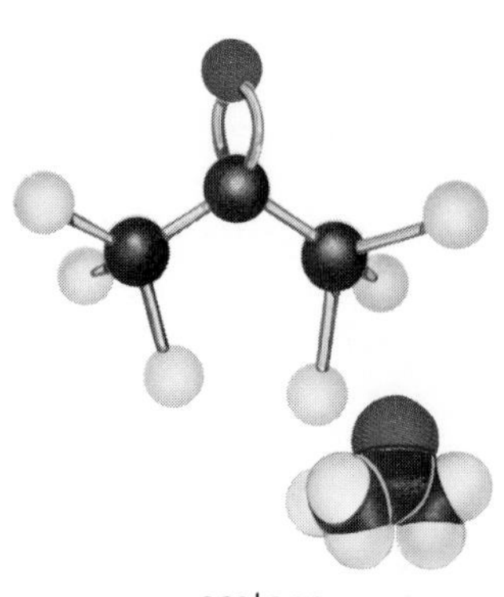
acetone

and that each of the three bond angles is approximately 120°. In the above, *X* could represent another carbon atom, a hydrogen atom, or an oxygen atom. This 120° angle is the orientation of three sp^2-orbitals. Consequently, it is sp^2 hybridization of the carbon atom that is chosen to rationalize the bonding in the carbonyl group. The three partially filled sp^2-orbitals form the three σ-orbitals and the partially filled *p*-orbital of the carbon atom forms the π-bond with a partially filled *p*-orbital of the oxygen atom. The formation of molecular orbitals involving the oxygen atom of the carbonyl group requires the use of two partially filled atomic orbitals of oxygen that have perpendicular ∞-fold axes of rotation. These orbitals could be the two partially filled *p*-orbitals of the ground state atom (He) $2s^2\ 2p_x^{\,2}\ 2p_y^{\,1}\ 2p_z^{\,1}$, or they could be the partially filled *p*-orbital and a partially filled hybrid sp^2-orbital, or they could be a partially filled *p*-orbital and a partially filled hybrid *sp*-orbital. The experimental evidence does not enable us to make a reasoned choice among these possibilities. In labeling the bonds below, we use the first option: two partially filled *p*-orbitals of the ground state atom. The identification of the atomic orbitals used in the formation of the molecular orbitals for acetaldehyde is

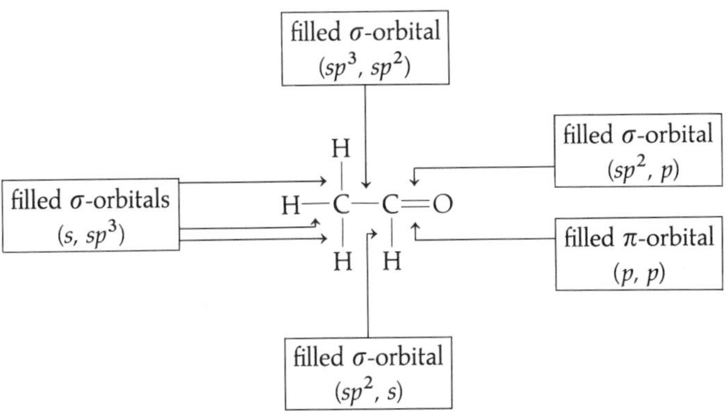

Acetaldehyde

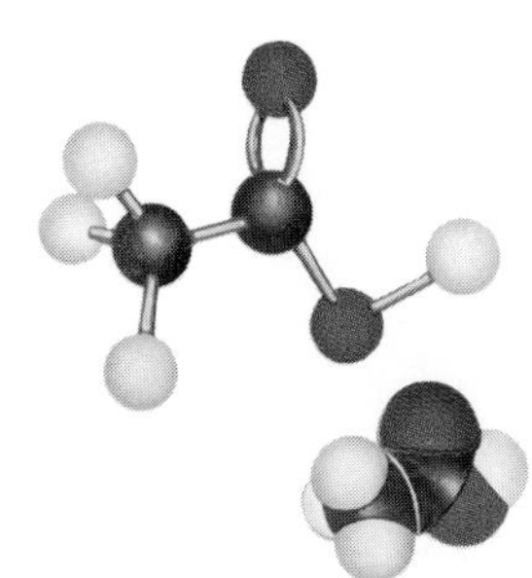

acetic acid

Future discussions will trade on the valence bond approach to the rationalization of the structures of covalent units, not only the structure of molecules but also the structure of polyatomic ions such as the ammonium ion, $NH_4^{(1+)}$, and the carbonate ion, $CO_3^{(2-)}$. Experimental bond angles are a key too whether hybridization is indicated and, if so, what type.

bond angle	type of hybridization
90°	none
109°	sp^3
120°	sp^2
180°	sp

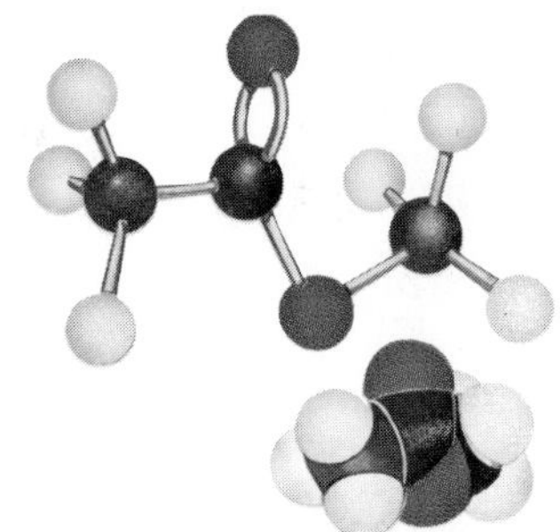

methyl acetate

Note that the four bonding sites of the carbon atom are rationalized in terms of <u>four</u> partially filled atomic orbitals and that there are three sets of hybrid orbitals.

Set I	sp^3	sp^3	sp^3	sp^3	(no *pi*-bonds formed)
Set II	sp^2	sp^2	sp^2	$2p_z$	(one *pi*-bond formed)
Set III	sp	sp	$2p_y$	$2p_z$	(two *pi*-bonds formed)

If one sp^3-orbital is used, all of Set I orbitals must be used; if one sp^2-orbital, then all of Set II; if one sp, then all of Set III.

These sets of atomic orbitals are also applicable to atoms, such as nitrogen and oxygen, that have fewer than four bonding sites. For those atoms, some of these atomic orbitals are filled by two electrons with antiparallel spins. Such pairs of electrons are referred to as **nonbonding electrons** or "**lone pair electrons.**" The choice of hybrid orbitals used is dictated by the experimental facts.

Carbon Dioxide

Linear Triatomic Molecules

Carbon dioxide and hydrogen cyanide are both linear molecules and both involve multiple bonding. Carbon is the central atom in both molecules.

	carbon dioxide	hydrogen cyanide
	CO_2	HCN
	O=C=O	HC≡N
	$\ddot{\underset{..}{O}}$::C::$\ddot{\underset{..}{O}}$	H:C:::N:
Electron count:		
from the carbon atom:	1 × 4 = 4	1 × 4 = 4
from the hydrogen atoms:		1 × 1 = 1
from the oxygen atoms:	2 × 6 = 12	
from the nitrogen atom:		1 × 5 = 5
total	16	10
Shown in Lewis structure:	16	10
	linear	linear
	nonpolar molecule	polar molecule

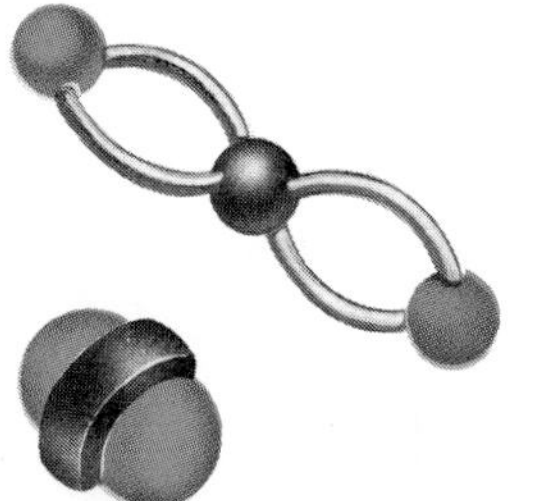

carbon dioxide

hydrogen cyanide

Linearity requires that the appropriate set of atomic orbitals for carbon, the central atom, is two partially filled *sp*-orbitals and two partially filled *p*-orbitals. The orientation of these orbitals has already been discussed in detail. The ground state partially filled atomic orbitals for the hydrogen atom (one *s*-orbital), for the oxygen atoms (two *p*-orbitals), and for the nitrogen atoms (three *p*-orbitals) are adequate to form the appropriate number of σ-orbitals and π-orbitals.

In carbon dioxide, each σ-orbital (either the one with the carbon atom and one oxygen atom or the one with the carbon atom and the other oxygen atom) can be rationalized as being formed by maximum overlap of a partially filled *sp*-orbital of the carbon atom and a partially filled *p*-orbital of the oxygen atom. The two π-orbitals (one with the carbon atom and one oxygen atom, the other with the carbon atom and the other oxgyen atom) are each formed by maximum side-to-side overlap of a partially filled *p*-orbital of the carbon atom and a partially filled *p*-orbital of an oxygen atom. (It would also be possible to rationalize the experimental facts using either sp^2 hybridization or *sp* hybridization for the oxygen atoms just as long as each oxygen atom has at least one partially filled *p*-orbital.)

In hydrogen cyanide, the two σ-orbitals (one with the hydrogen atom and the carbon atom and the other with the carbon atom and the nitrogen atom) are formed by the maximum overlap of the partially filled *s*-orbital of the hydrogen atom and one partially filled *sp*-orbital of the carbon atom and by the maximum overlap of the other partially filled *sp*-orbital of the carbon atom and a partially filled *p*-orbital of the nitrogen atom. The two π-orbitals (both with the carbon atom and the nitrogen atom) are formed in each case by maximum side-to-side overlap of a partially filled *p*-orbital of the carbon atom and a partially filled *p*-orbital of the nitrogen atom. Note that the nitrogen atom must have at least two partially filled *p*-orbitals to form the two *pi*-bonds. Either the ground state atom $1s^2\ 2s^2\ 2p_x^{\ 1}\ 2p_y^{\ 1}\ 2p_z$ or the *sp* hybridized atom will do the trick.

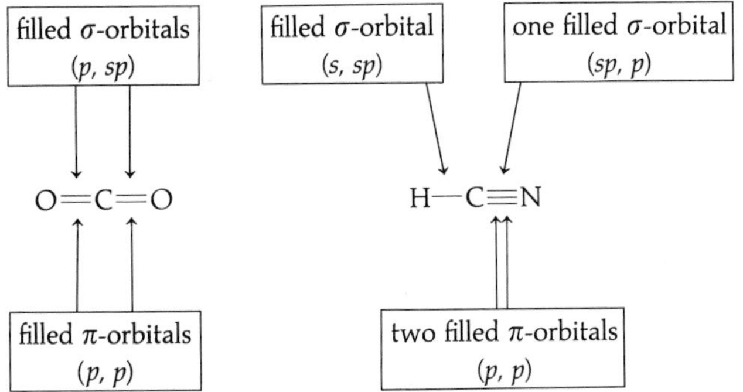

Hydrogen Cyanide

Polar and Nonpolar Bonds; Polar and Nonpolar Molecules

All of the bonds in carbon dioxide, hydrogen cyanide, and water are polar.

O=C=O   H—C≡N   H—O—H

The atoms directly bonded to each other have different electronegativity values and the electrons of the molecular orbitals are not, on the average, equally distributed with respect to the atoms of the two elements.

Polar Bonds
Polar Molecules
Nonpolar Molecules

The measure of the polarity of molecules is the dipole moment. The values of the dipole moments for these compounds in the gas phase at one atmosphere pressure and approximately room temperature are

carbon dioxide	0 debye
hydrogen cyanide	2.98
water	1.85

Hydrogen cyanide and water molecules are polar. Carbon dioxide molecules are not polar. All bonds in these three molecules are polar.

The polarity of molecules is discussed in terms of the response of molecules in the gas phase to an electrostatic field produced by two charged metal plates, one with a high positive charge and one with a high negative charge. In Figure 9-16, the nature of the electrostatic forces acting on two hypothetical molecules is diagrammed. One hypothetical molecule is an ellipsoid with a very small negative charge at one end and an equal positive charge at the other end. The whole molecule, as with all molecules, has a net charge of zero.

The other hypothetical molecule is also an ellipsoid but there are equal negative charges at each end of the molecule and a positive charge equal to the sum of the two negative charges in the middle. For both molecules, all negative charges are repelled by the negative plate and attracted by the positive plate. All positive charges are attracted by the negative plate and repelled by the positive plate. The forces acting on the molecules are shown as vectors (arrows indicating the direction of the

Figure 9-16

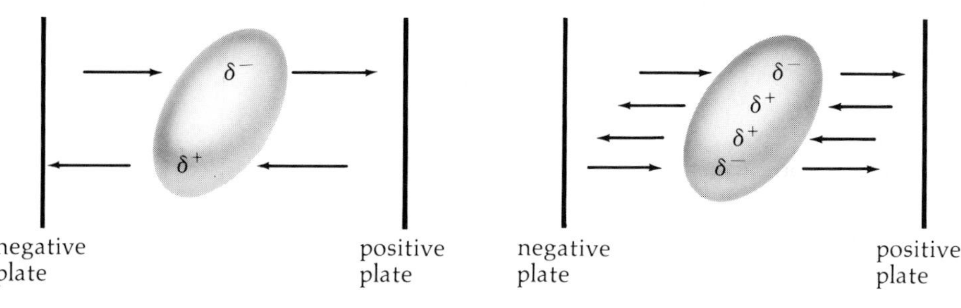

ELECTROSTATIC FORCES OF ATTRACTION AND REPULSION ACTING ON TWO HYPOTHETICAL MOLECULES.

forces). To show fine points such as the dependencies of the magnitudes of the forces on the distances from the charged plates is not attempted in these diagrams.

In Figure 9-16, on the left, the forces of attraction and the forces of repulsion move the negative end of the molecule toward the positive plate and move the positive end of the molecule toward the negative plate. Consequently, the molecule rotates. This hypothetical molecule is a polar molecule. All molecules of this type tend to become aligned by this rotation. Complete alignment is never attained since the molecules of the gas are in constant motion and the collision of molecules induces randomness in the orientation of the molecules. Hydrogen cyanide exhibits the behavior of this hypothetical molecule.

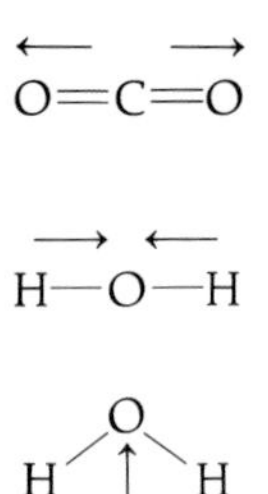

In Figure 9-16, on the right, the forces of attraction and repulsion would move both ends of the molecule toward the positive plate and move the center of the molecule toward the negative plate. The net result is that there is no movement of this molecule. Molecules of this type do not align with an electrostatic field. The molecule as a whole is nonpolar. The distribution of charges in the carbon dioxide molecule is that of this hypothetical molecule. The bonds are polar but the symmetry of the molecule nullifies the polarity of the bonds to give a nonpolar molecule. In the case of a linear molecule such as O=C=O, the balance of forces is fairly obvious. Carbon tetrachloride, with its four polar bonds, is also a nonpolar molecule. This nonpolarity is the consequence of the balance of forces due to the tetrahedral nature of the molecule. One of the evidences that water is not a linear molecule, H—O—H, is its dipole moment. If water were linear, its dipole moment would be zero.

In summary, the ball–stick–spring model was specified in terms of number of bonding sites and the orientation of these bonding sites. These specifications were imposed on the model in order that the model be consistent with known properties of compounds. The valence bond model is much more detailed. The valence bond model is based upon a model for the distribution of electons in atomic orbitals. The development of the atomic orbital model was strongly guided by existing knowledge of the properties of atoms, particularly the emission and absorption spectra of atoms. The valence bond approach manipulates atomic orbitals to attain a system of molecular orbitals that is consistent with known properties of molecules and is used to predict properties that have not been measured. Such predictions are always subject to experimental test.

A comprehensive treatment of the valence bond approach also utilizes *d*-orbitals and *f*-orbitals in the formation of molecular orbitals. For the elements discussed in this chapter, there are no partially filled *d*- and *f*-orbitals in the ground state atoms (and other low-energy state atoms) and these orbitals are not involved in the formation of molecular orbitals. *d*-Orbitals and *f*-orbitals are important in the consideration of compounds of elements that form more than four bonds. Such bond formations are characteristic of large atoms, particularly the atoms of the transition metals and the rare earths (the lanthanide series and the actinide series of elements).

All modern approaches to the nature of the chemical bond have to do with the balance of forces arising from the electrostatic attraction between electrons and nuclei, the electrostatic repulsion of nuclei and nuclei, the electrostatic repulsion of electrons and electrons, and the much weaker magnetic interactions of electrons.

Scramble Exercises

1. Hydrogen sulfide, H_2S, has an experimental hydrogen–sulfur–hydrogen bond angle of 93°.

- **(a)** Starting with the atomic number for sulfur, write a plausible Lewis electron dot structure for hydrogen sulfide.
- **(b)** On the basis of the valence bond approach, propose the molecular orbitals involved in the bonding within the hydrogen sulfide molecule and the atomic orbitals from which the molecular orbitals are formed.
- **(c)** Predict and give the basis of the prediction as to whether the H_2S molecule is polar or nonpolar.

2. The formula of methanol is CH_3OH. The experimental value for the hydrogen–carbon–hydrogen bond angles is 109.3° and the value for the hydrogen–oxygen–carbon bond angle is 108.9°.

- **(a)** Give a plausible Lewis dot structure for methanol.
- **(b)** On the basis of the valence bond approach, predict the nature of the carbon–oxygen molecular orbital and the atomic orbitals from which it is formed.

3. The formula of methyl amine is CH_3NH_2:

```
      H
      |      H
      |    /
  H — C — N
      |    \
      |      H
      H
```

The hydrogen–carbon–hydrogen bond angles are 109.5° and the hydrogen–nitrogen–hydrogen bond angle is 106°. The nitrogen in methyl amine has three bonding sites.

(a) Write a plausible Lewis electron dot structure for methyl amine.
(b) On the basis of the valence bond approach, predict the nature of the carbon–nitrogen molecular orbital and the atomic orbitals from which it is formed.

4. Formaldehyde, H_2CO, is the simplest molecule containing a carbonyl group. The molecule is planar and all bond angles are approximately 120°.

(a) Write out the ball–stick–spring structure and give a plausible Lewis electron dot structure for formaldehyde.
(b) On the basis of the valence bond model, identify all of the molecular orbitals in the molecule of formaldehyde and identify the atomic orbitals from which they are formed.

5. The formula for propylene is CH_3CHCH_2:

The carbon atoms have been numbered 1, 2, and 3 for identification. The bond angles at carbon-1 and also at carbon-2 are approximately 120°. The bond angles at carbon-3 are approximately 109°.

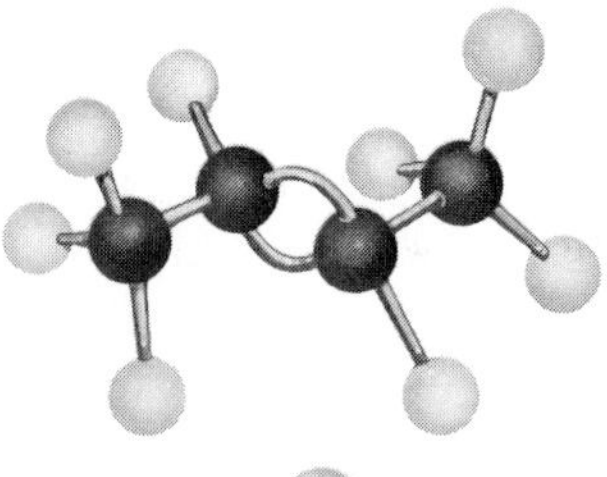

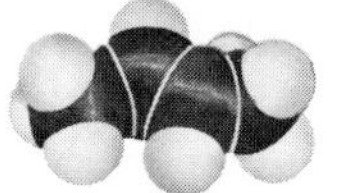

trans-2-butene

(a) Write out a Lewis electron dot structure for propylene.
(b) According to the valence bond approach, identify the appropriate type of hybrid atomic orbitals for each carbon atom and describe the molecular orbitals that constitute the bonding of carbon-1 to carbon-2.

6. From the standpoint of the valence bond approach, explore the nature of the double bond in

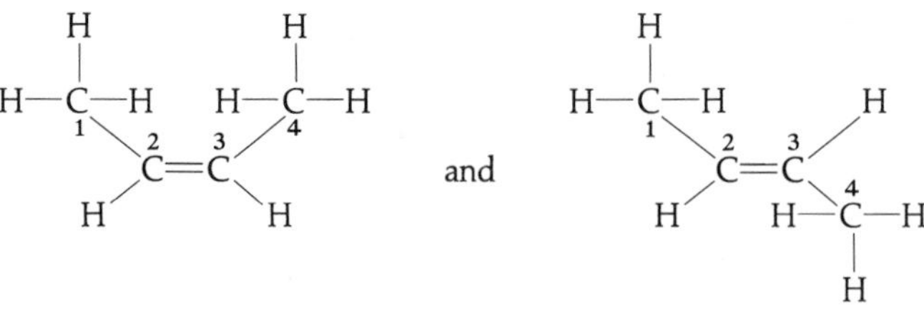

cis-2-butene   *trans*-2-butene

and the role of that double bond in preserving the identities of these two structural isomers. All bond angles at carbon-2 and at carbon-3 are approximately 120°. All bond angles at carbon-1 and at carbon-4 are approximately 109.5°.

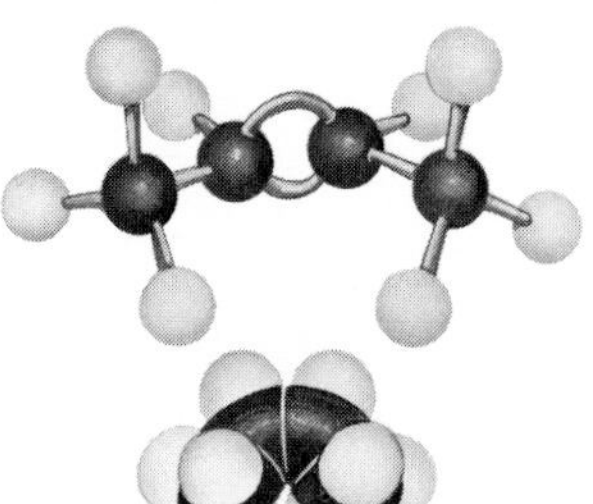

cis-2-butene

Additional exercises for Chapter 9 are given in the Appendix.

A Look Ahead

VERY LARGE MOLECULES

Although this chapter has dealt with small molecules—molecules of fewer than twelve atoms—the concepts introduced here are equally applicable to very large molecules—molecules that contain not only thousands of atoms but hundreds of thousands of atoms. There are simply more atoms to be bonded by *sigma*-bonds and more atoms to be bonded by both *sigma*- and *pi*-bonds. With only a *sigma*-bond, rotation occurs about the bond. With both a *sigma*-bond and a *pi*-bond, rotation is prohibited and planarity is imposed upon a group of atoms: at least four atoms if a carbonyl group,

$$\begin{matrix} X \\ & \searrow \\ & & C=O \\ & \nearrow \\ X \end{matrix}$$

is involved; at least six atoms if a carbon–carbon double bond

$$\begin{matrix} X & & & & X \\ & \searrow & & \nearrow & \\ & & C=C & & \\ & \nearrow & & \searrow & \\ X & & & & X \end{matrix}$$

is involved. These regions of planarity impose a certain amount of restraint on the configurations large molecules can assume. Such restraints are important factors in determining the properties of proteins and other biologically significant molecules.

Cyclic structures (ring structures), either with or without double bonds, also impose restricted motion on segments of a large molecule. These ring structures are important factors in the properties of starches, celluloses, and nucleic acids.

Molecules with long chains of atoms have considerable freedom as to how these chains coil or double back in loops. Polar groups along such chains interact with each other to hold segments of the molecule in juxtaposition if the charges are unlike or to keep segments of the molecule apart if the charges are like. Some large molecules are explored in Chapters 17 through 21. You now have adequate knowledge background to start working your way through these chapters.

10

COMPOUNDS OF SODIUM, MAGNESIUM, AND RELATED ELEMENTS

Chapter 10

The compounds of sodium and magnesium are invariably solids at room temperature. They all have high melting points, are frequently colorless, and are frequently soluble in water. The best known of these compounds is table salt, sodium chloride, NaCl. The symbol Na comes from the Latin *natrium* for sodium.

The elements sodium and magnesium are the first two members of the third period of the Periodic Table:

$$\text{sodium } (Z = 11) \ (\text{Ne})\ 3s^1 \qquad \text{magnesium } (Z = 12) \ (\text{Ne})\ 3s^2$$

These elements are metals and their surfaces, when clean, have the characteristic metallic sheen. Sodium and the other elements of the first column of the Periodic Table are known as the **alkali metals.** Magnesium and other elements of the second column are the **alkaline earths.** The white crystalline subtances characteristic of the alkali flats in arid regions of the West contain compounds of the alkali metals and the alkaline earths. Alkali metal elements and alkaline earth elements form ionic compounds—very seldom, if ever, covalent compounds.

Alkali Metals
Alkaline Earths
Ionic Compounds

Elements such as carbon, nitrogen, oxygen, and the halogens are nonmetals. The compounds of these nonmetals, which we discussed in previous chapters, are covalent compounds, usually polar covalent compounds. These nonmetals also form ionic compounds. Sodium chloride, NaCl, and magnesium chloride, $MgCl_2$, are ionic compounds of a metal and a nonmetal. In this chapter, we will deal with some of the compounds of the alkali metals and the alkaline earths. The properties of these compounds are quite different from the properties of the compounds we have been studying and we shall first look at some of these compounds—where they are found, some of their properties, and how they are used.

Chlorides

Magnesium sulfate, $MgSO_4$, a white solid, is sold in the retail market under the name epsom salts. It is frequently added to the hot water used to soak sore muscles and was at one time extensively used as a laxative. This compound may also be responsible in part for the distress that tourists encounter when they make extensive use of bottled mineral water in areas where the safety of the local drinking water is questioned.

Sulfates

Magnesium carbonate, $MgCO_3$, a white crystalline solid, frequently occurs with calcium carbonate, $CaCO_3$, a white crystalline compound of another alkaline earth, in very extensive mineral deposits known as dolomite and limestone. Limestone is primarily calcium carbonate; dolomite contains a higher proportion of magnesium carbonate. Both compounds have very limited solubility in water. Extensive beds of

Carbonates

limestone underlie the bluegrass country of Kentucky and are believed by some to have been a large factor in determining the soil condition upon which bluegrass and horses have thrived. Carlsbad Caverns, Mammoth Caves, and the Mark Twain Caves in the Missouri River bluffs are all cavities in limestone formations. The limestone was originally deposited through the action of marine organisms when these areas were below sea level. In a much later geological period, after these areas were above sea level, these caves were very slowly formed as the calcium carbonate dissolved in fresh water containing some dissolved carbon dioxide. Spectacular stalactites and stalagmites are the consequence of the redeposition of calcium carbonate by evaporation from dripping water that bears these dissolved materials. These processes are extremely slow: geologically significant time periods are very long.

Recrystallization processes occur when beds of limestone are subjected to conditions of extreme heat and pressure by geological processes, and the resultant material may emerge as marble. This mineral is still primarily calcium carbonate with some magnesium carbonate. It is a long long time span from the marine organisms to marble halls. The lime used in agriculture, particularly in regions where the soil tends to be acidic, is usually crushed limestone but it may also be crushed sea shells or crushed marble. The degree to which these materials are crushed determines how rapidly they react. The more surface available, the more rapidly crushed material reacts, and more frequent applications of lime are required.

The word "lime" properly refers to the product obtained when limestone or marble or sea shells are heated to a high temperature. The resulting mixture of calcium

Oxides

oxide and magnesium oxide is the lime. The carbon dioxide gas escapes into the air during the conversion:

$$CaCO_3(s) \xrightarrow[\text{heat}]{} \underset{\text{calcium oxide}}{CaO(s)} + CO_2(g) \qquad \Delta H = 180 \text{ kilojoules} \quad 43 \text{ kilocalories}$$

$$MgCO_3(s) \xrightarrow[\text{heat}]{} \underset{\text{magnesium oxide}}{MgO(s)} + CO_2(g) \qquad \Delta H = 117 \text{ kilojoules} \quad 28 \text{ kilocalories}$$

Calcium oxide and magnesium oxide, particularly the former, are important ingredients of cement and by their chemical reactions with water and various silicates (compounds of silicon) are largely responsible for the setting of concrete. Lime has also had a more glamorous role in society. Both oxides are extremely stable with respect to heat and they can be kept at high temperatures for long periods of time without appreciable deterioration. The "limelight" of stage and politics goes back to the days when a cylinder of lime was heated to a very high temperature by an oxyhydrogen flame to provide a high-intensity light source. To radiate light is a property of all hot solids. The higher the temperature, the whiter the light; the larger the surface, the more intense the light.

Magnesium oxide, MgO, also known as magnesia, is frequently used in refractory ceramics. Calcium oxide is not so used even though it holds up very well at high temperature. The reactivity of calcium oxide with water introduces problems in fabrication and handling. A slurry of magnesium oxide in water is the well-advertised milk of magnesia used as an antacid and a mild laxative.

Gypsum is a mineral that can be carved easily and is frequently made into desk ornaments and lamp bases. Although the pure calcium sulfate dihydrate, $CaSO_4 \cdot 2\,H_2O$, of which it is composed is white, gypsum frequently is shaded and banded by other trace compounds. The formation of these banded deposits is a fascinating geological and chemical story. Unusually compact, fine-textured gypsum is known as alabaster. Heated, gypsum loses water and becomes the powder known as plaster of Paris.

$$CaSO_4 \cdot 2\,H_2O(s) \longrightarrow CaSO_4(s) + 2\,H_2O(g)$$

On mixing with water, the dihydrate is again formed and the slurry sets into a solid mass. If you have worn a cast, it may have been calcium sulfate dihydrate.

Perhaps the best known compound of barium, another alkaline earth, is barium sulfate, $BaSO_4$. It is the white solid (incorrectly called barium) used as a slurry in water to define the digestive tract in X-ray studies. It is nontoxic and extremely insoluble in water and in dilute acid solutions such as those encountered in the stomach. The role of the barium sulfate in X-ray examinations is to absorb the X-rays and thus delineate the open pathway of the digestive tract and the contours of that pathway.

Gratuitous Information 10-1

SALT

Sodium chloride is essential for life and there have been societies in which salt has had more economic and political value than gold. It is still transported by backpack to remote settlements in the Himalayan mountains where no other mode of transportation—not even pack animals—is practical. Salt occurs in many places in the world as large mineral deposits known as halite. The rock salt used on streets to melt ice and in home freezers to make ice cream is usually crushed halite that is discolored by trace materials. Sodium chloride is also recovered from sea water by evaporation. One of Mahatma Gandhi's early political actions in India had to do with the British regulations on salt recovered from the sea. Today, salt is produced in increasing quantities as a by-product of the recovery of water from sea water for domestic use.

The salt that is carried inland by the wind at the seashore might be best described as wind-borne droplets of sea water or possibly small crystals of sodium chloride. The mechanism by which sea water is believed to get into the air involves sea foam. As a bubble rises, the water underneath the bubble also rises and thus acquires momentum in the upward direction. When the bubble bursts at the surface, this water from the underside of the bubble continues to rise and, according to high-speed photography, a small pinnacle of water forms. It is the very tip of this pinnacle that continues into the air as a droplet that can then be picked up and carried by the wind. During storms, sea spray, of course, causes much more salt water to be dispersed into the atmosphere.

Magnesium chloride, $MgCl_2$, is frequently an impurity in table salt. It is this compound that is responsible for salt becoming moist in humid weather and then caking when it dries out. The magnesium chloride absorbs the moisture and actually dissolves in the water. When the water evaporates later, the redeposited magnesium chloride, along with some dissolved sodium chloride, cements the sodium chloride crystals together.

Block salt is provided for domestic animals, particularly those that sweat, and used to attract wild animals into the range of camera and gun. Small birds such as white crown sparrows may be found enjoying the crumbs from these blocks of salt. Recently, there has been increased concern about the biological consequences of excess sodium chloride in human diets.

Any other inert compound containing atoms with heavy nuclei would absorb X-rays equally well.

Melting Points of Alkali Metal Compounds

The melting points and the solubilities in water of two compounds of each of the alkali metals are given in Table 10-1. Included in the table are the formulas of the compounds and the formulas of the ions in an aqueous solution of the compounds.

Table 10-1

COMPOUNDS OF ALKALI METALS

compound name, formula, color	melting point (°C)	solubility (grams per 100 grams water)	ions in aqueous solution	
lithium chloride $LiCl(s)$ white	614	64	Li^{1+} lithium ion colorless	Cl^{1-} chloride ion colorless
lithium nitrate $LiNO_3(s)$ white	264	90	Li^{1+}	NO_3^{1-} nitrate ion colorless
sodium chloride $NaCl(s)$ white	801	36	Na^{1+} sodium ion colorless	Cl^{1-}
sodium sulfate $Na_2SO_4(s)$ white	884	s (soluble)	Na^{1+}	SO_4^{2-} sulfate ion colorless
potassium chloride $KCl(s)$ white	776	34	K^{1+} potassium ion colorless	Cl^{1-}
potassium chromate $K_2CrO_4(s)$ yellow	968	63	K^{1+}	CrO_4^{2-} chromate ion yellow
rubidium chloride $RbCl(s)$ white	715	77	Rb^{1+} rubidium ion colorless	Cl^{1-}
rubidium sulfate $Rb_2SO_4(s)$ white	1060	42	Rb^{1+}	SO_4^{2-}
cesium chloride $CsCl(s)$ white	646	162	Cs^{1+} cesium ion colorless	Cl^{1-}
cesium bromide $CsBr(s)$ white	636	124	Cs^{1+}	Br^{1-} bromide ion colorless

In this table, one of the compounds for each element is the chloride, the compound of the alkali metal and chlorine. The other compound is a rather random choice to exhibit some of the common types of compounds other than the chloride.

Note the range of numerical values of the melting points of the compounds. Only one, lithium nitrate, 264 °C, has a melting point below 600 °C. The boiling points are, of course, higher. These high melting points and boiling points are in marked contrast to those of the covalent compounds we have been discussing. At room temperature and pressure, most of those covalent compounds have been gases or liquids, and many of those that are solids melt below 300 °C. Sucrose, table sugar, molecular formula $C_{12}H_{22}O_{11}$, has a melting point of 185 °C and this is a high value for a carbon–hydrogen–oxygen compound of this molecular weight.

Sodium chloride in a salt shaker is white. A large crystal of sodium chloride is colorless and transparent. The whiteness of the many small crystals of salt is the reflection of white light by the surfaces of randomly oriented crystals. With one exception, all of the compounds listed in Table 10-1 are colorless, but samples of the compounds as we see them in reagent bottles are white. With one exception, aqueous solutions of the compounds listed in Table 10-1 are colorless and transparent. The exception in both cases is potassium chromate. A large crystal of potassium chromate is yellow and transparent. Finely divided solid potassium chromate is yellow and opaque. An aqueous solution of potassium chromate is yellow and transparent. How yellow depends upon the concentration of the solution. The color is a property of the chromate ion. The pigment chrome yellow is lead chromate, $PbCrO_4$, a water-insoluble chromate. (Lead is a transition metal, not an alkali metal.)

Solubility in Water of Alkali Metal Compounds

The solubilities in water are expressed in terms of grams of the compound that dissolve in 100 grams of water. The solubilities are dependent upon temperature, and handbooks specify the temperature for each solubility tabulated. For the values quoted in Table 10-1, temperatures range from 0 °C to 25 °C. It is the order of magnitude of the solubilities that is significant for this discussion, not the specific values, and the specific temperatures are omitted in this table. Note that the solubilities range from less than 40 g per 100 g of water to more than 150 g per 100 g of water. The saturated solution of cesium chloride contains more grams of dissolved solid than grams of water. Essentially, all compounds of the alkali metals are quite soluble in water.

Ions in Aqueous Solutions

An aqueous solution of sodium chloride contains sodium ions, $Na^{\text{(1+)}}$(aq), and chloride ions, $Cl^{\text{(1−)}}$(aq). The magnitude of the charge on each ion equals the magnitude of the charge of the electron. The net charge of all the ions in the solution is zero. An aqueous solution of sodium sulfate contains sodium ions, $Na^{\text{(1+)}}$(aq), and sulfate ions, $SO_4^{\text{(2−)}}$(aq). In the aqueous solution of sodium chloride, the number of sodium ions equals the number of chloride ions. In the aqueous solution of sodium sulfate, the number of sodium ions is twice as great as the number of sulfate ions. The net charge for the solution is zero. Each sulfate ion, $SO_4^{\text{(2−)}}$, consists of a group of five atoms: one sulfur atom and four oxygen atoms covalently bonded together. This group of five atoms has within it two more electrons than the total electrons of the five atoms. Consequently, the ion has the charge of the two extra electrons. This charge of 2− is the charge of the ion as a whole.

Table 10-2

COMPOUNDS OF ALKALINE EARTHS

compound name, formula, color	melting point (°C)	solubility (grams per 100 grams water)	ions in aqueous solution	
beryllium chloride $BeCl_2(s)$ white	405	vs (very soluble)	Be^{2+} beryllium ion colorless	Cl^{1-}
beryllium sulfate $BeSO_4(s)$ white	d 550 (decomposes)	not available	Be^{2+}	SO_4^{2-}
magnesium chloride $MgCl_2(s)$ white	708	54	Mg^{2+} magnesium ion colorless	Cl^{1-}
magnesium sulfate $MgSO_4(s)$ white	d 1124 (decomposes)	26	Mg^{2+}	SO_4^{2-}
calcium chloride $CaCl_2(s)$ white	772	74	Ca^{2+} calcium ion colorless	Cl^{1-}
calcium sulfate $CaSO_4(s)$ white	1450	0.20	Ca^{2+}	SO_4^{2-}
strontium chloride $SrCl_2(s)$ white	873	54	Sr^{2+} strontium ion colorless	Cl^{1-}
strontium sulfate $SrSO_4(s)$ white	1605	0.011	Sr^{2+}	SO_4^{2-}
barium chloride $BaCl_2(s)$ white	1560	37	Ba^{2+} barium ion colorless	Cl^{1-}
barium sulfate $BaSO_4(s)$ white	1580	0.00023	Ba^{2+}	SO_4^{2-}
radium chloride $RaCl_2(s)$ white	1000	s (soluble)	Ra^{2+} radium ion colorless	Cl^{1-}
radium sulfate $RaSO_4(s)$ white	not available	0.000002	Ra^{2+}	SO_4^{2-}

METALLIC CONDUCTION

All metals are good conductors of electricity. Some are better conductors than others but all are good conductors of electricity in comparison to those substances that are classified as nonmetals. This conductivity is rationalized as the freedom of electrons to flow in the metal. If a difference in electrical potential (voltage) is applied to the two ends of a wire, an electric current flows in the wire and the magnitude of that current is directly proportional to the applied difference in potential. The proportionality constant is the resistance. The relation as it is usually expressed as an equation is

$$\begin{pmatrix}\text{difference in}\\ \text{potential}\end{pmatrix} = \begin{pmatrix}\text{electrical}\\ \text{current}\end{pmatrix} \times \begin{pmatrix}\text{electrical}\\ \text{resistance}\end{pmatrix}$$

$$V = I \times R$$

The units most frequently used to express these quantities are volts, amperes, and ohms:

$$\text{volts} = \text{amperes} \times \text{ohms}$$

The resistance of a piece of wire depends not only upon its composition but also upon its length and its cross-section. It should come as no great surprise that the longer the wire, the greater its resistance to the flow of current and that the smaller the cross-sectional area of the wire, the greater the resistance, all other things being equal.

All of the above has to do with the movement of electrons in a metal and is known as metallic conduction. The most available electrons are the *s* electrons.

Although we usually think of metallic conduction in connection with the solid phase, metallic conduction is also a property of metals in the liquid phase. Mercury is the only metallic element that is a liquid under room conditions. Liquid mercury is frequently used to make electrical contacts and is used in mercury switches to make and break circuits silently by simply tilting a small sealed container of mercury to close or open the space between two wires.

Instruments to measure differences in potential, the voltage drop between two electrical contacts, are readily available and so are instruments to measure current, the rate of flow of electricity. A current of one ampere flowing for a period of one second corresponds to the transfer during that one second of 6.24×10^{18} electrons. In lecture demonstration equipment, ordinary light bulbs are frequently used as a very crude indicator of electron flow. The light bulb is put into the electrical circuit by cutting one of the lead wires from the source of current and wiring in the socket for the light bulb. The lower the wattage of the bulb, the smaller the electron flow required to raise the temperature of the filament in the bulb sufficiently for the filament to glow.

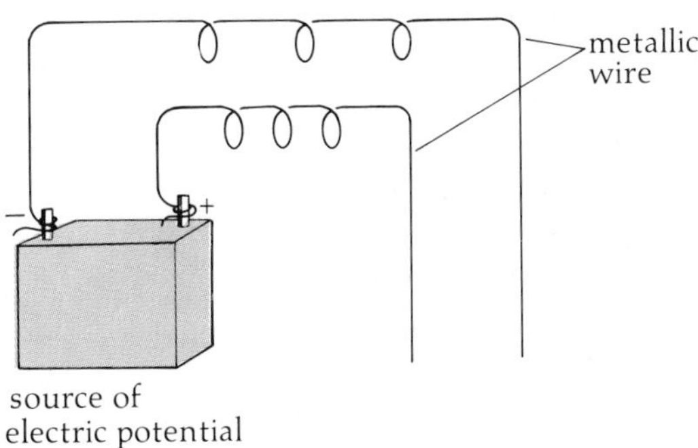

Incomplete Circuit (no electrical current)

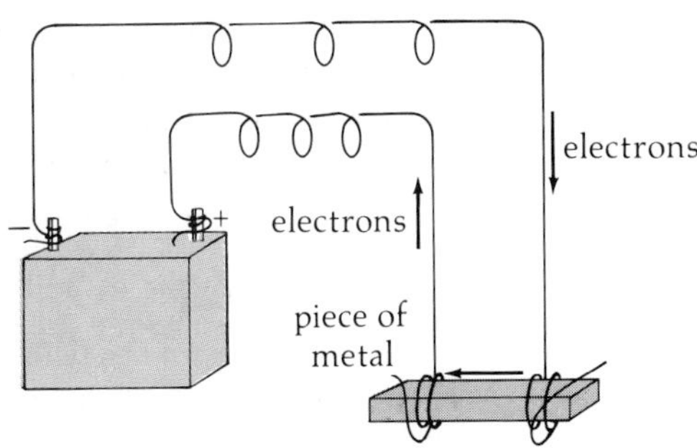

Complete Circuit with Metallic Conductor

Alkaline Earths

Table 10-2, Compounds of Alkaline Earths, is a companion table to Table 10-1, Compounds of Alkali Metals. The compounds of beryllium, the first member of the family, and the compounds of radium, the last member of the family, have not been studied as extensively as the compounds of the other members of the family and information concerning these compounds in handbooks is fragmentary. The compounds of both elements are toxic but for quite different reasons. In this second table, compounds are limited to the chlorides and the sulfates. All have melting points above 400 °C. One, beryllium sulfate, decomposes at 550 °C without melting. All are colorless compounds and multiple crystal samples are white. The chlorides are soluble in water; where values are available, more than 30 g of the compound dissolve per 100 g of water. The solubilities of the sulfates range from 26 g per 100 g water for $MgSO_4$ to 2×10^{-6} g per 100 g water for $RaSO_4$. Barium sulfate, which is used in X-ray studies of the intestinal tract in part because of its insolubility, has a solubility of 2.3×10^{-4} g per 100 g water (0.00023 g or 0.23 mg per 100 g of water).

Empirical Formulas of Ionic Compounds

Positive ions are known as **cations,** negative ions as **anions.** Table 10-3 is an extended but still very limited list of anions that pair with alkali metal and alkaline earth cations in forming crystals of the compounds. The empirical formula of each compound is made up of the smallest number of cations and anions that have a net charge of zero. For example, the empirical formula of magnesium phosphate is made up of the smallest number of magnesium ions, Mg^{2+}, and phosphate ions, PO_4^{3-}, that have a net charge of zero. The smallest "common denominator" of 2 and 3 is 6; therefore three Mg^{2+} and two PO_4^{3-} are required, and the empirical formula of magnesium phosphate is $Mg_3(PO_4)_2$.

All compounds made up of positive ions and negative ions are called salts, with the exception of acids and compounds containing hydroxide ions, OH^{1-}, and oxide ions, O^{2-}. The term "salt" in this generic sense of a very large class of compounds is an artifact of the development of the language of chemistry and is no longer used in

Table 10-3

SELECTED ANIONS

halide ions (colorless)		polyatomic anions (colorless)		
fluoride	F^{1-}	hydroxide ion	OH^{1-}	
chloride	Cl^{1-}	nitrate ion	NO_3^{1-}	
bromide	Br^{1-}	carbonate ion	CO_3^{2-}	
iodide	I^{1-}	phosphate ion	PO_4^{3-}	
oxygen family ions (colorless)		polyatomic anions (colored)		
oxide	O^{2-}	permanganate ion	MnO_4^{1-}	purple
sulfide	S^{2-}	chromate ion	CrO_4^{2-}	yellow
selenide	Se^{2-}			
telluride	Te^{2-}			

its original connotation as a product of acid–base reactions. Acid–base reactions are perceived today in terms of much more specific concepts. (Chapters 13 and 14.)

Evidences for Ionic Compounds

The evidence for ions is derived from the properties of the compounds we now call **ionic compounds.** Three types of properties are considered here: the electrical conductivity of the molten compounds, the electrical conductivity of aqueous solutions of the compounds, and the spatial distribution of atoms in crystals of the compounds.

Electrical Conductivity of the Liquid Phase

Above 801 °C, sodium chloride becomes a liquid and the liquid sodium chloride conducts an electric current. The test equipment can be very simple: a dish of molten sodium chloride, a source of electrical potential, two connecting wires with the socket for a light bulb inserted into one wire, several light bulbs of different wattages, and two electrodes that can be used to make the electrical contact into the high-temperature liquid. For molten sodium chloride, electrodes made of copper (melting point 1083 °C) or the copper wires themselves are adequate. For higher temperatures, platinum (m.p. 1769 °C) or graphite electrodes are commonly used. Graphite, a crystaline form of carbon, sublimes above 2000 °C without melting. The same equipment is also used to test aqueous solutions. Figure 10-1.

Electrical Conductivity of Aqueous Solutions

As shown by the response of the light bulb, both molten salt and the aqueous solution of salt complete the electrical circuit. In similar tests using water or the liquid phases of other covalent compounds, the light bulb gives no visible response. The conductivity of molten sodium chloride or the aqueous solution of sodium chloride is rationalized in terms of charged particles—the positive sodium ions, $Na^{(1+)}$, and the negative chloride ions, $Cl^{(1-)}$—that are free to move in the liquid phase in response to the electrostatic forces produced by the negative charge of one electrode and the positive charge of the other electrode. The sodium ions (the cations) move toward the negative electrode and the chloride ions (the anions) move toward the positive

Figure 10-1

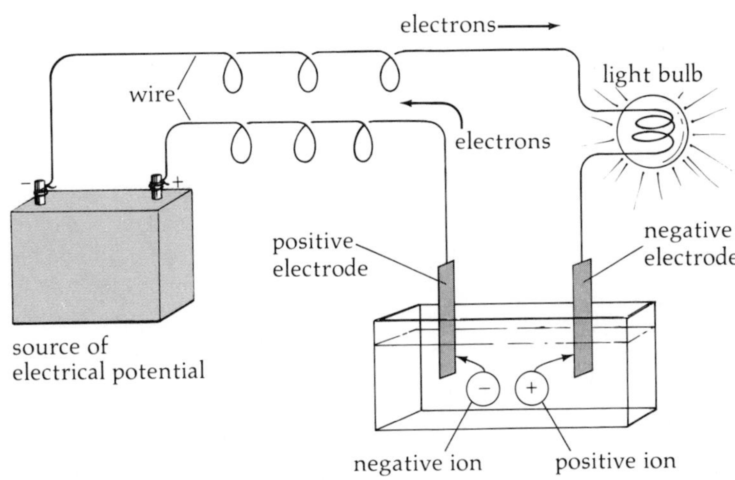

COMPLETE CIRCUIT WITH IONIC CONDUCTION.

electrode. At the surface of each electrode, a chemical reaction occurs. The chemical reaction at the negative electrode removes electrons from the electrode. The chemical reaction at the positive electrode surrenders electrons to the electrode. The products produced by the chemical reactions at each electrode can be identified and measured. The number of electrons removed from the negative electrode can be determined; also, the number of electrons transferred to the positive electrode can be measured. The two numbers are the same. Everything that is known about the conductivities of these liquid phases and the electrode reactions can be rationalized in terms of ions. Ions have to the chemists the same reality as molecules. The liquid phases of all alkali metal compounds and all alkaline earth compounds and all of the aqueous solutions of these compounds readily conduct electric currents. This conductivity is in sharp contrast to the lack of conductivity of the liquid phases of all covalent compounds and the aqueous solutions of most covalent compounds. There are some covalent compounds that react with water to produce ions and these solutions are consequently conductors of electricity. The conductivity of the liquid phase of a compound like beryllium sulfate, which decomposes before it melts, cannot be studied. The conductivity of the aqueous solution of a compound such as barium sulfate, which has very limited solubility in water, is very slight, and the detection of that conductivity requires a more sensitive indicator than the light bulb. The concern about handling electrical equipment and live wires under wet conditions relates to the properties of ionic materials dissolved in the water, not to the water itself.

Crystal Structure

By X-ray analysis, the positions of the nuclei of atoms in crystals can be determined with precision: the heavier the nuclei, the greater the precision. The X-ray analysis of sodium chloride crystals shows that each sodium nucleus is surrounded by six chlorine nuclei and that each chlorine nucleus is surrounded by six sodium nuclei. (Figure 10-2.) In all of the above, the distance between a sodium nucleus and an adjacent chlorine nucleus is 0.281 nanometers. The relation of the sodium nucleus to each of the six surrounding chlorine nuclei is exactly the same. There is no grouping, no pairing, to give an NaCl unit with a shorter sodium nucleus–chlorine nucleus distance. This evidence is not consistent with the concept of a crystal made up of molecules. It is consistent with the concept of a crystal made up of an orderly array of ions. For the solid phase, there can be no molecular formula for sodium chloride, only the **empirical formula,** NaCl, that expresses the relative numbers of atoms.

There is spectral evidence that the vapor phase of sodium chloride, obtained only at very high temperatures, exists as a mixture of free sodium ions, free chloride ions, ion pairs (NaCl units), and other small aggregates of sodium and chloride ions. Covalent compounds in the vapor phase exist entirely as molecules.

The sodium chloride crystal structure is a remarkably simple stacking of ions. This type of crystal structure does not hold for all 1:1 ionic compounds—one positive ion to one negative ion. The diameters of the ions are important in the balancing of the forces of attraction and repulsion to attain an array in which the net forces are zero. In some cases, the relative sizes of the ions result in spaces that are of sufficient magnitude that other ions or molecules may be built into the crystal. This accounts for the two molecules of water in the formula for gypsum, $CaSO_4 \cdot 2\ H_2O$, calcium sulfate dihydrate. The dot between the $CaSO_4$ and the $2\ H_2O$ simply indicates

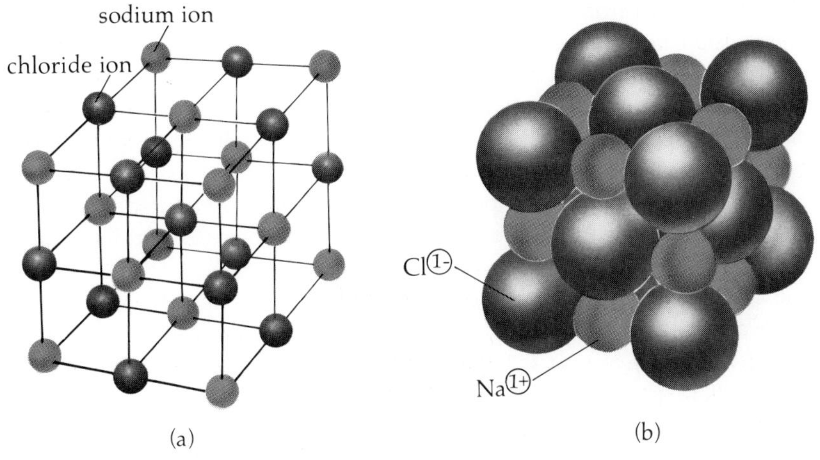

SODIUM CHLORIDE CRYSTAL LATTICE. (a) Distribution of sodium ions and chloride ions in space. Wires are used in three-dimensional models to support the balls. They are in no way a part of the crystal itself. (b) A space-filling model.

Hydrates

that, from the standpoint of relative abundance, the crystal has two molecules of water tucked away for every $CaSO_4$ unit. The formation of hydrates is not unusual for compounds made up of a small ion such as the calcium ion, Ca^{2+}, and a large ion such as the sulfate ion, SO_4^{2-}. The latter ion has to be large simply on the basis of the five atoms that are covalently bonded together in the ion. The magnesium sulfate that is sold under the name epsom salts is actually the heptahydrate $MgSO_4 \cdot 7\,H_2O$. Sodium chloride does not form hydrates. Calcium chloride forms a series of hydrates: $CaCl_2 \cdot H_2O$, $CaCl_2 \cdot 2\,H_2O$, and $CaCl_2 \cdot 6\,H_2O$. This capacity of calcium chloride to accommodate water in the crystal structure makes anhydrous calcium chloride, $CaCl_2$; calcium chloride monohydrate, $CaCl_2 \cdot H_2O$; and calcium chloride dihydrate, $CaCl_2 \cdot 2\,H_2O$, effective dehydrating agents, and they are frequently used to remove water vapor from small isolated volumes of air. For example,

$$\underset{\text{calcium chloride dihydrate}}{CaCl_2 \cdot 2\,H_2O(s)} + 4\,H_2O(g) \longrightarrow \underset{\text{calcium chloride hexahydrate}}{CaCl_2 \cdot 6\,H_2O(s)}$$

The numerical values for melting points and solubilities given in Tables 10-1 and 10-2 are for the anhydrous compounds. Many of the numerical values for the various hydrates are also given in handbooks. Frequently, hydrates decompose by the loss of water before they melt.

Interesting patterns of crystallization may be imposed by the shape of an ion. The carbonate ion, CO_3^{2-}, is a pinwheel of three oxygen atoms around a central carbon atom with the centers of all four nuclei in a plane (sp^2-hybridization). In calcite (calcium carbonate), $CaCO_3$, these pinwheels are arrayed in a very ordered fashion but in a quite different pattern from that of sodium chloride. It is the orientation of the ions that is responsible for crystals of calcite having properties that are dependent upon the direction of orientation of the crystal. A pure crystal of calcite is transparent to visible light but the velocity of light in the crystal depends upon the orientation of the crystal to the light waves. This seems like a rather esoteric property. It is, however, the property that made it possible to design optical instruments to study the transmission of polarized light by compounds—studies that contributed a great deal to our understanding of the orientation of atoms within molecules and ions and to our understanding of natural products and biological processes.

The distribution of forces within crystals of compounds with unequal numbers of positive and negative ions, such as magnesium chloride, $MgCl_2$ (a 1:2 crystal), and sodium sulfate, Na_2SO_4 (a 2:1 crystal), are more complicated than the distribution of forces in 1:1 crystals.

Solvents for Ionic Compounds

One of the best solvents for an ionic compound, an ionic crystal, is water; among the poorest solvents are the liquid hydrocarbons and ethers. These differences relate, at least in part, to the dielectric constants of these liquids. In discussing the force between two isolated charged particles, we have been concerned only with the magnitude of the charges and the distance separating the charges:

$$\text{force} = \frac{\text{(charge particle I)(charge particle II)}}{\text{(distance separation)}^2}$$

This relation applies to the ions in a crystal, but in solution the properties of the solvent that occupies the space between the ions must also be considered. The properties of the solvent come in as the dielectric constant of the solvent in the denominator of the force equation:

$$\text{force} = \frac{\text{(charge I)(charge II)}}{\text{(dielectric constant)(distance)}^2}$$

The dielectric constant of space (a vacuum) is exactly 1, hence the simplified relation in discussing isolated charges. The dielectric constants for all gases are slightly more, very slightly more, than 1: 1.00+. The dielectric constants for liquid hydrocarbons are of the order of 2; the value for ethanol, CH_3CH_2OH, is approximately 24; and the value for water is approximately 80 at room temperature, a little less at higher temperatures. This value for water is exceptionally high. There are a few known compounds that have higher dielectric constants, but these compounds are not readily available. The high dielectric constant of water decreases the forces between ions in aqueous solutions.

The solution of an ionic solid such as sodium chloride in water can be considered as the conversion of the crystalline solid made up of ions into independent hydrated ions separated by water:

$$NaCl(s) \longrightarrow Na^{1+}(aq) + Cl^{1-}(aq)$$

The reverse process is crystallization.

$$Na^{1+}(aq) + Cl^{1-}(aq) \longrightarrow NaCl(s)$$

A saturated solution exists when both processes are proceeding at equal rates so that the quantity of dissolved material remains unchanged by time:

$$\mathrm{NaCl(s)} \qquad \mathrm{Na^{1+}(aq) + Cl^{1-}(aq)}$$

The nature of the solvent comes into the discussion of solubility in two ways:

- the intervening solvent between two ions reduces the forces of attraction between the ions in aqueous solution, and
- the solvent molecules may interact with one or both of the ions.

Both of these factors contribute to the stability of the ions in solution and thus decrease the tendency for crystallization from the water solution to take place. Each of these factors will be discussed.

The force of attraction between a sodium ion and a chloride ion in air is given by the following:

$$\text{force in air} = \frac{(1+)(1-)}{1.00\ (\text{distance between the ions})^2}$$

The force of attraction between a sodium ion and a chloride ion in water is given by

$$\text{force in water} = \frac{(1+)(1-)}{80\ (\text{distance between the ions})^2}$$

Dielectric Constant of the Solvent

Sodium ions and chloride ions introduced into air lead to the formation of crystals of sodium chloride. Sodium ions and chloride ions introduced into water are much less likely to form crystals. The force between the ions in water is 1/80 of the force between ions, at the same distances, in air. A very high concentration of ions in water is required for crystals to form. Ethanol, dielectric constant 24, is a much less effective solvent for sodium chloride, or any other ionic compound, than water.

Water molecules are polar and tend to orient about ions in solution.

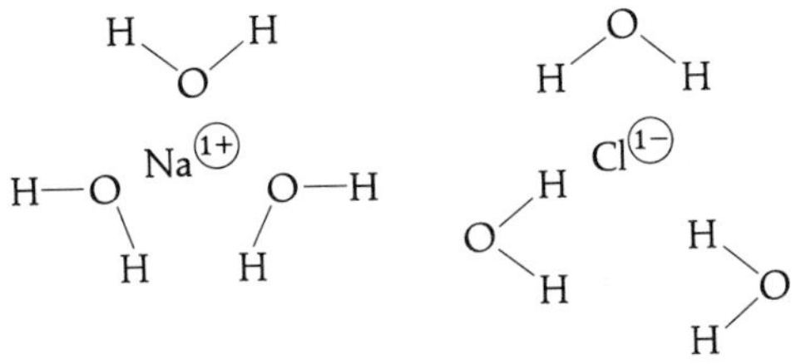

```
      H   H                      O
       \ /                     /   \
        O                     H     H

 H—O   Na(1+)   O—H        H      Cl(1-)
   \            /         /
    H          H         O          H
                          \          \
                           H          O
                                     /
                                    H
```

Hydration of Ions

For the cation, water molecules orient with the oxygen—the more electronegative of the two elements in water—nearest to the ion. For the anion, water molecules orient with the hydrogen—the more electropositive of the two elements—nearest to the ion. These processes are known as **hydration** or **solvation.** The fit of the wedge-shaped water molecules is clearly more effective for positive ions than for negative ions of the same diameter.

The magnitude of force between ions is, of course, related to the charges on the ions. According to the force equation, the force between a magnesium ion, Mg^{2+}, and a sulfate ion, SO_4^{2-}, is four times as great as the force between a sodium ion,

$Na^{(1+)}$, and a chloride ion, $Cl^{(1-)}$, at the same separation. The larger charges on ions stabilize the crystals. The larger charges on ions also increase the attraction of ions for polar molecules such as water.

Two empirical rules of solubility in water are that essentially all compounds of alkali metals are soluble in water and that all nitrates are soluble in water.

The remaining challenge in this chapter is to correlate the properties of the compounds and the ions of the alkali metals and the alkaline earths with the ground state electron distribution of the atoms. Starting with the atomic numbers, you can, of course, write out the notation for these electron distributions. For convenience, these and also the ground state electron distribution for the preceding inert gases are given in Table 10-4.

The ground state distribution of electrons of each alkali metal atom is the ground state distribution of an inert gas element with one additional electron in the next *s*-orbital. The ground state distribution of electrons of each alkaline earth atom is the ground state distribution of an inert gas element with two additional electrons in the next *s*-orbital. Example:

sodium (Ne) $3s^1$ magnesium (Ne) $3s^2$

Table 10-4

GROUND STATE ELECTRON DISTRIBUTION IN ATOMS OF THE ALKALI METALS AND THE ALKALINE EARTHS IN TERMS OF THE NEAREST INERT GAS

nearest inert gas		at. no.	alkali metals		at. no.	alkaline earths		at. no.
helium $1s^2$	He	2	lithium (He) $2s^1$	Li	3	beryllium (He) $2s^2$	Be	4
neon (He) $2s^2$ $2p^6$	Ne	10	sodium (Ne) $3s^1$	Na	11	magnesium (Ne) $3s^2$	Mg	12
argon (Ne) $3s^2$ $3p^6$	Ar	18	potassium (Ar) $4s^1$	K	19	calcium (Ar) $4s^2$	Ca	20
krypton (Ar) $4s^2$ $3d^{10}$ $4p^6$	Kr	36	rubidium (Kr) $5s^1$	Rb	37	strontium (Kr) $5s^2$	Sr	38
xenon (Kr) $5s^2$ $4d^{10}$ $5p^6$	Xe	54	cesium (Xe) $6s^1$	Cs	55	barium (Xe) $6s^2$	Ba	56
radon (Xe) $6s^2$ $4f^{14}$ $5d^{10}$ $6p^6$	Rn	86				radium (Rn) $7s^2$	Ra	88

Where does the valence bond approach applied to these ionic compounds lead? Sodium chloride is used as the test case:

Valence Bond Approach

sodium (Z = 11) (Ne) $3s^1$ chlorine (Z = 17) (Ne) $3s^2\ 3p_x{}^2\ 3p_y{}^2\ 3p_z{}^1$

The partially filled atomic orbitals available for the formation of a molecular orbital are the 3*s*-orbital of the sodium atom, and the $3p_z$-orbital of the chlorine atom. These are entirely consistent with the formation of a *sigma*-orbital for NaCl.

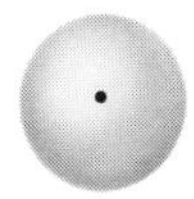

partially filled
3*s*-orbital
sodium atom

partially filled
3*p*-orbital
chlorine atom

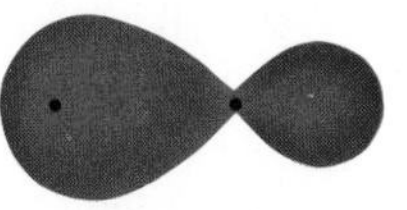

Na Cl
σ-orbital for NaCl

Gratuitous Information 10-2

QUARTZ SAND

There are a few compounds that have very high melting points even though they are not ionic compounds. One of these, quartz, is discussed in detail here as a contrast to ionic crystals and molecular crystals. The melting point of quartz is above 1600 °C. The most commonly encountered sand in the United States is made up of grains of quartz. Each grain is a molecule. (So you have seen individual molecules after all.) Break the grain with the blow of a hammer and each fragment is a molecule. Quartz has the empirical formula SiO_2. The four bonds of silicon are tetrahedrally oriented: silicon is sp^3 hybridized. Each bond of an atom of silicon is *sigma*-bonded to an oxygen atom that is in turn *sigma*-bonded to another silicon atom. The total grain is a three-dimensional covalent work. Each silicon atom is covalently bonded to four oxygen atoms. Each oxygen atom is covalently bonded to two silicon atoms. There are twice as many oxygen atoms as silicon atoms and the bonding extends throughout the grains. This type of solid is known as a **covalent** crystal.

Quartz is remarkably resistant to fracture, and when it does break, there are no preferred cleavage planes such as those found in **molecular** crystals and **ionic** crystals. Molecular crystal fractures occur between planes of molecules. Ionic crystal fractures occur between planes of ions. Covalent crystals have no fracture planes. The forces between molecules that hold molecules together in molecular crystals are much weaker than the electrostatic forces that hold ions together in ionic crystals. The forces between ions in ionic crystals are much weaker than the covalent bonds that hold atoms together in the covalent crystals of quartz. When quartz is fractured, covalent bonds are broken and the surfaces are curved. Such fractures are known as conchoidal (shell-like) fractures and are clearly evident on the surfaces of arrowheads and other Stone Age instruments shaped by chipping flint and obsidian, both forms of SiO_2. Large crystals of quartz formed in geological processes by crystallization from molten SiO_2 are frequently exhibited in museums and displayed at roadside stands in areas where these crystals are found. Pure silicon dioxide is colorless and can be as transparent as high-quality crystal glass. One form of natural quartz is known as rock crystal, others as agate, rose quartz, and opal. Quartz is our most abundant mineral; hence the great amounts of quartz sand formed by weathering. Quartz is an essential component of all commercial glass.

Melting points below 300 °C are characteristic of covalent compounds such as ethanol and sucrose. Melting points above 300 °C are characteristic of ionic compounds such as sodium chloride and also characteristic of very high molecular weight covalent substances such as silicon dioxide.

Certainly this σ-bond would be a very polar bond: the electronegativity of sodium is listed as 0.8 and that of chlorine as 3.5. This unusually large difference is consistent with the two electrons in the σ-orbital spending most of their time in the vicinity of the chlorine nucleus and thus creating a marked separation in charge. The limiting situation to this unequal distribution would be to consider the pair of electrons as primarily occupying the region around the chlorine nucleus. In this case, the chlorine atom would achieve essentially a full net charge of 1− and the sodium atom would achieve essentially a full net charge of 1+. The chlorine atom would have essentially a total of 18 electrons and a nuclear charge of 17+, and the sodium atom would have essentially a total of 10 electrons and a nuclear charge of 11+—a very polar σ-bond. This structure, predicted by the valence bond approach, corresponds closely to the "ion pair" unit for which there is experimental evidence in the high-energy vapor phase of sodium chloride. However, the structure does not directly relate to a crystal, in which every sodium nucleus is surrounded by six equidistant chlorine nuclei and every chlorine nucleus is surrounded by six equidistant sodium nuclei.

Electron Transfer Approach

The more usual approach to the structure of sodium chloride is to assume that the sodium ion has one less electron than the sodium atom and thus the charge of 1+ (11 protons and 10 electrons):

	electron structure	**Lewis structure**
sodium atom, Na	$1s^2\ 2s^2\ 2p^6\ 3s^1$	Na·
sodium ion, Na$^{(1+)}$	$1s^2\ 2s^2\ 2p^6$	(Na)$^{(1+)}$ or (:N̈a:)$^{(1+)}$

and that the chloride ion has one more electron than the chlorine atom and hence the charge of 1− (17 protons and 18 electrons):

	electron structure	**Lewis structure**
chlorine atom, Cl	$1s^2\ 2s^2\ 2p^6\ 3s^2\ 3p^5$	:C̈l·
chloride ion, Cl$^{(1-)}$	$1s^2\ 2s^2\ 2p^6\ 3s^2\ 3p^6$	(:C̈l:)$^{(1-)}$

The Lewis structures of the ions are consistent with the magic octet of the Lewis approach. The octet of electrons shown in the Lewis structure for the sodium ion is the octet of electrons of quantum level 2. The octet of electrons shown in the Lewis structure for the chloride ion is the octet of electrons of quantum level 3. The sodium ion has the electron structure of the neon atom but the sodium ion is quite different from the neon atom. The sodium ion has a charge of 1+; the neon atom has no charge. The nucleus of the sodium ion has 11 protons; the nucleus of the neon atom has 10 protons.

Sizes of Ions and Atoms

The radii of atoms and ions have been extensively studied through a variety of experimental measurements. The values determined depend to some degree on the experimental method used. The generally accepted values are given in Table 10-5 for the alkali metal atoms, the alkali metal ions, the alkaline earth atoms, the alkaline earth ions, the halogen atoms, the halide ions, and the noble gas atoms.

Note that the radius of the sodium ion is small with respect to the radius of the sodium atom and that the radius of the chloride ion is large with respect to the radius of the chlorine atom. In each case, the difference is rationalized in terms of the different numbers of electrons in the electron atmospheres of the atom and the ion of the element. Both the atom and the ion have the nuclear charge characteristic of the element.

Table 10-5

RADII OF ATOMS AND IONS: ALKALI METALS, ALKALINE EARTHS, HALOGENS, AND INERT GASES (IN NANOMETERS)

period	alkali metals		alkaline earths		halogens		inert gases	
1							helium	
							He	0.09
2	lithium		beryllium		fluorine		neon	
	Li	0.15	Be	0.11	F	0.07	Ne	0.11
	Li^{1+}	0.06	Be^{2+}	0.03	F^{1-}	0.14		
3	sodium		magnesium		chlorine		argon	
	Na	0.19	Mg	0.16	Cl	0.10	Ar	0.15
	Na^{1+}	0.10	Mg^{2+}	0.07	Cl^{1-}	0.18		
4	potassium		calcium		bromine		krypton	
	K	0.23	Ca	0.20	Br	0.11	Kr	0.17
	K^{1+}	0.13	Ca^{2+}	0.10	Br^{1-}	0.20		
5	rubidium		strontium		iodine		xenon	
	Rb	0.25	Sr	0.21	I	0.13	Xe	0.19
	Rb^{1+}	0.15	Sr^{2+}	0.11	I^{1-}	0.22		
6	cesium		barium				radon	
	Cs	0.27	Ba	0.22			Rn	0.22
	Cs^{1+}	0.17	Ba^{2+}	0.14				

	sodium atom	sodium ion	chlorine atom	chloride ion
protons	11	11	17	17
electrons	11	10	17	18
radius	0.19 nm	0.10 nm	0.10 nm	0.18 nm

The sodium atom has more electrons than the sodium ion and, consequently, the volume of the atom is greater than the volume of the ion. The chloride ion has more electrons than the chlorine atom and, consequently, the volume of the ion is greater than the volume of the atom. The radius of the chloride ion is 1.8 times the radius of the sodium ion. The relative sizes seem even more startling when spheres of these relative sizes are considered. The spheres drawn below to represent the two ions have the correct relative dimensions. See also Figure 10-2.

sodium ion, Na^{1+}

chloride ion, Cl^{1-}

In the undisturbed ion (such as an isolated ion), the center of this net positive or negative charge would be at the center of the atom and consequently coincident with the center of the nucleus. The properties of the ions can be treated in terms of the net charges of the ions. The small radius of the sodium ion allows its effective charge to closely approach the electron atmosphere of the chloride ion and, by its presence, distort the outer electron atmosphere of the chloride ion, pulling the electrons towards the sodium ion. This outer atmosphere is primarily the electrons of the 3*p*-orbitals. The distortion of the electron atmosphere of the sodium ion by the very large chloride ion would be much less pronounced. Under these distorted conditions, the centers of the effective charges are no longer coincident with the nuclei, and the situation approaches the formation of a *sigma*-bond, a very polar *sigma*-bond.

Percent Ionic Character of Covalent Bonds

With different atoms of similar electronegativities, there is considerable advantage to considering bonds as polar *sigma*-bonds, and such a bond is described as having a certain degree of ionic character. In this scheme, the sodium chloride, NaCl, bond is considered to have 70% ionic character; hydrogen fluoride, HF, 55% ionic character; and hydrogen chloride, HCl, 20% ionic character. Sodium chloride is usually considered as an ionic compound, hydrogen chloride as a polar covalent compound. The bond in a homonuclear diatomic molecule, such as diatomic chlorine, Cl_2, has, of course, 0% ionic character.

There is a tendency to worry about how and when the chlorine atom acquired the additional electron to become the chloride ion, how and when the sodium atom gave away the electron to become the sodium ion. These are interesting questions, but from the standpoint of the sodium chloride they are not very relevant. Most sodium atoms became sodium ions and most chlorine atoms became chloride ions some time in the far distant past of geological time. It is quite possible that most sodium ions and many chloride ions have never been sodium atoms and chlorine atoms. Ions of some elements have as much reality as atoms. The questions that have more relevance are: How can the sodium ion be given an electron to obtain an atom of sodium? and How can the electron be taken away from the chloride ion to obtain an atom of chlorine? These are exactly the reactions that occur at the electrodes in testing the electrical conductivity of molten sodium chloride.

Electrolysis of Molten Sodium Chloride

The sodium ions move toward the negative electrode. The negative electrode is negative due to the excess electrons supplied by the source of electrical potential. The sodium ion accepts an electron and thus removes an electron from the electrode:

$$Na^{1+} + e^{1-} \longrightarrow Na$$

sodium ion	electron obtained from the electrode	sodium atom

The chloride ions move toward the positive electrode. The positive electrode is positive due to the removal of electrons by the source of the electrical potential. The chloride ion gives up an electron to the electrode and, in the process, becomes a chlorine atom:

$$Cl^{1-} \longrightarrow Cl + e^{1-}$$

chloride ion	chlorine atom	electron transferred to the electrode

The net effect is that an electron has been transferred from the negative electrode to the positive electrode.

Both of these equations can be read in terms of moles:

$$\text{one mole sodium ions} + \text{one mole electrons} \longrightarrow \text{one mole sodium atoms}$$

$$\text{one mole chloride ions} \longrightarrow \text{one mole chlorine atoms} + \text{one mole electrons}$$

A mole of electrons is 6.023×10^{23} electrons and is frequently called a **faraday** in honor of Michael Faraday, an early investigator of electrical phenomena.

Chlorine atoms react to form diatomic molecules:

$$2\,Cl \longrightarrow Cl_2$$

chlorine atoms	diatomic molecule of chlorine

and in this sense diatomic chlorine is the ultimate product, along with metallic sodium, of the electrolysis of molten sodium chloride. The electrolysis of an aqueous solution of sodium chloride does not give metallic sodium. Instead, the reaction at the negative electrode involves water and the products are hydrogen gas, H_2, and hydroxide ions, OH^{1-}. Chlorine is the primary product at the positive electrode when concentrated solutions of sodium chloride are used. Expect more information about electron transfer reactions in Chapter 16.

Comparison of Ions and Atoms

Table 10-5 presents experimental information about the atoms and ions of the alkali metals, the alkaline earths, and the halogens that can be correlated with the position in the Periodic Table and the distribution of electrons in atoms and ions. Many of these correlations are self-evident and easily rationalized.

- The atoms of the alkali metals increase in radii from lithium to cesium: the atoms are quite similar in nature but increase in mass and in numbers of electrons.
- The ions of the alkali metals all have a charge of 1+ : in each case, the atom attains a noble gas structure by losing an *s* electron.
- The ions of the alkali metals increase in radii from lithium ion to cesium ion.
- An alkali metal ion has a smaller radius than the corresponding alkali metal atom: both have the same nuclear charge but the corresponding ion has fewer electrons.
- The atoms of alkaline earths increase in radii from beryllium to radium.
- The ions of the alkaline earths all have a charge of 2+.
- The ions of the alkaline earths increase in radii from beryllium ion to radium ion.
- An alkaline earth ion has a smaller radius than the corresponding alkali metal ion: both have the same number of electrons but the alkaline earth ion has the greater nuclear charge.

A very parallel set of statements could be made with respect to the atoms of the halogen family, the halide ions of the halogen family, and the atoms of the noble gas family. The only differences are that all of the halide ions have a charge of 1− and that a halide ion has a larger radius than the atom of the corresponding noble gas. The last is merely the consequence of the additional electron having to be accommodated in the ion without an increase in charge on the nucleus.

Isoelectronic Atoms and Ions

The oxide ion, O^{2-}, the fluoride ion, F^{1-}, the neon atom, Ne, the sodium ion, Na^{1+}, and the magnesium ion, Mg^{2+}, would seem to be a strange bag of particles but they have one characteristic in common. The electron atmosphere is in each case made up of ten electrons. These five particles are isoelectronic. Since the nuclear charge increases from 8+ for the oxide ion to 12+ for the magnesium ion, it is no surprise that their radii decrease throughout the series. The chemical properties of these five isoelectronic particles—four ions and one atom—are very different. The sulfide ion, S^{2-}, the chloride ion, Cl^{1-}, the argon atom, Ar, the potassium ion, K^{1+}, and the calcium ion, Ca^{2+}, constitute another group of **isoelectronic** particles, particles that contain the same number of electrons.

Scramble Exercises

1. Review: Starting with the atomic number for barium, $Z = 56$, write out the full notation for the distribution of electrons in a ground state atom of barium.

2. Write down the empirical formulas for several ionic compounds, such as barium chloride, lithium sulfate, potassium phosphate, and magnesium phospate. See Table 10-3 for the formulas of polyatomic anions. Always check that the net charge for the formula of the compound is zero.

3. Using Table 10-6 for the formulas of transition metal cations, write down the empirical formulas for the following ionic compounds: iron (III) chloride, iron (III) sulfate, chromium (III) sulfate, iron (II) phosphate, and chromium (III) phosphate.

4. Which of the following four compounds—LiF, LiI, CsF, and CsI—would be expected to have the highest percent ionic character for the bond between the two atoms? Rationalize your choice.

5. Rationalize the relative diameters of the bromine atom and the bromide ion.

6. Rationalize the relative diameters of the rubidium ion and the bromide ion.

7. The dielectric constant of ethanol is 24, the dielectric constant of water 80. Use this information to rationalize (a) the experimental fact that sodium chloride is less soluble in ethanol than in water and (b) the experimental fact that the addition of ethanol to a saturated solution of sodium chloride in water leads to the precipitation (the formation) of crystals of sodium chloride.

   Ethanol and water mix in all proportions, and the dielectric constant of ethanol–water mixtures lies somewhere between 24 and 80—depending upon the relative amounts of ethanol and water in the mixture.

8. In the electrolysis of molten sodium chloride, the products are sodium atoms and chlorine atoms. See the text for the electrode reactions. How many electrons would be involved, at the negative electrode, in the formation of 0.0100 moles of sodium atoms? At the same time, how many moles of chlorine atoms would be formed at the positive electrode? How many moles of diatomic chlorine molecules, Cl_2, would be formed at the positive electrode? How many moles of sodium chloride would be converted to the elements? How many grams of sodium chloride converted to the elements?

9. The melting point reported in Table 10-1 for lithium nitrate, 264 °C, is exceptionally low for an ionic compound. Endeavor to suggest a plausible explanation for this very low value.

Additional exercises for Chapter 10 are given in the Appendix.

A Look Ahead

Most of the elements are metals. In this chapter we have considered the alkali metals and the alkaline earths at the extreme left of the Periodic Table. Many substances you recognize as metals are transition elements, the adjacent ten columns of elements of the Periodic Table. These are the elements that we associate with *d*-orbitals, and much of their chemistry is attributed to electrons in *d*-orbitals in the same sense that the chemistry of the alkali metals and the alkaline earths is attributed to electrons in *s*-orbitals and the chemistry of the nonmetals attributed to electrons in *p*-orbitals.

Chemically speaking, the transition elements are much more versatile than the alkali metals and the alkaline earths. They form cations and also complex anions. For example, chromium forms the cation chromium(III), Cr^{3+}, and the complex anions chromate ion, CrO_4^{2-}, and the dichromate ion, $Cr_2O_7^{2-}$. Many of these elements form more than one cation. For example, iron forms both the iron(II) ion, Fe^{2+}, and the iron(III) ion, Fe^{3+}. The symbol Fe is derived from the Latin *ferrum*, for iron.

Many of the ions formed by the transition elements are colored—at least under some conditions. The aqueous solutions of all compounds of copper(II)—such as copper(II) sulfate, $CuSO_4$, and copper(II) chloride, $CuCl_2$—are blue or blue-green in color. Their crystalline hydrates, such as $CuSO_4 \cdot 5\ H_2O$ and $CuCl_2 \cdot 2\ H_2O$, are also blue and blue-green. Anhydrous copper(II) sulfate is a slightly gray-white solid. The blue color is associated with the ion $Cu(H_2O)_4^{2+}$—the copper(II) tetrahydrate ion, also called the tetraaquo copper(II) ion. The names and charges for a few cations of transition metals are given in Table 10-6.

Table 10-6

SELECTED TRANSITION METAL CATIONS

colorless ions		colored ions in aqueous solutions		
silver(I) ion	Ag^{1+}	iron(II) ion	Fe^{2+}	pale green
lead(II) ion	Pb^{2+}	iron(III) ion	Fe^{3+}	yellow-brown
zinc(II) ion	Zn^{2+}	copper(II) ion	Cu^{2+}	blue
mercury(II) ion	Hg^{2+}	nickel(II) ion	Ni^{2+}	green
		cobalt(II) ion	Co^{2+}	pink
		chromium(III) ion	Cr^{3+}	green

11

HYDROGEN AND OTHER GASES

Chapter 11

Gases are conceptually the simplest of the phases, even though they seem to be elusive from an experimental point of view. Almost all gases are colorless and these colorless gas samples must be confined in containers. It was the rationalization of the results of early experiments with gases that led to the basic concepts of atoms and molecules and to the evolution of a system of atomic weights, empirical formulas, molecular formulas, the concept of moles, and the concept of an absolute (kelvin) temperature. The historical development is a fascinating saga, but it is a meandering story of a search for consistent models that could rationalize a number of experimental observations. There were numerous false leads and great confusion in the use of terms as the vocabulary and the science developed simultaneously. In retrospect, we are inclined to marvel at the perseverance of individuals in the pursuit of understanding throughout the time period between the early study of pressure–volume relations by Boyle in the middle of the seventeenth century and the proposal by Avogadro early in the nineteenth century that equal volumes of gases at the same temperature and pressure contain the same number of molecules. It was this insight of Avogadro that led to the concept of the mole and the eventual experimental evaluation of the number 6.02×10^{23}, now known as Avogadro's number. With this understanding of the nature of gases, the elucidation of chemical change was off and running.

In this chapter we shall consider some of the physical properties of hydrogen gas and then compare the properties of other gases with those of hydrogen. A major interest in our exploration of gases is the evidence these properties give of attractive forces between molecules in the gaseous phase.

The gas phase is the natural habitat of small molecules at room conditions. Gases seem elusive to capture and measure but they can be confined to tanks, balloons, and cylinders with movable pistons. In the laboratory, leakproof glass equipment is easily constructed for work with gases from extremely low pressures to pressures that do not greatly exceed the pressure of the atmosphere. Metal equipment is required for work at high pressures and is frequently used for work at low pressures. With glass equipment, liquid mercury is frequently used to confine the gas and to move the gas from one part of the equipment to another. The presence of the liquid mercury, of course, introduces mercury vapor into the system by the evaporation of liquid mercury. The highest pressure mercury vapor can attain at room temperature is 2×10^{-4} torr. This is a very small pressure. The average atmospheric pressure at sea level is about 760 torr.

In this chapter, we shall explore the properties of hydrogen gas, H_2, and then expand the exploration to other gases. The text is brief; the background sections are numerous and, in some cases, long. It may be useful to read, or at least scan, all of the background sections before proceeding to the text and then returning to the background material as needed.

It can be shown experimentally that the pressure, *P*, of a sample of hydrogen gas is dependent upon

- the number of molecules in the sample,
- the volume to which the sample is confined, and
- the temperature of the gas.

11-1 *Background Information*

PRESSURE

Pressure is defined as the quotient of force divided by the area on which that force is acting:

$$\text{pressure} = \frac{\text{force}}{\text{area}}$$

Snowshoes distribute the force (largely gravitational force) over a greater area than ordinary footwear does. A two-hundred-pound woman wearing spike heels is in violation of some city building codes that restrict the magnitude of the force on a unit area. She is also a menace to the airtight but lightweight metal floors of pressurized cabins in commercial airplanes. Carpets in aircraft serve more than a decorative function.

The units of pressure can be any unit of force divided by any unit of area. In commerce in this country, pounds per square inch is frequently used. Note that pounds are being used here as a unit of force, not as a unit of mass. A force of one pound simply means the force of gravity acting on a mass of one pound in a specified gravitational field.

Scientists frequently measure the pressures of gases and liquids in terms of columns of liquid, particularly mercury: witness those bright, shiny columns of mercury that show up in the measurement of blood pressure. Pressure measurements using columns of liquids are based on the response of liquids to the gravitational force. All liquids flow towards the center of the earth to the extent permitted by their containers: liquids seek their own level. But other forces, such as those arising from the pressure of gases, may operate against the gravitational forces and two surfaces of the same liquid may stand at different levels.

The level of the liquid in a wide-diameter drinking straw is essentially the same as the level of the liquid in the glass until the pressure of the gas in the straw is either increased by blowing more air into the straw from the lungs or reduced by sucking out, swallowing, some of the air. The vertical distance between the level of the liquid in the glass and the level of the liquid in the straw is a measure of the difference of the gas pressures on the two liquid surfaces, one in the glass and one in the straw.

The sketches below represent two U-tubes. Both arms of both U-tubes are sealed at the top. Both U-tubes contain a liquid. Both arms of both U-tubes contain gas.

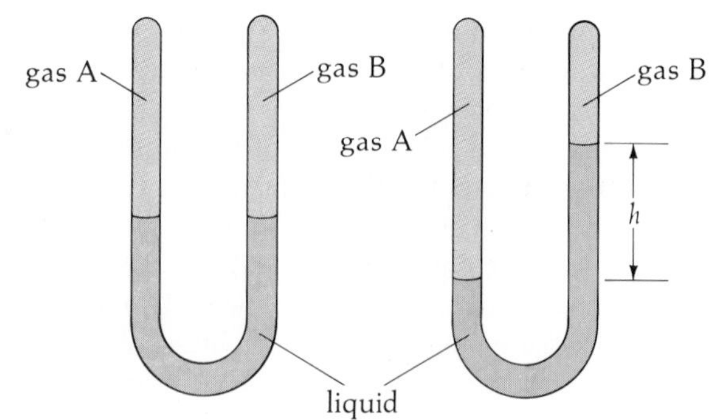

The pressure of hydrogen gas is said to be dependent upon the three independent variables: number of molecules, volume, and temperature. Each of these three variables can be independently varied. In varying any one of the three, the value of the pressure will be changed.

Relation of Pressure to Volume

The dependency of the pressure of hydrogen gas on volume alone can be studied by confining a sample of hydrogen gas so that no molecules can escape and carrying out the experiment at a constant temperature that can be achieved simply by working in a thermostated room or immersing the equipment in a thermostated liquid. It is no surprise that the pressure of the hydrogen gas increases as the sample of the gas

---

In the U-tube to the left, the vertical difference in height of the two liquid surfaces is zero: gas A and gas B are at the same pressure. In the other U-tube, the vertical difference in the levels of the two liquid surfaces, h, is a measure of the difference in pressure: gas A has the higher pressure.

If there is no gas in one arm, the vertical distance between the liquid levels in the two arms is a measure of pressure of the gas in the other arm. If this arm is not sealed but open to the atmosphere and the liquid is mercury, the U-tube constitutes a mercury barometer. Another, more common design of a mercury barometer consists of a straight tube, sealed at one end, standing open-end down in a dish of mercury. It is, of course, necessary to have no air in the tube. (Gratuitous Information 11-1) Mercury is the chosen liquid due to its very high density (13.6 g mL^{-1}), its low volatility (vapor pressure of 2×10^{-4} torr at room temperature), and its inertness to chemical change.

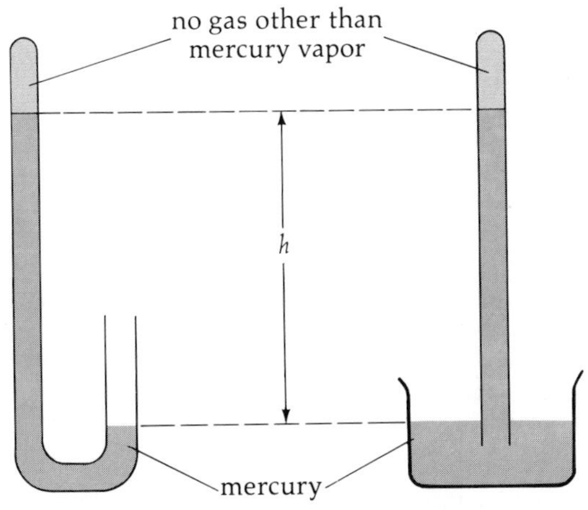

two forms of mercury barometers

The unit of pressure defined in terms of a column of mercury is called the "torr" in honor of an assistant of Galileo, Evangelista Torricelli, who first constructed barometers and lugged them up a mountainside to measure atmospheric pressure at higher elevations. By definition, the **torr** is the pressure required to support a column of mercury to a height of one millimeter. The complete definition also specifies the gravitational field and the temperature of the mercury.

At sea level, the normal variation of the height of the mercury in a barometer is within the range of 750 to 770 mm. The column of water in a barometer constructed with water (density of 1.0 g mL^{-1}) would stand at heights 13.6 times the heights of the column of mercury. Another unit of pressure is defined in terms of a **standard atmosphere.** The "atmosphere," a unit of pressure, is defined to be exactly 760 torr.

DIRECT PROPORTIONS

The simplest relation between two variables is a direct (first power) proportion. The cost of an order of ham sandwiches is directly proportional to the number of sandwiches. Written with the proportionality symbol "$\propto$," this statement is

$$\text{cost} \propto \text{number of sandwiches}$$

With the insertion of the cost per sandwich, this statement becomes an equality:

$$\text{cost} = \left[\frac{\text{cost}}{\text{per sandwich}}\right] \times \left[\frac{\text{number of}}{\text{sandwiches}}\right]$$

For an order of eight sandwiches at $3.00 per sandwich,

$$\text{cost} = 3\,\frac{\text{dollars}}{\cancel{\text{sandwich}}} \times 8\,\cancel{\text{sandwiches}}$$
$$= 24 \text{ dollars}$$

If y is directly proportional to x, the insertion of a proportionality constant, k, in $y \propto x$ yields an equality: $y = kx$. In the case of the ham sandwiches, the price per sandwich (3 dollars/sandwich) is the proportionality constant k.

If the cost is plotted against the number of sandwiches, the line is straight, the line passes through the origin (0, 0), and the slope of the line is 3 dollars/sandwich. (The slope is any vertical displacement divided by the corresponding horizontal displacement.)

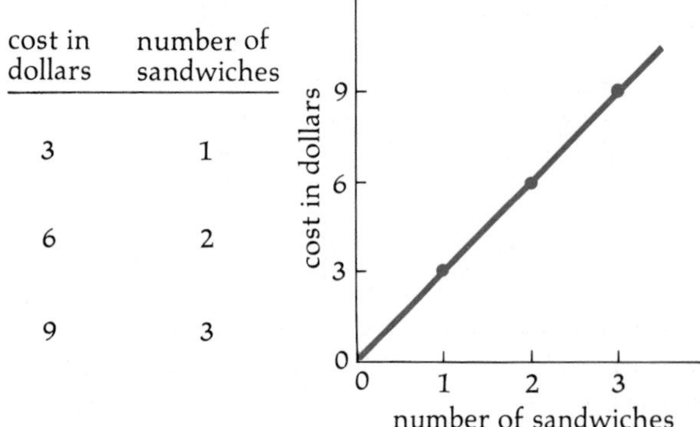

cost in dollars	number of sandwiches
3	1
6	2
9	3

If y is directly proportional to x, a plot of y vs x yields a straight line that passes through the origin and has a slope of k.

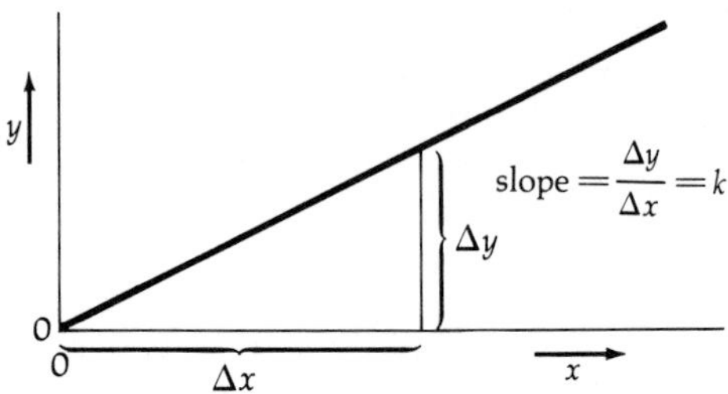

The proportionality constant, the slope, frequently has a physical significance that can be easily identified. For example, the mass of a liquid or solid is directly proportional to its volume:

$$\text{mass} \propto \text{volume}, \quad \text{and}$$
$$\text{mass} = k \text{ volume}$$

If the mass of a 10-milliliter sample of a liquid is 20 grams, then k must be equal to 2.0 g/mL:

$$20 \text{ g} = k \times 10 \text{ mL}$$
$$k = \frac{20 \text{ g}}{10 \text{ mL}} = 2.0 \text{ g/mL}$$

and k is nothing more than the mass in grams of a one-milliliter sample of the liquid. The proportionality constant, k, in

$$\text{mass} = k \text{ volume}$$

is known as density:

$$k = \frac{\text{mass}}{\text{volume}} = \text{density}$$

The units of mass and the units of volume are entirely a matter of choice. The numerical value of density depends upon the units chosen and the choice of units must be given with the numerical value of density.

In the background sections dealing with length, volume, force, and energy, we have used equations for direct proportions to convert units of length to other units of length, etc. For example:

$$\left[\begin{array}{c}\text{length}\\ \text{in meters}\end{array}\right] = \left[1 \times 10^{-3}\ \frac{\text{meters}}{\text{millimeters}}\right]\left[\begin{array}{c}\text{length in}\\ \text{millimeters}\end{array}\right]$$

In the above expressions the proportionality constant is

$$1 \times 10^{-3}\ \frac{\text{meters}}{\text{millimeters}}$$

Note that in the right hand side of this equation millimeters cancel millimeters and the units on both sides of the equation reduce to meters. This is as it must be. In an equality, the units on both sides of this equation must be the same.

In the equations for the direct proportion

$$\left[\begin{array}{c}\text{energy}\\ \text{in joules}\end{array}\right] = \left[4.18\ \frac{\text{joules}}{\text{calories}}\right]\left[\begin{array}{c}\text{energy}\\ \text{in calories}\end{array}\right]$$

the proportionality constant is

$$4.18\ \frac{\text{joules}}{\text{calories}}$$

and the units reduce to joules on both sides of the equation. This is the equation that defines the calorie.

Inside the back cover of this book is a listing of proportionality constants frequently used in chemical studies.

Gratuitous Information 11-1

CONSTRUCTION OF AN APPROXIMATE BAROMETER

To set up an approximate barometer is rather fun. Remove all jewelry and either wear thin plastic gloves or be very careful to start with clean, dry hands and wash your hands thoroughly as soon as the barometer is complete. Almost fill a clean, dry tube with clean, dry mercury. (Safety Note: Mercury is toxic, and care should be taken to keep it confined to closed containers whenever possible and to work over a tray to keep the mercury from getting scattered about. Mercury should not be heated except under very carefully specified conditions and then only by someone who fully understands the hazards involved.) Place a thumb over the open end and invert to the degree required for the air bubble to run the length of the tube to the closed end. (The large bubble collects small pockets of air from along the surface of the glass.) Completely fill tube with mercury and, with the assistance of your thumb, invert the tube, immerse the end of tube and thumb in dish of mercury, withdraw thumb, and clamp the tube in the vertical position.

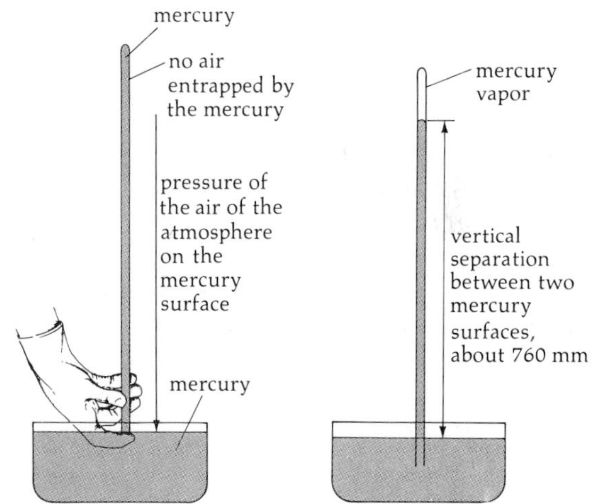

The product is not a precision barometer since there will be some residual air. To get some idea of the amount of air remaining in the tube, tip the column to reduce the maximum vertical height, open end still immersed in the pool of mercury. If the mercury can be brought to the top of the tube, the exclusion of air has been pretty successful. If the tube is tipped from the vertical position rather quickly, the mercury approaches the end of the tube rapidly. If air is present, the mercury bounces back on this cushion of air. A sharp metallic ring, rather like a light metal hammer striking a metal anvil, as the mercury and glass collide indicates very little air, but be careful—don't break the tube! Since the atmosphere supports a column of mercury to a height of about 760 mm at sea level, the length of the glass tube should be a little more than this.

Columns of mercury are inoperable as a means of measuring pressure in a space capsule. Why?

INVERSE PROPORTIONS

Another frequently encountered relation between two variables is an inverse proportion. For example, the time required for ants to remove some cake crumbs is inversely proportional to the number of ants:

$$\text{time} \propto \frac{1}{\text{ants}}$$

In making this statement, it is assumed that there are no sociological problems in the ant colony and that each ant does its fair share of the work. It is also assumed that there are no traffic problems and that the presence of more ants does not interfere with the progress of the work being done by any one ant. It is also assumed that only the ants are removing the cake crumbs.

Here again, this proportional relation can be stated as an equality by the introduction of a proportionality constant k:

$$\text{time} = k \times \frac{1}{\text{ants}} = \frac{k}{\text{ants}}$$

The equation can, of course, also be written in the form

$$\text{ants} \times \text{time} = k$$

If it is known how many minutes it takes a known number of ants to remove the cake crumbs, then the numerical value of k is known. For example, if it takes ten ants thirty minutes to remove the crumbs,

$$k = 10 \text{ ants} \times 30 \text{ minutes} = 300 \text{ ant minutes}$$

and a table of values could be set up to display this relation graphically.

number of ants	required time (minutes)	reciprocal ants: $\frac{1}{\text{ants}}$ or ants^{-1}
1	300	1.00
3	100	0.33
5	60	0.20
10	30	0.10
15	20	0.067
20	15	0.050
30	10	0.033
60	5	0.017
100	3	0.010
300	1	0.0033

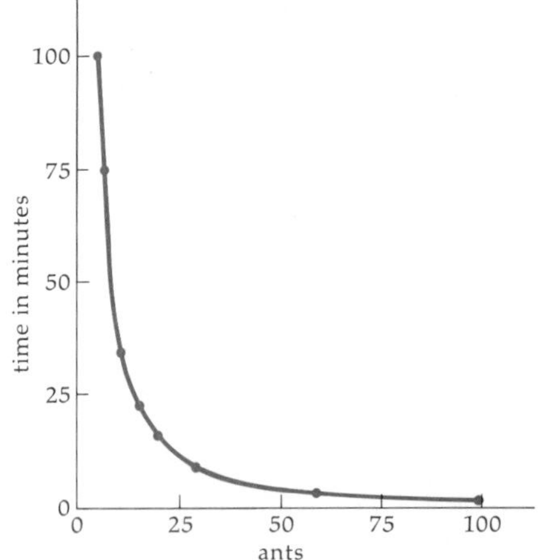

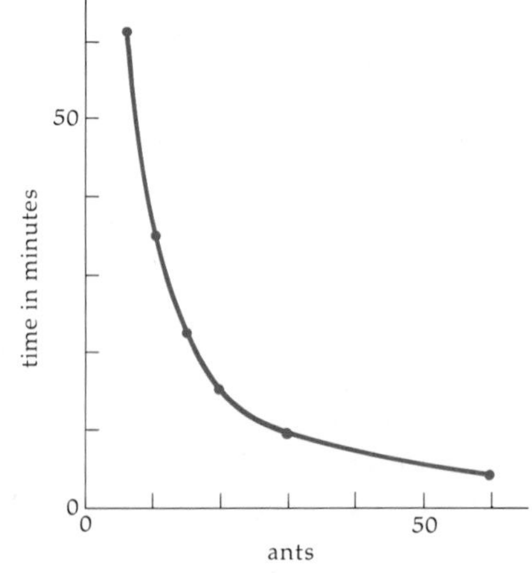

Although these graphs may seem to be quite different, they are in fact the same values plotted on different scales. Curves such as these have a characteristic shape and emphasize the inverse nature of the relation between the number of ants and the time required to do the job. Scientists, however, have a strong preference for straight lines, particularly straight lines that go through the origin, and would prefer to plot time against 1/ants (or to plot ants against 1/time).

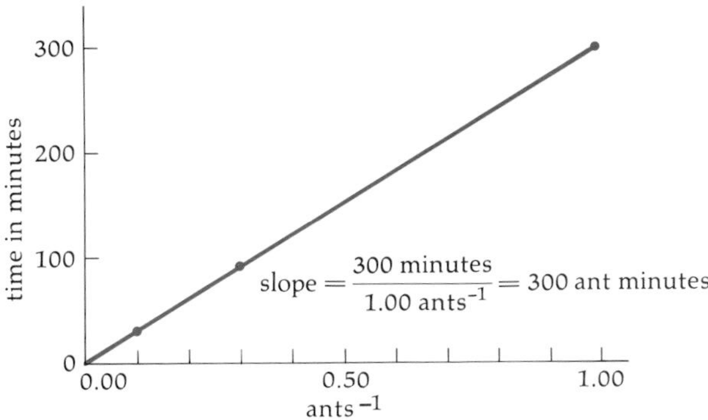

This is a neat trick. By an honest manipulation, the inverse relation between time and ants has been expressed as a direct proportion between time and (1/ants). This is a useful way to express the relation, although it must be admitted that a discussion of reciprocal ants (ants^{-1}) does seem a bit strange. The units of k, ant minutes, are comparable to the much more familiar unit man hours (person hours?).

Inverse proportion relations are extremely common: the number of sandwiches that can be prepared from a jar of peanut butter is inversely proportional to the amount of filling placed in each sandwich; the number of candy bars that can be purchased with a fixed sum of money is inversely proportional to the cost per candy bar; the length of time the world's petroleum resources will last is inversely proportional to the number of barrels used per day. Note that, in inverse proportions, the relation between the two variables is restrictive: as one increases, the other must decrease at the same rate.

In direct proportions, the relation between two variables is expansive: as one increases, the other must increase at the same rate.

is pushed into a smaller volume or that the pressure of the hydrogen gas decreases as the sample of the gas is allowed to occupy a larger volume. This is knowledge of common experience with gases to anyone who has pushed on the piston of a bicycle pump without letting gas escape from the pump. Measurements show that there is a very definite and simple relation between the pressure of the sample of hydrogen and the volume. If the volume is doubled, the pressure of the hydrogen is one-half the original value. If the volume is reduced to one-third its original value, the pressure is increased to three times its original value. This is an inverse relation: as one increases, the other decreases. But it is more than just an inverse relation. It is an inverse (first power) proportion and can be represented by the expresssion

$$\text{pressure} \propto \frac{1}{\text{volume}}$$

This expression is read: Pressure is inversely proportional to volume. It can also be read: Pressure is directly proportional to the fraction 1/volume or The pressure is directly proportional to the reciprocal of volume. The phrase "first power" indicates no exponents other than "1".

Relation of Pressure to Temperature

The relation of the pressure of the hydrogen gas to temperature can similarly be investigated by confining the sample of hydrogen gas to some arbitrary volume and simply changing the temperature of the room in which the work is done. Again, it is no surprise that an increase in temperature is accompanied by an increase in pressure. "Don't heat a closed container." Here again, there is a simple mathematical relation between the pressure of the hydrogen gas and the temperature of the hydrogen gas if the temperature is measured from an absolute zero of temperature in the same sense that all volumes are measured from an absolute zero for volume and all pressures are measured from an absolute zero of pressure. Temperatures on this absolute temperature scale, T, are known as **kelvin** temperatures and are related to temperature on the Celsius scale by the equation

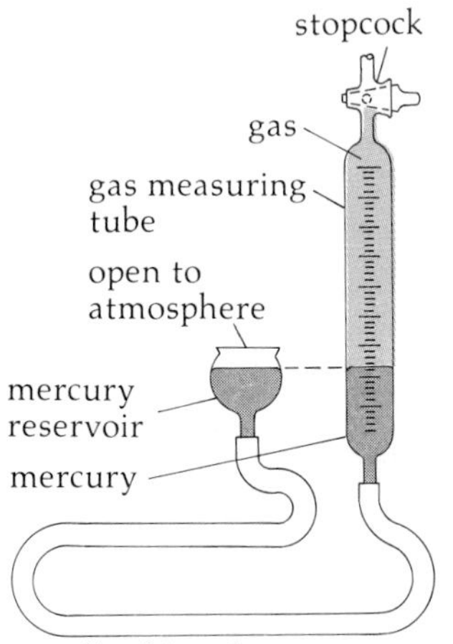

pressure of gas equal to the atmospheric pressure

$$\text{temperature kelvin} = T_K = 273.15 + \text{temperature Celsius}$$

In terms of temperature expressed on this kelvin temperature scale, the pressure of a sample of hydrogen gas is found to be directly proportional to the first power of the temperature. If the kelvin temperature of the sample of hydrogen gas is doubled, the pressure of that sample of gas is doubled providing the volume has remained unchanged. This is a direct relation: as one increases, the other increases. But it is more specific than just a direct relation. It is a direct (first power) proportion and can be represented by

$$\text{pressure} \propto \text{temperature kelvin}$$

This expression is read: Pressure is directly proportional to temperature on the kelvin scale.

Relation of Pressure to Number of Molecules

The relation of the pressure of hydrogen gas to the number of molecules of hydrogen can be investigated by simply adding hydrogen gas to a vessel of constant volume in a room at constant temperature. It is common experience that the pressure of the gas increases if more gas is added and that the pressure decreases if gas escapes. Measurements show that the relation of pressure to the number of molecules of hydrogen is a direct (first power) proportion:

$$\text{pressure} \propto \text{number of molecules}$$

Pressure is directly proportional to the number of molecules. The number of molecules involved in any sensible sample of hydrogen gas is so large that it is customary to express the number of molecules in terms of moles of molecules. The pressure of the hydrogen gas is directly proportional to the first power of the number of moles of hydrogen gas confined to a given volume at a given temperature:

$$\text{pressure} \propto \text{moles of hydrogen}$$

The number of moles of hydrogen gas, H_2, in any experiment is, of course, the mass of the sample in grams divided by 2.02 grams mole^{-1}.

The separate relations of the pressure of hydrogen gas to each of the three variables can be combined in a single expression:

$$\left(\frac{\text{pressure}}{\text{hydrogen gas}}\right) \propto \left(\frac{1}{\text{volume}}\right) \times \left(\frac{\text{temperature}}{\text{kelvin}}\right) \times \left(\frac{\text{moles}}{\text{hydrogen gas}}\right)$$

$$P \propto \frac{1}{V} \times T \times n$$

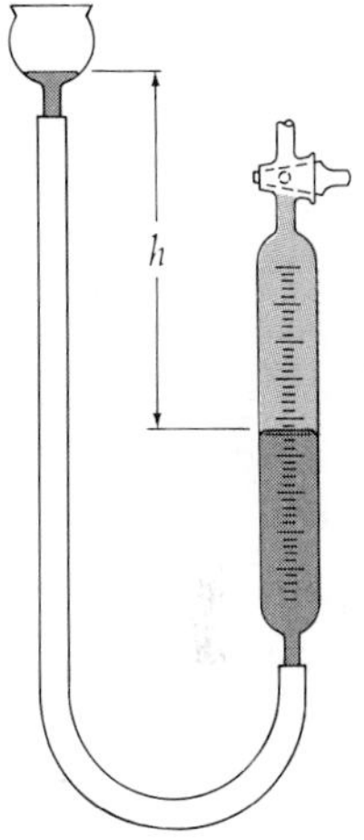

pressure of gas greater than the atmospheric pressure

The pressure of hydrogen gas is proportional to the product of the reciprocal of the volume of hydrogen gas, the kelvin temperature of the hydrogen gas, and the number of moles of hydrogen gas. The validity of this expression can be demonstrated by showing that each of the three separate expressions is encompassed in this statement. For example, if temperature and number of moles are constant, they are no longer variables and the expression reduces to

$$\text{pressure} \propto \frac{1}{\text{volume}}$$

Similar tests can be made for the relations pressure $\propto$ temperature kelvin and pressure $\propto$ moles of hydrogen gas.

As with all proportionality statements, the general expression can be converted to an equality by the introduction of a proportionality constant. Any symbol could be used to represent this proportionality constant but it has become customary to use the symbol R:

$PV = nRT$

$$P = R \times \frac{1}{V} \times T \times n \quad \text{or} \quad PV = nRT$$

The value of the proportionality constant, R, must be determined experimentally. To determine a value for R, it is only necessary to measure the four quantities mass, pressure, temperature, and volume for one sample of hydrogen gas and then calculate R:

$$R = \frac{PV}{nT}$$

For example, an experiment that showed that a 0.0271-gram sample of hydrogen gas occupied a volume of 324 milliliters at 27 °C and a pressure of 1.02 atmospheres is adequate to calculate the value of R:

Evaluation of R for Hydrogen

$$n = \text{moles hydrogen gas} = 0.0271 \text{ grams} \div 2.02 \frac{\text{grams}}{\text{mole}} = 0.0134 \text{ mole}$$

$$T = \text{temperature kelvin} = 273 + 27 = 300 \text{ K}$$

$$P = \text{pressure} = 1.02 \text{ atmospheres}$$

$$V = \text{volume} = 324 \text{ milliliters} \quad \text{or} \quad 0.324 \text{ liters}$$

$$R = \frac{1.02 \text{ atm} \times 0.324 \text{ liters}}{0.0134 \text{ moles} \times 300 \text{ K}} = 0.0822 \frac{\text{atm liters}}{\text{moles kelvin}}$$

Using the data from this one experiment, the value of R is 0.0822 atm liters moles^{-1} kelvin^{-1}.

THE ABSOLUTE ZERO TEMPERATURE

The approach to a concept of a lower limit for temperature, an absolute zero temperature, through gas measurements is a fascinating story in itself. It is a comparatively easy experimental matter to confine a sample of gas to a container under conditions such that the pressure of the gas is maintained at a constant value and to investigate the dependency of the volume of that gas sample on the temperature. Such an investigation of a particular sample of hydrogen gas at a particular pressure, P_1, yields data of the type given in the table below. The graphical presentation of the data, volume in milliliters (y axis) versus temperature in degrees Celsius (x axis), displays the dependency of volume on temperature in an even more striking manner.

Temperature (° C)	Volume (mL)
−50	82
0	100
50	118
100	137
150	155
200	173

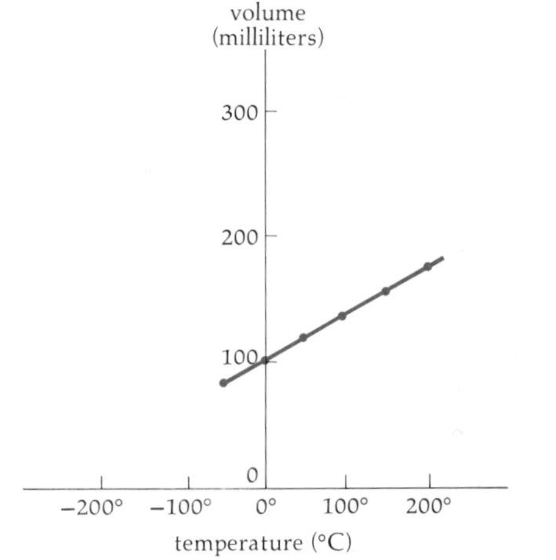

The volume increases as the temperature is increased, and a plot of the data gives a straight line (a linear relation). The relation is, however, not a direct proportion: the volume at 200 °C is not double the volume at 100 °C. The line does not go through the origin (0, 0).

Work with the same sample of gas at a pressure, P_2, which is maintained at a fixed higher value for this set of measurements, yields a similar set of data, but all values of volume for corresponding temperatures are lower and the graph of the data yields a straight line that lies below the first line.

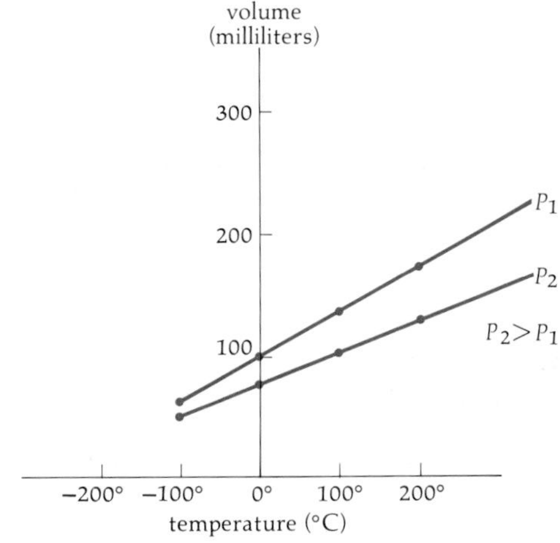

Investigations of many samples of gas (either hydrogen gas or some other gas or a mixture of gases) at the same pressure or some other pressure (the pressure must be kept at some low constant value for each set of measurements) yield similar sets of data and a linear graph is obtained for each sample at each pressure. Choices of sample sizes and pressures could lead to two or more lines being superimposed. In general, however, the positions of the lines present a fan pattern. The startling aspect of this graph is that all of the lines seem to be headed toward a common point. An extrapolation (extension) of these experimental lines to lower temperatures leads approximately to a common point of intersection at zero volume. In other words, all of these samples of gases have behaved at the pressure and temperature range studied as if there is a temperature at which all of these gas samples would cease to occupy volume. There

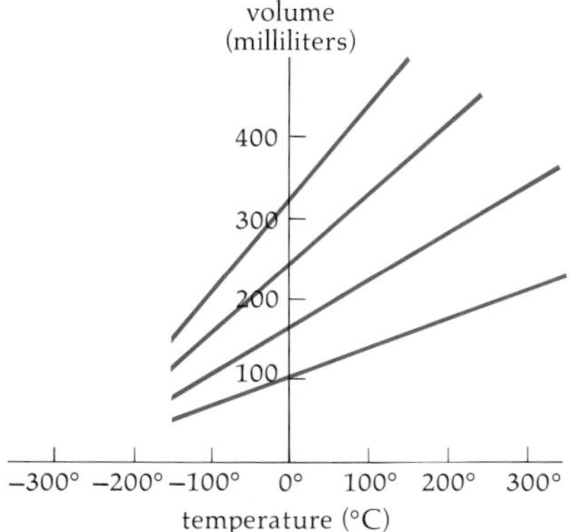

must be something very special about this temperature, and it would be much more meaningful to reckon all other temperatures from this unique temperature. The **absolute temperature scale,** also known as the **kelvin temperature scale,** takes this unique temperature as zero.

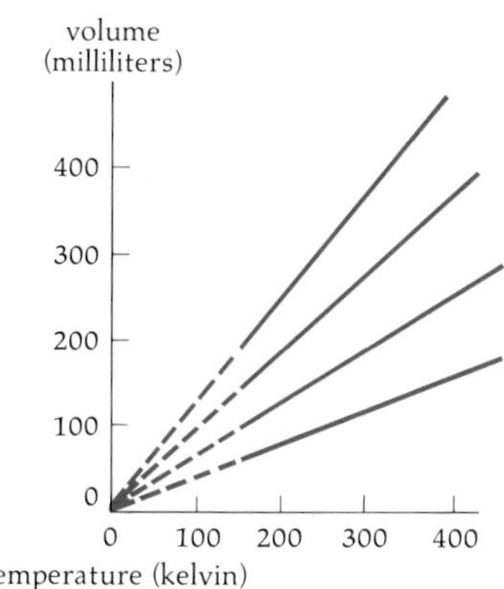

Extremely careful experimental work with gases at low pressures has located this unique temperature at $-273.15\ ^\circ C$ (Celsius). By defining this temperature as zero kelvin and continuing to use the Celsius degree as the unit of measurement, the kelvin scale is related to the Celsius scale by the relation

$$\text{kelvin temperature} = 273.15 + \text{degrees Celsius}$$
$$K = 273.15 + {}^\circ C$$

and the freezing point of water becomes 273.15 K, the boiling point 373.15 K. The units of the kelvin scale and the units of the Celsius scale are the same. The measurements simply start from different points. Note that, on the kelvin scale, the volumes of the samples of gas at constant pressure are directly proportional to the temperature: double the temperature and the volume is also doubled. The extrapolated lines pass through the origin (0 mL, 0 kelvin).

Two questions should be asked before leaving this development based on the behavior of gases. Do all gases exhibit the linear relation between volume and temperature under conditions of constant pressure and the rather limited temperature range covered? And what are the experimental facts concerning the behavior of gases in that low temperature range traversed by the extrapolation of gas behavior observed in the -50 to $200\ ^\circ C$ range? The answer to the first is a very definite negative. Many gases deviate markedly from the linear relation, particularly at the higher pressures. All gases deviate from the straight line behavior in the extrapolated region, particularly at the lower temperatures. The answer to the second is that all gases eventually condense to form liquids and even solids. This is a bit disappointing since it is exciting to contemplate the possibility of a sample shrinking to zero volume and then reappearing as the temperature is raised! This would have been the ultimate in disappearing acts.

Zero kelvin has not been attained experimentally and it is interesting to consider whether this absolute zero of temperature is attainable. There is no question that temperatures as low as 0.001 K have been attained, and research is underway to study the properties of materials near 0.000 01 K. The concept of an absolute zero temperature has importance to science that goes far beyond the properties of gases.

If we chose to use the volume in milliliters and the pressure in torr,

$$V = 324 \text{ mL}$$

$$P = 1.02 \cancel{\text{atm}} \times 760 \frac{\text{torr}}{\cancel{\text{atm}}} = 775 \text{ torr}$$

$$R = \frac{775 \text{ torr} \times 324 \text{ mL}}{0.0134 \text{ moles} \times 300 \text{ K}} = 6.25 \times 10^4 \frac{\text{torr mL}}{\text{moles kelvin}}$$

The proportionality constant has units and its numerical value is dependent upon the units chosen. In the development of the equation $PV = nRT$, commitment was made to moles and the kelvin temperature. The values given above are based on one set of data—not very precise data at that. A number of more precise measurements yields the following average values for R:

$$0.08205 \text{ liter atmosphere mole}^{-1} \text{ kelvin}^{-1}$$

$$6.240 \times 10^4 \text{ mL torr mole}^{-1} \text{ kelvin}^{-1}$$

Use of $PV = nRT$

This $PV = nRT$ relation is remarkably useful for hydrogen gas. To the degree that this relation is valid, if any three of P, V, n, and T are specified, the fourth can be calculated.

For example, the volume of one mole of hydrogen gas at 273 K and 1.00 atmosphere can be very readily calculated:

$$PV = nRT$$

$$V = \frac{nRT}{P}$$

If we choose to express R in liter atmosphere mole^{-1} kelvin^{-1}, then n must be expressed in moles ($n = 1.00$ mole), T must be expressed in kelvin ($T = 273$ kelvin), P must be expressed in atmospheres ($P = 1.00$ atm), and the calculations give the volume in liters.

$$V = \frac{1.00 \cancel{\text{mole}} \times 0.082 \dfrac{\text{liter atm}}{\cancel{\text{mole}} \cancel{\text{kelvin}}} \times 273 \cancel{\text{kelvin}}}{1.00 \text{ atm}}$$

$$= \frac{22.4 \text{ liter} \cancel{\text{atm}}}{1.00 \cancel{\text{atm}}} = 22.4 \text{ liters}$$

One liter is about a quart and 22.4 liters is more than 5 gallons. In a similar manner, the volume of a mole of hydrogen gas at 273 kelvin can be found for other pressures. See Table 11-1.

The volume of 0.0224 liters (22.4 milliliters) at 1000 atmospheres is about one-tenth of a cup. This is an extremely small volume into which one mole of hydrogen gas would have to be compressed.

Test $PV = nRT$ for Hydrogen Gas

But how valid is this $PV = nRT$ relation for hydrogen gas? The answer to that can only lie in experimental measurements made over a wide range of experimental conditions. This becomes a little messy and laborious due to the number of variables.

Table 11-1

VOLUMES CALCULATED FROM $PV = nRT$ FOR ONE MOLE HYDROGEN GAS AT 273 K

pressure (in atmospheres)	volume (in liters)
0.0010	2240
0.10	224
1.0	22.4
10	2.24
100	0.224
1000	0.0224

Testing the validity can be simplified by testing at one temperature at a time and by focusing upon the value of the fraction PV/RT for one mole of hydrogen gas.

According to the equation $PV = nRT$, the number of moles equals the fraction PV/RT:

$$n = \frac{PV}{RT}$$

For the special case of one mole of gas that we have elected to use, $n = 1.000$ and the fraction PV/RT would be expected to have the value $PV/RT = 1.000$ for all values of pressure and all values of temperature. This is the reference line, the broken line, in Figure 11-1. In this figure, experimental values of the ratio PV/RT for the one mole of hydrogen gas at 273 K are plotted against pressure for a wide range of pressures. According to the equation $PV = nRT$, regardless of the value of the pressure, the values of the fraction PV/RT would be expected to equal 1.000 for the one mole of hydrogen gas and all points would be expected to fall on the reference line. The deviation of the solid line from the broken reference line in Figure 11-1 is the deviation of hydrogen gas behavior at 273 K from the $PV = nRT$ relation. The solid line is called an **isotherm**—all measurements at the same temperature, in this case the 273 K isotherm.

Clearly, the behavior of hydrogen gas at 273 K does deviate from the reference line—not greatly at pressures less than 10 atmospheres (the extreme left of the graph) but very substantially at pressures above 100 atmospheres. At 1000 atmospheres (at the extreme right of the graph), the fraction PV/RT is 1.6, not 1.000. The value of the fraction PV/RT is 60% greater than the value predicted by the equation.

Hydrogen gas is less compressible at high pressures than would be expected on the basis of its compressibility at low pressures. The deviation is interpreted to be the consequence of the experimental volume being about 60% greater than the volume calculated using the equation $PV = nRT$. The numerical value 0.082 liter atm mole^{-1} kelvin^{-1} for R is a value determined from experimental measurements at one atmosphere or less than one atmosphere.

Figure 11-1

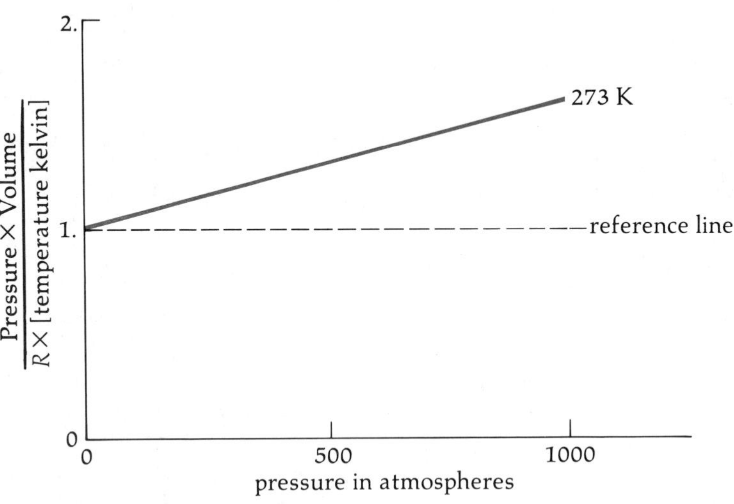

TEST OF $PV = nRT$ EQUATION FOR ONE MOLE OF HYDROGEN GAS AT 273 K AND VARIOUS PRESSURES.

Test $PV = nRT$ for Other Gases

Is the equation $PV = nRT$ applicable to other gases, such as helium gas, He; nitrogen gas, N_2; ammonia gas, NH_3; and methane gas, CH_4? One approach to answering this question would be to make a series of measurements for each gas and endeavor to evaluate the gas constant, R, for each gas. Another approach would be to use the numerical value of R obtained for hydrogen gas and test the validity of the equation using the experimental values of the fraction PV/RT for one mole of another gas at selected temperatures. Figure 11-2 gives the 200 K and the 473 K isotherms for nitrogen gas, N_2, and the same isotherms for methane gas, CH_4. Clearly, there is great diversity in behavior—not only between the behavior of the two gases but in the behavior of each gas at the different temperatures.

Clearly, however, all curves have one region of marked similarity: as the pressures approach zero at the extreme left of the graph, all values of PV/RT approach 1.000. The equation and the value of R obtained for hydrogen gas are also applicable for nitrogen and for methane at low pressures! Much experimental work is done at pressures less than 2 atmospheres, and in this range the $PV = nRT$ relation using the value of R derived for hydrogen gas is quite applicable.

Universal Gas Constant

Studies carried out with a great variety of gases show that all gases are described by the relation $PV = nRT$ at sufficiently low pressures and sufficiently high temperatures and that the same value of R is applicable for all gases in these low-pressure and high-temperature ranges. Consequently, R has become known as the "**universal gas constant.**" At pressures above a few atmospheres, the $PV = nRT$ relation does not precisely describe the behavior of any gas, including hydrogen gas.

Figure 11-2

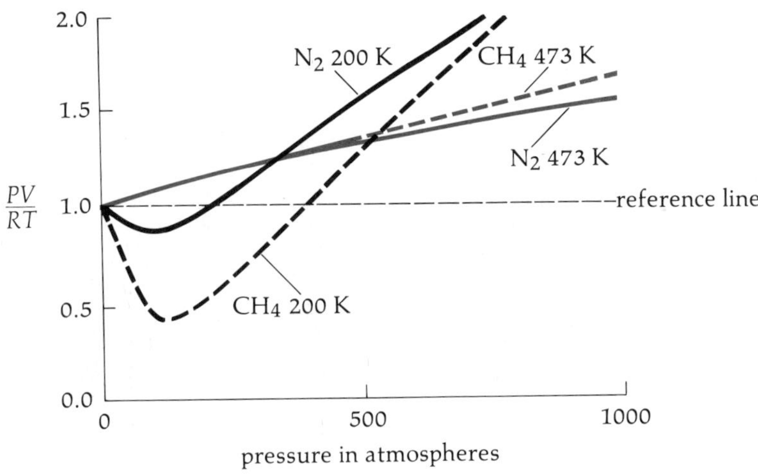

TEST OF THE $PV = nRT$ RELATION FOR NITROGEN GAS AND METHANE GAS. The 200 K and 473 K isotherms of experimental PV/RT fractions vs pressure for one mole of nitrogen gas, N_2, and also for one mole of methane gas, CH_4. The curves are a test of the validity of the $PV = nRT$ equation (using the numerical value of R derived for hydrogen gas) for nitrogen gas and methane gas.

Ideal Gas Equation

Ideal Gas Behavior

Real Gas Behavior

$PV = nRT$ describes the behavior of all gases in a limited low pressure range. This experimental evidence indicates that all gases in this limited pressure range have some characteristics that are independent of the identity of the molecules themselves and also independent of the individual masses of the molecules. This uniformity of behavior is not exhibited at high pressures where the molecules are forced to occupy a smaller volume and it is not true of molecules in either the liquid or the solid state. This behavior of gases under low pressures is in a sense an idealized state of affairs and $PV = nRT$ has become known as the **"ideal gas equation"** and "R" as the universal gas constant. A gas that conforms to the equation is said to "exhibit **ideal gas behavior.**" Real gases exhibit ideal gas behavior at low pressures.

A mixture of gases that do not react in the chemical sense is described equally well by the ideal gas equation so long as n, the number of moles used in the equation, is the total number of moles—the sum of the number of moles of molecules of each kind of polyatomic gas present plus the number of moles of atoms of each kind of monatomic gas present.

Universality of Gas Behavior

In order to rationalize this universality of gas behavior, a model is proposed for an ideal gas whose properties are independent of the identity of the units in the gas phase regardless of whether they are atoms of monatomic gases, such as helium, or molecules of polyatomic gases, such as H_2. For convenience in the following discussion of gases, the term molecules will be used to include atoms of monatomic gases as well as molecules of polyatomic gases. This model is known as the **kinetic molecular**

Kinetic Molecular Model for Ideal Gases

model for ideal gases. In this model, molecules of all gases are endowed with a number of properties:

- molecules in the gas are separated by space,
- molecules themselves do not take up space,
- all molecules are in motion,
- molecules do not attract other molecules and all collisions are perfectly elastic, and
- the average kinetic energy of the molecules is directly proportional to the kelvin temperature. (Some molecules have much higher kinetic energies, some much lower.)

Velocities of Molecules Having the Average Kinetic Energy

Using well-known relations from physics, it is comparatively easy to show that this model rationalizes a relation of the type $PV = cnT$, where c is a constant and T is the kelvin temperature. If the experimental value of R is accepted for c, it is also easy to evaluate the average kinetic energy of molecules at specific temperatures. If the identity of the molecule is known and consequently the mass of the molecule is

11-4 Background Information

MOMENTUM

Momentum is defined as the product of mass times velocity:

$$\text{momentum} = \text{mass} \times \text{velocity}$$

The magnitude of momentum is a measure of the tendency of an object to continue in motion with the same velocity. Using the values given in Table 11-2 (page 244) for the mass of a molecule of hydrogen gas and the velocity of a molecule of hydrogen gas having the average kinetic energy at 300 K, the momentum of this molecule along its line of motion is given by

$$\text{momentum} = (0.33 \times 10^{-23}\ \text{g}) \times \left(19.3 \times 10^{4}\ \frac{\text{cm}}{\text{sec}}\right)$$

$$= 6.37 \times 10^{-19}\ \frac{\text{g cm}}{\text{sec}}$$

or

$$= (0.33 \times 10^{-26}\ \text{kg}) \times \left(19.3 \times 10^{2}\ \frac{\text{m}}{\text{sec}}\right)$$

$$= 6.37 \times 10^{-24}\ \frac{\text{kg m}}{\text{sec}}$$

The small momentum of a molecule of hydrogen gas is the consequence of the small mass of a molecule of hydrogen gas. The velocity is quite impressive.

The units of momentum and the numerical value of momentum are entirely dependent upon the unit used to express mass and the units used to express velocity.

Momentum is a vector quantity: momentum has direction as well as magnitude. If a molecule has a head-on collision with a wall and the collision is perfectly elastic, the molecule returns along its initial path at the same speed but moving in the reverse direction. If the momentum preceding the collision is mv, the momentum following the collision is $-mv$, making a total change in momentum of $2mv$.

KINETIC ENERGY

Any object in motion possesses kinetic energy, and the equation

$$\text{kinetic energy} = \tfrac{1}{2}\,\text{mass} \times (\text{velocity})^2$$
$$KE = \tfrac{1}{2}mv^2$$

is probably one of the best known equations of physics. On collision, all of that kinetic energy has to go somewhere. Herein lie the hazards associated with automobiles in rapid motion.

If the mass is expressed in kilograms and the velocity in meters per second, the kinetic energy has the units kg m^2 sec^{-2}. To take a simple example, a mass of 4 kilograms moving with a velocity of 3 meters per second has a kinetic energy of 18 kg m^2 sec^{-2}.

$$KE = \frac{1}{2} \times 4\ \text{kg} \times \left(3\ \frac{\text{m}}{\text{sec}}\right)^2$$
$$= 18\ \text{kg m}^2\ \text{sec}^{-2}$$

This set of units is frequently encountered and 1 kg m^2 sec^{-2} is given the name joule, abbreviated J. The above kinetic energy can therefore be expressed as 18 joules.

The joule, a unit of energy, was introduced in Background Information 5-1, on energy. The approach at that time was the energy expended by a force acting through a distance:

$$\text{Energy} = \text{force} \times \text{distance}$$

and a force of one newton acting through a distance of one meter was defined as the expenditure of one joule of energy:

$$1\ \text{joule} = 1\ \text{newton} \times 1\ \text{meter}$$

We should be able to show that joule = newton × meter = kg meter^2 sec^{-2}. The newton is the force that gives the mass of one kilogram an acceleration of one meter per second squared. (See Background Information 4-2, on force.)

$$1\ \text{newton} = 1\ \text{kilogram} \times 1\frac{\text{meter}}{\text{second}^2}$$

$$1\ \text{joule} = 1\ \text{newton} \times 1\ \text{meter}$$

$$1\ \text{joule} = 1\ \text{kilogram} \times \frac{\text{meter}}{\text{second}^2} \times \text{meter}$$
$$= 1\ \text{kg m}^2\ \text{sec}^{-2}$$

The unit joule has not been extensively used in this country but there is now a concerted push to use this unit and we may find ourselves counting joules instead of calories. The calorie is now defined as 4.1840 joules.

Note that using the values for mass and velocity given in Table 11-2 (p. 244) gives the kinetic energy of a molecule having the average kinetic energy. For hydrogen gas at 300 K,

$$\underset{\text{per molecule}}{\text{KE}} = \frac{1}{2}\left(0.33 \times 10^{-26}\ \text{kg}\right)\left(19.3 \times 10^2\ \frac{\text{m}}{\text{sec}}\right)^2$$
$$= 6.1 \times 10^{-21}\ \text{kg m}^2\ \text{sec}^{-2}$$
$$= 6.1 \times 10^{-21}\ \text{joules}$$

and the kinetic energy of a mole of hydrogen gas molecules having the average kinetic energy at 300 K is

KE per mole

$$= 6.1 \times 10^{-21}\ \frac{\text{joules}}{\cancel{\text{molecule}}} \times 6.0 \times 10^{23}\ \frac{\cancel{\text{molecule}}}{\text{mole}}$$

$$= 3.66 \times 10^3\ \frac{\text{joules}}{\text{mole}}$$

Since

$$(\text{energy in joules}) = 4.184\ \frac{\text{joules}}{\text{calories}}\ (\text{energy in calories})$$

$$\text{KE per mole} = 3.66 \times 10^3\ \frac{\text{joules}}{\text{mole}} \div \frac{4.18\ \text{joules}}{\text{calories}}$$

$$= 8.76 \times 10^2\ \frac{\text{calories}}{\text{moles}}$$

known, it is a simple matter to calculate the velocity of the molecule that has the average kinetic energy. This last step, the calculation of the velocity, follows directly from the equation for kinetic energy:

$$\text{kinetic energy} = \tfrac{1}{2} \times \text{mass} \times (\text{velocity})^2$$
$$\text{KE} = \tfrac{1}{2}\, mv^2$$

The velocities in meters per second of the molecules that have the average kinetic energy are given in Table 11-2 for several gases at 300 K and 400 K. If the velocities in meters per second in Table 11-2 don't seem impressive, multiply them by 3600 seconds per hour to obtain the velocities in meters per hour. One kilometer is roughly equivalent to 0.6 miles and the slowest of these velocities, the value for the carbon dioxide molecules at 300 K, corresponds to almost 900 miles per hour; the largest, the value for hydrogen at 400 K, corresponds to almost 5000 miles per hour. In both cases, these are the velocities of the molecules that have the average kinetic energy. The velocities of many molecules are much greater and the velocities of many molecules are much smaller.

Note that the velocity of the molecule of hydrogen gas, which has the average kinetic energy at 300 K, is greater than the corresponding velocities of molecules of the other gases. Since all gases at the same temperature have the same average kinetic energy, it follows from the kinetic energy equation, $\text{KE} = \frac{1}{2}mv^2$, that the hydrogen molecule with the smallest mass has the greatest velocity.

If the average kinetic energy for gas molecules is directly proportional to the kelvin temperature, it is to be expected that the average kinetic energy will be greater at 400 K than at 300 K: 33% greater.

The kinetic molecular model for ideal gases relates to the behavior of gases in a great multiplicity of ways. In terms of a purely descriptive approach, some of these relations are

- the space between molecules allows the gases to be compressed or to be expanded;

Table 11-2

VELOCITIES OF MOLECULES WITH THE AVERAGE KINETIC ENERGY, AT 300 K AND 400 K

	mass of one molecule (in grams)	**velocities (in meters per second)**	
		at 300 K	**at 400 K**
hydrogen, H_2	0.33×10^{-23}	19.3×10^2	22.3×10^2
helium, He	0.66×10^{-23}	13.7×10^2	15.8×10^2
methane, CH_4	2.7×10^{-23}	6.7×10^2	7.8×10^2
oxygen, O_2	5.3×10^{-23}	4.8×10^2	5.6×10^2
carbon dioxide, CO_2	7.3×10^{-23}	4.1×10^2	4.8×10^2

- the assumed zero volume of the molecules would allow the volume of a sample of gas to be reduced to zero;
- the motion of molecules creates pressure by continuous impact on any surface;
- the frequency of impact on a unit area of the container wall and the momentum of each molecule at time of impact determine the magnitude of the pressure;
- the velocity of the molecule determines how frequently the molecule gets back to a surface to strike again and also determines the momentum at the time of impact;
- the momentum, mass times velocity, is directly proportional to the velocity;
- this double entry of velocity into the determination of the frequency of impacts with the wall and the momentum of the molecule at the time of impact makes pressure dependent upon velocity × velocity—upon $(\text{velocity})^2$;
- velocity squared relates directly to kinetic energy, $\frac{1}{2}$ mass × $(\text{velocity})^2$;
- average kinetic energy has been related directly to the absolute temperature by the model;
- the higher the temperature, the greater the average kinetic energy; the greater the velocity, the more frequent the collisions and the harder the molecules strike the surface; and
- the lack of attractive forces between molecules leads to perfectly elastic collisions and allows each molecule to proceed without restraint.

The kinetic molecular model for **ideal gases** is used as an approximation for **real gases** with emphasis on the differences between real gases and the model for the ideal gas. The interpretation of the graphs presented in Figure 11-2 is that the behavior of nitrogen gas and methane gas closely approximates the behavior of an ideal gas in a small pressure range at the extreme left of the graph (less than 10 atmospheres) but that, at higher pressures, the deviation of real gas behavior from ideal behavior is marked.

Positive Deviations from Ideal Gas Behavior

At extreme pressures, all known gases exhibit positive deviations: the experimental curves run above the ideal $PV/RT = 1$ reference line. For many gases, there is an intermediate range of pressures in which the gases exhibit negative deviations: the experimental curves run below the ideal PV/RT line. Actually, all real gases exhibit this region of negative deviation provided the isotherm is for a sufficiently low temperature.

Negative Deviations from Ideal Gas Behavior

Deviations of Real Gases from the Kinetic Molecular Model for Ideal Gases

A deviation that is first negative and then positive implies that real gases differ from ideal gases in at least two respects. Let us consider the positive deviations first at the extremely high pressures. One mole of an ideal gas at 300 K and 1.00 atmosphere would occupy a volume of

$$V = \frac{nRT}{P} = \frac{1.00\ \cancel{\text{mole}} \times 0.082\ \dfrac{\text{liter}\ \cancel{\text{atm}}}{\cancel{\text{mole}}\ \cancel{\text{kelvin}}} \times 300\ \cancel{\text{kelvin}}}{1.00\ \cancel{\text{atm}}}$$

$$= 24.6\ \text{liters}$$

One mole of an ideal gas at 300 K and 1000 atmosphere would occupy a volume of

$$V = \frac{1.00 \times 0.082 \times 300}{1000}$$
$$= 0.0246 \text{ liter} \quad \text{or} \quad 24.6 \text{ milliliters}$$

A volume of 25 liters is a bit more than a cubic foot; a volume of 25 milliliters is less than the volume of a demitasse cup. Consequently, the positive deviations of real gases are rationalized by attributing the positive deviation of the experimental value of PV/RT to the product PV being larger than the ideal simply because the actual volumes of the molecules themselves become significant when gases are compressed into this very small volume. Real molecules occupy volume, and this volume cannot be neglected at high pressures as is done in the model for an ideal gas.

The negative deviations exhibited by real gases at an intermediate pressure range are rationalized in terms of attractive forces between molecules. If these molecules really do attract each other, then the molecules approaching the walls of the container will be pulled back by the molecules not as close to the wall. The molecules do not collide with the wall as frequently and their momentum at time of impact with the wall is reduced. On both scores, the attractive forces between the molecules reduce the experimental pressure below the value to be expected for one mole of an ideal gas confined to the same volume at the same temperature.

In summary, positive deviations from ideal gas behavior are rationalized in terms of the real volumes of molecules; negative deviations are rationalized in terms of the attractive forces between real molecules. The first increases the experimental volume as compared to the ideal volume; the second decreases the experimental pressure as compared to the ideal pressure.

The concept of an ideal gas is intellectually intriguing. The use of the ideal gas equation, even as an approximation, is very convenient. The deviations from ideal gas behavior provide insight into the magnitude of the forces between molecules and the volumes of molecules. Both of these factors relate directly to the transition from the vapor (gas) phase to the liquid phase.

van der Waal's Forces

The forces between molecules of a gas are called van der Waals' forces. It is van der Waals' forces that hold molecules together in liquids and hold molecules together in molecular crystals (crystals made up of molecules). To designate these forces as van der Waals' forces says nothing about the nature of the forces. The name recognizes the individual, Johannes van der Waals, who recognized that such forces do exist. A gas that shows a large negative deviation from ideal behavior is at a temperature that is not far removed from the temperature at which the gas can be converted to the liquid phase either by pushing the molecules a little closer together by increasing the pressure or by lowering the temperature of the gas further to slow the molecules. The gas phase is known as the expanded phase, the liquid phase and the solid phase as condensed phases. A gas occupies a large volume in comparison to the volume of the same mass of a substance as a liquid or a solid at the same temperature and the same pressure.

The ideal gas equation, $PV = nRT$, and the values for the universal gas constant, R, derived by measurements on a real gas under favorable conditions of low pressures and temperatures far above the boiling point of the liquid, are extensively used for real gases. Ordinary laboratory experimental work is performed at pressures below 10 atmospheres—frequently at 1 atmosphere or below. In using the equation, the usual care must be taken with regard to the use of units. The units for both sides of an equation must always be the same. If R is used in liters atmospheres mole^{-1} kelvin^{-1}, the pressure must be in atmospheres and the volume must be in liters. If R is used in milliliters torr mole^{-1} kelvin^{-1}, the pressure must be in torr and the volume in milliliters.

Vibrational and Rotational Motion of Molecules

In this chapter, we have been concerned with the translational motion of molecules, the motion that translates the center of mass of the molecule from one place to another. Molecules are also involved in two other types of motion in which the position of the center of mass of the molecule does not change. One of these is vibrational motion and the other is rotational motion. In vibrational motion of a diatomic molecule, such as hydrogen, H_2, or hydrogen chloride, HCl, the two atoms move in unison along the line of centers of the two atoms to alternately extend and shorten the distance separating the two atoms.

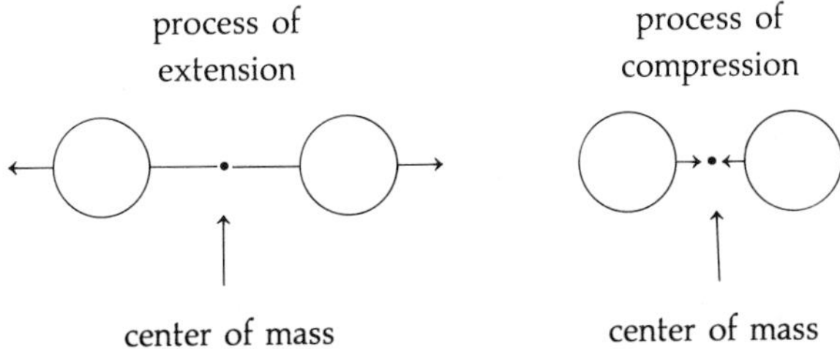

If the two atoms are of equal mass, the displacements of the two atoms are equal but in opposite directions. If the two atoms are of unequal mass, the displacement of the lighter atom is the greater and the center of mass remains unchanged. Vibrational motion of diatomic molecules was explored in detail in Chapter 8. One of the characteristics of vibrational motion is that it is quantized. In molecules more complicated than diatomic molecules, there are a number of modes of vibration.

In rotational motion, the center of mass remains unchanged and the molecule rotates about the center of mass. For a diatomic molecule such as hydrogen, rotational motion can be represented by the following diagram:

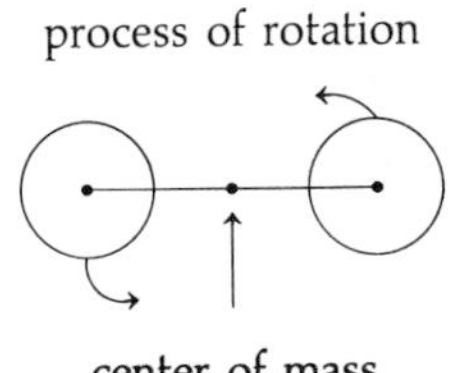

One of the characteristics of rotational motion is that rotational motion is quantized. All polyatomic molecules are simultaneously involved in translational motion, vibrational motion, and rotational motion. Atoms such as helium, He, are involved in translational motion only.

Scramble Exercises

1. Using $PV = nRT$, calculate the volume that one mole of hydrogen gas would occupy at a temperature of 300 K and a pressure of 1 atm. The temperature 300 K, 27 °C, corresponds to the temperature of a warm summer day. This calculation assumes ideal gas behavior for hydrogen gas at 27 °C and 1 atm.

2. Using $PV = nRT$, calculate the volume that one mole of hydrogen gas would occupy at a temperature of 300 K and a pressure of (a) 10 atm, (b) 100 atm, and (c) 1000 atm. These calculations also assume ideal gas behavior for hydrogen gas at these higher pressures. At these pressures, the experimental value of volume would be higher than the calculated value.

3. Hydrogen chloride gas, HCl, exhibits greater negative deviation from ideal gas behavior than hydrogen gas, H_2. Suggest plausible explanations of this difference.

4. Suggest the basis for using measurements of hydrogen gas at low pressures to evaluate R, the ideal gas constant.

5. It is an experimental fact that hydrogen gas escapes about four times as fast as oxygen gas through a small opening in the wall of a container under the same conditions of temperature and pressure of the confined gas. Propose a plausible explanation for this observed fact. The opening, though small, is very large in comparsion to the diameters of molecules. Table 11-2 has useful information.

6. The kinetic energy of a molecule of hydrogen gas having the average kinetic energy at 300 K was evaluated in Background Information 11-5, on kinetic energy. In a similar manner, use the values given in Table 11-2 to evaluate the average kinetic energy of molecules of methane, CH_4, and molecules of oxygen, O_2, at 300 K.

7. What is the relation of the average kinetic energy of molecules of oxygen in a sample of oxygen at 400 K to the average kinetic energy of molecules of oxygen in a sample of oxygen at 300 K?

8. Starting with $PV = nRT$, calculate the number of molecules of hydrogen, H_2, in a sample of hydrogen gas that occupies a volume of 10.0 milliliters at a pressure of 800 torr and 127 °C.

9. Starting with $PV = nRT$, calculate the number of atoms of helium, He, in a sample of helium gas that occupies a volume of 10.0 milliliters at a pressure of 800 torr and 127 °C.

10. Relate the answers you obtained in Exercises 8 and 9 to the proposal made by Avogadro early in the nineteenth century that equal volumes of gases at the same temperature and pressure contain the same number of molecules. Avogadro arrived at this concept through the study of the volumes of gases that react and the volumes of gas products formed when all the measurements of gas volumes were made at the same temperature and pressure. For example, (1) for all gases measured at 27 °C and one atmosphere pressure, 10 mL of hydrogen

gas and 10 mL of chlorine gas react to give 20 mL of the compound hydrogen chloride, and (2) for all gases measured at 127 °C and 0.50 atmosphere pressure, 10 mL of hydrogen gas reacts with 5 mL of oxygen gas to give 10 mL of water vapor (water in the gas phase). These relations may not seem so strange to us. We know the formulas for the molecules and can write equations. This was far from true for the state of science at the time of Avogadro. Relate these equations for the chemical reaction to statement (1) and statement (2) made above.

$$H_2(g) + Cl_2(g) \longrightarrow 2\,HCl(g)$$
$$2\,H_2(g) + O_2(g) \longrightarrow 2\,H_2O(g)$$

Additional exercises for Chapter 11 are given in the Appendix.

A Look Ahead

The next chapter deals with the liquid phase. At pressures of one atmosphere or less, the molar volume, the volume of one mole, of a liquid is seldom more than 1% of the molar volume of a gas at the same temperature and pressure. As compared to gases, liquids are essentially incompressible. When the pressure on an ideal gas is doubled, the volume is reduced by 50%. When the pressure on a liquid is doubled, the volume is reduced by less than 1%. Most liquids expand when heated but the degree of expansion is small as compared to the expansion of a gas. There is no simple relation, such as the $PV = nRT$ equation for gases, that relates the variables pressure, volume, temperature, and number of moles for one liquid, much less for all liquids.

In a sense, liquids can be considered as the extreme deviation from ideal gas behavior. The physical properties of liquids are determined by the volumes of molecules and the van der Waals' forces that bring and keep molecules together in the liquid phase.

Chapter 12 will have more meaning if you are able

- to compare the experimental behavior of real gases with the ideal behavior described by the equation $PV = nRT$,
- to discuss ideal gas behavior in terms of the model for an ideal gas, and
- to discuss real gas behavior in terms of deviations from the model for ideal gases.

In Chapter 12, the concept of van der Waals' forces is extended to physical properties of liquids and in Chapters 13 and 14 to chemical reactions of water with other polar molecules.

12

OIL, WATER, AND LIQUID SOLUTIONS

Chapter 12

The statement "Oil and water don't mix" is purely a matter of semantics. A liquid that does not mix with water is called an oil in the most general sense of that term. Naturally occurring oils, such as vegetable oils and petroleum, are mixtures of compounds. It is easier to start this exploration with a pure compound that has limited solubility in water, for example, benzene, C_6H_6.* When benzene and water are shaken together, droplets of benzene may be dispersed throughout the water or droplets of water may be dispersed throughout the benzene, depending upon the relative volumes of benzene and water being shaken. In either case, on standing, the dispersion separates into two layers. The top layer is largely benzene with a low concentration of dissolved water; the bottom layer is largely water with a low concentration of dissolved benzene. If the two liquids have been thoroughly mixed over a period of time, each layer is saturated at that temperature with the other compound. Benzene is only slightly miscible in water, water is only slightly miscible in benzene, and the two compounds are frequently said to be immiscible. If you wish to be a purist, no two liquids are absolutely immiscible.

Benzene and Water

In contrast to this, ethanol, CH_3CH_2OH, and water are completely miscible at room temperature. Water never becomes saturated with ethanol and ethanol never becomes saturated with water. There are never two separate liquid phases. Solutions of ethanol and water cover the entire range in concentration from essentially 100% ethanol to essentially 100% water. The rationalization for these marked differences in properties of benzene and ethanol with water lies in differences in the forces between molecules in liquids. If ethanol and water mix in all proportions, the attractive forces between an ethanol molecule and a water molecule must compare favorably with the forces of attraction between a molecule of ethanol and another molecule of ethanol and also with the forces of attraction between a molecule of water and another molecule of water. If this were not true, either the molecules of ethanol would draw together or the molecules of water would draw together and the mixing of ethanol and water would yield two liquid phases, two layers. The immiscibility of benzene and water is to be rationalized in terms of particularly strong forces of attraction between molecules of benzene or particularly strong forces of attraction between molecules of water in comparison to the forces of attraction between benzene molecules and water molecules.

Ethanol and Water

* CAUTION: Benzene vapor is toxic. Any hydrocarbon, such as an isomer of hexane, C_6H_{14}, can be used in demonstrations.

To understand the properties of pure liquids and solutions, it is necessary to understand the nature of van der Waals' forces between molecules. Much can be learned about the forces between like molecules by a consideration of the properties of pure compounds that are liquids.

Properties of Pure Liquid Compounds

Benzene

In the evaporation of benzene, the molecules of benzene must escape the van der Waals' forces of attraction of the liquid phase to enter the vapor phase in which the molecules are widely separated and comparatively free of van der Waals' forces. The volume of the vapor phase is extremely large in comparison to the volume of the same mass of the compound in the condensed phase at the same temperature and the same pressure, frequently a thousand times as great.

The escape of molecules of benzene from the liquid goes on continuously:

$$C_6H_6(l) \longrightarrow C_6H_6(g)$$

Molecules of benzene in the vapor phase also return to the liquid phase:

$$C_6H_6(g) \longrightarrow C_6H_6(l)$$

The rate of return to the liquid phase depends upon the concentration of the benzene molecules in the immediate vicinity of the liquid surface. If air currents carry away the benzene molecules, there will be very little return of benzene molecules to the liquid phase and, in time, all of the liquid will vaporize. Figure 12-1.

If vaporization takes place into a closed space, such as a bell jar or an inverted wide-mouth jar or an inverted beaker over the open dish of benzene, the concentration of molecules of benzene in the vapor phase builds up and the rate of condensation

Figure 12-1

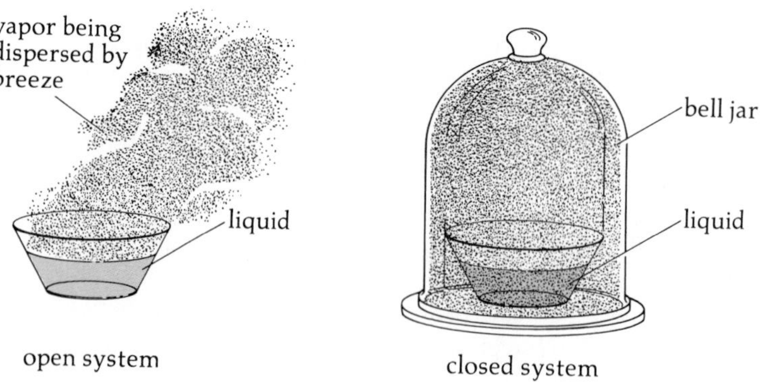

EVAPORATION OF LIQUID IN AN OPEN DISH AND IN A CLOSED SYSTEM. Dots represent molecules of the vapor.

(the rate of the return of benzene molecules to the liquid phase) increases. If the temperature remains constant, the rate of condensation of benzene molecules from the vapor phase eventually becomes equal to the rate of vaporization of benzene molecules from the liquid phase. When the rate of condensation is equal to the rate of vaporization, the liquid benzene and the vapor of benzene are said to be in **equilibrium** and the equation is written with two arrows to indicate this state of equilibrium:

$$C_6H_6(l) \rightleftharpoons C_6H_6(g)$$

Vapor Pressure

Under conditions of equilibrium, the pressure of the benzene vapor has reached a constant value and is called the **vapor pressure** of the benzene for that temperature. At 20 °C, the vapor pressure of benzene is 75 torr.

To measure the vapor pressure (the pressure of the vapor under equilibrium conditions) of a compound at a specified temperature is conceptually an easy matter. Connect a U-tube barometer (Background Information 11-1, on pressure) to a vessel that can be evacuated, pump out the air from the vessel until the levels of the mercury in the two arms of the barometer are equal, introduce the liquid for which the vapor pressure is to be determined into the vessel, and read the maximum pressure attained. (Figure 12-2). This is the vapor pressure. In practice, there are troublesome details such as airtight connections of the barometer and the pump to the vessel, an appropriate valve to isolate the system from the pump, an appropriate constant-temperature

Figure 12-2

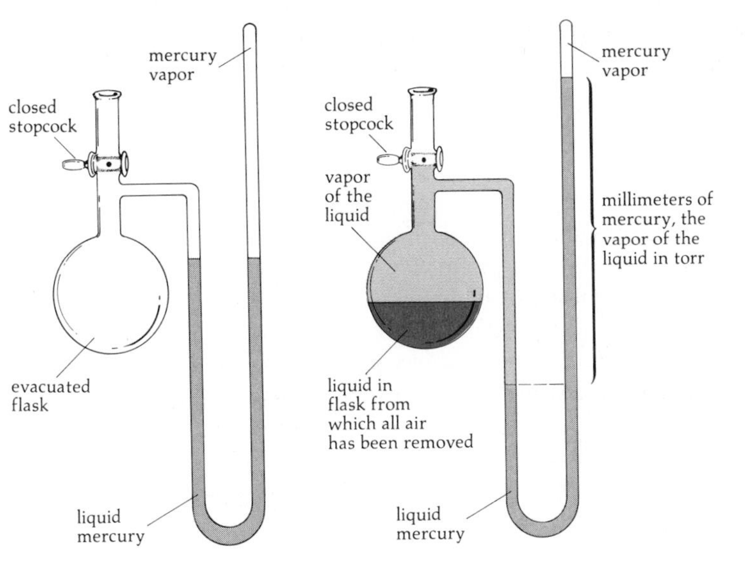

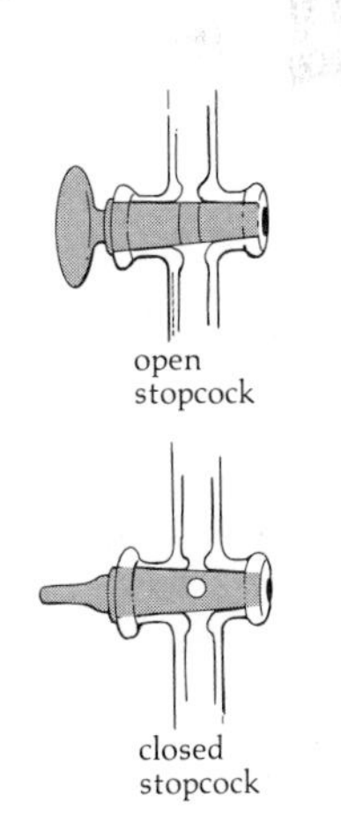

SCHEMATIC REPRESENTATION OF EQUIPMENT THAT COULD BE USED TO MEASURE THE VAPOR PRESSURE OF A LIQUID.

Table 12-1

VAPOR PRESSURES OF BENZENE AND WATER IN TORR

temperature (°C)	benzene	water
20	75	18
40	183	55
60	393	149
80	758	355
100	1338	760
120	2310	1489
140	3730	2711

room or thermostat, and an appropriate mechanism to introduce the liquid without introducing air.

At 20 °C, the vapor pressure of water is 18 torr. If on a rainy day the temperature of the atmosphere is 20 °C and the actual pressure due to water vapor in the air is 18 torr, the relative humidity of the atmosphere is 100%. Air conditioning units are designed to deliver air with a relative humidity of less than 60%.

The vapor pressures in torr for benzene and water are given in Table 12-1 for the temperature range 20 to 140 °C.

The graphs of the vapor pressures of benzene and water versus temperature are given in Figure 12-3.

Dependence of Vapor Pressure on Temperature

Clearly, the vapor pressure of benzene is greater than the vapor pressure of water at all temperatures. Benzene is the more volatile compound. The manner in which vapor pressure increases with temperature for the two compounds is similar. As the temperature increases, the vapor pressure increases but the relation is not a direct proportion: the lines are not straight and they do not go through the origin.

Boiling Point
Normal Boiling Point

The temperature at which the vapor pressure of a liquid equals one atmosphere, 760 torr, is the **boiling point** of that liquid, sometimes called the **normal boiling point** of the liquid: 80.1 °C for benzene and 100.0 °C for water. At these temperatures, the vapor of the pure liquid compound has sufficient pressure to sustain a bubble of the vapor within the liquid while the liquid itself is under a pressure of one atmosphere. These are the bubbles that are formed on the heated surface and rise to the surface of the liquid as the liquid boils. The temperature of liquid benzene has to be raised to 80.1 °C for the vapor pressure of benzene to reach 760 torr. The temperature of liquid water has to be raised to 100 °C for the vapor pressure of water to reach 760 torr. As you may remember, the temperature of boiling water under a pressure of one atmosphere was used to establish 100 °C. If the depth of the liquid is appreciable, then the actual temperature and the vapor pressure have to be slightly higher in order to be able to sustain the bubble of vapor under the added pressure due to liquid above the bubble. The vast volume of these bubbles in comparison to the volume of the liquid that disappears demonstrates the great difference in volume between the

Figure 12-3

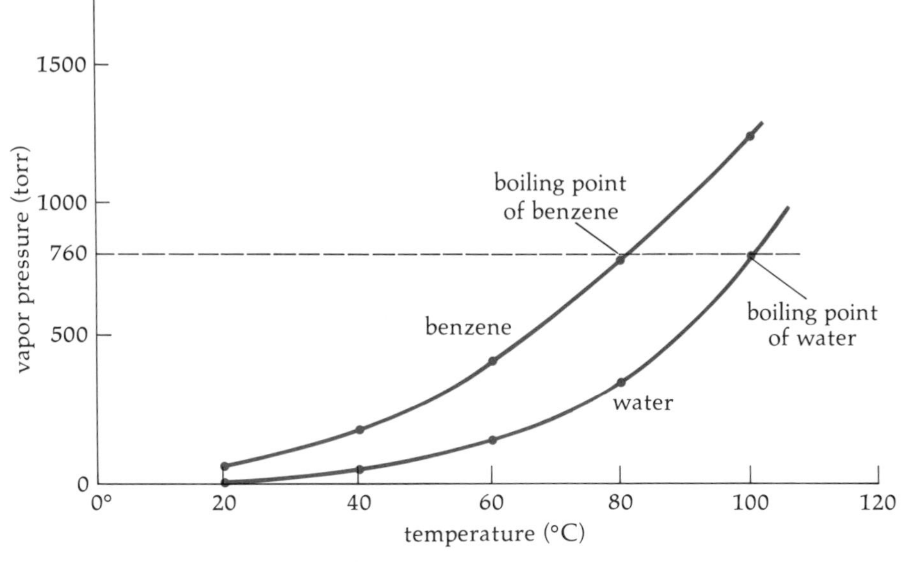

GRAPH OF VAPOR PRESSURE VS TEMPERATURE FOR BENZENE AND WATER.

n-pentane

expanded vapor phase and the condensed liquid phase. Those very small bubbles that appear on the walls of the container and sing in a most reassuring way when tap water is heated are merely the escape of dissolved air and, as anyone who likes a good cup of tea knows, are not a true indication of boiling. The bubbles of dissolved gas appear before the temperature of the water reaches its boiling point. If the gas pressure on the surface of the liquid is not one atmosphere but some value below one atmosphere, then the temperature at which a liquid boils is less than the normal boiling point and that temperature can be read from the vapor pressure–temperature graph for that compound. At an elevation of 10,000 feet, the boiling point of water is reduced to approximately 90 °C. The normal boiling points of compounds are a very convenient means of comparing in a qualitative way the ease with which molecules of a liquid escape to the vapor phase.

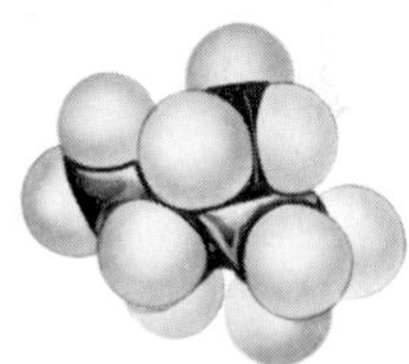

2-methylbutane

2,2-dimethylpropane

A number of factors influence the vapor pressure of liquids, and it is a challenge to identify them and explore them one at a time. Many of the readily available liquid compounds contain carbon and hydrogen and we begin this exploration with the three isomers of pentane, C_5H_{12} (molecular mass 72.2). Their structures and boiling points are given in Table 12-2. All bond angles within the molecules are the 109° tetrahedral angle (sp^3 hybridization of carbon).

The three isomers have quite different boiling points: one above room temperature, one just about room temperature on a warm summer day, and one well below room temperature. A space-filling model of the isomer that boils at 10 °C is practically a sphere. A space-filling model of the isomer that boils at 36 °C is some-

Normal Boiling Points of Isomers of Pentane

Table 12-2

NORMAL BOILING POINTS OF THE THREE STRUCTURAL ISOMERS OF PENTANE

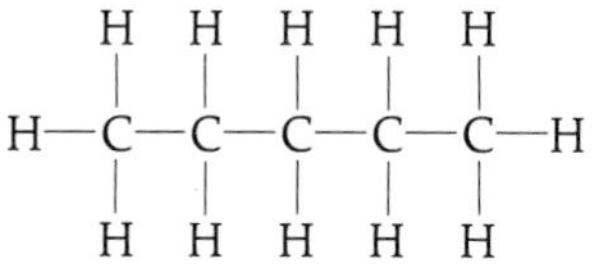

```
    H  H  H  H  H
    |  |  |  |  |
 H—C—C—C—C—C—H
    |  |  |  |  |
    H  H  H  H  H
```

n-pentane 36 °C

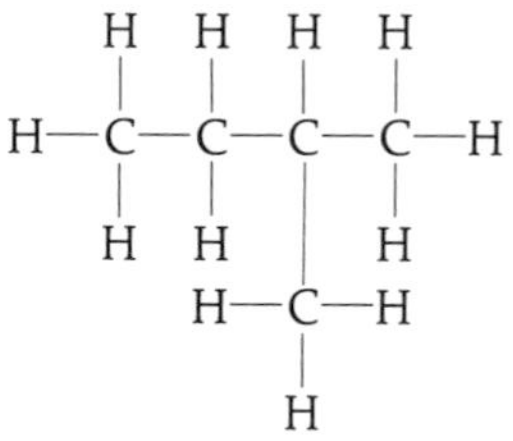

```
    H  H  H  H
    |  |  |  |
 H—C—C—C—C—H
    |  |  |  |
    H  H  |  H
       H—C—H
          |
          H
```

2-methylbutane 28 °C

```
          H
          |
       H—C—H
    H     |     H
    |     |     |
 H—C—C—C—H
    |     |     |
    H     |     H
       H—C—H
          |
          H
```

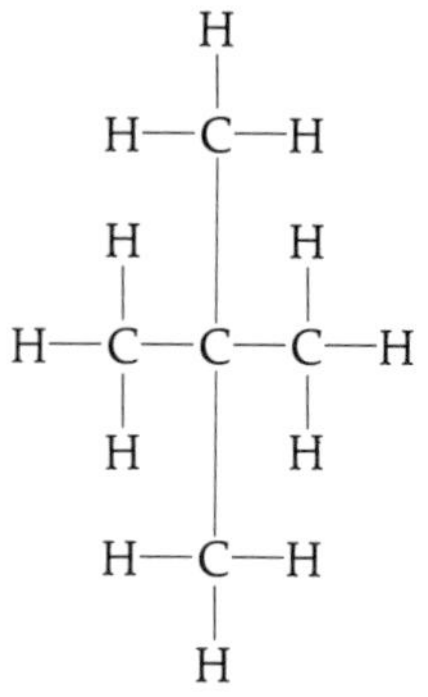

2,2-dimethylpropane 10 °C

what of a plump worm. It is the spherical isomer that escapes most readily from the liquid phase and, consequently, the liquid phase of that isomer has the lowest boiling point.

Normal Boiling Points of Straight-Chain Alkanes

The normal boiling points are given in Table 12-3 for the first seven members of the series of straight-chain, saturated hydrocarbons. Clearly the boiling point increases with the number of carbon atoms in the chain, also with the molecular mass. In all of these compounds, the bond angles are the 109° of the tetrahedral angle (sp^3 hybridization of carbon) with free rotation about all σ-bonds. In terms of space-filling models, all of these molecules have plump worm conformations. It is not possible to assess whether the decrease in volatility throughout the series is the consequence of an increase in molecular mass or an increase in the length of the wormlike structures, possibly both.

Table 12-3

NORMAL BOILING POINTS OF LOW MOLECULAR MASS, STRAIGHT-CHAIN ALKANES

		mol mass	**boiling point at 760 torr (°C)**
methane	CH_4	16	−162
ethane	CH_3CH_3	30	−89
propane	$CH_3CH_2CH_3$	44	−42
n-butane	$CH_3(CH_2)_2CH_3$	58	−1
n-pentane	$CH_3(CH_2)_3CH_3$	72	36
n-hexane	$CH_3(CH_2)_4CH_3$	86	69
n-heptane	$CH_3(CH_2)_5CH_3$	100	98

Table 12-4

NORMAL BOILING POINTS OF THE FOUR STRUCTURAL ISOMERS OF C_4H_9Cl

```
     H   H   H   H                    H   H   H   H
     |   |   |   |                    |   |   |   |
 H — C — C — C — C — Cl           H — C — C — C — C — H
     |   |   |   |                    |   |   |   |
     H   H   H   H                    H   H   |   H
                                              Cl
```

1-chlorobutane 78 °C | 2-chlorobutane 68°C

```
            H                               H
            |                               |
        H — C — H                       H — C — H
     H      |      H                 H      |      H
     |      |      |                 |      |      |
 H — C ──── C ──── C — Cl        H — C ──── C ──── C — H
     |      |      |                 |      |      |
     H      H      H                 H      Cl     H
```

1-chloro-2-methylpropane 68 °C | 2-chloro-2-methylpropane 52 °C

Normal Boiling Points of the Structural Isomers of C_4H_9Cl

The four monochlorobutanes, C_4H_9Cl, molecular mass 92.5, are also informative. (Table 12-4.) Their boiling points range from 52 °C for the most globby of the isomers to 78 °C for the straight-chain molecule. In visualizing space-filling models for these isomers, remember that the chlorine atom is large as compared to the carbon atom. The molecular masses are, of course, the same and the only difference lies in the arrangement of the atoms. The carbon–chlorine bond is polar—2.5 for carbon and 3.5 for chlorine on the electronegativity scale—and the chlorine atom has more mass than a CH_3 group: 35.5 as compared to 15. In comparison to the pentanes, the monochlorobutane molecules have slightly more mass than the pentanes—92.5 as compared to 72. In all cases, the boiling point of the monochlorobutane is greater than that of the corresponding isomeric pentane; for the closest structural matches, approximately 40 °C higher. This is a greater difference than would be expected for the difference in molecular weights and is attributed to the polar carbon–chlorine bond with electrostatic attraction between the appropriate regions of negative and positive partial charges within one molecule and the regions of negative and positive partial charges in other molecules.

Normal Boiling Points of the Structural Isomers of $C_4H_{10}O$

The isomers of $C_4H_{10}O$, mol wt 74, provide a greater variety of structures due to the formation of three ethers as well as four alcohols. (Table 12-5.) A number of correlations are evident. The normal boiling points of the alcohols are much higher than the normal boiling points of the ethers, at least 50 °C higher when comparisons are made between "straight-chain" isomers and also between branched-chain isomers. The boiling points of the alcohols are greater than those of the corresponding monochlorobutane by about 30 °C in spite of the smaller molecular weight of the alcohols: 74 as compared to 92.5. There must be something very special about the hydroxyl groups, —OH, of the alcohols. The values on the electronegativity scale for hydrogen, carbon, and oxygen are 2.1, 2.5, and 3.5, respectively. The uniqueness of the alcohols

Table 12-5

ISOMERIC STRUCTURES OF $C_4H_{10}O$ AND THEIR NORMAL BOILING POINTS

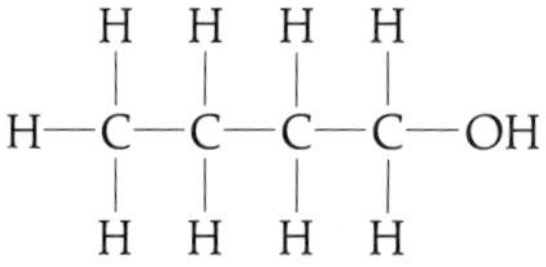

```
   H  H  H  H
   |  |  |  |
H— C— C— C— C—OH
   |  |  |  |
   H  H  H  H
```

1-butanol 117 °C

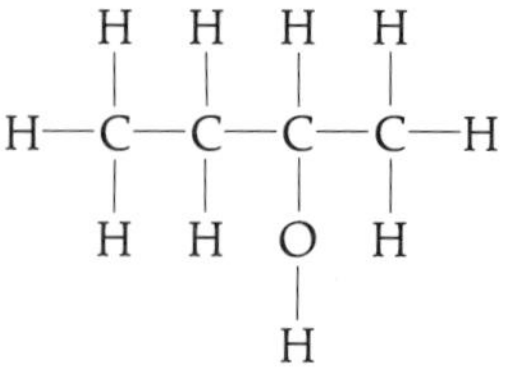

```
   H  H  H  H
   |  |  |  |
H— C— C— C— C—H
   |  |  |  |
   H  H  O  H
          |
          H
```

2-butanol 96 °C

```
          H
          |
       H— C—H
   H      |      H
   |      |      |
H— C——————C——————C—OH
   |      |      |
   H      H      H
```

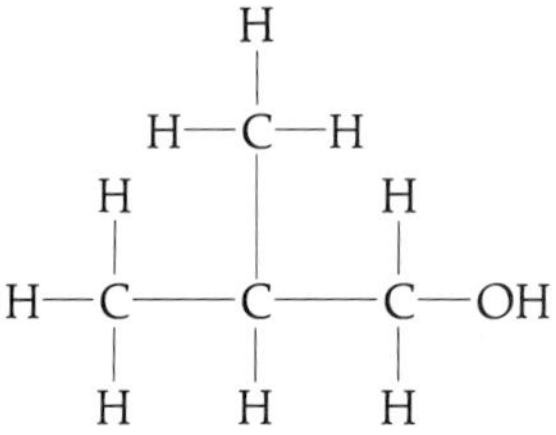

2-methyl-1-propanol 108 °C

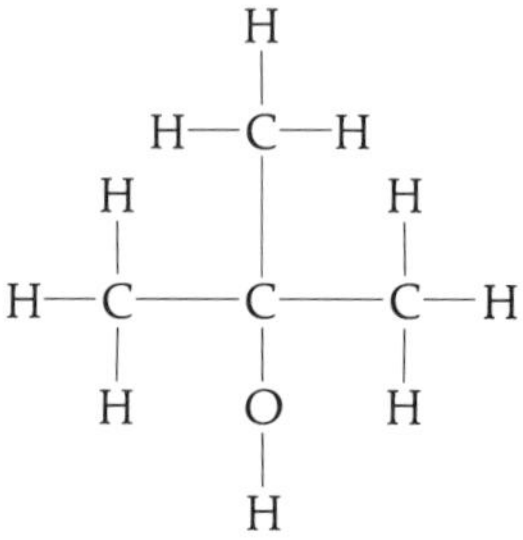

```
          H
          |
       H— C—H
   H      |      H
   |      |      |
H— C——————C——————C—H
   |      |      |
   H      O      H
          |
          H
```

2-methyl-2-propanol 82 °C

```
   H  H  H     H
   |  |  |     |
H— C— C— C— O— C—H
   |  |  |     |
   H  H  H     H
```

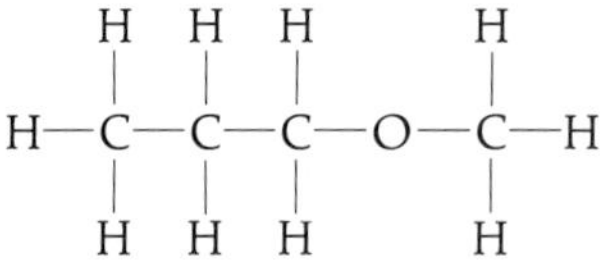

methyl *n*-propyl ether 39 °C

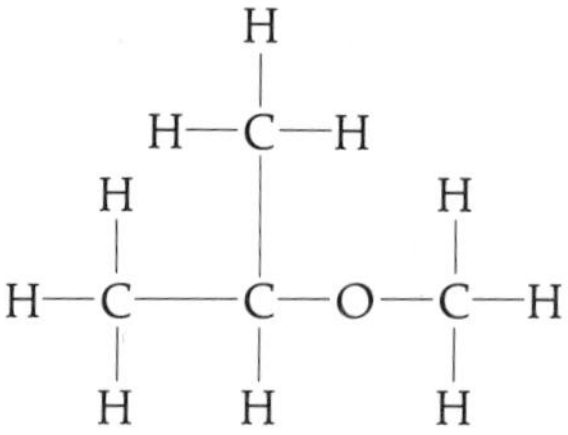

```
          H
          |
       H— C—H
   H      |      H
   |      |      |
H— C——————C— O— C—H
   |      |      |
   H      H      H
```

methyl isopropyl ether 32 °C

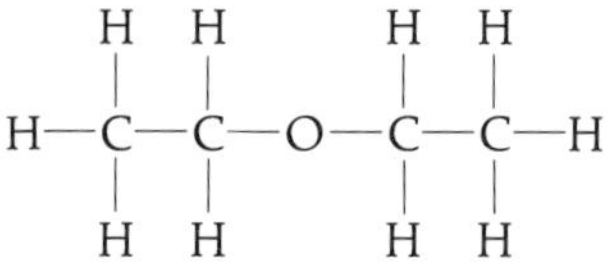

```
   H  H     H  H
   |  |     |  |
H— C— C— O— C— C—H
   |  |     |  |
   H  H     H  H
```

diethyl ether 35 °C

is believed to lie in the difference between the electronegativity of hydrogen and the electronegativity of oxygen. For its molecular weight of 18, water also has an exceptionally high boiling point. More about this later.

Dependencies of Vapor Pressure on Temperature

It is not surprising that the vapor pressure increases as the temperature is raised. More molecules have the momentum (mass × velocity) to escape from the liquid phase and more molecules in the vapor phase have the kinetic energy to escape being trapped by the liquid phase.

Another manifestation of the forces between molecules is the energy required to disperse the molecules of the condensed liquid phase to give the expanded vapor phase. The energy required to convert one mole of liquid benzene at 80 °C to benzene vapor at one atmosphere is approximately 8 kilocalories (32 kilojoules). The energy required to convert one mole of liquid water to water vapor at one atmosphere and 100 °C is about 10 kilocalories (42 kilojoules):

$$C_6H_6(l) \longrightarrow C_6H_6(g) \qquad \Delta H_{353\,K} = 8.1 \text{ kcal}$$
$$H_2O(l) \longrightarrow H_2O(g) \qquad \Delta H_{373\,K} = 9.7 \text{ kcal}$$

The symbol ΔH is again the change in enthalpy—in this case, also called the heat of vaporization. For both compounds, the same number of molecules, Avogadro's number of molecules, are delivered into the vapor phase at a pressure of one atmosphere but the temperatures are different—353 K and 373 K—so the average kinetic energies of the molecules in the vapor phase are different and the volumes occupied by the vapor are different.

The dependency of vapor pressure on temperature (Figure 12-1) is closely related to this heat of vaporization, and vapor pressure, p, can be expressed by the equation

$$\log_{10} p = \frac{-\Delta H_v}{2.303\,RT} + \text{constant}$$

where ΔH_v is the heat of vaporization, R is the universal gas constant, and T is the kelvin temperature. The other constant can be evaluated for each compound by substituting the heat of vaporization and any pair of experimental values for the vapor pressure and the temperature into the equation. This constant is characteristic of the compound: the numerical value for the constant also depends on the units used for the vapor pressure. In using this equation, care is taken to use consistent units for the heat of vaporization, ΔH_v, and the universal gas constant. If the heat of vaporization is expressed in kilocalories per mole, then the universal gas constant must be expressed in kilocalories per mole per kelvin. Although it may seem very strange, liter atmospheres and milliliter torrs are units of energy, and the universal gas constant can be expressed in terms of joules and also in terms of calories.

$$R = 8.31 \text{ joules mole}^{-1} \text{ kelvin}^{-1}$$
$$= 1.99 \text{ calories mole}^{-1} \text{ kelvin}^{-1}$$

It is not important that you use the equation but it is important that you know that there is a mathematical relation between the vapor pressure of a compound and the heat of vaporization of that compound and that the constant in the equation has to be experimentally determined for each compound.

Condensation is the reverse of vaporization and the heat of condensation is numerically the same as the heat of vaporization but the sign is reversed:

$$H_2O(g) \longrightarrow H_2O(l) \qquad \Delta H_{373\,K} = -9.7 \text{ kcal}$$

When 18 g of steam, $H_2O(g)$, condense to 18 g of liquid water, $H_2O(l)$, 9.7 kcal are released. This is the reason that steam at 100 °C produces such severe burns. For every 18 g of steam that condense on the skin, 9.7 kcal are transferred to the skin.

By comparison, 18 g of liquid water (less than 2 tablespoons) at 100 °C produce a very minor burn.

Le Chatelier's Principle

Le Chatelier, a nineteenth-century French scientist, reached a remarkably perceptive conclusion concerning the consequences of meddling with a system at equilibrium. Very freely paraphrased, his conclusion was that a system forced out of equilibrium readjusts in a manner that tends to undo the change it has just suffered. Raise the temperature of a system of water and water vapor at equilibrium and part of that added heat is used up by converting more liquid water to the vapor phase. In so doing, the temperature elevation is in part diminished. In a sense, Le Chatelier's principle might be called an observation of the natural perversity of a disturbed system. **Le Chatelier's principle** is not an exact relation given by a mathematical equation, and it is not a rationalization of how or why the change occurs. It is, however, a very useful indicator of the direction of the response to changes imposed on systems in equilibrium.

Viscosity

Other properties of liquids are also related to the forces between molecules. Some liquids flow like water and others are as slow as molasses. The property of liquids related to resistance to flow is quite familiar and most people have an intuitive feeling for viscosity born of long experience in transferring liquids. **Viscosity** is defined in terms of the force necessary to achieve certain specified flow conditions and is most frequently measured in a unit called a poise. The definition of viscosity is rather complicated and very difficult to apply in a direct fashion. For qualitative consideration, intuition is adequate, but it may lead to erroneous conclusions. There is a tendency to feel that a viscous liquid is also a dense liquid. A viscous oil, frequently called a heavy oil, does in fact float on water, a consequence of a volume of oil having a mass smaller than the mass of the same volume of water. Density and viscosity are quite different properties. Viscosity is related to motion of molecules with respect to other molecules. Density is the mass per unit volume. It is common experience that the viscosity of a liquid is decreased by an increase in temperature (hot pancake syrup pours more readily than cold pancake syrup), an inverse relation but not an inverse proportion.

Surface Tension

Another commonly observed property of a liquid is the tendency for a small volume of a liquid to assume a spherical shape. This is the consequence of the forces between the molecules bringing molecules close together. The actual shape of a drop on a dry surface is the result of the balance of (1) the forces between molecules of the liquid, (2) the forces between the surface and the molecules of the liquid, and (3) the force of gravity on the molecules. The forces between the molecules of the liquid tend to reduce the total surface of the drop to a minimum. The property is known as **surface tension.** It can be defined in two ways: as the force acting across a one-centimeter edge of surface or as the energy required to increase the surface area by one square centimeter. The two definitions are entirely equivalent although they seem to be quite different. A force acting through a distance is the expenditure of energy. The surface tension of water in contact with air is quite different from the surface tension of water in contact with benzene.

van der Waals' Forces

The forces between molecules in the liquid phase are believed to differ in degree but not in kind from the van der Waals' forces believed to be responsible for the negative deviation of real gases from ideal behavior. If the pressure is low enough and the temperature is high enough, all liquid compounds become gases—assuming, of course, that the compound does not decompose at the higher temperature. The forces between molecules are in all cases believed to be charge–charge interactions—also called dipole–dipole interactions. Polar bonds are an obvious source of charge–charge interaction. With an unequal distribution of charge over the molecule, two molecules may tend to pair or a large number of molecules may become a continuously shifting heel-to-toe routine that tends to hold the molecules together. Figure 12-4. This type of attractive force is believed to play a significant role with polar molecules, such as the monochlorinated saturated hydrocarbons, the alcohols, and water. To a somewhat smaller degree, it could also play a role with nonpolar molecules that contain polar bonds: carbon dioxide, O═C═O, and acetylene, H—C≡C—H. Both are linear molecules (*sp* hybridization) and the symmetry of the molecules leads to net permanent dipole moments of zero for the molecules as a whole in spite of markedly polar bonds between oxygen and carbon and much less polar bonds between hydrogen and carbon. There are charge distributions within the nonpolar molecules that can be a source of attractive forces with other molecules.

Forces between Polar Molecules

Forces between Polar Groups

A more challenging problem is to rationalize the attractive forces between atoms. The monatomic gases—He, Ne, Ar, Kr, and Xe—do form liquids under rather extreme conditions: high pressures and low temperatures. On the average, the distribution of electrons in an atom is such that the center of negative charge coincides with the nucleus of the atom and there is no separation of the center of negative charge and the center of positive charge in the atom. At any instant, however, the distribution of electrons may be such that the center of negative charge does not coincide with the nucleus. These fleeting inequalities in distribution of charge are thus fleeting dipole moments. They do not have staying power, but while they last, they can also disturb

Induced Dipoles

Figure 12-4

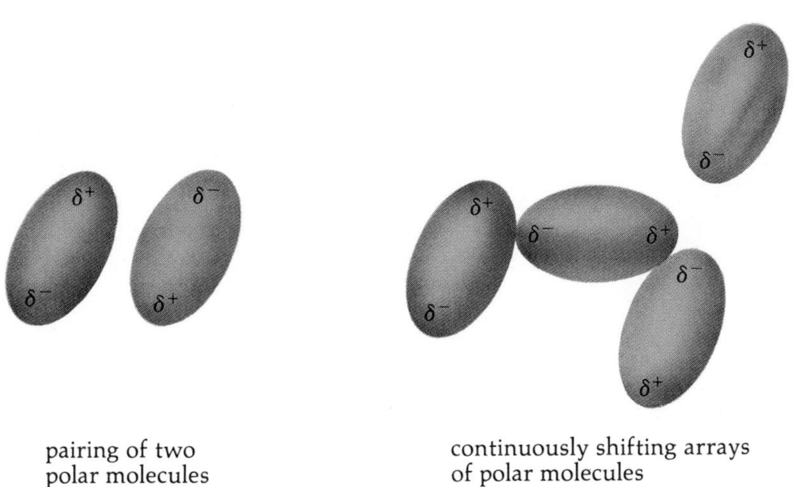

Interaction of Polar Molecules.

the electron atmospheres of adjacent atoms and the total mob effect is not negligible. Disruptions of the spherical symmetry of the electron atmosphere of an atom can be brought about by the collision of atoms in their random motions. This type of induced van der Waals' forces is known as **London** (Fritz, not the city) **forces.** The more delocalized the electron atmosphere is (as in the larger atoms), the more significant the London interactions become. This is consistent with the increase in boiling point with the increase in size of the noble gas elements.

London Forces

benzene, C_6H_6

The concept of **induced dipole moments** is also consistent with the high boiling point of benzene, C_6H_6 (mol wt 78), as compared to *n*-hexane, $CH_3(CH_2)_4CH_3$ (mol wt 86): 80 °C as compared to 69 °C. The benzene molecule is a rather special molecule with all the nuclei of the 12 atoms lying in a plane and all of the bond angles being 120°. It is frequently represented by either of the following structures:

```
          H                                H
          |                                |
          C                                C
       //   \                           /    \\
   H—C         C—H                  H—C        C—H
     ||        |          or          |        ||
   H—C         C—H                  H—C        C—H
       \    //                          \\   /
          C                                C
          |                                |
          H                                H
```

Each of the 12 *sigma*-bonds is considered to involve at least one sp^2 hybrid orbital of a carbon atom. All 12 of the ∞-fold axes of rotation of the σ-orbitals lie in the plane of the molecule. Each of the *pi*-bonds between adjacent carbon atoms is considered to arise from a partially filled unhybridized *p*-orbital of each of these two adjacent carbon atoms. For the six carbon atoms, there are a total of six *p*-orbitals and six electrons. The two structures drawn above have simply paired off the partially filled *p*-orbitals differently. The six carbon–carbon bond lengths are known to be equal, 0.140 nm, and there is no justification for an arbitrary pairing of the *p*-orbitals to give a sequence of alternating single and double bonds with different bond lengths.

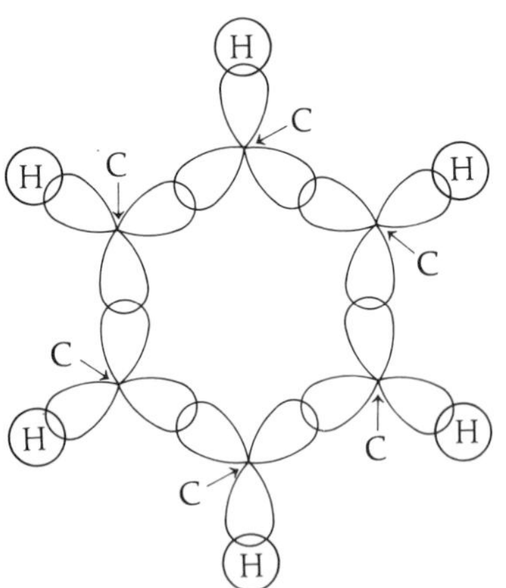

formation of the σ-orbitals of benzene

partially filled p-orbitals
of the six carbon atoms
of benzene

Delocalized Molecular Orbitals

The current approach is to consider a much more delocalized electron structure for benzene, with the six partially filled p-orbitals (for all of which the ∞-fold axis of rotation is perpendicular to the plane of the molecule) interacting around the ring to give two doughnut-type clouds of electron charge, one doughnut above the plane of the nuclei and one below the plane of the nuclei. The negative charge density of these two clouds is derived from the six electrons. This is a molecular orbital encompassing six nuclei; it is also a π-system formed by side-to-side overlap of six partially filled p-orbitals.

the six carbon atom nuclei of the benzene ring lie in a plane between the two "doughnuts" of high electron probability

The pictograph

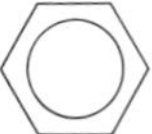

is frequently used as a symbol for benzene. Each corner is understood to represent a carbon atom and its attached hydrogen atom. The straight lines represent carbon–carbon *sigma*-bonds and the circle represents the delocalized π-cloud system of the six-carbon ring. It is this π-cloud of electrons that is particularly susceptible to London interactions and is believed to account for this relatively high boiling point of 80 °C for benzene. Space-filling models show the molecule to be a compact, thick disk. The dipole moment of benzene is zero.

Table 12-6

NORMAL BOILING POINTS AND DIELECTRIC CONSTANTS FOR ETHANOL AND OTHER TWO-CARBON COMPOUNDS

	molecular weight	normal boiling point (°C)	dipole moment (debye)
CH_3CH_2OH ethanol	46	78	1.69
CH_3CH_2Cl chloroethane	64	12	2.05
CH_3OCH_3 dimethyl ether	46	−23	1.30
CH_3CH_2SH ethanethiol	62	35	1.58
$CH_3CH_2NH_2$ ethylamine	45	−33	1.47

Table 12-6 provides evidence of the uniqueness of the boiling point of ethanol as compared to the boiling points of similar two-carbon compounds. At a pressure of one atmosphere, ethanol boils at 78 °C and chloroethane at 12 °C, in spite of the fact that the chloroethane molecule has a higher dipole moment and a higher molecular weight. With the exception of ethanethiol, other comparisons are equally dramatic.

Table 12-7 provides evidence of the uniqueness of the boiling point of water as compared to the boiling points of other low molecular weight compounds of

Table 12-7

NORMAL BOILING POINTS (°C) OF WATER AND OTHER SMALL MOLECULAR COMPOUNDS OF HYDROGEN

CH_4	NH_3	H_2O	HF
−161	−33	100	19
		H_2S	HCl
		−61	−85
		H_2Se	HBr
		81	−66
		H_2Te	HI
		130	−35

hydrogen. The pattern of Table 12-7 is that of the Periodic Table for the elements other than hydrogen. Most of the boiling points are below 0 °C. In the halogen family, the boiling points of the hydrogen halides increase with molecular weight, with the exception of hydrogen fluoride. In the oxygen family, the boiling points increase with molecular weight, with the exception of water. Water boils at 100 °C as compared to −61 °C for hydrogen sulfide, H_2S. The water molecule is more polar than the hydrogen sulfide molecule, but the magnitude of this difference cannot account for a difference of 161 degrees in boiling point. These abnormally high boiling points and a multiplicity of other evidences have led to the conclusion that hydrogen bonded to the small highly electronegative oxygen atom is responsible for another type of interaction between molecules in which the hydrogen acts as a bridge between two oxygen atoms. This interaction is called a **hydrogen bond.**

Hydrogen Bonds

The hydrogen bond is believed to arise from the very small size and the very simple structure of the hydrogen atom. In a *sigma*-bond, the two electrons of the bond are primarily involved in the charge distribution between the two atoms. This leaves the proton of the hydrogen atom comparatively unprotected. This is particularly true when the *sigma*-bond involves a highly electronegative atom such as an atom of fluorine, oxygen, or nitrogen. In this case, the charge distribution is displaced toward the electronegative atom and the proton of the hydrogen atom is exposed to a greater degree than in a bond with a less electronegative atom. This allows the proton to approach any other molecule very closely. The preferred approach would, of course, be to the region of the molecule that has a high negative charge density. The proton of the hydrogen atom of one molecule of water or alcohol approaches very close to the oxygen atom of another molecule of water or alcohol.

For hydrogen bonding between molecules of water, the three atoms involved, oxygen–hydrogen–oxygen, lie on a straight line and the interaction is believed to be entirely electrostatic, not a bond in the usual *sigma*-bond sense of sharing electrons. The energy of dissociation of a hydrogen bond in water is approximately 5 kcal $mole^{-1}$. As chemical bonds go, the "hydrogen bond" is not a strong bond. The energy required to break the hydrogen bond between two molecules of water is less than 10% of the energy required to break the *sigma*-bond between the hydrogen

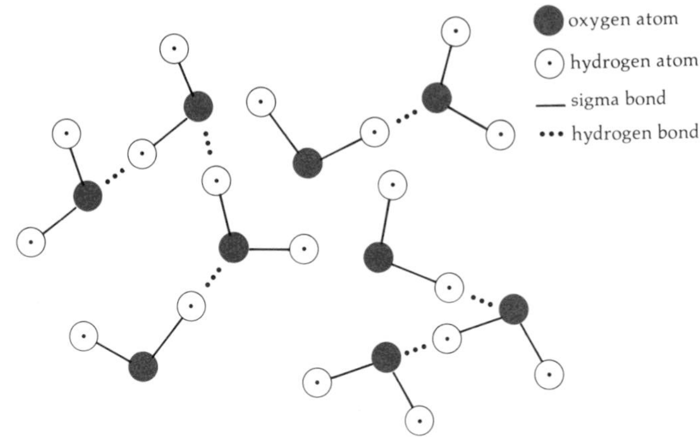

schematic two-dimensional representation of nine molecules of water grouped by hydrogen bonding into a dimer, a trimer, and a tetramer

atom and the oxygen atom in one molecule of water. The hydrogen bond between a hydrogen atom in one molecule and an oxygen atom in another molecule is longer than the σ-bonds between the oxygen atom and the hydrogen atom within a single molecule of water (the figure above is not exact). It is possible for hydrogen atoms of two molecules of water to form two hydrogen bonds with the oxygen atom of a third molecule of water but this is probably uncommon in liquid water. It is, however, believed to be the rule in ice and to determine the crystal properties of the solid phase of water. In ice, each oxygen atom is surrounded by four atoms of hydrogen—two σ-bonds and two hydrogen bonds—with tetrahedral orientation.

The extensive degree of hydrogen bonding that takes place in liquid water raises considerable uncertainty as to the nature of the molecular unit in liquid water. It is certainly not a single H_2O unit and may be best represented by $(H_2O)_n$, where n is unknown and probably variable. In the vapor phase, the molecule is definitely the simple H_2O unit.

Hydrogen bonding does not occur to the same degree with the corresponding sulfur compounds ethanethiol, CH_3CH_2SH, and hydrogen sulfide, H_2S. The sulfur atom is not as electronegative as the oxygen atom so the proton does not become as vulnerable. The sulfur atom is also a larger atom and the distance of nearest approach is larger.

In summary, van der Waals' forces are electrostatic forces and encompass

Summary of van der Waals' Forces

- dipole–dipole interaction of permanent dipoles,
- London forces—induced dipole interactions, and
- hydrogen bonding.

All molecules can be involved in interactions of the London type. London forces are, in general, the weakest of the van der Waals' forces and most significant in molecules with delocalized electron atmospheres. Hydrogen bonding, when the structures and proximity of molecules permit hydrogen bonding to occur, is the strongest of the van der Waals' forces.

Hydrogen bonds belong in the "count your blessings" list of every biological organism on this planet. We would not get along too well if water boiled at a temperature lower than the boiling point of hydrogen sulfide, lower than −61 °C. Hydrogen bonding is extremely important in determining how we are put together and how we function. (Chapters 17 through 20.)

Miscible and Immiscible Liquids

Returning to the problem of benzene and water, what types of compounds are oils? Liquid compounds that are quite unlike water. Liquid compounds that do not form strong hydrogen bonds with water. These include the hydrocarbons and the halogen compounds of the hydrocarbons. It is the hydrogen bonding of water molecules with water molecules that excludes molecules of other compounds and maintains the separate water phase. Ethanol does form hydrogen bonds and ethanol is completely miscible with water. Hydrogen atoms in water molecules hydrogen bond to oxygen atoms in ethanol; hydrogen atoms of the —OH group in ethanol hydrogen bond to the oxygen atoms in water. The term **hydrophobic,** water-hating, is frequently associated with compounds such as hydrocarbons that are immiscible with water. The term is misleading. It is not that the compounds repel water but that water attracts water strongly and, in so doing, excludes the other compound.

Low molecular weight carbon–hydrogen–oxygen compounds such as

ethanol, CH_3CH_2OH,

acetone, $(CH_3)_2CO$,

acetic acid, $CH_3C(=O)—OH$

and

the ethyl ester of acetic acid, $CH_3C(=O)—OCH_2CH_3$,

contain oxygen atoms that can hydrogen-bond with molecules of water and these compounds are very soluble in water. On the other hand, compounds such as

1-pentanol, $CH_3CH_2CH_2CH_2CH_2OH$,

3-pentanone, $(CH_3CH_2)_2CO$,

hexanoic acid, $CH_3CH_2CH_2CH_2CHC(=O)—OH$,

and

the ethyl ester of propanoic acid, $CH_3CH_2C(=O)—OCH_2CH_3$,

are only slightly soluble in water. That these compounds dissolve even slightly in water is the consequence of hydrogen bonding of water molecules to the electronegative oxygen atoms in these molecules. That these compounds are only slightly soluble in water is the consequence of the extensive hydrocarbon-like segments of these molecules. An exploration of the space-filling model of the 20-carbon "straight-chain" saturated hydrocarbon $C_{20}H_{42}$

$$CH_3CH_2CH_2CH_2CH_2CH_2CH_2CH_2CH_2CH_2CH_2CH_2CH_2CH_2CH_2CH_2CH_2CH_2CH_2CH_3$$

is helpful in understanding the many configurations of saturated hydrocarbon chains. The experimental bond angles are all 109.5°. A remarkable degree of flexibility is introduced into the chain by the 19 carbon–carbon *sigma*-bonds.

Three configurations of this molecule are given in Figure 12-5. Note that the three configurations given in this figure are for a single isomer of $C_{20}H_{42}$. They differ from each other only in rotations about carbon–carbon *sigma*-bonds. The sequence of bonding within the molecule remains unchanged and all bond angles remain the same. In Figure 12-6, the carbon chain is revealed for two of these configurations by the removal of one hydrogen atom from each carbon atom.

The extended chain requires a unique set of orientations about the carbon–carbon *sigma*-bonds and a bit of care is required to orient the carbon atoms to give the extended chain. Figure 12-7 displays this orientation of the carbon atoms. The probability of 20 carbon atoms assuming this unique orientation is extremely small. The most probable orientation of a carbon–carbon chain is some variation of a serpentine configuration.

Long hydrocarbon chains attached to carboxylate groups

$$-\overset{\overset{\displaystyle O}{\|}}{C}-O-$$

are essential components of animal fats, vegetable oils, and cell membranes (Chapter

Figure 12-5

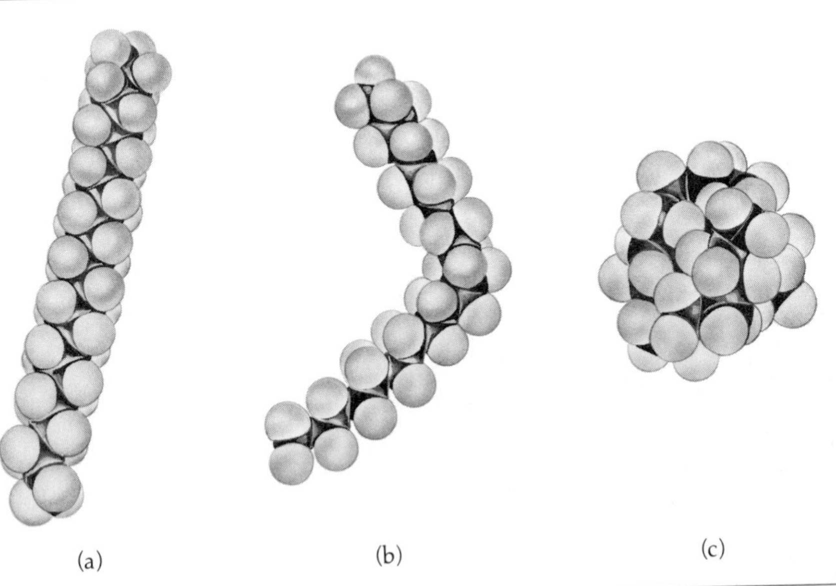

THREE CONFIGURATIONS OF THE "STRAIGHT-CHAIN" ISOMER, $C_{20}H_{42}$. (a) The fully extended chain; (b) a bent molecule that differs from the extended chain only by 180° rotation about one carbon–carbon *sigma*-bond near the center of the molecule; (c) a globby molecule that differs from the extended chain by varying degrees of rotation about the 19 carbon–carbon *sigma*-bonds.

20.) The properties of a compound such as stearic acid, $CH_3(CH_2)_{16}COOH$, a saturated fatty acid, are determined by the properties of hydrocarbon chains and the properties of the carboxylic acid group.

globby molecule
of stearic acid

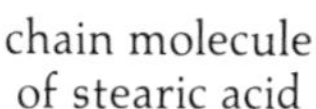

chain molecule
of stearic acid

Composition of Solutions

Concentration of Solutions

Since solutions are so much a part of our existence, it is useful to be able to express the concentration of a solution in numerical terms, and a great number of systems have developed as means of convenience. Instructions for many products to be used domestically may be stated in terms such as "Use two tablespoons to make one pint of solution by adding water." Such solutions can be prepared quite reproducibly, but the quantities used have nothing to do with quantities that are chemically relevant. Note that "two tablespoons per pint of solution" is actually a quotient:

$$\frac{\text{tablespoons of solute}}{\text{pints of solution}} = 2 \text{ tablespoons pint}^{-1}$$

The solution would have the same concentration if 6 tablespoons of solute were used to prepare three pints of solution: there would just be more of the solution.

Molarity

Concentrations that would be most useful to chemists should obviously be expressed in terms of moles of one or more of the substances that make up the solution. One of the most commonly used expressions of concentration of liquid solutions is molarity, M. **Molarity** is defined as the quotient of moles of solute divided by liters of solution. This is an extremely useful definition. Expressed as an equation,

Figure 12-6

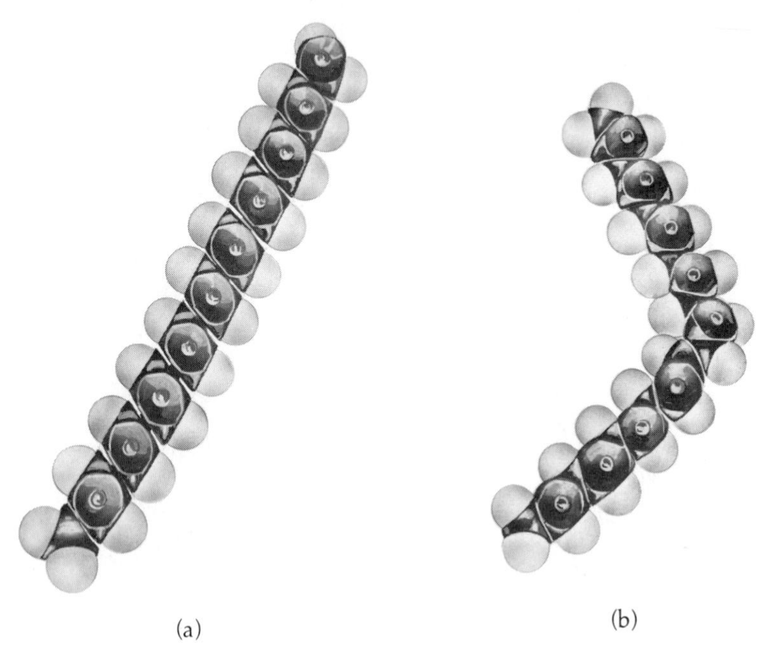

INCOMPLETE MODELS FOR THE "STRAIGHT CHAIN" ISOMER OF $C_{20}H_{42}$. (a) The extended chain (Figure 12-5a) with one hydrogen atom removed from each carbon atom; (b) the bent chain (Figure 12-5b) with one hydrogen atom removed from each carbon atom.

$$\text{molarity} = \frac{\text{moles of solute}}{\text{liters of solution}} = \text{M}$$

This equation involves only three quantities. If two of these are known, the third can be calculated. Don't underestimate the usefulness of this equation. Laboratories are filled with glass equipment calibrated in terms of liters or, more frequently, in terms of milliliters, and the experimental determination of moles of solute is no problem, assuming, of course, that the formula of that compound is known and the molecular weight can be found by adding the appropriate numbers of atomic weights.

For example, to prepare 300 milliliters of a 1.6 molar solution of sucrose in water, first find out how many moles of sucrose are needed. The quotient, moles of solute/liters of solution, is to equal 1.6 moles/liters and the volume of the solution is to be 0.300 liters.

$$\frac{x\text{ moles}}{0.300\text{ liters}} = 1.6\,\frac{\text{moles}}{\text{liters}}$$

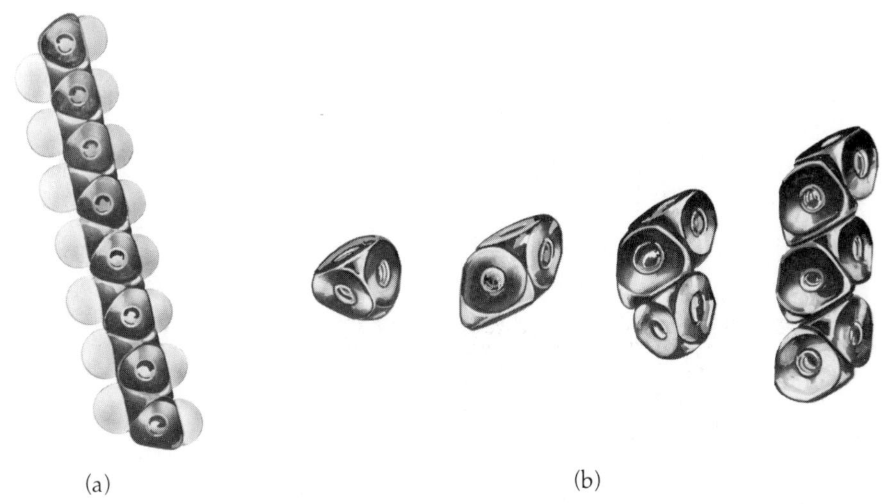

ORIENTATION OF THE CARBON ATOM SPACE-FILLING MODELS IN THE EXTENDED CHAIN. (a) An expanded segment of the extended chain with one hydrogen atom removed from each carbon atom (Figure 12-6a); (b) the orientation of carbon atoms models to give the extended carbon chain.

$$x \text{ moles} = 1.6 \frac{\text{moles}}{\cancel{\text{liters}}} \times 0.300 \cancel{\text{liters}}$$

$$= 0.48 \text{ moles}$$

The formula of sucrose is $C_{12}H_{22}O_{11}$, and one mole of sucrose, therefore, has a mass of 342 grams. Consequently, all that is needed is

$$342 \frac{\text{grams}}{\cancel{\text{mole}}} \times 0.48 \cancel{\text{moles}} = 164 \text{ grams}$$

to march off to the laboratory, weigh out 164 grams of sucrose, add water to dissolve the sucrose, and bring the total volume of the solution to 300 ml by continuing to add water. In using molarity, the emphasis is on the number of moles of the solute that would be distributed throughout a liter of solution.

Another method of expressing composition places the emphasis upon the fraction of the moles present that are of a particular compound. Very logically, this method of expressing concentration is called **mole fraction.** For a solution prepared by mixing 1.00 mole of ethanol with 2.00 moles of water, the mole fraction of ethanol is

Mole Fraction

$$\frac{1.00 \cancel{\text{moles}}}{3.00 \cancel{\text{moles}}} = 0.33$$

and the mole fraction of water is

$$\frac{2.00 \cancel{\text{moles}}}{3.00 \cancel{\text{moles}}} = 0.67$$

Note that the sum of the mole fractions for the solution is 1.00. (Mole fractions are always expressed as decimal quantities.)

In general terms for a solution made up of two compounds, compound A and compound B, the mole fraction of A, X_A, is given by the equation

$$\text{mole fraction of A} = \frac{\text{moles A}}{\text{total moles}} = \frac{\text{moles A}}{\text{moles A} + \text{moles B}} = X_A$$

and the mole fraction of B, X_B, in the same solution is given by the equation

$$\text{mole fraction of B} = \frac{\text{moles B}}{\text{total moles}} = \frac{\text{moles B}}{\text{moles A} + \text{moles B}} = X_B$$

Since the solution contains only two compounds, A and B, the fraction that isn't A has to be B and the sum of X_A and X_B must be the total solution:

$$X_A + X_B = 1$$

If the solution contains three compounds—A, B, and C—the sum of the three mole fractions must be 1.

In discussing the properties of solutions, the mole fractions are most informative since they give directly the relative number of molecules of each compound to the total number of molecules present. In a solution of benzene, C_6H_6, and diethyl ether, $CH_3CH_2OCH_2CH_3$, that has a mole fraction of 0.20 for benzene and 0.80 for diethyl ether, there are twenty molecules of benzene for every eighty molecules of diethyl ether. Two out of ten molecules are benzene; eight out of ten molecules are diethyl ether.

Note that the units of molarity are moles liters^{-1} but that mole fractions are fractions—moles divided by moles—and mole fractions have no units.

Ideal and Nonideal Solutions

In order to consider further the concept of ideal and nonideal solutions, we shall first consider a solution made up of toluene and *para*-xylene:

CH_3 CH_3

CH_3

methylbenzene, also called toluene

1,4-dimethylbenzene, also called *para*-xylene

Table 12-8

NORMAL BOILING POINTS AND VAPOR PRESSURES FOR TOLUENE AND *para*-XYLENE AT THREE SPECIFIED TEMPERATURES

	toluene	*para*-xylene
boiling point	111 °C	138 °C
vapor pressure		
30 °C	32 torr	11 torr
60 °C	123	50
90 °C	358	166

At room temperature, both are colorless liquids. Their boiling points and vapor pressures at three temperatures are given in Table 12-8. There are two other isomers of xylene that differ from *para*-xylene by the relative positions of the two methyl groups on the benzene ring, but for simplicity this 1,4-isomer will be referred to as xylene in the following discussion.

We shall arbitrarily choose a solution that has the composition expressed in mole fractions

$$\underset{\text{liquid}}{X_{\text{toluene}}} = 0.20$$

$$\underset{\text{liquid}}{X_{\text{xylene}}} = 0.80$$

and also arbitrarily choose to work at 30 °C. This solution is in a closed flask from which all air has been removed.

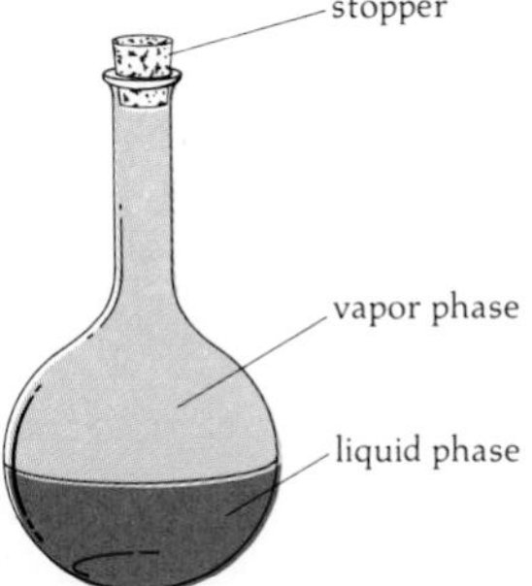

Composition of the Vapor Phase

Question: What is the equilibrium pressure of toluene in the vapor phase? of xylene in the vapor phase? The answers are that we do not know without making an experimental determination. Ideally, we might expect the pressure of the vapor of toluene, P, to be directly proportional to the mole fraction of toluene in the liquid phase:

$$\underset{\text{vapor}}{P_{\text{toluene}}} \propto \underset{\text{liquid}}{X_{\text{toluene}}}$$

and the proportionality constant to be the vapor pressure of pure toluene at 30 °C:

$$\underset{\text{vapor}}{P_{\text{toluene}}} = 32 \text{ torr} \times \underset{\text{liquid}}{X_{\text{toluene}}} \quad \text{at } 30\ ^\circ\text{C}$$

The proposal that the vapor pressure of toluene is directly proportional to the mole fraction of toluene in the liquid phase is equivalent to the assumption that, whatever molecules of toluene do in pure toluene, molecules of toluene continue to do the same thing in the solution of toluene and xylene, but only part of the molecules reaching the surface of the solution are toluene. In this particular case, only two out of ten molecules are toluene:

$$\underset{\text{vapor}}{P_{\text{toluene}}} = 32 \text{ torr} \times 0.20 = 6.4 \text{ torr}$$

For this particular solution at 30 °C, the expected pressure of toluene in the vapor phase is 6.4 torr. Similarly, we would expect the pressure of xylene to be directly proportional to the mole fraction of xylene in the liquid phase:

$$\underset{\text{vapor}}{P_{\text{xylene}}} \propto \underset{\text{liquid}}{X_{\text{xylene}}}$$

and the proportionality constant to be the vapor pressure of pure xylene at 30 °C.

$$\underset{\text{vapor}}{P_{\text{xylene}}} = 11 \text{ torr} \times X_{\text{xylene}} \quad \text{at } 30\ ^\circ\text{C.}$$

For this particular solution, the expected pressure of xylene is 8.8 torr.

$$\underset{\text{vapor}}{P_{\text{xylene}}} = 11 \text{ torr} \times 0.80 = 8.8 \text{ torr}$$

If the experimental values of these pressures are 6.4 torr for toluene and 8.8 torr for xylene, the solution is said to be ideal. If the experimental values are different, the solution is said to be nonideal. Due to the similarity in molecular structures of these two compounds, a solution of toluene and xylene would be expected to be ideal or very close to ideal. For some nonideal solutions, the experimental values are greater than the ideal values. For others, the experimental values are smaller than the ideal values.

Using the calculated ideal values of pressure, the expected total pressure in the vapor phase is 15.2 torr:

$$\underset{\text{vapor}}{P_{\text{total}}} = 6.4 \text{ torr} + 8.8 \text{ torr} = 15.2 \text{ torr at } 30\ ^\circ\text{C}$$

and the expected composition of the vapor phase expressed in mole fractions is

$$\underset{\text{vapor phase}}{X_{\text{toluene}}} = \frac{6.4 \text{ torr}}{15.2 \text{ torr}} = 0.42 \text{ at } 30\ ^\circ\text{C}$$

$$\underset{\text{vapor phase}}{X_{\text{xylene}}} = \frac{8.8 \text{ torr}}{15.2 \text{ torr}} = 0.58 \text{ at } 30\ ^\circ\text{C}$$

Note that the expected mole fraction of toluene in the vapor phase (0.58) is greater than the mole fraction of toluene in the liquid phase (0.20). This is not surprising. Toluene is more volatile than xylene. Pure toluene has the higher vapor pressures and the lower boiling point. The vapor phase in equilibrium with a solution is always richer in the more volatile component than the liquid phase of the solution.

If the temperature of the flask and its contents is raised to 60 °C and time is allowed for equilibrium to establish, the expected pressure of toluene in the vapor phase becomes 24.6 torr and the expected pressure of the xylene becomes 40.0 torr.

$$P_{\text{toluene vapor}} = 123 \text{ torr} \times 0.20 = 24.6 \text{ torr} \quad \text{at } 60\ ^\circ\text{C}$$

$$P_{\text{xylene vapor}} = 50 \text{ torr} \times 0.80 = 40.0 \text{ torr} \quad \text{at } 60\ ^\circ\text{C}$$

For equilibrium at 90 °C, the expected pressures for toluene and xylene in the vapor phase are 71.2 and 132.8 torr.

The relationship for solutions that we have just proposed and used

Raoult's Law

$$\begin{pmatrix}\text{pressure of compound A}\\ \text{in the vapor phase in}\\ \text{equilibrium with the}\\ \text{liquid solution}\end{pmatrix} = \begin{pmatrix}\text{vapor pressure of}\\ \text{pure compound A}\end{pmatrix} \times \begin{pmatrix}\text{mole fraction of}\\ \text{compound A}\\ \text{in the solution}\end{pmatrix}$$

$$P_{\text{A}} = P^0_{\text{A}} \times X_{\text{A}}$$

is known as Raoult's law in honor of the French scientist who first proposed it.

Figure 12-8

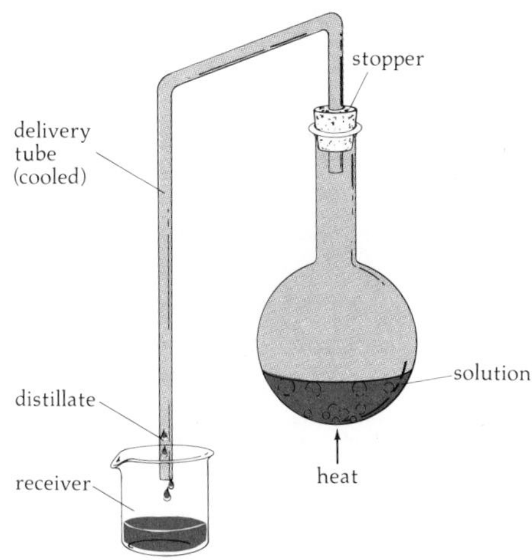

SCHEMATIC STILL.

12-1 *Gratuitous Information*

STILLS AND THE DISTILLATION PROCESS

The following discussion is based upon our previous exploration of the composition of the vapor phase in equilibrium with a solution of toluene and xylene. To distill this solution, the stopper in the flask could be replaced by another stopper through which the end of a bent glass tube has been inserted and the flask heated. Figure 12-8.

As the temperature of the solution rises, the rate of evaporation of both toluene and xylene increases and the pressures of both compounds in the vapor phase increase. Eventually, the sum of the pressure of toluence vapor and the pressure of xylene vapor in the flask equals atmosphere pressure and the liquid boils. As the solution boils, the mixture of toluene and xylene vapor pushes the air out of the flask and the delivery tube as the vapor flows through the flask and the tube. If the tube is appropriately cooled, the vapor condenses and the distillate drips into some appropriate receiver. In practice, the solution would be distilled in well-designed equipment with thermometers in place, an electrical heater, a cold water jacket built around the vertical portion of the delivery tube, and a connected receiver. The details of the equipment are not important to this discussion. What is important is that the composition of the liquid collected (the distillate) from the condensation of the vapor, is different from the composition of the solution being boiled and that the distillate is richer in the more volatile component.

Those high-rise structures so characteristic of petroleum refineries are complex distillation equipment to separate crude petroleum into fractions, some of which are made up of very volatile hydrocarbons, others of less volatile hydrocarbons. Gasoline is one of the more volatile fractions produced. The actual composition of gasoline is balanced to give a more volatile mixture for winter driving conditions than for summer driving. For ideal solutions and most nonideal solutions, each time the process of distillation is repeated, the distillate becomes richer in the more volatile compounds than the solution that is being distilled. The great height of petroleum refinery distillation equipment is to allow for multiple cycles of vaporization and condensation. Ethanol is more volatile than water and the stills of mountain moonshine fame are crude contraptions to distill over a solution more concentrated in alcohol than the fermentation mixture. Not only does distillation improve the alcoholic content of the product, it also separates the product from nonvolatile substances present in the fermentation mixture.

If the properties of a solution are consistent with **Raoult's law,** the solution is said to be an ideal solution. If the properties of a solution are not consistent with Raoult's law, the solution is said to be nonideal. What are the other properties of ideal solutions of two liquid compounds? The two liquids are miscible in all proportions; the heat of mixing the two liquids is zero, and the volume of the solution must equal the sum of the volumes of the two liquids before they are mixed. In all liquids there are, of course, attractive forces between the molecules. In an ideal solution, the attractive forces between molecules of A and molecules of B must be very similar to the attractive forces between molecules of A and other molecules of A and also very similar to the attractive forces between molecules of B and other molecules of B. This equality of attraction among all the molecules is most probable for compounds that have molecules of similar mass and similar structure.

Scramble Exercises

1. Glycerol is a colorless, high-viscosity liquid that decomposes at about 290 °C without boiling at a pressure of one atmosphere. Glycerol is completely miscible with water. Suggest a plausible explanation of these properties.

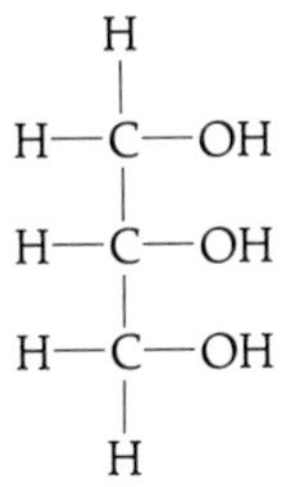

```
     H
     |
  H—C—OH
     |
  H—C—OH
     |
  H—C—OH
     |
     H
```

glycerol

2. How many grams of glycerol (see the above exercise for the formula of glycerol) are dissolved in 100 ml of a 0.100-molar solution of glycerol in water?

3. At 20 °C, the vapor pressure of pure benzene, C_6H_6, is 75 torr and the vapor pressure of pure toluene is 23 torr. Which is the more volatile compound, benzene or toluene?

4. A solution of benzene and toluene does in fact very closely approach ideal behavior.

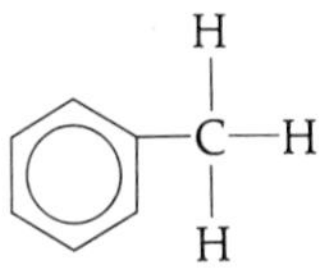

C_6H_6 benzene and $C_6H_5CH_3$ toluene

Assuming ideal behavior and equilibrium between the vapor phase and the liquid phase at 20 °C, calculate the pressure of benzene and the pressure of toluene in the vapor phase if the mole fraction of toluene in the liquid phase is 0.40. Also calculate the mole fraction of toluene and the mole fraction of benzene in the vapor phase. See Exercise 3 for additional information.

5. What is the mole fraction of ethanol, C_2H_5OH, in a solution prepared by dissolving 10.0 grams of ethanol in 10.0 grams of water?

6. For the two isomers of xylene other than *para*-xylene, 1,4-dimethylbenzene, propose the relative positions of the two methyl groups on the benzene ring.

Additional exercises for Chapter 12 are given in the Appendix.

A Look Ahead

In this chapter, we have been discussing covalent compounds. Aqueous solutions of these compounds, if in fact they dissolve appreciably in water, do not conduct an electric current. In the next chapter, we shall explore aqueous solutions of covalent compounds that react with water to give aqueous solutions that do conduct—aqueous solutions that contain ions. In a sense, these reactions might be considered as hydrogen bonding that goes beyond hydrogen bonding to a complete transfer of the nucleus of the hydrogen atom, the proton, to a water molecule. For example, hydrogen chloride gas reacts with water to give hydronium ions, H_3O^{1+}, and chloride ions, Cl^{1-}, in aqueous solutions.

$$HCl(g) + H_2O(l) \longrightarrow \underset{\text{hydronium ion}}{H_3O^{1+}(aq)} + Cl^{1-}(aq)$$

$$H\!:\!\ddot{\underset{\cdot\cdot}{Cl}}\!: + :\!\ddot{\underset{\underset{H}{\cdot\cdot}}{O}}\!:\!H \longrightarrow \left[H\!:\!\overset{\overset{H}{\cdot\cdot}}{\underset{\underset{H}{\cdot\cdot}}{O}}\!:\right]^{1+} + \left[:\!\ddot{\underset{\cdot\cdot}{Cl}}\!:\right]^{1-}$$

The resultant solution is hydrochloric acid, an acid that is essential to the digestion of the food we eat. You do, indeed, have an acid stomach.

13

VINEGAR AND OTHER SOUR SOLUTIONS: THE BRÖNSTED CONCEPT OF ACIDS

Chapter 13

The Brönsted Concept of Acids

A Note to the Reader

Acid–base phenomena are complex and also very significant to many chemical systems, including biological systems and the environment within which they flourish or perish. There are several concepts of acids and bases. The approach here is to select one concept, the **Brönsted concept** of acids and bases, and explore acid–base phenomena in terms of that concept.

Lots of numbers are used in the presentation. These are there to give reality to the phenomena being discussed. By the time you have completed the chapter, you should be able to

- write the chemical equations for the reactions of acids with water,
- formulate the K_A expressions,
- compare the relative strengths of acids given either the K_A values or the pK_A values,
- sketch and interpret the percent abundance–pH diagrams given either the relevant K_A values or pK_A values, and
- discuss buffers in terms of percent abundance–pH diagrams.

It would be nice if you could convert a K_A value into the corresponding pK_A value, but you can get along pretty well by just approximating pK_A. For example, if K_A is 3.4×10^{-8}, pK_A does equal 7.47, but to be able to recognize that pK_A must be between 7 and 8 is adequate. There is an extensive background section on logarithms to the base 10. Use it to the extent you can use it profitably. You may already be very proficient in the use of logarithms but it is also quite possible that you have never encountered them before. Most scientific calculators, even the inexpensive ones, can be used to evaluate the logarithm of a number.

Acetic Acid

Vinegar is an aqueous solution of acetic acid, CH_3COOH, and possibly a few other things, the identity of which depends upon how that particular sample of vinegar was made. Apple cider and other fruit juices contain a number of molecules and ions that may remain or be transformed into other molecules and ions during the fermentation process. White vinegar may be prepared by dissolving acetic acid in water.

Pure acetic acid is a colorless liquid that freezes at 16.6 °C and boils under a pressure of one atmosphere at 118 °C. Spectacular displays of white crystals are frequently formed during the freezing process, and for this reason pure acetic is frequently called glacial acetic acid. The odor associated with pickles is to a large degree dominated by the odor of acetic acid molecules that escape into the vapor phase. The formula as it is usually written—CH_3COOH—might seem to indicate that the two oxygens are bonded together and that the compound is a peroxide. Not true: —COOH represents the carboxylic acid group in which one oxygen atom is doubly bonded (a *sigma* and a *pi*-bond) to the carbon atom. The other oxygen atom is singly bonded (a *sigma*-bond) to the same carbon atom and also singly bonded (a *sigma*-bond) to the hydrogen atom. (Build the ball-stick-spring model.)

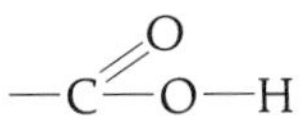

$$-C(=O)-O-H \qquad H-\underset{H}{\overset{H}{C}}-\overset{O}{\overset{\|}{C}}-OH \qquad H{:}\underset{H}{\overset{H}{C}}{:}\overset{O:}{C}{:}O{:}H$$

carboxylic acid group 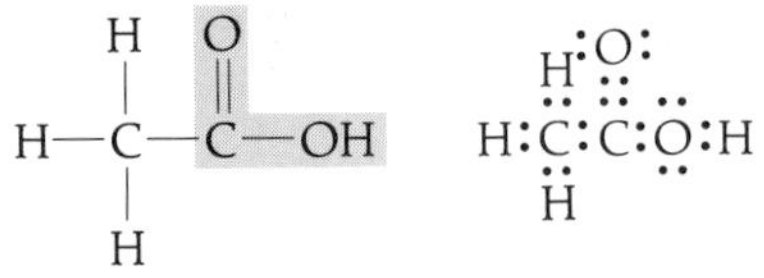 — acetic acid molecule

The three bond angles at the carbon atom of the carboxylic acid group are approximately 120° and the four atoms in the acetic acid molecule nearest to the double bond are coplanar (sp^2 hybridization for the atom of carbon in the carboxylic acid group). It is the properties of the carboxylic acid group that make acetic acid and many other carboxylic acids attractive as food additives. The carboxylic acid group is characteristic of organic acids, of which there are a great number.

Reaction with Water

The boiling point of the acetic acid indicates that it is a covalent compound made up of molecules. The fact that water and acetic acid are miscible in all proportions indicates that there are strong attractive forces between molecules of acetic acid and molecules of water. This is to be expected since both molecules contain polar hydroxyl groups,

$$-\overset{\delta-}{O}-\overset{\delta+}{H}$$

and hydrogen bonding is to be expected. The presence of the double bond to oxygen in the carboxylic acid group makes the hydrogen–oxygen bond in the acid more polar than the hydrogen–oxygen bond in the water molecule and it is the hydrogen bonding shown below that is of primary importance.

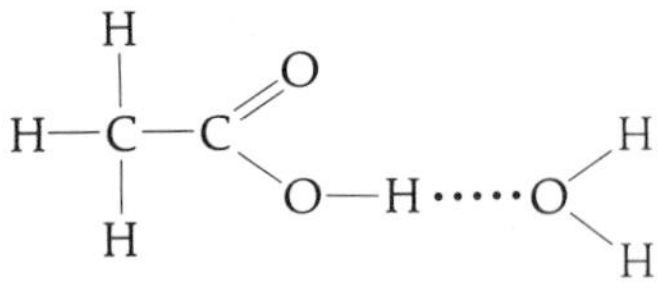

$$H-\underset{H}{\overset{H}{C}}-C(=O)-O-H\cdots O(H)(H)$$

The situation is more complicated than normal hydrogen bonding. An aqueous solution of acetic acid conducts electricity—not as well as an aqueous solution of an ionic compound such as sodium chloride, but it does conduct. In comparison, although ethanol, CH_3CH_2OH, and water mix in all proportions, the resultant solutions do not conduct.

Hydronium Ion

Some of the acetic acid molecules react with water by the transfer of the proton of the carboxylic acid group to a water molecule. The two ions that are the product of the reaction are the **acetate ion,** CH_3COO^{1-}, and the **hydronium ion,** H_3O^{1+}:

$$\underset{\text{acetic acid molecule}}{CH_3COOH(aq)} + H_2O(l) \longrightarrow \underset{\text{hydronium ion}}{H_3O^{1+}(aq)} + \underset{\text{acetate ion}}{CH_3COO^{1-}(aq)}$$

```
 H:O:                            H (1+)      H:O:  (1-)
 .. .. ..           ..           ..           .. ..  ..
H:C:C:O:H  +  H:O:   ⟶    H:O:     +  H:C:C:O:
 ..    ..           ..           ..           ..     ..
 H                  H            H            H
```

Brönsted acid Brönsted base

This reaction might be thought of as the consequence of hydrogen bonding that has gone too far. Note that the transfer involves only the proton, the nucleus of the hydrogen atom. Both electrons of the oxygen–hydrogen *sigma*-bond in the acetic acid molecule remain with the oxygen atom. This accounts for the negative charge of the acetate ion and the positive charge of the hydronium ion. It is the hydronium ion, $H_3O^{(1+)}$, that is responsible for the sour taste of vinegar and other acidic aqueous solutions. It is also the hydronium ion that is responsible for the properties of vinegar as a food preservative.

Brönsted Concept

Acids and bases as classes of compounds can be approached in a number of ways. In this discussion, we shall follow the approach known as the **Brönsted concept** of acids and bases. According to this concept, the acetic acid molecule, by donating the proton, is said to be acting as a **Brönsted acid** and the water molecule, by accepting the proton, is said to be acting as a **Brönsted base.** Any molecule or ion that donates a proton is called a Brönsted acid and any molecule or ion that accepts a proton is called a Brönsted base.

The acetate ions, $CH_3COO^{(1-)}$, are proton "hungry" and the reverse reaction between the acetate ions and the hydronium ions takes place continuously (note the

$$CH_3COOH(aq) + H_2O(l) \longleftarrow H_3O^{(1+)}(aq) + CH_3COO^{(1-)}(aq)$$

reversed arrow) at the same time that the forward reaction between the acetic acid molecules and the water takes place continuously. In this reverse reaction, the hydronium ion, by donating a proton, is said to be a Brönsted acid and the acetate ion, by accepting a proton, is said to be a Brönsted base. Don't forget that an acetic acid solution contains great numbers of molecules and ions. As some molecules of acetic acid are reacting with water, some acetate ions are reacting with hydronium ions.

$$H_3O^{(1+)}(aq) + CH_3COO^{(1-)}(aq) \longrightarrow H_2O(l) + CH_3COOH(aq)$$

hydronium ion acetate ion water acetic acid

Brönsted acid Brönsted base

When the two rates are equal, the rate at which acetic acid molecules disappear by reaction with water is equal to the rate at which acetic acid molecules appear by the reaction of acetate ions with hydronium ions. Consequently, the number of acetic acid molecules in the solution remains constant. Also, the number of acetate ions in the solution remains constant and the number of hydronium ions in the solution remains constant. This is the equilibrium situation that is established instantaneously in any aqueous solution prepared by adding acetic acid to water (or to any aqueous solution) and by maintaining that solution at a contant temperature. At **equilibrium,** the two rates are equal and opposite. At equilibrium, the concentration of the acetic acid molecules, CH_3COOH, the concentration of the acetate ions, $CH_3COO^{(1-)}$, and the concentration of the hydronium ion, $H_3O^{(1+)}$, remain unchanged although the acid and its conjugate are both continuously reacting.

The state of equilibrium, with equal and opposite rates, is indicated by the two arrows in the chemical equation:

Conjugate Pair

conjugate pair

$$CH_3COOH(aq) + H_2O(l) \rightleftharpoons H_3O^{(1+)}(aq) + CH_3COO^{(1-)}(aq)$$

Brönsted acid	Brönsted base	Brönsted acid	Brönsted base

conjugate pair

In the Brönsted terminology, the acetic acid molecule and the acetate ion are called a **conjugate acid–base pairs**; the water molecule and the hydronium ion are also a Brönsted **conjugate acid–base pair.** By the transfer of a proton, a Brönsted acid becomes its conjugate base, and a Brönsted base, by the acceptance of a proton, becomes its conjugate acid.

In commercial vinegar solutions, the total concentration of the acetic acid is about 0.7 molar (0.7 moles acetic acid dissolved per liter of solution). This concentration expresses the moles of acetic acid that would be added to water to make a liter of solution. The 0.7 figure accounts for the sum of the moles of acetic acid molecules and the moles of acetate ion present in the solution at any instant. In a 0.7 molar aqueous solution of acetic acid at room temperature, almost 1% of the acetic acid molecules added exists as acetate ions approximately 99% of the acetic acid added exists as the molecules of acetic acid. At lower total concentrations, the percentage that exists as acetate ions is greater. At high temperatures, the percentage that exists as acetate ions is also greater. It is important to keep in mind that what is an acetic acid molecule one instant is an acetate ion the next instant.

Hydrogen Chloride

Another aqueous solution that is definitely sour is called hydrochloric acid. Hydrogen chloride, HCl, is a colorless gas under normal room conditions. Its boiling point is −85 °C under a pressure of 1 atmosphere. The reaction of dissolved hydrogen chloride with water is represented by the equation

Reaction with Water

$$HCl(aq) + H_2O(l) \longrightarrow H_3O^{(1+)}(aq) + Cl^{(1-)}(aq)$$

hydrogen chloride molecule — chloride ion

conjugate pair

$$H\!:\!\ddot{\underset{..}{Cl}}\!: \;+\; H\!:\!\ddot{\underset{\textstyle H}{\underset{..}{O}}}\!: \longrightarrow \left[H\!:\!\overset{\textstyle H}{\ddot{\underset{\textstyle H}{\underset{..}{O}}}}\!:\right]^{(1+)}(aq) + \left[:\!\ddot{\underset{..}{Cl}}\!:\right]^{(1-)}(aq)$$

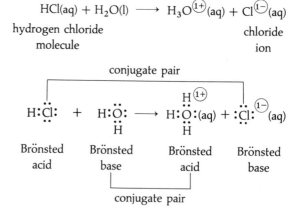

Brönsted acid	Brönsted base	Brönsted acid	Brönsted base

conjugate pair

Hydrogen chloride is reacting as a Brönsted acid and water is reacting as a Brönsted base. The HCl molecule and the Cl^{1-} ion are a conjugate pair. The reverse reaction, in which the hydronium ion, H_3O^{1+}, reacts as an acid and the chloride ion, Cl^{1-}, reacts as a base, is indicated by the reverse arrow:

Conjugate Pairs

$$HCl(aq) + H_2O(l) \longleftarrow H_3O^{1+}(aq) + Cl^{1-}(aq)$$

The equilibrium situation, in which the two rates are equal and opposite and, consequently, all concentrations have steady values, is represented by the two arrows:

$$\underset{\text{acid}}{HCl(aq)} + \underset{\text{base}}{H_2O(l)} \rightleftharpoons \underset{\text{acid}}{H_3O^{1+}(aq)} + \underset{\text{base}}{Cl^{1-}(aq)}$$

(conjugate pair: HCl and Cl^{1-}; conjugate pair: H_2O and H_3O^{1+})

It is the aqueous solution of hydrogen chloride that is known as **hydrochloric acid.** This is the acid in your stomach that facilitates the digestion of the food you eat.

In an aqueous solution of hydrochloric acid prepared to be 0.7 molar (0.7 moles HCl dissolved in water to give one liter of solution), almost 100% of the hydrogen

13-1 *Gratuitous Information*

pH AND THE THINGS WE EAT AND DRINK

We like the taste of hydronium ion. Listed below are a number of things we eat and drink with their normal pH ranges. The smaller the pH, the greater the concentration of the hydronium ion in the liquid phase. Aqueous solutions that have a pH less than 7.00 are said to be acidic; more than 7.00, basic.

Food	pH
limes	1.8–2.0*
soft drinks	2.0–4.0
lemons	2.2–2.4
wines	2.8–3.8
jams and jellies	2.8–4.0
apples	2.9–3.3
grapefruit	3.0–3.3
pickles, sour	3.0–3.4
strawberries	3.0–3.5
oranges	3.0–4.0
rhubarb	3.1–3.2
blackberries	3.2–3.6
raspberries	3.2–3.6
cherries	3.2–4.0
sauerkraut	3.4–3.6
peaches	3.4–3.6
grapes	3.5–4.5
olives	3.6–3.8
pears	3.6–4.0
tomatoes	4.0–4.4
beers	4.0–5.0
bananas	4.5–4.7
pumpkin	4.8–5.2
cheese	4.8–6.4
carrots	4.9–5.3
beets	4.9–5.5
beans	5.0–6.0
bread, white	5.0–6.0
cabbage	5.2–5.4
asparagus	5.4–5.8
potatoes	5.6–6.0
peas	5.8–6.4
tuna	5.9–6.1
corn	6.0–6.5
salmon	6.1–6.3
oysters	6.1–6.6
water, drinking	6.5–8.0
eggs, fresh	7.6–8.0

* Values listed here are taken from *CRC Handbook of Chemistry and Physics*, ed. Robert C. Weast, Melvin J. Astle, and William H. Beyer, 64th ed. (Boca Raton, Fla.: CRC Press, 1983), p. D-151.

Taste depends not only upon the pH but also upon the type and concentration of sugar present. Sweet and sour pork is just exactly that: a high concentration of sugar and a high concentration of hydronium ion.

chloride molecules reacts with water to give hydronium ions and chloride ions. At equilibrium, the concentration of the hydrogen chloride molecule in the solution is very small, almost zero.

As a consequence of the differences in capacity of CH_3COOH molecules and HCl molecules to transfer protons to water molecules, acetic acid is called a **weak acid** and hydrochloric acid is called a **strong acid.**

Hydronium ions, $H_3O^{(1+)}$, are common to all aqueous solutions and it has be-

13-1 *Background Information*

LOGARITHMS TO THE BASE 10

Any number can be expressed as 10 raised to the appropriate power:

$$1000 = 10^3 \qquad 0.24 = 10^{-0.62}$$
$$240 = 10^{2.38} \qquad 0.10 = 10^{-1}$$
$$100 = 10^2 \qquad 0.024 = 10^{-1.62}$$
$$24 = 10^{1.38} \qquad 0.010 = 10^{-2}$$
$$10 = 10^1 \qquad 0.0024 = 10^{-2.62}$$
$$2.4 = 10^{0.38} \qquad 0.0010 = 10^{-3}$$
$$1 = 10^\circ$$

In each case, the exponent is said to be the logarithm of the number. Three is the logarithm of 1000, 2.38 is the logarithm of 240, and -2.62 is the logarithm of 0.0024.

Stated in general terms, any number, N, can be expressed as 10 raised to some power, X:

$$N = 10^X$$

and the logarithm of the number N is the exponent X:

$$\log N = X$$
$$\log 1000 = 3$$
$$\log 240 = 2.38$$
$$\log 0.0024 = -2.62$$

Since these are all based on the power of 10, the more complete statements are

$$\log_{10} N = X$$
$$\log_{10} 1000 = 3$$
$$\log_{10} 240 = 2.38$$
$$\log_{10} 0.0024 = -2.62$$

Note that, in $1000 = 10^3$, the 3 is a whole number—an exact integer.

Since the logarithm of 10 is 1:

$$\log 10 = 1$$

and the logarithm of 1 is 0:

$$\log 1 = 0$$

the logarithm of any number between 1 and 10 must have a value between 0 and 1. For example, the logarithm of 2.4 is 0.38:

$$2.4 = 10^{0.38}$$
$$\log 2.4 = 0.38$$

The value 0.38 is found in a log table. See Table 13-1.

Find 2 in the "number" column at the extreme left of the table. Find 4 in the "number" row at the top of the table. Use these to locate the 38 in the body of the table in much the same way you use the code to find a city on a road map. The 38 is a decimal quantity and is known as the mantissa of the logarithm. In a similar manner, check yourself on the use of the table with the following additional examples:

$$\log 7.6 = 0.88$$
$$\log 1.8 = 0.26$$

Numbers such as 240 and 0.0024 should be changed immediately to

$$240 = 2.4 \times 10^2 \quad \text{and}$$
$$0.0024 = 2.4 \times 10^{-3}$$

$$\log 240 = \log (2.4 \times 10^2)$$
$$= \log 2.4 + \log 10^2$$
$$= 0.38 + 2 = 2.38$$

come the practice to focus on the molar concentration of the hydronium ions in aqueous solutions. The concentrations that are most frequently of interest range between 1×10^{-1} moles per liter and 1×10^{-13} moles per liter. These concentrations are quite small and aqueous solutions are frequently discussed in terms of another quantity that is related to the molar concentration of the hydronium ion. This quantity is pH (read as the two letters *p* and *h*). The relation between the molar concentration of hydronium ion and pH is indicated by the small table on page 290. *pH*

$$\log 0.0024 = \log (2.4 \times 10^{-3})$$
$$= \log 2.4 + \log 10^{-3}$$
$$= 0.38 + (-3) = -2.62$$

Expressed in general terms,

$$\log (ab) = \log a + \log b$$

This is, of course, completely consistent with the general rule that

$$x^a \times x^b = x^{a+b}$$
$$2.4 = 10^{0.38}$$
$$240 = 2.4 \times 10^2$$
$$= 10^{0.38} \times 10^2 = 10^{2.38}$$
$$\log 240 = \log 2.4 + \log 10^2$$
$$= 0.38 + 2 = 2.38$$

Use the following to check yourself out on the evaluation of logarithms:

$$\log 6.0 \times 10^{23} = 23.78$$
$$\log 1.8 \times 10^{-5} = -4.74$$

The logarithm table used here is a two-place table. Most log tables are either four- or five-place tables. These are used in exactly the same fashion. The values given in the body of these tables are again decimal quantities corresponding to the logarithms of quantities between one and ten. The values are, however, expressed to a larger number of figures.

$$\log 2.436 = 0.3868$$
$$\log 2.4365 = 0.38677$$

Many calculators—including hand calculators—have a key to evaluate logarithms to the base 10. These give more figures than is justified by the experimental values frequently fed into the calculator. As a rough guide, retain in the mantissa (the decimal portion) the same number of figures present in the experimental number.

log 2.4 = 0.38~~02112417~~

6
log 243. = 2.38~~5~~606274

60
log 2.432 = 0.38~~59~~63571

log 2432.6 = 3.3860707~~02~~

Natural Logarithms Logarithms to the base 10 are known as common logarithms. Another system uses the base $e = 2.718 \ldots$

$$\log_e 2.4 = 0.88$$
$$2.4 = e^{0.88} = (2.718)^{0.88}$$

It is customary to use ln to represent $\log_e$ and to reserve log for the base 10.

$$\ln 2.4 = 0.88 \qquad 2.4 = e^{0.88}$$
$$\log 2.4 = 0.38 \qquad 2.4 = 10^{0.38}$$
$$\ln 2.4 = 2.303 \log 2.4$$

These natural logarithms (base *e*) arise in many theoretical developments and are therefore very significant in the sciences—including mathematics.

molar concentration of hydronium ion (in moles per liter)	pH
1.0×10^{-1} or 0.10	1.0
1.0×10^{-2} or 0.010	2.0
1.0×10^{-3} or 0.0010	3.0
1.0×10^{-4} or 0.00010	4.0
etc. to	
1.0×10^{-13} or 0.000 000 000 000 10	13.0

If you are at liberty to add reagents to an aqueous solution, the pH of that solution can be controlled at will within the range pH = 1 to pH = 13 by selecting an appropriate reagent and adding the appropriate amount. More about this later.

It is relatively easy to experimentally determine the pH of an aqueous solution. This can be done by means of an instrument known as a pH meter. (What else would you call it?) When the sensing elements (the electrodes) are immersed in the solution, a pH meter reads the pH directly. The pH can also be determined approximately

Acid–Base Indicators

through the use of water-soluble Brönsted acid–base conjugate pairs where the acid

Table 13-1

LOGARITHMS OF NUMBERS BETWEEN 1 AND 10

number	0	1	2	3	4	5	6	7	8	9
1	00	04	08	11	15	18	20	23	26	28
2	30	32	34	36	38	40	41	43	45	46
3	48	49	51	52	53	54	56	57	58	59
4	60	61	62	63	64	65	66	67	68	69
5	70	71	72	72	73	74	75	76	76	77
6	78	79	79	80	81	81	82	83	83	84
7	85	85	86	86	87	88	88	89	89	90
8	90	91	91	92	92	93	93	94	94	95
9	95	96	96	97	97	98	98	99	99	99*

* Log 9.9 lies between 0.99 and 1.00 and is closer to 1.00 than it is to 0.99.

and its conjugate base have different colors and there is consequently a color change as the acid is converted to the base or the base is converted to the acid.

conjugate pair

$$\underset{\text{color A}}{\underset{\text{acid}}{\text{Brönsted}}} + H_2O \rightleftharpoons H_3O^{1+} + \underset{\text{color B}}{\underset{\text{base}}{\text{Brönsted}}}$$

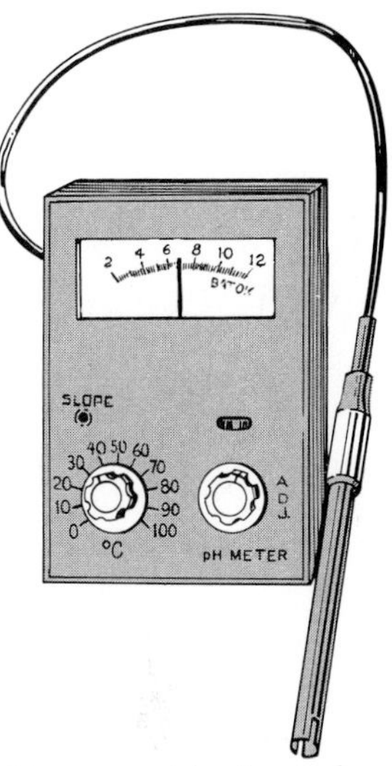

pH meter with electrodes protected by plastic shield

These acid–base pairs that undergo a color change as one is transformed into the other are known as **acid–base indicators.** Many of the flower and fruit pigments are Brönsted acids. (You can design and test your own kitchen experiments.) For red cabbage to stay red in a salad you need an acidic (low pH) salad dressing. For each indicator, there is a specific pH range in which the color change is most apparent. Indicators can, therefore, be used to identify these pH ranges.

We now return to a more detailed discussion of the acetic acid–acetate ion conjugate pair equilibrium.

conjugate pair

$$CH_3COOH + H_2O \rightleftharpoons H_3O^{1+} + CH_3COO^{1-}$$

conjugate pair

In this, we shall focus upon the relative amounts of CH_3COOH and CH_3COO^{1-} in aqueous solutions of controlled hydronium ion concentration and, consequently, controlled pH. The aqueous solution whose pH is to be controlled could be prepared by dissolving a specific number of moles of acetic acid, CH_3COOH, in water to give a specific volume of solution, or the aqueous solution could be prepared by dissolving the same number of moles of sodium acetate, CH_3COONa, a solid, in water to give the same volume of solution. Or the solution could be prepared by dissolving both acetic acid and sodium acetate in water to give the same volume of solution so long as the sum of the moles of acetic acid and the moles of sodium acetate used in the preparation of this solution equals the moles of acetic acid or the moles of sodium acetate used singly in preparing the other solutions. The point here is that the solution we are going to be discussing could be made up in a number of ways. The sodium acetate is a white, water-soluble, ionic compound and consequently provides an easy mechanism for introducing acetate ion into the solution. The sodium ion, although present in the solution, is in no way involved with the reactions that are to be discussed. The significant characteristics of the solution to be discussed are (1) that the total moles of acetic acid molecules plus the total moles of acetate ion present in the solution have some specific value and (2) that the total volume of the solution remains the same or essentially the same:

$$\text{moles } CH_3COOH + \text{moles } CH_3COO^{1-} = \text{specific value}$$

At this time we shall not worry about how we control the hydronium ion concentration of this solution, except to say that it will be done by adding water-soluble reagents that do not contain either acetic acid molecules or acetate ion and that the volumes of the reagents added will not significantly change the total volume of the solution.

pH and the Percent Abundance of Acetic Acid and Its Conjugate Base

When the pH of the solution is adjusted to 2.0 (hydronium concentration equals 1.0×10^{-2} moles per liter), it can be determined that 99.8% of the total moles of the conjugate pair are acetic acid molecules and only 0.2% are acetate ions.

$$100 \times \frac{\text{moles } CH_3COOH}{\text{moles } CH_3COOH + \text{moles } CH_3COO^{(1-)}} = 99.8\%$$

$$100 \times \frac{\text{moles } CH_3COO^{(1-)}}{\text{moles } CH_3COOH + \text{moles } CH_3COO^{(1-)}} = 0.2\%$$

When the pH of the same solution is adjusted to 9.0 (hydronium concentration 1.0×10^{-9} moles per liter), it can be determined that 0.01% of the total moles of the conjugate pair is acetic acid molecules and 99.99% are acetate ion:

$$100 \times \frac{\text{moles } CH_3COOH}{\text{total moles conjugate pair}} = 0.01\%$$

$$100 \times \frac{\text{moles } CH_3COO^{(1-)}}{\text{total moles conjugate pair}} = 99.99\%$$

At pH's less than 2.0 (hydronium ion concentrations more than 1.0×10^{-2} moles per liter), almost all (more than 99.8%) of this conjugate pair is present as the conjugate acid, and at pH's greater than 9.0 (hydronium ion concentrations less than 1.0×10^{-9} moles per liter), almost all of the conjugate pair (more than 99.99%) is present as the conjugate base.

It is logical that if there is a relatively high concentration of hydronium ion, $H_3O^{(1+)}$, present in the solution, there is a relatively high probability that an acetate ion, $CH_3COO^{(1-)}$, can bump into a hydronium ion and pick up a proton to become the conjugate acid, CH_3COOH. This is reflected in the 99.8% abundance of the acid molecules at a pH of 2.0. At relatively low concentrations of hydronium ion, there is a reduced probability for the acetate ion to bump into hydronium ions, and consequently the concentration of the conjugate base builds up. This is reflected in the 99.99% abundance of the acetate ion at a pH of 9.0.

Percent Abundance vs pH for Acetic Acid Aqueous Solutions

The relatively high abundance of the conjugate acid at equilibrium for solutions with relatively high concentrations of hydronium ion and the relatively high concentration of the conjugate base with relatively low concentrations of hydronium ion is characteristic of Brönsted acid–base conjugate pairs in general. The specific numerical values at a given temperature are characteristic of the specific conjugate pair in much the same way that the numerical value of a melting point is a characteristic of the specific substance.

All of this can be displayed graphically for the specific conjugate pair acetic acid–acetate ion in a plot of percent abundance against pH. Figure 13-1.

Note that, at all pH's,

$$\begin{matrix}\text{\% abundance} \\ \text{acetic acid molecule}\end{matrix} + \begin{matrix}\text{\% abundance} \\ \text{acetate ion}\end{matrix} = 100\%$$

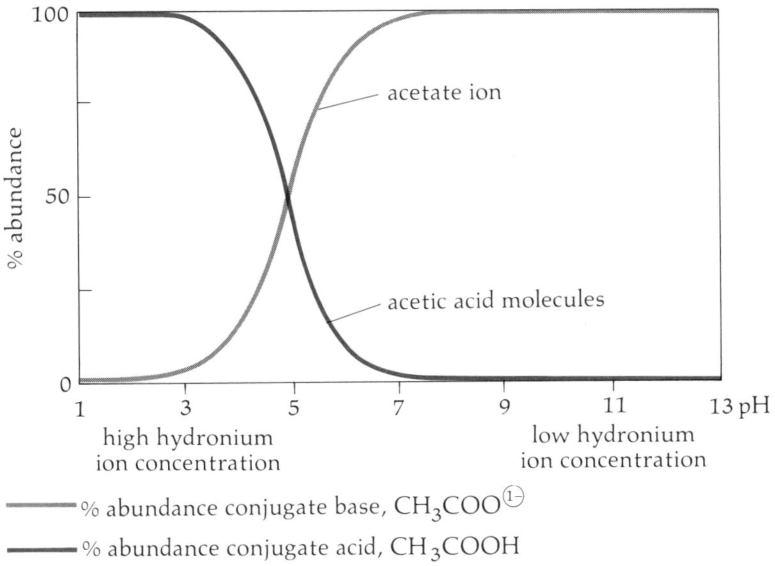

PERCENT ABUNDANCE VS pH CURVES FOR AQUEOUS SOLUTIONS OF ACETIC ACID AND ACETATE ION AT 25 °C.

Fifty Percent Crossover

and that the two percent abundance curves cross at the pH where 50% of the conjugate pair added exists as acetic acid molecules and 50% of the conjugate pair added exists as acetate ion. The pH at which this crossover occurs is a characteristic property of the conjugate pair and we shall pay particular attention to these **fifty percent crossovers** in percent abundance vs pH graphs. For the acetic acid–acetate ion conjugate pair, this pH is 4.75. At a pH of 4.75,

$$\text{molar concentration of the acetic acid molecules} = \text{molar concentration of the acetate ion}$$

For convenience, square brackets, [ ], are commonly used to represent molar concentration and the above equality can be written

$$[CH_3COOH] = [CH_3COO^{1-}]$$
$$[\text{conjugate acid}] = [\text{conjugate base}]$$

In the vicinity of the fifty percent crossover in the percent abundance vs pH graph, the percent abundances of the conjugate pair change very markedly with very small changes in pH. Specific values are given in Table 13-2 for the acetic acid–acetate ion conjugate pair in aqueous solution at 25 °C.

There is a simple mathematical relation that describes the equilibrium relations between an acid and its conjugate base. This relation for the acetic acid–acetate ion

Table 13-2

PERCENT ABUNDANCES OF ACETIC ACID MOLECULES AND ACETATE IONS IN AQUEOUS SOLUTIONS OF SPECIFIED pH AT 25 °C

	percent abundance	
pH	**CH_3COOH**	**$CH_3COO^{(1-)}$**
4.50	64	36
4.60	59	41
4.70	51	49
4.75	50	50
4.80	48	52
4.90	43	57
5.00	36	64

equilibrium is formulated below.

The equilibrium: $CH_3COOH + H_2O \rightleftarrows H_3O^{(1+)} + CH_3COO^{(1-)}$

The relation, known as the K_A (read as two separate letters) for acetic acid, is

K_A Equilibrium Constant for Acetic Acid

$$\underset{\text{acetic acid}}{K_A} = \frac{\begin{pmatrix}\text{molar concentration}\\ \text{hydronium ion}\end{pmatrix} \times \begin{pmatrix}\text{molar concentration}\\ \text{acetate ion}\end{pmatrix}}{\begin{pmatrix}\text{molar concentration}\\ \text{acetic acid molecules}\end{pmatrix}}$$

Using the square brackets to represent molar concentration, the expression becomes

$$\underset{\text{acetic acid}}{K_A} = \frac{[H_3O^{(1+)}] \times [CH_3COO^{(1-)}]}{[CH_3COOH]}$$

The experimental numerical value at 25 °C of K_A is 1.76×10^{-5}. The value K_A has to be determined from experimental measurements, but once it is determined it has a constant value for the temperature at which it was determined and is ideally independent of all other variables. All of the numerical values that have been given earlier in this chapter for acetic acid could be (and were) calculated from this numerical value of K_A for acetic acid. The letter K arises because this expression ideally has a specific value, a constant value for each temperature. The A denotes that the expression is based upon the reaction of the acid with water. (This is an ideal equation in the sense that $PV = nRT$ is an ideal equation. Later in this chapter, we explore the relation of real systems to this equation. It is a useful relation and, in the meantime, it will be taken as valid under all conditions.)

The percent abundance diagram for aqueous solutions of hydrochloric acid and chloride ions displays very different phenomena. Figure 13-2. If the diagram were correctly drawn, the percent abundance curve for the chloride ion would be so close to the 100% line, it would seem to coincide with that line and the hydrogen chloride curve would seem to coincide with the 0% abundance line. At all pH's, dilute aqueous solutions of hydrochloric acid contain essentially no HCl molecules. (Note the restriction: dilute solutions.)

Percent Abundance vs pH for Aqueous Solutions of Hydrogen Chloride

$$\underset{\text{Brönsted acid}}{HCl} + H_2O \rightleftharpoons H_3O^{(1+)} + \underset{\text{Brönsted base}}{Cl^{(1-)}}$$

The K_A equilibrium expression for hydrochloric acid, comparable to the one for acetic acid, has such a high numerical value for K_A it is said to approach infinity:

K_A for Hydrochloric Acid

$$\underset{\text{hydrochloric acid}}{K_A} = \frac{[H_3O^{(1+)}][Cl^{(1-)}]}{[HCl]} \longrightarrow \infty \quad \text{at } 25\ ^\circ C$$

This is consistent with the statement that the molar concentration of hydrogen chloride molecules, [HCl], approaches zero. With this quantity in the denominator, the

Figure 13-2

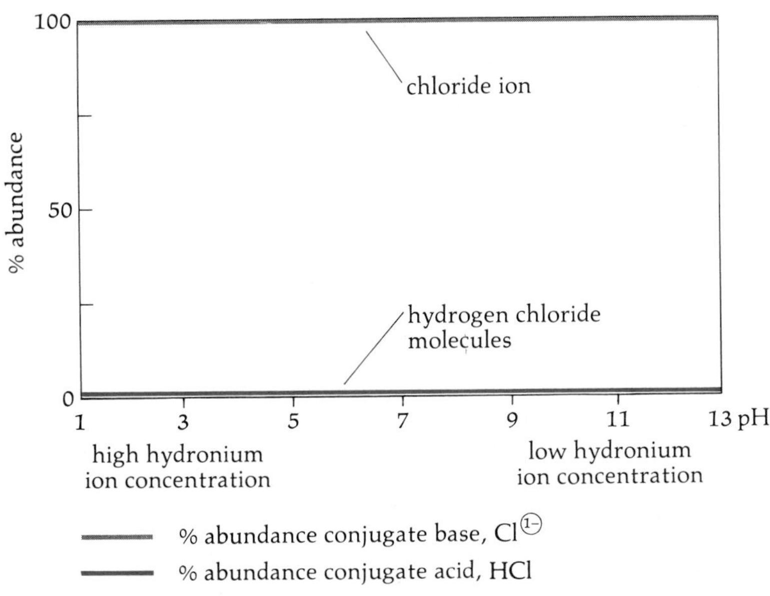

PERCENT ABUNDANCE VS pH CURVES FOR AQUEOUS SOLUTIONS OF HYDROCHLORIC ACID AND CHLORIDE ION AT 25 °C.

fraction must approach infinity as [HCl] approaches 0:

$$\frac{[H_3O^{1+}][Cl^{1-}]}{[HCl]} \longrightarrow \infty$$

Hydrochloric acid is said to be a strong acid. The percent abundance curves do not cross.

K_A for Nitric Acid

Nitric acid is another inorganic acid that reacts with water. It is also a strong acid:

$$\underset{\text{nitric acid}}{HNO_3} + H_2O \rightleftarrows H_3O^{1+} + \underset{\text{nitrate ion}}{NO_3^{1-}}$$

$$\underset{\text{nitric acid}}{K_A} = \frac{[H_3O^{1+}][NO_3^{1+}]}{[HNO_3]} \longrightarrow \infty \quad \text{at } 25\ ^\circ\text{C}$$

An aqueous solution of hydrochloric acid or nitric acid is a ready source of hydronium ion and these acids are frequently added to other solutions, such as the acetic acid–acetate ion solutions, to increase the concentration of the hydronium ion and thus to decrease the pH of the solution.

We have been using the quantity pH and know the relation of pH to the molar concentration of the hydronium ion. For example,

$$\text{for pH} = 2.0 \qquad [H_3O^{1+}] = 1.0 \times 10^{-2} \text{ moles per liter}$$

and

$$\text{for pH} = 9.0 \qquad [H_3O^{1+}] = 1.0 \times 10^{-9} \text{ moles per liter}$$

Definition of pH

As pH is used here, **pH is a defined quantity.** pH is defined to be the negative logarithm to the base 10 of the molar concentration of the hydronium ion. By definition,

$$\text{pH} = -\log_{10} [H_3O^{1+}]$$

where the square brackets indicate the concentration of the hydronium ion in moles of hydronium ion per liter of solution. pH is the negative logarithm of the hydronium ion concentration only because we choose to say that it is. The great advantage of using logarithms is that extremely small concentrations of hydronium ion can be discussed without writing out those troublesome decimal quantities or using expressions containing negative powers of 10.

Note that the pH increases as the molar concentration of the hydronium ion decreases. This is simply a consequence of the manner in which pH is defined. For example, if $[H_3O^{1+}] = 1.00 \times 10^{-2}$ moles liter^{-1},

$$\begin{aligned} \text{pH} &= -\log (1.00 \times 10^{-2}) \\ &= -(\log 1.00 + \log 10^{-2}) \\ &= -(0.00 + (-2)) = 2.00 \end{aligned}$$

hydronium ion concentration (in moles liter^{-1})	pH
$[H_3O^{1+}]$	$-\log_{10} [H_3O^{1+}]$
1.00×10^{-1}	1.00
1.00×10^{-2}	2.00
1.00×10^{-3}	3.00
1.00×10^{-4}	4.00
etc.	etc.

If $[H_3O^{1+}] = 1.00 \times 10^{-9}$ moles liter^{-1},

$$\begin{aligned} pH &= -\log (1.00 \times 10^{-9}) \\ &= -(\log 1.00 + \log 10^{-9}) \\ &= -(0.00 + (-9)) = 9.00 \end{aligned}$$

Many crucial biological systems are restricted to a narrow pH range, 6.5–7.5, and several to the extremely narrow range 7.3–7.5. The pH ranges for some human biological materials*:

blood	7.3–7.5	duodenal contents	4.8–8.2
spinal fluid	7.3–7.5	feces	4.6–8.4
saliva	6.5–7.5	urine	4.8–8.4
milk	6.6–7.6	intracellular fluids	
bile	6.8–7.0	muscle	6.1
gastric contents	1–3	liver	6.9

pK_A is another defined quantity. pK_A is defined to be the negative logarithm of K_A:

$$pK_A = -\log K_A$$

where K_A, the equilibrium constant for the acid, is defined to be *pK_A*

$$K_A = \frac{[H_3O^{1+}] \times [\text{conjugate base}]}{[\text{conjugate acid}]}$$

for the equilibrium

$$\text{conjugate acid} + H_2O \rightleftarrows H_3O^{1+} + \text{conjugate base}$$

* Values listed here are taken from *CRC Handbook of Chemistry and Physics*, ed. Robert C. Weast, Melvin J. Astle, and William H. Beyer, 64th ed. (Boca Raton, Fla.: CRC Press, 1983), p. D-151.

For acetic acid, a weak acid, at 25 °C,

$$K_A = 1.76 \times 10^{-5}$$

and therefore

$$\begin{aligned} pK_A &= -\log_{10}(1.76 \times 10^{-5}) \\ &= -(\log 1.76 + \log 10^{-5}) \\ &= -(0.25 + (-5)) \\ &= -(-4.75) = 4.75 \end{aligned}$$

Note that the above relations can be used to show that the fifty percent crossover of the percent abundance vs pH curves for acetic acid is at pH = pK_A.

At the crossover, where

$$[CH_3COOH] = [CH_3COO^{1-}]$$

the expression

$$K_A = \frac{[H_3O^{1+}][CH_3COO^{1-}]}{[CH_3COOH]} = 1.76 \times 10^{-5}$$

reduces to

$$K_A = [H_3O^{1+}] = 1.76 \times 10^{-5}$$

and

$$pK_A = pH = 4.75$$

Other Carboxylic Acids

Acetic acid is, of course, a specific example for a weak acid. The relation is completely general for weak acids.

Another example, lactic acid, a compound in sour milk, has a K_A of 1.4×10^{-4} and a pK_A of 3.85 at 25 °C. The reaction with water is

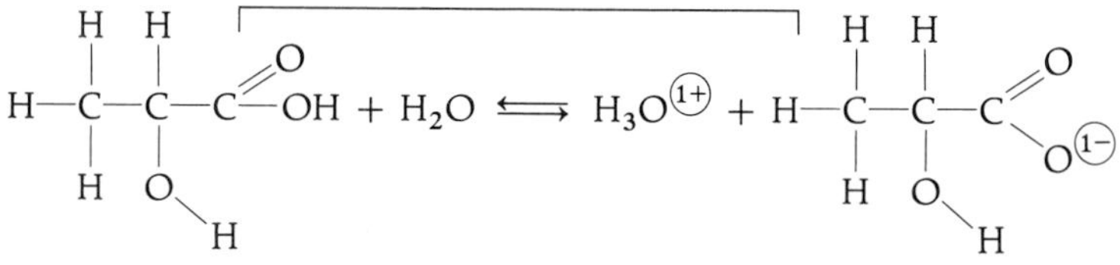

$$\underset{\text{lactic acid}}{CH_3{-}CH(OH){-}C(=O){-}OH} + H_2O \rightleftharpoons H_3O^{1+} + \underset{\text{lactate ion}}{CH_3{-}CH(OH){-}C(=O){-}O^{1-}}$$

$$K_A = \frac{[H_3O^{1+}][\text{lactate ion}]}{[\text{lactic acid}]} = 1.4 \times 10^{-4} \quad \text{at } 25\ °C$$

The numerical value of K_A for lactic acid indicates that lactic acid is a stronger acid than acetic acid: 1.4×10^{-4} is larger than 1.76×10^{-5} (0.00014 > 0.0000176). The larger value of K_A is the consequence of the more extensive reaction of the conjugate acid (the equilibrium concentration of which is in the denominator of the K_A expression) to give the conjugate base (equilibrium concentration in the numerator) and the hydronium ion (equilibrium concentration also in the numerator).

Given the numerical value of pK_A, 3.85, it is quite easy to sketch a set of approximate percent abundance vs pH curves for lactic acid and its conjugate base. The key point, the fifty percent crossover, falls at pH = 3.85. Simply locate this point on the graph and sketch in curves similar to those for acetic acid. For the conjugate pair, at each pH, % acid + % base = 100%. The axes must be labeled and the curves must be labeled. Figure 13-3.

These percent abundance vs pH curves are very helpful in that you know immediately that, at pH's less than 3.85, [lactic acid] > [lactate ion] and that the lower the pH, the more pronounced the inequality. You also know that, at pH's greater than 3.85, [lactate ion] > [lactic acid] and that the higher the pH, the more pronounced the inequality. At pH's over 9, for example, the concentration of the acid molecule is extremely small.

Phosphoric Acid—a Polyprotic Acid

Many organic acids and a number of inorganic acids are essential to biological processes in both animals and plants. The average adult, in good health, produces about six liters of 0.1 molar hydrochloric acid per day. It is this strong acid that is responsible for the low pH, 1 to 3, of the gastric contents in humans. Two other inorganic acids are of particular biological interest: phosphoric acid, H_3PO_4, and carbonic acid, H_2CO_3. Conjugate pairs derived from these acids are instrumental in the control of the pH of the cells and the pH of blood. The nucleic acids, DNA and RNA, are esters of phosphoric acid. (Chapter 19.)

Figure 13-3

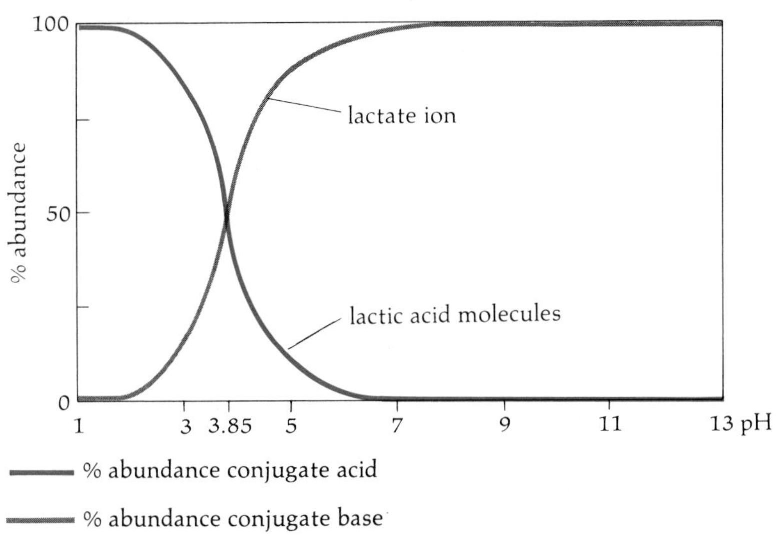

PERCENT ABUNDANCE VS pH CURVES FOR AQUEOUS SOLUTIONS OF LACTIC ACID AND LACTATE ION AT 25 °C.

There are two phosphoric acids: *meta*-phosphoric acid, HPO_3, and *ortho*-phosphoric acid, H_3PO_4:

HPO_3 H—O—P(=O)=O H:Ö:P(:O:)(:O:)

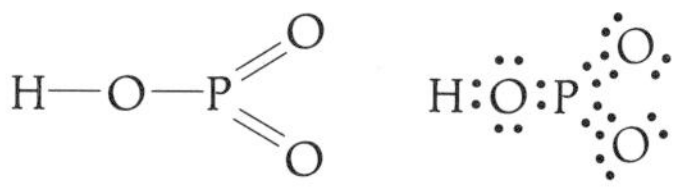

meta-phosphoric acid

H_3PO_4 H—O—P(=O)(—O—H)—O—H

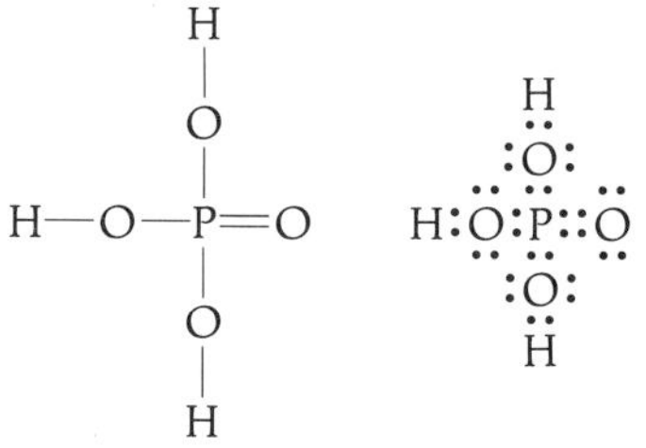

ortho-phosphoric acid

Meta-phosphoric acid in the presence of water becomes *ortho*-phosphoric acid:

$$HPO_3 + H_2O \longrightarrow H_3PO_4$$

meta-phosphoric acid *ortho*-phosphoric acid

and it is the *ortho*-phosphoric acid that is of interest in biological systems. *Ortho*-phosphoric acid is also used in colas and other soft drinks. H_3PO_4 is commonly called phosphoric acid without designating that it is the *ortho* compound. Note that, in the Lewis structures given for the phosphoric acids, the electron count for phosphorus is 10—five pairs of electrons. Phosphorus is a third-row element with available $3d$-orbitals.

Although nitrogen and phosphorus are in the same family of the Periodic Table, and two nitric acids, HNO_3 and H_3NO_4, might be predicted, only the *meta* compound, HNO_3, is known experimentally. This is rationalized on the basis of the nitrogen atom being too small to accommodate the four oxygen atoms required for the *ortho* compound and also the nitrogen atom being able to accommodate only eight electrons in the highest principal quantum level of the ground state atom. Phosphorus can accommodate more than the octet in the third principal quantum level. It was pointed out earlier that the Lewis octet concept was most applicable to small atomic number elements.

Phosphoric acid, the *ortho* compound, is an interesting compound in that it contains three hydrogen atoms bonded through oxygen to the central phosphorus atom. Each of these hydrogen atoms is acidic—meaning each proton can be transferred to a molecule of water with the formation of the phosphate ion, PO_4^{3-}. It is helpful to consider the reaction of phosphoric acid with water in three steps:

Reaction 1

$$H_3PO_4 + H_2O \longrightarrow H_3O^{1+} + H_2PO_4^{1-}$$

dihydrogen phosphate ion

Reaction 2

$$H_2PO_4^{1-} + H_2O \longrightarrow H_3O^{1+} + HPO_4^{2-}$$

monohydrogen phosphate ion

Reaction 3

$$HPO_4^{2-} + H_2O \longrightarrow H_3O^{1+} + PO_4^{3-}$$

Each of these reactions is reversible, and in all aqueous solutions containing phosphoric acid there are three equilibria involving the three Brönsted acid–base conjugate pairs:

Conjugate Pairs

Equilibrium 1

$$\underset{\text{acid}}{H_3PO_4} + H_2O \rightleftarrows H_3O^{1+} + \underset{\text{base}}{H_2PO_4^{1-}}$$

Equilibrium 2

$$\underset{\text{acid}}{H_2PO_4^{1-}} + H_2O \rightleftarrows H_3O^{1+} + \underset{\text{base}}{HPO_4^{2-}}$$

Equilibrium 3

$$\underset{\text{acid}}{HPO_4^{2-}} + H_2O \rightleftarrows H_3O^{1+} + \underset{\text{base}}{PO_4^{3-}}$$

In the first equilibrium above, H_3PO_4 and $H_2PO_4^{1-}$ are a conjugate pair in which $H_2PO_4^{1-}$ is the base. In the second equilibrium, $H_2PO_4^{1-}$ and HPO_4^{2-} are a conjugate pair in which $H_2PO_4^{1-}$ is the acid. In a very similar manner the HPO_4^{2-} ion also has different roles in the second and third equilibria.

An aqueous solution that contains any one of the three ions, $H_2PO_4^{1-}$, HPO_4^{2-}, or PO_4^{3-}, also contains the other two—but the relative abundances of the three ions are quite different and it is the interdependencies of these relative abundances at various pH's that is of particular interest.

For a given temperature, each of these equilibria is described by an equilibrium constant expression:

K_{A1}, K_{A2}, and K_{A3}

Equilibrium 1

$$K_{A1} = \frac{[H_3O^{1+}][H_2PO_4^{1-}]}{[H_3PO_4]} = 8 \times 10^{-3} \qquad pK_{A1} = 2.1 \quad 25\ °C$$

Equilibrium 2

$$K_{A2} = \frac{[H_3O^{1+}][HPO_4^{2-}]}{[H_2PO_4^{1-}]} = 6 \times 10^{-8} \qquad pK_{A2} = 7.2 \quad 25\ ^\circ C$$

Equilibrium 3

$$K_{A3} = \frac{[H_3O^{1+}][PO_4^{3-}]}{[HPO_4^{2-}]} = 2 \times 10^{-13} \qquad pK_{A3} = 12.7 \quad 25\ ^\circ C$$

The percent abundance vs pH curves for the three Brönsted acids of phosphoric acid are given in Figure 13-4 (page 305). It is important that you understand that diagram. The intervening material, which is quite detailed, is intended to assist in attaining that understanding. You may wish to refer to that figure as you work through these materials.

The numerical values of the three K_A's clearly show that H_3PO_4 is a much stronger acid than $H_2PO_4^{1-}$ and $H_2PO_4^{1-}$ is a much stronger acid than HPO_4^{2-}: $8 \times 10^{-3} > 6 \times 10^{-8} > 2 \times 10^{-13}$. This could have been anticipated. In the transfer of the first proton, the positive proton must escape from the H_3PO_4 molecule, leaving behind the $H_2PO_4^{1-}$ ion with its single negative charge. In the transfer of the second proton, the proton must escape from the $H_2PO_4^{1-}$ ion, leaving behind a doubly charged negative ion, HPO_4^{2-}. In the transfer of the third proton, that proton must escape from the HPO_4^{2-} ion, leaving behind a triply charged negative ion, PO_4^{3-}. The first transfer is the easiest transfer to bring off: H_3PO_4 is a stronger acid than $H_2PO_4^{1-}$. Similarly, $H_2PO_4^{1-}$ is a stronger acid than HPO_4^{2-}. A comparison of the K_A values for the three acids with the K_A value for acetic acid (1.76×10^{-5}) shows that H_3PO_4 is a stronger acid than acetic acid ($8 \times 10^{-3} > 1.76 \times 10^{-5}$), that $H_2PO_4^{1-}$ is a weaker acid than acetic acid ($6 \times 10^{-8} < 1.76 \times 10^{-5}$), and that HPO_4^{2-} is an extremely weak acid in comparison to acetic acid ($2 \times 10^{-13} \ll 1.76 \times 10^{-5}$). These comparisons may be more impressive if you write the K_A's as decimal quantities.

The marked interdependencies of the abundances of the H_3PO_4 molecule and its three ions can be demonstrated by selecting various pH values, such as 1, 7, and 13, and substituting the corresponding hydronium concentration into each of the three K_A expressions.

For pH = 1, the hydronium ion concentration equals 1.0×10^{-1} moles per liter:

$$K_{A1} = \frac{1.0 \times 10^{-1} \times [H_2PO_4^{1-}]}{[H_3PO_4]} = 8 \times 10^{-3}$$

and Ratio 1 becomes

$$\frac{[H_2PO_4^{1-}]}{[H_3PO_4]} = \frac{8 \times 10^{-3}}{1.0 \times 10^{-1}} = 8 \times 10^{-2} = 0.08$$

Similarly, from K_{A2} and K_{A3},

$$\text{Ratio 2} \qquad \frac{[HPO_4^{2-}]}{[H_2PO_4^{1+}]} = 6 \times 10^{-7} = 0.000\,000\,6$$

$$\text{Ratio 3} \qquad \frac{[PO_4^{3-}]}{[HPO_4^{2-}]} = 2 \times 10^{-12} = 0.000\,000\,000\,002$$

At a pH of 1.0, the H_3PO_4 molecule concentration is large in comparison to the $H_2PO_4^{1-}$ ion concentration (Ratio 1):

$$[H_3PO_4] > [H_2PO_4^{1-}]$$

At a pH of 1.0, the $H_2PO_4^{1-}$ ion concentration is much greater than the HPO_4^{2-} ion concentration (Ratio 2):

$$[H_2PO_4^{1-}] \gg [HPO_4^{2-}]$$

At a pH of 1.0, the HPO_4^{2-} ion concentration is much greater than the PO_4^{3-} ion concentration (Ratio 3):

$$[HPO_4^{2-}] \ggg [PO_4^{3-}]$$

The relative concentrations of all four species at pH = 1 are qualitatively expressed by

$$[H_3PO_4] > [H_2PO_4^{1-}] \gg [HPO_4^{2-}] \ggg [PO_4^{3-}]$$

At pH = 1, the H_3PO_4 molecule is the predominant species present in the solution.

The three ratios for the three pH values (1, 7, and 13) are given below in the usual exponential notation and also in the decimal notation. Use the values just discussed for a pH value of 1 to orient to this table.

		pH		
		1	**7**	**13**
Ratio 1	$\frac{[H_2PO_4^{1-}]}{[H_3PO_4]}$	8×10^{-2} 0.08	8×10^{4} 80000.	8×10^{10} 80000000000.
Ratio 2	$\frac{[HPO_4^{2-}]}{[H_2PO_4^{1-}]}$	6×10^{-7} 0.0000006	6×10^{-1} 0.6	6×10^{5} 600000.
Ratio 3	$\frac{[PO_4^{3-}]}{[HPO_4^{2-}]}$	2×10^{-13} 0.0000000000002	2×10^{-6} 0.000002	2 2

At a pH of 7, the concentration of the H_3PO_4 molecule is much less than the concentration of the $H_2PO_4^{1-}$ ion (Ratio 1):

$$[H_3PO_4] \ll [H_2PO_4^{1-}]$$

At pH 7, the concentration of the $H_2PO_4^{1-}$ ion is very slightly greater than the concentration of the HPO_4^{2-} ion (Ratio 2), but note that the concentrations of the two ions are almost equal:

$$[H_2PO_4^{1-}] > [HPO_4^{2-}]$$

At pH 7, the concentration of the HPO_4^{1-} ion is much greater than the concentration of the PO_4^{3-} ion (Ratio 3):

$$[HPO_4^{2-}] \gg [PO_4^{3-}]$$

The relative concentrations of the four species at pH = 7 are qualitatively expressed as follows:

$$[H_3PO_4] \ll \underbrace{[H_2PO_4^{1-}] > [HPO_4^{2-}]}_{\text{not greatly different}} \gg [PO_4^{3-}]$$

At pH 7, the predominant species are $H_2PO_4^{1-}$ and HPO_4^{2-}.

At a pH of 13, the concentration of the H_3PO_4 molecule is much much less than the concentration of the $H_2PO_4^{2-}$ ion (Ratio 1):

$$[H_3PO_4] \lll [H_2PO_4^{1-}]$$

At pH 13, the concentration of the $H_2PO_4^{1-}$ ion is much less than the concentration of the HPO_4^{2-} ion (Ratio 2):

$$[H_2PO_4^{1-}] \ll [HPO_4^{2-}]$$

At pH 13, the concentration of the HPO_4^{2-} ion is less than the concentration of the PO_4^{3-} ion (Ratio 3), but note that it is only one half as great:

$$[HPO_4^{2-}] < [PO_4^{3-}]$$

The relative concentrations of the four species at pH = 13 are qualitatively expressed as follows:

$$[H_3PO_4] \lll [H_2PO_4^{1-}] \ll \underbrace{[HPO_4^{2-}] < [PO_4^{3-}]}_{\text{not greatly different}}$$

At pH 13, the predominant species are the PO_4^{3-} ion and the HPO_4^{2-} ion.

The relations of predominant species to pH of aqueous solution of H_3PO_4 and its cation are summarized in the table on the next page.

It should, however, be clear that, at all pH values, all four species are present—some at extremely low concentrations. At all pH values,

$$\text{moles } H_3PO_4 + \text{moles } H_2PO_4^{1-} + \text{moles } HPO_4^{2-} + \text{moles } PO_4^{3-} = \text{fixed value}$$

The magnitude of the fixed value depends upon how the solution was prepared.

Percent Abundance vs pH Curves for Aqueous Solutions of Phosphoric Acid

The percent abundance–pH graph for phosphoric acid can be rather easily sketched. There are three pK_A's and, consequently, three fifty percent crossover points.

crossover 1	$[H_3PO_4] = [H_2PO_4^{1-}]$	$pH = pK_{A1} = 2.1$
crossover 2	$[H_2PO_4^{1-}] = [HPO_4^{2-}]$	$pH = pK_{A2} = 7.2$
crossover 3	$[HPO_4^{2-}] = [PO_4^{3-}]$	$pH = pK_{A3} = 12.7$

	pH		
	1	**7**	**13**
$[H_3O^{1+}]$	1×10^{-1}	1×10^{-7}	1×10^{-13} moles/liter
predominant species	H_3PO_4	$H_2PO_4^{1-}$ and HPO_4^{2-}	PO_4^{3-} and HPO_4^{2-}
	high hydronium ion concentration		low hydronium ion concentration

The pH of cola beverages is achieved by the use of phosphoric acid and phosphate ion mixtures to give a pH between 2 and 3. Consequently, these beverages contain primarily the conjugate pair H_3PO_4 and $H_2PO_4^{1-}$. On the other hand, biological materials in the pH range slightly above 7 would contain primarily the conjugate pair $H_2PO_4^{1-}$ and HPO_4^{2-}.

Figure 13-4

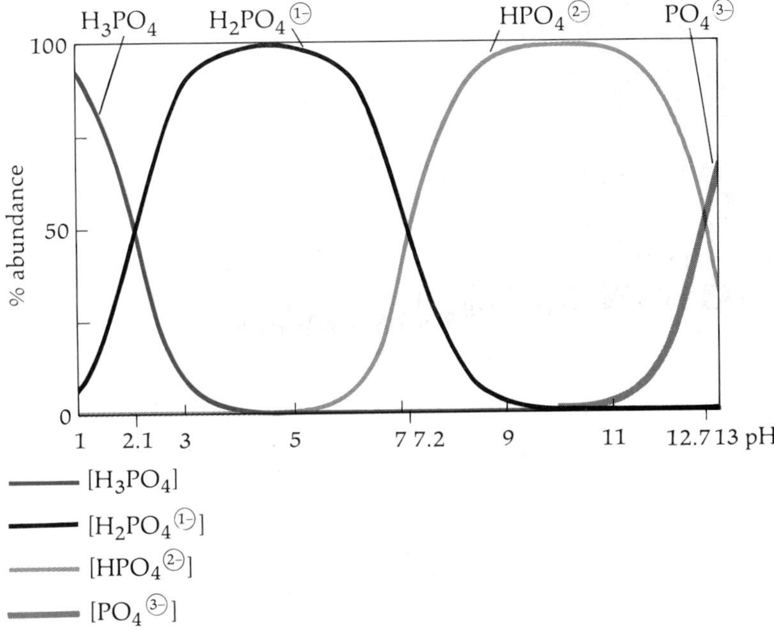

PERCENT ABUNDANCE VS pH CURVES FOR AQUEOUS SOLUTIONS OF PHOSPHORIC ACID AND THE THREE PHOSPHATE IONS, 25 °C.

This percent abundance vs pH curve may seem quite complicated but it is in fact very easy to construct, and it does display just about anything anyone might care to know about phosphoric acid in aqueous solutions. In order to construct a graph, write the various species across the top in order of decreasing strength as acids and use the three pK_A's to mark in the three crossovers.

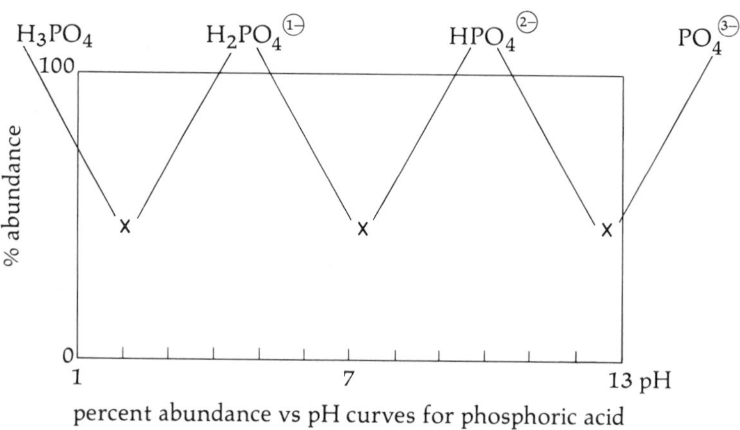

percent abundance vs pH curves for phosphoric acid

At pH 2.1, the crossover involves the conjugate pair for pK_{A1}: H_3PO_4 and $H_2PO_4^{1-}$. At pH 7.2, the crossover involves the conjugate pair for pK_{A2}: $H_2PO_4^{1-}$ and HPO_4^{2-}. At pH 12.7, the crossover involves the conjugate pair for pK_{A3}: HPO_4^{2-} and PO_4^{3-}. At each PH, % abundance H_3PO_4 + % abundance $H_2PO_4^{1-}$ + % abundance HPO_4^{2-} + % abundance PO_4^{3-} = 100%. The curve for $H_2PO_4^{1-}$ acting as a base rises to a maximum of almost 100 percent following the first crossover and then, acting as an acid, descends through the second crossover. Similarly, the curve for HPO_4^{2-} acting as an acid rises to a maximum of almost 100 percent following the second crossover and then, acting as a base, descends through the third crossover.

Carbonic Acid

The story for carbonic acid, H_2CO_3, is less complicated than that for phosphoric acid in that there are only two protons involved. Carbonic acid is a diprotic acid; phosphoric acid is a triprotic acid.

Equilibrium 1

$$H_2CO_3 + H_2O \rightleftharpoons H_3O^{1+} + HCO_3^{1-}$$

carbonic acid molecules — hydrogen carbonate ion

$$K_{A1} = \frac{[H_3O^{1+}][HCO_3^{1-}]}{[H_2CO_3]} = 4.3 \times 10^{-7} \qquad pK_{A1} = 6.37 \quad 25\ ^\circ C$$

Equilibrium 2

$$HCO_3^{1-} + H_2O \rightleftharpoons H_3O^{1+} + CO_3^{2-}$$

hydrogen carbonate ion — carbonate ion

$$K_{A2} = \frac{[H_3O^{1+}][CO_3^{2-}]}{[HCO_3^{1-}]} = 5.6 \times 10^{-11} \qquad pK_{A2} = 10.25 \quad 25\ °C$$

In Scramble Exercise 4, you are asked to sketch the percent abundance vs pH diagram for carbonic acid.

Buffer Solutions

The point has been made repeatedly that the percent abundance vs pH curves are very steep in the region where the pH equals a pK_A or approximately equals a pK_A. In this region, a large change in the relative abundances of the acid and its conjugate base is associated with only a small change in pH. Numerical values were given in Table 13-2 for the acetic acid–acetate ion conjugate pair for pH's at and near the value 4.75. Solutions with a pH close to 50% abundance crossover are called **buffer solutions** and deserve detailed consideration. Assume that you have a solution at 25 °C in which the concentration of the acetic acid molecule is equal to the concentration of the acetate ion:

$$[CH_3COOH] = [CH_3COO^{1-}]$$

The pH of this solution is 4.75. To this solution add some hydrochloric acid—H_3O^{1+} and Cl^{1-} ions in water. The chloride ion remains as chloride ion and is of no concern to the topic being discussed. The added hydronium ion, H_3O^{1+}, would be expected to reduce the pH. In actual fact, the pH of the solution may change very little. Much of the added hydronium ion reacts with the acetate ion, decreasing the acetate ion concentration, $[CH_3COO^{1-}]$, and increasing the concentration of the acetic acid molecules, $[CH_3COOH]$.

$$\underset{\text{acid}}{H_3O^{1+}} + \underset{\text{base}}{CH_3COO^{1-}} \longrightarrow CH_3COOH + H_2O$$

The degree to which this reaction occurs is described by the K_A expression for acetic acid:

$$K_A = \frac{[H_3O^{1+}][CH_3COO^{1-}]}{[CH_3COOH]} = 1.76 \times 10^{-5}$$

When the hydrochloric acid is added, the increased concentration of hydronium ion leads the expression

$$\frac{[H_3O^{1+}][CH_3COO^{1-}]}{[CH_3COOH]}$$

to have an instantaneous value greater than 1.76×10^{-5} and the concentrations shift until the *new* values for the three concentrations give a value for the expression

$$\frac{[H_3O^{1+}][CH_3COO^{1-}]}{[CH_3COOH]}$$

that does again equal 1.76×10^{-5}. During this period of adjustment, the hydronium ion concentration (numerator) decreases, the acetate ion concentration (numerator) decreases, and the concentration of the acetic acid molecule (denominator) increases. The combination of these changes leads again to the 1.76×10^{-5} value of K_A.

If the solution started with a substantial concentration of both acetate ion and acetic acid molecules, the value of the ratio $[CH_3COO^{1-}]/[CH_3COOH]$ changes very little. For example, if each was 0.200 moles per liter,

$$\text{the ratio} \quad \frac{[CH_3COOH]}{[CH_3COO^{1-}]} = \frac{0.200}{0.200} = 1.00$$

is changed very little by reducing the acetate ion concentration by 0.001 moles per liter and increasing the acetic acid molecule concentration by 0.001 moles per liter:

$$\frac{[CH_3COO^{1-}]}{[CH_3COOH]} = \frac{0.200 - 0.001}{0.200 + 0.001} = \frac{0.199}{0.201} = 0.99$$

The new hydronium ion concentration can be found from the K_A expression

$$K_A = [H_3O^{1+}] \times \frac{[CH_3COO^{1-}]}{[CH_3COOH]} = 1.76 \times 10^{-5}$$

$$[H_3O^{1+}] = \frac{1.76}{0.99} \times 10^{-5} = 1.78 \times 10^{-5} \text{ moles liter}^{-1}$$

$$pH = -\log(1.78 \times 10^{-5}) = 4.75$$

The resultant pH of the solution would still be 4.75. The pH has, of course, decreased, but the decrease is so small it is not evident in the above calculation. If the moles of hydrochloric acid required to bring about the above change had been added to the same volume of distilled water, the pH of that solution would have become approximately 3.00.

To carry the same example further, if the moles of hydrochloric acid changed the acetate ion concentration to 0.190 moles/liter and the concentration of the acetic acid molecules to 0.210 moles/liter, the hydronium ion concentration would again be calculated from the equilibrium constant relation

$$[H_3O^{1+}]\frac{0.190}{0.210} = 1.76 \times 10^{-5}$$

$$[H_3O^{1+}] = 1.76 \times 10^{-5} \times \frac{0.210}{0.190} = 1.95 \times 10^{-5} \text{ moles liter}^{-1}$$

and the pH would become 4.71. The same number of moles of hydrochloric acid added to the same volume of distilled water would lead to a solution with a pH of approximately 2.0.

To carry the example one step further, the addition of more moles of hydrochloric acid could lead to an acetate ion concentration of 0.100 moles per liter and an acetic acid molecule concentration of 0.300 moles per liter. The corresponding hydronium ion concentration would become 5.28×10^{-5} mole per liter and the pH of the solution would become 4.28. The same number of moles of hydrochloric acid added to the same volume of distilled water would lead to a solution with a pH of approximately 1.

All of the above can be said in general terms very simply. The solution containing acetic acid molecules and acetate ion at about equal concentrations (about 50% abundance) resists changing pH when an acid, such as hydrochloric acid, is added. The solution is said to be buffered at a pH of 4.75.

When a base, such as sodium hydroxide (an ionic compound), is added to water or an aqueous solution, it tends to reduce the concentration of the hydronium ion and consequently increases the pH of the solution. For sodium hydroxide, it is the hydroxide ion, OH^{1-}, that is the base. The sodium ion, Na^{1+}, does not react with any of the ions or molecules in this solution and it is not a factor in this discussion. The sodium ion simply remains in solution. The hydroxide ion is a Brönsted base and accepts protons from a Brönsted acid such as the hydronium ion:

$$\underset{\text{acid}}{H_3O^{1+}} + \underset{\text{base}}{OH^{1-}} \longrightarrow H_2O + H_2O$$

When sodium hydroxide solution is added to an aqueous solution buffered at a pH of 4.75 with the conjugate pair acetic acid molecules and acetate ions, the hydroxide ion can also react with the acetic acid molecule:

$$\underset{\text{acid}}{CH_3COOH} + \underset{\text{base}}{OH^{1-}} \longrightarrow H_2O + CH_3COO^{1-}$$

After the addition of the sodium hydroxide solution, the concentrations of the three ions and molecules (H_3O^{1+}, CH_3COO^{1-}, and CH_3COOH) again adjust to satisfy the K_A expression

$$\frac{[H_3O^{1+}][CH_3COO^{1-}]}{[CH_3COOH]} = 1.76 \times 10^{-5}$$

The addition of hydroxide ion to the solution does decrease the hydronium ion concentration and consequently raises the pH, but the changes are very small in comparison to the changes that would be obtained by adding the same number of moles of sodium hydroxide to the same volume of pure water. The acetic acid–acetate ion solution is buffered with respect to the addition of a base as well as buffered with respect to the addition of an acid.

The phosphoric acid–phosphate ions percent abundance vs pH curves show that phosphoric acid and/or its ions can be used to produce solutions buffered at pH's in the vicinity of 2.1, 7.2, and 12.7 at 25 °C. The relevant conjugate pairs are

for a pH of 2.1: H_3PO_4 and $H_2PO_4^{1-}$

for a pH of 7.2: $H_2PO_4^{1-}$ and HPO_4^{2-}

for a pH of 12.7: HPO_4^{2-} and PO_4^{3-}

Another way of discussing buffer action is simply to note that the addition of an acid to a solution of some specific pH tends to decrease the pH but that the phenomena in the solution follow the percent abundance vs pH curves. If a conjugate pair is present in the solution and the curves are steep, a lot of the added acid will be used up in reacting with the conjugate base in the solution. Similarly, the addition of a base tends to increase the pH, but the phenomena in the solution again follow the percent abundance vs pH curves with the percent abundance of the conjugate base increasing and the percent abundance of the conjugate acid decreasing. The addition of acid "drives" the system along the percent abundance curves to the left of the pK_A value; the addition of base, to the right.

By selecting the relative quantities of the conjugate acid and the conjugate base, a solution can be buffered at a preselected pH near a 50% abundance vs pH crossover as well as at the pH equal to this pK_A. For example, if the concentration of the acetic acid molecule is 0.080 moles per liter and the concentration of the acetate ion is 0.100 moles per liter, the solution is buffered at a pH of 4.85:

$$\frac{[H_3O^{\textcircled{1+}}](0.100)}{(0.080)} = 1.76 \times 10^{-5}$$

$$[H_3O^{\textcircled{1+}}] = 1.41 \times 10^{-5}$$

$$pH = -\log(1.41 \times 10^{-5}) = 4.85$$

Summary Table of K_A and pK_A Values

Up to this point we have discussed two organic acids—acetic acid and lactic acid—and several inorganic acids—hydrochloric acid, nitric acid, phosphoric acid, and carbonic acid. Empirical knowledge (experimentally determined knowledge) concerning the strengths of these acids and the identities of their conjugate bases is summarized in Table 13-3. In this table, the acids are listed in order of decreasing strength—the

Table 13-3

K_A AND pK_A VALUES FOR SOME BRÖNSTED ACIDS AT 25 °C

	K_A	pK_A	acid	conjugate base
strongest acids ↑	approaches $+\infty$	approaches $-\infty$	HCl	$Cl^{\textcircled{1-}}$
	approaches $+\infty$	approaches $-\infty$	HNO_3	$NO_3^{\textcircled{1-}}$
increasing strength as acids	1×10^{-2}	2	H_3PO_4	$H_2PO_4^{\textcircled{1-}}$
	1.4×10^{-4}	3.9	lactic acid	lactate ion
	1.8×10^{-5}	4.8	CH_3COOH	$CH_3COO^{\textcircled{1-}}$
	4.3×10^{-7}	6.4	H_2CO_3	$HCO_3^{\textcircled{1-}}$
	6.8×10^{-8}	7.2	$H_2PO_4^{\textcircled{1-}}$	$HPO_4^{\textcircled{2-}}$
	5.6×10^{-11}	10.3	$HCO_3^{\textcircled{1-}}$	$CO_3^{\textcircled{2-}}$
weakest acids	2×10^{-11}	12.7	$HPO_4^{\textcircled{2-}}$	$PO_4^{\textcircled{3-}}$

strongest acids at the top, the weakest acids at the bottom. The K_A and the pK_A values quantitatively express the strength of each acid in aqueous solutions, and comparisons of strengths of acids can be made in terms of these experimentally determined values: the smaller the K_A, the weaker the acid. Terms such as weak and strong are relative terms—weak with respect to what? strong with respect to what?—and have no meaning unless the references for comparison are stated. Since all K_A values are based upon the equilibrium

$$\text{acid} + H_2O \rightleftharpoons H_3O^{(1+)} + \text{base}$$

the reference of comparison implicit in K_A and pK_A values is water.

There are a great number of other organic acids that contain the carboxylic acid group

$$-C(=O)-OH$$

Scramble Exercise 1 is concerned with several of these. As will be seen by their respective K_A's, they have different strengths as acids—but not very different. In addition, there are organic acids other than carboxylic acids. These will not be discussed here. Inorganic acids are quite diverse in structure and also in strength. A number of these inorganic acids will be introduced. In their exploration, it will be helpful to note where each acid fits into Table 13-3.

The hydrogen compounds of the members of the oxygen family and the members of the halogen family are acids.

hydrogen oxide	H_2O	hydrogen fluoride	HF
hydrogen sulfide	H_2S	hydrogen chloride	HCl
hydrogen selinide	H_2Se	hydrogen bromide	HBr
hydrogen telluride	H_2Te	hydrogen iodide	HI

Hydrogen Halides

All of the hydrogen halides are strong acids in their reaction with water (that is, K_A approaches $+\infty$), with the exception of hydrogen fluoride. The K_A for HF is 3.5×10^{-4} and the pK_A is 3.45 at 25 °C.

Hydrogen Compounds of Oxygen Family Elements

All of the hydrogen compounds of the oxygen family are weak acids. It may seem strange to consider water as an acid. In the reactions with acetic acid and hydrochloric acid, water reacted as a base in accepting a proton to become the conjugate acid, $H_3O^{(1+)}$. When water reacts as an acid, it is the proton donor and becomes the hydroxide ion, $OH^{(1-)}$, its conjugate base:

$$\underset{\text{acid}}{H_2O} + \text{base} \longrightarrow \text{acid} + \underset{\text{base}}{OH^{(1-)}}$$

This reaction of water as an acid is a very common role for water and it will be discussed in considerable detail in the next chapter.

The compound H_2S is a diprotic acid:

Equilibrium 1

$$H_2S + H_2O \rightleftarrows H_3O^{1+} + HS^{1-}$$

hydrogen sulfide ion

$$K_{A1} = 9.1 \times 10^{-8} \qquad pK_{A1} = 7.0 \quad 25\ °C$$

Equilibrium 2

$$HS^{1-} + H_2O \rightleftarrows H_3O^{1+} + S^{2-}$$

hydrogen disulfide molecule

sulfide ion

$$K_{A2} = 1.1 \times 10^{-12} \qquad pK_{A2} = 12.0 \quad 25\ °C$$

Note that both H_2S and HS^{1-} are very weak acids in their reaction with water as the base and that the hydrogen sulfide ion, HS^{1-}, is a particularly weak acid.

Other Inorganic Acids

Seven common inorganic oxygen acids are given below in the pattern of the positions in the Periodic Table of the elements other than hydrogen and oxygen:

H_3BO_3	H_2CO_3	HNO_3		
	H_4SiO_4	H_3PO_4	H_2SO_4	$HClO_4$

The molecules of the top row of compounds are isoelectronic.

	H_3BO_3	**H_2CO_3**	**HNO_3**
	boric acid	**carbonic acid**	**nitric acid**
hydrogen atoms	3 electrons	2 electrons	1 electron
central atom	5	6	7
oxygen atoms	24	24	24
total	32	32	32

The molecules of the bottom row of compounds are also isoelectronic.

	H_4SiO_4	H_3PO_4	H_2SO_4	$HClO_4$
	silicic acid	**phosphoric acid**	**sulfuric acid**	**perchloric acid**
hydrogen atoms	4 electrons	3 electrons	2 electrons	1 electron
central atom	14	15	16	17
oxygen atoms	32	32	32	32
total	50	50	50	50

The anions BO_3^{3-}, CO_3^{2-}, and NO_3^{1-} are also isoelectronic with 32 electrons each. The removal of the protons, the hydrogen nuclei, does not change the number of electrons. Of these 32 electrons, only 24 electrons are in the second quantum levels of their respective atoms and only 24 electrons need to be considered in the Lewis structures. There are a number of experimental facts to be accommodated. Each of the three anions is a planar with bond angles of 120° about the central atom. In each ion, the three bond distances are known to be equal. These bond distances in each case are shorter than the usual single bond distances for the two elements involved but longer than the usual double bond length. Presumably, whatever picture would be adequate for one ion would be adequate for all three ions. This is, however, not necessarily true. The three ions would have a different distribution of charge due to the differences in charge on the nuclei of the central atom: +5 for boron, +6 for carbon, and +7 for nitrogen. The larger the charge, the stronger the force pulling these 24 electrons toward the central atom. This is consistent with the HNO_3 being the strongest of these acids: the pair of electrons involved in the hydrogen–oxygen bond is pulled toward the central structure.

The four anions—SiO_4^{4-}, PO_4^{3-}, SO_4^{2-}, and ClO_4^{1-}—are isoelectronic and could conceivably have the same electronic structures with 50 electrons each. They do differ, however, in the nuclear charges on the central atoms: +14 for silicon, +15 for phosphorus, +16 for sulfur, and +17 for chlorine. The anions are known to be tetrahedral, with the four oxygen atoms at the corners of the tetrahedron that has the other atom at the center of the tetrahedron.

In all of these oxygen acids, the protons are believed to be bonded to an oxygen that is in turn bonded to the central atom. All of the anions are quite well known in compounds such as sodium borate, Na_3BO_3, and calcium silicate, Ca_2SiO_4. The pure acid compounds themselves are infrequently encountered even in laboratories where acidic solutions are frequently used. Pure silicic acid may be unattainable. It is an extremely weak acid with a reported K_{A1} of 2×10^{-10}. There is some question as to the reliability of this value, but there would be no doubt H_4SiO_4 is an extremely weak acid.

Pure phosphoric acid, H_3PO_4, is a colorless solid with a melting point of 42.4 °C. Very few chemists have ever seen the solid; in general they work with its aqueous solution. It is commercially marketed as an 85% by weight H_3PO_4 aqueous solution. Pure sulfuric acid is a colorless liquid freezing at 10.4 °C. It forms a number of hydrates, and here again chemists are accustomed to working with its aqueous solutions. It is frequently commercially marketed as a 98% by weight H_2SO_4 aqueous solution. It is a diprotic acid:

Equilibrium 1

$$\underset{\text{sulfuric acid molecules}}{H_2SO_4(aq)} + H_2O(l) \rightleftharpoons H_3O^{1+}(aq) + \underset{\text{hydrogen sulfate ions}}{HSO_4^{1-}(aq)}$$

$$K_{A1} = \frac{[H_3O^{1+}(aq)][HSO_4^{1-}(aq)]}{[H_2SO_4(aq)]} \longrightarrow +\infty$$

Equilibrium 2

$$\underset{\text{hydrogen sulfate ions}}{HSO_4^{1-}} + H_2O \rightleftharpoons H_3O^{1+}(aq) + \underset{\text{sulfate ions}}{SO_4^{2-}(aq)}$$

$$K_{A2} = \frac{[H_3O^{1+}(aq)][SO_4^{2-}(aq)]}{[HSO_4^{1-}(aq)]} \simeq 1 \times 10^{-2} \qquad pK_A \simeq 2.0$$

The H_2SO_4 molecule is an extremely strong acid and essentially no H_2SO_4 molecules remain in dilute aqueous solutions as molecules. The hydrogen sulfate ion, HSO_4^{1-}, is in itself a much stronger acid than acetic acid. The fifty percent crossover on the percent abundance vs pH diagram for the HSO_4^{1-}–SO_4^{2-} conjugate pair comes at a pH of 2 and, consequently, in aqueous solutions at pH's greater than 2, the SO_4^{2-} ion is the predominant species—at least in those solutions that are sufficiently dilute for the equilibrium constants to be applicable. More about this later.

Very closely related to H_4SiO_4, H_3PO_4, and H_2SO_4 is a series of more complicated molecules and anions. The simplest of these molecules are $H_6Si_2O_7$, $H_4P_2O_7$, and $H_2S_2O_7$. Each of these can be thought of as the consequence of the elimination of a molecule of water as two molecules of the acids react. For example,

$$H_3PO_4 + H_3PO_4 \longrightarrow H_4P_2O_7 + H_2O$$

In each case, the two "central atoms" become bonded through an oxygen atom. $H_4P_2O_7$ is known as pyrophosphoric acid since it is formed by pyrolysis of phosphoric acid (the decomposition of phosphoric acid by heating).

```
      H     H                  O     O                   O     O
      |     |                  ‖     ‖                   ‖     ‖
      O     O            H—O—P—O—P—O—H            H—O—S—O—S—O—H
      |     |                  |     |                   ‖     ‖
H—O—Si—O—Si—O—H                O     O                   O     O
      |     |                  |     |
      O     O                  H     H
      |     |
      H     H
```

There is a large array of related anions of these acids. The anions having the largest charges—$Si_2O_7^{6-}$, $P_2O_7^{4-}$, and $S_2O_7^{2-}$—have been identified in crystalline solids but not in aqueous solutions. The bonding through oxygen is very common with these third-row elements. The pyrophosphate linkages are important in biosynthetic processes and are probably best known in ATP and ADP: adenosine 5′-triphosphate and adenosine 5′-diphosphate. The pyrosilicate linkages are the bases of the structures of many silicate rocks and their formation is responsible for the setting of cement.

Assessment of the Validity of K_A Values

How reliable are K_A expressions? Are they applicable in general or are they ideal expressions that have limited applicability to real solutions? In all cases, K_A values are experimentally determined quantities and consequently must be determined using specific solutions under specific conditions. How well do the K_A values apply over wide concentration ranges? Solutions of acetic acid can range in concentration from essentially 100% acetic acid to essentially 100% water. Solutions of phosphoric acid commonly used in the laboratory range from 85% by weight H_3PO_4, in the concentrated phosphoric acid of commerce, to solutions that are so dilute they approach pure water. The behavior found here is very parallel to the behavior encountered with gases. At low pressures, where gas molecules are separated by a great deal of space, the ideal gas equation, $PV = nRT$, describes rather well the experimental behavior exhibited by the gas. At low concentrations, where the ions and molecules involved in the K_A expression are separated by a great deal of water, the K_A expressions describe rather well the phenomena that occur in the solution.

In an aqueous solution, there are a number of interactions: water molecules with water molecules, water molecules with the solute molecules, water molecules with the positive ions, water molecules with the negative ions, solute molecules with the solute molecules, each positive ion with all other positive ions and with all negative ions, each negative ion with all other negative ions and all positive ions. The interactions between ions are more significant than the interactions between molecules due to the strong charge–charge interactions of ions in addition to the usual van der Waals type of interactions between molecules.

How dilute must solutions be for the K_A expressions to be reasonably descriptive of experimental reality? Concentrations of about 0.1 molar are frequently considered as an upper limit for K_A expressions to be valid, and even here the deviations between a K_A expression and experimental reality may be significant. (This accounts for our consideration of pH in this chapter being limited to the range 1 to 13.) The separation between molecules of solute in 0.1 molar solutions corresponds roughly to the separation between molecules in a gas at 2 atmospheres pressure.

In a 0.010 molar solution of acetic acid at 25 °C, the K_A expression

$$CH_3COOH(aq) + H_2O(l) \rightleftharpoons H_3O^{1+}(aq) + CH_3COO^{1-}(aq)$$

$$K_A = \frac{[H_3O^{1+}(aq)][CH_3COO^{1-}(aq)]}{[CH_3COOH(aq)]}$$

describes the experimental phenomena rather well, but in a 1.0 molar solution of acetic acid, deviations from the ideal equilibrium expression may be significant. The total ionic environment in the solution is of prime importance. In the same sense that deviations from ideal gas behavior may be negative or positive, deviations from ideal K_A expressions may be both negative and positive. Negative deviations arise from attractive forces. Positive deviations arise from the very limited amount of free solvent between the ions.

Aqueous solutions are quite complex. In the first place, water in the liquid phase does not exist as simple H_2O units. The molecule of water is polar, extensive hydrogen bonding occurs and there may be no units smaller than $(H_2O)_2$ or $(H_2O)_3$. If liquid water itself is not made up of single H_2O units, what is the justification of writing equations using H_2O as the molecule and producing a specific ion H_3O^{1+}?

Validity of the H_3O^{1+} Ion

The justification is simplicity. H_3O^{1+} is an approximation to a complex situation. Even in water itself there is a very rapid rate of exchange of protons among molecules of water and there is no reason to believe that a specific unit H_3O^{1+} has a continued existence. The charge and the very small size of the proton make it improbable that the proton can wander about unattached. If it is attached, an otherwise unshared pair of electrons on an oxygen atom, a lone pair of electrons on the oxygen atom, is the most probable site. The number of molecules of water that are bound together through hydrogen bonding is both variable and uncertain. It might be that units such as $H_5O_2^{1+}$ or $H_7O_3^{1+}$ or $H_9O_4^{1+}$ are more suitable than H_3O^{1+}; they are also more troublesome to write than H_3O^{1+}. The unit H_3O^{1+} is therefore symbolic of a type of phenomenon. In many cases, the symbol H^{1+}, the hydrogen ion, is used instead. In water solution, the H^{1+} symbol is understood to represent a more complex but unspecified phenomenon involving water. Using the symbol H^{1+}, the reaction

$$CH_3COOH + H_2O \rightleftharpoons H_3O^{1+} + CH_3COO^{1-}$$

is written

$$CH_3COOH \rightleftharpoons H^{1+} + CH_3COO^{1-}$$

$$\frac{[H^{1+}][CH_3COO^{1-}]}{[CH_3COOH]} = K_i$$

where K_i is called the **ionization constant** or the dissociation constant. It is entirely equivalent to the K_A expression and has the identical numerical value. The reaction with water to give the hydronium ion provides a rationalization of formation of ions when the covalent compound dissolves in water. The equation for ionization written without showing the water molecule is simple but more mysterious. Tables of K_A and pK_A values are frequently listed in handbooks under "ionization constants" of acids.

It is our practice in this book to use H_3O^{1+} in general and $H^{1+}(aq)$ in writing oxidation–reduction equations in Chapter 16.

One further question: Why doesn't a K_A expression such as

$$K_A = \frac{[H_3O^{1+}][CH_3COO^{1-}]}{[CH_3COOH]}$$

include the molar concentration of water? The simplest answer is that the K_A expression is only useful at low concentrations of the solute. The solution is almost entirely water and the characteristics of most of the water are unchanged by the solute. Water itself is not a variable and nothing is to be gained by including water in this type of an equilibrium expression. In many other equilibria, where the concentration of water is a variable, water is included in the equilibrium expression.

Scramble Exercises

1. There are a great number of carboxylic acids that are very similar to acetic acid. Note that the strengths of the acids listed are different but not greatly different. Which is the strongest? the weakest?

		K_A	pK_A
formic acid	$HCOOH$	1.77×10^{-4}	3.75
acetic acid	CH_3COOH	1.76×10^{-5}	4.75
propionic acid	CH_3CH_2COOH	1.34×10^{-5}	4.87
n-butyric acid	$CH_3CH_2CH_2COOH$	1.54×10^{-5}	4.81
isobutyric acid	$(CH_3)_2CHCOOH$	1.44×10^{-5}	4.84
benzoic acid	C_6H_5COOH	6.46×10^{-5}	4.19

Be sure you can write the chemical equation for the reaction of any one of these acids with water. The name of the negative ion in each case can be arrived at by dropping the *-ic* suffix of the acid and adding *-ate*: acetic acid, acetate ion. *n*-Butyric acid is present in rancid butter and is largely responsible for its unpleasant qualities.

2. There are three chlorine-containing acids that are closely related to acetic acid:

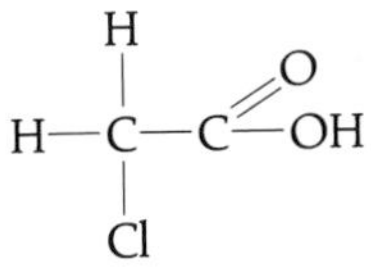

H—C(H)(Cl)—C(=O)—OH $ClCH_2COOH$ $K_A = 1.4 \times 10^{-3}$ $pK_A = 2.85$

chloroacetic acid

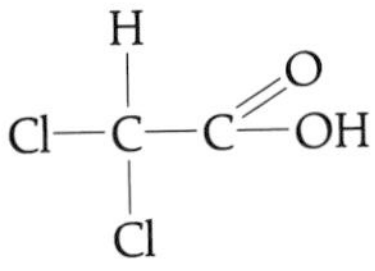

Cl—C(H)(Cl)—C(=O)—OH $Cl_2CHCOOH$ $K_A = 3.3 \times 10^{-2}$ $pK_A = 1.5$

dichloroacetic acid

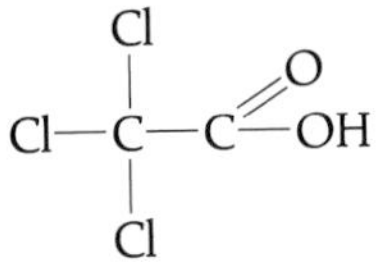

Cl—C(Cl)(Cl)—C(=O)—OH Cl_3CCOOH $K_A = 2 \times 10^{-1}$ $pK_A = 0.7$

trichloroacetic acid

Compare the strengths of these acids with each other and with the strength of acetic acid. Propose a plausible explanation for relative strengths of these acids. (How could the substitution of chlorine atoms for hydrogen atoms in the acetic acid molecule change the ease with which the proton of the carboxylic acid group can be donated to a water molecule?)

3. Write the equation for the reaction of one of these chloroacetic acids with water. Set up the corresponding equilibrium constant expression (the K_A expression). Sketch a percent abundance vs pH diagram for the aqueous solutions of this acid and its conjugate base for the pH range 1 to 13. Identify the curves. At a pH of 10, what is the predominant ion or molecule? At a pH of 1, what is the predominant ion or molecule?

4. Using the information given in the text of the chapter, sketch the percent abundance vs pH diagram for carbonic acid.

5. Oxalic acid is a diprotic acid that is found in oxalis plants and in rhubarb.

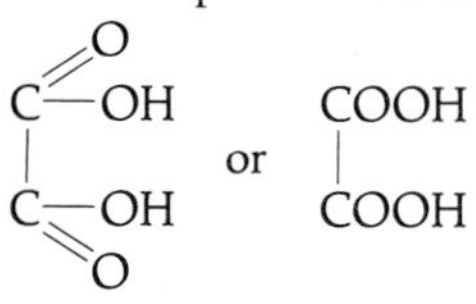

$$\begin{array}{c} C(=O)-OH \\ | \\ C(=O)-OH \end{array} \quad \text{or} \quad \begin{array}{c} COOH \\ | \\ COOH \end{array} \qquad \begin{array}{ll} K_{A1} = 5.9 \times 10^{-2} & pK_{A1} = 1.2 \\ K_{A2} = 6.4 \times 10^{-5} & pK_{A2} = 4.2 \end{array}$$

Write the equation for the first reaction of this acid with water and set up the K_{A1} expression.

Write the equation for the second reaction with water and set up the K_{A2} expression.

Give the formulas for the two sets of conjugate pairs involved in the above two reactions.

Sketch and label the percent abundance vs pH diagram for the oxalic acid–oxalate ion aqueous solutions for the pH range 1 to 13. In what pH range would the hydrogen oxalate ions

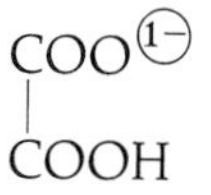

$$\begin{array}{c} COO^{1-} \\ | \\ COOH \end{array}$$

be the predominant species?

6. Citric acid is a triprotic carboxylic acid

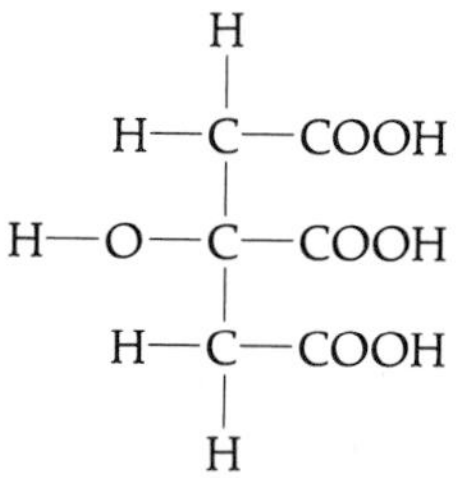

$$\begin{array}{c} H \\ | \\ H-C-COOH \\ | \\ H-O-C-COOH \\ | \\ H-C-COOH \\ | \\ H \end{array}$$

with three K_A's—8.4×10^{-4}, 1.8×10^{-5}, and 4.0×10^{-7}—and, of course, three pK_A's—3.08, 4.7, and 6.40. Assuming that the most reactive proton is on the central carboxylic acid group, write the three chemical equations with water corresponding to K_{A1}, K_{A2}, and K_{A3}. The values given above are for 18 °C.

7. Hydrogen sulfide, H_2S, is a gas at room conditions and is the compound responsible for the odor associated with rotten eggs and other decomposing protein. The formation of this compound from biological material is responsible for a significant amount of the sulfur that gets into the atmosphere. Hydrogen sulfide has only limited solubility in water and is a diprotic acid. Try your hand at any or all of the following:

Write the equation for the reaction of the first proton with water.

Set up the K_{A1} expression in terms of concentrations. The experimental value for K_{A1} is 9.1×10^{-8} at room temperature.

Compare the strength of H_2S as an acid with the strength of acetic acid, CH_3COOH, and with the strength of carbonic acid, H_2CO_3.

Try your hand at demonstrating that the pK_{A1} corresponding to a K_{A1} of 9.1×10^{-8} is a little more than 7.

Write the chemical equation for the reaction of the hydrogen sulfide ion, HS^{1-}, with water.

Set up the K_A expression for this reaction. This expression is, of course, the K_{A2} expression for the diprotic acid hydrogen sulfide, H_2S. The numerical value at 25 °C is 1.1×10^{-12}.

Sketch the percent abundance–pH diagram for the hydrogen sulfide equilibria in aqueous solutions over the pH range 1 to 13.

In what pH range is the sulfide ion, S^{2-}, the predominant species?

8. Within an aqueous solution made up by dissolving acetic acid and sodium acetate in water, the usual, equilibrium exists:

$$CH_3COOH(aq) + H_2O(l) \rightleftarrows H_3O^{1+}(aq) + CH_3COO^{1-}(aq)$$

Discuss in terms of Le Chatelier the manner in which this equilibrium responds to the addition of a few drops of hydrochloric acid—containing hydronium ions, H_3O^{1+}, and chloride ions, Cl^{1-}—to the aqueous solution.

Additional exercises for Chapter 13 are given in the Appendix.

A Look Ahead

In this chapter, we have explored the reactions of Brönsted acids with the solvent water. In these reactions, water molecules react as a Brönsted base in the formation of the conjugate acid of water, the hydronium ion, H_3O^{1+}:

$$CH_3COOH + H_2O \longrightarrow H_3O^{1+} + CH_3COO^{1-}$$

In the next chapter, we explore the reactions of Brönsted bases with the solvent water. In these reactions, water molecules react as Brönsted acids in the formation of the conjugate base of water: the hydroxide ion, OH^{1-}. Examples:

$$CH_3COO^{1-} + H_2O \longrightarrow OH^{1-} + CH_3COOH$$

Brönsted base — acetic acid molecule

$$NH_3 + H_2O \longrightarrow OH^{1-} + NH_4^{1+}$$

Brönsted base — ammonium ion

These reactions raise interesting questions as to the significance of the molecules of our most common solvent reacting both as a base and as an acid.

14

AMMONIA AND RELATED COMPOUNDS: THE BRÖNSTED CONCEPT OF BASES

Chapter 14

There are two parts to this chapter. Part I deals with ammonia and related compounds as bases in aqueous solutions. The approach is parallel to the previous chapter dealing with aqueous solutions of acids. Look for these parallels. In this, as in so many of the topics that we have dealt with before, the concepts seem very complex at the beginning. Later they reduce to a few comparatively simple relations. The first part is also highly relevant to later discussions of proteins (Chapter 17) and nucleic acids (Chapter 19). Part II has to do with the role of ammonia and related compounds in the world food supply.

Part I Ammonia and Related Compounds as Brönsted Bases

Ammonia

Ammonia, NH_3, is a colorless gas of characteristic odor associated with smelling salts, household ammonia, and a wide variety of cleaning agents. In all of these, the odor is due to NH_3 molecules that have escaped from these aqueous solutions. Ammonia is a Brönsted base. Its conjugate acid is the ammonium ion, $NH_4^{(1+)}$:

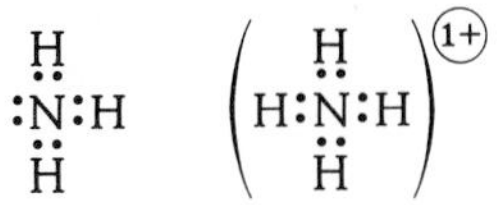

```
 H          /  H  \(1+)
:N:H       (H:N:H )
 H          \  H  /
```

The Three Ethylamines

The parentheses are used here to indicate that the 1+ charge is associated with the entire ion—not with one particular hydrogen atom. Amines, such as the methylamines, are also Brönsted bases, and any other amine could equally well be explored as a base in this chapter in place of ammonia. In every case, it is the unshared pair of electrons, the lone pair electrons, on the nitrogen atom that accepts the proton in the formation of the conjugate acid.

```
  H              CH3             CH3
  |              |               |
 :N—CH3         :N—CH3          :N—CH3
  |              |               |
  H              H               CH3
methylamine    dimethylamine    trimethylamine
```

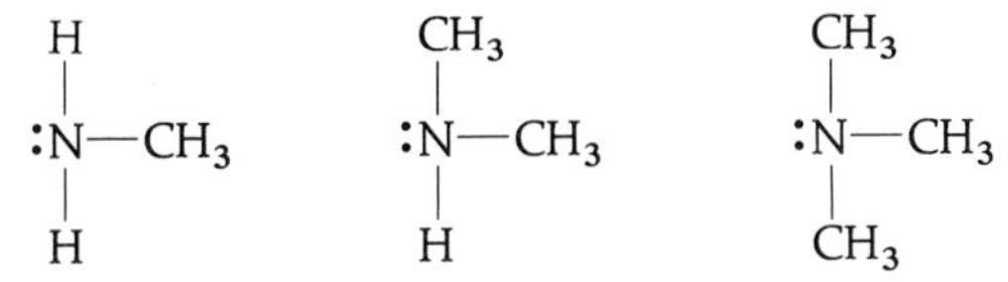

Much of the chemistry of the proteins and the nucleic acids is the chemistry of amine nitrogens in those very complex molecules.

First, a few words about ammonia and its relations to the hydrogen compounds of its first-row neighbors in the Periodic Table.

Table 14-1

PHYSICAL PROPERTIES OF COMPOUNDS ISOELECTRONIC WITH AMMONIA

	H H:C:H H **methane**	H H:N: H **ammonia**	H:O: H **water**	H:F: **hydrogen fluoride**
molecular weight	16	17	18	20
boiling point	−164 °C	−34 °C	100 °C	20 °C
dipole moment	0	1.47 D	1.85 D	1.82 D
bond angle	109.5°	107.3°	104.5°	
bond length	0.109 nm	0.101 nm	0.096 nm	0.100 nm
dielectric constants of liquid*	~2	~22	~80	~84

* For different temperatures.

Compounds Isoelectronic with NH_3

The ammonia molecule is isoelectronic with methane, water, and hydrogen fluoride. All of these molecules have a total of ten electrons, only eight of which are shown in the Lewis structures. The molecular weights increase only slightly through the series and the bond lengths decrease very slightly through the series (Table 14-1). The striking differences in normal boiling points (−164 to 100 °C) indicate the tremendous differences in attractive forces between molecules of the various compounds. The rationalization of such marked differences in attractive forces would be expected to lie in electrostatic phenomena: polar bonds, dipole moments, and hydrogen bonds. This is in keeping with the electronegatives of the five elements: 2.1 for hydrogen, 2.5 for carbon, 3.0 for nitrogen, 3.5 for oxygen, and 4.0 for fluorine. The carbon–hydrogen bonds would be the least polar. The tetrahedral symmetry of methane would, in any case, lead to a nonpolar molecule. In methane, there are also no unshared pairs of electrons essential to the acceptance of a proton. Methane with its boiling point of −164 °C is by far the most volatile of the compounds. The distribution of atoms in the other molecules leads to polar molecules and also allows for the close approaches of protons of other molecules essential for hydrogen bonding. Hydrogen fluoride can form long chains through hydrogen bonding but is somewhat limited, by its single proton per molecule, in forming more complicated networks of HF units.

The 107.3° and 104.5° bond angles of ammonia and water exceed the 90° angles predicted by the use of *p*-orbitals of the nitrogen ($1s^2\ 2s^2\ 2p_x{}^1\ 2p_y{}^1\ 2p_z{}^1$) and oxygen ($1s^2\ 2s^2\ 2p_x{}^2\ 2p_y{}^1\ 2p_z{}^1$) atoms. Both bond angles more nearly approach the 109.5° of the tetrahedral distribution rationalized in terms of four sp^3 hybrid orbitals. In keeping with the valence bond approach to these tetrahedral angles, the atomic orbitals of nitrogen involved in forming the bonds in ammonia are said to have considerable sp^3 character. Likewise, the atomic orbitals of oxygen involved in bonding in water are said to have some sp^3 character.

The ammonia molecule is a rather flat pyramid with the nitrogen at the apex. One of the modes of vibration of this molecule has provoked considerable interest. The molecule inverts, in the sense that an umbrella may turn inside out in a windstorm. In vibrational motion of the molecule, the center of mass of the molecule is considered as fixed. The nitrogen atom moves along the threefold axis of the pyramid toward the plane of the three hydrogen atoms. At the same time, the three hydrogen atoms move in the opposite direction. There is an instant in which the four atoms lie in a single plane with hydrogen–nitrogen–hydrogen bond angles of 120°. As the motion continues with the nitrogen atom on the other side of the plane of the three hydrogen atoms, the 107.3° bond angles are resumed and the inversion of the molecule is complete. The frequency of this inversion is characteristic of the molecule and is more reliable as a basis of measuring time than the motion of the earth. An ammonia clock developed by the National Bureau of Standards has an accuracy of three parts per billion. The frequency of the inversion vibration is a little more than 2.3×10^{10} vibrations per second. There are a number of other modes of vibration of the ammonia molecule. This inversion mode is of particular interest since there are so few molecules that invert.

Ammonia gas is very soluble in water: approximately 90 grams of ammonia per 100 grams of cold water. Molecules of both compounds are polar and hydrogen atoms of molecules of both compounds can participate in hydrogen bonding. There is also a chemical reaction between molecules of water and molecules of ammonia. Both molecules are capable of reacting as acids by donating protons and both molecules are capable of reacting as bases by accepting protons. In the reactions discussed in Chapter 13, the molecule of water reacted as a Brönsted base in the acceptance of a proton to give the hydronium ion, H_3O^{1+}. In the reaction of ammonia with water, water is the stronger acid. The reaction observed is the transfer of a proton from the water molecule to the ammonia molecule:

$$NH_3(aq) + H_2O(l) \longrightarrow \underset{\text{hydroxide ion}}{OH^{1-}(aq)} + \underset{\text{ammonium ion}}{NH_4^{1+}(aq)}$$

The Ammonia–Ammonium Ion Conjugate Pair

conjugate pair

$$\underset{\text{base}}{H{:}\overset{H}{\underset{H}{\ddot{N}}}{:}(aq)} + \underset{\text{acid}}{H{:}\overset{H}{\ddot{\underset{\cdot\cdot}{O}}}{:}(l)} \longrightarrow \underset{\text{base}}{:\ddot{\underset{\cdot\cdot}{O}}{:}H^{1-}(aq)} + \underset{\text{acid}}{H{:}\overset{H}{\underset{H}{\ddot{N}}}{:}H^{1+}(aq)}$$

conjugate pair

The name of the ammonium ion, a cation, utilizes the *-ium* ending characteristic of so many of the metals such as sodium and magnesium that also form positive ions, Na^{1+} and Mg^{2+}. The NH_4^{1+} ion has the tetrahedral symmetry characteristic of sp^3

hybridization of the central atom. The above reaction is reversible and a state of equilibrium is established in any aqueous solution that is at constant temperature and in

$$NH_3(aq) + H_2O(l) \rightleftharpoons OH^{1-}(aq) + NH_4^{1+}(aq)$$

a closed container so that there is not a continuous loss of ammonia gas. If there is a space above the liquid in the container, there must also be an equilibrium between dissolved ammonia molecules and ammonia molecules in the gas phase:

$$NH_3(aq) \rightleftharpoons NH_3(g)$$

The ammonia–water equilibrium is a typical Brönsted acid–base equilibrium

$$NH_3(aq) + H_2O(l) \rightleftharpoons OH^{1-}(aq) + NH_4^{1+}(aq)$$

K_B and pK_B for NH_3

and the degree of reaction at equilibrium is described in the usual fashion by an equilibrium constant that, in this case, is designated as the K_B

$$\underset{\text{ammonia}}{K_B} = \frac{\begin{pmatrix}\text{molar concentration}\\ \text{hydroxide ions}\end{pmatrix} \times \begin{pmatrix}\text{molar concentration}\\ \text{ammonium ions}\end{pmatrix}}{\begin{pmatrix}\text{molar concentration}\\ \text{ammonia molecules}\end{pmatrix}}$$

Using square brackets for molar concentrations

$$\underset{\text{ammonia}}{K_B} = \frac{[OH^{1-}][NH_4^{1+}]}{[NH_3(aq)]} = 1.77 \times 10^{-5} \qquad pK_B = 4.75 \quad 25\ ^\circ C$$

The experimentally determined numerical value of K_B of less than one is the consequence of the low concentrations of the ammonium ions and the hydroxide ions (both in the numerator) in comparison to the concentration of the molecules of ammonia (in the denominator). The ammonia reacts as a base. The numerical value of K_B is a measure of the strength of ammonia as a Brönsted base in its reaction with water. Note that the K_A expressions are formulated in terms of the reaction of the acid with molecules of water and that the K_B expressions are formulated in terms of the reaction of the base with molecules of water.

It is entirely fortuitous that the numerical value of the K_A for acetic acid (1.76×10^{-5}) and the numerical value of K_B for ammonia (1.77×10^{-5}) are so nearly the same—probably are the same within the experimental error in determining the numerical values for these equilibrium constants. It is pure chance that the strength of ammonia as a Brönsted base in water is essentially the same as the strength of acetic acid as a Brönsted acid in water.

The above discussion has focused upon the reaction of the ammonia molecule as a base with the solvent water. It is also informative to look at the related reaction of its conjugate acid, the ammonium ion, NH_4^{1+}, with the solvent water:

$$NH_4^{1+}(aq) + H_2O(l) \longrightarrow H_3O^{1+}(aq) + NH_3(aq)$$

and at the equilibrium established at constant temperature in a closed container:

$$NH_4^{1+}(aq) + H_2O(l) \rightleftarrows H_3O^{1+}(aq) + NH_3(aq)$$

This equilibrium is written in the form appropriate to K_A expressions:

K_A and pK_A for NH_4^{1+}

$$\underset{\text{ammonium ion}}{K_A} = \frac{[H_3O^{1+}][NH_3]}{[NH_4^{1+}]} = 5.65 \times 10^{-10} \qquad pK_A = 9.25 \quad 25\ ^\circ C$$

The experimentally determined numerical values of K_A and pK_A are a measure of the strength of the ammonium ion, NH_4^{1+}, as an acid in water, and these values place the ammonium ion as an acid between the dihydrogen phosphate ion, $H_2PO_4^{1-}$, and the hydrogen carbonate ion, HCO_3^{1-}, in Table 13-3. The ammonium ion is a weaker acid than $H_2PO_4^{1-}$ and a stronger acid than HCO_3^{1-}.

Any aqueous solution that contains ammonia molecules also contains ammonium ions. Any aqueous solution that contains ammonium ions also contains ammonia molecules. Another equilibrium in the aqueous solution can be explored by the addition of the two chemical expressions

$$NH_3(aq) + H_2O(l) \rightleftarrows OH^{1-}(aq) + NH_4^{1+}(aq)$$

and

$$NH_4^{1+} + H_2O \rightleftarrows H_3O^{1+} + NH_3$$

The net sum of these two chemical equations is

$$H_2O(l) + H_2O(l) \rightleftarrows H_3O^{1+}(aq) + OH^{1-}(aq)$$

This is the equilibrium for the reaction of water with water. It should not be a great surprise that, if water is both a Brönsted acid and a Brönsted base, this equilibrium is attained any time and any place liquid water is present. One molecule of water acts as the acid and the other molecule of water acts as the base. Note that, in one case, H_2O and OH^{1-} are the conjugate pair and, in the other case, H_2O and H_3O^{1+} are the conjugate pair.

The ammonia–ammonium ion equilibrium in aqueous solution can be used to explore equilibria between water and its ions in all aqueous solutions. By the multiplication of the K_B expression for ammonia, the conjugate base,

Product of $K_A \times K_B$

$$\underset{\text{ammonia}}{K_B} = \frac{[OH^{1-}][NH_4^{1+}]}{[NH_3]} = 1.77 \times 10^{-5} \qquad pK_B = 4.75 \quad 25\ ^\circ C$$

and the K_A expression for the ammonium ion, the conjugate acid,

$$\underset{\text{ammonium ion}}{K_A} = \frac{[H_3O^{1+}][NH_3]}{[NH_4^{1+}]} = 5.65 \times 10^{-10} \qquad pK_A = 9.25 \quad 25\ ^\circ C$$

a very significant relation is obtained:

$$K_A \times K_B = \frac{[H_3O^{(1+)}][NH_3]}{[NH_4^{(1+)}]} \times \frac{[OH^{(1-)}][NH_4^{(1+)}]}{[NH_3]} = 5.65 \times 10^{-10} \times 1.77 \times 10^{-5} \quad 25\ ^\circ C$$

This simplifies to

$$K_A \times K_B = [H_3O^{(1+)}][OH^{(1-)}] = 1.00 \times 10^{-14} \quad 25\ ^\circ C$$

Although this last expression involving hydronium ion and hydroxide ion has been arrived at here through the consideration of a particular conjugate acid–base pair, the ammonia–ammonium ion system in aqueous solution, the expression is entirely general and is applicable to all aqueous systems at 25 °C. Actually, there are two fundamental relations stated in this one expression:

$$K_A \times K_B = 1.00 \times 10^{-14} \quad 25\ ^\circ C$$

and

$$[H_3O^{(1+)}][OH^{(1-)}] = 1.00 \times 10^{-14} \quad 25\ ^\circ C$$

For every conjugate acid–base pair in aqueous solution, K_A is inversely proportional to K_B

$$K_A \propto 1/K_B$$

and, at 25 °C, the proportionality constant is 1.00×10^{-14}. For water or any aqueous solution, the molar concentration of the hydronium ion is inversely proportional to the molar concentration of the hydroxide ion

$$[H_3O^{(1+)}] \propto \frac{1}{[OH^{(1-)}]}$$

and the proportionality is again 1.00×10^{-14} at 25 °C.

Ion Product of Water K_W

For water and aqueous solutions, the product of the molar concentration of the hydronium ion $H_3O^{(1+)}$ and the molar concentration of the hydroxide ion $OH^{(1-)}$ is known as the "ion product of water," K_W:

$$K_W = [H_3O^{(1+)}][OH^{(1-)}] = 1.00 \times 10^{-14} \quad 25\ ^\circ C$$

In the same sense that pH and pK_A are defined by the equations

$$pH = -\log_{10} [H_3O^{(1+)}] \qquad \text{and} \qquad pK_A = -\log_{10} K_A$$

pOH and pK_B are defined by the equations

$$pOH = -\log_{10} [OH^{(1-)}] \qquad \text{and} \qquad pK_B = -\log_{10} K_B$$

$pK_A + pK_B$

It follows from $K_A \times K_B = 1.00 \times 10^{-14}$ at 25 °C for a conjugate pair that $pK_A + pK_B = 14.00$ at 25 °C and from $[H_3O^{(1+)}][OH^{(1-)}] = 1.00 \times 10^{-14}$ that pH +

pOH = 14.00 at 25 °C. For example,

$$-\log (K_A \times K_B) = -\log (1.00 \times 10^{-14})$$
$$-\log K_A - \log K_B = 14.00$$
$$pK_A + pK_B = 14.00 \quad 25\ °C$$

The significance of the last relation is indicated by the following table of the corresponding values for pH, molar concentrations of hydronium ion, molar concentrations of hydroxide ion, and pOH.

pH + pOH

pH	$[H_3O^{1+}]$ moles liter^{-1}	$[OH^{1-}]$ moles liter^{-1}	pOH	
1.00	1.0×10^{-1}	1.0×10^{-13}	13.00	↑ acid
3.00	1.0×10^{-3}	1.0×10^{-11}	11.00	solutions
5.00	1.0×10^{-5}	1.0×10^{-9}	9.00	
7.00	1.0×10^{-7}	1.0×10^{-7}	7.00	neutral
9.00	1.0×10^{-9}	1.0×10^{-5}	5.00	basic
11.00	1.0×10^{-11}	1.0×10^{-3}	3.00	solutions
13.00	1.0×10^{-13}	1.0×10^{-1}	1.00	↓

In pure water without added solutes, the only source of the hydronium ion and the hydroxide ion is the reaction of water molecules with water molecules:

$$H_2O(l) + H_2O(l) \longrightarrow H_3O^{1+}(aq) + OH^{1-}(aq)$$

In pure water, the concentration of the hydronium ion has to equal the concentration of the hydroxide ion,

$$[H_3O^{1+}] = [OH^{1-}] = 1.00 \times 10^{-7} \text{ moles per liter}$$

and both concentrations must be equal to 1.00×10^{-7} moles per liter at 25 °C. The pH of pure water at 25 °C is 7.00. This situation is taken as the pivot point in discussing aqueous solutions. Any aqueous solution in which the concentrations of these two ions are equal is called a **neutral aqueous solution,** regardless of what else may be present in the solution. An aqueous solution that has a higher concentration of hydronium ion than hydroxide ion is said to be acidic and an aqueous solution that has a lower concentration of hydronium ion than hydroxide ion is said to be basic. At 25 °C, an acidic aqueous solution has a pH of less than 7.00; a basic solution has a pH of more than 7.00.

Neutral Aqueous Solutions

At other temperatures, a neutral aqueous solution must still have equal concentrations of hydronium ion and hydroxide ion, but the numerical value will no longer be 1.00×10^{-7} moles per liter and the pH of the neutral water solution will no longer be 7.00. The K_W values for water at other temperatures are known. Note that the

Dependence of pH on Temperature

value of K_W and the value of pH for pure water at our normal body temperature of 37 °C differ significantly from the corresponding values at 25 °C.

°C	K_W	pK_W	pH neutral solution
0	1.14×10^{-15}	14.94	7.47
10	2.95×10^{-15}	14.53	7.27
20	6.76×10^{-15}	14.17	7.09
25	1.00×10^{-14}	14.00	7.00
30	1.48×10^{-14}	13.83	6.92
35	2.08×10^{-14}	13.68	6.84
40	2.95×10^{-14}	13.53	6.77

We return now to a further consideration of the ammonia–ammonium ion equilibrium in water.

Percent Abundance NH_3 and NH_4^{1+}

In setting up the percent abundance vs pH diagram for aqueous solutions of ammonia and its conjugate acid, it is convenient to also write in the pOH scale. Figure 14-1. The crossover of the two percent abundance curves at 50% abundance of each

Figure 14-1

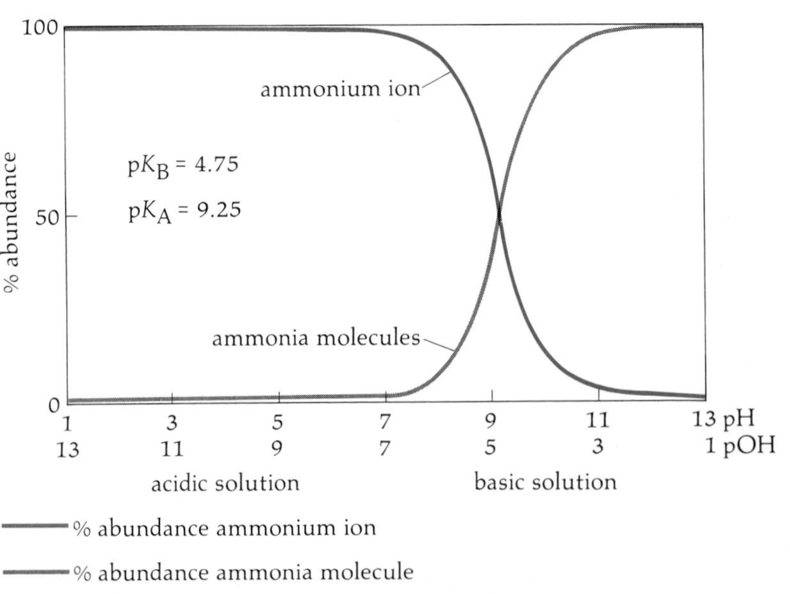

PERCENT ABUNDANCE VS pH DIAGRAM FOR THE AMMONIA–AMMONIUM ION CONJUGATE PAIR IN AQUEOUS SOLUTIONS AT 25 °C.

must come at pH = pK_A or pOH = pK_B. As would be expected, in a strongly acidic solution there is a great deal of ammonium ion, $NH_4^{(1+)}$, and relatively few ammonia molecules, NH_3. If there are lots of hydronium ions, $H_3O^{(1+)}$, then ammonia molecules have a greater chance of accepting a proton to become ammonium ions. At pH's less than 9.25, the ammonium ion, $NH_4^{(1+)}$, is the predominant species; at pH's more than 9.25, the ammonia molecule, NH_3, is predominant.

In drawing the above diagram, it has again been assumed that the pH of an aqueous solution can be controlled at will. In the next chapter, we shall finally face into the problem of how this can be brought about experimentally.

The 50% crossover of the ammonia–ammonium ion percent abundance curve is at pH 9.25 and the ammonia–ammonium ion conjugate pair can be used to buffer an aqueous solution at pH 9.25. A convenient source of ammonium ion is the ionic compound ammonium chloride, NH_4Cl. It is a water-soluble, white, crystalline compound and the chloride ion, $Cl^{(1-)}$, in no way enters into the equilibria that control the pH of the solution.

The Chloride Ion as a Base

We now have the background to rationalize the above statement concerning the chloride ion. The chloride ion is the conjugate base of hydrogen chloride, a very strong acid. K_A approaches ∞:

$$HCl(aq) + H_2O(l) \rightleftarrows H_3O^{(1+)}(aq) + Cl^{(1-)}(aq)$$

$$K_A = \frac{[H_3O^{(1+)}][Cl^{(1-)}]}{[HCl]} \longrightarrow \infty \quad 25\ ^\circ C$$

If K_A for HCl approaches infinity, K_B for $Cl^{(1-)}$, its conjugate base, approaches zero:

$$Cl^{(1-)}(aq) + H_2O(l) \rightleftarrows OH^{(1-)}(aq) + HCl(aq)$$

$$K_B = \frac{[OH^{(1-)}][HCl]}{[Cl^{(1-)}]} \longrightarrow 0$$

This follows from $K_A \times K_B = 1.00 \times 10^{-14}$. If K_A for HCl approaches infinity, K_B for its conjugate base must approach zero. The chloride ion, $Cl^{(1-)}$, has very little tendency to accept protons and is not a significant base. For a conjugate pair, if K_A approaches infinity, pK_A approaches negative infinity and pK_B approaches positive infinity.

Now that we have the relations

$$K_A \times K_B = 1.00 \times 10^{-14} \quad 25\ ^\circ C$$

and

$$pK_A + pK_B = 14.00 \quad 25\ ^\circ C$$

Tabulation of pK_A and pK_B Values

we could expand Table 13-3 to include for each acid the K_B and pK_B values for its conjugate base. This is the structure of Table 14-2 with the exception that all K_A values, all K_B values, and some of the conjugate pairs have been deleted to simplify the presentation and to make room for additional conjugate pairs such as the ammonium ion–ammonia molecule conjugate pair. To orient yourself to Table 14-2, locate this acid–base pair and note (1) that $pK_A + pK_B = 14.0$ for all pairs, (2) that the

Table 14-2

pK_A AND pK_B VALUES FOR SOME CONJUGATE ACID–BASE PAIRS IN AQUEOUS SOLUTIONS AT 25 °C

the acids		the conjugate bases	
pK_A	**formula**	**formula**	**pK_B**
approaches $-\infty$	HCl	Cl^{1-}	approaches $+\infty$
−1.7	H_3O^{1+}	H_2O	15.7
2	H_3PO_4	$H_2PO_4^{1-}$	12.
2.4	$^{1+}NH_3CH_2COOH$ glycine cation	$^{1+}NH_3CH_2COO^{1-}$ glycine zwitterion	11.6
4.8	CH_3COOH	CH_3COO^{1-}	9.2
6.4	H_2CO_3	HCO_3^{1-}	7.6
7.2	$H_2PO_4^{1-}$	HPO_4^{2-}	6.8
9.2	NH_4^{1+} ammonium ion	NH_3 ammonia	4.8
9.8	$^{1+}NH_3CH_2COO^{1-}$ glycine zwitterion	$NH_2CH_2COO^{1-}$ glycine anion	4.2
10.3	HCO_3^{1-}	CO_3^{2-}	3.7
10.7	$CH_3CH_2NH_3^{1+}$ ethylammonium ion	$CH_3CH_2NH_2$ ethylamine	3.3
12.7	HPO_4^{2-}	PO_4^{3-}	1.3
15.7	H_2O	OH^{1-}	−1.7
approaches $+\infty$	OH^{1-}	O^{2-}	approaches $-\infty$

(Left arrow, pointing up: increasing strength as an acid. Right arrow, pointing down: increasing strength as a base.)

ammonium ion is a weaker acid than the dihydrogen phosphate ion, and (3) that the ammonia molecule is a stronger base than the monohydrogen phosphate ion. Other conjugate pairs, such as those involving the zwitterion of glycine, are discussed later in this chapter.

As interesting as aqueous solutions of ammonia are, we would not devote this much attention to them if it were not for the commercial and biological significance of related compounds in which the nitrogen is bonded directly to one or more carbon atoms. The simplest of these are the alkyl amines, such as methylamine, CH_3NH_2; dimethylamine, $(CH_3)_2NH$; and trimethylamine, $(CH_3)_3N$. They have characteristic odors but the odor is more closely related to that of decomposing fish than it is to the "clean" odor of ammonia. The low molecular weight compounds are gases and are soluble in water but their volatility and solubility decrease with the length and the number of alkyl groups (carbon–hydrogen chains). All amines are Brönsted bases and all form alkyl ammonium ions. For example, ethylamine reacts with water to form ethylammonium ion:

The Three Ethylamines

$$\underset{\text{ethylamine}}{CH_3CH_2NH_2(aq)} + H_2O \longrightarrow OH^{1-}(aq) + \underset{\text{ethylammonium ion}}{CH_3CH_2NH_3^{1+}(aq)}$$

The experimental value of K_B for the equilibrium established is 5×10^{-4} at 25 °C, $pK_B = 3.3$.

$$\underset{\text{ethylamine}}{K_B} = \frac{[OH^{1-}][CH_3CH_2NH_3^{1+}]}{[CH_3CH_2NH_2]} = 5 \times 10^{-4} \qquad pK_B = 3.3 \quad 25\ °C$$

Ethylamine is a stronger base than ammonia itself. The conjugate acid, the ethylammonium ion, $CH_3CH_2NH_3^{1+}$, with its K_A of 2×10^{-11} and pK_A of 10.7, is a weaker acid than the ammonium ion, NH_4^{1+}:

$$CH_3CH_2NH_3^{1+}(aq) + H_2O \rightleftharpoons H_3O^{1+}(aq) + CH_3CH_2NH_2(aq)$$

$$\underset{\text{ethylammonium ion}}{K_A} = \frac{[H_3O^{1+}][CH_3CH_2NH_2]}{[CH_3CH_2NH_3^{1+}]} = 2 \times 10^{-11} \qquad pK_A = 10.7 \quad 25\ °C$$

$$K_B \times K_A = 5 \times 10^{-4} \times 2 \times 10^{-11} = 1.0 \times 10^{-14} \quad 25\ °C$$

and

$$pK_B + pK_A = 3.3 + 10.7 = 14.0 \quad 25\ °C.$$

Note the manner in which the ethylamine–ethylammonium ion conjugate pair is entered in Table 14-2:

10.7	$\mathbf{CH_3CH_2NH_3^{1+}}$	$CH_3CH_2NH_2$	3.3

Also note that the ethylammonium ion is one of the weaker acids and that ethylamine is one of the stronger bases entered in Table 14-2.

K_A, pK_A, K_B, and pK_B expressions can, of course, be formulated for any conjugate acid–base pair in aqueous solution. For each conjugate pair at a specified temperature only one of the four values (K_A, pK_A, K_B, and pK_B) needs to be determined experimentally: any of the other three can be calculated from that value. Table 14-3 summarizes the four values for the ammonia–ammonium ion conjugate pair and the four values for each of the three ethylamine–ethylammonium ion conjugate pairs. For these four conjugate pairs, note that (1) all of the ammonium ion acids are very weak

Table 14-3

K AND pK VALUES FOR THE CONJUGATE ACID–BASE PAIR FOR AMMONIA AND THE THREE ETHYLAMINES IN AQUEOUS SOLUTION, 25 °C

the conjugate acids			the bases		
K_A	**pK_A**			K_B	pK_B
5×10^{-10}	**9.3**	**NH_4^{1+} ammonium ion**	NH_3 ammonia	2×10^{-5}	4.7
2×10^{-11}	**10.7**	**$CH_3CH_2NH_3^{1+}$ ethylammonium ion**	$CH_3CH_2NH_2$ ethylamine	5×10^{-4}	3.3
2×10^{-11}	**10.7**	**$(CH_3CH_2)_3NH^{1+}$ triethylammonium ion**	$(CH_3CH_2)_3N$ triethylamine	5×10^{-4}	3.3
1×10^{-11}	**11.0**	**$(CH_3CH_2)_2NH_2^{1+}$ diethylammonium ion**	$(CH_3CH_2)_2NH$ diethylamine	1×10^{-3}	3.0

acids, (2) the ammonium ion is the strongest acid of the four ammonium ions included in the table, and (3) ammonia is the weakest base of the four nitrogen bases.

Handbooks present information such as that given in Tables 14-2 and 14-3 in a two-column format: pK_A and the formula of the acid. These are the two columns printed in boldface type in Tables 14-2 and 14-3.

The ammonia molecule and the ammonium ion have a host of relatives that are essential to both plant and animal life as it exists on this planet. These include the proteins and the nucleic acids. The proteins either constitute or are an important constituent of enzymes, hormones, skin, hair, tendons, muscles, and blood. The nucleic acids are essential to the synthesis of all cellular proteins and are the agents by which genetic makeup is stored and transmitted by the genes. The active components of viruses are nucleic acids. All protein molecules and all nucleic acid molecules are macromolecules containing a great number of atoms—in some cases more than 100 million atoms. This is a seemingly impossible complex array of atoms in molecules. If we are to exist, such large ions or molecules are inevitable, since neither ionic compounds with small ions nor covalent compounds with small molecules could have the flexibility and the "stay-togetherness" of tissues of plants and animals. On the other hand, the processes by which materials are moved about and converted into the proper compounds to provide the structures, functions, and energy balances of a biological system place very severe restrictions on the characteristics of the molecules and ions in the aqueous medium of biological systems. The forces between molecules, between ions, and between molecules and ions become extremely important. Ion–ion interactions, ion-dipole interactions, dipole-dipole interactions, and hydrogen bonding are all extremely important in maintaining dispersed systems in the aqueous medium and in creating the appropriate grouping of molecules and ions so that essential chemical reactions go on. There is a great deal of order in these incredibly complex ions and molecules. This order arises from the repeated use of a limited number of groups of atoms. Both proteins and nucleic acids will be explored in later chapters. The next section of this chapter provides some of the background for those explorations.

Groups of atoms that repeatedly occur in the proteins are exemplified by glycine, NH_2CH_2COOH:

Glycine

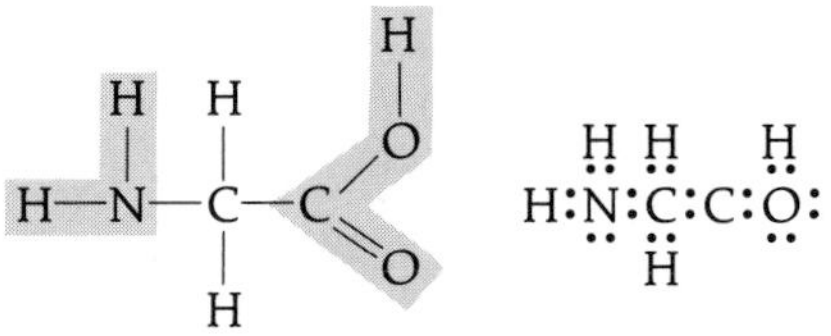

This is an interesting compound in that it is both a nitrogen base (very similar to an alkyl amine) and a carboxylic acid (very similar to acetic acid). Compounds of this type are known as amino acids. The discussion of glycine will seem very complex unless you can focus on where the action is. There are two sites of action. One is the amine group, $—NH_2$, and the other is the carboxylic acid group, —COOH. The amine group, a base, will, in acidic solutions, become the conjugate acid:

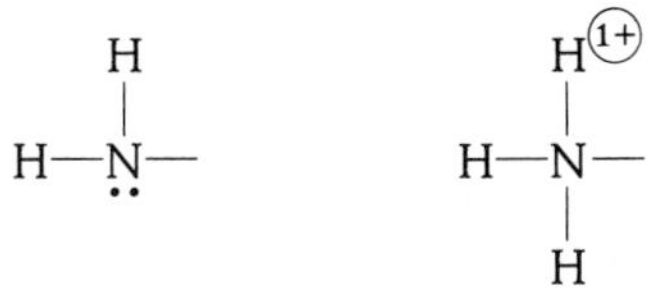

base	conjugate acid
amine group	ammonium ion group

The carboxylic acid group, an acid, will, in basic solutions, become the conjugate base:

acid	conjugate base
carboxylic acid group	carboxylate ion group

The Ions of Glycine

In aqueous solutions, there are three ions of glycine:

(1) The completely protonated cation:

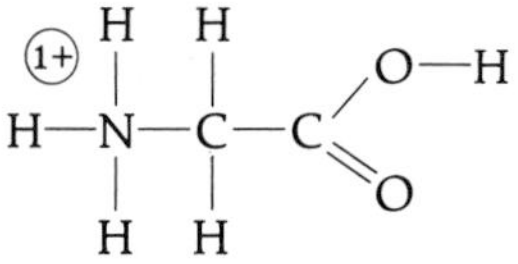

predominant in low pH solutions (highly acidic solutions)

This ion is a diprotic acid: one proton of the ammonium ion group and the proton of the carboxylic acid group can be transferred to Brönsted bases.

(2) The completely deprotonated anion:

$$\mathrm{H{-}\ddot{N}(H){-}C(H)(H){-}C(=O){-}O^{1-}}$$

predominant in high pH solutions (low concentration of hydronium ion)

The amine group is a base and the carboxylate ion is a base.

(3) The partially protonated ion:

$$\mathrm{{}^{1+}H{-}N(H)(H){-}C(H)(H){-}C(=O){-}O^{1-}}$$

predominant in middle-range pH's

The Zwitterion

This ion is both an acid (a proton of the ammonium ion group) and a base (the carboxylate ion). It is an unusual ion. As a whole, the net charge is zero. In that sense it is a molecule, but it is not an ordinary polar molecule. The magnitude of the charge at the ammonium ion group is the full charge of a proton and the magnitude of the charge at the carboxylate ion is the full charge of an electron. This structure is a double ion. Such a structure is known as a dipolar ion or as a zwitterion (from the German *zwitter*, for both).

The ion $^{1+}NH_3CH_2COOH$ is a diprotic acid. The two equilibria with water and the two K_A's are

Equilibrium 1

The carboxylic acid group is a stronger acid than the ammonium ion group. The first reaction involves the —COOH group.

$$\underset{\text{cation}}{{}^{1+}\mathrm{NH_3CH_2COOH}} + \mathrm{H_2O} \rightleftharpoons \mathrm{H_3O^{1+}} + \underset{\text{zwitterion}}{{}^{1+}\mathrm{NH_3CH_2COO^{1-}}}$$

$$K_{A1} = \frac{[\mathrm{H_3O^{1+}}][{}^{1+}\mathrm{NH_3CH_2COO^{1-}}]}{[{}^{1+}\mathrm{NH_3CH_2COOH}]} = 4.5 \times 10^{-3} \qquad pK_{A1} = 2.35 \quad 25\ ^\circ\mathrm{C}$$

Note the position of the acid–base pair

$^{1+}NH_3CH_2COOH$	$^{1+}NH_3CH_2COO^{1-}$
fully protonated glycine cation	glycine zwitterion

in Table 14-2. The glycine cation is clearly a stronger acid than acetic acid. Note that, in the comparison made in the preceding sentence, both acid–base pairs involve the reaction of the carboxylic acid group, —COOH, to give its conjugate base, the carboxylate group, $—COO^{1-}$.

Equilibrium 2

$$^{\oplus}NH_3CH_2COO^{\ominus} + H_2O \rightleftharpoons H_3O^{\oplus} + NH_2CH_2COO^{\ominus}$$

zwitterion — deprotonated anion

$$K_{A2} = \frac{[H_3O^{(1+)}][NH_2CH_2COO^{(1-)}]}{[^{(1+)}NH_3CH_2COO^{(1-)}]} = 1.7 \times 10^{-10} \qquad pK_{A2} = 9.78 \quad 25°C$$

Note the position of this acid–base pair

$^{(1+)}NH_3CH_2COO^{(1-)}$ glycine zwitterion — $NH_2CH_2COO^{(1-)}$ deprotonated glycine anion

in Table 14-2. The glycine zwitterion is a slightly weaker acid than the ammonium ion. Note that in the comparison made in the preceding sentence, both acid–base pairs involve the reaction of the ammonium ion group, $—NH_3^{(1+)}$ or $NH_4^{(1+)}$, to give the conjugate base, the amine group, $—NH_2$ or NH_3.

Percent Abundance vs pH

Figure 14-2 gives the relation of the percent abundance curves for the ions of glycine to pH in aqueous solutions of glycine at 25 °C. At the first 50% abundance crossover,

$$[^{(1+)}NH_3CH_2COOH] = [^{(1+)}NH_3CH_2COO^{(1-)}]$$
$$[H_3O^{(1+)}] = 4.5 \times 10^{-3} \text{ moles liters}^{-1}$$
$$pH = 2.35$$

At the second 50% abundance crossover in Figure 14-2,

$$[^{(1+)}NH_3CH_2COO^{(1-)}] = [NH_2CH_2COO^{(1-)}]$$
$$[H_3O^{(1+)}] = 1.7 \times 10^{-10} \text{ moles liters}^{-1}$$
$$pH = 9.78$$

At the pH of biological cells, it is the zwitterions of glycine that are significant. There is also evidence that even crystals of glycine are made up of the zwitterion $^{(1+)}NH_3CH_2COO^{(1-)}$, not the molecule NH_2CH_2COOH. Glycine is a white solid that decomposes at 260 °C without melting and the compound is much more soluble in water than in less polar solvents such as ethanol and diethyl ether. These are properties of a solid stabilized by ion–ion interaction rather than the much weaker van der Waals' forces.

α-Amino Acids

There is a great variety of amino acids that differ from glycine only in the replacement by other groups of one or both of the hydrogen atoms on the alpha carbon, the carbon adjacent to the carboxylic acid group. For example, alanine is glycine with

Figure 14-2

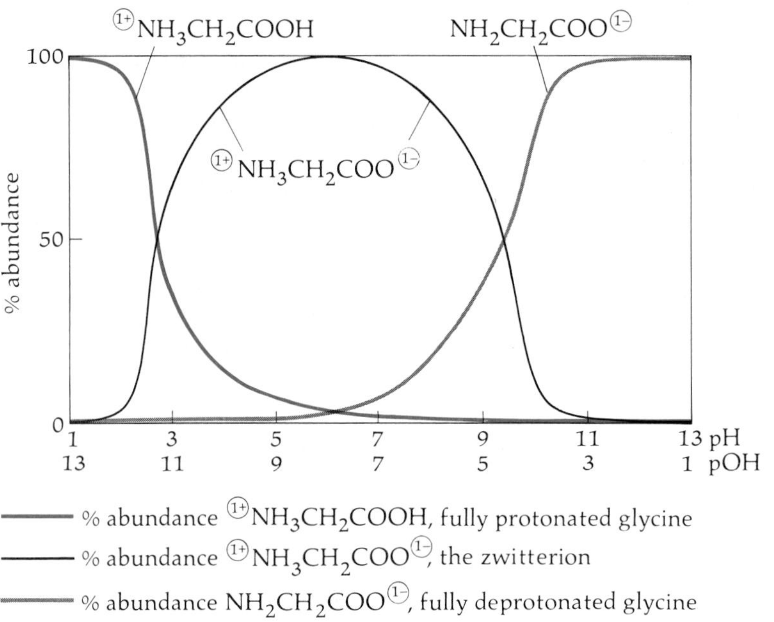

PERCENT ABUNDANCE VS pH DIAGRAM FOR THE IONS OF GLYCINE IN AQUEOUS SOLUTION AT 25 °C. The diagram has been distorted to show the crossover of the percent abundance of $^{(1+)}NH_3CH_2COOH$ and the percent abundance of $NH_2CH_2COO^{(1-)}$ at the isoelectric point, pH = 6.06, where the concentration of the positive ion equals the concentration of the negative ion. Drawn to scale, the curves for these two ions could not be distinguished from the 0% abundance line. Note that this crossover is not a 50% abundance crossover and the ions involved are not a conjugate pair.

one hydrogen atom on the α-carbon replaced by a methyl group:

$$\underset{\text{alanine}}{\underset{\displaystyle CH_3}{NH_2\overset{}{C}HCOOH}} \qquad {}^{(1+)}NH_3\underset{\displaystyle CH_3}{C}HCOO^{(1-)} \qquad {}^{(1+)}H{-}\underset{H}{\overset{H}{N}}{-}\underset{H-\underset{H}{C}-H}{\overset{H}{C}}{-}\overset{O}{\overset{\|}{C}}{-}O^{(1-)}$$

alanine zwitterion

Here again, the properties of the solid indicate that the solid is composed of the zwitterion, and the properties of aqueous solutions indicate that, in the middle pH range, it is the zwitterion that is the predominant ion. At 25 °C, pK_{A1} and pK_{A2} are

2.35 and 9.87. The corresponding values for glycine are 2.35 and 9.78. Consequently, the pH–percent abundance curves for alanine are almost identical to those of glycine. (Figure 14-2).

One of the interesting properties of zwitterions is that, in aqueous solution, these ions do not contribute to the electrical conductivity of the solution. With its net charge of zero, the zwitterion does not move toward either the positive electrode or the negative electrode. The hydronium ion concentration at which the zwitterion is at a maximum is known as the isoelectric point. The pH of the **isoelectric point** for glycine is 6.06 and for alanine, 6.11. Attention is called to the isoelectric point for glycine in the note that accompanies Figure 14-2.

Isoelectric Point

Alanine

Biologically active alanine has an isomer that is its mirror image. This type of isomerism is known as **stereoisomerism.** Until you build ball and stick models, you may not believe that a mirror image structure could not be superimposable and is, therefore, an isomer—not an identity. If there are stereoisomers of a compound, biological systems are frequently highly selective in the synthesis and use of one isomer and the rejection of the other isomer. An understanding of nonsuperimposable mirror images is fundamental to understanding how each of us is structured and how we operate at the molecular level.

Mirror Images

We use mirror images daily, but most of us have probably never thought much about mirror images. Two exercises may help.

Exercise 1: Stand in front of a mirror holding a comb in one hand, reach out with your free hand "to clasp the free hand of your image," memorize the image you see, and retain this image in your memory. Still "clasping the free hand of the image," turn beside the mirror to face the same direction as "the image," which is, of course, no longer there. Where is your hand with the comb in relation to the hand with the comb in "the image"? The image that another person sees of you is not the image you see of yourself when you look in a mirror. The image you see of yourself when you look at a photograph is not the image you see of yourself when you look in a mirror—unless the picture was reversed in printing.

Exercise 2: Where necessary, build the ball and stick models to answer each of the following four questions.

(a) If you built a ball and stick model of methane, CH_4, and then built a second ball and stick model that is the mirror image of the first model, would the two models be superimposable?

(b) If you built a ball and stick model of CH_3Cl and a second model that is the mirror image of the first model, would the two models be superimposable?

(c) If you built a ball and stick model of CH_2ClBr and a second model that is its mirror image, would the two models be superimposable? If you have difficulty building the mirror image, actually work with a mirror and match in a model what you see in the mirror.

(d) If you built a model of CHClBrI and also its mirror image, would the two structures be superimposable?

The answer to questions a, b, and c is, in each case, yes. The answer to question d is no. If you missed one of these, go back and endeavor to find where you went wrong. If your models in d were superimposable, you built the identity, not the mirror image. Interchange any two balls in one model and you will have the mirror image. Place the two models on a surface and turn them to achieve the mirror image orientation. The molecule CHClBrI is said to be **chiral.** The molecule has a mirror image that is not superimposable. The general characteristic of the molecule CHClBrI is that the tetrahedral carbon is attached to four different groups, and that carbon atom is said to be dissymetric. This is the characteristic of the α-carbon in alanine but not in glycine. Alanine is chiral; glycine is not. Molecules can have a nonsuperimposable mirror image for reasons other than having four different groups attached to a tetrahedral carbon atom, but these structures need not concern us now.

Chiral Compounds

Barring some misadventure, your left hand is a mirror image of your right hand and your right hand is a mirror image of your left hand. Your hands are chiral. The word chiral is derived from the Greek *cheir,* for hands.

The two stereoisomers of alanine are

Stereoisomers of Alanine

COO^{1-}, C, H, CH_3, NH_3^{1+} — biologically active alanine

COO^{1-}, C, H, CH_3, NH_3^{1+} — mirror image

The solid lines indicate that the σ-bonds lie in the plane of the paper. The broken line indicates that the σ-bond is below the plane of the paper. The solid wedge indicates that the σ-bond is above the plane of the paper. Free rotation about these σ-bonds allows the groups $—CH_3$, $—NH_3^{1+}$, and $—COO^{1-}$ to be turned to appropriate positions in comparing the two structures and we do not need to concern ourselves with the details of those groups.

It is important that you understand the concept of stereoisomerism. It is not important that you remember which spatial orientation is the biologically active alanine. There is a very precise nomenclature to deal with this very subtle type of isomerism. This nomenclature need not concern us.

One of the physical properties of chiral compounds is that each isomer is optically active in rotating plane-polarized light as the polarized light passes through a solution of the isomer. If one isomer rotates the plane of polarized light to the right, the mirror image of that isomer rotates the plane of polarized light to the left. If a solution contains both isomers at the same concentrations, the solution is optically inactive. What the molecules of one isomer do to plane-polarized light, the molecules of the other isomer undo.

Polarimeters, the instruments used to study optical activity, have two sets of calcite prisms ($CaCO_3$). A light source produces unpolarized light. The light transmitted by the first set of prisms is plane-polarized. The second set of prisms identifies the new position of the plane of the light after it passes through the solution. The optical property of calcite crystals, upon which the design of the sets of calcite prisms is based, is the consequence of the ordered array of the planar carbonate ions, CO_3^{2-}, in crystals of calcite.

With the exception of glycine, all the biologically significant *alpha*-amino acids are chiral and all have the same spatial orientation at the *alpha*-carbon as biologically active alanine.

Another chiral compound of biological origin, lactic acid, was discussed in Chapter 13 as an acid. The structures of the two stereoisomers are

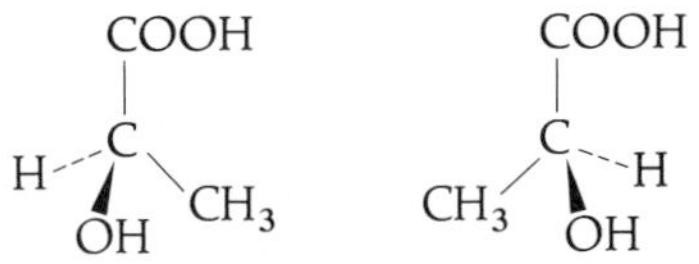

COOH | H–C(–OH)–CH_3 COOH | CH_3–C(–OH)–H

The isomer on the left is found in sour milk. The isomer on the right is a product of the fermentation of glucose.

Nitrogen Bases as Building Blocks of RNA and DNA

In the discussion of nucleic acids (Chapter 19), you will discover that two nitrogen bases are important building blocks of both RNA and DNA. Both of these nitrogen bases are ring compounds:

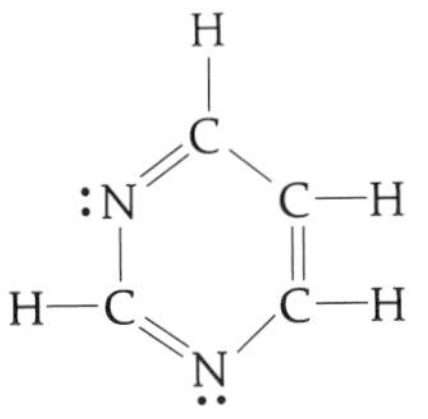

pyrimidine 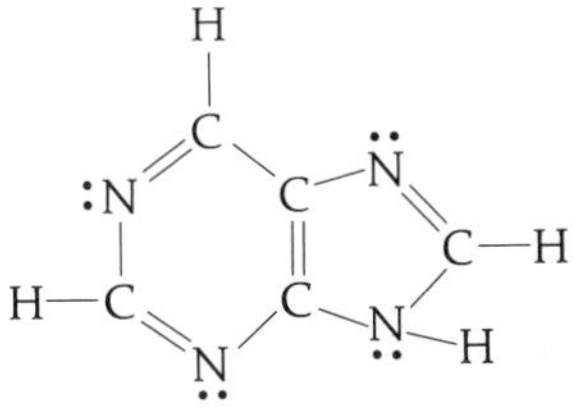 purine

also written

pyrimidine 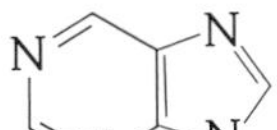 purine

It is the unshared pair of electrons of the nitrogen atoms, as in ammonia,

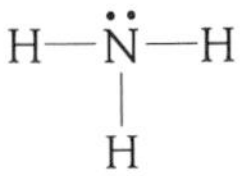

$$\mathrm{H{-}\ddot{N}{-}H} \atop \mathrm{|\;H}$$

that makes the compounds Brönsted bases. Each nitrogen atom with its unshared pair of electrons has the potential to accept a proton. The more schematic representation of the structures of the molecules is frequently used for simplicity. In this representation, the end of a straight line indicates a carbon atom unless the symbol for some other element appears at that position. Double bonds are indicated in the usual fashion. All structures are understood to include whatever hydrogen atoms are necessary to complete the structures: four bonding sites for carbon atoms and three bonding sites for nitrogen atoms. All of the atoms of pyrimidine lie in a plane. Purine also approaches being planar but does not quite make it. In either case, these ring structures introduce flat, rather rigid groups of atoms, "platelets of atoms," into the structure of any molecule in which they are incorporated. Three derivatives of pyrimidine, known as pyrimidine bases, and two derivatives of purine, known as purine bases, are key units in the structures of the ribonucleic acids (RNA) and the deoxyribonucleic acids (DNA). You may be familiar with their names: uracil, thymine, cytosine, guanine, and adenine. The ring structures of these nitrogen bases impose restrictions on the spatial configurations open to the very large nucleic acid molecules and also provide structural capabilities that are thought to be the basis for the storage and transmission of genetic information.

Part II Role of Ammonia and Related Compounds in the World Food Supply

Nitrogen Fertilizers

An increasing world population has accelerated the demand for nitrogen fertilizers. These include any compound of nitrogen that can be assimilated by plants. Traditionally, nitrogen fertilizers were supplied by the distribution of biological waste materials over the land. Quantities of such materials are limited and labor costs have been high. Mined sodium nitrate, $NaNO_3$, and a by-product of the steel industry, ammonium sulfate, $(NH_4)_2SO_4$, have been extensively used, but the production and distribution of ammonia, NH_3, and urea, $(NH_2)_2CO$, became economically feasible and these commercially synthesized materials now dominate the nitrogen fertilizer field.

Fixation of Nitrogen

Stability of the N_2 Molecule

There is no shortage of nitrogen. Eighty percent of the molecules in air are diatomic nitrogen molecules. The problem lies in the fixation of nitrogen, the conversion of the diatomic nitrogen of the air into chemical combination so that plants can deal with it. Ammonium compounds and nitrates are reasonably stable and in most cases there is no problem of their tending to decompose into the elements, so it would seem that it should be comparatively easy to get nitrogen from the atmosphere to react to form these compounds. Nitrogen gas is, however, notorious as an inert gas, and nitrogen gas is frequently used when it is desirable to maintain chemicals in an inert atmosphere. This is a consequence of the great stability of the diatomic nitrogen molecule. It requires 945 kilojoules (226 kcal)

$$N_2(g) \longrightarrow 2\ N(g) \qquad \Delta H = 945 \text{ kilojoules}$$

to dissociate one mole of diatomic nitrogen molecules into nitrogen atoms, as compared to 435 kilojoules (104 kcal) for the dissociation of one mole of diatomic hydrogen molecules into atoms and 497 kilojoules (110 kcal) for the dissociation of one mole of diatomic oxygen molecules into atoms:

$$H_2(g) \longrightarrow 2\ H(g) \qquad \Delta H = 435 \text{ kilojoules}$$

$$O_2(g) \longrightarrow 2\ O(g) \qquad \Delta H = 497 \text{ kilojoules}$$

Role of Electrical Storms

In order for diatomic nitrogen molecules to react, they have to be activated in some manner. During electrical storms, lightning provides this energy and oxides of nitrogen are produced in the atmosphere during storms. The most probable product is nitrogen monoxide, although a number of other oxides of nitrogen are also formed.

$$N_2(g) + O_2(g) \longrightarrow 2\ NO(g)$$

Following a series of reactions with more oxygen of the air and with water, dilute nitric acid, HNO_3, solutions rain upon the landscape. Throughout the ages, the total contribution of electrical storms to the fixed nitrogen of the world has been significant.

Role of Bacteria

One of the most intriguing methods of fixing nitrogen has been mastered by a variety of bacteria. The best known and managed are cooperative ventures involving bacteria and legumes such as peas, beans, and alfalfa. Enormous quantities of nitrogen are fixed by these bacteria. Several hundred species of nonleguminous plants also team up with bacteria to fix nitrogen, and serious research efforts are under way to promote the participation of other agricultural crops in symbiotic ventures with nitrogen-fixing bacteria.

In the fixation of nitrogen by bacteria and other microorganisms, the enzymatic reactions proceed under normal atmospheric conditions. These systems have been extensively studied and, although they are only partially understood, it is clear that enzymes (proteins) and other groups of atoms, including iron and molybdenum, are involved. There are very close parallels between these compounds and the hemoglobins of higher animals. In the symbiotic activity of legumes and bacteria, the legumes presumably provide some component required by an enzyme of the bacteria but not produced by the bacteria.

Nonsymbiotic nitrogen fixation occurs in a number of microorganisms including blue-green algae, and this is one of the reasons blue-green algae are considered attractive potential sources of protein.

Synthesis of Ammonia Gas from N_2 and H_2

Until other means are developed to meet the world food requirements, we are dependent upon the commercial production of ammonia. Well over fifty percent of the commercial nitrogen fertilizers used in the United States arise from the synthesis of ammonia from nitrogen gas and hydrogen gas. This is an exothermic reaction: heat is released during the reaction.

$$N_2(g) + 3H_2(g) \longrightarrow 2\,NH_3(g) \qquad \Delta H = -92 \text{ kilojoules } (-22 \text{ kcal})$$

Catalysts

In the Haber process, an appropriate molecular mixture of hydrogen gas and nitrogen gas flows over a solid catalyst under conditions of high temperature and pressure. A large fraction of the ammonia produced is removed, more hydrogen gas and nitrogen gas are added, and the reaction mixture is recirculated over the catalyst in a continuous process. All reactions between molecules proceed more rapidly under conditions that bring the molecules closer together and thus increase the contact between the reacting molecules. By adsorbing gases, a solid catalyst brings molecules closer together on its surface and, in the process of adsorption, also modifies the distribution of electrons in the molecules of the reactants. The adsorption of these gases on a catalyst increases the probability of breaking old bonds and forming new bonds. The nature of chemical change keeps the game fair. A catalyst always accelerates the rate of decomposition of the products, the reverse reaction, to the same degree it accelerates the forward reaction. If a reaction mixture is allowed to come to equilibrium, the composition of the equilibrium mixture is the same regardless of whether a catalyst is or is not present. The difference lies in how long it takes for the mixture to come to equilibrium. A reaction that might require hours, days, or even years to reach equilibrium without a catalyst may attain equilibrium at the same temperature almost instantaneously in the presence of an appropriate catalyst. By using a catalyst, you do not get something for nothing; you just get it faster. The catalysts that have been found to be effective in the reaction of N_2 with H_2 have been mixtures of some of the transition metals and transition metal oxides. These are the elements that belong to that block of ten elements in the middle of the Periodic Table. The chemistry of these transition metals frequently involves either their *d*-orbital electrons or their capacity to accommodate electrons from other atoms in unoccupied *d*-orbitals. The most commonly used catalyst seems to be a mixture of iron, iron oxide, and molybdenum. The fixation of nitrogen has extremely high economic, political, and social value and a great deal of research has been directed to the elucidation of the series of reactions that take place at or on the surface of the catalyst. It is difficult to study these surface reactions, and the mechanism of the reaction is not fully disclosed and presumably not fully understood.

The ideal catalyst would make it possible to attain rapid conversion near room temperature. The solid catalysts that have been developed for the ammonia synthesis are not that efficient and the reaction is carried out at an elevated temperature. The high temperature accelerates the forward reaction and it also accelerates the reverse reaction, the decomposition of ammonia into diatomic nitrogen and diatomic hydrogen. In this case, however, the yield is, in part, a losing game: the increase in temperature accelerates the reverse reaction, the decomposition-of-NH_3 reaction, to a greater degree than it accelerates the forward reaction. When equilibrium is established at the higher temperature, the yield of ammonia is less than it would have been if equilibrium had been established at a lower temperature. This result is entirely in keeping with the principle of Le Chatelier: starting with an equilibrium mixture at the lower temperature, the addition of heat to raise the temperature shifts the equilibrium to tend to use up that heat. In this case, it is the decomposition of ammonia that requires heat. The choice of the temperature to be used is an economic trade-off between rate of conversion and the fraction of the reactants converted.

Use of a Flowing System

Why the use of the flowing system? This is purely a matter of economics. The maximum yield per batch would be obtained if the reaction mixture remained in the reaction chamber until equilibrium was established. This delay, however, ties up expensive equipment, and the unit cost of production is reduced by a continous flow process in which at least part of the product is removed before the mixture is returned to the reaction chamber containing the catalyst. This process minimizes the back reaction and consequently increases the production of ammonia per unit time. The reaction is also carried out under high pressure—several atmospheres. For this particular reaction, four molecules of reactants yield two

$$N_2 + 3\,H_2 \rightleftarrows 2\,NH_3$$

molecules of products: an increase in pressure favors the production of ammonia. According to the principle of Le Chatelier, the compression of an equilibrium mixture into a smaller volume will shift the equilibrium to relieve the increased pressure in part by reaction to form fewer molecules—four molecules to form two molecules in this particular case. Higher pressures also increase the rate of reaction. The molecules are closer together: molecules adsorb more readily upon the surface of the catalyst and the reaction proceeds more readily.

Another part of the industrial production of ammonia is the production of the reaction mixture in the first place. It should be a clean mixture made up of three hydrogen molecules for every nitrogen molecule. This can be brought about in a very cleverly designed high-temperature process using air and steam with hydrocarbons or coal. The equations for the process using methane as the hydrocarbon are

Hydrocarbons as a Source of H_2

(1) two reactions of the oxygen of the air with the hydrocarbon that give hydrogen gas:

$$2\ CH_4(g) + O_2(g) \longrightarrow 2\ CO(g) + 4\ H_2(g)$$
$$CH_4(g) + O_2(g) \longrightarrow CO_2(g) + 2\ H_2(g)$$

Note that carbon monoxide is also a product in the first and carbon dioxide is also a product in the second reaction.

(2) two reactions of steam that give hydrogen gas—one with the hydrocarbon:

$$CH_4(g) + H_2O(g) \longrightarrow CO(g) + 3\ H_2(g)$$

and one with the carbon monoxide formed above:

$$CO(g) + H_2O \longrightarrow CO_2(g) + H_2(g)$$

By the appropriate choice of catalyst, pressure, temperature, and relative quantities of air and steam, the reactive mixture can be made to yield the desired proportions of nitrogen molecules left over from the air and hydrogen molecules formed by the four reactions.

The carbon dioxide gas and any excess of steam are removed by passing the resultant mixture through a drying tower containing solid calcium oxide (an ionic compound). The trivial name for CaO is lime.

$$CaO(s) + CO_2(g) \longrightarrow CaCO_3(s)$$
$$CaO(s) + H_2O(g) \longrightarrow Ca(OH)_2(s)$$

(The next questions are, of course: Can you regenerate the calcium oxide? Can you find a use for calcium carbonate and/or calcium hydroxide? Can you persuade someone to pay you for these unwanted products? What do you have to pay to get rid of these products?)

The noble gases, argon in particular, introduced with the air into the circulating reaction mixture can be recovered and marketed. With the continous addition of air and the continuous removal of ammonia, carbon dioxide, and water vapor, the noble gases collect in the reaction mixture to the point that it is economically attractive to separate them by low-temperature distillation and market them as commercial products.

Synthesis of Urea

The network of practices and possibilities extends in a seemingly endless fashion. Ammonia is converted to urea, $(NH_2)_2CO$, a low-melting, white, crystalline, water-soluble compound, by reaction with carbon dioxide under pressure and at elevated temperatures:

$$2\ NH_3(g) + CO_2(g) \longrightarrow \begin{matrix} H-N-H \\ \quad\diagdown \\ \quad\quad C{=}O(s) \\ \quad\diagup \\ H-N-H \end{matrix} + H_2O(g)$$

Urea is primarily used as a fertilizer, but it is also used as a feed supplement for cattle! Note the similarity of the formula of urea and the formula of carbonic acid:

$$\begin{matrix} H-O \\ & \diagdown \\ & & C=O \\ & \diagup \\ H-O \end{matrix}$$

Note that this reaction uses some of the carbon dioxide produced by previous reactions. Urea is a dibasic compound: each of the nitrogen atoms has an unshared pair of electrons and the potential to accept a proton. It is, however, a very weak base and aqueous solutions of urea are essentially neutral. (Urea is also formed in the biological degradation of proteins and large quantities are excreted in the urine.)

Synthesis of Ammonium Nitrate

Ammonium nitrate, NH_4NO_3, a white, crystalline, water-soluble, ionic compound, is also an attractive fertilizer since a high proportion of the mass of the compound is nitrogen. This reduces the transportation and handling charges to deliver a hundred pounds of fixed nitrogen to the customer. It is produced in a single operation through a multiplicity of reactions using ammonia and air to supply the oxygen. Here again, the reactions are controlled by a careful choice of catalysts, pressure, and temperature. Some of the reactions are

$$4\ NH_3(g) + 5\ O_2(g) \longrightarrow 4\ NO(g) + 6\ H_2O(g)$$
$$2\ NO(g) + O_2(g) \longrightarrow 2\ NO_2(g)$$
$$3\ NO_2(g) + H_2O(g) \longrightarrow 2\ HNO_3(g) + NO(g)$$
$$NH_3(g) + HNO_3(g) \longrightarrow NH_4NO_3(s)$$

The enthusiasm for the large-scale use of ammonium nitrate as a fertilizer is tempered by the fact that ammonium nitrate has been the chief component in some of the world's most disastrous explosions. Small quantities of ammonium nitrate are extremely difficult to detonate. Large samples have, however, been set off by other explosions or intense heat. The nitrogen reverts to the very stable diatomic molecules and the energy released becomes kinetic energy of the products (high temperature):

$$2\ NH_4NO_3(s) \longrightarrow 2\ N_2(g) + O_2(g) + 4\ H_2O(g) \qquad \Delta H = -105 \text{ kilojoules } (-25 \text{ kcal})$$

The very rapid increase in the moles of gases and the expansion of the gaseous products is the explosion.

What becomes of all of this nitrogen that is incorporated as amines and related compounds in biological systems? Much of it moves through the food chain to other organisms with the end product in animal metabolism being urea excreted in the urine. In the destruction of biological materials by fire, much of the nitrogen reverts to diatomic nitrogen. There is a great deal to be said for the use of biological materials as fertilizers, but the supply does not match or even approach the food production requirements of the current world population.

Scramble Exercises

1. Many of the properties of ammonia, the amines, and amino acids are believed to be a consequence of the unshared pair of electrons on the nitrogen atom.

Write out the Lewis electron structures for each of the ions and molecules involved in the reaction of methylamine and water:

$$CH_3NH_2(g) + H_2O(l) \rightleftarrows OH^{1-}(aq) + CH_3NH_4^{1+}(aq)$$

Note that it is this unshared pair of electrons on the nitrogen atom that makes it possible for ammonia and the amines to act as bases.

2. Set up the K_B expression for the above equilibrium and sketch the percent abundance diagram for the CH_3NH_2–$CH_3NH_3^{1+}$ conjugate pair in aqueous solutions. At 25 °C, $K_B = 5.6 \times 10^{-4}$ and $pK_B = 3.35$. Include both the pH scale and the pOH scale on the graph.

3. Complete the equation of the reaction of methylammonium ion with water:

$$CH_3NH_3^{1+} + H_2O \rightleftarrows H_3O^{1+} + \underline{\qquad}$$

and set up the corresponding K_A expression. Using the values of K_B and pK_B given above for methylamine, evaluate K_A and pK_A for this methylammonium ion.

Use Tables 14-2 and 14-3 to compare the Brönsted acid strength of the methylammonium ion, $CH_3NH_3^{1+}$, with the Brönsted acid strength of the ethylammonium ion, $CH_3CH_2NH_3^{1+}$, and also with the ammonium ion, NH_4^{1+}.

4. Write out the Lewis electron structure for pyrimidine and justify the statement that this compound is a nitrogen base.

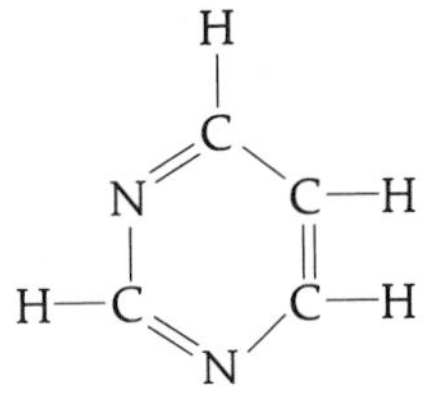

```
        H
        |
        C
      //  \
     N     C—H
     |     ||
   H—C     C—H
      \\  /
        N
```

5. Write out the Lewis electron structure for

```
        H
        |
        C
      /  \\
     N     C—H
     ||    |
   H—C     C—H
      \  //
        N
```

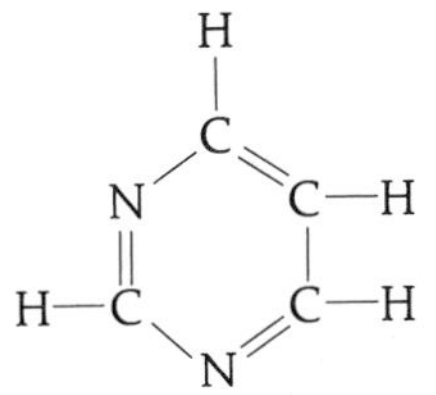

Is this the same structure written in Exercise 4?

It is an experimental fact that, in pyrimidine, all four carbon–nitrogen bond lengths are the same and both carbon–carbon bond lengths are the same. What could this imply about the structure of pyrimidine? The carbon–nitrogen bond lengths are shorter than the usual carbon–nitrogen single bond length and longer than the usual carbon–nitrogen double bond length. The carbon–carbon bond lengths are shorter than the usual single bond length and longer than the usual double bond length.

6. One of the methods of separating materials in aqueous solution is to use a source of direct current and two electrodes to apply an electrical potential that induces positive ions to move toward the negative electrode and negative ions to move toward the positive electrode. For the amino acid alanine, what would be the response of each of the following ions and molecules to an applied potential? This process of separation is known as electrophoresis and is extensively used to separate biological materials.

$$H_2N-\underset{CH_3}{\overset{H}{C}}-COOH \qquad \overset{(1+)}{H_3N}-\underset{CH_3}{\overset{H}{C}}-COOH$$

$$\overset{(1+)}{H_3N}-\underset{CH_3}{\overset{H}{C}}-COO^{(1-)} \qquad H_2N-\underset{CH_3}{\overset{H}{C}}-COO^{(1-)}$$

Which of these forms would predominate in solutions of low pH?

7. The two K's for the alanine ammonium ion at 25 °C are

$$\overset{(1+)}{H}-\underset{H}{\overset{H}{N}}-\underset{CH_3}{\overset{H}{C}}-COOH$$

alanine ammonium ion

$K_{A1} = 4.5 \times 10^{-3}$ $\quad pK_{A1} = 2.35$
(for the—COOH group)

$K_{A2} = 1.3 \times 10^{-10}$ $\quad pK_{A2} = 9.87$
(for the $\overset{(1+)}{H}-\underset{H}{\overset{H}{N}}-$ group)

Set up the two K_A expressions. Which of the two Brönsted acids is the stronger acid? Sketch and label the percent abundance diagram for the aqueous solutions of alanine at 25 °C.

In what pH range is

$$\overset{(1+)}{H}-\underset{H}{\overset{H}{N}}-\underset{CH_3}{\overset{H}{C}}-COO^{(1-)}$$

the predominant ion?

Estimate the isoelectric point for alanine.

8. Carboxylic acids such as acetic acid react with the alcohols such as ethanol

$$CH_3\overset{O}{\overset{\|}{C}}-OH + \underset{\text{ethanol}}{HOCH_2CH_3} \longrightarrow \underset{\substack{\text{ethyl acetate}\\ \text{(an ester)}}}{CH_3\overset{O}{\overset{\|}{C}}-OCH_2CH_3} + H_2O$$

and also react with amines such as ethylamine:

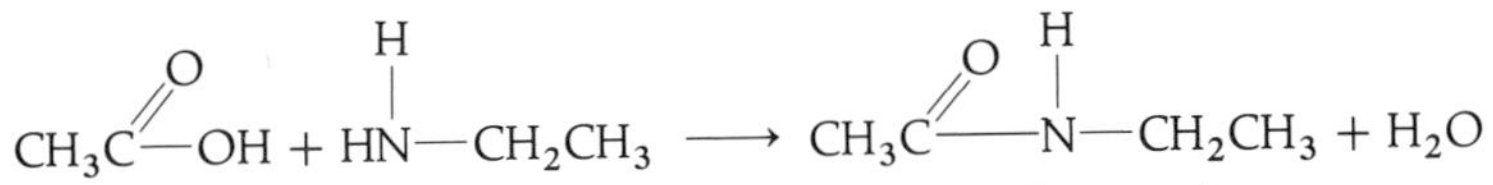

$$CH_3\overset{\overset{O}{/\!/}}{C}-OH + \underset{\text{ethylamine}}{H\overset{\overset{H}{|}}{N}-CH_2CH_3} \longrightarrow \underset{\substack{N\text{-ethylacetamide} \\ \text{(an amide)}}}{CH_3\overset{\overset{O}{/\!/}}{C}-\overset{\overset{H}{|}}{N}-CH_2CH_3} + H_2O$$

Point out the similarity between these two reactions. By the use of ethanol containing the oxygen-18 isotope, it has been possible to demonstrate that the water eliminated in the first reaction does not contain the oxygen-18 isotope. (The name *N*-ethylacetamide is based on the name acetamide, for CH_3CONH_2, in which the suffix *-amide* has replaced the suffix *-ic* of acetic acid. The *N*- simply indicates that the ethyl group is attached through the nitrogen atom.) Note that the reaction of a carboxylic acid and an alcohol forms an ester; the reaction of a carboxylic acid with an amine forms an amide.

9. Which of the following compounds, all derivatives of methane, have two stereoisomers?

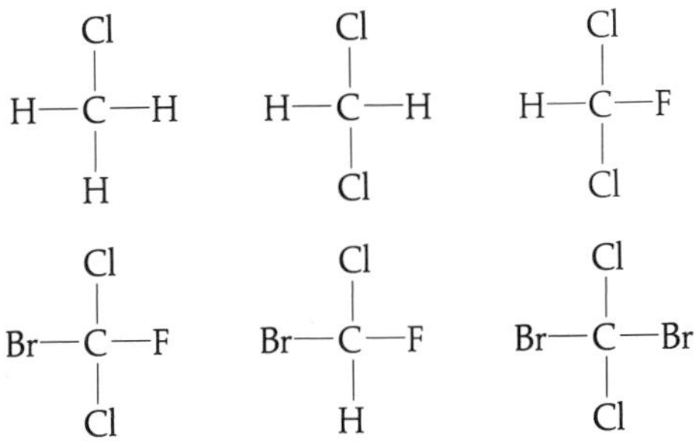

$$H-\overset{\overset{Cl}{|}}{\underset{\underset{H}{|}}{C}}-H \qquad H-\overset{\overset{Cl}{|}}{\underset{\underset{Cl}{|}}{C}}-H \qquad H-\overset{\overset{Cl}{|}}{\underset{\underset{Cl}{|}}{C}}-F$$

$$Br-\overset{\overset{Cl}{|}}{\underset{\underset{Cl}{|}}{C}}-F \qquad Br-\overset{\overset{Cl}{|}}{\underset{\underset{H}{|}}{C}}-F \qquad Br-\overset{\overset{Cl}{|}}{\underset{\underset{Cl}{|}}{C}}-Br$$

If you lack confidence in your answer, build the models.

10. The boiling point of liquid ammonia under a pressure of one atmosphere is $-33\ ^\circ C$. At a lower temperature or under a higher pressure, liquid ammonia can be used as a solvent in much the same way that liquid water can be used as a solvent. The ammonia molecule is potentially both a Brönsted acid and a Brönsted base in exactly the same sense that water is both a Brönsted acid and a Brönsted base. Write out the equilibrium expression for liquid ammonia parallel to the equilibrium expression for liquid water:

$$H_2O(l) + H_2O(l) \rightleftarrows H_3O^{\textcircled{1+}}(aq) + OH^{\textcircled{1-}}(aq)$$

On the basis of that equation, propose a definition for a neutral liquid ammonia solution. There are many ionic compounds, such as sodium amide, $NaNH_2$, that are very comparable to compounds such as sodium hydroxide, NaOH.

Additional exercises for Chapter 14 are given in the Appendix.

A Look Ahead

In the next chapter, we explore reactions of ions in aqueous solutions. We have just devoted two chapters to the exploration of the reactions of Brönsted acid–base conjugate pairs with liquid water and we would like to explore other types of reactions of ions in aqueous solutions without being concerned about the reactions of Brönsted acids and Brönsted bases. This is a luxury we never have. The introduction of either a Brönsted acid or a Brönsted base to water or to an aqueous solution leads to the formation of its conjugate and also leads to changes in the concentrations of the hydronium ions and hydroxide ions. Consequently, in Chapter 15, Brönsted acids and bases will be ever present in the exploration of aqueous solutions. From the stand-point of developing a sense of security with your understanding of acid–base phenomena, this continued experience with acids and bases in Chapter 15 is excellent.

Once you have acquired a familiarity with the phenomena and an understanding of the concepts, the initial overpowering complexity is to a large degree resolved and the concepts themselves may even begin to seem rather obvious. For example, in Chapter 13, the focus was upon the reactions of Brönsted acids with liquid water and the relative quantities of the Brönsted acid and its conjugate base that exist together in equilibrium in aqueous solutions. In this chapter, the initial focus has been upon the reactions of Brönsted bases with liquid water and the relative quantities of the Brönsted base and its conjugate acid that exist together in equilibrium in aqueous solutions. The two chapters approach a conjugate pair equilibrium from two directions, one from the reaction of the conjugate acid with the solvent water and the other from the reaction of the conjugate base with the solvent water. The end result is the same, irrespective of the route.

Why so much emphasis on aqueous solutions? Water is the most readily available solvent on this planet and all biological organisms are primarily aqueous systems.

15

REACTIONS OF IONS IN AQUEOUS SOLUTIONS

Chapter 15

Water is a unique compound and the chemistry of aqueous solutions is both complex and subtle with many reactions in aqueous solution involving the water molecule and the ions of water as reactants. The unique characteristics of water arise from the smallness of the molecule H_2O, the nonlinearity of the molecule, the polarity of the hydrogen–oxygen bond, and the small size of the hydrogen atom. Water molecules are polar. Water molecules form strong hydrogen bonds with other molecules of water and with molecules and ions of other compounds containing highly electronegative atoms such as oxygen and nitrogen. Water molecules react as Brönsted acids and also as Brönsted bases. Liquid water has an exceptionally high boiling point for such a low molecular weight compound. Liquid water has an exceptionally high dielectric constant. Water is an exceptionally good solvent for ionic compounds. All biological life on this planet is dependent upon the unique characteristics of water molecules and liquid water.

Water

In the consideration of aqueous solutions, we must be ever mindful of three ions that are closely related to the water molecule, H_2O. These ions are the hydronium ion, H_3O^{1+}; the hydroxide ion, OH^{1-}; and the oxide ion, O^{2-}. Two of these, the hydronium ion and the hydroxide ion, you already know a great deal about. The approach we take now is to explore the relation of these ions with water molecules through the treatment of the hydronium ion, H_3O^{1+}, as a triprotic acid in much the same way we approached *ortho*-phosphoric acid as a triprotic acid. In this approach, the three Brönsted acids are H_3O^{1+}, H_2O, and OH^{1-}. The three conjugate acid–base pairs are

The Ions of Water

Hydronium Ion as a Triprotic Acid

$$H_3O^{1+} \quad \text{and} \quad H_2O$$
$$H_2O \quad \text{and} \quad OH^{1-}$$
$$OH^{1-} \quad \text{and} \quad O^{2-}$$

The three reactions with water are

Reaction 1

$$\underset{\text{acid}}{H_3O^{1+}(aq)} + H_2O(l) \longrightarrow H_3O^{1+}(aq) + \underset{\text{base}}{H_2O(l)}$$

Reaction 2

$$\underset{\text{acid}}{H_2O(l)} + H_2O(l) \longrightarrow H_3O^{1+}(aq) + \underset{\text{base}}{OH^{1-}(aq)}$$

Reaction 3

$$OH^{1-}(aq) + H_2O(l) \longrightarrow H_3O^{1+}(aq) + O^{2-}(aq)$$

acid — base

In the same sense that the $H_2PO_4^{1-}$ ion is both a Brönsted base and a Brönsted acid, the water molecule is both a Brönsted base and a Brönsted acid. In the same sense that the HPO_4^{2-} ion is both a Brönsted base and a Brönsted acid, the OH^{1-} ion is both a Brönsted base and a Brönsted acid.

The three equilibria, the three K_A expressions, the three K_A values, and the three pK_A values at 25 °C are

Equilibrium 1 (for the acid–base pair H_3O^{1+} and H_2O)

$$H_3O^{1+}(aq) + H_2O(l) \rightleftarrows H_3O^{1+}(aq) + H_2O(l)$$

$$\underset{\text{hydronium ion}}{K_A} = \frac{\cancel{[H_3O^{1+}]}[H_2O]}{\cancel{[H_3O^{1+}]}} = 5.5 \times 10^{1} \qquad \underset{\text{hydronium ion}}{pK_A} = -1.7$$

Equilibrium 2 (for the acid–base pair H_2O and OH^{1-})

$$H_2O(l) + H_2O(l) \rightleftarrows H_3O^{1+}(aq) + OH^{1-}(aq)$$

$$\underset{\text{water}}{K_A} = \frac{[H_3O^{1+}][OH^{1-}]}{[H_2O]} = 1.8 \times 10^{-16} \qquad \underset{\text{water}}{pK_A} = 15.7$$

Equilibrium 3 (for the acid–base pair OH^{1-} and O^{2-})

$$OH^{1-}(aq) + H_2O(l) \rightleftarrows H_3O^{1+}(aq) + O^{2-}(aq)$$

$$\underset{\text{hydroxide ion}}{K_A} = \frac{[H_3O^{1+}][O^{2-}]}{[OH^{1-}]} \text{ approaches } 0 \qquad \underset{\text{hydroxide ion}}{pK_A} \text{ approaches } +\infty$$

Gratuitous Information 15-1 discusses the origin of these K_A values.

The summary of the above information concerning the three acid–base pairs is given in Table 15-1. The form used is that of Table 14-2.

As was to be expected, the hydronium ion, H_3O^{1+} is a stronger acid than the water molecule, H_2O, and the water molecule, H_2O, is a stronger acid than the hydroxide ion, OH^{1-}. It is easier for the proton to transfer from H_3O^{1+}, leaving behind a neutral molecule of water, than for a proton to transfer from a molecule of water, leaving behind a negatively charged hydroxide ion. It is easier for the proton to transfer from a molecule of water, leaving behind a negatively charged hydroxide ion, than for a proton to transfer from a hydroxide ion, leaving behind the oxide ion with its double negative charge. The magnitude of the pK_A values indicates that the hydronium ion, H_3O^{1+}, is a much stronger acid than the water molecule and that the

Table 15-1

PROPERTIES OF THE TRIPROTIC ACID, $H_3O^{(1+)}$

	the acids		the conjugate bases		
	pK_A	**formula**	formula	pK_B	
increasing acid strength ↑	**−1.7**	$\mathbf{H_3O^{(1+)}}$	H_2O	15.7	increasing base strength ↓
	15.7	$\mathbf{H_2O}$	$OH^{(1-)}$	−1.7	
	approaches +∞	$\mathbf{OH^{(1-)}}$	$O^{(2-)}$	approaches −∞	

water molecule is a much stronger acid than the hydroxide ion, $OH^{(1-)}$. The corollary of the above is that the oxide ion, $O^{(2-)}$, is a much stronger base than the hydroxide ion, $OH^{(1-)}$, and the hydroxide ion, $OH^{(1-)}$, is a much stronger base than the water molecule. To compare the strengths of these Brönsted acids and other Brönsted acids, refer to Tables 13-3 and 14-2. It is, of course, not necessary to include the pK_B values in Tables 13-3, 14-2, 14-3, and 15-1. The relation of the pK_A value and the pK_B value for a Broñsted conjugate pair at 25 °C is well known:

General Table of Brönsted Acid Strengths

$$pK_A + pK_B = 14.00 \quad 25\ ^\circ C$$

and it is now common practice for chemists—particularly biochemists—to work from tables that omit the pK_B values. These tables are compilations of experimentally determined properties of Brönsted acid–base pairs and take the form of the two columns printed in boldface type in Tables 14-2, 14-3, and 15-1.

Only a very few acids—such as hydrochloric acid, HCl; nitric acid, HNO_3; and sulfuric acid, H_2SO_4—are stronger than the hydronium ion, $H_3O^{(1+)}$. The oxide ion, $O^{(2-)}$, is by far the strongest base we have discussed and the statement is frequently made that there are no oxide ions in aqueous solutions. A more defensible statement might be that the concentration of the oxide ion is not sufficiently high for the oxide ion to have been experimentally detected in aqueous solutions.

In our exploration of the reactions of ions in aqueous solution, we shall consider three categories of reactions: Brönsted acid–base reactions, precipitation reactions, and oxidation–reduction reactions. The exploration of acid–base reactions in this chapter summarizes and extends the development of acids and bases in Chapters 13 and 14. The exploration of precipitation reactions addresses combinations of ions to form a solid phase that separates from the liquid phase of the aqueous solution. In the initial approach to precipitation reactions, all ions and experimental conditions are chosen so that the complication of Brönsted acid–base phenomena may be avoided. The exploration of oxidation–reduction reactions is deferred until Chapter 16, Oxygen Gas, the Molecular Orbital Approach to Diatomic Molecules, and Oxidation–Reduction Reactions.

K_A VALUES FOR THE TRIPROTIC ACID, $H_3O^{(1+)}$

In the text, we have presented the three Brönsted acid equilibria for the triprotic acid $H_3O^{(1+)}$ with water, the three K_A expressions, the three K_A values, and the three pK_A values. We need these three pK_A values to place the three acid–base pairs

$$\begin{array}{ccc} H_3O^{(1+)} & \text{and} & H_2O \\ H_2O & \text{and} & OH^{(1-)} \\ OH^{(1-)} & \text{and} & O^{(2-)} \end{array}$$

in tables such as Table 14-2, which compares Brönsted acid strengths quantitatively through the pK_A values characteristic of the various Brönsted acids.

For the comparisons to be valid, the same conventions (practices) must be used in formulating and evaluating all pK_A expressions. As a model, let us take the equilibrium for the transfer of the first proton of *ortho*-phosphoric acid to water:

$$H_3PO_4 + H_2O \rightleftarrows H_3O^{(1+)} + H_2PO_4^{(1-)}$$

$$K_A = \frac{[H_3O^{(1+)}] \times [H_2PO_4^{(1-)}]}{[H_3PO_4]}$$

Historically, the K_A expression for the dissociation of phosphoric acid was considered to be

$$H_3PO_4 \rightleftarrows H^{(1+)} + H_2PO_4^{(1-)}$$

$$K_A = \frac{[H^{(1+)}][H_2PO_4^{(1-)}]}{[H_3PO_4]}$$

and the use of this form of the K_A expression has persisted. Today, it would be more consistent with the chemical equation to express K_A in the following fashion:

$$K_A = \frac{[H_3O^{(1+)}][H_2PO_4^{(1-)}]}{[H_3PO_4][H_2O]}$$

It is, however, a great convenience to use the molar concentration of water, $[H_2O]$, in the denominator as 1, even though the molar concentration of water is not one mole per liter.

The molar concentration of water in a dilute aqueous solution is approximately 55 moles per liter. A dilute solution is mostly water and a liter of pure water at 25 °C has a mass of a little less than 1000 grams. The mass of water in a liter of a dilute solution may be a little less than the mass of water in a liter of pure water but, as an approximation, we can take the mass of water in a liter of a dilute aqueous solution to be approximately 1000 g. This mass of water is approximately 55 moles (1000 g ÷ 18 grams per mole). The molar concentration of water in a dilute aqueous solution is approximately 55 moles per liter.

In the evaluation of K_A for H_3PO_4, the molar concentration of water in the denominator was used as 1 rather than as 55 moles/liter. This saves a good deal of trouble and does not invalidate the comparison of K_A and pK_A values if we consistently use $[H_2O]$ as 1 in the evaluation of all K_A values.

How do we maneuver to obtain comparable K_A and pK_A values for the hydronium ion as an acid?

$$\underset{\text{acid}}{H_3O^{(1+)}} + H_2O \rightleftarrows H_3O^{(1+)} + H_2O$$

This seems like an exercise in stupidity. The two sides of the chemical equation are identical. But let us press on and see what comes of it.

$$\underset{\text{hydronium ion}}{K_A} = \frac{[H_3O^{(1+)}][H_2O]}{[H_3O^{(1+)}][H_2O]}$$

A solution has only one molar concentration of hydronium ions, so the $[H_3O^{(1+)}]$ in the numerator and the $[H_3O^{(1+)}]$ in the denominator are identities. A solution has only one molar concentration of water: for a dilute aqueous solution, that value is 55 moles per liter. But to be consistent with our practice in setting up K_A expressions in general, we use the $[H_2O]$ in the denominator as 1 and the $[H_2O]$ in the numerator as 55 moles per liter. So much for consistency.

$$\underset{\text{hydronium ion}}{K_A} = \frac{[H_3O^{1+}][H_2O]}{[H_3O^{1+}][H_2O]} = \frac{55}{1} = 55 = 5.5 \times 10^1$$

Note that this K_A value is greater than 1. The hydronium ion, H_3O^{1+}, is quite a strong acid as compared to most Brönsted acids we have discussed. See Table 14-2. Only a few acids—such as HCl, HNO_3, and H_2SO_4—are stronger. Since the K_A for hydronium ion is greater than 1, pK_A is a negative number:

$$\underset{\text{hydronium ion}}{pK_A} = -\log K_A = -\log 5.5 \times 10^1 = -\log 5.5 - \log 10^1 = -0.7 - 1 = -1.7$$

(The pK_A value for HCl approaches $-\infty$.)

What about the diprotic acid, H_2O?

$$H_2O + H_2O \rightleftharpoons H_3O^{1+} + OH^{1-}$$

$$\underset{\text{water}}{K_A} = \frac{[H_3O^{1+}][OH^{1-}]}{[H_2O][H_2O]}$$

We already know the value of the numerator. At 25 °C, $K_W = 1.00 \times 10^{-14}$. What about the two molar concentrations of water, $[H_2O]$, in the denominator? To be consistent with the K_A values of other Brönsted acids, one $[H_2O]$ must be used as 1 and the other as 55 moles/liter.

$$\underset{\text{water}}{K_A} = \frac{1.00 \times 10^{-14}}{1 \times 55} = 1.8 \times 10^{-16}$$

and

$$pK_A = 15.7$$

In comparison to other Brönsted acids we have discussed, water is a very weak Brönsted acid. See Table 14-2.

What about the monoprotic acid, the hydroxide ion, OH^{1-}? Clearly, this "acid" is going to be much weaker than H_2O:

$$OH^{1-} + H_2O \rightleftharpoons H_3O^{1+} + O^{2-}$$

$$\underset{\text{hydroxide ion}}{K_A} = \frac{[H_3O^{1+}][O^{2-}]}{[OH^{1-}][H_2O]}$$

Following the usual K_A convention, $[H_2O]$ in the denominator is used as 1:

$$\underset{\text{hydroxide ion}}{K_A} = \frac{[H_3O^{1+}][O^{2-}]}{[OH^{1-}]}$$

The experimental properties of aqueous solutions lead to the conclusion that the presence of oxide ions is undetectable and the concentration of oxide ions must approach zero. Therefore,

$$\underset{\text{hydroxide ion}}{K_A} \longrightarrow 0 \qquad \underset{\text{hydroxide ion}}{pK_A} \longrightarrow \infty$$

approaches zero. If K_A approaches zero, then pK_A, the negative logarithm of K_A, approaches positive infinity. The hydroxide ion in aqueous solutions is an extremely weak acid.

In the above, we have maneuvered to generate a set of K_A and pK_A values that are useful in making comparisons with other Brönsted acids and consistent with our empirical knowledge of the properties of aqueous solutions. It is these values that are given in comprehensive tables such as Table 14-2.

Brönsted Acid–Base Reactions

One of the first questions to be addressed about any solute added to water is whether the added molecules or ions are Brönsted acids or bases. If they are, one characteristic of the aqueous solution at equilibrium is the equilibria established by the reactions of those Brönsted acids and bases with water.

pH Control Using a Strong Acid

In Chapters 13 and 14, it was assumed that the pH of an aqueous solution can be varied between 1 and 13 at will, but we avoided the issue of how to bring about the change experimentally. A convenient way to decrease the pH (increase the hydronium ion concentration) of an aqueous solution is to add an aqueous solution of hydrochloric acid, HCl. Hydrochloric acid is a very strong acid (K_A approaches infinity and pK_A approaches negative infinity) and consequently its conjugate base, the chloride ion, Cl^{1-}, is a very weak base. To add a dilute solution of hydrochloric acid is to add water molecules, H_2O; hydronium ion H_3O^{1+}; and chloride ions, Cl^{1-}. (See Figure 13-2 for the percent abundance vs pH curves for hydrochloric acid.) Since the chloride ion is a very weak Brönsted base, its addition to a solution contributes to the ionic environment in the solution but is not a direct factor in the determination of the hydronium ion concentration of a solution. If the concentration of the hydrochloric acid solution added is sufficiently high and the volume added is sufficiently large, the hydronium ion concentration can be increased to 1×10^{-1} moles/liter or greater and the pH of the solution reduced to a value of 1 or less. More acid will be required if the solution to which the hydrochloric acid is being added was initially buffered.

pH Control Using a Weak Acid

Acids other than hydrochloric acid could be used to reduce the pH. Any strong acid, such as nitric acid (K_A approaches infinity and pK_A approaches negative infinity) would be equally effective. To add an aqueous solution of nitric acid, HNO_3, is to add primarily water molecules, H_2O; hydronium ions, H_3O^{1+}; and nitrate ions, NO_3^{1-}. The nitrate ion is an extremely weak base and its addition to an aqueous solution is not a factor in determining the pH of the solution. On the other hand, a weak acid, such as acetic acid ($K_A = 1.76 \times 10^{-5}$, $pK_A = 4.75$) cannot be used as effectively as a strong acid. To add an acetic acid solution is to add primarily water molecules, H_2O; acetic acid molecules, CH_3COOH; and much smaller quantities of hydronium ions, H_3O^{1+}, and acetate ions, CH_3COO^{1-}. In <u>all</u> aqueous solutions to which either acetic acid molecules or acetate ions are added, an equilibrium is established:

$$CH_3COOH(aq) + H_2O(l) \rightleftharpoons H_3O^{1+}(aq) + CH_3COO^{1-}(aq)$$

and the relative concentrations of the acetate ion and the acetic acid molecule conform to in the K_A expression:

$$K_A = \frac{[H_3O^{1+}] \times [CH_3COO^{1-}]}{[CH_3COOH]} = 1.76 \times 10^{-5} \quad 25\ ^\circ C$$

At low pH's, the concentration of the acetate ion is extremely low in comparison to the concentration of the acetic acid molecule. For pH–percent abundance curves, see Figure 13-1. Even though the concentration of the acetic acid solution added is high and a large volume is added, the hydronium ion concentration will not be increased above 1×10^{-3} moles/liter and the pH will not be reduced below 3.

pH Control Using a Strong Base

A convenient way to increase the pH (decrease the hydronium ion concentration) of an aqueous solution is to add either solid sodium hydroxide, NaOH, or an aqueous solution of sodium hydroxide. Sodium hydroxide is an ionic compound. The crystal lattice of the solid contains equal numbers of sodium ions, Na^{1+}, and hydroxide ions, OH^{1-}, and its aqueous solutions can be treated simply as a solution of sodium ions and hydroxide ions. The sodium ion, an alkali metal ion, is hydrated in the sense that all cations (positive ions) are surrounded by a shell of polar water molecules. The presence of the sodium ion contributes to the ionic environment in the solution but is not a direct factor in determining the hydronium ion concentration of a solution. Note that the reaction of the hydronium ion with the added hydroxide ion is the reverse reaction of the equation written for the water equilibrium:

$$H_2O(l) + H_2O(l) \rightleftharpoons H_3O^{1+}(aq) + OH^{1-}(aq)$$

If the concentration of the sodium hydroxide solution added is sufficiently high and the volume added is sufficiently great, the hydronium ion concentration of any solution can be reduced to 1×10^{-13} moles/liter or less and the pH of the solution is thus increased to 13 or greater. More sodium hydroxide solution will be required if the solution to which it is being added was initially buffered. The addition of any other alkali metal hydroxide, such as potassium hydroxide, KOH, is as effective in increasing the pH of the solution as the addition of sodium hydroxide. To add an aqueous solution of potassium hydroxide is to add primarily water molecules; potassium ions, K^{1+}; and hydroxide ions, OH^{1-}. The hydroxide ion is a very strong Brönsted base. Note the position of the water–hydroxide ion acid–base pair in Table 14-2.

The oxide ion, O^{2-}, is a stronger base than the hydroxide ion, OH^{1-}. Sodium oxide, Na_2O, is an ionic solid and is soluble in water. Why have we not suggested the addition of sodium oxide as a convenient way to increase the pH (decrease the hydronium ion concentration) of an aqueous solution? The addition of sodium oxide would be very effective. Not only does the oxide ion react with hydronium ion:

$$O^{2-}(aq) + H_3O^{1+}(aq) \longrightarrow H_2O(l) + OH^{1-}(aq)$$

but the oxide ion also reacts with water:

$$O^{2-}(aq) + H_2O \longrightarrow OH^{1-}(aq) + OH^{1-}(aq)$$

The K_B expression for the equilibrium is

$$O^{2-}(aq) + H_2O(l) \rightleftharpoons OH^{1-}(aq) + OH^{1-}(aq)$$

$$\underset{\text{oxide ion}}{K_B} = \frac{[OH^{1-}][OH^{1-}]}{[O^{-2}]}$$

K_B approaches ∞ and the oxide ion concentration approaches zero. The consequence of the addition of solid sodium oxide to an aqueous solution is equivalent to the consequence of the addition of sodium hydroxide solution and it is much more convenient to work with an aqueous solution of sodium hydroxide than to work with solid sodium oxide.

pH Control Using a Weak Base

The addition of an aqueous solution of a weak base such as ammonia, NH_3 ($K_B = 1.76 \times 10^{-5}$) can also be used to increase the pH of a solution. To add an aqueous solution of ammonia, NH_3, is to add primarily water molecules, H_2O; ammonia molecules, NH_3; and much smaller quantities of ammonium ion, $NH_4^{(1+)}$; and hydroxide ions, $OH^{(1-)}$. (See Percent Abundance–pH diagram for Ammonia–Ammonium Ion, Figure 14-1.) The reduction of the hydronium ion concentration of the solution to which the ammonia solution is added can be considered as the consequence of the reaction of the hydronium ion with some of the added hydroxide ion

$$H_3O^{(1+)}(aq) + OH^{(1-)}(aq) \longrightarrow H_2O(l) + H_2O(l)$$

or as the consequence of the reaction of the hydronium ion with ammonia molecules

$$H_3O^{(1+)}(aq) + NH_3(aq) \longrightarrow NH_4^{(1+)}(aq) + H_2O(l)$$

to give ammonium ions. In either case, the equilibrium

$$NH_3(aq) + H_2O(l) \rightleftarrows NH_4^{(1+)}(aq) + OH^{(1-)}(aq)$$

is established in any solution to which either ammonia molecules, NH_3, or ammonium ions, $NH_4^{(1+)}$, have been added and this equilibrium is described by the K_B expression for ammonia:

$$K_B = \frac{[NH_4^{(1+)}] \times [OH^{(1-)}]}{[NH_3]} = 1.76 \times 10^{-5} \quad 25\ ^\circ C$$

Even though the concentration of the ammonia solution added is high and a large volume is added, the hydronium ion concentration will not be reduced to a value below 1×10^{-11} moles/liter. Adding an ammonia solution will not increase the pH of the solution above 11.

pH of Aqueous Solutions of Sodium Chloride

Pure water has a pH of 7.00 at 25 °C. An aqueous solution may have a pH other than 7.00 if one or more of the ions or molecules added reacts with water. An aqueous solution of sodium chloride, NaCl, has a pH of essentially 7. The sodium ion, $Na^{(1+)}$, does not react with water to give either hydronium ions or hydroxide ions. The chloride ion, the conjugate base of HCl (a very strong Brönsted acid), is an extremely weak Brönsted base and the reaction of chloride ion with water is not significant:

$$Cl^{(1-)}(aq) + H_2O(l) \rightleftarrows HCl(aq) + OH^{(1-)}(aq)$$

$$K_B \text{ approaches } 0$$

The presence of the sodium ions and the chloride ions in the water does not disturb the equilibrium

$$H_2O + H_2O \rightleftarrows H_3O^{(1+)} + OH^{(1-)}$$

nor the concentrations of the ions of water in pure water ($[H_3O^{(1+)}] = [OH^{(1-)}] = 1.00 \times 10^{-7}$ moles/liter at 25 °C).

An aqueous solution of ammonium chloride, NH_4Cl, is acidic—pH less than 7. As the white, crystalline, ionic solid dissolves in water, the ammonium ion, NH_4^{1+}, and the chloride ion, Cl^{1-}, are dispersed through the water and the ammonium ion, NH_4^{1+}, a stronger Brönsted acid than water (Table 14-2), reacts with water:

pH of Aqueous Solutions of Ammonion Chloride

$$NH_4^{1+}(aq) + H_2O(l) \longrightarrow H_3O^{1+}(aq) + NH_3(aq)$$

increasing the hydronium ion concentration in the solution and consequently yielding a solution that is acidic. At equilibrium, the two chemical equilibria of particular significance in the discussion of the pH of the solution are

$$NH_4^{1+}(aq) + H_2O(l) \rightleftharpoons H_3O^{1+}(aq) + NH_3(aq)$$

and

$$H_2O(l) + H_2O(l) \rightleftharpoons H_3O^{1+}(aq) + OH^{1-}(aq)$$

At equilibrium, the concentrations of ions and molecules in the solution at 25 °C satisfy the two quantitative relations

$$\underset{\text{ammonium ion}}{K_A} = \frac{[H_3O^{1+}] \times [NH_3]}{[NH_4^{1+}]} = 5.65 \times 10^{-10}$$

and

$$\underset{\text{water}}{K_W} = [H_3O^{1+}] \times [OH^{1-}] = 1.00 \times 10^{-14}$$

At equilibrium, the hydronium ion concentration exceeds the hydroxide ion concentration in the ammonium chloride solution. The molar hydronium ion concentration, $[H_3O^{1+}]$, is greater than 1.00×10^{-7} and the pH of the solution is less than 7.00.

On the other hand, an aqueous solution of sodium acetate, CH_3COONa, is basic. (The concentration of the hydronium ion is less than the concentration of the hydroxide ion.) As sodium acetate, a white solid, dissolves in water, the sodium ions, Na^{1+}, and the acetate ions, CH_3COO^{1-} (from the crystal lattice) are dispersed in water. The acetate ion is a stronger Brönsted base than water (Table 14-2) and reacts with water molecules to give hydroxide ions:

pH of Aqueous Solutions of Sodium Acetate

$$CH_3COO^{1-}(aq) + H_2O(l) \longrightarrow OH^{1-}(aq) + CH_3COOH(aq)$$

At equilibrium, the two equilibria of particular significance in the discussion of the pH of the solution are

$$CH_3COO^{1-}(aq) + H_2O(l) \rightleftharpoons OH^{1-}(aq) + CH_3COOH(aq)$$

and

$$H_2O(l) + H_2O(l) \rightleftharpoons H_3O^{1+}(aq) + OH^{1-}(aq)$$

At 25 °C, the concentrations of molecules and ions in solution satisfy the quantitative relations

$$\underset{\text{acetate ion}}{K_B} = \frac{[OH^{1-}] \times [CH_3COOH]}{[CH_3COO^{1-}]} = 5.6 \times 10^{-10}$$

and

$$K_W = [H_3O^{1+}] \times [OH^{1-}] = 1.00 \times 10^{-14}$$

At equilibrium, the hydroxide concentration is greater than the hydronium ion concentration, the hydronium ion is less than 1×10^{-7} moles per liter, and the pH of the solution is greater than 7.

An aqueous solution of iron(III) chloride is acidic: $[H_3O^{1+}] > [OH^{1-}]$, $[H_3O^{1+}] > 1.00 \times 10^{-7}$ moles/liter, pH < 7.00. To discuss this gets us into a property of the iron(III) ion, Fe^{3+} (one of many transition metal ions) that we have yet to explore.

pH of Aqueous Solutions of Iron(III) Chloride

The cations of the alkali metals and the alkaline earths are solvated (hydrated) in the usual sense that positive ions are surrounded by an ill-defined shell of polar water molecules. The properties of transition metal compounds indicate that their cations bind water molecules in a much more specific sense and that much of the chemistry of transition metal cations, such as the iron(III) ion, is derived from these more complex ions. The addition of iron(III) chloride, $FeCl_3$ (also called ferric chloride), to soil increases the acidity—lowers the pH—of the water in the soil. This is of horticultural significance since acid soils are essential to the culture of plants such as blueberries and many evergreens. Part of the properties of aqueous solutions of $FeCl_3$ are attributed to the formation of a series of coordination complexes:

$Fe(H_2O)_6^{3+}$	hexaaquoiron(III) ion
$Fe(OH)(H_2O)_5^{2+}$	hydroxopentaaquoiron(III) ion
$Fe(OH)_2(H_2O)_4^{1+}$	dihydroxotetraaquoiron(III) ion
$Fe(OH)_3(H_2O)_3$	trihydroxotriaquoiron(III)

In each case, a total of six molecules or ions are coordinated with the iron(III) ion, and the iron(III) ion is said to have a coordination number of 6.

Why is the aqueous solution of iron(III) chloride, $FeCl_3$, acidic? What is the source of the hydronium ion? The hexaaquoiron(III) ion is a Brönsted acid and reacts with water by the transfer of a proton from one of the molecules of water of the complex to a molecule of liquid water:

$$Fe(H_2O)_6^{3+}(aq) + H_2O(l) \longrightarrow H_3O^{1+}(aq) + Fe(OH)(H_2O)_5^{2+}(aq)$$

The hydroxide ion, OH^{1-}, remains a part of the coordination complex and the charge of the conjugate base, $Fe(OH)(H_2O)_5^{2+}$, is 2+ rather than 3+. This ion is both a Brönsted acid and a Brönsted base. Its reaction as a Brönsted acid generates more hydronium ion:

$$Fe(OH)(H_2O)_5^{2+}(aq) + H_2O(l) \longrightarrow H_3O^{1+}(aq) + Fe(OH)_2(H_2O)_4^{1+}(aq)$$

The hexaaquoiron(III) ion, $Fe(H_2O)_6^{3+}$, is a well-defined structure with the six molecules of water uniformly distributed about the iron(III) ion. Each molecule of water is bonded through a *sigma*-bond to the iron, utilizing a pair of the unshared electrons of the water molecule to fill a *sigma*-orbital that encompasses the oxygen nucleus and the iron nucleus. A covalent bond formed through the use of a pair of electrons from one atom, rather than the use of one electron from each of two atoms, is designated as a **coordinate covalent bond.** Complex ions and molecules held together by coordinate covalent bonds are designated as **coordination complexes.** The three complex ions of iron(III) and the complex molecule of iron(III) listed previously are coordination complexes of iron(III).

Coordination Complexes of Transition Metal Ions

The transition metal elements form thousands, perhaps millions, of coordination complex ions and molecules. Since the formation of the *sigma*-bonds in complex ions and molecules utilizes pairs of unshared electrons of the coordinating molecules or ions, all Brönsted bases are candidates for the formation of coordination complexes with transition metal ions that have "unfilled *d*-orbitals." The system of nomenclature that has developed to name these complexes is certainly not our concern, but it is your right to discover that coordination complexes do exist and you may be amused to observe how a few of the coordination complexes are named. In the examples given, notice the combination of prefixes such as *hexa-*, *penta-*, *tetra-*, *tri-*, *hydroxo-*, and *aquo-* in the names of the coordination complexes.

Precipitation Reactions

Precipitation of Barium Chromate

When a few drops of a colorless 0.01 M barium chloride, $BaCl_2$, aqueous solution is added to a few milliliters of a pale yellow 0.01 M potassium chromate, K_2CrO_4, aqueous solution in a test tube, the solution instantaneously becomes opaque. A pale yellow precipitate, a pale yellow solid, slowly settles on the bottom of the test tube and the aqueous solution slowly becomes a transparent yellow liquid again. The solid is barium chromate, $BaCrO_4$, an ionic compound. Finely divided particles of the solid, distributed throughout the solution, scatter the light, and it is these suspended particles that are responsible for the opaqueness of the reaction mixture. In spite of the high density of the barium chromate, 4.5 grams per cubic centimeter as compared to 1.0 grams per cubic centimeter for water, the precipitate may be so finely divided that it settles very slowly in the gravitational field of the earth. The aqueous solution of the ionic compound barium chloride contains barium ions, $Ba^{2+}(aq)$, and chloride ions, Cl^{1-}. Both are colorless. The aqueous solution of the ionic compound potassium chromate, K_2CrO_4, contains potassium ions, $K^{1+}(aq)$, and chromate ions, $CrO_4^{2-}(aq)$. The potassium ion is colorless and the chromate ion is yellow. (Chrom is derived from the Greek *chroma*, for color.) The chemical reaction is very straightforward:

$$Ba^{2+}(aq) + CrO_4^{2-}(aq) \longrightarrow BaCrO_4(s)$$

Barium ion	chromate ion	barium chromate
colorless	yellow	yellow

On the addition of the barium chloride solution, the resultant solution is supersaturated with respect to barium chromate. Precipitation of solid barium chromate proceeds rapidly, with the reduction of the concentration of the barium ion and the concentration of the chromate ion, until the rate of precipitation of barium chromate equals the rate at which the newly formed solid barium chromate dissolves and an equilibrium between the solid and its ions in solution has been established:

$$BaCrO_4(s) \rightleftarrows Ba^{2+}(aq) + CrO_4^{2-}(aq)$$

Further precipitation of the chromate ion can be achieved by the continued drop-by-drop addition of the barium chloride solution. After each addition, an opaque cloud of finely divided barium chromate forms where the two solutions mix, and precipitation continues until equilibrium is again established. As more of the chromate ion is removed from the solution by precipitation, the yellow color of the chromate ion in the aqueous solution becomes less intense. Eventually the solution is colorless and the addition of the colorless barium chloride solution no longer produces a visible change. Note that potassium chloride, KCl, does not precipitate. All potassium compounds are too soluble in water to precipitate in experiments of this type using very dilute solutions.

Very similar phenomena occur when a few drops of 0.01 M barium chloride solution is added to a few milliliters of a colorless solution of 0.01 M sodium sulfate, Na_2SO_4. In this case, the precipitate is a white ionic compound, barium sulfate, $BaSO_4$. The reaction is

Precipitation of Barium Sulfate

$$\underset{\text{colorless}}{Ba^{2+}(aq)} + \underset{\substack{\text{sulfate ion}\\\text{colorless}}}{SO_4^{2-}(aq)} \longrightarrow \underset{\substack{\text{barium sulfate}\\\text{white}}}{BaSO_4(s)}$$

and the equilibrium expression for the saturated solution is

$$BaSO_4(s) \rightleftarrows Ba^{2+}(aq) + SO_4^{2-}(aq)$$

Note that sodium chloride, NaCl, does not precipitate. All sodium compounds are too soluble in water to precipitate in experiments of this type using very dilute solutions.

Solubility Product Constants

One of the characteristics of a saturated solution of an ionic compound is that the molar concentrations of the cations and the anions in the saturated solution are coupled by an equilibrium constant expression known as the **solubility product constant** for that ionic compound. For barium sulfate, the solubility product constant expression takes the form

$$\underset{BaSO_4}{K_{sp}} = [Ba^{2+}] \times [SO_4^{2-}]$$

where the square brackets indicate molar concentrations. At 25 °C, the experimental value of this constant is 1×10^{-10} moles²/liters²:

$$\underset{BaSO_4}{K_{sp}} = 1.0 \times 10^{-10} \text{ moles}^2 \text{ liters}^{-2} \quad 25\ °C.$$

In a saturated solution of barium sulfate, the molar concentration of the barium ion is inversely proportional to the molar concentration of the sulfate ion:

$$[Ba^{2+}] \propto \frac{1}{[SO_4^{2-}]}$$

At 25 °C, the proportionality constant, the solubility product constant (K_{sp}), is 1.0×10^{-10} moles^2/liters^2:

$$[Ba^{2+}] = 1.0 \times 10^{-10} \frac{1}{[SO_4^{2-}]}$$

If a saturated solution of barium sulfate is prepared by shaking solid barium sulfate, $BaSO_4$, with water, the concentration of the barium ion and the sulfate ion are the same and, at 25 °C, each is 1.0×10^{-5} moles per liter. In the experiments described here, in which the 0.01 M barium chloride solution is added dropwise to several milliliters of 0.01 M sodium sulfate solution, the sulfate ion concentration at equilibrium is going to be much greater than the concentration of the barium ion when the first precipitate is formed. Using the K_{sp} expression (the solubility product constant expression)

$$[Ba^{2+}(aq)] \times [SO_4^{2-}(aq)] = 1.0 \times 10^{-10} \text{ moles}^2 \text{ liters}^{-2} \quad 25\ °C$$

we can set up pairs of values for the two concentrations that satisfy the solubility product constant expression.

CONCENTRATIONS OF IONS THAT SATISFY K_{sp} FOR BARIUM SULFATE AT 25 °C

sulfate ion	**barium ion**
1.0×10^{-2} moles/liter	1.0×10^{-8} moles/liter
1.0×10^{-3}	1.0×10^{-7}
1.0×10^{-4}	1.0×10^{-6}
⋮	⋮
1.0×10^{-8}	1.0×10^{-2}

Experimentally, it would be quite a challenge to make the concentration of the barium ion and the concentration of the sulfate ion at equilibrium come out at 1.0×10^{-5} moles per liter by adding one reagent to another.

For barium chromate,

$$K_{sp} = [Ba^{2+}] \times [CrO_4^{2-}] = 2.0 \times 10^{-10} \text{ moles}^2 \text{ liters}^{-2} \quad 25\ °C.$$
$$BaCrO_4$$

At 25 °C, barium chromate is more soluble in water in terms of moles per liter than barium sulfate ($2 \times 10^{-10} > 1 \times 10^{-10}$).

Precipitation of Silver Halides

The addition of a few drops of 0.01 M aqueous solution of sodium chloride, NaCl, to a few milliliters of 0.01 M aqueous solution of silver nitrate, $AgNO_3$, precipitates silver chloride, AgCl:

$$Ag^{1+}(aq) + Cl^{1-}(aq) \longrightarrow AgCl(s)$$

silver ion colorless	colorless	silver chloride white

Note that sodium nitrate, $NaNO_3$, does not precipitate. The addition of 0.01 M aqueous solution of potassium bromide, KBr, to 0.01 M aqueous solution of silver nitrate, $AgNO_3$, yields a precipitate of silver bromide, AgBr:

$$\underset{\text{colorless}}{Ag^{1+}(aq)} + \underset{\substack{\text{bromide ion}\\\text{colorless}}}{Br^{1-}(aq)} \longrightarrow \underset{\substack{\text{silver bromide}\\\text{pale yellow}}}{AgBr(s)}$$

Note that potassium nitrate, KNO_3, does not precipitate. The addition of 0.01 M aqueous solution of potassium iodide, KI, to 0.01 M aqueous solution of silver nitrate, $AgNO_3$, yields a precipitate of silver iodide, AgI:

$$\underset{\text{colorless}}{Ag^{1+}(aq)} + \underset{\substack{\text{iodide ion}\\\text{colorless}}}{I^{1-}(aq)} \longrightarrow \underset{\substack{\text{silver iodide}\\\text{yellow}}}{AgI(s)}$$

The solubility product constant expressions for the equilibria established in saturated solutions of these three silver halides and their experimental values at 25 °C are

$$\underset{AgCl}{K_{sp}} = [Ag^{1+}] \times [Cl^{1-}] = 2 \times 10^{-10} \text{ moles}^2 \text{ liters}^{-2}$$

$$\underset{AgBr}{K_{sp}} = [Ag^{1+}] \times [Br^{1-}] = 8 \times 10^{-13} \text{ moles}^2 \text{ liters}^{-2}$$

$$\underset{AgI}{K_{sp}} = [Ag^{1+}] \times [I^{1-}] = 2 \times 10^{-16} \text{ moles}^2 \text{ liters}^{-2}$$

In this series of silver halides, silver chloride is the most soluble in water, silver iodide the least soluble. Silver bromide and silver iodide are less soluble in water in terms of moles per liter than barium sulfate and barium chromate.

All of the precipitates discussed above are 1 to 1 ionic compounds—one cation to one anion. The following series of precipitation reactions involve the formation of 2 to 1 or 1 to 2 ionic compounds—two cations to one anion or one cation to two anions. For these, the K_{sp} expression involves the square of the molar concentration of one of the ions.

Silver chromate, Ag_2CrO_4, a red solid, is precipitated by mixing 0.01 M aqueous solutions of sodium chromate, Na_2CrO_4, and silver nitrate, $AgNO_3$:

Precipitation of Silver Chromate

$$\underset{\text{colorless}}{2\,Ag^{1+}(aq)} + \underset{\text{yellow}}{CrO_4^{2-}(aq)} \longrightarrow \underset{\substack{\text{silver chromate}\\\text{dark red}}}{Ag_2CrO_4(s)}$$

The equilibrium expression for the saturated aqueous solution:

$$Ag_2CrO_4(s) \rightleftharpoons 2\,Ag^{1+}(aq) + CrO_4^{2-}(aq)$$

The solubility product constant expression:

$$\underset{Ag_2CrO_4}{K_{sp}} = [Ag^{1+}]^2 \times [CrO_4^{2-}] = 9 \times 10^{-12} \text{ moles}^3 \text{ liters}^{-3} \quad 25\text{ °C}.$$

The addition of a few drops of 0.01-M aqueous solution of potassium iodide, KI, to a few milliliters of 0.01 M aqueous solution of mercury(II) nitrate, $Hg(NO_3)_2$, gives a bright red precipitate of mercury(II) iodide, HgI_2:

Precipitation of Mercury(II) Iodide

$$\underset{\substack{\text{mercury(II) ion}\\ \text{colorless}}}{Hg^{+2}(aq)} + \underset{\text{colorless}}{2\,I^{1-}(aq)} \longrightarrow \underset{\substack{\text{mercury(II) iodide}\\ \text{red}}}{HgI_2(s)}$$

The equilibrium expression for a saturated solution:

$$HgI_2(s) \rightleftarrows Hg^{2+}(aq) + 2\,I^{1-}(aq)$$

The solubility product constant expression:

$$\underset{HgI_2}{K_{sp}} = [Hg^{2+}] \times [I^{1-}]^2 = 3 \times 10^{-29} \text{ moles}^3 \text{ liters}^{-3} \quad 25\ ^\circ C$$

Note that mercury(II) iodide is much less soluble than silver chromate ($3 \times 10^{-29} \ll 9 \times 10^{-10}$).

A 0.01 M aqueous solution of sodium hydroxide, NaOH, added to a 0.01 M aqueous solution of magnesium chloride, $MgCl_2$, gives a white precipitate of magnesium hydroxide, $Mg(OH)_2$:

Precipitation of Magnesium Hydroxide

$$Mg^{2+}(aq) + 2\,OH^{1-}(aq) \longrightarrow \underset{\substack{\text{magnesium hydroxide}\\ \text{white}}}{Mg(OH)_2(s)}$$

$$Mg(OH)_2(s) \rightleftarrows Mg^{2+}(aq) + 2\,OH^{1-}(aq)$$

$$\underset{Mg(OH)_2}{K_{sp}} = [Mg^{2+}] \times [OH^{1-}]^2 = 1 \times 10^{-11} \text{ moles}^3 \text{ liters}^{-3} \quad 25\ ^\circ C$$

A 0.01 M aqueous solution of sodium hydroxide, NaOH, added to a 0.01 M aqueous solution of ferric chloride, $FeCl_3$, gives a brown, voluminous, jellylike precipitate of iron(III) hydroxide that is extremely slow to settle. The traditional form in which the chemical equation, the chemical equilibrium expression, and the corresponding solubility product constant have been written are

Precipitation of the Hydroxide of Iron(III) Ion

$$\underset{\substack{\text{iron(III)}\\ \text{ion}}}{Fe^{3+}(aq)} + 3\,OH^{1-}(aq) \longrightarrow \underset{\substack{\text{iron(III) hydroxide}\\ \text{brown}}}{Fe(OH)_3(s)}$$

$$Fe(OH)_3(s) \rightleftarrows Fe^{3+}(aq) + 3\,OH^{1-}(aq)$$

$$\underset{Fe(OH)_3}{K_{sp}} = [Fe^{3+}] \times [OH^{1-}]^3 = 1 \times 10^{-36} \text{ moles}^4 \text{ liters}^{-4} \quad 25\ ^\circ C$$

This series of expressions is of questionable validity since the Fe^{3+} ion is not a predominant ion in aqueous solutions.

The experimental evidence of the formation of the solid is quite clear, but the interpretation of the formation as being the combination of ions to give an ionic compound of limited solubility in water may be false. The interpretation in terms of the complex ions $Fe(OH)(H_2O)_5^{2+}$ and $Fe(OH)_2(H_2O)_4^{1+}$ is quite different and rationalizes the incorporation of water to give the gelatinous precipitate. The equation written here is for the hydroxopentaaquoiron(III) ion. A very similar equation could be written for the dihydroxotetraaquoiron(III) ion:

$$Fe(OH)(H_2O)_5^{2+} + 2\ OH^{1-}(aq) \longrightarrow 2\ H_2O(l) + \underset{\substack{\text{trihydroxotriaquoiron(III)} \\ \text{brown}}}{Fe(OH)_3(H_2O)_3(s)}$$

In the above reaction, the complex ion acts as a Brönsted acid and two protons, one from each of two aquo groups, are transferred to hydroxide ions. The product, $Fe(OH)_3(H_2O)_3$, is a neutral covalent compound—not an ionic compound. The product is also capable of extensive hydrogen bonding with water molecules. Herein may be the basis of the jellylike form of the precipitate.

Effect of Brönsted Acids and Coordination Complexes on Precipitation Reactions

In the exploration of precipitation reactions in aqueous solutions, we have described nine experiments in which an aqueous solution of an ionic compound is added to an aqueous solution of another ionic compound. With the exception of the ninth experiment, the experiment using the aqueous solution of iron(III) chloride, care was taken to select ionic compounds and experimental conditions to avoid complications due to Brönsted acid reactions and complications due to the formation of coordination complexes. We shall now seek some of the complications that interfere with these and other precipitation reactions.

In the first eight experiments, an ionic compound was precipitated from the mixture of the two solutions. For these eight experiments, the ions brought together and the precipitates formed are listed below.

	ions				precipitate
I	K^{1+}	CrO_4^{2-}	Ba^{2+}	Cl^{1-}	$BaCrO_4$
II	Na^{1+}	SO_4^{2-}	Ba^{2+}	Cl^{1-}	$BaSO_4$
III	Na^{1+}	Cl^{1-}	Ag^{1+}	NO_3^{1-}	$AgCl$
IV	K^{1+}	Br^{1-}	Ag^{1+}	NO_3^{1-}	$AgBr$
V	K^{1+}	I^{1-}	Ag^{1+}	NO_3^{1-}	AgI
VI	Na^{1+}	CrO_4^{2-}	Ag^{1+}	NO_3^{1-}	Ag_2CrO_4
VII	K^{1+}	I^{1-}	Hg^{2+}	NO_3^{1-}	HgI_2
VIII	Na^{1+}	OH^{1-}	Mg^{2+}	Cl^{1-}	$Mg(OH)_2$

In each of the first five cases, a 1 to 1 ionic compound precipitated whenever the product of the molar concentration of its cation and the molar concentration of its anion in the combined solution exceeded the solubility product constant for the ionic compound. In each of the last three experiments, a 2 to 1 or a 1 to 2 ionic compound precipitated whenever the product of the molar concentrations of the ions raised to the appropriate powers exceeded the solubility product constant for the ionic compound.

The Effect of Low pH

None of the ions introduced in the eight experiments are Brönsted acids; none have protons that could be transferred. All of the anions are Brönsted bases and could accept protons, but all of the anions, with the exception of the chromate ion, CrO_4^{2-}, and the hydroxide ion, OH^{1-}, are the conjugate bases of strong acids such as nitric acid and the hydrogen halides. Consequently, all of the anions, except the chromate ion and the hydroxide ion, are extremely weak Brönsted bases.

Question: Would the results of the eight experiments have been the same if the hydronium ion concentration had been made 0.01 moles per liter by the addition of nitric acid to each of the original solutions? An approximate experimental test of this could be made by adding 0.1 M nitric acid drop by drop to the contents of each of the eight test tubes containing the precipitates and mixing the contents of the test tube by stirring with a glass rod.

Question: Does the precipitate in any of the eight test tubes dissolve? The experimental answer is yes. If sufficient volume of the acid is added, the precipitates dissolve in three of the test tubes. You may enjoy predicting which ones. The experimental answer is Experiment I, $BaCrO_4$; Experiment VI, Ag_2CrO_4; and Experiment VIII, $Mg(OH)_2$. The equilibria in these three test tubes before the nitric acid is added are

$$\text{I} \quad BaCrO_4(s) \rightleftarrows Ba^{2+}(aq) + CrO_4^{2-}(aq)$$

$$\text{VI} \quad Ag_2CrO_4(s) \rightleftarrows 2\,Ag^{1+}(aq) + CrO_4^{2-}(aq)$$

$$\text{VIII} \quad Mg(OH)_2(s) \rightleftarrows Mg^{2+}(aq) + 2\,OH^{1-}(aq)$$

In each case, the added hydronium ion reacts with the anion:

$$CrO_4^{2-}(aq) + H_3O^{1+} \longrightarrow H_2O(l) + HCrO_4^{1-}(aq)$$

$$OH^{1-}(aq) + H_3O^{1+}(aq) \longrightarrow H_2O(l) + H_2O(l)$$

The concentration of the anion is reduced. The solution is no longer saturated. The solid dissolves until the product of the concentrations of the ions of the precipitate again matches the numerical value of the solubility product constant and equilibrium between the solid and its ions is again established—or until all the solid dissolves. Barium chromate, $BaCrO_4$, and silver chromate, Ag_2CrO_4, would not have precipitated in the first place if the original solution had been acidic. Likewise, magnesium hydroxide, $Mg(OH)_2$, would not have been precipitated from an acidic solution. If the H_3O^{1+} concentration is 1×10^{-2} moles per liter in the original solution, the hydroxide concentration would be much too small (1×10^{-12} moles per liter, calculated from K_W) to produce a supersaturated solution of magnesium hydroxide, $Mg(OH)_2$:

$$\underset{Mg(OH)_2}{K_{sp}} = [Mg^{2+}] \times [OH^{1-}]^2 = 1 \times 10^{-11}$$

If $[OH^{1-}] = 1 \times 10^{-12}$, then at equilibrium,

$$[Mg^{2+}] \times [1 \times 10^{-12}]^2 = 1 \times 10^{-11} \text{ moles}^3/\text{liters}^3$$

$$[Mg^{2+}] = \frac{1 \times 10^{-11}}{1 \times 10^{-24}} = 1 \times 10^{13} \text{ moles/liter}$$

This concentration of magnesium ion is physically impossible. The mass and volume of 1×10^{13} moles of a compound of magnesium would require a truck (or trucks) to move it and certainly could not be dissolved in water to give one liter of solution.

[Note: the iron(III) hydroxide also dissolves on the addition of a strong acid and would not be formed in the first place from acidic solutions.]

The Effect of the Formation of Coordination Complexes

None of the alkali metal ions and the alkaline earth ions form coordination complexes. Mercury and silver are transition metal elements. Mercury(II) ion, Hg^{2+}, and silver ion, Ag^{1+}, would be expected to form coordination complexes. For simplicity in the following discussion, *aquo*-complexes are ignored.

In Experiment VII, red mercury(II) iodide, HgI_2, was precipitated by the addition of a few drops of 0.01 M potassium iodide, KI, to a few milliliters of an aqueous solution of 0.01 M mercury(II) nitrate, $Hg(NO_3)_2$:

$$Hg^{2+}(aq) + 2\, I^{1-}(aq) \longrightarrow \underset{\text{red}}{HgI_2(s)}$$

If the experiment had been carried out by the addition of a few drops of mercury(II) nitrate solution to a few milliliters of the potassium iodide solution, the red precipitate of mercury(II) iodide, HgI_2, would not have appeared or at least would have appeared and then disappeared again very quickly. On the addition of a larger volume of the mercury(II) nitrate solution, the red precipitate would have been formed. In carrying out a precipitation reaction, one does not expect to have to be concerned about which reagent is added or the relative volumes used so long as enough is added for the ion concentrations to exceed the solubility product constant for the ionic compound being precipitated. The K_{sp}, at 25 °C, for HgI_2 is 3×10^{-24}. It is a very insoluble compound and there should be no difficulty in exceeding the solubility product constant. The reason for the rather surprising "now you have it, now you don't" phenomena in precipitating HgI_2 is that mercury(II) ion forms a colorless coordination complex anion, HgI_4^{2-} (coordination number 4 for the mercury(II) ion). The name of this ion is tetraiodomercurate ion, but let us not bother about that.

HgI_4^{2-}

$$Hg^{2+}(aq) + 2\, I^{1-}(aq) \longrightarrow \underset{\text{red}}{HgI_2(s)}$$

$$Hg^{2+}(aq) + 4\, I^{1-}(aq) \longrightarrow \underset{\text{colorless}}{HgI_4^{2-}(aq)}$$

With an excess of iodide ion, the colorless complex ion, HgI_4^{2-}, is obtained in solution. With an excess of mercury(II) ion, the red precipitate, HgI_2, is obtained.

$Ag(NH_3)_2^{1+}$

Silver ion, Ag^{1+}, forms a coordination complex with ammonia, $Ag(NH_3)_2^{1+}$, and the silver chloride precipitate can be dissolved in an aqueous solution of ammonia:

$$\underset{\text{white}}{AgCl(s)} + \underset{\text{colorless}}{2\, NH_3(aq)} \longrightarrow \underset{\text{colorless}}{Ag(NH_3)_2^{1+}(aq)} + Cl^{1-}(aq)$$

The name of this coordination complex is diamminesilver ion. In Experiment III, an aqueous solution of sodium chloride was added to an aqueous solution of silver nitrate to precipitate silver chloride. If either solution had contained an adequate concentration of ammonia molecules, silver chloride would not have precipitated due to the tie-up of the silver ion, Ag^{1+}, in the coordination complex, $Ag(NH_3)_2^{1+}$ (coordination number 2 for silver ion).

Any chemical reaction that decreases the concentration of either ion of an ionic compound, which might otherwise be precipitated from an aqueous solution, may prevent precipitation. Examples from the preceding discussion:

- Conversion of CrO_4^{2-} to $HCrO_4^{1-}$ prevents the precipitation of $BaCrO_4$ and Ag_2CrO_4.
- Conversion of OH^{1-} to H_2O prevents the precipitation of $Mg(OH)_2$.
- Conversion of Hg^{2+} to HgI_4^{2-} prevents the precipitation of HgI_2.
- Conversion of Ag^{1+} to $Ag(NH_3)_2^{1+}$ prevents the precipitation of AgCl.

Gratuitous Information 15-2

TRANSITION METAL IONS ARE ACIDS, TOO

You are aware that an unshared pair of electrons is characteristic of a Brönsted base and also aware that an unshared pair of electrons of molecules and ions such as NH_3, H_2O, OH^{1-}, and I^{1-} is considered to be instrumental in the formation of coordination complexes such as $Ag(NH_3)_2^{1+}$, $Cu(NH_3)_4^{2+}$, HgI_4^{2-}, and $Fe(H_2O)_6$.

In the early 1920's, at about the same time Brönsted proposed the Brönsted concept of acids and bases, G. N. Lewis (of Lewis electron dot structure fame) proposed a more general concept of acids and bases. A **Lewis acid** is a molecule or ion that can accept a share in a pair of electrons and a **Lewis base** is a molecule or ion that can donate a share in a pair of electrons. In the Lewis concept, the reaction

$$\underset{\text{acid}}{Ag^{1+}(aq)} + \underset{\text{base}}{2\,NH_3(aq)} \longrightarrow Ag(NH_3)_2^{1+}(aq)$$

is a Lewis acid–base reaction. The silver ion, $Ag^{1+}(aq)$, is a Lewis acid and the ammonia molecule is a Lewis base. Transition metal ions have readily available "unfilled" *d*-orbitals. As the principal quantum levels become larger, the separation between energy levels is smaller and consequently "unoccupied" orbitals are more readily available to accept a share in a pair of electrons—to form a coordinate covalent bond.

The Brönsted concept of acid–base pairs is built entirely upon the transfer of a proton and, in aqueous solution, the most frequently encountered strong acid is the hydronium ion, H_3O^{1+}. How does the hydronium ion fit into the Lewis concept of an acid? In the reaction

$$H_3O^{1+}(aq) + NH_3(aq) \longrightarrow NH_4^{1+} + H_2O$$

how can the hydronium ion, H_3O^{1+}, be a Lewis acid? How does the hydronium ion accept a share in a pair of electrons of the ammonia molecule?

$$\left(\begin{array}{c} \cdot\cdot \\ H:\!O\!:H \\ H \end{array}\right)^{1+} + \begin{array}{c} H \\ :\!N\!:H \\ H \end{array} \longrightarrow \begin{array}{c} \cdot\cdot \\ H:\!O\!: \\ H \end{array} + \left(\begin{array}{c} H \\ H:\!N\!:H \\ H \end{array}\right)^{1+}$$

The hydrogen–oxygen bonds in the hydronium ion are, of course, polar bonds with each hydrogen atom having a partial positive charge. As a proton of a hydronium ion encounters the region of high electron density of the unshared pair of electrons of the nitrogen atom, the proton may become incorporated in a *sigma*-bond with the nitrogen atom of the ammonia molecule and cease to be incorporated in a *sigma*-bond with the oxygen atom of the hydronium ion. In the Lewis concept of acids, the transfer of a proton is a simultaneous acceptance by a proton of a share in a pair of electrons of the base and the release of the pair of electrons of its previous *sigma*-bond in the parent molecule or ion. All Brönsted bases are, of course, Lewis bases. The Lewis concept of acids and bases encompasses all Brönsted acids and bases and extends the concept of acids to include molecules and ions excluded by the Brönsted concept.

Scramble Exercises

1. The following series of experiments are carried out. All solutions are colorless aqueous solutions. In at least one of these experiments, a white precipitate is formed and in at least one of these experiments no precipitate is formed.
   (a) 0.01 M sodium chloride, NaCl, is added to 0.01 M potassium nitrate, KNO_3.
   (b) 0.01 M potassium chloride, KCl, is added to 0.01 M barium nitrate, $Ba(NO_3)_2$.
   (c) 0.01 M magnesium sulfate, $MgSO_4$, is added to 0.01 M barium nitrate, $Ba(NO_3)_2$.

   Identify an experiment in which a precipitate forms and write the appropriate equation for the reaction. Decisions depend entirely upon empirical knowledge acquired in this chapter or upon previously acquired empirical knowledge.

   Identify an experiment in which no reaction occurs and state the reason for your choice.

2. When an aqueous solution of sodium carbonate, Na_2CO_3, is added to an aqueous solution of barium chloride, $BaCl_2$, a white precipitate is obtained. Predict the formula and the name of the compound precipitated. Write an appropriate equation for the precipitation reaction.

3. Solid barium carbonate, $BaCO_3$, has a very small solubility in water but dissolves readily in hydrochloric acid or nitric acid. Suggest a plausible explanation for the solubility of barium carbonate in the acids and write an appropriate equation or equations.

4. The solubility product constants, at 25 °C, are given in this chapter for silver chromate, Ag_2CrO_4; magnesium hydroxide, $Mg(OH)_2$; and mercury(II) iodide, HgI_2. On the basis of these solubility product constants, compare the solubility in water of the three compounds. All three compounds are either 2 to 1 or 1 to 2 ionic compounds and, consequently, it is fair to compare the numerical values of their solubility product constants.

5. Copper(II) hydroxide, $Cu(OH)_2$, is a blue-green solid that is very slightly soluble in water. Copper(II) ion, Cu^{2+}, forms a tetraamminecopper(II) ion, $Cu(NH_3)_4{}^{2+}$, that has a most attractive royal blue color. Solid copper hydroxide dissolves readily in an aqueous solution of ammonia, NH_3. Suggest a plausible explanation for the solubility of copper(II) hydroxide in ammonia water and include a plausible equation or equations. The same reaction is involved when household ammonia, an aqueous solution of ammonia, is used to remove the blue stains that sometimes develop under dripping faucets in buildings that have copper water pipes. The blue substance deposited in the bathtub is possibly a combination of $CuCO_3$ and $Cu(OH)_2$.

6. When an aqueous solution of sodium chloride, NaCl, is added to solid silver chromate, Ag_2CrO_4, and the mixture stirred, the red solid disappears and a

white precipitate forms in its place. Suggest a plausible explanation. The solubility of silver chromate in water is very small.

7. Review: In terms of the valence bond model,
   (1) identify each of the bonds in the acetic acid molecule as a *sigma*-bond or a *pi*-bond,
   (2) identify the type of hybridization involved by each carbon atom in the acetic acid molecule, and
   (3) rationalize the experimental fact that the two carbon atoms and the two oxygen atoms of the acetic acid molecule lie in the same plane.

Additional exercises for Chapter 15 are given in the Appendix.

A Look Ahead

In Chapter 16, Oxygen Gas, the Molecular Orbital Approach to Diatomic Molecules, and Oxidation–Reduction Reactions, we shall at long last explore the molecular orbital, MO, approach to chemical bonding. This approach encompasses shared pairs of electrons in molecular orbitals but it is not limited to shared pairs of electrons in the sense that Lewis structures and the valence bond, VB, approach are limited to shared pairs of electrons. Scramble Exercise 7 in this chapter is an excellent review of the valence bond approach to one molecule.

We shall also explore reactions that involve the transfer of electrons from one atom, molecule, or ion to another atom, molecule or ion. For example, the conversion of an atom of sodium, Na, to a sodium ion, $Na^{(1+)}$, involves the transfer of an electron, $e^{(1-)}$:

$$Na \longrightarrow Na^{(1+)} + \underset{\text{electron}}{e^{(1-)}}$$

The equation written is only half of the story. Some atom, molecule, ion, or electrode must accept that electron. The conversion reaction of a sodium ion, $Na^{(1+)}$, to become a sodium atom, Na, involves the acceptance of an electron:

$$Na^{(1+)} + e^{(1-)} \longrightarrow Na$$

Here again, the equation written is only half of the story. Some atom, molecule, ion, or electrode must supply that electron.

For reasons that are purely historical, these reactions are known as oxidation–reduction reactions, in spite of the fact that oxygen is not necessarily involved in oxidation–reduction reactions.

16

OXYGEN GAS, THE MOLECULAR ORBITAL APPROACH TO DIATOMIC MOLECULES, AND OXIDATION–REDUCTION REACTIONS

Chapter 16

Within several kilometers of the surface of the earth, both directions, atoms of oxygen are more abundant than the atoms of any other element. Many of the minerals that constitute the solid crust are oxygen-containing compounds such as oxides, carbonates, phosphates, and silicates. In terms of numbers of atoms, the seas and the ice caps are 33% oxygen and the atmosphere is approximately 20% oxygen. In terms of the total universe, it is estimated that atoms of oxygen account for only 0.03% of all atoms, with the three most abundant being atoms of hydrogen (94%), helium, and neon.

Distribution of the Elements

Almost all life as we know it is dependent upon being able to take in diatomic oxygen, O_2, and give off carbon dioxide, CO_2, through natural processes. The availability of diatomic oxygen in the atmosphere is absolutely essential to our welfare. During photosynthesis, plants also take in carbon dioxide and return diatomic oxygen to the atmosphere.

Diatomic oxygen, O_2, reacts with all elements, with the exception of some of the rare gases, and with many compounds in reactions such as the combustion of compounds of carbon, hydrogen, and oxygen discussed in Chapter 4. The natural processes of decomposition of biological materials also utilize diatomic oxygen and produce carbon dioxide. All of these oxidation reactions continuously deplete the diatomic oxygen content of the atmosphere. Recharging the diatomic oxygen content of the atmosphere is entirely credited to the photosynthetic processes of green plants—hence the incorporation of green belts in urban planning.

Photosynthesis

In the photosynthetic reactions of plants, molecules of a variety of compounds essential to plants are formed in addition to diatomic oxygen. These include carbohydrates (Chapter 18) and other compounds that lead to the formation of proteins (Chapter 17), and fats (Chapter 20). The primary reactants are water and carbon dioxide. The following overall equation is written for the photosynthesis of glucose, a simple carbohydrate:

$$6\,CO_2(g) + 6\,H_2O(l) \xrightarrow{\text{light}} \underset{\text{glucose}}{C_6H_{12}O_6(aq)} + 6\,O_2(g)$$

The concept of photosynthesis is simple. For the synthesis to take place, bonds within the reactants water and carbon dioxide must be broken and bonds formed to give the products. The energy required to break the bonds in the reactants is greater than the energy released in forming bonds. The additional energy is supplied by the photons of light absorbed by the green chlorophyll of the plant. A complex series of reactions is involved in the conversion of carbon dioxide and water into glucose and diatomic oxygen. The detailed mechanisms of photosynthetic reactions are only partially understood. *Respiration* Respiration, reactions in which diatomic oxygen is a reactant and carbon dioxide is an end product, occurs continously in living plants as well as in living animals. In the presence of light of the appropriate frequency, photosynthesis and respiration proceed simultaneously in plants. If sufficient numbers of photons are absorbed, the rate of photosynthesis exceeds the rate of respiration.

In this chapter we shall explore

- the molecular orbital (MO) approach to oxygen and other diatomic molecules.
- the reaction of diatomic oxygen with elements, and
- a generalized concept of oxidation reactions that do not necessarily involve oxygen.

Molecular Orbital Approach to the Structure of Diatomic Oxygen and Other Diatomic Molecules

In the discussion of the distribution of electrons in diatomic molecules in Chapter 7, it was pointed out that both the conventional Lewis structure approach and the valence bond (VB) approach are based upon the concept of shared pairs of electrons, and both approaches are incapable of dealing with odd-electron molecules or even-electron molecules that are paramagnetic. Consequently, we turn to another model to rationalize the paramagnetic properties of diatomic oxygen and the nature of the bonding in odd-electron molecules. That model is known as the molecular orbital (MO) approach. Some concepts and much of the terminology of the molecular orbital approach are common to the valence bond approach, and close attention to the differences between the two approaches is essential. *Quick Review of Valence Bond Approach* First, a quick review of the nature of chemical bonding, the conventional Lewis structure of O_2, and the valence bond approach to O_2. If you need a more detailed review, refer to the presentation in Chapter 7.

- All chemical bonding in diatomic molecules is dependent on the distribution of the electron atmospheres of the two atoms about the two nuclei with a concentration of negative charge between the two nuclei being effective in bonding the two atoms together.
- atomic number of oxygen: $Z = 8$
- conventional Lewis structure for O_2 is $\ddot{\underset{\cdot\cdot}{O}}::\ddot{\underset{\cdot\cdot}{O}}$
- pseudo Lewis structure for O_2 is $:\dot{\underset{\cdot\cdot}{O}}:\dot{\underset{\cdot\cdot}{O}}:$ with unpaired electrons to account for its paramagnetism
- ground state distribution of electrons in atomic orbitals for the oxygen atom: $1s^2\ 2s^2\ 2p_x{}^2\ 2p_y{}^1\ 2p_z{}^1$

- the 90% electron probability representation of the partially filled $2p_y$ and $2p_z$ atomic orbitals of two oxygen atoms.

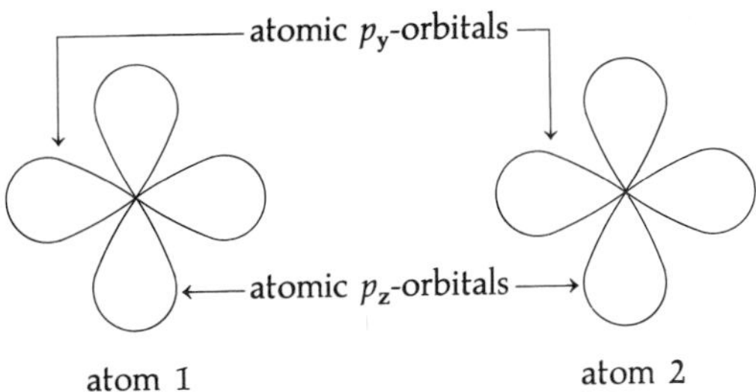

The filled $2p_x$-orbital extends above and below the plane of the paper and is not shown in this diagram.

- the 90% electron probability representation for the filled σ-orbital for the O_2 molecule formed by the end-to-end overlap of the partially filled $2p_y$-orbitals of the two oxygen atoms and the filled π-orbital for the O_2 molecule formed by the side-to-side overlap of the partially filled $2p_z$-orbitals of the two oxygen atoms:

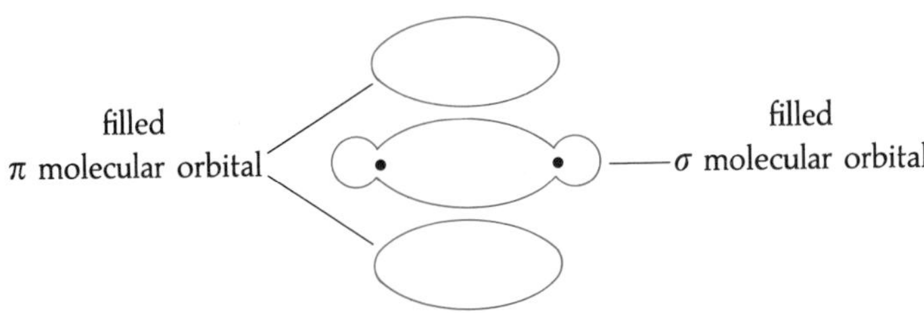

Note that, in the valence bond approach to O_2, four electrons of four partially filled atomic orbitals (designated here as the $2p_y$ electron and the $2p_z$ electron of each atom) have been assigned to two molecular orbitals, a *sigma*-orbital and a *pi*-orbital. Each molecular orbital is filled by two electrons having paired spins and each of these filled molecular orbitals contributes to the stability of the molecule. The other twelve electrons in filled atomic orbitals are ignored.

Molecular Orbital Approach

In the molecular orbital approach, all of the electrons of both atoms (a total of sixteen electrons in the case of O_2) are assigned to quantized energy states of molecules called molecular orbitals. Figure 16-1 presents the schematic energy level diagram for the ten lowest energy molecular orbitals of homonuclear diatomic molecules. This schematic energy sequence of molecular orbitals is to diatomic molecules what the schematic energy sequence of atomic orbitals, given in Figure 6-4, is to atoms, and the two diagrams are used in exactly the same way to assign electrons to molecular orbitals for ground state homonuclear diatomic molecules (Figure 16-1) and to assign electrons to atomic orbitals for ground state atoms (Figure 6-4). For both diagrams, all energies are negative and the lowest energy state is at the bottom of the diagram. The lowest energy molecular orbital, the most stable molecular orbital for an electron, is the σ_{1s} (read sigma-one-s) molecular orbital. In both diagrams, Figure 16-1 and

Figure 16-1

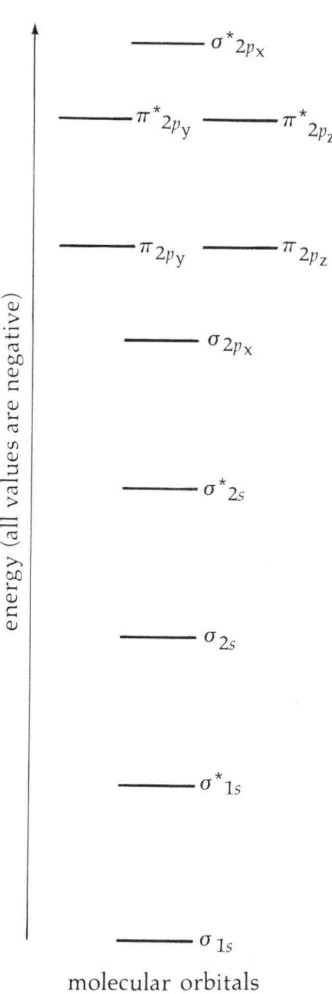

SCHEMATIC ENERGY LEVEL DIAGRAM FOR MOLECULAR ORBITALS OF HOMONUCLEAR DIATOMIC MOLECULES.

Figure 6-4, the orbitals are represented by short horizontal lines and each orbital can accommodate a maximum of two electrons with paired electron spins.

Quantized Energy States of Molecular Orbitals

The appropriate MO distribution of the 16 electrons in a ground state diatomic molecule of oxygen is found using the construction principle (*Aufbau Prinzip*) by assigning electrons according to the sequence of energy levels, according to rules equivalent to the Pauli exclusion principle (no two electrons are denoted by exactly the same quantum numbers), and Hund's rule (all degenerate energy levels are partially filled before a second electron with antiparallel spin is introduced). This distribution

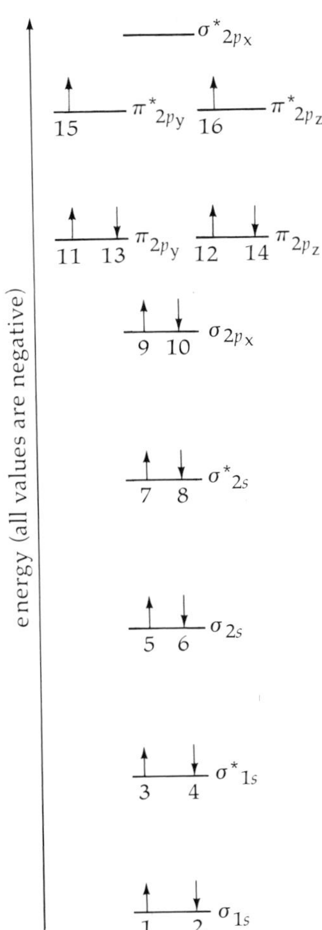

SCHEMATIC ENERGY LEVEL DIAGRAM FOR MOLECULAR ORBITALS OF A HOMONUCLEAR DIATOMIC MOLECULE WITH THE 16 ELECTRONS FOR THE O_2 MOLECULE FILLED IN. Each arrow represents one electron and the numerals indicate one order in which the electrons might be added to the energy level diagram.

of electrons is displayed in Figure 16-2 for the ground state diatomic molecule of oxygen, where each electron is represented by a vertical arrow in the usual fashion. The numerals under the arrows indicate one sequence in which the electrons could be assigned. Note that, with the degenerate (same energy) molecular orbitals π_{2p_y} and π_{2p_z} and again with the degenerate orbitals $\pi^*_{2p_y}$ and $\pi^*_{2p_z}$, one electron is assigned to each of the molecular orbitals having the same energy before a second electron is assigned to either orbital.

According to this molecular orbital approach, the ground state molecule of oxygen has two partially filled molecular orbitals: the $\pi^*_{2p_y}$ (read pi-star-two-p-y) molecular orbital and the $\pi^*_{2p_z}$ molecular orbital. The presence of these two unpaired electrons is consistent with the paramagnetic properties of diatomic oxygen. Here, at last, we have a model that is consistent with the paramagnetic properties of O_2. To understand further the significance of the distribution of electrons in the molecular orbitals of diatomic oxygen and other homonuclear diatomic molecules, it is necessary to explore how the molecular orbital diagram, Figure 16-1, is derived from atomic orbitals and the atomic orbital diagram. Figure 6-4.

In the MO approach, the wave functions of all molecular orbitals are derived by the combination of the wave functions for atomic orbitals for the two atoms. This we shall explore in some detail. It is not at all necessary for you to learn how to start with the atomic orbital diagram, Figure 6-4, and produce the structure of the molecular orbital diagram, Figure 16-1, but you should acquire a general concept of how molecular orbitals relate to atomic orbitals. The presentation is long and somewhat tedious. As we proceed, note the repetition in the approach and focus on the application of the molecular orbital approach to specific molecules and ions, such as H_2, H_2^{1+}, H_2^{1-}, He_2, O_2, and N_2, as they come into the discussion. The σ- and π-orbitals generated by the molecular orbital approach are called **bonding molecular orbitals.** One electron or two paired electrons in either a σ or π bonding orbital contribute to the electron density between the two nuclei and hence to the stability of the diatomic molecule. A pair of electrons in a σ or π molecular orbital contributes more stability to the molecule than one electron in the same molecular orbital. A filled bonding molecular orbital in the molecular orbital approach is equivalent to a filled molecular orbital in the valence bond approach. The σ^* and π^* molecular orbitals of the MO approach are called **antibonding molecular orbitals.** Antibonding molecular orbitals are unique to the molecular orbital approach: they have no counterpart in the valence bond approach. One electron or two paired electrons in either a σ^* or π^* antibonding orbital diminish the stability of the diatomic molecule, with a pair of electrons in an antibonding orbital being more destabilizing than one electron in the same antibonding orbital.

Bonding Orbitals σ and π

Antibonding Orbitals σ^* and π^*

The wave function for the σ_{1s} molecular orbital for a diatomic homonuclear molecule is derived, as indicated by the notation σ_{1s}, from the sum of the wave function for the 1s atomic orbital of one atom and the wave function for the 1s atomic orbital of the second atom:

Derivation of Bonding sigma-Orbitals from s Atomic Orbitals

$\psi_{\sigma_{1s}}$	=	ψ_{1s}	+	ψ_{1s}
wave function for the σ_{1s} molecular orbital		wave function for the 1s atomic orbital for atom 1		wave function for the 1s atomic orbital for atom 2

It is again the square of the wave function that gives the electron probability distribution that is the orbital. The shapes of the bonding σ_{1s}-orbitals are ellipsoids, identical with those obtained in the valence bond approach by combining two 1s atomic orbitals. The σ_{1s}-orbital has an ∞-fold axis of symmetry through the two nuclei.

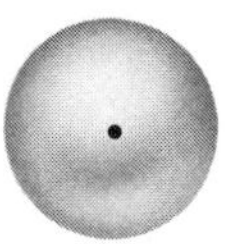
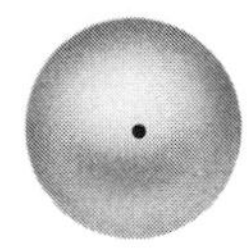

the 90% probability representation of the 1s-orbitals of the two separate atoms

the 90% probability representation of the σ_{1s}-orbital of the molecule

The wave function for the σ^*_{1s} molecular orbital for a diatomic molecule is derived from the difference of the wave function for the 1s atomic orbital of one atom and the wave function for the 1s atomic orbital of the second atom.

Derivation of Antibonding sigma-Orbitals from s Atomic Orbitals

$\psi_{\sigma^*_{1s}}$	=	ψ_{1s}	−	ψ_{1s}
wave function for the σ^*_{1s} molecular orbital		wave function for the 1s atomic orbital for atom 1		wave function for the 1s atomic orbital for atom 2

The shape of the molecular σ^*_{1s} molecular orbital, derived from the square of the $\psi_{\sigma^*_{1s}}$ wave function, compared to the shapes of the 1s atomic orbitals of the two separate atoms is indicated below in the usual 90% probability sketches.

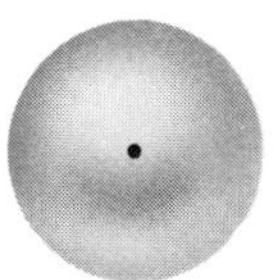
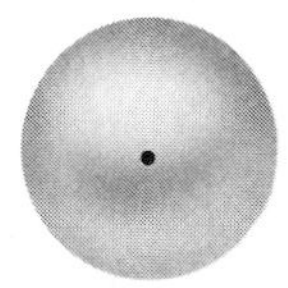

the 90% probability representation for the 1s-orbitals of the two separate atoms

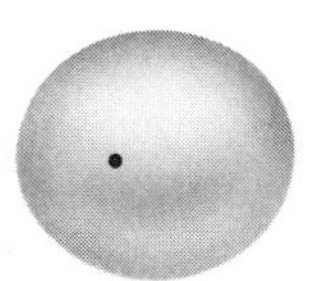

the 90% probability representation for the σ^*_{1s}-orbital of the molecule

This σ^*_{1s} molecular orbital is quite different from any of the molecular orbitals we have discussed before. There are two regions of high electron probability—one off center to one nucleus and the other off center with respect to the other nucleus. The σ^*_{1s} molecular orbital has again an ∞-fold axis of symmetry. The unique characteristic of this molecular orbital is that the region of high probability for electrons is not primarily between the two nuclei but beyond both nuclei at the two ends of the molecule. Since it is high electron density between two nuclei that provides the attraction to offset the repulsion between the two positive nuclei, electrons in the σ^*_{1s} molecular orbital do not contribute to the stability of the molecule. Instead, electrons in the σ^*_{1s} molecular orbital destabilize the bond between the two nuclei and the σ^*_{1s} molecular orbital is called an antibonding orbital. Antibonding molecular orbitals are an essential part of the molecular orbital approach. Note that there are five antibonding (starred) molecular orbitals and five bonding molecular orbitals in the schematic energy level diagram, Figure 16-1. The 10 orbitals accommodate a maximum of 20 electrons.

Relative Energies of σ and σ^* Orbitals

A comparison of the energy associated with the σ_{1s} molecular orbital (a bonding orbital) and the energy associated with the σ^*_{1s} molecular orbital (an antibonding

Figure 16-3

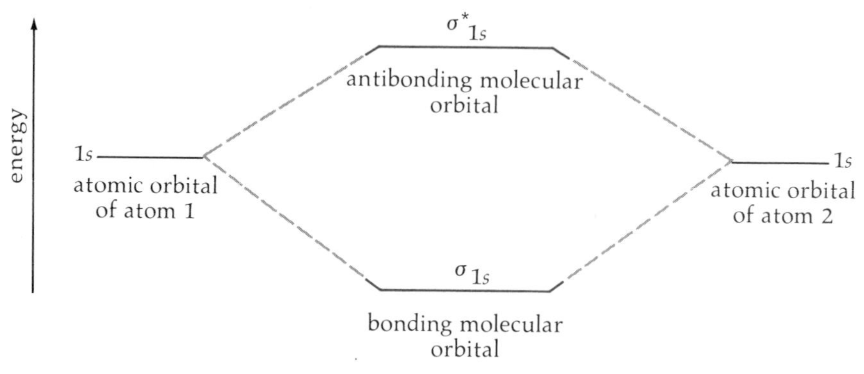

RELATIVE ENERGIES ASSOCIATED WITH THE σ_{1s} AND σ^*_{1s} MOLECULAR ORBITALS AND THE TWO 1*s* ATOMIC ORBITALS FROM WHICH THEY ARE DERIVED.

orbital) relative to the energy associated with the 1*s* atomic orbital of the two separate atoms is given in Figure 16-3. All of these energy relations are derived from the wave functions for atomic orbitals, which are themselves expressions that satisfy energy relations for the atoms.

Note that the energy state associated with the σ_{1s} molecular orbital of the molecule is lower (more stable) than the energy state associated with the 1*s* atomic orbitals of the two separate atoms and that the energy state associated with the σ^*_{1s} molecular orbital of the molecule is higher (less stable) than the energy state associated with the 1*s* atomic orbitals of the two separate atoms.

The bit of the schematic energy level diagram for molecular orbitals of homonuclear diatomic molecules given in Figure 16-3 is sufficient to consider the diatomic hydrogen, H_2, ground state molecule and the diatomic helium, He_2, ground state molecule and also the ground states of the diatomic hydrogen ions, $H_2^{(1+)}$ and $H_2^{(1-)}$, and the diatomic helium $He_2^{(1+)}$ ion. These diatomic units have numbers of electrons ranging from one for the $H_2^{(1+)}$ ion to four for the He_2 molecule.

We know that hydrogen gas under room conditions is diatomic and helium gas is monatomic. The molecular orbital model is entirely consistent with these experimental facts. The two electrons of the diatomic hydrogen molecule fill the σ_{1s} molecular orbital. According to the above molecular orbital energy level diagram, the energy associated with the σ_{1s} molecular orbital is lower than the energy associated with the 1*s* atomic orbitals of the two separate atoms. The H_2 molecule is more stable than the two separate atoms.

The Hydrogen Molecule

The Helium Molecule

According to Figure 16-3, the four electrons of a ground state diatomic helium molecule would be assigned first to fill the σ_{1s} molecular orbital and then to fill the σ^*_{1s} molecular orbital. The net result is that the sum of the energy associated with a filled σ_{1s} bonding orbital and the energy associated with a filled σ^*_{1s} antibonding orbital is essentially the same as the energy associated with the two separate helium atoms. The diatomic helium molecule is not more stable than the separate helium atoms and the molecular orbital approach predicts that helium is monatomic. What the filled σ_{1s} bonding orbital would contribute to the stability of the diatomic helium molecule, the filled σ^*_{1s} antibonding orbital would cancel out.

The Diatomic Ions $H_2^{(1+)}$, $H_2^{(1-)}$, and $He_2^{(1+)}$

In the molecular orbital approach to the $H_2^{(1+)}$ ion, its one electron is assigned to the σ_{1s} molecular orbital. This is a one-electron bond that contributes stability to the ion but it is not nearly as stable as the two-electron bond of the H_2 molecule. The unit $H_2^{(1+)}$ occurs in hydrogen gas electrical discharge tubes and has been extensively studied. It is frequently called the hydrogen molecule ion. (Of course, the $H_2^{(1-)}$ ion is also a hydrogen molecule ion.)

One Electron Bond in $H_2^{(1-)}$ and $He_2^{(1+)}$

According to Figure 16-3, two of the three electrons of the $H_2^{(1-)}$ ion, and also of the $He_2^{(1+)}$ ion, would be assigned to fill the σ_{1s} molecular orbital and the third electron to partially fill the σ^*_{1s}-orbital. These ions would be expected to have a bond strength similar to the bond strength of the $H_2^{(1+)}$ ion. The bonding orbital has one more electron than the antibonding orbital. Both ions are considered to have an effective one-electron bond.

The story is very much the same for the generation of the wave functions of the σ_{2s} molecular orbital and the σ^*_{2s} molecular orbital from the combination of the wave functions of the 2*s* atomic orbitals of the two atoms. See Figure 16-1 for the position of σ_{2s} and σ^*_{2s} orbitals in the schematic energy level diagram.

The three 2*p*-orbitals (p_x, p_y, and p_z) of one atom and the three 2*p*-orbitals of the other atom generate six molecular orbitals: three bonding orbitals—σ_{2p_x}, π_{2p_y}, and π_{2p_z}—which can accommodate a maximum of six electrons, and three antibonding orbitals—$\sigma^*_{2p_x}$, $\pi^*_{2p_y}$, and $\pi^*_{2p_z}$—which can also accommodate a maximum of six electrons. The section of the schematic energy level diagram encompassing the molecular orbitals generated from atomic orbitals $2p_x$, $2p_y$, and $2p_z$ for two atoms is given in Figure 16-4.

Formation of Molecular Orbitals Using p Atomic Orbitals

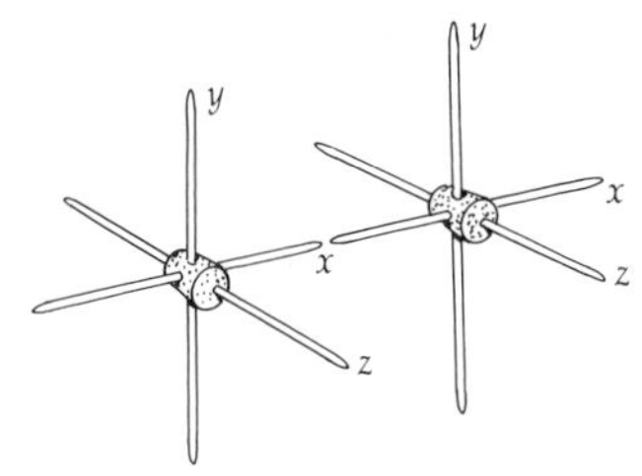

This figure as a whole is complex, but the parts of which it is composed are straightforward. Each part will be discussed in detail. Take care to identify each part on the diagram as it is discussed. The diagram is labeled for a specific orientation of the atomic orbitals of the two atoms. This orientation is for the ∞-fold axis of rotation for the $2p_x$-orbital of atom 1 to be the continuation of the ∞-fold axis of rotation of the $2p_x$-orbital of atom 2, for the ∞-fold axis of rotation of the $2p_y$-orbital of atom 1 to be parallel to the ∞-fold axis of rotation of the $2p_y$-orbital of atom 2, and for the ∞-fold axis of rotation of the $2p_z$-orbital of atom 1 to be parallel to the ∞-fold axis of rotation of the $2p_z$-orbital of atom 2. We focus first on the interaction of the wave functions of the $2p_x$ atomic orbitals of the two homonuclear atoms to give the wave functions for the σ_{2p_x} molecular orbital, a bonding orbital:

$$\underset{\text{molecule}}{\psi_{\sigma^*_{2p_x}}} = \underset{\text{atom 1}}{\psi_{2p_x}} + \underset{\text{atom 2}}{\psi_{2p_x}}$$

Derivation of σ Orbitals from p Atomic Orbitals

This is exactly the same process as the formation of a σ-orbital in the valence bond approach by end-to-end overlap of two *p*-orbitals:

the $2p_x$-orbitals of the two separate atoms   the σ_{2p_x}-orbital of the molecule

Figure 16-4

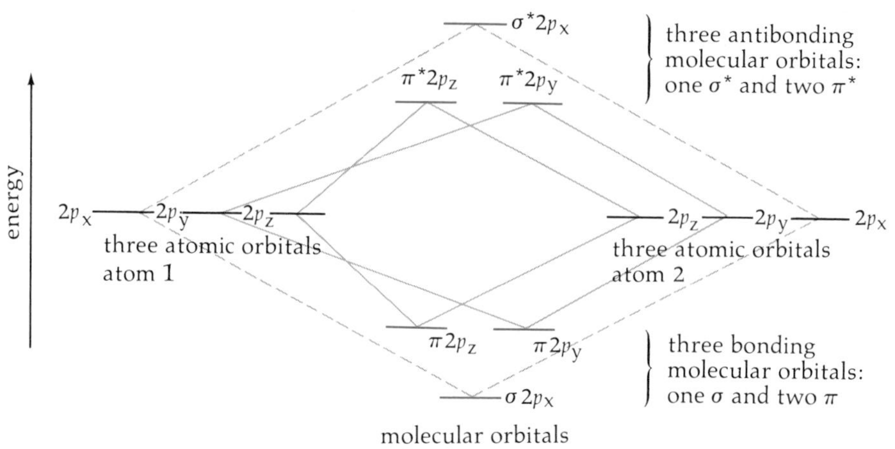

RELATIVE ENERGIES ASSOCIATED WITH THE MOLECULAR ORBITALS GENERATED FROM THE $2p_x$, $2p_y$, AND $2p_z$ ATOMIC ORBITALS OF TWO HOMONUCLEAR ATOMS.

The energy state associated with this bonding σ_{2p_x} molecular orbital is lower (more stable) than the energy associated with the $2p_x$ atomic orbitals of the two separate atoms, Figure 16-4. The σ_{2p_x}-orbital is a bonding orbital.

The wave function for the $\sigma^*_{2p_x}$ molecular orbital, an antibonding orbital, is derived from the difference between the wave functions for the $2p_x$ atomic orbitals:

$$\underset{\text{molecule}}{\psi_{\sigma^*_{2p_x}}} = \underset{\text{atom 1}}{\psi_{2p_x}} - \underset{\text{atom 2}}{\psi_{2p_x}}$$

Derivation of σ^* Orbitals from p Atomic Orbitals

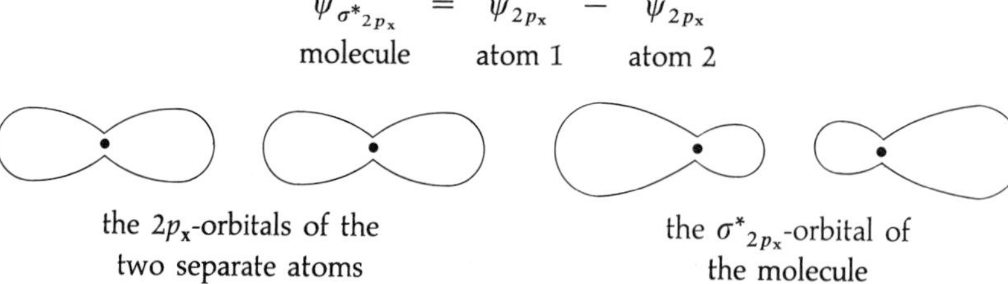

As with all antibonding molecular orbitals, the greater part of the negative charge density is not between the two nuclei, and the energy state associated with this antibonding molecular orbital is higher (less stable) than the energy state associated with the $2p_x$ atomic orbitals of the two separate atoms, Figure 16-4. Note that the 90% distribution representation sketched above for the $\sigma^*_{2p_x}$ molecular orbital represents one molecular orbital—not two molecular orbitals. See Figure 16-1 for the positions of the σ_{2p_x} molecular orbital and the $\sigma^*_{2p_x}$ molecular orbital in the schematic energy level diagram.

The wave functions for the π_{2p_y}-orbital and the $\pi^*_{2p_y}$-orbital of the molecule are generated by the combination of the wave function for the $2p_y$-orbital of atom

1 with the wave function for the $2p_y$-orbital of atom 2 in the side-by-side orientation. Pictorially, the combination to give the bonding π-orbital in the molecular orbital approach is exactly the same as the combination to give the π molecular orbital in the valence bond approach:

Derivation of π Orbitals from p Atomic Orbitals

$$\underset{\text{molecule}}{\psi_{\pi_{2p_y}}} = \underset{\text{atom 1}}{\psi_{2p_y}} + \underset{\text{atom 2}}{\psi_{2p_y}}$$

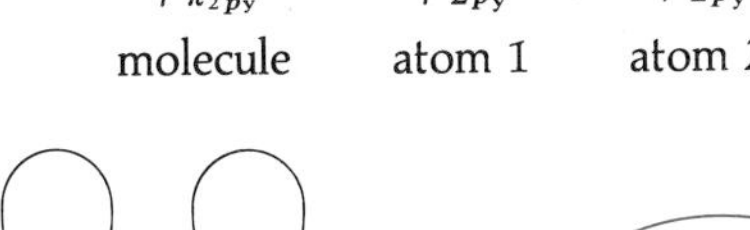

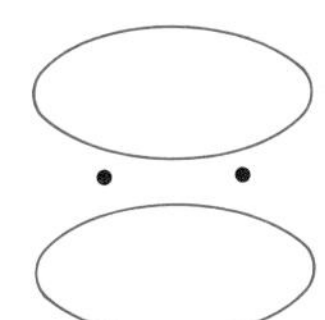

the $2p_y$-orbitals of the two atoms

the π_{2p_y}-orbital of the molecule

The energy state associated with the bonding π_{2p_y} molecular orbital is less (more stable) than the energy state associated with the $2p_y$ atomic orbitals of the two separate atoms. (Figure 16-4.)

Derivation of π^* Orbitals from p Atomic Orbitals

The wave function for the antibonding $\pi^*{}_{2p_y}$ molecular orbital is derived (as are all antibonding molecular orbitals) from the difference of the wave functions of two atomic orbitals. In this case, it is the difference between the $2p_y$ atomic orbitals of the two atoms:

$$\underset{\text{molecule}}{\psi_{\pi^*{}_{2p_y}}} = \underset{\text{one atom}}{\psi_{2p_y}} - \underset{\text{the other atom}}{\psi_{2p_y}}$$

the $2p_y$-orbitals of the two separate atoms

the $\pi^*{}_{2p_y}$-orbital of the molecule

Clearly, the distribution of electron charge in this $\pi^*{}_{2p_y}$ molecular orbital is counterproductive in forming a chemical bond. The energy state associated with this antibonding orbital is greater (less stable) than the energy state associated with the $2p_y$ atomic orbitals of the two separate atoms. (Figure 16-4). The $\pi^*{}_{2p_y}$ molecular orbital is an antibonding orbital. Note the relation of the energy state associated with the $\pi^*{}_{2p_y}$ molecular orbital to the energy state associated with the $\sigma^*{}_{2p_x}$ molecular orbital (Figure 16-1.)

Relative Energies of σ and σ^*, and π and π^* Orbitals

The generation of the π_{2p_z} and the $\pi^*{}_{2p_z}$ molecular orbitals using the $2p_z$ atomic orbitals of the two atoms is the same as that delineated for the formation of the π_{2p_y} and the $\pi^*{}_{2p_y}$ molecular orbitals using the $2p_y$ atomic orbitals for the two atoms. The energies associated with the π_{2p_y} and the π_{2p_z} molecular orbitals are identical and these two orbitals are degenerate. The energies associated with the $\pi^*_{2p_y}$ and the $\pi^*_{2p_z}$ molecular orbitals are identical and these two orbitals are degenerate. Again, see Figure 16-1 to see the relation of these orbitals to the general scheme of molecular orbitals. The orientation in space of the π_{2p_z} and the $\pi^*_{2p_z}$ molecular orbitals is, of course, not the same as the orientation of the π_{2p_y} and $\pi^*_{2p_y}$ molecular orbitals. In terms of the sketches given for the π_{2p_y} and the $\pi^*_{2p_y}$ molecular orbitals, the regions of high electron probability of the π_{2p_z} and $\pi^*_{2p_z}$ molecular orbitals are above and below the plane of the paper.

With this much background, we can now return to the discussion of the distribution of the 16 electrons of diatomic oxygen in the ground state molecule as presented in Figure 16-2.

The Diatomic Oxygen Molecule

The bonding contribution of the pair of electrons in the σ_{1s} molecular orbital is cancelled by the antibonding character of the pair of electrons in the σ^*_{1s} molecular orbital. The bonding contribution of the pair of electrons in the σ_{2s} molecular orbital is cancelled by the antibonding character of the pair of electrons in the σ^*_{2s} molecular orbital. Consequently, the first eight electrons assigned to the four lowest molecular orbitals in the schematic energy level diagram are ineffective in bonding the two atoms of oxygen in the diatomic molecule—four electrons in bonding orbitals and four electrons in the corresponding antibonding orbitals.

The next six electrons are assigned to bonding molecular orbitals: paired electrons to the σ_{2p_x} molecular orbital, paired electrons to the π_{2p_y} molecular orbital, and paired electrons to the π_{2p_z} molecular orbital.

The last two electrons are assigned to two degenerate antibonding molecular orbitals: one to the $\pi^*_{2p_y}$-orbital and one to the $\pi^*_{2p_z}$-orbital. In the MO approach, it is the degeneracy of these two orbitals that accounts for two unpaired electrons and the paramagnetic property of diatomic oxygen, and it is the antibonding nature of these two orbitals that accounts for the dissociation energy of diatomic oxygen

$$O_2 \longrightarrow O + O \qquad \Delta H = 498 \text{ kilojoules per mole } O_2$$

being so much lower than the dissociation energy of diatomic nitrogen. With only 14 electrons, the $\pi^*_{2p_y}$-orbital and the $\pi^*_{2p_z}$-orbital of diatomic nitrogen are not occupied.

$$N_2 \longrightarrow N + N \qquad \Delta H = 950 \text{ kilojoules per mole } N_2$$

The Double Bond in Diatomic Oxygen

According to the MO approach, the bonding in O_2 is said to be equivalent to a double bond

$$\frac{10 \text{ electrons in bonding orbitals} - 6 \text{ electrons in antibonding orbitals}}{2}$$

The Triple Bond in Diatomic Nitrogen

and the bonding in N_2 is equivalent to a triple bond [(10 − 4)/2]. The triple bond for diatomic nitrogen is consistent with the Lewis structure and also the valence bond approach.

Of the two primary components of air, diatomic oxygen is by far the more reactive. Reactions of diatomic oxygen with both elements and compounds are very common and we are dependent upon photosynthesis to continuously replenish the world supply of diatomic oxygen. Reactions of diatomic nitrogen with elements and compounds are very rare and tremendous investments in equipment and energy are made annually to convert diatomic nitrogen into nitrogen fertilizers to promote the growth of plants. According to the MO approach to the structure of these two diatomic molecules, the differences in the reactivity of O_2 and N_2 molecules are attributed to the two unpaired electrons in antibonding orbitals in the oxygen molecule—one in the $\pi^*_{2p_y}$ molecular orbital and one in the $\pi^*_{2p_z}$ molecular orbital—and the lack of electrons in these antibonding orbitals in nitrogen molecules. Nitrogen molecules are diamagnetic and have the exceptionally high dissociation energy.

Figure 6-4 (p. 109) is schematic in that it displays the sequence of the atomic orbitals according to energy for all elements. For atoms of all elements, the lowest energy is associated with the 1*s* atomic orbital, but the numerical value of that energy is a characteristic of the element. For example, the energy associated with the 1*s* atomic orbital of oxygen is lower than the energy associated with the 1*s* atomic orbital of nitrogen (due to the larger nuclear charge of +8 for the oxygen atom as compared to the nuclear charge of +7 for the nitrogen atom). Likewise, Figure 16-1 is schematic since it is derived from the wave functions of the atomic orbitals. For all homonuclear diatomic molecules, the σ_{1s} molecular orbital has the lowest energy, but the numerical value of the energy is again characteristic of the element. Figure 16-1 is also schematic in that the separations between energy levels have been distorted for ease of presentation. As energy increases, the separation between energy states becomes smaller and smaller.

Heteronuclear Diatomic Molecules

In Figure 16-3, for homonuclear diatomic molecules, the energy associated with the 1*s* atomic orbital for atom 1 is the same as the energy associated with the 1*s* atomic orbital of atom 2. In Figure 16-4, for homonuclear diatomic molecules, the energy associated with the 2*p* atomic orbitals of atom 1 is the same as the energy associated with the 2*p* atomic orbitals of atom 2. For heteronuclear diatomic molecules such as NO and CO, the energies associated with the 1*s*, the 2*s*, and the 2*p* atomic orbitals of the different elements are different. Consequently, Figure 16-3 and Figure 16-4 are not strictly applicable to heteronuclear diatomic molecules. However, for molecules such as NO and CO, where the atomic numbers of the two elements involved are not greatly different, the energy level diagram of Figure 16-1 can be used as an approximation. Figure 16-1 is not applicable to molecules such as HCl. Scramble Exercises 2 and 3 are an exploration of the molecular orbital approach to carbon monoxide, CO; nitrogen monoxide (also called nitric oxide), NO; and the cyanide ion, CN^{1-}.

Excited Diatomic Molecules

Figure 16-1 is applicable to excited diatomic molecules in the same sense that Figure 6-4 is applicable to excited atoms. In an excited state, one or more electrons may occupy higher molecular orbitals, leaving vacancies in molecular orbitals that are filled in the ground state molecule. The absorption spectra of ground state molecules and the emission spectra of excited state molecules is a characteristic of the energy separations between molecular orbitals in the same sense that the absorption spectrum

Emission Spectra of Excited Diatomic Molecules

of ground state atoms and the emission spectrum of excited atoms is a characteristic of the energy separation between atomic orbitals. However, both the absorption spectrum and the emission spectrum of molecules are complicated by simultaneous transitions between vibrational energy states of molecules and simultaneous transitions between rotational energy states of molecules. Atoms absorb or emit photons of specific energies—narrow bands; molecules absorb or emit photons over a wide range of energies—broad bands. The broad bands encompass photons of energies corresponding to the sum of the change in electronic energy states plus the much smaller change in vibrational energy states plus the change in the much, much smaller rotational energy states.

All of the preceding discussion of the molecular orbital approach to chemical bonding has been limited to diatomic molecules. In principle at least, the approach can be extended to larger molecules. In practice, the wave functions for molecular orbitals become increasingly complex as the number of atoms in the molecule is increased, and approximations must be introduced. The great contribution of the molecular orbital approach is the concept of quantized energy states of molecules determined by the distribution of electrons in molecular orbitals. The great contribution of the valence bond approach to chemical bonding is the concept of directed bonds. MO is an approach to understanding energy changes. VB is an approach to an understanding of the three-dimensional distribution of atoms in molecules and ions. Chemists use the approach that is helpful in addressing the problem at hand—and, in many cases, use both approaches in addressing the same problem.

Reactions of Diatomic Oxygen with Elements

Very Reactive Metals

Calcium, an alkaline earth metal, reacts with oxygen gas, O_2, to form calcium oxide, CaO, a white, crystalline, ionic compound that melts at 2580 °C:

$$2\,Ca(s) + O_2(g) \longrightarrow \underset{\text{calcium oxide}}{2\,CaO(s)}$$

At room temperature, the metallic sheen of a freshly cut surface of calcium quickly becomes clouded in air by the formation of a coating of calcium oxide and the calcium is partially protected from further reaction with oxygen. At elevated temperatures, the reaction with oxygen proceeds rapidly with the evolution of heat and light. To protect calcium from reaction with the oxygen of the air, the metal is stored under oil.

Relatively Unreactive Metals

The alkali metals and the alkaline earth metals react vigorously with diatomic oxygen. Other metals are not as reactive and there is a cluster of ten metals in the Periodic Table that are particularly unreactive. These elements are designated as the platinum metals, the coin metals, and mercury.

			29 COPPER **Cu** 63.54	
44 RUTHENIUM **Ru** 101.07	45 RHODIUM **Rh** 102.90	46 PALLADIUM **Pd** 106.42	47 SILVER **Ag** 107.87	
76 OSMIUM **Os** 190.2	77 IRIDIUM **Ir** 192.22	78 PLATINUM **Pt** 195.08	79 GOLD **Au** 196.97	80 MERCURY **Hg** 200.59
platinum metals			coin metals	mercury

The atoms of all of these elements have *d*-orbital electrons and the elements 76 → 80 also have *f*-orbital electrons. Several of these elements have unusual properties. Copper and gold are colored and mercury is the only metal that is a liquid at room temperature. The economic value of this cluster of elements is related to their metallic luster, their high electrical conductivity, their limited abundance, their limited chemical reactivity, and, in a few cases, their colors. The tarnishes that slowly form on copper and silver are usually the oxides CuO and Ag_2O. Under some conditions, the sulfides may also be formed. The reactivities of other metals with diatomic oxygen lie between those of calcium and this cluster of ten metals.

Reactive Nonmetals

Sulfur, a nonmetal, exists in two crystalline forms. Both are yellow and both melt below 120 °C. The element is extensively used in commercial chemical processes and it is frequently shipped by rail in gondolas, cars of the type used to transport coal. Heated well above the melting point, sulfur vapor burns in air with the evolution of heat and light:

$$S(g) + O_2(g) \longrightarrow \underset{\text{sulfur dioxide}}{SO_2(g)}$$

The product, sulfur dioxide, is a gas at room temperature and is thus clearly a covalent compound. In general, the oxides of nonmetals are covalent compounds and, under room conditions, the oxides of nonmetals exist as gases, as liquids, or as solids that have low melting points. One striking exception to this is silicon dioxide (commonly known as quartz), which has the empirical formula SiO_2 and melts above 1600 °C. Silicon dioxide forms a molecular crystal with each silicon atom *sigma*-bonded to four atoms of oxygen and each oxygen atom *sigma*-bonded to two atoms of oxygen. (Gratuitous Information 10-2 and Figure 21-2.)

Oxidation–Reduction Reactions

The oxidation of calcium with diatomic oxygen and the oxidation of sulfur with diatomic oxygen will be used to explore two general concepts of **oxidation–reduction reactions,** frequently called **redox reactions.**

Since the products of the oxidation of calcium with diatomic oxygen are two ions, the calcium ion Ca^{2+} and the oxide ion O^{2-}, the reaction of calcium with oxygen can be analyzed in terms of two half-reactions. One half-reaction has to do with the oxidation of calcium atoms, Ca, to calcium ions, Ca^{2+}, and the other half-reaction has to do with the reduction of the diatomic oxygen molecules, O_2, to oxide ions, O^{2-}.

Half-Reaction of Oxidation

In the oxidation of calcium atoms to calcium ions, electrons are also a product of the half-reaction:

$$Ca \longrightarrow Ca^{2+} + 2\,e^{1-} \qquad \text{half-reaction of oxidation}$$

This is consistent with the electron structure of the calcium atom, (Ar) $4s^2$, and the electron structure of the calcium ion, (Ar). Note that the net charges on the two sides of this equation are balanced (zero for this particular half-reaction).

In the conversion of diatomic oxygen, O_2, to two oxide ions, O^{2-}, four electrons are reactants:

Half-Reaction of Reduction

$$O_2 + 4\,e^{1-} \longrightarrow 2\,O^{2-} \qquad \text{half-reaction of reduction}$$

The source of these electrons is the half-reaction of oxidation of calcium atoms. Note again that the net charges on the two sides of this equation for the half-reaction of reduction are balanced (-4 on both sides for this particular half-reaction of reduction).

The Total Reaction

The equation for the total reaction of oxidation–reduction is the sum of the equation for the half-reaction of oxidation and the equation for the half-reaction of reduction—with appropriate adjustment of quantities to eliminate electrons from both sides of the equation for the total reaction:

$$\begin{array}{rcl} 2 \times (Ca & \longrightarrow & Ca^{2+} + 2\,e^{1-}) \\ O_2 + 4\,e^{1-} & \longrightarrow & 2\,O^{2-} \\ \hline 2\,Ca + O_2 & \longrightarrow & 2\,Ca^{2+} + 2\,O^{2-} \end{array}$$

Rewritten to show phases, the equation for the total reaction is

$$2\,Ca(s) + O_2(s) \longrightarrow 2\,CaO(s)$$

In this reaction, calcium metal, Ca, has been oxidized to calcium ion, Ca^{2+}, by diatomic oxygen gas. Diatomic oxygen gas has been reduced to oxide ions, O^{2-}, by metallic calcium. In being oxidized, calcium atoms have lost electrons. In being reduced, diatomic oxygen molecules have gained electrons. This oxidation–reduction reaction has involved the transfer of electrons from calcium atoms to diatomic oxygen molecules. The reaction of sulfur with diatomic oxygen forms a covalent product, and another approach will be made to the discussion of this reaction later.

OZONE

The air taken into the cabins of planes flying at 35,000 feet contains sufficient ozone to cause eye irritation and headaches, and commercial airlines use catalysts to convert ozone to diatomic oxygen. Ozone, O_3, is a triatomic oxygen molecule. It is a blue gas and has a characteristic odor that is frequently evident around electric generators, electric motors, and other equipment that produces electrostatic discharges.

A molecule of ozone is formed by the dissociation of a diatomic oxygen molecule into oxygen atoms:

$$O_2(g) \longrightarrow 2\,O(g)$$

and the addition of an atom of oxygen to a diatomic molecule of oxygen:

$$O(g) + O_2(g) \longrightarrow \underset{\text{ozone}}{O_3(g)}$$

The first reaction is endothermic and utilizes the energy of the electrostatic discharge or the energy of photons of ultraviolet light. The second reaction is exothermic and the O_3 molecule is "too hot" to stay together. The only ozone molecules that survive are the O_3 molecules that transfer part of their energy to another molecule—any other molecule—by collision before the O_3 unit flies apart in the normal process of vibration.

Ozone gas is a powerful oxidizing agent and significantly reduces the life time of rubber, many fabrics, and many dyes. Ozone, in an acidic aqueous solution, is an exceptionally strong oxidizing agent:

$$O_3 + 2\,H^{1+} + 2\,e^{1-} \longrightarrow O_2 + H_2O$$

half-reaction of reduction

In France and some other European countries, ozone is used in place of diatomic chlorine in the purification of water. Chlorine is also an oxidizing agent.

The ozone in the stratosphere and the upper atmosphere absorbs ultraviolet light of wavelengths shorter than 300 nanometers and thus protects the earth from very high energy ultraviolet light destructive to cells of both plants and animals. In recent years, there has been considerable concern about the destruction of this protective layer of ozone brought about by the use of fully halogenated chlorofluoroalkanes as the propellants in sprays cans. These compounds are either one-carbon or two-carbon compounds in which all the hydrogen atoms of methane or ethane are replaced by atoms of chlorine and atoms of fluorine. As a group, these compounds, commonly referred to as fluorocarbons, are ideal for consumer product use. Their physical properties are excellent for use as propellants. They are noncombustible and are extremely stable in the environment—consequently not a fire hazard and nontoxic. It is exactly that stability that leads to the ozone problem. The molecules do not degrade in our environment and some of the molecules diffuse eventually into the stratosphere where high-energy ultraviolet photons from the sun lead to their dissociation into free radicals (odd electron atoms or groups of atoms). These free radicals react with ozone. The free radical chemistry of the stratosphere is extremely complex and it is still not clear to what degree the fluorocarbons would destroy the ozone layer and to what degree the increased radiation reaching the earth would increase the incidence of skin cancer and otherwise negatively affect life on this planet. There is a significant time lag between the time the little button on the spray can is pushed and the time the fluorocarbons reach the stratosphere. Consequently, it was deemed prudent to regulate the use of fluorocarbons in spray cans in this country and to immediately discourage their use in spray cans throughout the world. Some of the propellants currently being used in spray cans are combustible and thus a fire hazard. The regulation of the use of the fluorocarbons in spray cans may be the first regulation in this country based upon the predictions of a model that had not been experimentally tested. There is now some experimental evidence that oxides of nitrogen, NO_x, discharged into the atmosphere from furnaces and internal combustion engines, may have an effect on the protective ozone layer in the upper atmosphere that is equal to or greater than that of the fluorocarbons.

O_2 and O_3 are allotropic forms of oxygen. Many elements occur in more than one form—frequently in different crystalline forms. Diamond and graphite are allotropic forms of the element carbon.

Generalized Concepts of Oxidation–Reduction Reactions

In a very similar fashion, calcium burns in chlorine gas, Cl_2, to give calcium chloride, $CaCl_2$, a white, crystalline, ionic compound that melts at 772 °C:

A Redox Reaction without Oxygen

$$Ca(s) + Cl_2(g) \longrightarrow CaCl_2(s)$$

Here again, equations can be written for the half-reaction converting calcium atoms to calcium ions:

$$Ca \longrightarrow Ca^{2+} + 2\,e^{1-}$$

and also for the half-reaction converting diatomic chlorine, Cl_2, into chloride ions, Cl^{1-}:

16-2 *Gratuitous Information*

FLAMES, BIOLUMINESCENCE, AND CHEMILUMINESCENCE

Flames produced in the complete combustion of compounds of carbon and hydrogen (hydrocarbons) and also of compounds of carbon, hydrogen, and oxygen (such as alcohols and ethers) have a characteristic blue color. Spectrographic studies of the emission spectra of these flames reveal a number of bands not only in the visible region of the spectrum but also in the ultraviolet and the infrared regions of the electromagnetic spectrum. The most prominent bands in the visible region of the spectrum are attributed to the emission of excited C_2 molecules formed in the high-temperature portion of the flame. The system of bands arises from transitions of electrons in excited C_2 molecules from high-energy molecular orbitals to unfilled lower-energy molecular orbitals in much the same way that the emission lines of helium and neon gases in electric discharge tubes arise from the transition of electrons from high-energy atomic orbitals to lower-energy atomic orbitals.

The emission spectrum of the diatomic C_2 molecule is, however, more complex. In the transition of a high energy state molecule to a lower energy state molecule, two changes in the molecule may be involved in addition to the transition of an electron from one molecular orbital to a lower molecular orbital. These are a change from one quantized vibrational level to another and/or a change from one quantized rotational level to another. The energy of the photon emitted is equal to the sum of the energy change in the electron transition, the energy change in the vibrational transition, and the energy change in the rotational transition. Consequently, the number of lines that appear in the emission spectrum of the diatomic molecule is increased by these quantized vibrational and rotational transitions. The energy changes associated with vibrational transitions are much less than the energy changes associated with the electron transitions, and the energy changes associated with rotational transitions are much less than the energy changes associated with vibrational transitions. It is the energy of the electron transition between the molecular orbitals of C_2 that places the emission spectrum of C_2 in the visible region of the spectrum. It is the energies of the vibrational and rotational transitions that generate clusters of closely spaced lines in the emission spectrum of the diatomic molecule. The experimental reality is that the capabilities of the equipment may not resolve these closely spaced lines. And a cluster of these closely spaced lines may appear as a broad line or band, and the emission spectrum of a diatomic molecule is frequently referred to as a band spectrum—rather than as a line spectrum.

The complexity of the combustion process is indicated by the emission spectra of other molecules, such as C_3 and CH, in other regions of the total spectrum of the flame. Even the emission bands of the CN molecule may be detected in the ultraviolet region of the spectrum if combustion takes place with air. All of these rather strange molecules react with oxygen in the formation of the final products of complete combustion: H_2O and CO_2. In the case of CN, oxides of nitrogen are also formed.

$$Cl_2 + 2\ e^{1-} \longrightarrow 2\ Cl^{1-}$$

This is again an electron transfer reaction. Calcium atoms donate the electrons, and diatomic chlorine molecules accept the electrons. The sum of the equations for the two half-reactions is

$$Ca + Cl_2 \longrightarrow Ca^{2+} + 2\ Cl^{1-}$$

or, rewritten to show phases,

$$Ca(s) + Cl_2(g) \longrightarrow CaCl_2(s)$$

Very similar reactions involving the transfer of electrons were explored in Chapter 5 in a discussion of sodium chloride.

Oxidation reactions are also key to the emission of photons by those biological organisms that glow in the dark. Our emotional reaction to this bioluminescence can be very strong. We take pleasure in the glow of protozoans in tropical seas and the flicker of fireflies on a summer evening. We find the glow of fox fire on our shoes and the hooves of horses to be eerie in a passage through damp decaying vegetation on the floor of a forest, and we shrink from the glow of a dead fish. Conditioned by the emission of light from high-energy sources like incandescent solids, flames, and electric discharge tubes, we find "cold light" to be strange indeed. Actually, the phenomenon is quite common but the intensity of the light emitted can frequently be very low and consequently infrequently observed.

The emission of photons by fireflies has been extensively studied and is discussed in terms of the oxidation of luciferin by diatomic oxygen in the presence of an enzyme, luciferase. The oxidation being referred to here is not the complete high-temperature oxidation of luciferin to carbon dioxide and water but the very small changes in molecular structure at ambient temperatures in which two electrons per molecule are transferred from the luciferin molecule to diatomic oxygen (lost electrons oxidation). The molecular orbital sources of the electrons transferred result in the products of oxidation being excited molecules with vacancies in molecular orbitals of lower energy than other occupied molecular orbitals. The energy changes associated with the conversion of these excited states to the ground state determine the energy of photons emitted.

Don't rush off to find the formula of luciferin and the formula of its oxidation product. There are many luciferins. Even for fireflies, the formula of luciferin is species specific. The term luciferin is used to designate any compound that emits visible light during enzymatic oxidation by diatomic oxygen in a biological system usually in the temperature range of 20 to 30 °C. The structures of a number of luciferins have been determined and some have been synthesized. Their molecular weights are in the general range of 200 to 400 and there is variability in the grouping of atoms within the molecules and even variability in the elements that make up the various luciferins.

Fox fire is due to the bioluminescence of fungi and bacteria that inhabit the decaying vegetation and not to the decaying material itself. Likewise, the bioluminescence of decaying meat and fish is attributed to bacteria. There are, of course, live fish that have highly developed bioluminescence systems that are thought to be significant factors in their life styles and in their survival.

In addition to bioluminescence, there are other oxidation reactions that emit "cold light" completely independent of biological systems or enzymes. Examples of this phenomenon, chemiluminescence, are considered by most chemists to be intriguing laboratory curiosities. However, you may carry a "light stick" in your car.

It is now customary to consider all electron transfer reactions as oxidation–reduction reactions regardless of whether oxygen is or is not a reactant. For the reaction of calcium metal with chlorine gas, the calcium atoms are oxidized to calcium ions by diatomic chlorine molecules and diatomic chlorine molecules are reduced to chloride ions by calcium atoms.

Redox Reactions Defined in Terms of Electron Transfer

In this generalized concept of **oxidation–reduction reactions** as **electron transfer reactions,** oxidation involves the loss of electrons and reduction involves the gain of electrons:

$$Ca \longrightarrow Ca^{2+} + 2\,e^{1-} \qquad \text{half-reaction of oxidation}$$
$$Cl_2 + 2\,e^{1-} \longrightarrow 2\,Cl^{1-} \qquad \text{half-reaction of reduction}$$

The concept of redox reactions as electron transfer reactions is straightforward. If you find the terminology elusive, join the many generations of students who have preceded you in resorting to "LEO the GERman" or "LEO the lion went GER." Loss of Electrons Oxidation. Gain of Electrons Reduction. In the process of acting as an oxidizing agent, the oxidizing agent is reduced. (Chlorine, the oxiding agent, is reduced to chloride ion.) In the process of acting as a reducing agent, the reducing agent is oxidized. (Calcium, the reducing agent, is oxidized to calcium ion.) This extension of the term oxidation to chemical reactions that do not involve oxygen is an example of the perpetuation of a term to encompass concepts quite different from the initial meaning of the term.

Question: Is the treatment of redox reactions in terms of two half-reactions with a transfer of electrons from one half to the other half an intellectual exercise, or is there experimental evidence that these half-reactions can be separated and the transfer of electrons from one to the other demonstrated?

Answer: There is abundant experimental evidence to demonstrate that half-reactions can take place at different sites with transfer of electrons from one site to the other through a conducting wire. Such an experimental demonstration could be carried out for the reaction of calcium and chlorine gas. However, the experiment would have to be carried out at a temperature above the melting point of calcium chloride and provision would have to be made to control the very reactive chlorine gas. We shall discuss instead another redox reaction, the oxidation of metallic copper, Cu, by silver ion, Ag^{1+}. This reaction can be carried out at room temperature using aqueous solutions. The redox reaction itself can be demonstrated by placing a freshly cleaned piece of copper in a dilute aqueous solution of silver nitrate. Gradually, the colorless aqueous solution becomes blue as copper(II) ions, Cu^{2+}, are formed, and the yellow copper surface becomes covered by a coating of metallic silver. The equations for the half-reactions are

The Copper–Silver Ion Electrochemical Cell

half-reaction of oxidation $\qquad Cu(s) \longrightarrow \underset{\substack{\text{copper(II) ion} \\ \text{blue}}}{Cu^{2+}(aq)} + 2\,e^{1-}$

half-reaction of reduction $\qquad \underset{\substack{\text{silver ion} \\ \text{colorless}}}{Ag^{1+}(aq)} + e^{1-} \longrightarrow Ag(s)$

total reaction $$Cu(s) + 2\,Ag^{1+}(aq) \longrightarrow Cu^{2+}(aq) + 2\,Ag(s)$$

Note the adjustment of quantities to eliminate electrons from the equation for the total reaction.

This reaction, which can be carried out in a beaker, gives some indication of the products and the half-reactions but it provides no evidence of the transfer of electrons. To demonstrate the flow of electrons, the metallic copper must not come in contact with the aqueous solution of silver nitrate. A reaction vessel with two compartments is required. But the design of the reaction vessel must be such that ions can move from one compartment to the other to maintain a net ionic charge of zero in all small volumes of the solutions. One way to structure such a reaction vessel is to use a V-shaped reaction vessel to which a warm aqueous solution of gelatin and potassium nitrate is added in sufficient quantity to isolate the two side arms. On cooling, the gel sets and forms a barrier to the flow of liquid from one arm of the vessel to the other.

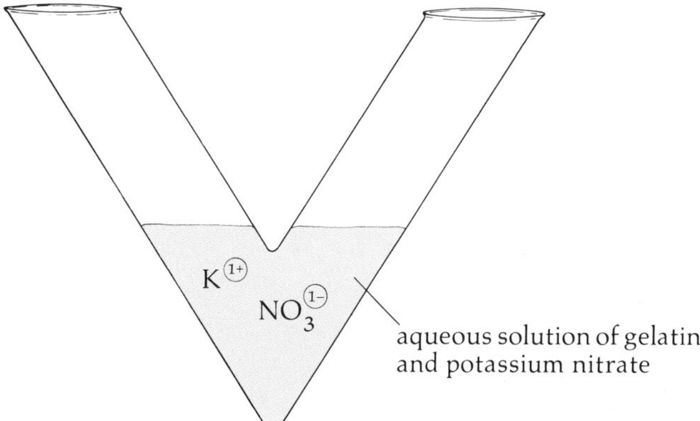

A dilute aqueous solution of copper(II) nitrate (blue) is added to the left-hand arm and about an equal volume of diluted aqueous silver nitrate solution (colorless) is added to the right-hand arm.

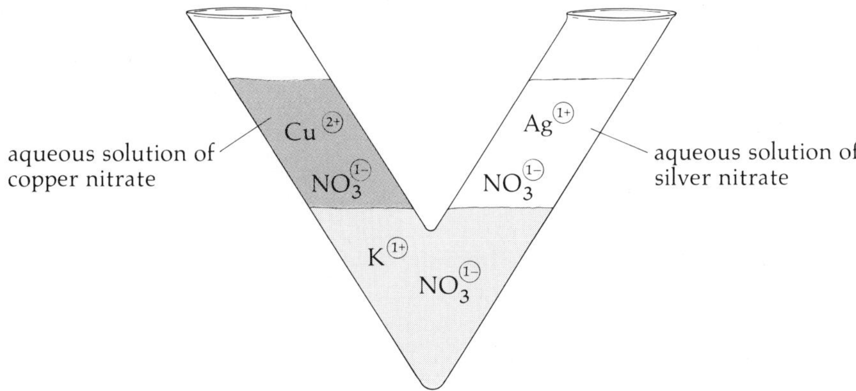

A freshly cleaned copper electrode (a strip of copper metal) that has been connected to an electrical lead (a piece of wire) about 20 cm long is mounted so that part of the copper electrode is immersed in the aqueous solution of copper(II) nitrate. In a similar manner, a freshly cleaned silver electrode connected to an electrical lead is mounted so that part of the silver dips into the silver nitrate solution. If the free ends of the electrical leads are touched together, the electrical circuit is completed and electrons would be able to flow through the wire from the copper electrode to the silver electrode or from the silver electrode to the copper electrode, but there is nothing to indicate the flow of electrons. Instead of shorting this electrochemical cell by touching the wires together, the two leads can be connected to a detector. The instrument usually chosen indicates the direction the electrons tend to flow and also measures the electric potential (the voltage) produced by the electrochemical cell. Such an indicator shows that the flow of electrons in the wire is from the copper electrode to the silver electrode and that the electrical potential for this cell is of the order of 0.5 volts. The exact potential depends upon the temperature of the cell and the concentrations of the solutions.

The Reactions

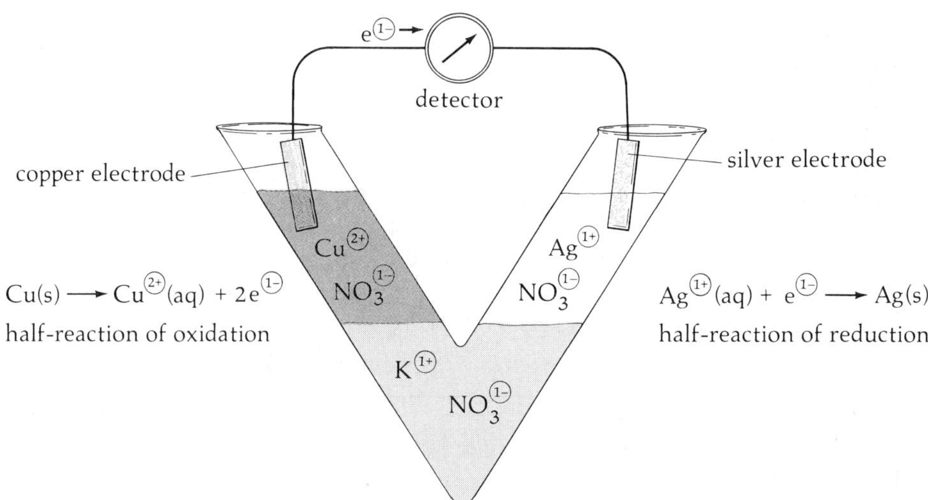

In this electrochemical cell, a number of processes proceed simultaneously:

Simultaneous Processes

- Atoms of the copper electrode become copper(II) ions, leaving the extra electrons on the copper electrode. This introduces copper(II) ion into the aqueous solution around the copper electrode and makes the copper electrode negative. As the reaction proceeds, the solution around the copper electrode becomes increasingly blue.
- Silver ions pick up electrons from the silver electrode and are deposited as silver atoms on the silver electrode. This decreases the concentration of silver ions in the aqueous solution around the silver electrode and makes the silver electrode positive. As the reaction proceeds, the solution around the electrode continues to be colorless.

- Electrons flow through the wire from the copper electrode to the silver electrode.
- All ions move in all of the aqueous media (the copper(II) nitrate solution, the potassium nitrate gel, and the silver nitrate solution) to balance off the increase in positive ion concentration in the region of the copper electrode and the decrease in positive ion concentration in the region of the silver electrode. All positive ions move toward the silver electrode and all negative ions move toward the copper electrode to the degree necessary for the net ionic charge in all small volumes of the solutions to be maintained at zero. This movement of ions is known as the **ionic conductance** of the cell.

Electrochemical cells of this type not only support the concept of half-reactions and the transfer of electrons but also provide a mechanism for comparing the relative tendencies of metallic copper to become copper ion:

$$Cu \longrightarrow Cu^{2+} + 2\ e^{1-}$$

and metallic silver to become silver ion:

$$Ag \longrightarrow Ag^{1+} + e^{1-}$$

If metallic silver had the greater tendency to give away electrons and become silver ion, all of the reactions would have been reversed and the silver electrode would have become the negative electrode.

The magnitude of the electrical potential produced by the electrochemical cell is a quantitative measure of the relative tendencies of the two metals to become positive ions by the transfer of electrons, a quantitative measure of the relative tendencies of the half-reactions of oxidation to occur. By measuring the electric potentials of a number of electrochemical cells, it is possible to build up tables that do for half-reactions of either oxidation or reduction what Tables 13-3, 14-2, and 14-3 do for Brönsted acids or bases. All compare the tendencies of a particular type of reaction to occur.

Batteries

In electrochemical cells, chemical energy is converted to electrical energy and it is electrochemical cells that are the basis of batteries and dry cells. As current is withdrawn from a battery, the initial reactants are converted to products in the two half-reactions and the battery is said to run down. If the half-reactions are reversible and none of the products have escaped, the battery can be recharged by applying an electrical potential to the electrodes that reverses the flow of electrons and reverses the chemical half-reactions at both electrodes. In the process of recharging, the initial chemical composition of the cell is restored. The goal of the engineer is to design and produce batteries that are long-lasting, rechargeable, light in weight, small in volume, safe, and inexpensive through an ingenious choice of half-reactions of oxidation and reduction and the development of a clever design for the interior and the packaging of the battery.

Redox Reactions That Cannot Be Explored Experimentally in Terms of Half-Reactions

Many redox reactions cannot be experimentally approached in terms of two half-reactions. In the oxidation of sulfur by diatomic oxygen, there is a single product: SO_2, a covalent molecule. To endeavor to discuss the reaction of diatomic oxygen with sulfur in terms of half-reactions and the transfer of electrons has no meaning.

Oxidation States of Sulfur

A rather arbitrary system of oxidation numbers has developed as an aid to the classification of the oxidation states of elements in molecules and ions. Table 16-1 classifies some sulfur-containing molecules and ions according to the oxidation number of the sulfur in the various molecules and ions.

To convert SO_2 (oxidation number +4) to H_2SO_3 (oxidation number also +4) does not involve a redox reaction:

$$SO_2 + H_2O \longrightarrow H_2SO_3$$

To convert SO_2 (oxidation number +4) to SO_3 (oxidation number +6) requires an oxidizing agent such as diatomic oxygen:

$$2\ SO_2 + O_2 \longrightarrow 2\ SO_3$$

In any redox reaction, the total increase in oxidation number must equal the total decrease in oxidation number.

Oxidation States of Nitrogen

Table 16-2 classifies a number of nitrogen-containing molecules and ions according to the oxidation number of nitrogen in various molecules or ions.

In general, it is the nonmetals other than the inert gases that form these extensive arrays of oxygen- and hydrogen-containing molecules and ions.

Most oxidation numbers are integers, some are zero, some are positive, and some are negative. Oxidation numbers are not ionic charges, although the rules for counting oxidation numbers have been defined in such a way that the oxidation number of an element in a simple ion such as copper(II) ion, Cu^{2+}, is numerically equal to the charge on the ion. Take care not to get trapped into thinking of oxidation numbers as actual charges on atoms in general.

A set of rules is used to count oxidation numbers. Some of these rules:

Counting Oxidation Numbers

- All free elements—such as Cu, Ag, O_2, Cl_2, and Ne—are assigned an oxidation number of zero regardless of whether the molecule is monoatomic or polyatomic.
- The oxygen atom is assigned an oxidation number −2 in all compounds except peroxides. In peroxides, such as hydrogen peroxide, H_2O_2, the oxygen atom is assigned an oxidation number of −1.

Table 16-1

OXIDATION STATES OF SULFUR IN COMMON SULFUR-CONTAINING MOLECULES AND IONS

oxidation number sulfur	formulas of molecules and ions			
+6	SO_3	H_2SO_4	HSO_4^{1-}	SO_4^{2-}
+4	SO_2	H_2SO_3	HSO_3^{1-}	SO_3^{2-}
0	S (usually S_8 in solid sulfur)			
−2	H_2S	HS^{1-}	S^{2-}	

Table 16-2

OXIDATION STATES OF NITROGEN IN COMMON NITROGEN-CONTAINING MOLECULES AND IONS

oxidation number nitrogen	formulas of molecules and ions		
+5	N_2O_5	HNO_3	NO_3^{1-}
+4	NO_2	N_2O_4	
+3	N_2O_3	HNO_2	NO_2^{1-}
+2	NO		
+1	N_2O		
0	N_2		
−3	NH_3	NH_4^{1+}	

- Hydrogen atoms are assigned an oxidation number of +1 in all compounds except hydrides. In hydrides, such as sodium hydride, NaH, hydrogen is assigned an oxidation number of −1.
- The sum of the oxidation numbers of all of the atoms in molecules of compounds must equal zero. For example, in sulfurous acid, H_2SO_3, the oxidation number for the sulfur atom must be +4.

$$2 \times (+1) + S + 3 \times (-2) = 0$$
$$2 + S - 6 = 0$$
$$S = +4$$

- The sum of the oxidation numbers of all atoms in an ion must equal the charge on the ion. For example, the oxidation number of the nitrogen atom in the nitrate ion, NO_3^{1-}, is +5.

$$N + 3(-2) = -1$$
$$N + (-6) = -1$$
$$N = +5$$

The formulas of some compounds seem to indicate oxidation numbers that are not integers—for example, the oxide of iron, Fe_3O_4, formed upon burning iron, in the combustion of iron, at high temperatures.

$$3 \times Fe + 4 \times (-2) = 0$$
$$3\ Fe = 8$$
$$Fe = 2\tfrac{2}{3}$$

There are two other oxides of iron: FeO and Fe_2O_3. The oxidation number of iron in the first is +2 and in the second +3. Another rationalization of the formula F_3O_4

is that the oxidation number of one atom of iron is +2 and the oxidation number of two atoms of iron is +3.

$$1 \times (+2) + 2 \times (+3) + 4 \times (-2) = 0$$
$$2 \quad + \quad 6 \quad - \quad 8 \quad = 0$$

In a redox reaction, some atoms undergo an increase in oxidation number (oxidation) and other atoms undergo a decrease in oxidation number (reduction). In the reaction

$$2\,S + O_2 \longrightarrow 2\,SO_2$$

the transition in oxidation number of sulfur is from 0 to +4 (oxidation) per atom of sulfur and the transition of oxidation number for oxygen is from 0 to −2 per atom of oxygen (reduction). The sulfur is oxidized and the oxygen is reduced. For the total reaction, the net change in oxidation numbers is zero.

In the discussion of the reactions of ions in aqueous solutions in Chapter 15, it was pointed out that, in addition to the two types of reactions discussed there—Brönsted acid–base reactions and precipitation reactions—a third type of reaction of ions in aqueous solutions involved oxidation–reduction reactions. In Brönsted acid–base reactions and in combination of ions to form a precipitate, there are no changes in oxidation states. Much of the chemistry of the elements and their compounds involves changes in oxidation numbers, and the class of reactions known as **oxidation–reduction** reactions or **redox** reactions is extremely large.

Many of the redox reactions that occur in aqueous solutions can be treated in terms of half-reactions and the transfer of electrons. One of these, the reaction of a piece of metallic copper in an aqueous solution of silver nitrate, has been discussed already in considerable detail in this chapter.

Rusting of Iron

A series of reactions involved in the rusting of iron to form hydrated iron(III) oxide, $Fe_2O_3 \cdot x\,H_2O$, will now be discussed. Part of these reactions are redox reactions, some are not. Frequently, these reactions proceed simultaneously. In this exploration, coordination complexes of the ions of iron will be ignored. There are enough complications without them.

Metallic iron does not react with dry diatomic oxygen. It does react with diatomic oxygen in the presence of water. The first reaction is the oxidation of iron to iron(II) ion, Fe^{2+}, and the reduction of diatomic oxygen to hydroxide ion:

$2 \times (Fe(s)$	$\longrightarrow$	$Fe^{2+}(aq) + 2\,e^{1-})$	oxidation
$O_2(aq) + 2\,H_2O(l) + 4\,e^{1-}$	$\longrightarrow$	$4\,OH^{1-}(aq)$	reduction
$2\,Fe(s) + O_2(aq) + 2\,H_2O(l)$	$\longrightarrow$	$4\,OH^{1-}(aq) + 2\,Fe^{2+}(aq)$	total

Note: The half-reaction of reduction

$$O_2 + 2\,H_2O + 4\,e^{1-} \longrightarrow 4\,OH^{1-}$$

may not seem quite so complex if you think of it in two parts—the reduction of diatomic oxygen, O_2, to oxide ions, O^{2-}:

$$O_2 + 4\,e^{1-} \longrightarrow 2\,O^{2-} \qquad \text{reduction}$$

and the Brönsted acid–base reaction of oxide ion, O^{2-}, with water:

$$\underset{\text{base}}{O^{2-}} + \underset{\text{acid}}{H_2O} \longrightarrow OH^{1-} + OH^{1-}$$

Note that, in this reaction, the conjugate acid of the oxide ion is the hydroxide ion, OH^{1-}, and the conjugate base of water is also the hydroxide ion. The oxide ion, $[:\ddot{O}:]^{2-}$, is an extremely strong Brönsted base, and the oxide ion is never a product in the presence of water. Consequently, it is reasonable to expect the product of the reduction of diatomic oxygen in the presence of water to be hydroxide ions, not oxide ions. Note that the sum of these two reactions is $O_2 + 2H_2O + 4e^{1-} \longrightarrow 4\,OH^{1-}$.

Gratuitous Information 16-3

RELATION OF OXIDES OF METALS AND NONMETALS TO BRÖNSTED ACIDS AND BASES

If the oxide of a metal, such as calcium oxide, CaO, dissolves in water, the oxide ion, O^{2-}, reacts with water

$$\underset{\text{base}}{O^{2-}} + \underset{\text{acid}}{H_2O} \longrightarrow \underset{\text{base}}{OH^{1-}} + \underset{\text{acid}}{OH^{1-}}$$

and the pH of the aqueous solution is more than 7.00. (Oxides of metals such as tin and silver do not dissolve in water and the pH of the water remains unchanged.)

In the above reaction, the oxide ion, a Brönsted base, accepts a proton to become the hydroxide ion, OH^{1-}, the conjugate acid. It seems a bit strange to call the hydroxide ion a Brönsted acid since we are so accustomed to thinking of the hydroxide ion as a Brönsted base. However, the hydroxide ion does have a proton that can be transferred to a strong Brönsted base. The problem is that there are very few Brönsted bases strong enough for this transfer to occur to a significant degree and we cease to think of the hydroxide ion as a Brönsted acid. An aqueous solution of an oxide of a metal has a pH that is greater than 7.

If the oxides of nonmetals such as sulfur dioxide, SO_2, and phosphorus pentaoxide, P_2O_5, dissolve in water, these oxides react with water to form Brönsted acids

$$SO_2 + H_2O \longrightarrow \underset{\text{sulfurous acid}}{H_2SO_3}$$

$$\underset{\text{acid}}{H_2SO_3} + \underset{\text{base}}{H_2O} \longrightarrow \underset{\text{acid}}{H_3O^{1+}} + \underset{\text{base}}{HSO_3^{1-}}$$

$$P_2O_5 + 3\,H_2O \longrightarrow \underset{\text{phosphoric acid}}{2\,H_3PO_4}$$

$$\underset{\text{acid}}{H_3PO_4} + \underset{\text{base}}{H_2O} \longrightarrow \underset{\text{acid}}{H_3O^{1+}} + \underset{\text{base}}{H_2PO_4^{1-}}$$

and the pH of the aqueous solution is less than 7.00.

The chemistry of metals and compounds of metals is in general quite different from the chemistry of nonmetals and compounds of nonmetals. Much of the science of chemistry has to do with the investigation of these differences in properties and with the development of models to rationalize these properties.

Iron(II) ion, Fe^{2+}, is oxidized further to iron(III) ion, Fe^{3+}, by dissolved oxygen in water:

$$4 \times (Fe^{2+} \longrightarrow Fe^{3+} + e^{1-}) \quad \text{oxidation}$$
$$O_2 + 2\,H_2O + 4\,e^{1-} \longrightarrow 4\,OH^{1-} \quad \text{reduction}$$
$$4\,Fe^{2+} + O_2 + 2\,H_2O \longrightarrow 4\,OH^{1-} + 4\,Fe^{3+} \quad \text{total}$$

(Always check these equations with respect to number of atoms and with respect to net charges.)

The hydroxide of iron(II), $Fe(OH)_2$, and the hydroxide of iron(III), $Fe(OH)_3$, have very limited solubilities in water and both ions of iron are readily precipitated in the presence of hydroxide ion—if the ion concentrations are sufficiently high.

$$Fe^{2+}(aq) + 2\,OH^{1-}(aq) \longrightarrow Fe(OH)_2(s)$$
$$Fe^{3+}(aq) + 3\,OH^{1-}(aq) \longrightarrow Fe(OH)_3(s)$$

Both of these precipitation reactions are reversible and the precipitation of iron(II) hydroxide

$$Fe^{2+}(aq) + 2\,OH^{1-}(aq) \rightleftharpoons Fe(OH)_2(s)$$

does not prevent the further oxidation of iron(II) ion, Fe^{2+}.

The complete dehydration of iron(III) hydroxide leads to the formation of iron(III) oxide, Fe_2O_3, a red solid:

$$2\,Fe(OH)_3(s) \longrightarrow \underset{\text{red}}{Fe_2O_3(s)} + 3\,H_2O$$

In the presence of water, the dehydration is not complete, and iron rust is a red-brown to yellow-brown solid of indefinite composition, $Fe_2O_3 \cdot x\,H_2O$. The rusting of iron is a complex process involving a number of simultaneous reactions.

Question: Why is rusting of the iron of an automobile so much more of a problem whenever salt, NaCl, is present? Neither sodium ions nor chloride ions are directly involved in the chemical reactions. The answer is that the sodium chloride plays the same role in the oxidation of iron and the reduction of oxygen that the potassium nitrate, in the gelatin, played in the electrochemical cell used to study the oxidation of metallic copper and the reduction of silver ion. The two half-reactions

Role of Salt in Rusting of Iron

$$Fe \longrightarrow Fe^{2+} + 2\,e^{1-}$$

and

$$2\,H_2O + O_2 + 4\,e^{1-} \longrightarrow 4\,OH^{1-}$$

can take place at different sites as long as there is a metal conductor for the transfer of the electrons and there are ions that are free to move and maintain a net ionic charge in all small volumes of the water. The metal of the car is the metallic conductor that shorts many sites of oxidation and many sites of reduction wherever the protective coating of paint is broken. The sodium ions and the chloride ions of the dissolved salt contribute to the ionic conductance that is essential to the completion of the electrochemical cells.

There are a tremendous number of oxidation–reduction reactions. All atoms, all molecules, and all ions become involved in redox reactions. The contents of the cells of plants and animals are electrochemically balanced. It is energy derived from redox reactions that keeps us warm and enables us to move and think.

Scramble Exercises

1. Use the schematic MO energy level diagram for homonuclear diatomic molecules, Figure 16-1, as a basis to comment on the distribution of electrons in molecular orbitals (according to the MO approach) and the probable dissociation energy as compared to the dissociation energies of O_2 and N_2 for

(a) ground state diatomic fluorine, F_2, and
(b) ground state diatomic neon, Ne_2.

2. Although the schematic MO energy level diagram for homonuclear diatomic molecules given in Figure 16-1 was developed for homonuclear diatomic molecules, it can also be used as an approximation for heteronuclear diatomic molecules as long as the atomic numbers of the two elements are approximately equal.

Use Figure 16-1 to comment on the distribution of electrons in molecular orbitals for ground state molecules of carbon monoxide, CO. What homonuclear diatomic molecule is isoelectronic with carbon monoxide?

3. See the introductory statement to Exercise 2. Use Figure 16-1 to speculate on the bonding and the comparable stability of NO molecules, CN molecules, and cyanide ions, $CN^{(1-)}$. The NO molecule and the cyanide ion are well known. Cyanogen, $(CN)_2$, is a well-known molecule at room temperature. CN has been identified in flames during the combustion of hydrocarbons in air.

4. Hydronium ion, $H_3O^{(1+)}$, is an oxidizing agent. The half-reaction of reduction is

$$2\ H_3O^{(1+)}(aq) + 2\ e^{(1-)} \longrightarrow H_2(g) + 2\ H_2O(l)$$

For simplicity in writing equations for redox reactions, the hydrogen ion, $H^{(1+)}$, is usually used in place of the hydronium ion and the equation for the half-reaction of reduction is written

$$2\ H^{(1+)}(aq) + 2\ e^{(1-)} \longrightarrow H_2(g) \qquad \text{reduction}$$

Metallic zinc, Zn, "dissolves" in 6 M hydrochloric acid with the evolution of hydrogen gas, H_2, and the formation of zinc ion, $Zn^{(2+)}$, which remains in solution with the chloride ion in the water of the hydrochloric acid solution. Write the equation for the half-reaction of oxidation of zinc to zinc ion, and combine the two equations to give the total equation for the redox reaction.

5. Metallic zinc also dissolves in 6 M acetic acid with the evolution of hydrogen gas, H_2, and the formation of zinc ion, $Zn^{(2+)}$. The chemical reaction is exactly the same as the reaction of metallic zinc with 6 M hydrochloric acid (Exercise 4) but the reaction using 6 M acetic acid is much slower than the reaction using 6 M hydrochloric acid. Suggest a plausible explanation.

6. Metallic magnesium, Mg, dissolves in 6 M hydrochloric acid and also in 6 M acetic acid with the evolution of hydrogen gas, H_2, and the formation of magnesium ion, Mg^{2+}, which remains in solution. Write the equations for the half-reaction of oxidation of magnesium, the half-reaction of reduction of hydronium ion (or hydrogen ion), and the total redox reaction.

   The reaction of magnesium with each acid is much more rapid than the reaction of zinc with the same acid. Suggest a plausible explanation.

7. Metallic copper, Cu, does not react with either 6 M hydrochloric acid or 6 M acetic acid. Suggest a plausible explanation.

8. Metallic copper does react with 6 M nitric acid. Copper(II) ion, Cu^{2+}, with its characteristic blue color, becomes apparent and a colorless gas, which is not hydrogen gas, is evolved. This colorless gas is nitrogen monoxide, NO.

   The half-reaction of oxidation is

$$Cu \longrightarrow Cu^{2+} + 2\,e^{1-}$$

   It is the nitrate ion, NO_3^{1-}, that is the oxidizing agent in the acidic solution. The oxidation number of nitrogen in the nitrate ion is +5 and the oxidation number of nitrogen in the nitrogen monoxide molecule, NO, is +2. All other oxidation numbers remain unchanged. Hydronium ion (written here as hydrogen ion for simplicity) is also involved in the half-reaction of reduction:

$$H^{1+} + NO_3^{1-} + e^{1-} \longrightarrow NO + H_2O$$

   The nitrate ion is a strong oxidizing agent in an acidic solution. (Copper does not react with an aqueous solution of sodium nitrate. Its aqueous solution contains nitrate ions, NO_3^{1-}, and sodium ions, Na^{1+}, but the pH of the solution is approximately 7.)

   One way to balance the equation for this half-reaction of reduction is to balance by atoms

$$4\,H^{1+} + 1\,NO_3^{1-} + \underline{\quad\quad}\,e^{1-} \longrightarrow 1\,NO + 2\,H_2O$$

   and then select the appropriate number of electrons to attain the same net charge on both sides of the equation for the half-reaction of reduction. Check the number of electrons chosen against the change in oxidation number for nitrogen: +5 to +2. Complete the equation for the half-reaction of reduction and write the equation for the total reaction. Just in case you need help, a total of six electrons must be transferred to eliminate electrons in combining the equations for the two half-reactions.

   As soon as nitrogen monoxide molecules come in contact with diatomic oxygen molecules in the air, another oxidation–reduction reaction yields a single product: nitrogen dioxide, NO_2. Nitrogen dioxide is a brown gas (covalent compound). Write the equation for this redox reaction.

9. If you have done Exercise 8, try your hand and mind at writing appropriate equations for dissolving metallic silver, Ag (another coin metal), in 6 M nitric acid. Assume that the product of oxidation is silver ions, Ag^{1+}, and the product of reduction is again nitrogen monoxide, NO.

10. The sulfur dioxide gas, SO_2, released into the atmosphere during the combustion of sulfur-containing fuels may end up as sulfuric acid, H_2SO_4, a strong diprotic acid, in acid rain. The predominant ion of sulfur in a dilute solution of sulfuric acid is the sulfate ion, SO_4^{2-}.

The series of reactions involved in the conversion of SO_2 to the sulfate ion depends upon the experimental conditions, such as temperature and the concentration of the various reactants. Two possible sequences of reactions are

(1) the oxidation of SO_2 to SO_3 by diatomic oxygen, O_2, followed by the dissolving of SO_3 in water to give the ions of sulfuric acid; and

(2) the dissolving of SO_2 in water to give sulfurous acid, H_2SO_3, and its ions followed by the oxidation of sulfurous acid by diatomic oxygen, O_2, to the ions of sulfuric acid.

The oxidation of SO_2 to SO_3 is a slow reaction, and under existing experimental conditions in the atmosphere, the second series of reactions is the more probable. The equation discussed below is for the overall conversion of SO_2 to SO_4^{2-} and is the sum of the two reactions that occur in path 1 and also in path 2.

Show that the oxidation number of sulfur in the sulfate ion, SO_4^{2-}, is +6 and complete the following half-reaction of oxidation of sulfur dioxide to sulfate ion, SO_4^{2-}.

$$\_\_SO_2(g) + \_\_H_2O(l) \longrightarrow \_\_SO_4^{2-}(aq) + \_\_e^{1-} + \_\_H^{1+}(aq)$$

(Complete the charge balance as well as the atom balance in this half-reaction of oxidation.)

Atmospheric oxygen, O_2, is the oxidizing agent. Complete the following half-reaction of reduction

$$\_\_H^{1+} \ \_\_O_2 + \_\_e^{1-} \longrightarrow \_\_H_2O$$

and combine the two half-reactions to obtain the total reaction for a transfer of a total of four electrons. In the final equation, if a molecule or an ion is both a reactant and a product, reduce the number of molecules or ions on both sides of the equation to eliminate that recurring molecule or ion on one side of the equation. In the equation for the total reaction, the right-hand side of the equation is $2\ SO_4^{2-} + 4\ H^{1+}$. Note that this corresponds to the ions of two molecules of sulfuric acid, H_2SO_4. Sulfuric acid is a strong acid, and it is correct for the ions of sulfuric acid to be given as the products.

11. Any oxide of nitrogen that is introduced into the atmosphere may end up in acid rain as the nitrate ion, NO_3^{1-}, of nitric acid due to oxidation of the oxide of nitrogen by diatomic oxygen. The skeleton of the equation for the half-reaction of oxidation of nitrogen dioxide, NO_2, is

$$\_\_H_2O + \_\_NO_2 \longrightarrow \_\_NO_3^{1-} + \_\_e^{1-} + \_\_H^{1+} \qquad \text{oxidation}$$

The skeleton of the equation for the half-reaction of reduction of diatomic oxygen, O_2, is

$$\_\_H^{1+} + \_\_O_2 + \_\_e^{1-} \longrightarrow \_\_H_2O \qquad \text{reduction}$$

Balance these equations for the two half-reactions and combine to give the total reaction.

In the equation for the total reaction, the right-hand side of the equation is $\rightarrow 4\ NO_3^{1-} + 4\ H^{1+}$. Note that these are the ions corresponding to four molecules of nitric acid, HNO_3. Nitric acid is a very strong acid, and it is correct for the ions of nitric acid to be the final products of the reaction.

In the discussion of atmospheric pollution, the oxides of nitrogen in general are frequently listed as NO_x without endeavoring to specify exactly which oxides of nitrogen are present.

Additional exercises for Chapter 16 are given in the Appendix.

A Look Ahead

First, a look back. In this chapter, we have completed our exploration of chemical concepts and brought together many of the concepts we have explored before:

- the molecular orbital approach to chemical bonding and the properties of molecules and ions;
- the relation of the molecular orbital approach to the valence bond approach to chemical bonding;
- a generalized concept of oxidation reactions that does not necessarily involve reactions with oxygen;
- the use of chemical change to produce an electrical current; and
- chemical changes that involve both redox reactions and Brönsted acid–base reactions.

Chapter 16 is a demanding chapter. If you have managed to work your way through this chapter, or even parts of this chapter, you have accomplished a great deal.

With Chapter 17, Proteins, we begin the exploration of very large molecules synthesized and used by biological systems and the exploration of very large molecules synthesized and marketed by the chemical industries. The biological systems got there first. Synthetic polymers are the product of this century.

The large molecules that will be explored contain thousands of atoms of a very few nonmetals and are characterized by long chains of atoms arranged in repeating patterns. Such molecules are called polymers. The natural polymers include the proteins, the starches, the celluloses, and the nucleic acids. The synthetic polymers include the nylons, the polyethylenes, the polybutadienes, and the polyesters. The natural polymers are biodegradable. The synthetic polymers are in general not biodegradable.

The five chapters dealing with these large molecules introduce chemical systems that are biologically and economically of great significance. The chemical systems may be new to you, but the principles involved in their exploration have been developed in the preceding chapters. Our concern in the exploration of these chemical systems is broad concepts—not details that must be mastered by those who work with large molecules.

Chapter 22, the last chapter in this book, is an exploration of nuclear reactions—reactions involving changes in the nuclei of atoms. Chemical reactions involve changes in the electron atmospheres of atoms—not changes in the nuclei of atoms.

17
PROTEINS

Chapter 17

Muscles, tendons, gelatin, egg albumin, enzymes, hemoglobin, wool, silk, hair, nails, and skin seem an unlikely collection to have much in common from a chemical point of view. Yet all are entirely, or largely, made up of a class of compounds known as proteins. Approximately 75% of the dry mass of both plants and animals is proteins. Protein molecules vary in size. The smallest are the protein hormones (such as the protein portion of snake venom) with molecular weights of a few hundred. Intermediate sizes include molecules like insulin with molecular weights of a few thousand, and the largest are the fibrous proteins (the molecules that make up tendons) with molecular weights running into the hundred thousands. In spite of this great diversity of size of the molecules and the great diversity of properties of the substances, proteins all arise from repetitions of simple units and the structural restrictions on the freedom of these units to move with respect to each other.

All proteins are made up of long chains of nitrogen atoms alternating with pairs of carbon atoms:

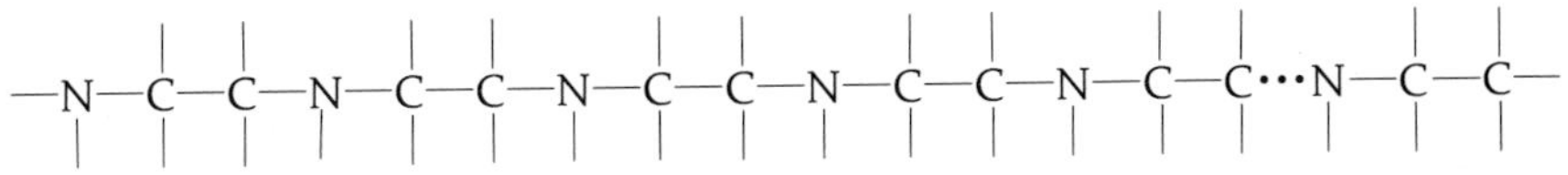

The Protein Backbone

$$-\text{N}-\text{C}-\text{C}-\text{N}-\text{C}-\text{C}-\text{N}-\text{C}-\text{C}-\text{N}-\text{C}-\text{C}-\text{N}-\text{C}-\text{C}\cdots\text{N}-\text{C}-\text{C}-$$

This representation of the chain is, of course, no more complete than the carbon–carbon chain is for a saturated hydrocarbon. More details of the structure are to come. This sequence of atoms beginning with a nitrogen atom and ending with a pair of carbon atoms is the carbon–nitrogen backbone of the protein chain or protein strand. The terminal carbon atom is a part of a carboxylic acid group, $-\text{COOH}$; or its conjugate base, the carboxylate ion group, $-\text{COO}^{1-}$; or, in some cases, the amide group

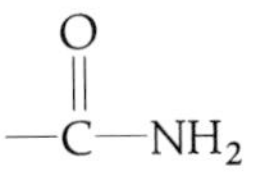

$$-\overset{\overset{\displaystyle\text{O}}{\|}}{\text{C}}-\text{NH}_2$$

The third bonding sites of all nitrogen atoms in the chain are satisfied by hydrogen atoms:

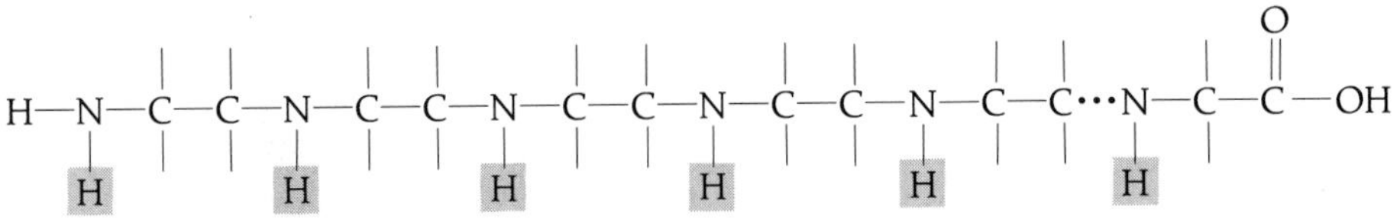

$$\text{H}-\underset{\text{H}}{\text{N}}-\text{C}-\text{C}-\underset{\text{H}}{\text{N}}-\text{C}-\text{C}-\underset{\text{H}}{\text{N}}-\text{C}-\text{C}-\underset{\text{H}}{\text{N}}-\text{C}-\text{C}-\underset{\text{H}}{\text{N}}-\text{C}-\text{C}\cdots\underset{\text{H}}{\text{N}}-\text{C}-\overset{\overset{\displaystyle\text{O}}{\|}}{\text{C}}-\text{OH}$$

One carbon atom of each pair of carbon atoms carries a double bond to an oxygen atom:

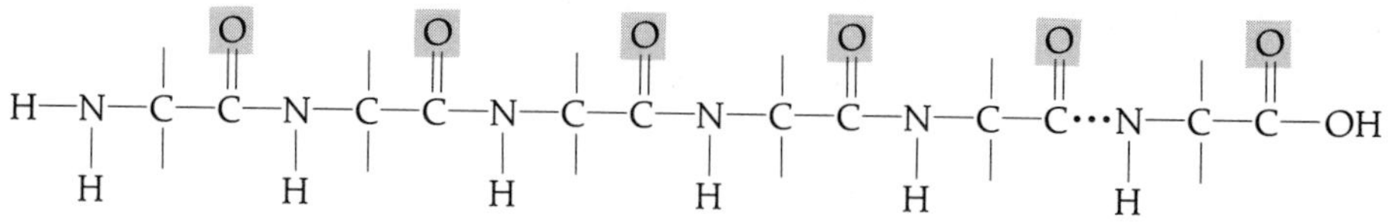

$$\mathrm{H{-}N(H){-}C{-}C({=}O){-}N(H){-}C{-}C({=}O){-}N(H){-}C{-}C({=}O){-}N(H){-}C{-}C({=}O){-}N(H){-}C{-}C({=}O)\cdots N(H){-}C{-}C({=}O){-}OH}$$

This presence of the carbonyl group

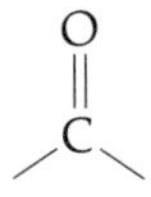

$$\mathrm{{>}C{=}O}$$

next to the amine group

$$\mathrm{{>}N{-}H}$$

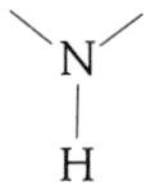

changes the properties of the amine group in many ways, as we shall see. The differences in protein molecules depend entirely upon the lengths of these chains and the atoms or groups of atoms that satisfy the remaining two bonding sites for the carbon atoms (one carbon atom in each pair of carbon atoms). In general, each of these carbon atoms carries a hydrogen atom and another group of atoms designated at the moment as R_1, R_2, . . .

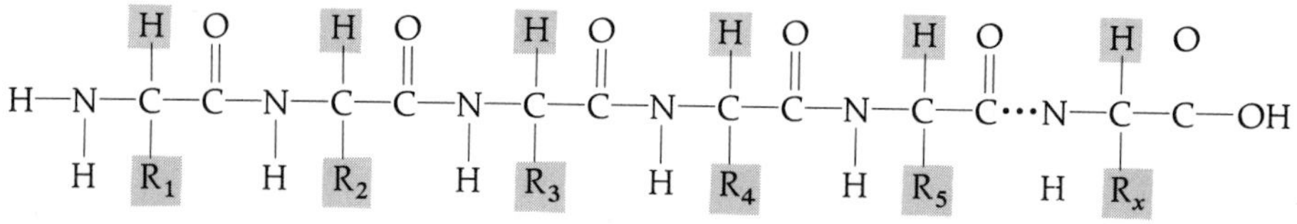

$$\mathrm{H{-}N(H){-}CH(R_1){-}C({=}O){-}N(H){-}CH(R_2){-}C({=}O){-}N(H){-}CH(R_3){-}C({=}O){-}N(H){-}CH(R_4){-}C({=}O){-}N(H){-}CH(R_5){-}C({=}O)\cdots N(H){-}CH(R_x){-}C({=}O){-}OH}$$

With the exception of the identities of the R groups (called side chains), proteins are monotonous repetitions of a rather simple unit. On hydrolysis, reaction with water at either high or low pH, proteins are broken into units called residues. These residues are α-amino acids or their conjugate bases:

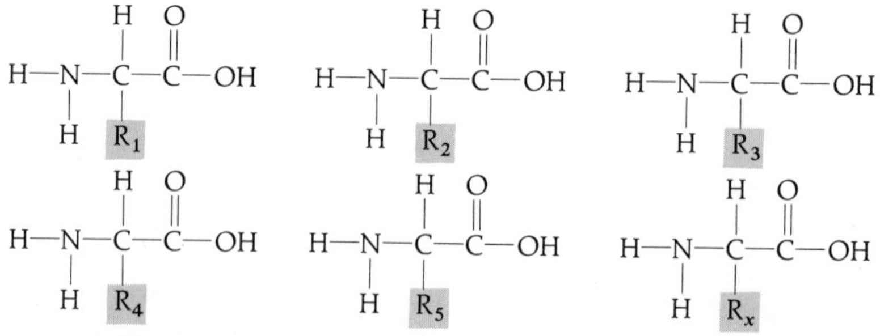

$$\mathrm{H{-}N(H){-}CH(R_1){-}C({=}O){-}OH} \qquad \mathrm{H{-}N(H){-}CH(R_2){-}C({=}O){-}OH} \qquad \mathrm{H{-}N(H){-}CH(R_3){-}C({=}O){-}OH}$$

$$\mathrm{H{-}N(H){-}CH(R_4){-}C({=}O){-}OH} \qquad \mathrm{H{-}N(H){-}CH(R_5){-}C({=}O){-}OH} \qquad \mathrm{H{-}N(H){-}CH(R_x){-}C({=}O){-}OH}$$

Amino Acid Residues

Table 17-1 gives a list of the common "R" groups of naturally occuring amino acids grouped according to the type of functional group these side chains have.

The carbon chain atoms in carboxylic acids are named after letters of the Greek alphabet: *alpha, beta, gamma,* etc., lettered "alphabetically" from the carboxyl group.

$$\underset{\gamma}{CH_3}\underset{\beta}{CH_2}\underset{\alpha}{CH_2}COOH$$

In the *alpha*-amino acids, the amine group is attached to the *alpha*-carbon. *Beta-, gamma-,* and other amino acids are also well known but they are of lesser biological significance. We shall refer to α-amino acids simply as amino acids.

One amino acid, alanine, was discussed in some detail in Chapter 14, page 340.

$$\begin{array}{ccccccccc} & & & & \text{H} & & \text{O} & & \\ & & & & | & & \| & & \\ \text{H} & - & \text{N} & - & \text{C} & - & \text{C} & - & \text{OH} \\ & & | & & | & & & & \\ & & \text{H} & & \text{CH}_3 & & & & \end{array} \qquad \text{alanine}$$

In alanine, the R group is the methyl group. This amino acid exists as the zwitterion

Zwitterions

$$\begin{array}{ccccccccc} & & ^{(1+)}\text{H} & & \text{H} & & \text{O} & & \\ & & | & & | & & \| & & \\ \text{H} & - & \text{N} & - & \text{C} & - & \text{C} & - & \text{O}^{(1-)} \\ & & | & & | & & & & \\ & & \text{H} & & \text{CH}_3 & & & & \end{array} \qquad \text{zwitterion of alanine}$$

in crystals of alanine and in aqueous solutions of alanine at the isoelectric point. At pH's greater than the isoelectric point, it becomes the deprotonated ion.

$$\begin{array}{ccccccccc} & & & & \text{H} & & \text{O} & & \\ & & & & | & & \| & & \\ \text{H} & - & \text{N} & - & \text{C} & - & \text{C} & - & \text{O}^{(1-)} \\ & & | & & | & & & & \\ & & \text{H} & & \text{CH}_3 & & & & \end{array} \qquad \text{deprotonated ion}$$

and at pH's lower than the isoelectric point, alanine becomes the completely protonated ion:

$$\begin{array}{ccccccccc} & & ^{(1+)}\text{H} & & \text{H} & & \text{O} & & \\ & & | & & | & & \| & & \\ \text{H} & - & \text{N} & - & \text{C} & - & \text{C} & - & \text{OH} \\ & & | & & | & & & & \\ & & \text{H} & & \text{CH}_3 & & & & \end{array} \qquad \text{completely protonated ion}$$

At no time does alanine exist in any major fraction as the neutral molecule.

$$\begin{array}{ccccccccc} & & & & \text{H} & & \text{O} & & \\ & & & & | & & \| & & \\ \text{H} & - & \text{N} & - & \text{C} & - & \text{C} & - & \text{OH} \\ & & | & & | & & & & \\ & & \text{H} & & \text{CH}_3 & & & & \end{array} \qquad \text{nonexistent neutral molecule}$$

To write alanine as the neutral molecule is a fiction, but it is a convenient fiction so long as we all know that it is a fiction and understand in what way it is a fiction. The same set of circumstances applies to all the other amino acids. We will write them all as neutral molecules in this chapter, acknowledging the fiction involved.

Table 17-1

R GROUPS OF α-AMINO ACIDS COMMONLY FOUND IN PROTEINS

the amino acid	abbreviation	R	pH isoelectric point
aliphatic R group			
glycine	Gly	H	6.0
alanine	Ala	CH_3	6.0
valine*	Val	$(CH_3)_2CH$	6.0
leucine*	Leu	$(CH_3)_2CHCH_2$	6.0
isoleucine*	Ile	$CH_3CH_2CHCH_3$	6.0
hydroxy-containing R group			
serine	Ser	CH_2OH	5.7
threonine*	Thr	CH_3CHOH	6.5
acidic R groups and their amides			
aspartic acid	Asp	CH_2COOH	3.0
asparagine	Asn	CH_2CONH_2	5.4
glutamic acid	Glu	CH_2CH_2COOH	3.1
glutamine	Gln	$CH_2CH_2CONH_2$	5.7
basic R groups			
lysine*	Lys	$CH_2(CH_2)_3NH_2$	9.5
hydroxylysine	Hyl	$CH_2CH_2CHOHCH_2NH_2$	9.2

* Must be a part of the human diet. The other amino acids can be synthesized in our bodies.
† The imino acids are not a simple R group and will be discussed later in the text. These amino acids are called <u>imino</u> acids rather than amino acids to emphasize the fact that the α-amino group is part of a ring structure.

the amino acid	abbreviation	R	pH isoelectric point
basic R groups *(continued)*			
histidine*	His	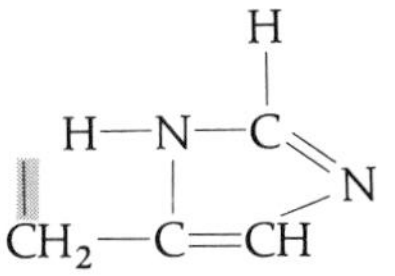 H–N–CH=N–CH=C–CH_2 (imidazole ring: CH_2—C=CH, H—N—C(H)=N)	7.6
arginine	Arg	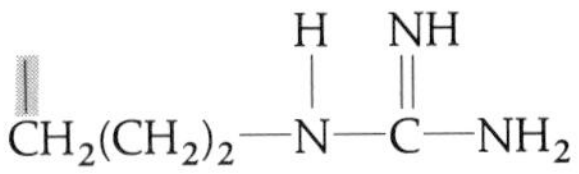 $CH_2(CH_2)_2$—N(H)—C(=NH)—NH_2	10.8
aromatic R groups			
phenylalanine*	Phe	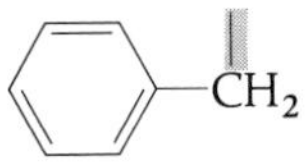 C_6H_5—CH_2	5.9
tyrosine	Tyr	 HO—C_6H_4—CH_2	5.6
tryptophan*	Trp	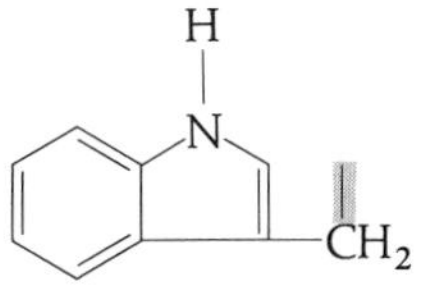 indole (N—H)—CH_2	5.9
sulfur-containing R groups			
cysteine	Cys	CH_2SH	5.0
methionine*	Met	$CH_2CH_2SCH_3$	5.7
imino acids†			
proline	Pro	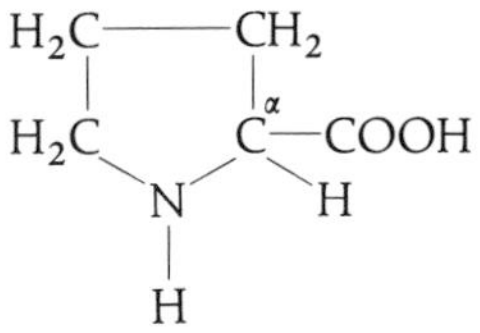 H_2C—CH_2; H_2C, C^{α}(H)—COOH; N—H (ring)	6.3
hydroxyproline	Hyp	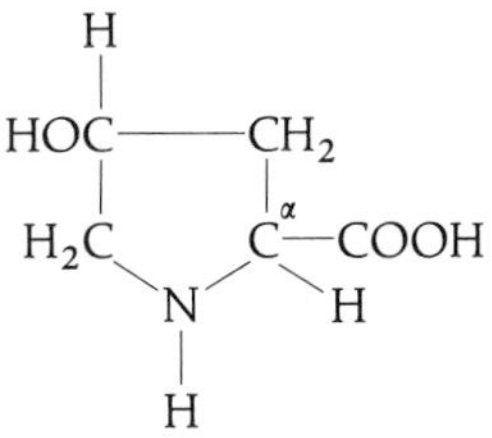 HOC(H)—CH_2; H_2C, C^{α}(H)—COOH; N—H (ring)	5.8

The simplest amino acid is glycine (R = H). Under appropriate experimental conditions, a molecule of glycine reacts with a molecule of alanine to give either of two products, glycine–alanine or alanine–glycine:

$$
\mathrm{H}-\underset{\underset{\displaystyle\mathrm{H}}{|}}{\mathrm{N}}-\overset{\overset{\displaystyle\mathrm{H}}{|}}{\underset{\underset{\displaystyle\mathrm{H}}{|}}{\mathrm{C}}}-\overset{\overset{\displaystyle\mathrm{O}}{\|}}{\mathrm{C}}-\mathrm{O}-\mathrm{H}
+
\mathrm{H}-\underset{\underset{\displaystyle\mathrm{H}}{|}}{\mathrm{N}}-\overset{\overset{\displaystyle\mathrm{H}}{|}}{\underset{\underset{\displaystyle\mathrm{CH_3}}{|}}{\mathrm{C}}}-\overset{\overset{\displaystyle\mathrm{O}}{\|}}{\mathrm{C}}-\mathrm{O}-\mathrm{H}
\longrightarrow
\mathrm{H}-\underset{\underset{\displaystyle\mathrm{H}}{|}}{\mathrm{N}}-\overset{\overset{\displaystyle\mathrm{H}}{|}}{\underset{\underset{\displaystyle\mathrm{H}}{|}}{\mathrm{C}}}-\overset{\overset{\displaystyle\mathrm{O}}{\|}}{\mathrm{C}}-\underset{\underset{\displaystyle\mathrm{H}}{|}}{\mathrm{N}}-\overset{\overset{\displaystyle\mathrm{H}}{|}}{\underset{\underset{\displaystyle\mathrm{CH_3}}{|}}{\mathrm{C}}}-\overset{\overset{\displaystyle\mathrm{O}}{\|}}{\mathrm{C}}-\mathrm{OH}+\mathrm{H_2O}
$$

glycine (Gly) + alanine (Ala) → glycine–alanine (Gly–Ala)

$$
\mathrm{H}-\underset{\underset{\displaystyle\mathrm{H}}{|}}{\mathrm{N}}-\overset{\overset{\displaystyle\mathrm{H}}{|}}{\underset{\underset{\displaystyle\mathrm{CH_3}}{|}}{\mathrm{C}}}-\overset{\overset{\displaystyle\mathrm{O}}{\|}}{\mathrm{C}}-\mathrm{O}-\mathrm{H}
+
\mathrm{H}-\underset{\underset{\displaystyle\mathrm{H}}{|}}{\mathrm{N}}-\overset{\overset{\displaystyle\mathrm{H}}{|}}{\underset{\underset{\displaystyle\mathrm{H}}{|}}{\mathrm{C}}}-\overset{\overset{\displaystyle\mathrm{O}}{\|}}{\mathrm{C}}-\mathrm{O}-\mathrm{H}
\longrightarrow
\mathrm{H}-\underset{\underset{\displaystyle\mathrm{H}}{|}}{\mathrm{N}}-\overset{\overset{\displaystyle\mathrm{H}}{|}}{\underset{\underset{\displaystyle\mathrm{CH_3}}{|}}{\mathrm{C}}}-\overset{\overset{\displaystyle\mathrm{O}}{\|}}{\mathrm{C}}-\underset{\underset{\displaystyle\mathrm{H}}{|}}{\mathrm{N}}-\overset{\overset{\displaystyle\mathrm{H}}{|}}{\underset{\underset{\displaystyle\mathrm{H}}{|}}{\mathrm{C}}}-\overset{\overset{\displaystyle\mathrm{O}}{\|}}{\mathrm{C}}-\mathrm{OH}+\mathrm{H_2O}
$$

alanine (Ala) + glycine (Gly) → alanine–glycine (Ala–Gly)

Di- and Tripeptides

The new carbon–nitrogen bond is, in both cases, marked with red. This linkage is known as the **peptide linkage.** Molecules made from two amino acids are called dipeptides.

A molecule of alanine can react with another molecule of alanine to give the dipeptide alanine–alanine and a molecule of glycine can react with another molecule of glycine to give the dipeptide glycine–glycine. Consequently a mixture of glycine and alanine would, of course, give all four possible dipeptides mentioned above.

A molecule of either amino acid can react with any of the above dimers to form tripeptides. Illustrating the process with glycine–alanine and alanine,

$$
\underset{\text{Gly–Ala}}{\text{glycine–alanine}} + \underset{\text{Ala}}{\text{alanine}} = \begin{cases} \text{glycine–alanine–alanine} \\ \text{Gly–Ala–Ala} \\ \text{or} \\ \text{alanine–glycine–alanine} \\ \text{Ala–Gly–Ala} \end{cases}
$$

Each trimer contains two peptide linkages. Natural proteins may contain 300 or more peptide linkages. Proteins are **polypeptides.**

Chiral Amino Acids

Look back at the structure of the analine molecule and you will see that the α-carbon carries four different groups; thus optical isomerism is possible at the α-carbon. (Chapter 14, page 340.) All the amino acid residues in naturally occurring proteins are, with very few exceptions, of one particular configuration. Glycine, of course, is not chiral (optically active) and does not have the possibility of optical isomers, but all the other amino acids do. It is one of the mysteries of the evolution of life that, in the course of that evolution, nature chose one particular isomer nearly exclusively over the other possibility.

The variety of properties exhibited by proteins is enormous. Some protein molecules are involved in the structure of hard, rigid materials (fingernails); elastic materials (the walls of veins and arteries, ligaments and muscles); tough, relatively unstretchable materials (skin and tendons); and flexible, water-soluble materials (the globular proteins—hemoglobin, antibodies, egg albumin, and enzymes). How this simple repetition of amino acids can lead to such differences in properties is one of the fascinations of protein chemistry. It is also interesting to observe that, in the explanation of the properties of these often large and complicated molecules, extensive use is made of the same ball and stick models and space-filling models with which we started this text. The models are useful for the simplest molecules (methane) and for very complex molecules (proteins). We will use many pictures of ball and stick and space-filling models in this chapter. It would be nice if every student had a model set big enough to build even a fraction of a protein molecule, but that is prohibitively expensive. We urge you, however, to look at a fully constructed model of a protein if the opportunity ever arises.

When dealing with complicated structures with a variety of possible spatial organizations, the usual tendency in science is to divide the problem up into smaller, more comprehensible units. Exactly so in the case of proteins: the problem of trying to rationalize the whole structure of a protein is broken down into trying to understand four aspects of the problem, designated as the primary structure, the secondary structure, the tertiary structure, and the quaternary structure. You will note that these words are just a fancy way of saying the "first, second, third, and fourth" levels of organization.

PRIMARY STRUCTURE

Our discussion has dealt with the basic building blocks of the protein chain, the amino acids. The length of the chain and the sequence of amino acids in it determine what is called the primary structure of a protein, and all the other properties of a protein depend ultimately on this primary structure.

Amino Acid Sequences

One of the major achievements of modern chemistry is that the sequence of amino acids in a protein, even a rather large one, can be determined exactly. In most cases (the proteins responsible for immunity are a major exception), a specific protein such as insulin or myoglobin in a given species always has the same exact number of amino acids in the chain in the same exact order. Figure 17-1 shows the amino acid sequence in human insulin, a protein that is actually two polypeptide chains joined

Figure 17-1

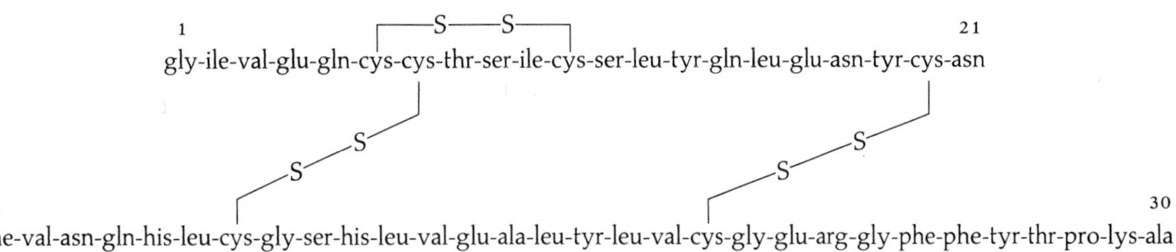

THE AMINO ACID SEQUENCE IN HUMAN INSULIN, A PROTEIN THAT IS TWO POLYPEPTIDE CHAINS WITH DISULFIDE BRIDGES. The amino acids labelled 1 are the amino terminal ends and the ones labelled 21 and 30 are the carboxylic acid terminal ends of the two chains.

through bridges of sulfur atoms (the disulfide bridge). The sequence of amino acids for the protein performing the same function differs slightly from species to species. For example, the primary structures of myoglobin (the oxygen-carrying protein in muscle) for the whale and the horse are given in Figure 17-2 with the similarities

Figure 17-2

1 10

Whale: val—leu—ser—glu—gly—glu—trp—gln—leu—val—leu—his—val—trp—ala—
Horse: gly—leu—ser—asp—gly—glu—trp—gln—gln—val—leu—asn—val—trp—gly—

20 30

lys—val—glu—ala—asp—val—ala—gly—his—gly—gln—asp—ile—leu—ile—arg—leu—
lys—val—glu—ala—asp—ile—ala—gly—his—gly—gln—glu—val—leu—ile—arg—leu—

40

phe—lys—ser—his—pro—glu—thr—leu—glu—lys—phe—asp—arg—phe—lys—his—leu—
phe—thr—gly—his—pro—glu—thr—leu—glu—lys—phe—asp—lys—phe—lys—his—leu—

50 60

lys—thr—glu—ala—glu—met—lys—ala—ser—glu—asp—leu—lys—lys—his—gly—val—
lys—thr—glu—ala—glu—met—lys—ala—ser—glu—asp—leu—lys—lys—his—gly—thr—

70 80

thr—val—leu—thr—ala—leu—gly—ala—ile—leu—lys—lys—lys—gly—his—his—glu—
val—val—leu—thr—ala—leu—gly—gly—ile—leu—lys—lys—lys—gly—his—his—glu—

90 100

ala—glu—leu—lys—pro—leu—ala—gln—ser—his—ala—thr—lys—his—lys—ile—pro—
ala—glu—leu—lys—pro—leu—ala—gln—ser—his—ala—thr—lys—his—lys—ile—pro—

110

ile—lys—tyr—leu—glu—phe—ile—ser—glu—ala—ile—ile—his—val—leu—his—ser—
ile—lys—tyr—leu—glu—phe—ile—ser—asp—ala—ile—ile—his—val—leu—his—ser—

120 130

arg—his—pro—gly—asn—phe—gly—ala—asp—ala—gln—gly—ala—met—asn—lys—ala—
lys—his—pro—gly—asn—phe—gly—ala—asp—ala—gln—gly—ala—met—thr—lys—ala—

140 150

leu—glu—leu—phe—arg—lys—asp—ile—ala—ala—lys—tyr—lys—glu—leu—gly—try—
leu—glu—leu—phe—arg—asn—asp—ile—ala—ala—lys—tyr—lys—glu—leu—gly—tyr—

153

gln—gly
gln—gly

THE AMINO ACID SEQUENCES IN MYOGLOBIN FROM THE WHALE AND FROM THE HORSE. Identical sequences are shaded.

highlighted. It is a general rule, developed after careful study, that the further apart in evolution two species are, the more different the amino acid sequence of a similar protein will be. With the many identical sequences of amino acids in whale and horse myoglobin, we suspect the animals might be rather closely related biologically, and, of course, both are mammals.

It has become customary to discuss the secondary, tertiary, and quaternary structures as if they were independent of the primary structure. However, the primary structure both makes possible and requires all other levels of organization.

SECONDARY STRUCTURE

Secondary structure has come to be identified with the configuration about the peptide bonds and how the hydrogen bonds of the protein molecules are arranged. Our first concern will be with the configuration of the atoms involved in the peptide bond. Experimental values for bond angles and bond lengths are given in Figure 17-3.

The Planar Peptide Bond

One observation is that the angles ∠CCO and ∠OCN and ∠CNC are near 120°. In ammonia and the amines, the bond angles with nitrogen at the apex are expected to be a little less than 109°. In the peptide linkage, the bond angles with nitrogen at the apex are more nearly 120°—some larger, some smaller. As you may recall, 120° bond angles are characteristic of ethylene, CH_2CH_2, a planar molecule. The important atoms in the peptide bond, for the moment, are the atoms encompassed by the center circle:

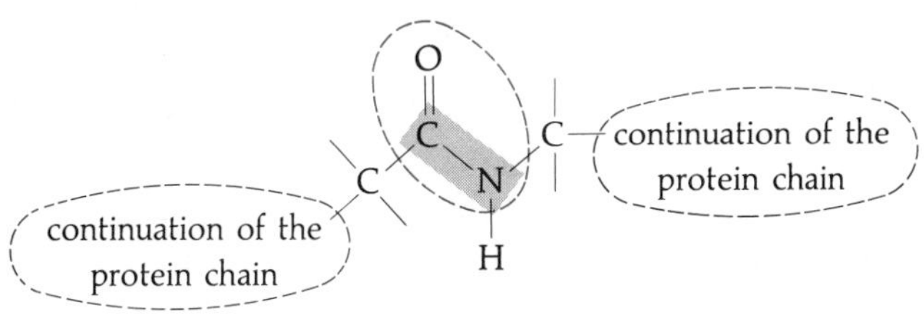

Figure 17-3

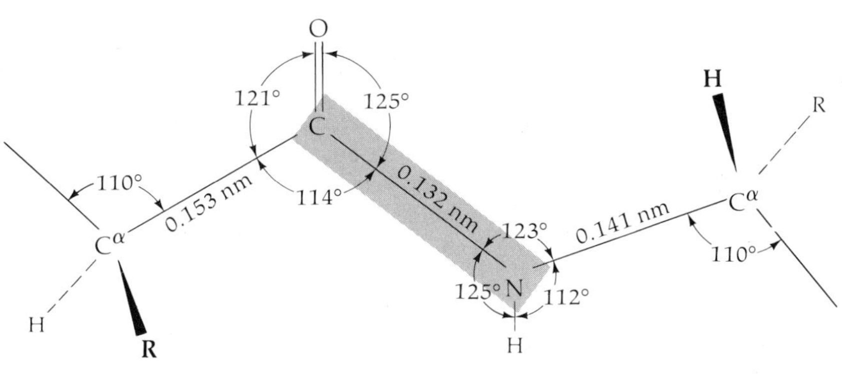

BOND LENGTHS AND ANGLES FOR A SINGLE PEPTIDE UNIT.

If we look closely at the electronegativities of these atoms, we note that oxygen is much more electronegative than carbon, and the carbon–oxygen double bond is quite polar with the oxygen atom carrying the partial negative charge and the carbon atom carrying the partial positive charge:

Now note that the nitrogen atom, if considered to be sp^3 hybridized, carries a pair of electrons (a "lone pair") in one of the sp^3 orbitals that is not involved in bonding. It would not be surprising if these electrons were attracted to the partial positive charge on the carbon. One often indicates this situation using arrows, where the electrons in the double bond tend to be attracted to the oxygen atom and the "lone pair" electrons on nitrogen tend to be attracted to the carbon atom.

The net result of this displacement of electrons is to produce electron density distributions that are described by the diagram below. The dotted line means that the C—N bond is slightly more than a single bond and the C—O bond is slightly less than a double bond. Another way of stating this concept is to say that the C—N and C—O bonds both have partial double bond character.

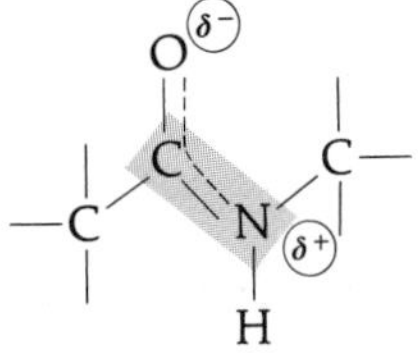

It is not possible to build this idea into a ball and stick molecule, although the valence bond model can be modified to accommodate the concept. What is important for us here is that the partial double bond character rationalizes the planarity of the six atoms most closely related to the peptide bond, and the six atoms inside this now larger center circle would lie in the same plane. (Figure 17-3 and Figure 17-4.)

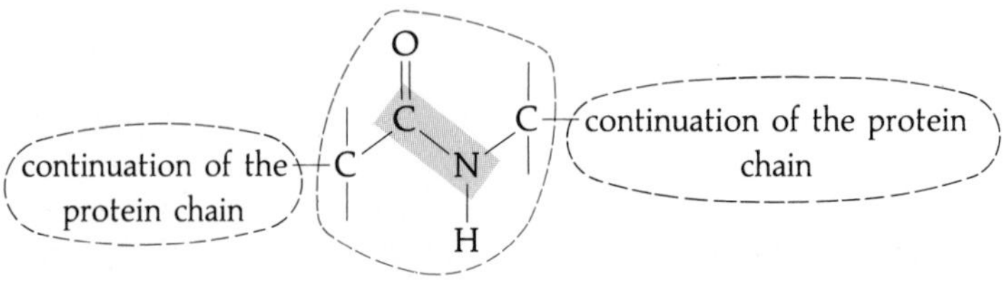

Figure 17-4

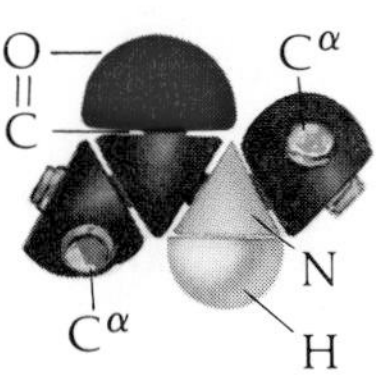

space-filling model of peptide diagrammed in Figure 17-3

(a)

repeating unit .723 nm

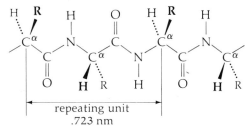

larger piece of a protein chain drawn in its most extended form. The only free rotation about bonds in the chain is about the *sigma*-bonds to the α-carbon atom.

(b)

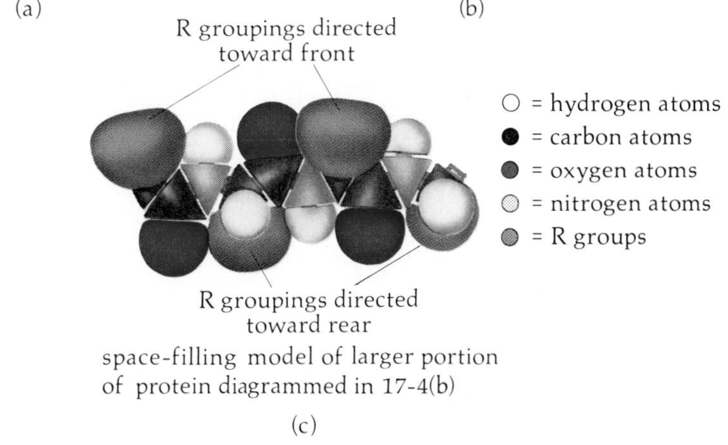

space-filling model of larger portion of protein diagrammed in 17-4(b)

(c)

SOME MODELS OF THE *trans* PEPTIDE BOND.

cis and trans Peptide Bonds

This planarity means that the peptide bond is capable of *cis* and *trans* isomers. We have shown the *trans* isomer in all cases above; the *cis* isomer is drawn below.

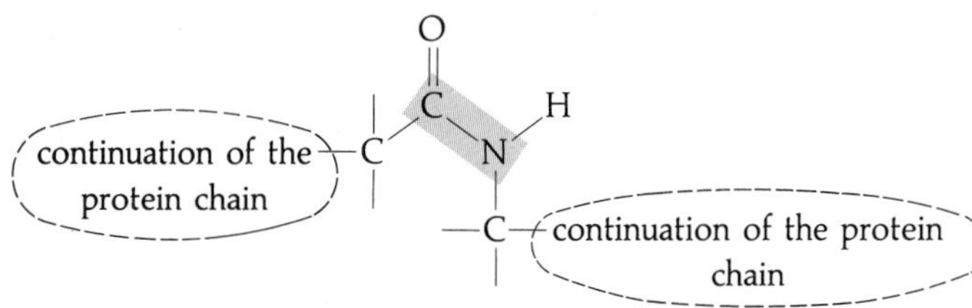

Note that the polypeptide chain continues across the peptide bond in the *trans* isomer and continues on the same side of the peptide bond in the *cis* isomer. Both isomers are possible, but in the vast majority of cases, nature has chosen the *trans* isomer for proteins; the *cis* isomer is very rare indeed, but not unknown, in protein strands.

Another consequence of the partial double bond is that the nitrogen atom of the peptide no longer has an unshared pair of electrons. This nitrogen atom is no longer a Brönsted base.

Hydrogen Bonding in Proteins

The other major feature of the secondary structure is hydrogen bonds. Note from the structure of the protein chain that the oxygen atom doubly bonded to carbon is electronegative and there is a hydrogen atom on the electronegative nitrogen atom, so all the conditions are met for a strong hydrogen bond between the carbonyl (keto) oxygen atom

$$-\overset{\overset{\displaystyle O}{\|}}{C}-$$

and the amide hydrogen atom.

$$-\underset{\underset{\displaystyle H}{|}}{N}-$$

The "keto–amide" hydrogen bond occurs whenever the appropriate portions of the same chain or of two separate chains are side by side.

This type of hydrogen bonding is very important in the structure of proteins. Folds in the protein chain (or rotations about single bonds in the structure) maximize the number of hydrogen bonds. There are several ways to achieve this result. We'll take a look at some of the most important and common configurations. Our first concern will be the α-structure, the helix. (A helix as used here means a structure like an old-fashioned door spring.) Next, we will discuss the β-structure, or the pleated sheet structure, and then the γ-helix characteristic of connective tissue. Our final concern will be the so-called random configuration of the globular proteins. An overall view of these structures can be obtained from the figures presented throughout the chapter.

The α-Helix

In the *alpha*-helix, the protein chain winds about a central axis and the keto–amide hydrogen bonds are essentially parallel to that helix axis. (Figure 17-5). The diagrams deserve careful examination. Try to follow the carbon–carbon–nitrogen atom backbone chain in both the ball and stick diagram and in the space-filling model, and then attempt to locate the keto–amide hydrogen bonds indicated on the diagram. The hydrogen bonds are intramolecular—between atoms of the same chain—the

$$>N-H \quad \text{and} \quad O{=}C<$$

Figure 17-5

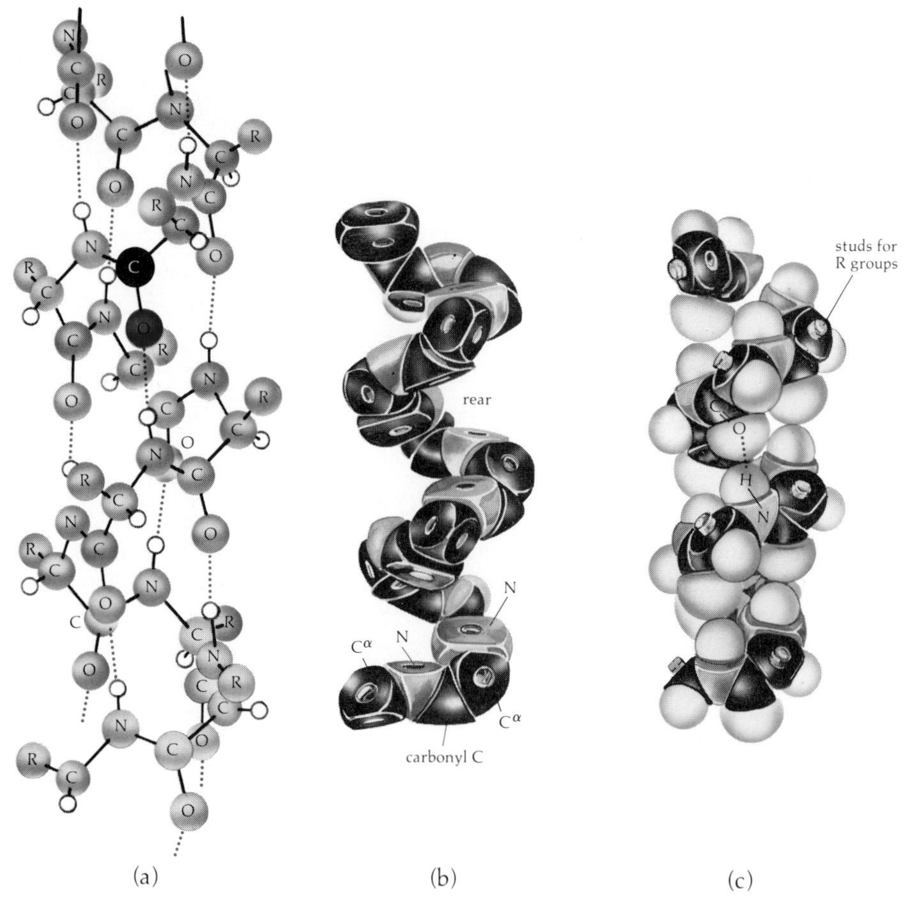

THE *alpha*-HELIX CONFORMATION OF A POLYPEPTIDE CHAIN. (a) A drawing of a ball and stick model. The small unidentified circles represent hydrogen atoms. In this model, the carbonyl double bond is shown with a single line for simplicity. The hydrogen bonds are indicated with dotted lines. This is a very difficult picture to follow because some atoms and some bonds are obscured by other atoms and bonds. If you work at it you can find the $\cdots$N—C—C—N—C—C$\cdots$ backbone. (b) A picture of a space-filling model of the backbone of the α-helix. All the H and O atoms have been removed so you can follow the

$$-\mathrm{N}-\mathrm{C}-\mathrm{C}-\mathrm{N}-\mathrm{C}-\mathrm{C}-$$

pattern of the chain through the α-helix. (c) The same space-filling model with hydrogen and oxygen atoms added. The R groups are, however, still missing. They would attach at the small studs or bosses showing in the picture. Note the hydrogen bond.

Figure 17-6

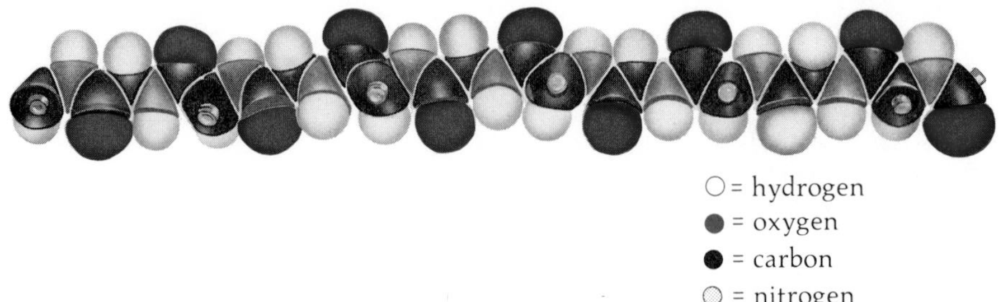

THE α-HELICAL SEGMENT IN FIGURE 17-5 SHOWN HERE FULLY EXTENDED. The studs for the R groups are still visible.

groups brought into position by the rotation of the chain about that central axis. This hydrogen bonding is an important component of many proteins; in hair (keratin), the protein is almost entirely in the α-helical form. Please note also the helix "extended" in Figure 17-6.

The β Pleated Sheet

The β-structure, or "pleated sheet" structure, is characteristic of silk. (Figure 17-7.) In this structure, two or more protein chains run alongside one another but with the nitrogen terminal end of one chain corresponding to the carbon terminal end of the other—the so-called antiparallel alignment. The keto–amide hydrogen bonds are perpendicular to the direction of the chain and link two chains together: the keto–amide hydrogen bonds are inter molecular. The reason this structure is called a "pleated sheet" is fairly obvious from the structural drawing but not so obvious from the picture of the space-filling model. This structure does not accommodate bulky side chains (R groups); β-structures are limited to proteins containing large amounts of simple amino acids: alanine, glycine, and serine, where the R groups are $-CH_3$, H, and $-CH_2OH$.

The γ (Collagen) Helix

The collagen helix (γ-helix) is more complicated yet. Its structure is the result of a large proportion of *imino* acids (proline and hydroxyproline) in the chain. We have indicated in Table 17-1 the complete structures of proline and hydroxyproline. The carboxyl group

$$-\overset{\overset{\displaystyle O}{\|}}{C}-OH$$

of the previous amino acid in the protein chain attaches to the nitrogen atom of the proline or hydroxyproline (with the elimination of a water molecule) and the chain continues through the group of the imino acid,

$$-\overset{\overset{\displaystyle O}{\|}}{C}-OH$$

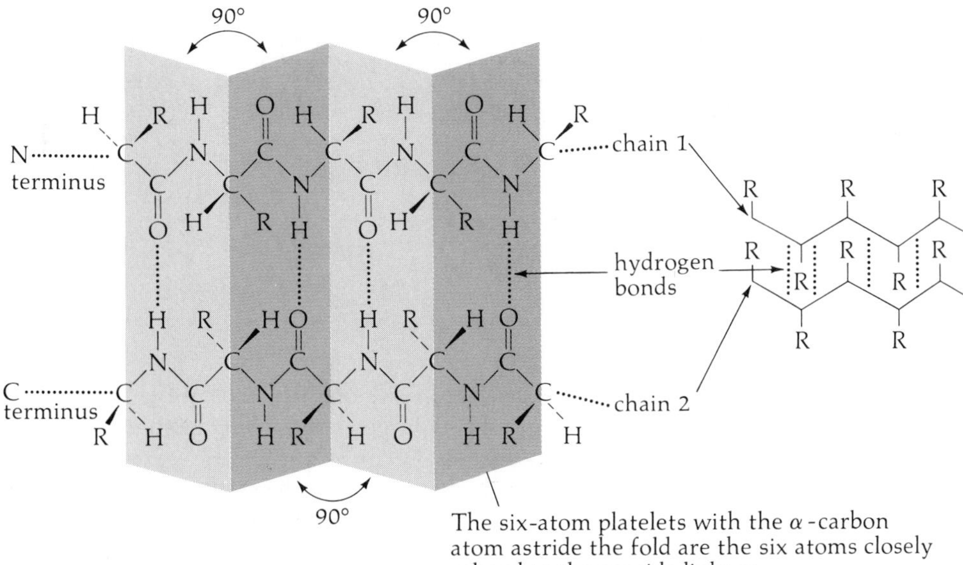

THE β-PLEATED SHEET STRUCTURE. (a) The drawing depicts the hydrogen bonding of the pleated sheet configuration between two peptide segments aligned in an antiparallel orientation. (b) The space-filling model shows a group of interlocking pleated sheets characteristic of many fibrous proteins, including silk. The section shown consists of 44 residues and 7 polypeptide strands. The hooks represent points of attachment for hydrogen bonds.

reacting with the amino group on the next amino acid in the chain. Imino acids, because of the ring structures, put bends or kinks in the chains.

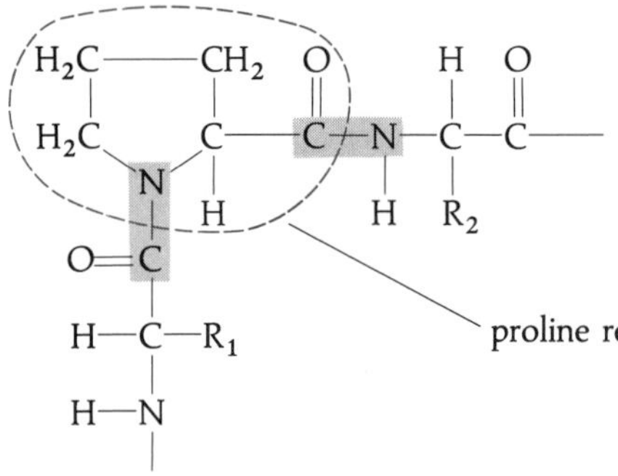

proline residue

The collagen structure is believed to consist of three separate chains wound about each other in a "triple helix" with the hydrogen bonds being intermolecular and perpendicular to the central helix axis. Figure 17-8 gives a schematic representation of the collagen structure. Most of the details are missing. Compare the collagen γ-helix presentations in Figure 17-8 with the α-helix presentations in Figure 17-5.

The globular proteins—those that circulate in the blood, for example—are typically smaller in molecular weight than the few mentioned above and much more soluble in water. For quite some time, it was considered that these proteins had a completely random structure. However, with the increased sophistication of modern structural analyses, it has become clear that these structures are very well defined by the secondary structures and by other interactions in the proteins that give rise to what are known as tertiary structures.

TERTIARY STRUCTURE

Interactions That Contribute to Tertiary Structure

Tertiary structure is determined by hydrophobic interactions (interactions between the R groups and the solvent), charge–charge interactions, and van der Waals' interactions between elements of the protein chain. The term "hydrophobic," introduced in Chapter 12, page 269, is used to describe the fact that side chains containing lots of $—CH_2—$ groups (alanine, valine, leucine) tend to be in contact with themselves rather than with the aqueous medium in which the protein is dissolved. Folding of the protein chain tends to maximize the number of "hydrophobic" (water-hating) R groups on the inside of the protein and the number of "hydrophilic" (water-loving) R groups, such as serine (R = $—CH_2OH$), on the outside of the protein in aqueous solutions such as blood. Charge–charge interactions refer to the interaction of positively and negatively charged groups on the protein, such as $—COO^{(1-)}$ and $—NH_3^{(1+)}$, with each other and with the solvent. The van der Waals' interactions, dipole–dipole interactions including both permanent and induced dipoles, are most evident in the hydrocarbon-like parts of the chains.

The high degree of order these interactions can impose on what seems to be just a random arrangement of the protein chains is shown in Figures 17-9 and 17-10. What is perhaps more interesting is that these exotic convolutions of the chain seem to be important for function—especially for antibodies and enzymes.

Figure 17-8

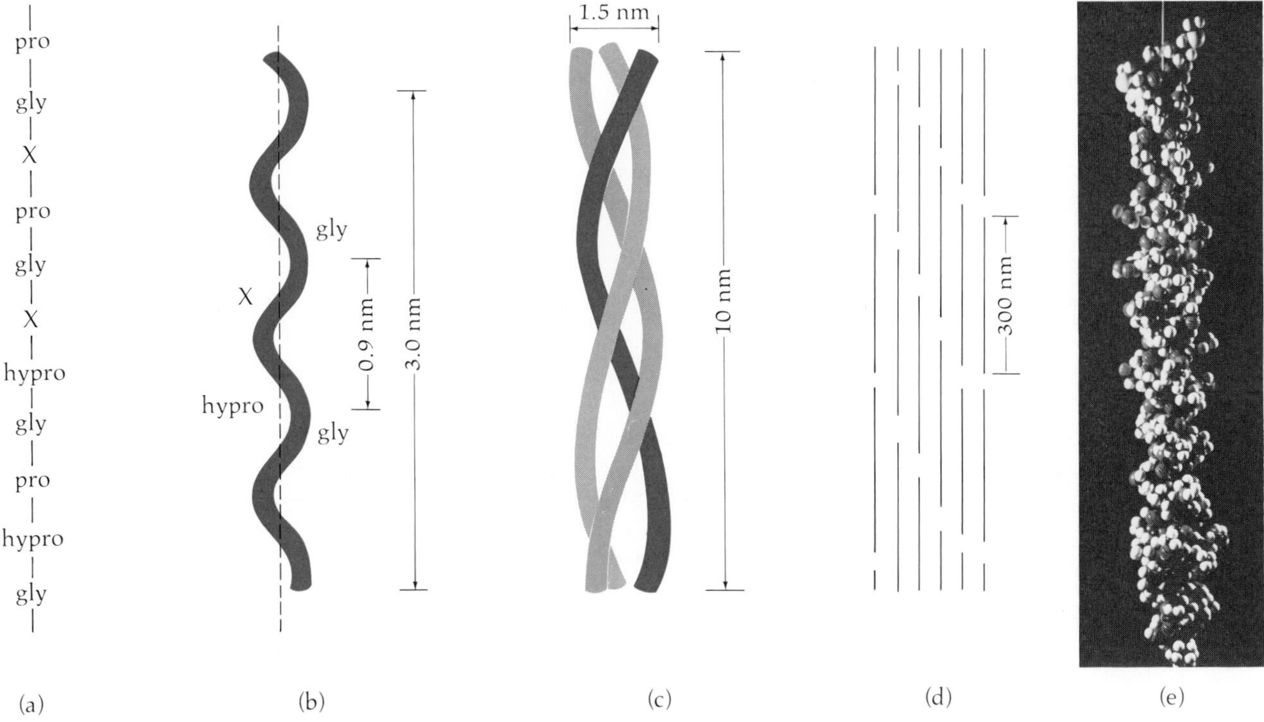

A SCHEMATIC REPRESENTATION OF THE COLLAGEN STRUCTURE. (a) Amino acid sequence of unit chain; every third residue is glycine; X represents any one of the usual amino acids. (b) Left-handed helical conformation of individual chain. (The helix is not an α-helix.) (c) Triple-helix arrangement of collagen; also called tropocollagen; three left-handed helixes coiled around each other with a right-handed twist [see the space-filling model (e)]. (d) Ordered packing of tropocollagen molecules in a collagen fibril of connective tissue; in some fibrils the adjacent molecules are covalently cross-lined; a very rigid macro structure. (e) A photograph of a space-filling model of the collagen triple helix.

Protein chains can also have covalent cross-links in them. The amino acid cysteine has the structure

$$\begin{array}{c} \text{H} \\ | \\ \text{H}_2\text{N}-\text{C}-\text{COOH} \\ | \\ \text{CH}_2 \\ | \\ \text{SH} \end{array}$$

Figure 17-9

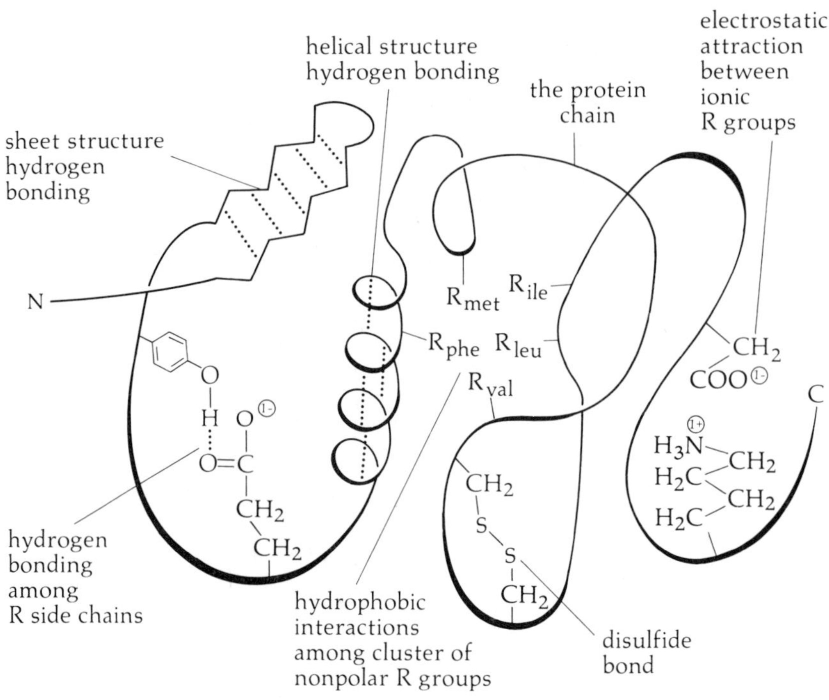

A Diagrammatic Representation of Stabilizing Forces in Globular Proteins. Only one of each type of interaction is shown. The frequency of occurrence of each type varies from protein to protein. The sheet and helical structures, and disulfide bonds, can all be absent, all present, or any combination can be present. The other types of interactions are found in all globular proteins.

The Disulfide Bond

Two of these amino acids, in different parts of the protein chain, can react with each other under oxidizing conditions to form a covalent bond: a disulfide bridge, $—CH_2—S—S—CH_2—$. (Refer to Figure 17-1.) When this occurs, two parts of the chain are tied together, which will produce a loop in the chain, or a link between two separate chains. This disulfide link has a very important effect on the conformation of the chain or chains involved and is considered either part of the secondary or part of the tertiary structure. Under reducing conditions in the laboratory, this bond can be broken, which will often "denature" the protein (change the protein from its native, active configuration to an inactive one). Other methods of denaturation include changing the temperature, pH, or concentration and types of ions in the protein solution or adding a "denaturing agent," such as urea, to the solution. Sometimes denaturation is reversible; sometimes it is not (like frying an egg).

QUATERNARY STRUCTURE

Quaternary structure deals with the interactions between fully formed protein chains. In hemoglobin, for example, the fully active molecule consists of four protein

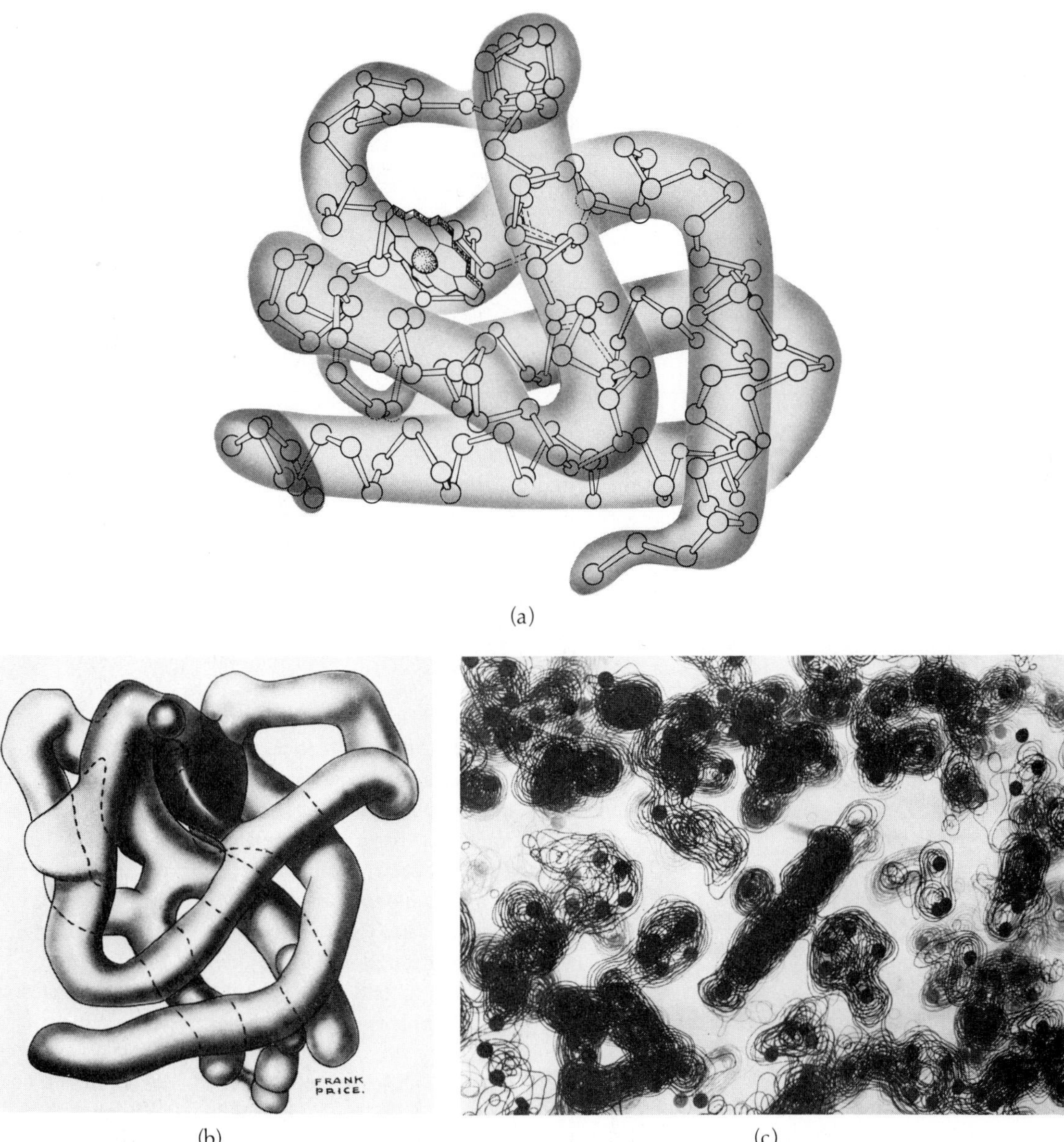

(a)

(b)

(c)

Some Representations of the Three-Dimensional Structure of Myoglobin.
(a) is a drawing to emphasize the helical segments of the chain; no other features are shown. (b) is a clay model simply showing the path of the polypeptide chain and the location of the heme group, the dark disc near the top center. Again, no structural details are shown. (c) is the electron density map obtained from X-ray diffraction experiments. This map is the basis for the other models.

Figure 17-11

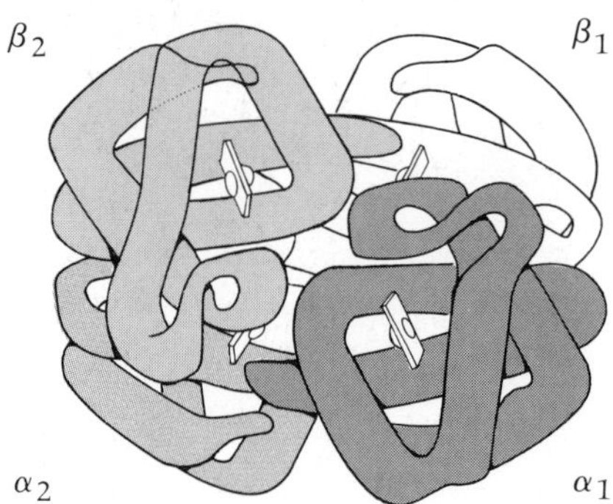

THE FOUR SEPARATE CHAINS OF HEMOGLOBIN ASSEMBLED INTO THE FUNCTIONING PROTEIN. The positions of the α_1, α_2, β_1, and β_2 chains, as well as the position of the heme groups, represented by the cardboard squares, are indicated.

17-1 *Gratuitous Information*

METABOLISM OF PROTEINS

The metabolism of proteins by mammals yields water, carbon dioxide, and urea

$$H_2N-\overset{\overset{\displaystyle O}{\|}}{C}-NH_2$$

Urea is a white, crystalline solid that is soluble in water, and urea is excreted in the urine. Note that urea is the result of only partial oxidation of proteins. Further oxidation would produce oxides of nitrogen, such as NO, NO_2, or the nitrate ion NO_3^{1-}. The metabolism of proteins is complicated and depends on species. In sea birds, for example, one end product of protein metabolism is guanidine

$$H_2N-\overset{\overset{\displaystyle NH}{\|}}{C}-NH_2$$

This compound is a major constituent of sea bird droppings, given the polite name "guano." Vast deposits of guano exist where sea birds have been nesting for long times— places like the Pacific coast of Chile. Over time, further oxidation of guanidine produces the nitrate ion, and "Chilean nitrates" have played a major role in history as a source of nitrate ion both for fertilizer and for explosives.

It is impossible to write a general equation for the oxidation of a protein, partly because the end product of oxidation varies so much among species but more so because there is such a wide variety of R groups in proteins. One can, however, give a rough average figure of 1 gram of protein providing about 20 kilojoules (5 kcal) of energy in humans. One gram of protein + oxygen → oxidation products + 2×10^4 joules.

More important than metabolism is the role of proteins in the diet to provide the "essential" amino acids, those that are required by a certain species but cannot be synthesized by that species. A general rule: the higher up the food chain an organism is, the more dependent that organism becomes on external sources for essential amino acids. We have marked the essential amino acids for humans in Table 17-1. From the number of them, we could predict that humans are near the top of the food chain. A simple bacterium like *E. coli* can synthesize all the amino acids it needs.

chains: two of one kind (α-chains) and two of another (β-chains). The α and β used here simply designate two different types of protein chains and have nothing to do with previous usage of the letters α and β. The complete structure is therefore $\alpha_2\beta_2$, and the quaternary structure refers to the details of the interaction between these chains. (See Figure 17-11 for a schematic representation of this situation in hemoglobin.) Hemoglobin also incorporates a "prosthetic" group, the heme, which is essential for its function. Many enzymes have these nonprotein portions. Another well-known case is chlorophyll, the green pigment in plants. The diversity and complexity of protein structure rationalizes the wide variety of properties that are seen in nature.

Interactions That Contribute to Quaternary Structure

Proteins do more, however, than just have structural roles in the body; they also have functions (specific purposes). The class of proteins called enzymes, for example, is necessary to catalyze the multitude of chemical reactions that go on in the body: metabolism (conversion of food to other products and energy), synthesis of proteins and nucleic acids (see Chapter 19), and nerve transmission, to name just a few. Other functioning proteins include hormones (insulin, for example) and the antibodies, to name only two other classes. Enzyme proteins are characterized by their "active site," the region of the protein where catalytic activity is believed to

Enzymes

Gratuitous Information 17-2

RESTYLING YOUR HAIR

Some people have straight hair and want curled hair, some have curled hair and want straight hair, and some want a different curl from what they have.

The major protein constituent of hair is keratin, which has a very high percentage of α-helical structure. It also has a number of cystine residues, which make disulfide cross-links possible. We'll examine here the chemistry behind modifying hair with an iron and by means of a "permanent wave" preparation.

What keeps keratin in the α-helical structure are hydrogen bonds and disulfide cross-links. We can intervene in the formation of hydrogen bonds and cross-links by the keratin molecules in hair and thus change the amount of curl. If the hair is moistened, there is an attraction between the water molecule and the hydrogen bonding sites in keratin and intramolecular hydrogen bonds are replaced, in part, by hydrogen bonds to water. Hair becomes limp when wet. If you now bend your hair around a curved object and apply heat to remove the water molecules, the intramolecular hydrogen bonds reform differently from their natural way of doing so, and you have put an artificial wave in your hair. As is well known, this type of wave vanishes next time hair becomes wet. The artificial wave is not permanent.

The process of creating a permanent wave depends on the existence of disulfide cross-links in keratin. As the hair is moistened and held in place appropriately by rollers or pins, the disulfide cross-links are first removed and then re-formed with the hair in its new position. The new hydrogen bonds formed on drying are held tightly in place by a network of disulfide cross-links. The only way to change a truly permanent wave is to cut the hair off.

The rather strong (and often unpleasant) odors associated with permanent wave preparations are the result of the chemical reagents necessary to do the job. Disulfide cross-links are reduced (removed) with compounds called thiols (sulfur compounds much like alcohols), which are famous for their powerful odors. (The defensive odor of skunks features several of these compounds.) Reformation of the cross-links is an oxidation step, and, while the chemicals involved are not as pungent as the thiols, they do have strong characteristic odors.

The tanning of leathers involves, among other things, the introduction of permanent artificial cross-links into the protein collagen, the major constituent of hides. Even stronger and less pleasant odors have long been associated with the tanning process.

be centered. The configuration of the enzyme often involves a "cleft," maintained by the secondary and tertiary structures of the protein, into which the substrate (the molecule that is acted on by the enzyme) fits, often almost exactly. Once the substrate is bound to the enzyme, a chemical reaction takes place, often involving the side chains (R groups) of the enzyme, the solvent, and the substrate. Then the substrate (now changed by the chemical reaction) is released from the enzyme, which is then ready to receive a new substrate molecule. Figure 17-12 gives some idea of this process in the specific case of the enzyme lysozyme, which hydrolyzes carbohydrates.

Studies on the structure and function of proteins and nucleic acids (Chapter 19) are two of the most productive and interesting areas of chemistry today. Because of their biological importance, these molecules are also studied by biologists, biochemists, and physicists. The area of study that is developing from the interaction of these four more traditional fields is called molecular biology, and it is certain to become a more and more important part of our attempts to understand normal biological mechanisms and disease.

Figure 17-12

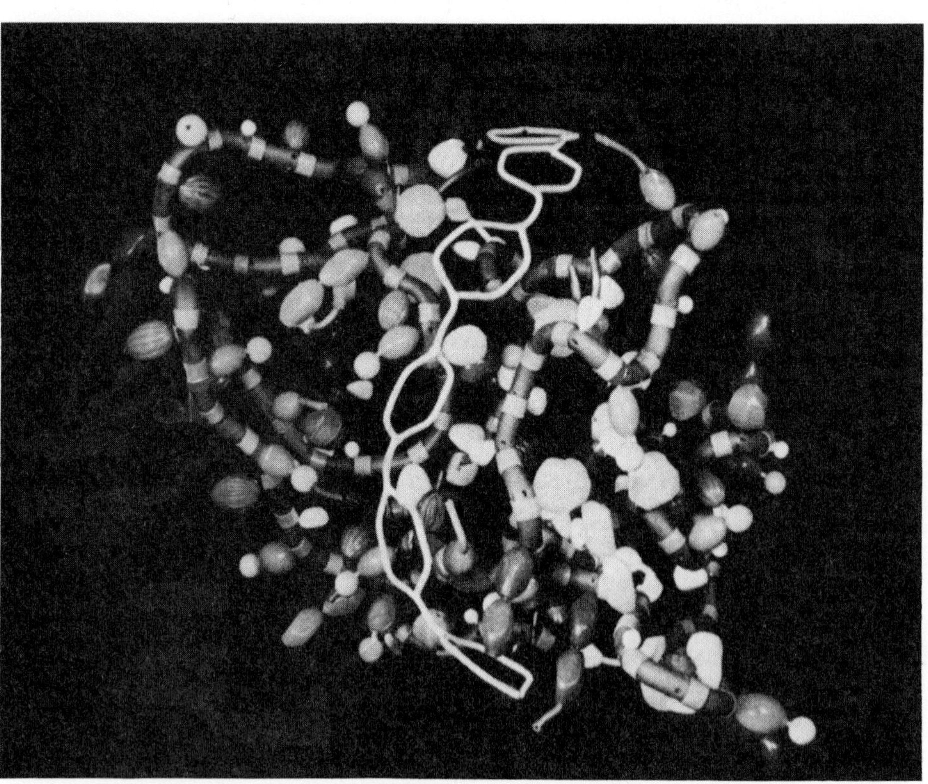

A SCHEMATIC REPRESENTATION OF THE BONDING OF A SUBSTRATE (A CARBOHYDRATE) TO AN ENZYME (LYSOZYME). The carbohydrate substrate, with its six-membered ring structure, is represented by the wire-frame model in the diagram.

Scramble Exercises

1. Write out the full structure for the dipeptide leu–lys.

2. Draw the full structure of the dipeptide proline–proline. Take care. Table 17-1 gave the full structural formula of proline (not just an R group). Use that full structural formula when you write out the structure of the dipeptide. Assume a pH range of 6 to 8.

3. Write out several possible polypeptide sequences using the amino acids abbreviated as Ala, Gly, Ser, Tyr, Asp, Leu, and Val. The first one might be in the order just given—Ala–Gly–Ser–Tyr–Asp–Leu–Val.

4. Write out one of those sequences above using the full structures (all the carbon, nitrogen, hydrogen, and oxygen atoms and all the bonds).

5. Those of you with the requisite drawing skills might wish to attempt drawing that sequence in Exercise 4 as (a) an α-helical segment and (b) a pleated sheet (β) form. Remember it takes two chains side by side to make a pleated sheet. Several students who like tinker toys might also wish to pool model sets to get enough atoms to make models of these structures.

6. Write out the full structures of the *cis* and *trans* isomers around the peptide bond between Ala and Ser in the structure Gly–Ala–Ser–Leu. Again, a pooling of model sets may also be helpful.

Additional exercises for Chapter 17 are given in the Appendix.

A Look Ahead

This is the first in a sequence of chapters that deal with large molecules. Part of our purpose is to show that the rather special properties of large molecules are the natural consequence of their construction from rather simple repeating units. In this chapter, we have shown how proteins are chains of repeating amino acid residues. The next chapter shows how the properties of carbohydrates (starches and celluloses) result from the buildup of a chain with repeating sugar units.

18

SUGARS, STARCH, GLYCOGEN, AND CELLULOSE: THE CARBOHYDRATES

Chapter 18

Green plants are efficient in converting carbon dioxide and water into a great variety of compounds. Among them are those classified as carbohydrates.

$$n\,CO_2 + y\,H_2O \xrightarrow[\text{green plants}]{\text{sunlight}} C_n(H_2O)_y + n\,O_2$$

The process is known as photosynthesis and takes its energy from the sun. (The word carbohydrate and the formula $C_n(H_2O)_y$ have no structural significance; they simply indicate the relative number of atoms in a molecule and explain, since there are apparently a number of H_2O units for a number of C atoms, why this class of compounds is known as carbohydrates.) Sugars, starches, and celluloses are all carbohydrates.

Photosynthesis

Photosynthetic processes are our primary source of food substances and were the primary source of the materials that were converted over long periods of time into the fossil fuels. Chemists could start with carbon dioxide and water and, under suitably controlled conditions, synthesize a great variety of compounds of carbon, hydrogen, and oxygen, some of which would be carbohydrates. However, with the current state of the art, we are much better at producing fertilizers and modifying plant products to meet our purposes than we are at synthesizing carbohydrates.

Isomerism in Sugars

In this discussion of carbohydrates, we shall start with one specific sugar: glucose, with the formula $C_6H_{12}O_6$. The six-carbon sugars are called **hexoses.** Glucose is an **aldohexose,** a six-carbon sugar that contains an aldehyde group

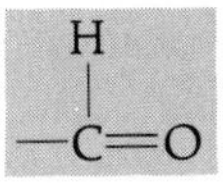

$$\begin{array}{c} H \\ | \\ -C{=}O \end{array}$$

D-Glucose, a Six-Carbon Sugar

There are 16 straight-chain (i.e., containing no ring structures) isomeric hexoses. One of these 16 isomers is the uniquely significant compound known as D-glucose, or dextrose. The representation of structural isomers of the hexoses on a flat surface presents some problems. Our concern here is how the isomers arise rather than the conventions used to represent the open chain isomeric hexoses.

Two representations of D-glucose are given below:

$$
\begin{array}{c}
H{-}{}^{1}C{=}O \\
| \\
H{-}{}^{2}C{-}O{-}H \\
| \\
H{-}O{-}{}^{3}C{-}H \\
| \\
H{-}{}^{4}C{-}O{-}H \\
| \\
H{-}{}^{5}C{-}O{-}H \\
| \\
H{-}{}^{6}C{-}O{-}H \\
| \\
H
\end{array}
\qquad\qquad
\begin{array}{c}
CHO \\
H{-}{+}{-}OH \\
HO{-}{+}{-}H \\
H{-}{+}{-}OH \\
H{-}{+}{-}OH \\
CH_2OH
\end{array}
$$

The structure at the left displays the aldehyde group, the hydroxyl groups, and the system of numbering the six carbon atoms in the chain. All bonds, with the exception of the carbon–oxygen double bond in the carbonyl group, are single bonds, and consequently rotation is possible about all of these bonds.

If two identical ball and stick models are made (please get out your model set and build along with this discussion) of this array of atoms, it becomes readily apparent that an exchange of the hydrogen atom with the hydroxyl group at carbon-2 (the second carbon atom) on one of the models gives a molecule that is no longer identical (superimposable) with the unmodified model. There are, therefore, two isomers that depend only upon the relative positions of the hydrogen atom and the hydroxyl group at carbon-2. These two isomers are evidence of a lack of symmetry at this position, and the carbon-2 carbon atom is said to be a dissymmetric carbon atom. This situation always arises when four different atoms or groups of atoms are bonded to a carbon atom. (Refer to amino acids in Chapter 14, page 335.)

In a very similar manner, it is easy to demonstrate that, again, the relative positions of the hydrogen atom and the hydroxyl group at carbon-3 determine two isomers. In other words, carbon-3 is also a dissymmetric carbon. Carbon-4 and

18-1 *Gratuitous Information*

SUGAR IN URINE

Glucose is the major sugar that circulates in the blood stream, and the oxidation of glucose in muscle is the major energy source that allows us to get from place to place. Diabetes is a disease (usually a lack of the hormone insulin produced in the pancreas) that prevents normal metabolism of glucose, and, as a result, the concentration of glucose can become abnormally high in the blood stream. When that happens, the kidneys, which are our blood-purifying organs, are unable to handle the load and some glucose appears in the urine. The preliminary test for diabetes is the "sugar in the urine." The test for the presence of glucose in urine depends upon the presence of the aldehyde group, which is a good reducing agent. The oxidizing agent used is one that produces a color when it is reduced by the aldehyde group of glucose. "Test tapes" used by diabetics are simply pieces of paper impregnated with the proper reagent to develop a color in the presence of an aldehyde.

carbon-5 are also dissymmetric, but carbon-6 is symmetric. Carbon-6 is not attached to four different atoms or groups of atoms. An exchange of a hydrogen atom with the hydroxyl group at carbon-6 on a model does not change the identity of the structure. The exchange at carbon-6 is simply equivalent to a rotation about the carbon–carbon single bond (σ-bond) between carbon-5 and carbon-6. It is the four dissymmetric carbon atoms—carbon-2, carbon-3, carbon-4, and carbon-5—that account for the sixteen straight-chain isomeric aldohexoses: two isomers at each of four atoms, or $2 \times 2 \times 2 \times 2 = 16$.

The structure at the right on page 440 uses a conventional representation employed by carbohydrate chemists. This symbolism is known as a Fisher projection. Our only concern is the recognition that the D-glucose, so important to biological systems, is one specific isomer of the aldohexoses, not just any aldohexose.

Cyclic Structures of Sugars

There are two cyclic isomers of D-glucose. These two isomers are designated as α-D-glucose (*alpha*-D-glucose) and β-D-glucose (*beta*-D-glucose). In an aqueous solution of D-glucose, the three isomers exist in equilibrium:

$$\underset{\text{(cyclic)}}{\alpha\text{-D-glucose}} \rightleftarrows \underset{\text{(straight chain)}}{\text{D-glucose}} \rightleftarrows \underset{\text{(cyclic)}}{\beta\text{-D-glucose}}$$

Partial structures are given below to focus attention on the relation of the cyclic structures to the straight chain. The complete structures of α-D-glucoses and β-D-glucose are given in Figure 18-1.

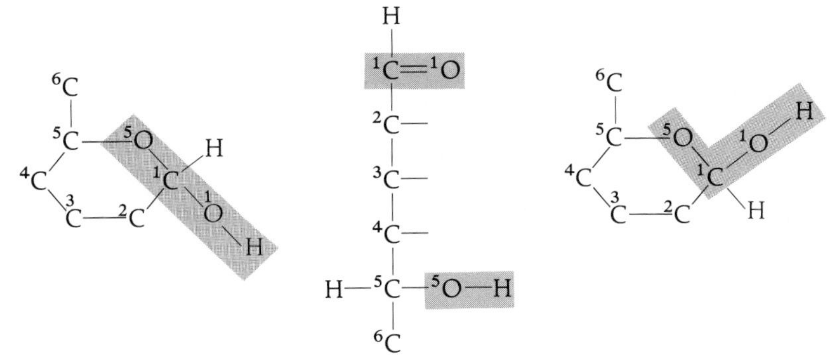

α-D-glucose
(partial structure)

D-glucose
(partial structure)

β-D-glucose
(partial structure)

The bonding at carbon-2, carbon-3, carbon-4, and carbon-6 is the same in all three isomers. The ring structure involves an ether linkage

$$\overset{5}{C}-\overset{5}{O}-\overset{1}{C}$$

Isomerism in Cyclic Sugars

Trace back the position of these numbered atoms to their position in the open D-glucose chain. In the cyclic structure, carbon-1 is dissymetric and the only difference between α-D-glucose and β-D-glucose is the relative position of the hydrogen atom and the hydroxyl group at carbon-1. This seems like a very small difference, but biologically it is an extremely important difference, as we shall see. Starches are polymers of α-D-glucose and the celluloses are polymers of β-D-glucose. The starches are digestible by humans, the celluloses are not. In forming the ring, the *alpha* and *beta* isomers result from the two possible arrangements of the hydrogen atom and the hydroxyl group at carbon-1. In terms of the ball and stick model, the orientation of the hydroxyl group with respect to the hydrogen atom at carbon-1 could be thought of as the consequence of which spring of the double bond opens up in forming the new bonds to oxygen and hydrogen. In terms of orbitals, it can be thought of as the consequence of the manner in which the hybrid sp^2-orbitals (planar) of carbon-1 change over to hybrid sp^3-orbitals (tetrahedral) in the cyclic structure. The sp^2 hybrid orbitals cannot be involved in a chiral center. The change from sp^2 to sp^3 hybridization introduces a new chiral carbon atom with two possible isomers.

An aqueous solution of D-glucose contains an equilibrium mixture of all three forms. At equilibrium, the concentration of the β-D-glucose is greater than the concentration of the α-D-glucose and the concentration of the α-D-glucose is much greater than the concentration of the straight-chain D-glucose. If one form is removed by chemical reaction, the equilibrium shifts to maintain the same relative distribution.

The complete formulas for the three isomeric D-glucoses are given in Figure 18-1. Ball and stick models of these structures indicate that the rings are not planar. X-ray studies confirm this prediction and show the rings to be bent into a manner commonly designated as a chair conformation. Figure 18-2.

The Chair Form

Fructose, the sugar associated with fruits, also has the molecular formula $C_6H_{12}O_6$. Fructose is a hexose but it does not contain the aldehyde group: it is not an **aldohexose.** Straight-chain fructose is instead a **ketohexose.** In glucose, the carbonyl group contains carbon-1; in fructose, the carbonyl group contains carbon-2. There are eight possible straight-chain ketohexoses. In all of these, the carbonyl group contains carbon-2. The D-fructose isomer of the ketohexoses is commonly found in

D-Fructose, a Five-Carbon Sugar

Figure 18-1

H—C=O
|
H—C—OH
|
HO—C—H
|
H—C—OH
|
H—C—OH
|
H—C—OH
|
H

α-D-glucose   D-glucose   β-D-glucose

STRUCTURES OF THE THREE ISOMERS OF D-GLUCOSE.

Figure 18-2

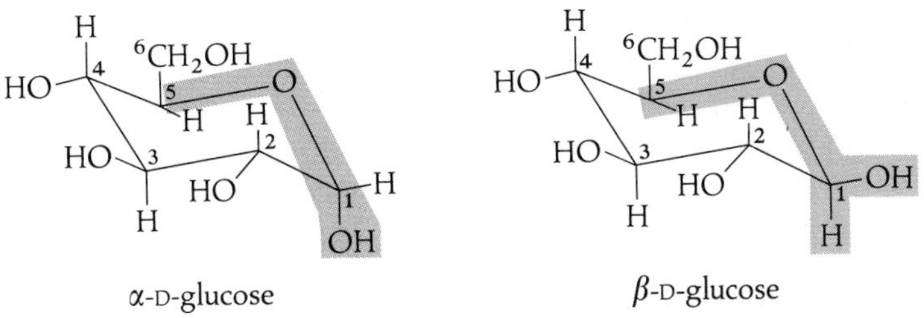

THE CHAIR FORMS OF α-D-GLUCOSE AND β-D-GLUCOSE.

fruits. The straight-chain compound exists in equilibrium with two cyclic isomers, α-D-fructose and β-D-fructose. Figure 18-3. Note that the straight-chain structure contains only three dissymmetric carbon atoms: carbon-3, carbon-4, and carbon-5. This accounts for there being eight straight-chain isomers for ketohexoses—$2 \times 2 \times 2 = 8$—instead of the sixteen for the aldohexoses. The cyclic structures contain a five-atom instead of a six-atom ring and α-D-fructose and β-D-fructose are the two chiral isomers arising from the dissymmetry at carbon-2 in the ring structures.

During the transitions from the straight chain to cyclic D-fructose structures, the bonding at carbon-1, carbon-3, carbon-4, and carbon-6 remains unchanged. The formation of this five-member cyclic ether of a ketohexose is entirely parallel to the formation of the six-member cyclic ether of the aldohexose. The difference arises exclusively from the position of the carbonyl group in the ketohexose and the aldohexose.

Figure 18-3

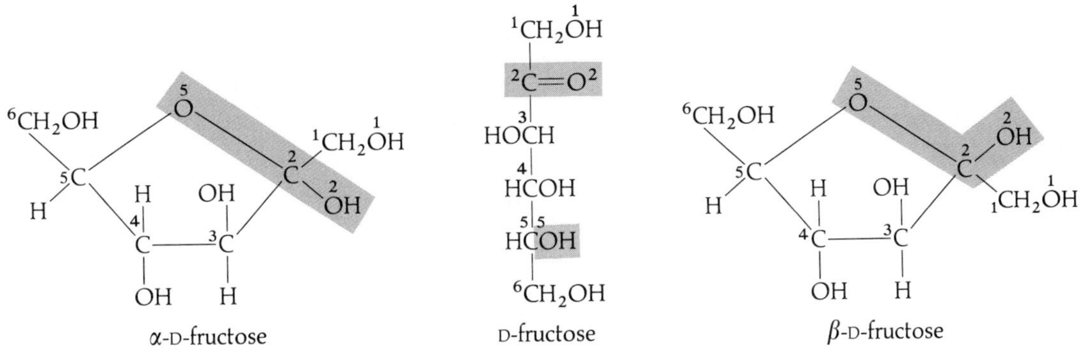

THE THREE ISOMERIC STRUCTURES OF D-FRUCTOSE. Oxygen atoms 1, 2, and 5 are numbered in all three structures.

There is a great multiplicity of sugars, not only in the sense of the isomers of the hexoses but also in terms of the length of the carbon chain, which may be as short as four atoms or as long as eight. In all cases, there is either an aldehyde or a ketone group and a number of alcohol groups. These rather low molecular weight compounds are soluble in water and insoluble in nonpolar solvents such as benzene. Many of them crystallize to give rather hard crystals. The marked solubilities of these compounds in water and the hardness of the crystals are rationalized in terms of extensive hydrogen bonding with molecules of water in solution and with other molecules of the sugar in the crystal.

Phosphate Esters of Sugars

Since these compounds are alcohols, they can react with acids to form esters. One biologically important class of esters is those with phosphoric acid, one of which is α-D-glucose-1-phosphate. The number of protons on the phosphate group and consequently the charge on the ion depends, of course, on the pH of the solution. In a more acidic solution, there would be more protons attached.

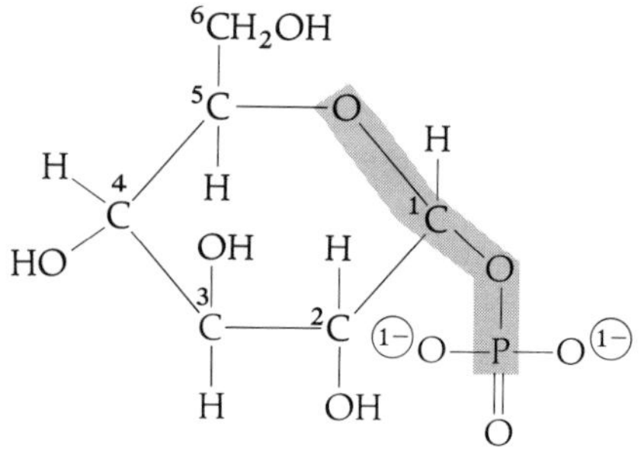

α-D-glucose-1-phosphate ion,
also called α-D-glucose-1-P

Note the similarity of this structure to the HPO_4^{2-} ion of phosphoric acid:

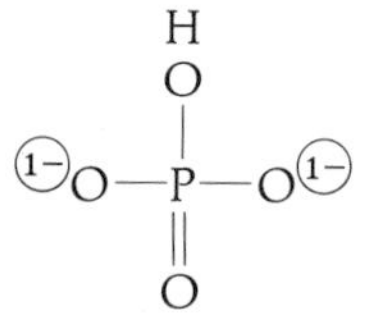

Since this phosphoric acid ester of glucose plays such an important role in biological reactions, it is sometimes referred to as the mobilized form of glucose.

Sucrose (Table Sugar)

Sucrose, $C_{12}H_{22}O_{11}$, the common table sugar, is a disaccharide made up of a double ring structure that yields, on hydrolysis with one molecule of water, one molecule of D-glucose and one molecule of D-fructose. Figure 18-4.

To orient yourself to a structure such as this requires some patience. It is comparatively easy to recognize the six-member ring—five carbon atoms and one oxygen atom—of glucose at the left and the five-member ring—four carbon atoms and one oxygen atom—of the fructose at the right. It is a great deal more troublesome to trace through the details of identifying that they are α-D-glucose and β-D-fructose. The fructose is particularly difficult since the fructose ring (Fig. 18-3) has been turned

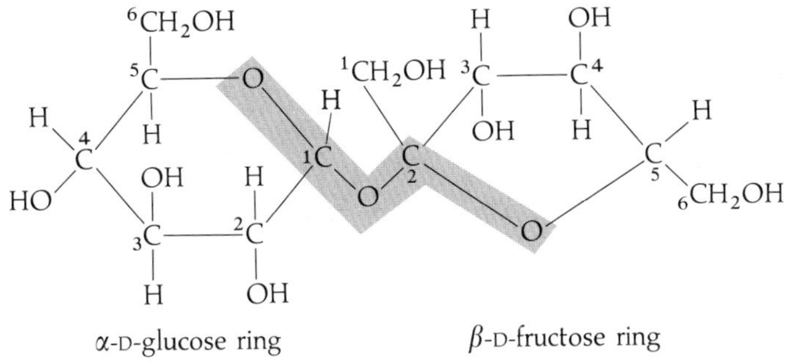

Figure 18-4

$$\alpha\text{-D-glucose ring} \qquad \beta\text{-D-fructose ring}$$

THE DOUBLE RING STRUCTURE OF SUCROSE, A DISACCHARIDE. The α-D-glucose and β-D-fructose rings are joined by an ether linkage at carbon-1 of the α-D-glucose ring and at carbon-2 of the β-D-fructose ring.

180° and inverted to bring carbon-2 of the fructose in position with carbon-1 of the glucose molecule. On hydrolysis, sucrose yields an equilibrium mixture of the three D-glucose isomers and three D-fructose isomers:

$$\text{sucrose} + H_2O \xrightarrow{\text{hydrolysis}} \text{D-glucose} + \text{D-fructose}$$

Sucrose is the principle sugar found in beets and cane. The degree to which sucrose withstands hydrolysis during refining processes and cooking processes is clear evidence of the stability of this disaccharide to hydrolysis to give the two monosaccharides glucose and fructose. In the kitchen, in a kettle, this hydrolysis reaction can be catalyzed by the addition of an acid such as acetic acid or citric acid from fruits, and the rate can, of course, be increased by elevating the temperature. This process is familiar to all who have made jams or jellies. In biological systems an enzyme, invertase, catalyzes the hydrolysis of sucrose to D-glucose and D-fructose and the reaction proceeds readily at body temperature. In the process of making honey, bees provide an appropriate enzyme and honey is a 50/50 mixture of the D-glucoses and the D-fructoses. Chocolate-covered cherries contain a D-fructose/D-glucose mixture formed after the chocolate is applied to a solid mixture containing sucrose and an appropriate enzyme.

Combustion of Sucrose

If the energy requirements of the body justify it, sucrose is eventually oxidized to carbon dioxide and water by a long series of reactions involving enzymes and co-enzymes. The total energy produced is the same as the heat released by complete combustion:

$$C_{12}H_{22}O_{11} + 12\,O_2 \longrightarrow 12\,CO_2 + 11\,H_2O$$
$$\Delta H = -5600 \text{ kilojoules } (-1350 \text{ kcal}) \text{ per mole of sucrose}$$

This figures out to about 17 kilojoules (4 kcal) per gram of sucrose. If the energy requirements do not justify immediate consumption, the hexoses are converted to glycogen, which becomes a readily available source of energy in the blood stream and in tissues throughout the body.

Lactose

Lactose, the sugar found in milk, is a disaccharide of D-glucose and D-galactose. The latter monosaccharide is also a hexose and is identical to D-glucose with the exception of the exchange of the hydrogen atom and the hydroxyl group at carbon-4. At birth, normal mammals have an enzyme, lactase, that promotes the hydrolysis of lactose. In many individuals the production of this enzyme descreases with age, and lactose may not be tolerated, much less digested, by many adults. This is particularly true for individuals whose ancestors come from regions of the world where milk and milk products are not a normal part of the diet for adults.

Plants specialize in forming very large molecules that can be hydrolyzed into monosaccharides. If only repeating units of α-D-glucose are involved, the polysaccharide is a starch. If only repeating units of β-D-glucose are involved, the polysaccharide is a cellulose. The hydroxyl group at carbon-1 and the hydroxyl group at carbon-4 are involved in polymer formation by ether linkages between adjacent rings. The following very schematic representation will be used for α-D-glucose and β-D-glucose.

α-D-glucose
(<u>very</u> schematic representation)

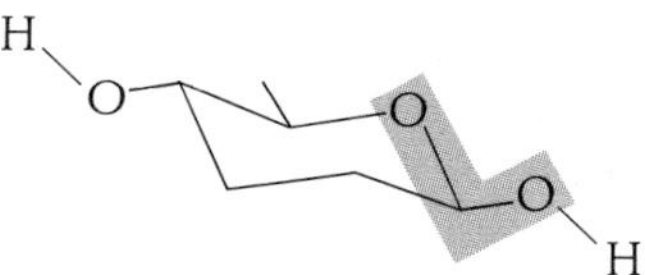

β-D-glucose
(<u>very</u> schematic representation)

Note the orientation of the hydroxyl groups at carbon-1. See Figure 18-2 for more complete structures of the monomers.

Figures 18-5 and 18-6 give a very schematic representation of a segment of a starch molecule and a segment of a cellulose molecule.

Starch
Cellulose

It is indeed remarkable that plants, through their enzyme systems, assemble units of α-D-glucose into the starch stored in their stems and seeds, particularly the seeds, and assemble β-D-glucose units into the cellulose that forms the supporting structure of the plant and its cells walls. Starch and cellulose are very similar in their structures but very different in their functions. The difference is underscored by the fact that enzyme systems of humans readily convert starches into glucose units but are incapable of converting the celluloses into glucose units. On the other hand, many animals such as cows flourish on a diet of grass. This apparently is, however, not a characteristic of the cow's enzyme systems but of the enzyme systems of microorganisms that inhabit the multiple stomachs of the cow. In the digestion of starch, enzymes called phosphorylases convert the end glucose units into α-D-glucose-1-phosphate ion and metabolism is on its way. Other enzymatic systems may break the long chains of starch into shorter segments. This process begins in the mouth with the enzyme salivary amylase during the act of chewing food.

Figure 18-5

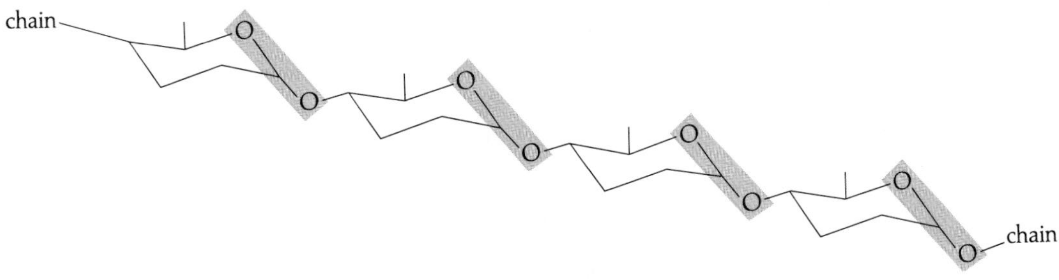

STARCH. Schematic representation of the ether linkages between the carbon-1 atom of one α-D-glucose ring and the carbon-4 atom of the next α-D-glucose ring in a segment of an amylose starch molecule. These continuous chain starches are known as amyloses and may include as many as 100,000 glucose units.

Branched-Chain Polysaccharides

In natural systems, neither the starches nor the celluloses are limited to continuous chains of glucose units bonded 1 → 4. Branching occurs by some glucose unit being bonded to three other glucose units, each of which is a part of a chain of glucose units. In the case of starch, as indicated in Figure 18-7, the primary bonding is 1 → 4 with an occasional branching by 1 → 6 bonding. Diversity in branching leads to the great diversity in natural starches. These branched-chain starches are known as amylopectins, some of which are estimated to contain more than a million glucose units.

Natural starch occurs in granules ranging in diameter from 2 to 100 microns. These granules are insoluble in water but swell in hot water during cooking into small balloons that finally burst. On cooling, the cooked starch forms an opaque paste.

Most natural starches contain 70 to 80% amylopectin (branched chain) and 20 to 30% amylose (straight chain), and the paste obtained on cooking reflects the properties of both types of molecules. A hot amylopectin solution yields a clear, nongelling paste. A hot amylose starch solution sets to an opaque, rigid gel. Both sets

Figure 18-6

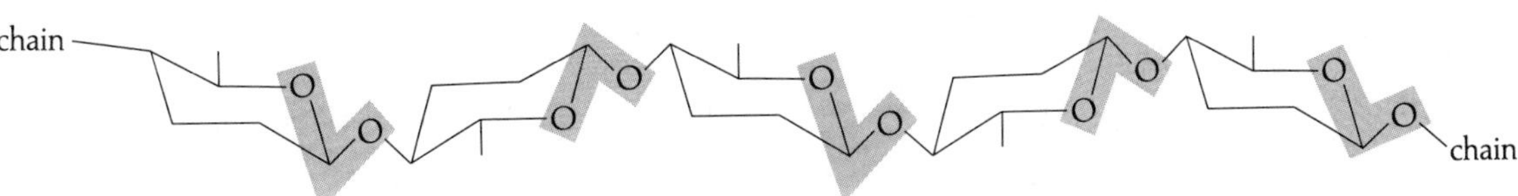

CELLULOSE. Schematic representation of repeated units of β-D-glucose bonded at carbon-1 and carbon-4 through an ether linkage. Note that alternate rings have been inverted to preserve the bond angles at the ether linkage between adjacent rings.

Figure 18-7

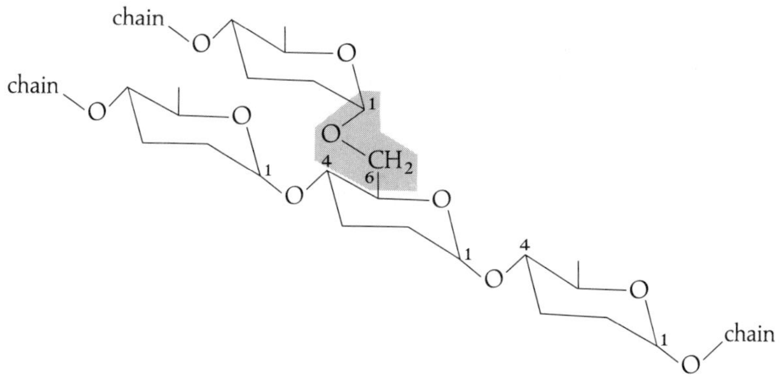

SCHEMATIC REPRESENTATION OF A SEGMENT OF AMYLOPECTIN, A BRANCHED-CHAIN STARCH.

of properties are attractive in the preparation of food products, and strains of corn have been developed that produce a high yield of amylose starch and other strains developed that produce a high yield of amylopectin starch.

Cornstarch—Uses in Everyday Life

Processed cornstarch is big business in the United States, and great skill has been developed not only in modifying chain lengths but also in controlling the cross-linking of one chain to another in order to have materials that impart desirable characteristics to jellies, jams, pastries, ice cream, soups, puddings, gelled desserts, package mixes, gum confections, and candies. In many of the above, it is the physical properties of the product that are of prime importance. It is now common practice to chemically introduce some cross-linking between chains by treating the natural starch with a reagent that does not destroy the granules and that can be easily washed out

18-2 Gratuitous Information

"THIS WEATHER TAKES THE STARCH OUT OF ME"

Anyone who has done much ironing is aware of the use of an aqueous solution of starch to help in "finishing" cotton and similar garments. The clothes to be ironed are dipped in the starch solution or the starch solution is sprayed on the clothes and the heat of the iron removes the water. A network of starch molecules linked by hydrogen bonds between themselves and between the starch molecules and the cellulose molecules of the fabric (if it's cotton) is left behind, which serves to stiffen the fabric and remove wrinkles. As the garment is worn, the wearer perspires and water molecules reenter the starch-fiber network. New hydrogen bonds between the starch molecules and the celluloses and the water molecules form and the fabric loses its stiffness, becoming limp and wrinkled. The analogy is used all the time as, on hot muggy days, one often hears, "This weather takes the starch out of me."

of the starch later. It has been found that one cross-link to another chain at intervals of 1000 to 2000 glucose units introduces significant differences in the physical properties of food products. One area of interest is the production of edible films, for soluble, digestible food packaging. Materials that are edible and protect sensitive foods from the oxygen and the moisture of the air are the ultimate in biodegradable packaging. (Ice cream cones are perhaps our most popular biodegradable dishes.) Corn syrup, another starch product, is an aqueous solution of partially hydrolyzed cornstarch that contains a great deal of glucose along with a wide variety of glucose chain molecules that have low molecular weights in comparison with the original starches.

Artificial Sweeteners

It is fairly easy to compare the relative sweetness of the various sugars—and other compounds that are not sugars at all from the chemical point of view—by matching the taste of aqueous solutions of the various sugars at various concentrations of the solutions. D-glucose is only about half as sweet as sucrose (the beet or cane sugar of the market-place), but D-fructose is about one and a half times as sweet. Lactose is only about a fifth as sweet as sucrose. On the other hand, a derivative of benzene commonly known as saccharin is judged to be more than 500 times as sweet as sucrose.

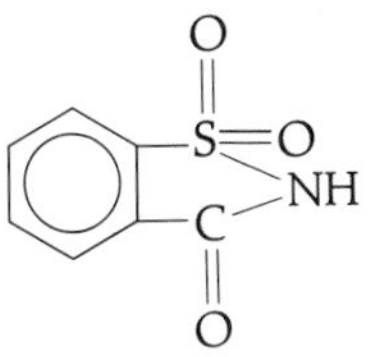

saccharin

Aspartame, or Nutrasweet, perhaps best known for its use in soft drinks, is a dipeptide (see Chapter 17, page 418) and is about 160 times as sweet as sucrose.

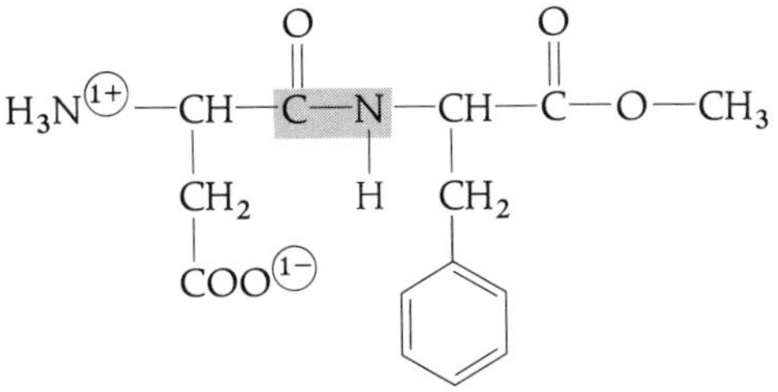

aspartame

The relation of taste to chemical structure is still a part of the great unknown.

Glycogen

Glycogen, referred to earlier in this chapter in connection with a holding pattern of glucose as an energy reserve in blood and animal tissues, is a polyglucose very similar to the amylopectin starches produced by plants. The α-D-glucose 1 → 4 chain is usually highly branched, with 1 → 6 branches about every tenth glucose unit in the 1 → 4 chain. This high degree of branching is attributed to a "glycogen branching enzyme," which is believed to "snip off" (cleave) di- and triglucose fragments of a 1 → 4 chain and attach them at the 6 position on the same or another 1 → 4 chain. Glycogen molecules may build up to quite large units containing as many as a million glucose units. One of the ways glucose units move back into the production of energy is through the conversion of the end glucose unit of 1 → 4 chains to α-D-glucose-6-P, a mobilized form of glucose, by the action of phosphylases. (See page 444 for the structure of α-D-glucose-1-P. Here we are discussing the use of α-D-glucose-6-P.)

Much of the beauty of our natural world is closely related to those cellulose 1 → 4 chains of β-D-glucose molecules and the infrequent 1 → 6 branching that occurs along those chains. Plants rise above the ground on graceful stems that bend with the wind, the ice, fruit, birds, and the actions of small children. The seas and oceans of the world were explored in wooden ships with sails made of cotton and the pioneers went west in wooden wagons pulled by animals sustained by the leaves and stems of plants. We enjoy both the feel and the visual beauty of highly polished wood and fabrics made from cotton and flax fibers.

Degradation of Cellulose in the Environment

The durability of cellulose is impressive, but those 1 → 4 and 1 → 6 linkages do yield to the enzyme systems of microorganisms and cellulose is biodegradable. The enzymes that participate in the formation of a chemical bond also participate in the breaking of that bond. Biodegradation is particularly evident under warm and humid conditions, but "dry rot" does occur with the aid of microorganisms under comparatively dry conditions.

There is considerable interest in the conversion of cellulose into usable food for people. This conversion can be brought off by appropriate organisms or by appropriate enzymes obtained from organisms and by synthetic catalysts modeled on natural enzyme systems. With an increasing world population, the conversion of whole plants into available food energy becomes increasingly attractive.

By the chemical treatment of cellulose, changes have been brought about that change the surface properties of natural fibers, leading to fabrics that may be "no-iron" and much easier to take care of. More extensive chemical changes have led to a variety of cellulose acetates (rayon) that can be extruded into fibers or films or cast into hard, tough, and durable items used as components in the manufacture of automobiles, appliances, sporting equipment, and toys.

Scramble Exercises

All of the following relate to the structure of those carbohydrates that are important in nucleic acids.

1. Write out the general formula for the aldopentoses, $C_5H_{10}O_5$, as a straight chain.

2. Identify the dissymmetric carbon atoms and determine the number of straight-chain isomers. One of these, known as D-ribose, is biologically significant.

3. Write out a reasonable ring structure for D-ribose. You have, of course, no way to predict the actual orientations of the hydrogen atoms and hydroxyl groups. The ether linkage connects carbon-1 and carbon-4. Does the possibility of *alpha* and *beta* structures arise on the basis of the relative hydrogen–hydroxyl orientation at carbon-1?

4. Write out a reasonable structure for ribose-3-phosphate ion using the ring structure of ribose.

5. Write out a reasonable structure for ribose-5-phosphate ion using the ring structure of ribose.

6. At equilibrium, an aqueous solution of D-ribose always contains a mixture of α-D-ribose, β-D-ribose, and the "straight-chain" D-ribose. Rationalize the fact that the relative concentrations of the three D-riboses remain unchanged following a reaction that has selectively converted part of the α-D-ribose into some other compound.

   You may wish to check your predictions for the structures of these compounds with the actual structures given in Chapter 19.

Additional exercises for Chapter 18 are given in the Appendix.

A Look Ahead

The next chapter discusses the nucleic acids, the molecules responsible for carrying genetic information. Sugars are a vital part of these molecules, and the scramble exercises for this chapter are directed toward the five-carbon sugars (pentoses) essential to the structures of the nucleic acids. Nucleic acids are, once again, very large molecules—in fact, some of the largest known. These molecules have very complex structures and serve complex purposes. However, the basic structure is once again the repetition of rather simple units.

19

DNA AND RNA: THE MOLECULES OF INHERITANCE

Chapter 19

The extraordinary variety of living things has held the interest of the human race since ancient times. The mystery of how these diverse creatures recreate themselves and organize their body parts into organs—and how those organs function—remains as fascinating today as it ever was. Perhaps the most exciting scientific development of the twentieth century is our progress in understanding the molecular basis on which the replication, development, and function of living things rest. In this chapter, we explore a simple but reasonably up-to-date view of how a single molecule, DNA, deoxyribonucleic acid, can carry all the necessary information for the creation of an organism; how that molecule can be replicated, how the "message" of the DNA molecule can be transcribed by another molecule (RNA, ribonucleic acid), and how the genetic message encoded in the DNA molecule can be translated into a functional molecule of a protein. While the details of the process are fascinating and complicated, the basic ideas are simple and by now so firmly imbedded in the structure of modern molecular biology that what is presented here is often called "the central dogma of molecular biology." Like most dogmas, it is an oversimplification and details are more and more being called into question.

Biological Importance of DNA and RNA

This chapter attempts to answer the questions (1) What is the DNA molecule and what properties of the molecule allow it to be replicated (duplicated) during cell division? (2) How can the information stored in the DNA molecule be "read" (transcribed) by and transferred to other molecules? and (3) How does the message on the DNA molecule allow for the synthesis of a protein molecule with a specific molecular weight and a particular sequence of amino acids? There are many more interesting questions to be asked and answered—an obvious one is Why does your DNA molecule make you a human being and not, say, a frog? That question will not be answered in detail here. What makes this area of scientific endeavor so interesting is that new developments appear almost daily. What was once an unanswerable question is now general knowledge. Progress is both rapid and exciting.

Nucleic acids get their name because they are acidic molecules found in the nuclei (centers) of biological cells. The terms "deoxyribonucleic acid" and "ribonucleic acid" come from the type of sugar molecule incorporated into the structure.

The Building Blocks of DNA

DNA molecules are very large molecules, made up of three types of building blocks. Once again, the story of how this information was obtained and how the building blocks fit together is one of the great sagas of modern science. We present here only the results of years of work. The building blocks are phosphoric acid (a triprotic acid), a sugar (deoxyribose in DNA), and a set of four organic molecules that contain carbon, hydrogen, oxygen, and nitrogen. Nearly all the atoms in these four molecules lie in a plane and each of these four molecules is a base in the acid–base sense. These four molecules are called the nitrogen bases. (See Chapter 14, page 333, for a discussion of nitrogen bases.)

The structures of these building blocks are

(1) phosphoric acid, H_3PO_4, a tetrahedral molecule with the phosphorus atom in the center and the four oxygen atoms at the corners of a regular tetrahedron:

Phosphoric Acid

$$\begin{array}{c} HO \\ \diagdown \\ HO\cdots P{=}O \\ \diagup \\ HO \end{array}$$

(2) deoxyribose, a five-carbon sugar (with one oxygen atom less than ribose), usually represented in the closed cyclic form:

Deoxyribose, a Five-Carbon Sugar

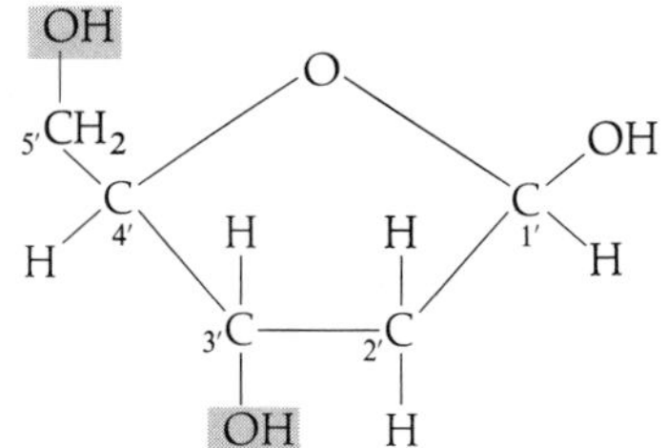

the structure of deoxyribose

Chapter 18 (pages 437–51) has more details on the structure of sugars. (The carbon atoms in the molecule are numbered as shown and are usually given primes to distinguish them from the carbon atoms in the nitrogen bases.)

(3) the nitrogen bases. The structures of the third type of building blocks, the nitrogen bases, will be dealt with in a later paragraph.

Phosphoric acid is an acid and the sugars have hydroxyl, —OH, groups so it is not surprising that the acid and the alcohols form esters. These esters make up the backbone of the long-chain DNA molecules. Figure 19-1. The alcohol groups attached to the 3′ and 5′ carbon atoms of the sugar are involved in the formation of the ester linkages.

The esters that we have discussed in Chapter 3 involved the product formed by the reaction of a carboxylic acid with an alcohol:

$$\underset{\text{carboxylic acid}}{R{-}\overset{\overset{\displaystyle O}{\|}}{C}{-}O{-}H} + \underset{\text{alcohol}}{H{-}O{-}\overset{\overset{\displaystyle H}{|}}{\underset{\underset{\displaystyle H}{|}}{C}}{-}R'} \longrightarrow \underset{\text{ester}}{R{-}\overset{\overset{\displaystyle O}{\|}}{C}{-}O{-}\overset{\overset{\displaystyle H}{|}}{\underset{\underset{\displaystyle H}{|}}{C}}{-}R'} + H_2O$$

The Phosphate Ester

The ester involved in the backbone of the DNA molecule was formed by the reaction

Figure 19-1

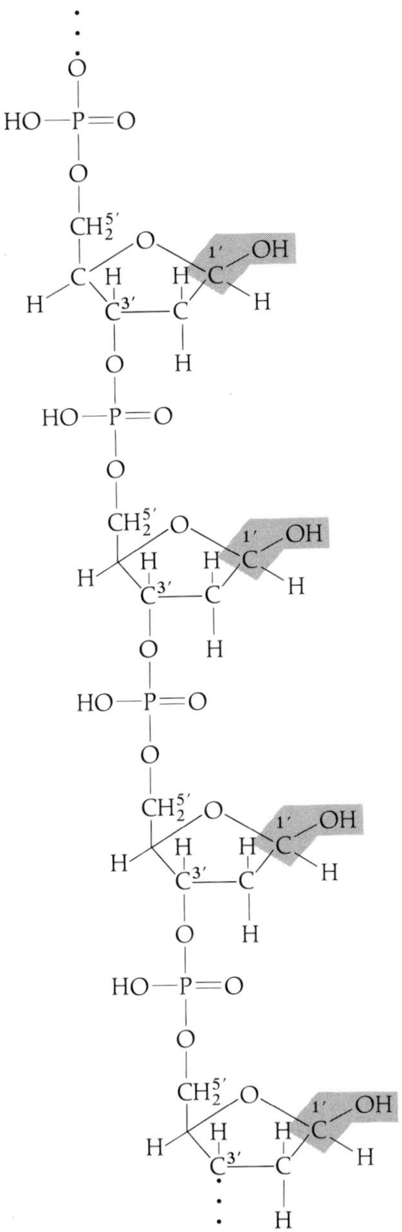

THE PHOSPHATE–SUGAR BACKBONE OF THE DNA MOLECULE. The nitrogen bases are attached to the carbon-1′ atoms through amine linkages.

of phosphoric acid, a triprotic acid, and an alcohol:

$$\mathrm{H{-}O{-}\overset{\displaystyle O}{\overset{\|}{\underset{\underset{\displaystyle H}{|}}{\underset{\displaystyle O}{\underset{|}{P}}}}}{-}OH} + \mathrm{H{-}O{-}\overset{\displaystyle H}{\overset{|}{\underset{\underset{\displaystyle H}{|}}{C}}}{-}R''} \longrightarrow \mathrm{HO{-}\overset{\displaystyle O}{\overset{\|}{\underset{\underset{\displaystyle H}{|}}{\underset{\displaystyle O}{\underset{|}{P}}}}}{-}O{-}\overset{\displaystyle H}{\overset{|}{\underset{\underset{\displaystyle H}{|}}{C}}}{-}R''} + H_2O$$

phosphoric acid (triprotic acid) — an ester that is also a diprotic acid

A second ester linkage can also be formed with another molecule of an alcohol:

$$\mathrm{R'''{-}\overset{\displaystyle H}{\overset{|}{\underset{\underset{\displaystyle H}{|}}{C}}}{-}O{-}H} + \mathrm{H{-}O{-}\overset{\displaystyle O}{\overset{\|}{\underset{\underset{\displaystyle H}{|}}{\underset{\displaystyle O}{\underset{|}{P}}}}}{-}O{-}\overset{\displaystyle H}{\overset{|}{\underset{\underset{\displaystyle H}{|}}{C}}}{-}R''} \longrightarrow \mathrm{R'''{-}\overset{\displaystyle H}{\overset{|}{\underset{\underset{\displaystyle H}{|}}{C}}}{-}O{-}\overset{\displaystyle O}{\overset{\|}{\underset{\underset{\displaystyle H}{|}}{\underset{\displaystyle O}{\underset{|}{P}}}}}{-}O{-}\overset{\displaystyle H}{\overset{|}{\underset{\underset{\displaystyle H}{|}}{C}}}{-}R''} + H_2O$$

alcohol — diprotic acid — a diester that is also a monoprotic acid

DNA as a Polyanion

The remaining proton of the phosphoric acid group is a strong acid. In the nucleus of the cell, where DNA is found, the phosphoric acid group is entirely deprotonated. (Refer to the percent abundance–pH curve of phosphoric acid, Figure 13-4, page 305. K_A for this acid is roughly equivalent to K_{A_1} for phosphoric acid.) The backbone of the DNA molecule carries many negative charges; these negative charges are balanced by positive ions, particularly Mg^{2+} ions, associated with the chain in close but mobile proximity. The DNA molecule is actually an ion (a very large polyanion, to be more precise), but it is almost never referred to that way. We will stay with common usage and call it a molecule throughout this chapter.

The third structural element in DNA molecules is the nitrogen bases. Except for an occasional substitution or modification, the bases are

The Nitrogen Bases

Adenine (A)	Cytosine (C)
Thymine (T)	Guanine (G)

The structures are given in Figure 19-2. The ring structures containing double bonds impose a high degree of planarity on the molecules; most of the atoms in the bases lie in a single plane. The bases attach to the sugar part of the –sugar–phosphate–sugar–phosphate–backbone at the carbon-1′ position through a *sigma*-bond between the carbon-1′ atom of the sugar and the nitrogen atom of the base (marked with an asterisk) in each structure. A water molecule is eliminated in the process. See Figure 19-3. None of the three —OH groups of deoxyribose remain as free —OH groups in the DNA structure.

The nitrogen bases are capable of forming strong hydrogen bonds with each other. Figure 19-4 shows the bases oriented for the formation of hydrogen bonds between Adenine and Thymine and between Guanine and Cytosine. As the figure shows, the G—C pair is capable of forming three hydrogen bonds and the A—T pair is capable of forming two hydrogen bonds.

Figure 19-2

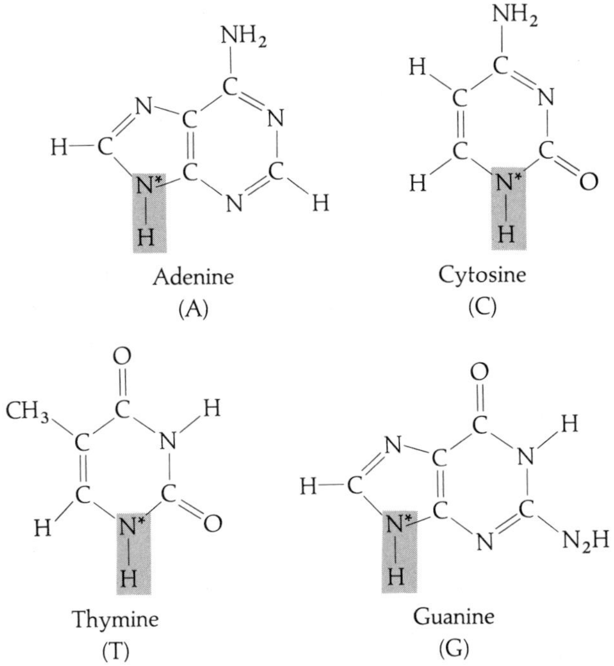

Adenine
(A)

Cytosine
(C)

Thymine
(T)

Guanine
(G)

THE FOUR NITROGEN BASES OF DNA.

Hydrogen Bonding between the Bases

Other possibilities for hydrogen bonds between the nitrogen bases exist but the ones shown are the strongest and are believed to impart important structural characteristics to the DNA molecule. It is often said that G pairs only with C (or C with G) and A pairs only with T (or T with A).

A Schematic Diagram of DNA

It is one of the great feats of deduction and induction in modern science that combined the known properties of the molecules (sugar, phosphoric acid, and the four nitrogen bases)—along with a very meager amount of structural information obtained from physical measurements—to produce, as James Watson and Francis Crick did, the "double helix" model for the structure of DNA. A schematic view of the proposed structure can be likened to a ladder. The sugar–phosphate chain forms the rails, and the bases—hydrogen-bonded to each other—form the steps. A very short sequence is indicated here:

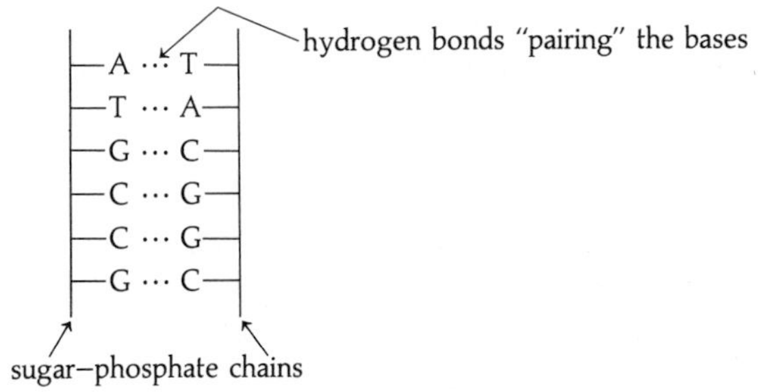

```
|—A ··· T—|   hydrogen bonds "pairing" the bases
|—T ··· A—|
|—G ··· C—|
|—C ··· G—|
|—C ··· G—|
|—G ··· C—|
sugar–phosphate chains
```

Figure 19-3

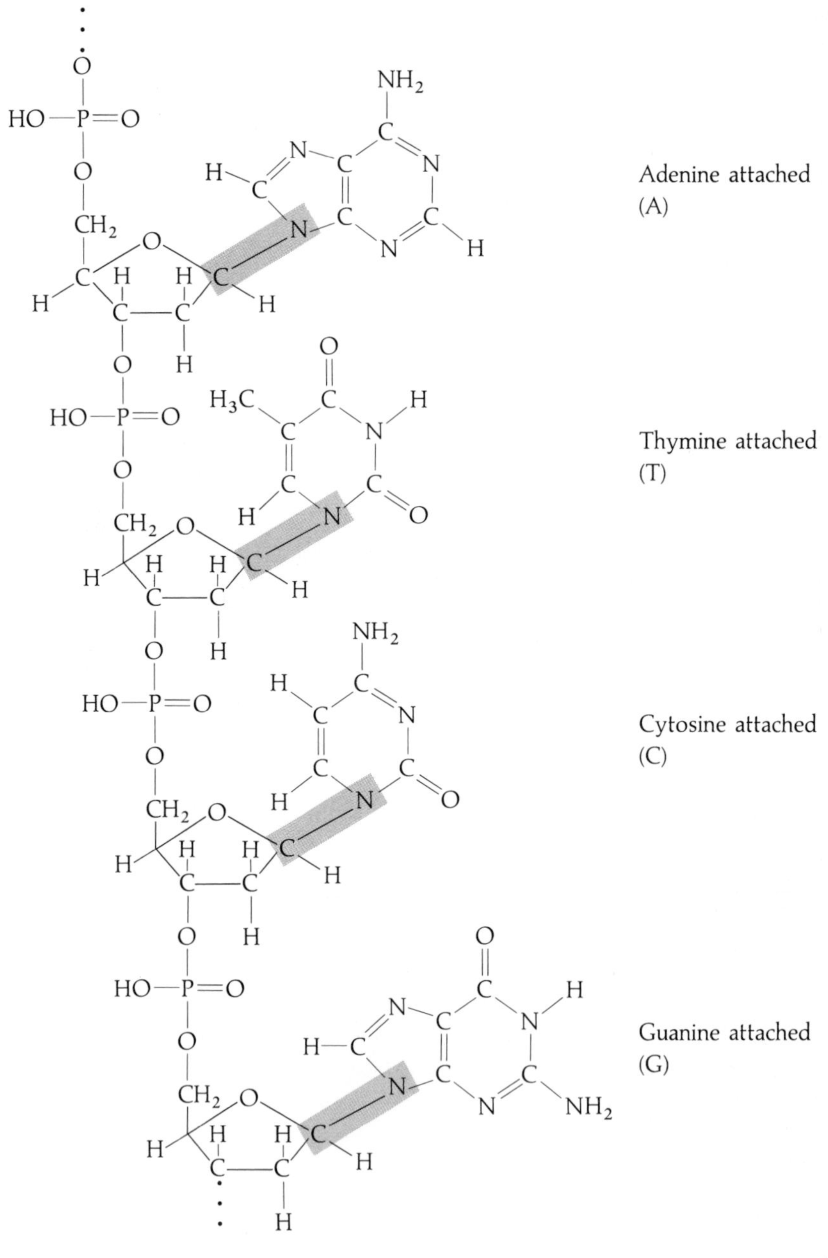

Adenine attached (A)

Thymine attached (T)

Cytosine attached (C)

Guanine attached (G)

A Brief Portion of a DNA Chain Showing the –Phosphate–Sugar–Phosphate–Sugar– Backbone with One of Each of The Four Nitrogen Bases Attached Through an Amine Linkage.

Figure 19-4

Guanine ::::::::: Cytosine
(three hydrogen bonds)

≈ 1.1 nm

(a)

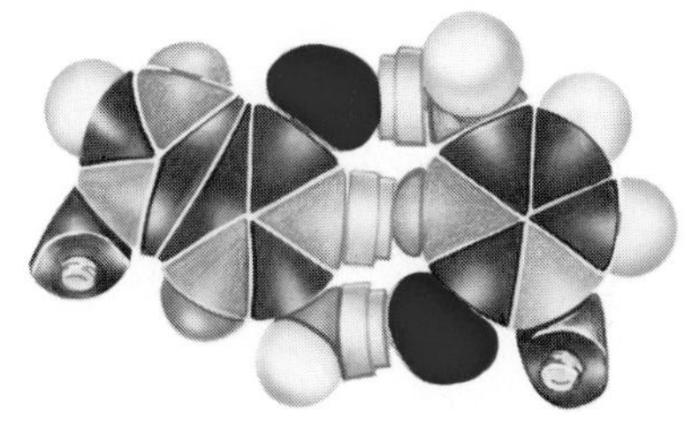

Adenine :::::::: Thymine
(two hydrogen bonds)

≈ 1.1 nm

(b)

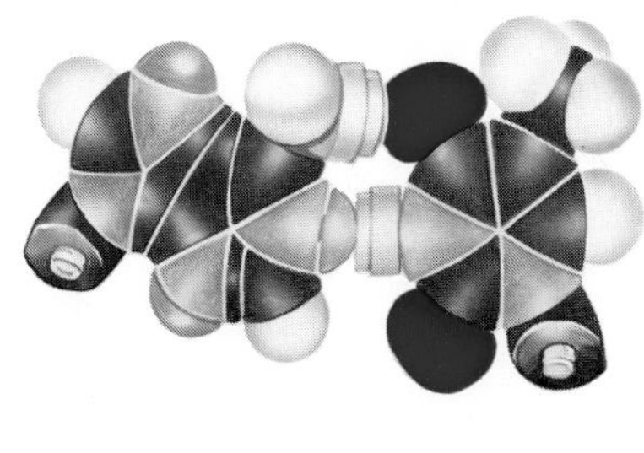

SKETCHES AND SPACE-FILLING MODELS OF THE HYDROGEN BONDS BETWEEN THE BASE PAIRS IN DNA. (a) Guanine–Cytosine (G—C) base pairs; (b) Adenine–Thymine (A—T) base pairs. The hydrogen atoms involved in the hydrogen bonds are represented by discs.

The feature that is missing from the "ladder schematic" is that the structure is helical, not extended in a straight line. One needs a ladder designed by Salvador Dali, capable of being twisted about a central axis much like a spiral staircase. (But remember, in a spiral staircase the inner rail doesn't spiral, whereas in DNA both do.) An outline drawing of this structure is given in Figure 19-5.

Models of the DNA Double Helix

A ball and stick model or a space-filling model representing all the atoms is very helpful to study but one big enough to show all the features is hard to handle and expensive to construct. Two photographs of such models are given in Figure 19-6.

The total molecular weight of a typical DNA molecule can be as high as 1×10^7. (This is large even in comparison with the largest protein molecules.) It is a very long molecule, yet in the nucleus of a cell it is very tightly packed, the extraordinary length of the molecule providing the flexibility required for the molecule to be folded back on itself many times. Lots more can be said about the details of

Figure 19-5

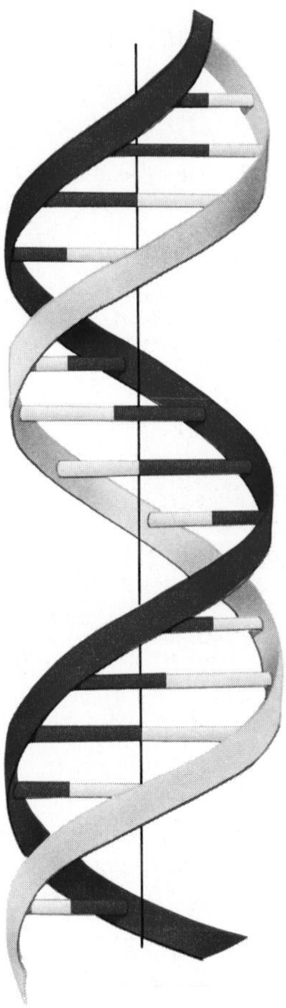

A VERY SCHEMATIC REPRESENTATION OF THE DNA DOUBLE HELIX, AS PROPOSED BY WATSON AND CRICK. The ribbons represent the –sugar–phosphate–sugar– backbone chain, and the horizontal bars represent the hydrogen-bonded base pairs that hold the two strands together. The central axis of the helix is denoted by the vertical line.

the structure, but for the purpose of explaining the function of the molecule, we will revert to the "ladder schematic." You must realize, however, that the schematic bears about as much resemblance to the actual structure of the molecule as a stick figure does to a human being.

Storage of Genetic Information

It is the length of the DNA molecule (number of base pairs) and the sequence of bases that contain the genetic information so, fundamentally, what makes you different from a frog is that your DNA molecule is bigger (mammalian DNA is bigger than amphibian DNA, which is in turn bigger than bacterial DNA) and the sequence

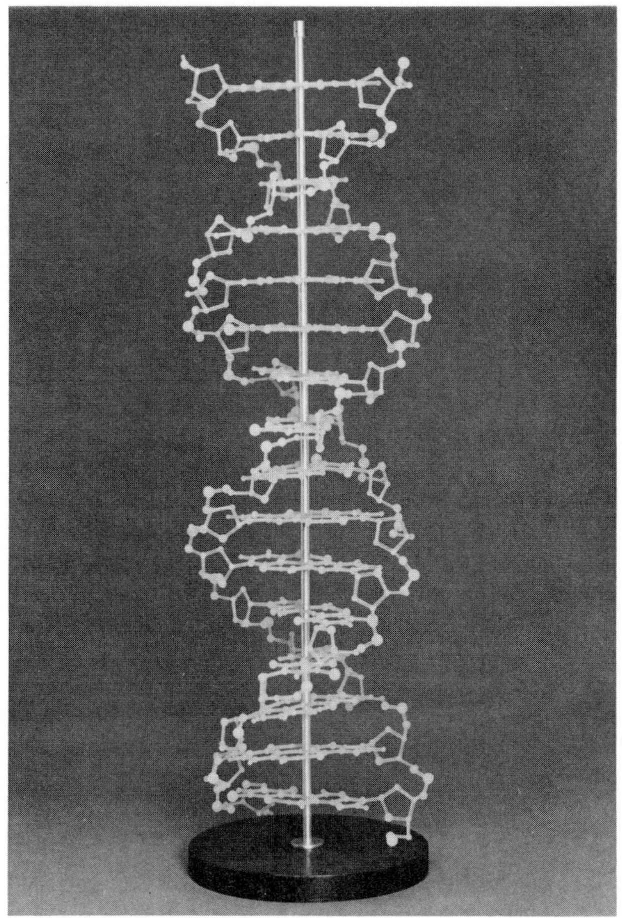

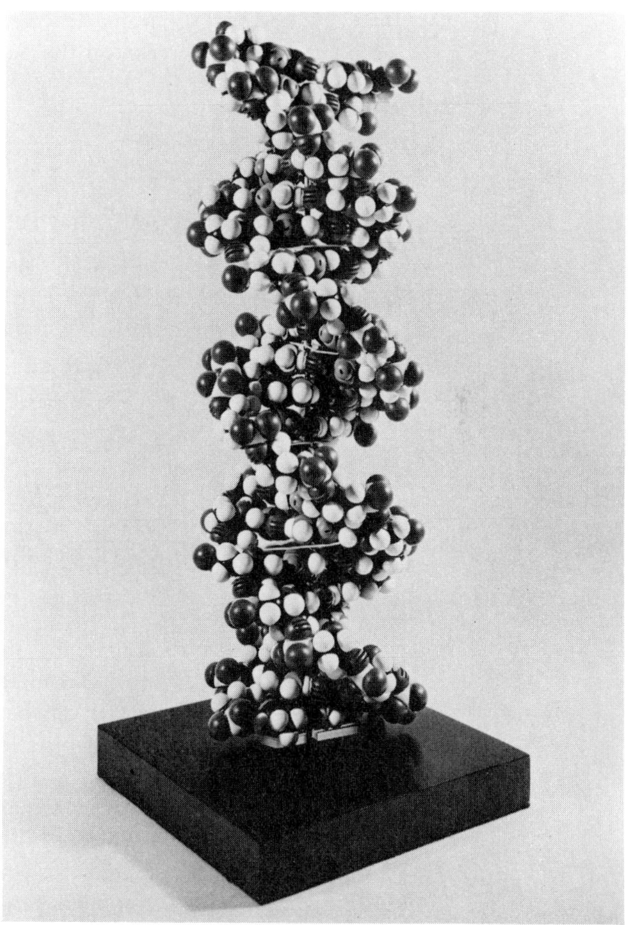

TWO MODELS OF THE DNA STRUCTURE. (a) A plastic ball and stick model of a segment of a DNA molecule. (b) A space-filling model of a segment of a DNA molecule.

of bases in your DNA is different from that in a frog. Your DNA molecule is also different, although in much finer detail, from your friends', from your parents', and from your siblings' unless you are an identical twin.

It is your DNA molecule, along with your biological and personality development that comes later, that makes you unique. One of the more remarkable things about modern science is that it is now possible to give the base sequence for quite large DNA molecules or parts of molecules. The sequence for a particular fragment of *E. coli* (a bacterium) DNA is given in Table 19-1.

Table 19-1

PARTIAL SEQUENCE OF THE BASES IN *E. coli* DNA (ONE STRAND). This sequence of 650 bases starts at the 5′-phosphate end of the chain and runs to the 3′-phosphate end. The breaks on the page are for convenience only. This is one continuous sequence.

5′—C G G G A A A G C G C A T A A A C T G G A G G A A T A A G C A G C A A A A C G C A C A A A C C G	50
T A A C C A A A C G C G C A A T T T A T T T A A A A A G G G A C T A G A C A G A G G G G T G G G A A	100
G T C C G T A T T A T C C A C C C C C G C A A C G G C G C T A A G C G C C C G T A G C T C A G C T G	150
G A T A G A G C G C T G C C T C C G G A G G C A G A G G T C T C A G G T T C G A A T C C T G T C G A	200
G C G C G C C A T T T A G T C C C G G C G C T T G A G C T G C G G T G G T A G T A A T A C C G C G T	250
A A C A A G A T T T G T A G T G G T G G C T A T A G C T C A G T T G G T A G A G C C C T G G A T T G	300
T G A T T C C A G T T G T C G T G G G T T C G A A T C C C A T T A G C C A C C C C A T T A T T A G A	350
A G T T G T G A C A A T G C G A A G G T G G C G G A A T T G G T A G A C G C G C T A G C T T C A G G	400
T G T T A G T G T C C T T A C G G A C G T G G G G G T T C A A G T C C C C C C C C T C G C A C C A C	450
G A C T T T A A A G A A T T G A A C T A A A A A T T C A A A A A G C A G T A T T T C G G C G A G T A	500
G C G C A G C T T G G T A G C G C A A C T G G T T T G G G A C C A G T G G G T C G G A G G T T C G A	550
A T C C T C T C T C G C C G A C C A A T T T T G A A C C C C G C T T C G G C G G G G T T T T T T G T	600
T T T C T G T G C A T T T C G T C A C C C T C C C T T C G C A A T A A A C G C C C G T A A T A A—3′	650

It's boring reading, much like a telephone book. The wonder of the phone book is that everyone in it can talk to everyone else; and what's remarkable about the DNA sequence is that it can be known, that it is known, and that the sequence confers certain properties on the DNA molecule, which in turn leads to certain properties of the organism.

Replication

Since the base sequence of a given DNA molecule is specific, it is relatively straightforward to see how an exact copy of a DNA molecule might be made. Exact copies are required if a cell is to be able to divide to give two identical cells. Look at the very simplified model of a ten-base pair sequence in a DNA molecule given in Figure 19-7. This diagram is further simplified by replacing the sugar–phosphate chains with straight lines—so our simplified model now is

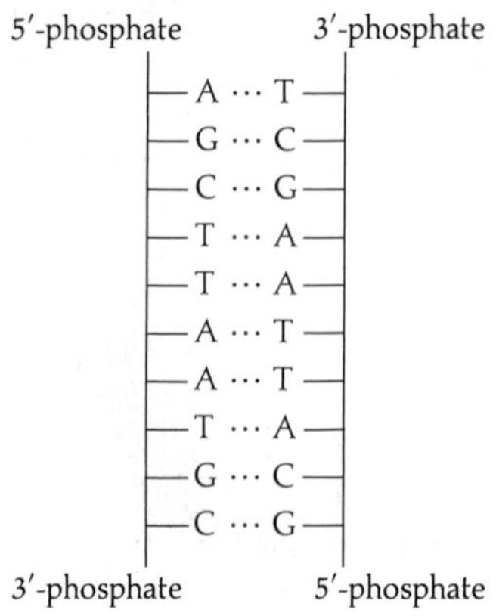

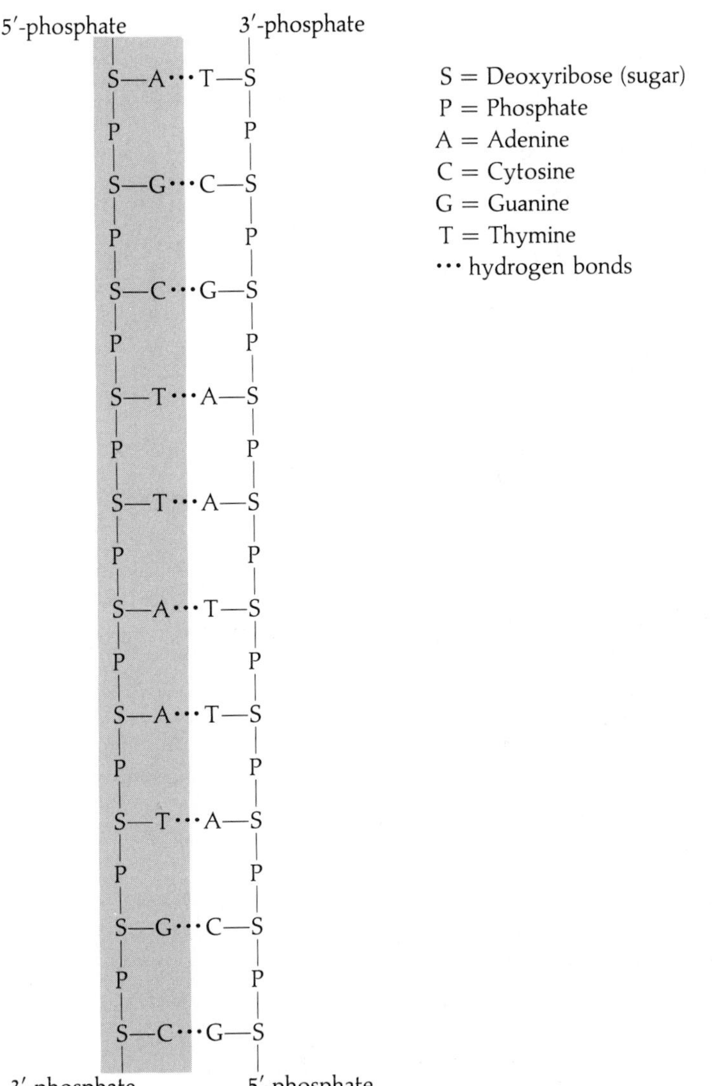

A SCHEMATIC REPRESENTATION OF THE DNA MOLECULE. Note that the two strands are complementary to each other because of the rules of base pairing, and that the two strands are "antiparallel." The strand on the left starts at the 5′ carbon atom of a sugar molecule and ends on the 3′ carbon atom of a sugar molecule. The strand on the right does just the opposite.

In the cell, action of some proteins can separate the strands (sometimes more dramatically called unzipping the strands) to give two single strands.

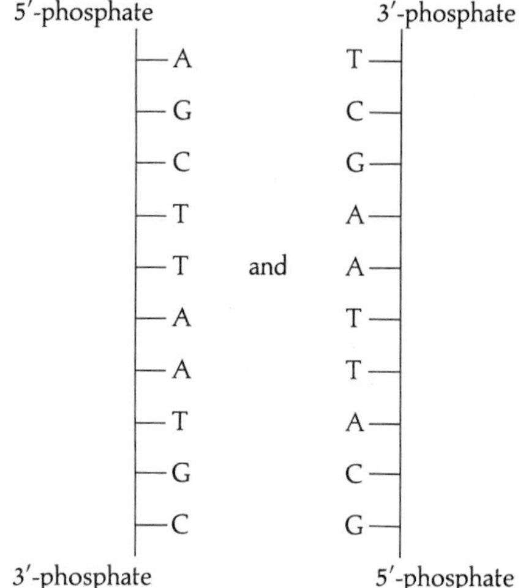

Now all that needs to be done is to assemble the bases, sugars, and phosphoric acid molecules to form the complement to each single strand. In biological systems this process, too, is accomplished with the aid of proteins.

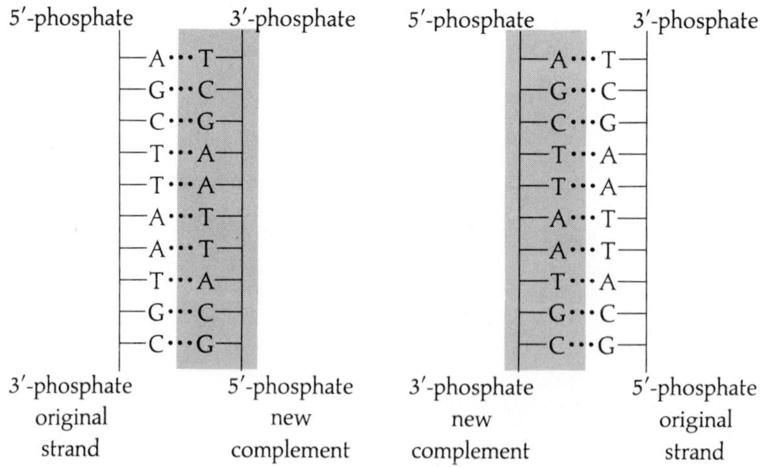

We have now constructed two exact copies of the original molecule, one strand of the original molecule being in each copy. The impossibility of assembling all the parts at once for a molecule of molecular weight 1×10^7 means that the actual process in the cell must involve only a small segment of the whole molecule at one time and the process must proceed down the length of the DNA chain rapidly. We indicate that process in an even more simplified model as

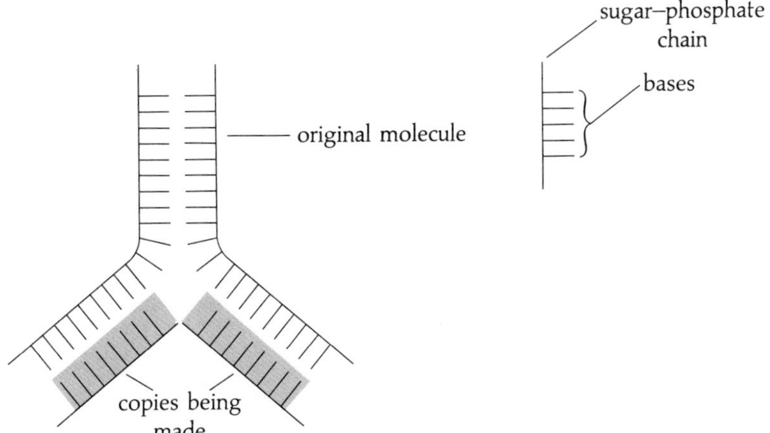

Those who have seen motion pictures of cell division in biology classes or at a science museum or on TV realize that the whole process takes very little time—only a few seconds.

Gratuitous Information 19-1

HUMAN INSULIN FROM A BACTERIUM

Modern experimentation with a DNA molecule now allows us to modify the genetic material of microorganisms. A lot has been said about the possible dangers of this new knowledge, but so far, and we hope forever, the applications have been for good. One of the first examples was modifying the bacterium *E. coli* so that it can produce human insulin. How this is done is a fascinating story. A virus that infects *E. coli* has, like many such "plasmids," a closed circular DNA molecule, represented here by a double circle:

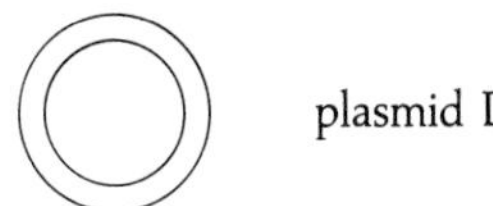

plasmid DNA

Enzymes are available that split this DNA into uneven ended fragments:

plasmid with piece deleted
Note the uneven ends; these are often called "sticky" ends.

The sequence of DNA that codes for human insulin is known; it can be isolated from human DNA or it can be synthesized, again with the complementary "sticky" ends.

sequence of DNA that codes for human insulin—with "sticky" ends

By reacting this piece of human DNA with the plasmid DNA fragment and the appropriate enzymes (enzymes that tie DNA fragments together are called "ligases"), we can incorporate the human insulin fragment into the plasmid DNA.

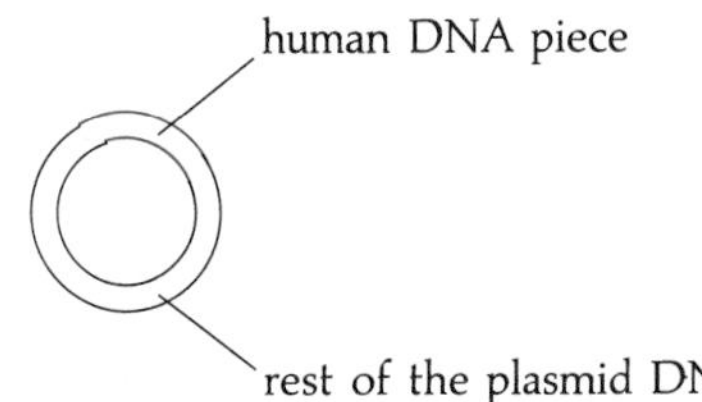

Now, when the modified plasmid DNA becomes incorporated into the *E. coli* DNA, the *E. coli* makes not only the cell products required by its own DNA and by the plasmid DNA, but also human insulin. Since the *E. coli* can be grown in a tank, production of human insulin becomes much simpler than the difficult process of isolating insulin from blood. True human insulin has become readily available and can be used instead of horse or pig insulin.

Biological cells do more than just replicate; they develop (change form and function) and they manufacture products necessary for the organism. The instructions for the development and function of the cell are also coded in the DNA molecule in the sequence of the base pairs. The questions then are how is the message read (transcribed) and how is the message expressed (translated). These concerns led to the consideration of the other types of nucleic acids present in the cell. There are now known to be many but we restrict our attention to only two: messenger ribonucleic acid (mRNA) and transfer ribonucleic acid (tRNA). Note first that these molecules are ribonucleic acids (RNA's), not deoxyribonucleic acids (DNA's). That means that the sugar part of the backbone of the molecule is ribose, not deoxyribose.

Messenger RNA

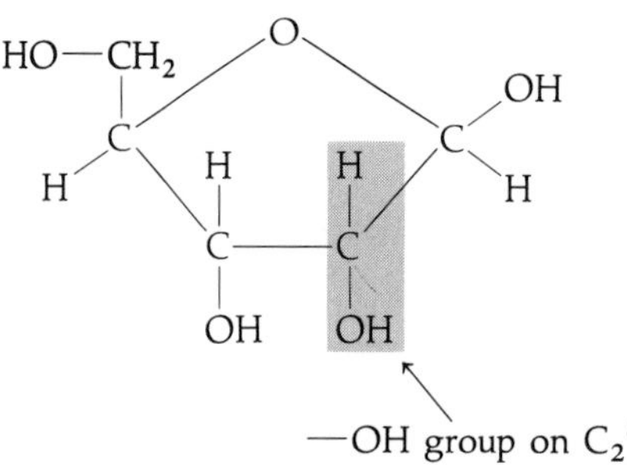

—OH group on C_2'

ribose with a hydrogen atom and a hydroxyl group at carbon-2'

deoxyribose with two hydrogen atoms at carbon-2'

The RNA's are singly stranded molecules while the DNA's are double strands. The extra —OH group on the RNA's makes them more soluble in the aqueous part of the cell (the cytoplasm) than DNA and prevents the formation of a double helical structure. In cells with nuclei (eukaryotes), the DNA molecule stays in the nucleus while the RNA's go back and forth between the nucleus and the cytoplasm. The other major difference between DNA and RNA is that, in RNA's, the base Thymine (T) is replaced by a similar base, Uracil (U).

Thymine

Uracil

Transcription of the Message

The point of attachment to the sugar molecule is indicated by an asterisk as before. A segment of a DNA molecule contains a message and a specific messenger RNA molecule can be produced to correspond with it. For our short sequence, for example, the corresponding mRNA molecule would be as shown in Figure 19-8. However, it must be understood that mRNA's are big molecules, too. A short ten-base pair fragment would be treated by the cell as a waste product and broken down and its parts used over again. Cells are both very efficient and ruthless.

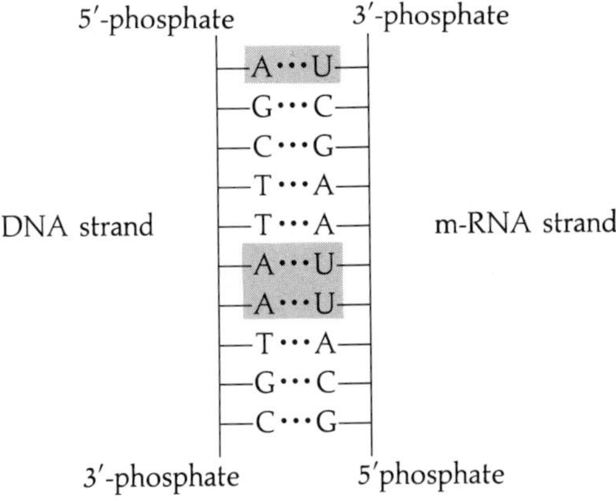

A SECTION OF THE DNA MOLECULE WITH A CORRESPONDING MESSENGER RNA STRAND BOUND TO IT. Note that A (adenine) in DNA is now complementary to U (uracil) in the RNA strand.

The message of part of the DNA molecule has now been transcribed to its specific messenger RNA, which, with its added solubility (the result of the extra —OH group on the sugar and the replacement of T by U) in the cytoplasm, can get its message to the region of the cell where it can be used. There are many such places in the cell, but we will focus on only one: the ribosome, a cell body responsible for the manufacture of proteins. We have said the information of the DNA molecule is contained in the sequence of bases, and in Chapter 17 we have said that the properties of proteins depend upon the specific sequence of amino acids. Perhaps your mind has already made the obvious connection that the sequence of bases must be the code for the sequence of amino acids in the proteins. The question is how.

Transfer RNA

The current explanation for this process is the result of what was called at the time "cracking the genetic code" and involves the third class of nucleic acids: transfer RNA, or tRNA. These tRNA molecules transfer amino acids to the messenger RNA. By comparison with DNA or mRNA, tRNA molecules are small, about 100 bases in length. They also have a very interesting shape, called a cloverleaf, although, as the detailed model shows, the three-dimensional shape differs from any ordinary cloverleaf. (See Figure 19-9.)

Translation of the Genetic Code

The tRNA's have the sugar–phosphate backbone of the RNA's but they contain all five of the bases mentioned above plus a variety of modified bases that seem to be important for maintaining the correct functioning geometry. (Gratuitous Information 19-2 has the structures of some of these modified bases.) One region of the tRNA is designed to read (translate) the message on the mRNA.

The message is encoded in the sequence of bases, and they are read in individual sequences of three—the so-called triplet code. For example, the three bases UUU in a row on the messenger RNA would be matched by a sequence AAA on the

Figure 19-9

(a) (b) (c)

THE STRUCTURE OF PHENYLALANINE tRNA. (a) and (b) are two-dimensional representations of the sequence of bases; (c) is an attempt to indicate the three-dimensional structure. Note how this single-stranded molecule doubles back on itself to form loops. Letters other than A, G, C, T, and U represent modified and unusual bases found in tRNA. Some of the structures are given in Gratuitous Information 19-2.

transfer RNA. Each triplet codon on the mRNA is matched by a specific triplet anticodon on the tRNA. The particular tRNA that has the anticodon in the correct place in its sequence of bases also has, at another part of its structure, a specific place for binding to a particular amino acid. Thus each amino acid has its own tRNA, and that tRNA recognizes a particular triplet of bases on the mRNA. Some amino acids actually have more than one tRNA. The diagrams given as Figure 19-9 are for the phenylalanine tRNA, the first tRNA molecule to have its full structure determined in detail. The function of the transfer RNA's is to bring the amino acids to the messenger RNA in the sequence specified by the triplets (codons) on the mRNA molecule. We can simplify the tRNA structure to a schematic or stick figure as well. One way of doing that is to use the following diagram.

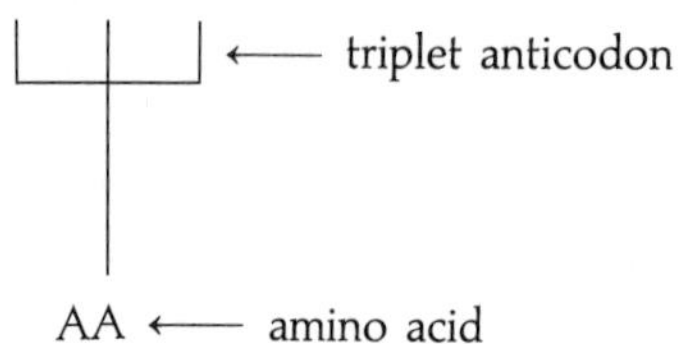

SOME OF THE MODIFIED BASES FOUND IN tRNA

Modifications involving changes of atoms are indicated by shading. An asterisk (*) represents the point of attachment to the ribose $C_1{}'$ atom.

Derived from Uracil (U)

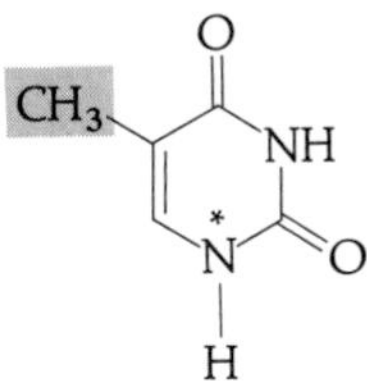

ribothymidine (T)

dihydrouridine (D)

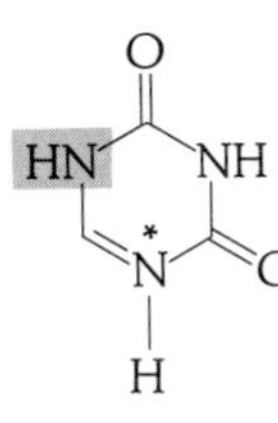

pseudouridine

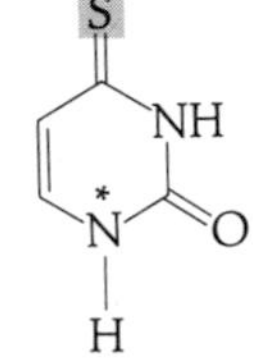

4-thiouridine (S^4U)

Derived from Cytosine (C)

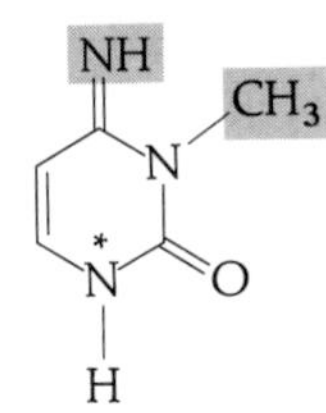

3-methylcytidine (m^3C)

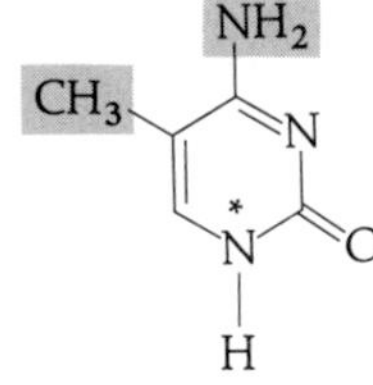

5-methylcytidine (m^5C)

Derived from Adenine (A)

inosine (I)

N^6-methyladenosine (m^6A)

N^6-isopentenyladenosine (i^6A)

Derived from Guanine (G)

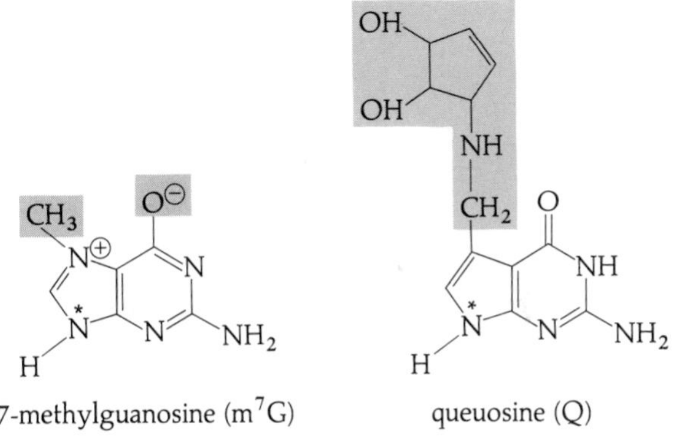

7-methylguanosine (m^7G)

queuosine (Q)

wyosine (Y)

Figure 19-10

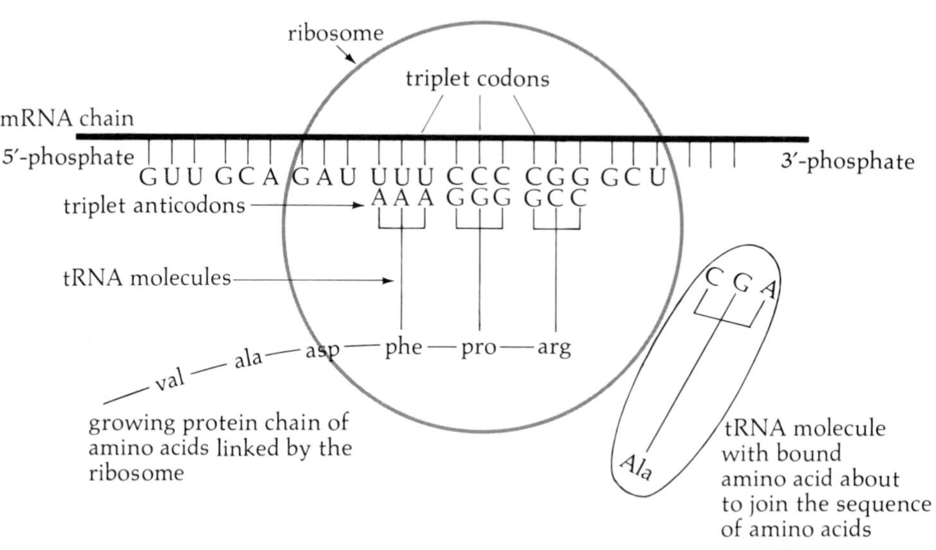

PROTEIN SYNTHESIS IN THE RIBOSOME REPRESENTED VERY SCHEMATICALLY.

The long vertical line represents the tRNA molecule with a triplet of bases—the anticodon—(three short vertical lines) at the top and a specific amino acid (AA) bound at the bottom.

Synthesis of Proteins

The whole process of protein synthesis in the ribosome is schematically diagrammed in Figure 19-10. The ribosome, represented by the large circle, moves along the mRNA strand from the 5′-phosphate end to the 3′-phosphate end. tRNA molecules with their bound amino acids are brought to the mRNA strand in the order specified by the triplet codons on the mRNA strand. Enzymes in the ribosome join the amino acids together by forming the peptide links, and a protein molecule with a specific length and a specific sequence of amino acids is produced.

The Genetic Code

Every possible triplet on the messenger RNA has significance in terms of synthesizing a particular protein. The genetic code, as we know it, is shown in Figure 19-11. The figure perhaps needs some explanation. The code reads from the 5′ end toward the 3′ end of the messenger RNA, and with four possible bases at each position in the triplet, there are 64 possibilities. These are tabulated in rows with the first base named, the second considered a variable, and the third base one of the four possibilities. Each triplet and the corresponding amino acid are given in the body of the table. Note that there are redundancies in the code; a particular amino acid has more than one codon (triplet sequence) that will code for it. (Serine is designated by the codons UCU, UCC, UCA, UCG, AGU, and AGC, for example.)

To help with the use of the genetic code table for mRNA, consider the tRNA molecule with its bound amino acid required to join the others on the mRNA strand in Figure 19-10. The next codon on the mRNA strand is GCU. In Figure 19-11 we

Figure 19-11

base at 5′ end of codon ↓	middle base of codon				base at 3′ end of codon ↓	
	U	C	A	G		
U	UUU–phe	UCU–ser	UAU–tyr	UGU–cys	U	
	UUC–phe	UCC–ser	UAC–tyr	UGC–cys	C	
	UUA–leu	UCA–ser	UAA	UGA	A	UAA, UGA, UAG } termination
	UUG–leu	UCG–ser	UAG	UGG–trp	G	
C	CUU–leu	CCU–pro	CAU–his	CGU–arg	U	
	CUC–leu	CCC–pro	CAC–his	CGC–arg	C	
	CUA–leu	CCA–pro	CAA–gln	CGA–arg	A	
	CUG–leu	CCG–pro	CAG–gln	CGG–arg	G	
A	AUU–ile	ACU–thr	AAU–asn	AGU–ser	U	
	AUC–ile	ACC–thr	AAC–asn	AGC–ser	C	
	AUA–ile	ACA–thr	AAA–lys	AGA–arg	A	
	AUG–met (and initiation)	ACG–thr	AAG–lys	AGG–arg	G	
G	GUU–val	GCU–ala	GAU–asp	GGU–gly	U	
	GUC–val	GCC–ala	GAC–asp	GGC–gly	C	
	GUA–val	GCA–ala	GAA–glu	GGA–gly	A	
	GUG–val	GCG–ala	GAG–glu	GGG–gly	G	

A Summary of the 64 Triplet Codons of the Genetic Code on Messenger RNA and Their Known Coding Assignments. There is much more complication to the whole picture than is shown in any of the diagrams or in this table. The order in which the sequence is read is given at the top of the diagram.

look in the far left column for G, the first base of the codons that are on the bottom sets of four rows. Next, look in the far right column: all the triplets containing U as the last base are in the top row across this set (GUU, GCU, GAU, and GGU). The codon of interest is the second one, GCU, which the table tells us is the codon for alanine. Consequently, the anticodon shown in the tRNA molecule in Figure 19-10 is CGA.

The tRNA molecule for alanine binds alanine at one portion of the molecule and has the anticodon, the sequence of complementary bases CGA, at the position where it binds to the mRNA strand.

One question that might have passed through your mind during this description is: How does the apparatus know where to start and where to stop? The genetic code of mRNA contains specific start and stop codons for this purpose (UAA, UAG, and UGA). You might also ask the question Does the whole process ever make mistakes? The answer is yes: either the DNA or the mRNA can be damaged, leading to rearrangements of the triplet codes, or the transcription or translation process can work incorrectly. However, the cell mechanism is subtle enough to recognize most mistakes and simply destroys them and uses the pieces over again. Some mistakes, however, may not be destroyed and may lead to new cell products that are viable in the organism and may become permanent features of the organism—mutations. The disease sickle cell anemia, for example, is the result of the single substitution of the amino acid valine for glutamic àcid in hemoglobin (at position 6 in the β-chain). From the codons, that would mean a simple substitution of a Uracil for an Adenine in the codon GUA for valine vs GAA for glutamic acid. In spite of the difficulties sickle cell anemia causes its sufferers, it presumably has survived in the human population because it gives resistance to malaria.

Mutations

Summary

Research into the nature and function of DNA and RNA molecules continues at a very rapid pace. The story becomes both more complicated and more interesting as more knowledge becomes available. The portion of the saga we have discussed here, however, can be summarized fairly simply and briefly. The DNA molecule carries the genetic information in the sequence of bases along its length. Cell replication involves the splitting of the double-stranded DNA molecule into two single strands and the construction of a complement to each strand, a process that produces two identical copies of the original molecule.

Information coded in the DNA molecule is read (transcribed) by complementary mRNA molecules. The genetic code is the sequence of triplets of bases on this messenger molecule. The translation of that code into useful cell products, such as protein, is accomplished by transfer RNA molecules, specific for their purposes.

Scramble Exercises

The text gives the base sequence (Table 19-1) for a portion of one strand of the *E. coli* DNA molecule. Use a portion of that DNA sequence (say, 301–324 inclusive: 24 bases) and

(1) write out the base sequence of the complementary DNA strand,
(2) write out the base sequence of the complementary messenger RNA strand,

(3) determine and write out what the amino acid sequence coded for by this region of the *E. coli* DNA would be using the code given in Figure 19-10.

Additional exercises for Chapter 19 are given in the Appendix.

A Look Ahead

The remaining class of large molecules of biological importance to be discussed in these chapters is the fats and lipids, which are the subject of Chapter 20. These molecules are in some ways simpler than the proteins, carbohydrates, and nucleic acids discussed so far. However, the assembly of many of these molecules into functional structural entities, such as the biological membrane, has a fascination all its own. That topic will be the concluding section of Chapter 20.

20

ANIMAL FATS, VEGETABLE OILS, LIPIDS, AND MEMBRANES

Chapter 20

The term fat as both a noun and an adjective has diverse connotations in our language, ranging from "the best and the richest" to "slothful and stupid." In some societies and at various moments in history, to be fat is a symbol of high economic status, of being able to afford the best. In this country and at this time, being excessively fat is considered a threat to health. Chemically speaking, animal fats and vegetable oils are triesters. One of the triesters found in beef tallow is stearin:

Triesters

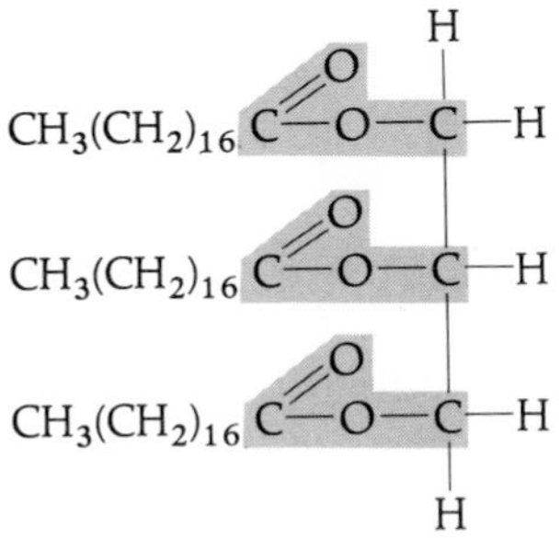

$$
\begin{array}{l}
\qquad\qquad\qquad\qquad\quad \mathrm{H} \\
\qquad\qquad\qquad\qquad\quad | \\
\mathrm{CH_3(CH_2)_{16}C(=O)-O-C-H} \\
\qquad\qquad\qquad\qquad\quad | \\
\mathrm{CH_3(CH_2)_{16}C(=O)-O-C-H} \\
\qquad\qquad\qquad\qquad\quad | \\
\mathrm{CH_3(CH_2)_{16}C(=O)-O-C-H} \\
\qquad\qquad\qquad\qquad\quad | \\
\qquad\qquad\qquad\qquad\quad \mathrm{H}
\end{array}
$$

Note the three ester linkages:

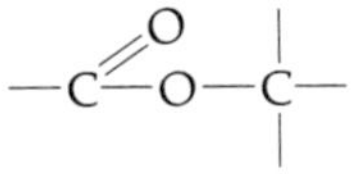

$$\mathrm{-C(=O)-O-\overset{|}{\underset{|}{C}}-}$$

Under the appropriate conditions, the above stearin reacts with three molecules of water to give three molecules of stearic acid and one molecule of glycerol:

Hydrolysis of Triesters

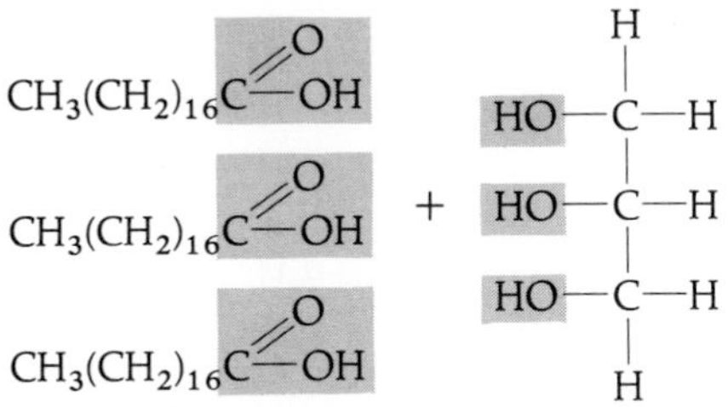

$$
\begin{array}{lcl}
 & & \quad\ \mathrm{H} \\
\mathrm{CH_3(CH_2)_{16}C(=O)-OH} & & \mathrm{HO-\overset{|}{C}-H} \\
\mathrm{CH_3(CH_2)_{16}C(=O)-OH} & + & \mathrm{HO-\overset{|}{C}-H} \\
\mathrm{CH_3(CH_2)_{16}C(=O)-OH} & & \mathrm{HO-\overset{|}{\underset{|}{C}}-H} \\
 & & \quad\ \mathrm{H} \\
\text{stearic acid} & & \text{glycerol}
\end{array}
$$

Glycerol is an alcohol with three hydroxyl —OH groups. Stearic acid

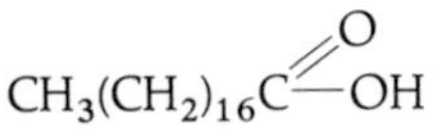

$$\mathrm{CH_3(CH_2)_{16}C(=O)-OH}$$

is a member of a group of carboxylic acids called the fatty acids, obviously from their isolation from fats. There are many fatty acids that can be obtained from the hydrolysis of fats or oils that occur in nature.

In the chapter on solutions (Chapter 12), the term oil was used to denote any substance that does not mix with water. In this chapter, we are restricting our attention to a particular group of molecules that is a subset of the more general class. The term "oil" in this chapter refers to **triglyceride molecules of natural origin.** The motor oil in your car is usually a petroleum compound, a fairly high molecular weight hydrocarbon, and is not a triglyceride.

Animal fats and vegetable oils, the triglycerides, are often represented by the general formula

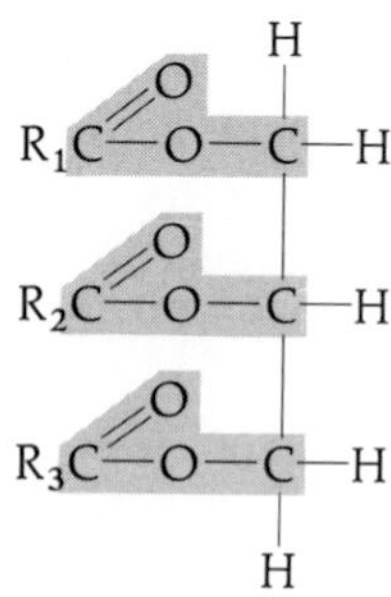

where R_1, R_2, and R_3 represent the appropriate alkyl groups (the appropriate carbon–carbon chains).

Fatty Acids

Some fatty acids, particularly those obtained on the hydrolysis of animal fats, have saturated carbon–carbon chains. Others, particularly those obtained on the hydrolysis of vegetable oils, have unsaturated carbon–carbon chains. Linoleic acid

$$CH_3(CH_2)_3(CH_2CH{=}CH)_2(CH_2)_7C({=}O){-}OH$$

is an unsaturated 18-carbon fatty acid. Stearic acid is a saturated 18-carbon fatty acid. Note the similarities and the differences in these two fatty acids:

$$\begin{array}{ccccccccccccccccccc}
 & H & H & H & H & H & H & H & H & H & H & H & H & H & H & H & H & H & O \\
 & | & | & | & | & | & | & | & | & | & | & | & | & | & | & | & | & | & \| \\
H- & C- & C- & C- & C- & C- & C= & C- & C- & C= & C- & C- & C- & C- & C- & C- & C- & C- & C-OH \\
 & | & | & | & | & | & & & | & & & | & | & | & | & | & | & | & \\
 & H & H & H & H & H & & & H & & & H & H & H & H & H & H & H &
\end{array}$$

linoleic acid

$$\begin{array}{ccccccccccccccccccc}
 & H & H & H & H & H & H & H & H & H & H & H & H & H & H & H & H & H & O \\
 & | & | & | & | & | & | & | & | & | & | & | & | & | & | & | & | & | & \| \\
H- & C- & C- & C- & C- & C- & C- & C- & C- & C- & C- & C- & C- & C- & C- & C- & C- & C- & C-OH \\
 & | & | & | & | & | & | & | & | & | & | & | & | & | & | & | & | & | & \\
 & H & H & H & H & H & H & H & H & H & H & H & H & H & H & H & H & H &
\end{array}$$

stearic acid

Conformation of Fatty Acids

A space-filling model of a saturated fatty acid stretched out full length looks very much like a caterpillar, while one or more double bonds in the fatty acid produce bends or kinks in the hydrocarbon chain (see Figures 20-1 and 20-2). However, the

Figure 20-1

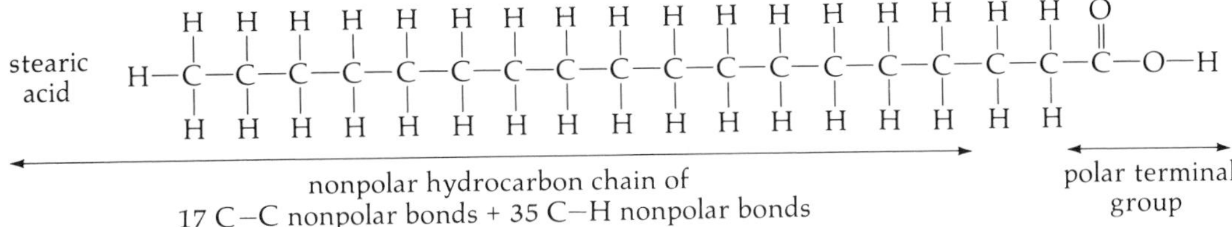

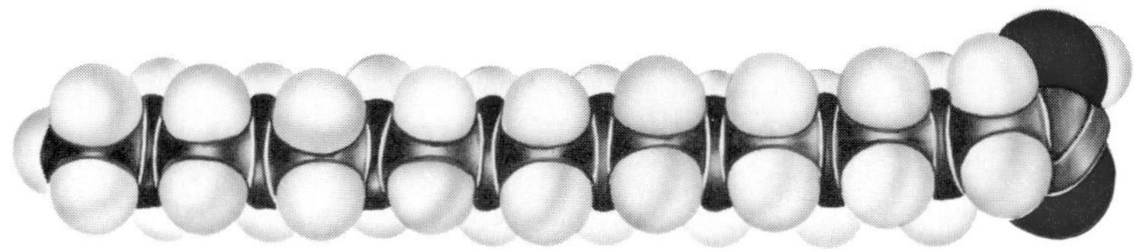

A SPACE-FILLING MODEL OF STEARIC ACID.

hydrocarbon chain is almost never fully stretched out under normal conditions. Instead, the chains arrange themselves in a globular fashion. Freedom of rotation about the great number of carbon–carbon single bonds in the R groups as well as about the carbon–oxygen single bonds leads to a much less ordered array, with the more polar carbon–oxygen regions of the molecule buried within the center of the molecule.

A molecule of a triglyceride presents a nonpolar exterior. Fats are not readily soluble in water. In general, biological systems are water systems, but water tends to be excluded from regions that have a high concentration of triglycerides. These regions are said to be hydrophobic, water-hating. The chemistry of hydrophobic

Figure 20-2

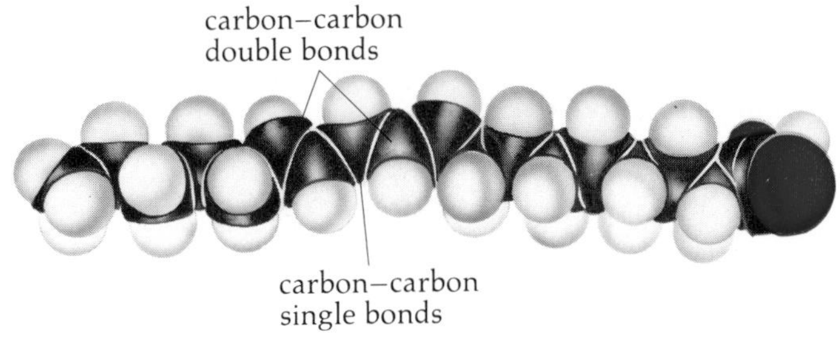

A SPACE-FILLING MODEL OF LINOLEIC ACID.
$CH_3CH_2CH_2CH_2CH_2CH{=}CHCH_2CH{=}CHCH_2CH_2CH_2CH_2CH_2CH_2CH_2COOH$

regions, such as the brain, is quite different from the chemistry of the hydrophilic, the water-loving, regions of other biological organs.

When triglycerides or similar molecules are placed in nonaqueous solvents such as hydrocarbons, the polar groups of several molecules tend to associate on the inside

Micelles

of a cluster of molecules, leaving the hydrocarbon chains on the outside. In aqueous environments, the hydrocarbon chains of several molecules tend to cluster together on the inside and the polar groups on the outside. If we represent the three nonpolar chains as a line (hydrocarbon chains) and the polar group on the end as a circle, the molecule of the fat could be represented as ————O. Several of these triglyceride molecules clustered together in a nonaqueous environment could be represented as

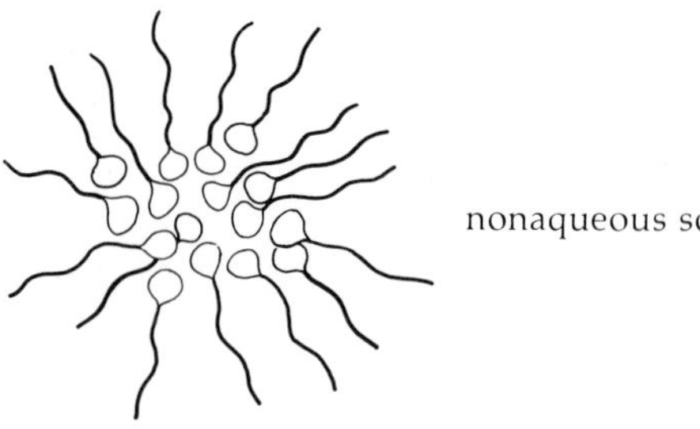

or in an aqueous environment as

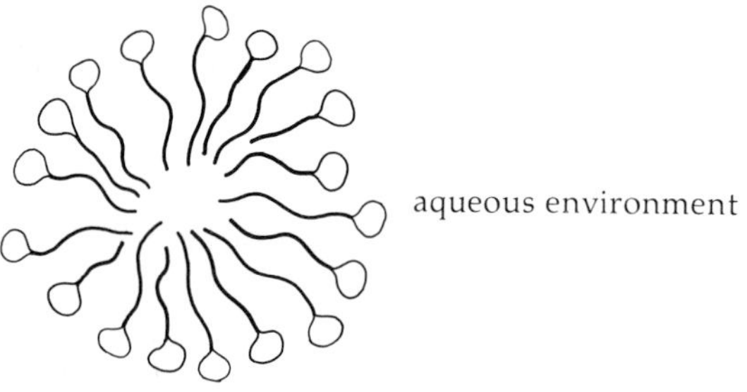

The above drawings are two-dimensional representations of three-dimensional phenomena. The clusters above are known as micelles and are responsible for the cloudy or milky appearance of fats dispersed in water (milk, salad dressings).

Another possible arrangement of these molecules is the formation of a bilayer, a double layer of molecules oriented like this:

Bilayers

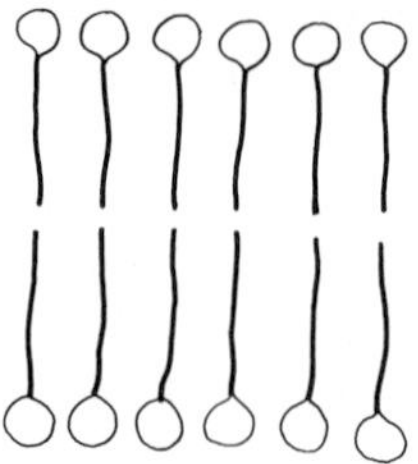

and one can imagine this structure forming a closed sphere, with some of the solvent trapped inside.

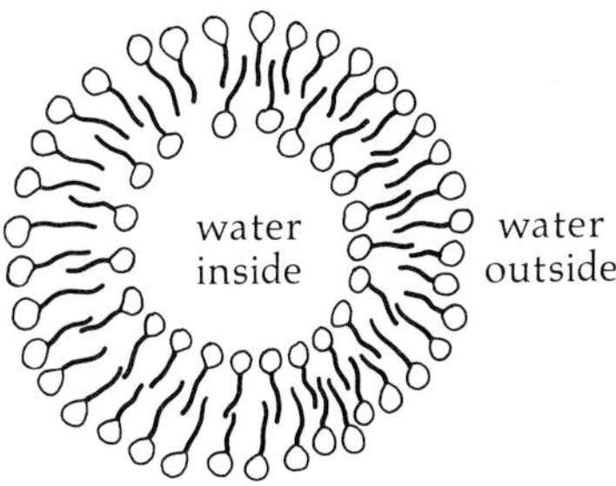

This kind of a structure suggests itself as a possible start on a model for the biological cell membrane. We will have more about that later.

The hydrolysis of triglycerides is represented by the general equation

$$
\begin{array}{l}
R_1C(=O)\text{—O—}\overset{\displaystyle H}{\underset{|}{\overset{|}{C}}}\text{—H} + \text{H—O—H} \\
R_2C(=O)\text{—O—}\overset{|}{\underset{|}{C}}\text{—H} + \text{H—O—H} \\
R_3C(=O)\text{—O—}\underset{\displaystyle H}{\overset{|}{\underset{|}{C}}}\text{—H} + \text{H—O—H} \\
\quad\text{triglyceride}
\end{array}
\xrightarrow{\text{hydrolysis}}
\begin{array}{l}
R_1C(=O)\text{—OH} \\
R_2C(=O)\text{—OH} \\
R_3C(=O)\text{—OH} \\
\text{fatty acids}
\end{array}
+
\begin{array}{l}
\text{HO—}\overset{\displaystyle H}{\underset{|}{\overset{|}{C}}}\text{—H} \\
\text{HO—}\overset{|}{\underset{|}{C}}\text{—H} \\
\text{HO—}\underset{\displaystyle H}{\overset{|}{\underset{|}{C}}}\text{—H} \\
\quad\text{glycerol}
\end{array}
$$

In the above equation, the three R groups may be three different groups or two R groups may be identical or all three of the R groups may be identical. In stearin, the three groups are identical.

Digestion of Fats

Fats and oils are hydrolyzed entirely or in part during the digestive processes in the small intestine and absorbed through the intestinal wall. The degree to which fats and oils survive the cooking processes of boiling, roasting, baking, and frying is clear evidence that triglycerides are quite resistant to hydrolysis, yet these reactions of hydrolysis go on quite smoothly at body temperature during the limited time the digestive mixture is in the small intestine. This striking difference in reactivity is attributed to the action of several enzymes (glycerol ester hydrolases) known as the lipases. Note that both words, hydrolases and lipases, are plural. Several enzymes of

20-1 *Gratuitous Information*

SOAPS AND DETERGENTS

Historically, animal fat has been important for the preparation of soap. Soaps are made by heating fats with a rather concentrated aqueous solution of sodium hydroxide (lye) in what is called the saponification ("soap-making") reaction.

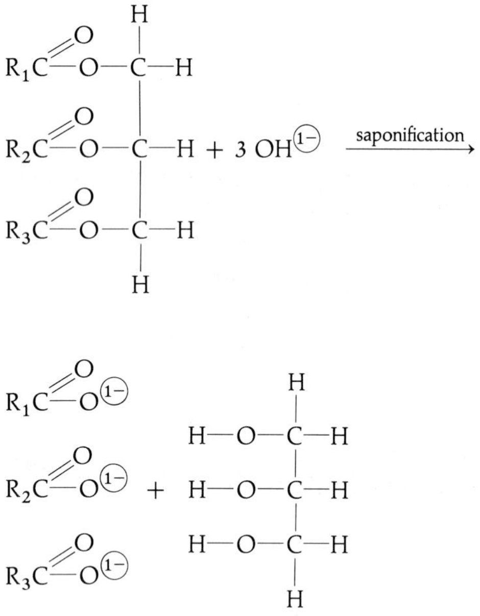

$$
\begin{array}{l}
\qquad\qquad\quad\;\; H \\
\qquad\qquad\quad\;\; | \\
R_1C(=O)-O-C-H \\
\qquad\qquad\quad\;\; | \\
R_2C(=O)-O-C-H \; + \; 3\,OH^{1-} \xrightarrow{\text{saponification}} \\
\qquad\qquad\quad\;\; | \\
R_3C(=O)-O-C-H \\
\qquad\qquad\quad\;\; | \\
\qquad\qquad\quad\;\; H
\end{array}
$$

$$
\begin{array}{lcl}
 & & \quad\;\;\; H \\
R_1C(=O)O^{1-} & & \quad\;\;\; | \\
 & & H-O-C-H \\
R_2C(=O)O^{1-} & + & H-O-C-H \\
 & & H-O-C-H \\
R_3C(=O)O^{1-} & & \quad\;\;\; | \\
 & & \quad\;\;\; H
\end{array}
$$

There are, of course, sodium ions in the solution. Assuming that excess lye has not been used, equal numbers of sodium ions and

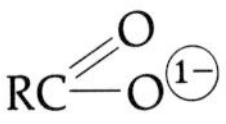

$$RC(=O)O^{1-}$$

ions make up the solid soap. Soaps are ionic compounds involving the sodium ion and therefore soaps are fairly soluble in water.

In washing, the removal of water-soluble substances does not require a soap; the removal of substances that are not soluble in water may be brought about with the aid of a soap or a detergent.

It is the anions of the soap, such as the stearate ions,

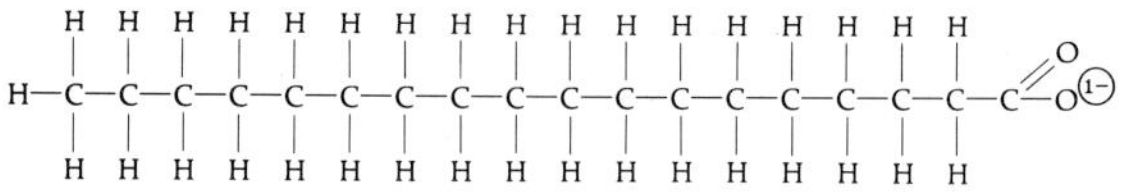

$$H-\overset{H}{\underset{H}{C}}-\overset{H}{\underset{H}{C}}-\overset{H}{\underset{H}{C}}-\overset{H}{\underset{H}{C}}-\overset{H}{\underset{H}{C}}-\overset{H}{\underset{H}{C}}-\overset{H}{\underset{H}{C}}-\overset{H}{\underset{H}{C}}-\overset{H}{\underset{H}{C}}-\overset{H}{\underset{H}{C}}-\overset{H}{\underset{H}{C}}-\overset{H}{\underset{H}{C}}-\overset{H}{\underset{H}{C}}-\overset{H}{\underset{H}{C}}-\overset{H}{\underset{H}{C}}-\overset{H}{\underset{H}{C}}-\overset{H}{\underset{H}{C}}-C(=O)O^{1-}$$

that are responsible for the cleaning action of a soap. The alkyl-chain part of the anion is nonpolar and tends to dissolve in nonpolar substances. The carboxylate ion

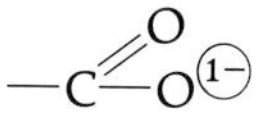

$$-C(=O)O^{1-}$$

portion of the anion is attracted to water by the polarity of the carbon–oxygen bonds and by the charge on the ion. When soapy water makes contact with a droplet of oil, the anion of the soap bridges the boundary between the oil and the water with the carbon–carbon "tail" in the oil and the carboxylate ion

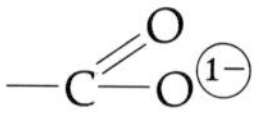

$$-C(=O)O^{1-}$$

"head" in the water. (See Figure 20-3.) Mechanical agitation breaks up the oil into droplets and the anion of the soap maintains an emulsion by preventing the coalescence of the oil droplets. The presence of the soap also lowers the surface tension of the water, and the soapy water is effective in penetrating the spaces between the fibers of cloth and small crevices in the fibers. Compounds that lower surface tension in this way are called surfactants.

In an earlier age, it was common practice to make soap from tallow, bacon grease, or other fats. During the colonial period, lye was prepared by leaching wood ashes with water and allowing the solution to evaporate from a wooden trough to yield a sufficiently concentrated solution. Apparently, there were two standard methods of testing the concentration of the solution. The solution was deemed sufficiently concentrated if a fresh egg floated on it or the edges of a chicken feather curled after the feather was swished through the solution.

All surfactants are very similar in their action. A surfactant is a large molecule or ion with a nonpolar end and either a polar or an ionic end. Soap is only one example of a surfactant. Detergents are surfactants that have a synthetic group replacing the carboxylate portion of the natural soaps.

$$\mathrm{H{-}\underset{H}{\overset{H}{C}}{-}\underset{H}{\overset{H}{C}}{-}\underset{H}{\overset{H}{C}}{-}\underset{H}{\overset{H}{C}}{-}\underset{H}{\overset{H}{C}}{-}\underset{H}{\overset{H}{C}}{-}\underset{H}{\overset{H}{C}}{-}\underset{H}{\overset{H}{C}}{-}\underset{H}{\overset{H}{C}}{-}\underset{H}{\overset{H}{C}}{-}\underset{H}{\overset{H}{C}}{-}\underset{H}{\overset{H}{C}}{-}\underset{H}{\overset{H}{C}}{-}\underset{H}{\overset{H}{C}}{-}\underset{H}{\overset{H}{C}}{-}\underset{H}{\overset{H}{C}}{-}\underset{H}{\overset{H}{C}}{-}\underset{H}{\overset{H}{C}}{-}O{-}\underset{O}{\overset{O}{S}}{-}O^{1-}}$$

a sulfate ester ion

One very troublesome problem arises in the use of soaps. If the water contains alkaline earth ions, such as calcium ion, Ca^{2+}, the anion of the soap is precipitated to give a curdy solid that is difficult to remove by rinsing.

$$2\ CH_3(CH_2)_{16}C(=O){-}O^{1-}(aq) + Ca^{2+}(aq)$$

$$\longrightarrow (CH_3(CH_2)_{16}C(=O){-}O)_2Ca(s)$$

This curd is particularly difficult to deal with in automatic washing machines. During the spin cycle, the clothes act as a filter and trap the curd as the water moves through the clothes to the exterior of the drum. Tattletale gray is inevitable. The great success of automatic washing machines was dependent upon the development of detergents that are not precipitated by hard water (water containing calcium or other alkaline earth ions.)

Figure 20-3

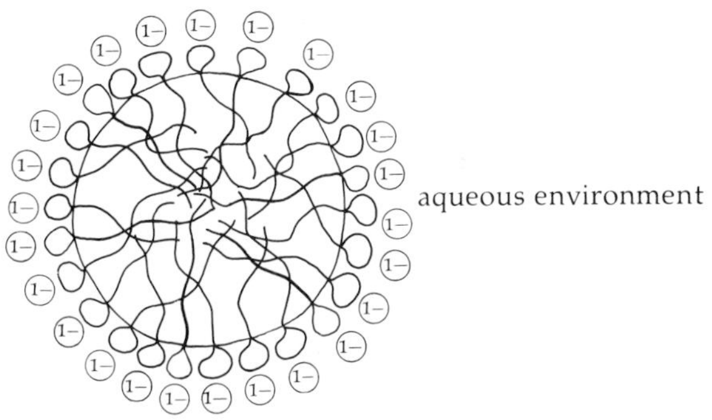

OIL DROP WITH SOAP ANIONS IMBEDDED "TAIL FIRST" AND THE POLAR CARBOXYLATE GROUPS ($-COO^{1-}$) POINTING OUTWARD.

high specificity are involved. All three ester linkages do not react at the same rate and a series of diglycerides may be formed:

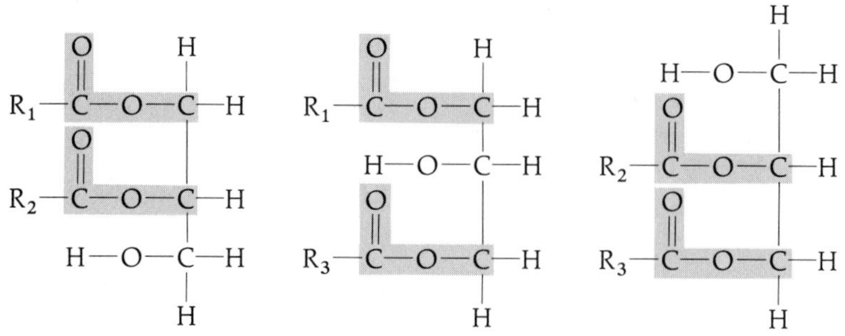

```
    O      H                O      H                        H
    ‖      |                ‖      |                        |
R1—C—O—C—H            R1—C—O—C—H                H—O—C—H
    O      |                       |                O       |
    ‖      |                H—O—C—H                ‖       |
R2—C—O—C—H                O      |            R2—C—O—C—H
           |                ‖      |                O       |
     H—O—C—H          R3—C—O—C—H                ‖       |
           |                       |            R3—C—O—C—H
           H                       H                        |
                                                            H
```

the diglycerides

Some hydrolases are selective in promoting hydrolysis at an end carbon—the *alpha*-carbons of the glycerol—and other enzymes are specific for hydrolysis at the central position—the *beta*-carbon of the glycerol.

A series of monoglycerides are also formed:

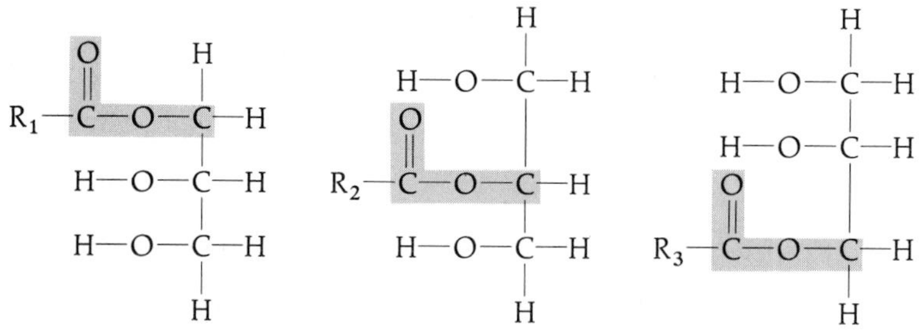

```
                                   H                        H
    O      H                       |                        |
    ‖      |                H—O—C—H                H—O—C—H
R1—C—O—C—H                O      |                        |
           |                ‖      |                H—O—C—H
     H—O—C—H          R2—C—O—C—H                O       |
           |                       |                ‖       |
     H—O—C—H                H—O—C—H          R3—C—O—C—H
           |                       |                        |
           H                       H                        H
```

the monoglycerides

20-2 *Gratuitous Information*

RANCIDITY IN FATS

The unsaturated fats and oils have double bonds in them and these bonds are the points of easiest attack in the process of oxidation. This vulnerability makes them "more digestible" than saturated fats, but also more susceptible to spoilage. When a compound such as linoleic acid is partially oxidized at one of the double bonds, one of the products can be the six-carbon acid caproic acid:

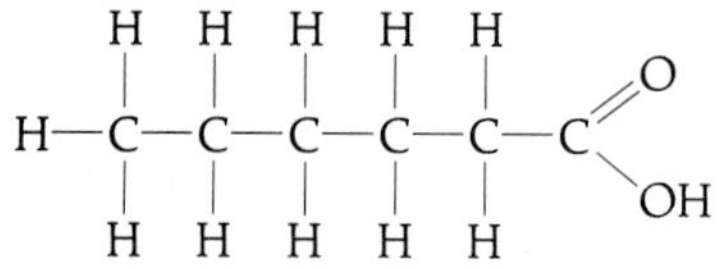

```
    H  H  H  H  H
    |  |  |  |  |    //O
H—C—C—C—C—C—C
    |  |  |  |  |    \
    H  H  H  H  H     OH
```

Carboxylic acids of this length have very strong and often very unpleasant odors; this one (caproic acid) is partly responsible for the characteristic odor of goats. Other unpleasant materials produced by partial oxidation of the double bonds include aldehydes and peroxides. As fats containing double bonds age, they tend to accumulate these materials and become rancid. To prevent rancidity, unsaturated fats are often reacted with hydrogen gas under pressure (linoleic acid + 2 H_2 → stearic acid) to produce compounds that are sold as "hydrogenated" vegetables oils, which are now solids.

It is these monoglycerides and diglycerides that pass through the intestinal wall and are reconverted to the triglycerides on the other side. One of the characteristics of an enzyme is that it facilitates both the forward reaction and the reverse reaction. Consequently, the enzymes that are responsible for the hydrolysis of a fat may also be responsible for the restructuring of the fat after diffusion through the intestinal wall. The reconstituted triglycerides are transported by the lymphatic system to the blood.

Caloric Content of Fats

If the energy requirements of the body justify it, the triglycerides are oxidized by a series of chemical reactions to water and carbon dioxide. These reactions occur in a variety of tissues: the liver, kidneys, and heart. A number of enzymes are involved and a number of intermediate compounds are formed in the process but the heat evolved for the overall reaction is the same as that produced by the direct oxidation to carbon dioxide and water. For stearin, the change in enthalpy is given by

$$2\ C_{57}H_{110}O_6 + 163\ O_2 \longrightarrow 114\ CO_2 + 110\ H_2O$$
$$\Delta H = -7.5 \times 10^4 \text{ kilojoules } (-18{,}000 \text{ kcal})$$

This figures out to approximately 42 kilojoules (10 kcal) per gram. Triglycerides in excess of the energy requirements are deposited within cells with a corresponding increase in volume. It is the objective of every reducing diet to reduce the volume of these cells.

Saturated and Unsaturated Fats

Triglycerides are roughly classified as oils, fats, and waxes on the basis of the ease with which the compound can be poured or spread. The ease of spreading depends in part upon the length of the alkyl chains of the fatty acids but depends primarily upon the degree of saturation (the absence of double bonds in the alkyl chains) and the temperature. The greater degree of saturation corresponds to a more "solid" triglyceride. A triglyceride may, of course, be the ester of both saturated and unsaturated fatty acids.

A number of fatty acids are obtained on the hydrolysis of natural fats and oils. Most of these contain a long "straight chain" of carbon atoms, although a few are branched and a few contain small rings of carbon atoms. There are very clearly a number of strong biological preferences. Nearly all of the fatty acids contain an even number of carbon atoms. This fact has strong implications for the manner in which they are originally formed and, if the same enzymes are involved, for the manner in which they are degraded to carbon dioxide and water. Animals such as horses, cows, and people utilize the fats and oils contained in their diet and also synthesize triglycerides from carbohydrates. The triglycerides synthesized contain a very high proportion of saturated fatty acids. Plants and cold-blooded animals, on the other hand, produce primarily triglycerides of unsaturated fatty acids. Consequently, fats extracted from plants and seeds are usually oils. Perhaps the most interesting biological preference is that the *cis* conformation at each double bond is synthesized almost exclusively.

Shortenings

At one time, lard, the cooking fat obtained from pork, was considered the most desirable shortening, and the dilution of lard with the less expensive vegetable oils was discouraged by more or less rigorously enforced regulations. At least one court case involving a charge of adulteration of lard was thrown out of court when it was demonstrated that hogs fed soybeans have the capacity to incorporate the unsaturated fatty acids of the seed in their own fat.

In more recent years, public preference for lard has been diverted to vegetable shortenings largely through the advertising of those companies that market them. Such advertising is based upon studies that indicate that saturated fatty acids are involved in the formation of an important precursor of cholesterol, which in turn may deposit on the walls of blood vessels, particularly in the blood vessels of individuals who restrict their physical activity. Animal fats also tend to have cholesterol itself as a component of the mixture. On the other hand, unsaturated vegetable oils often are hydrogenated to obtain a solid product. The phrase "partially hydrogenated soybean oil" that currently appears on the label of a popular margarine is a euphemism for a significant conversion of carbon–carbon double bonds to carbon–carbon single bonds by the chemical addition of hydrogen to the unsaturated soybean oil. Insofar as the public can determine, this margarine may contain an even higher proportion of saturated fatty acids than butter. However, vegetable oil margarines do not contain cholesterol, and it might even be possible to find some that have not been hydrogenated. The phrase "all natural," which is so popular these days, deserves to be looked at carefully. Look at the labels to see if the product has been hydrogenated.

Lipids

The substances that can be extracted from biological samples with an organic solvent that does not mix with water, such as diethyl ether, are known as lipids. The lipids include—in addition to the oils, fats, and waxes—the phospholipids, the sphingolipids, the cerebrosides, the steroids, and the terpenes. These are indeed a mixed bag of compounds and they range from long-chain alcohols with an amine group (sphingolipids) to polycyclic (complex ring) compounds (steroids).

Phospholipids

The close relation of the phospholipids to fats is shown by the general formula

```
                      H
            O         |
           //         |
       R1C—O—C—H
            O         |
           //         |
       R2C—O—C—H
          O           |
          ||          |
   R3O—P—O—C—H
          |           |
          O           H
          |
          H
```

general formula of phospholipids (also called phosphatides)

Both the phospholipids and the fats are triesters of glycerol. All phospholipids are esters of phosphoric acid. There are a great number of phospholipids since R_1, R_2, R_3 can represent a wide variety of groups. One example is shown in Figure 20-4. Note the relation of the phospholipids to phosphoric acid:

Figure 20-4

$$\overbrace{(CH_3)_3\overset{\oplus}{N}CH_2}^{R_3}CH_2O\overset{\overset{O}{\|}}{\underset{\underset{O^{\ominus}}{|}}{P}}OCH_2-\overset{\overset{H}{|}}{\underset{|}{C}}-O\overset{\overset{O}{\|}}{C}\overbrace{CH_2CH_2CH_2CH_2CH_2CH_2CH_2CH{=}CHCH_2CH_2CH_2CH_2CH_2CH_2CH_2CH_3}^{R_2}$$

$$CH_2O\underset{\underset{O}{\|}}{C}\underbrace{CH_2CH_2CH_2CH_2CH_2CH_2CH_2CH_2CH_2CH_2CH_2CH_2CH_2CH_2CH_2CH_3}_{R_1}$$

(ionic) polar hydrophilic "head" — nonpolar hydrophobic "tail"

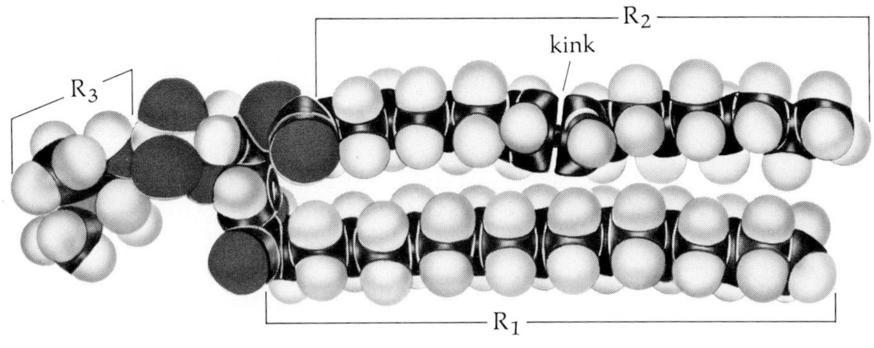

STRUCTURE AND SPACE-FILLING MODEL OF A TYPICAL PHOSPHOLIPID STRUCTURE. Note the kink produced by the *cis* double bond.

$$H-O-\overset{\overset{O}{\|}}{\underset{\underset{\underset{H}{|}}{O}}{|}}{P}-O-H$$

This same type of phosphate ester linkage

$$R-O-\overset{\overset{O}{\|}}{\underset{\underset{\underset{H}{|}}{O}}{|}}{P}-O-R$$

is an essential part of the nucleic acids. Phospholipids are esters in the same way that the phosphate–sugar linkages in DNA and RNA are esters.

Figure 20-5

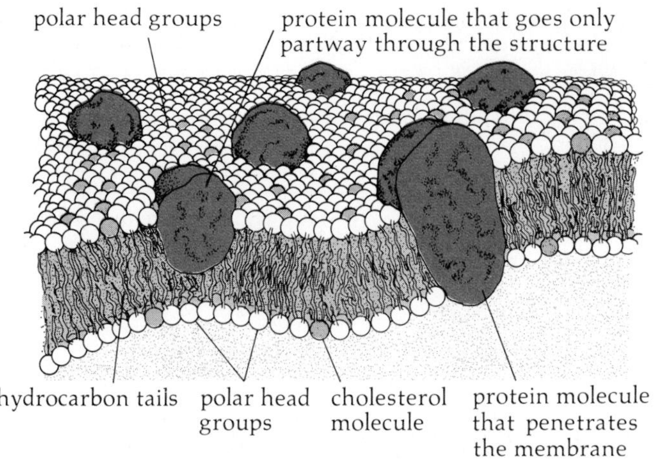

THE FLUID MOSAIC MODEL OF MEMBRANE STRUCTURE ACCORDING TO SINGER AND NICOLSON. The basic structure is the lipid bilayers. The large globs represent proteins, some of which are only on the surface; some of which go partway through the membrane, and some of which go all the way through the membrane. Cholesterol molecules are also incorporated in the lipid bilayer.

Structure of Biological Membranes

One of the most important features of a biological cell is its membrane, the structure that separates it from its environment, defining a cell interior that is different in composition and properties from the exterior. In addition, the structure of the membrane must allow for material to pass in and out of the cell to enable the cell to function. The proposal of the "fluid mosaic" model for the cell membrane, by S. J. Singer and G. L. Nicolson in 1972, was a major step forward in our attempts to understand these important and interesting structures. The lipid bilayer is the basic building block of the membrane, with protein molecules and cholesterol molecules imbedded in it. The elements of the structure are relatively free to move around while keeping the same general appearance—thus the term "fluid mosaic." Figures 20-5 and 20-6 should be explored carefully to get the idea of this model for the structure of the membrane. The functions of the membrane are believed to reside in the phospholipid bilayer (for structure), the cholesterol molecules (for fluidity), and the protein molecules (for transport of material across the membrane and for cell–cell interactions).

Scramble Exercises

1. Peanuts contain roughly 40% fat, 25% protein, 20% carbohydrates, and 15% water before processing. On complete combustion, the fuel value of peanuts ranges from 5 to 6 kilocalories per gram. One of the oils in peanuts is the

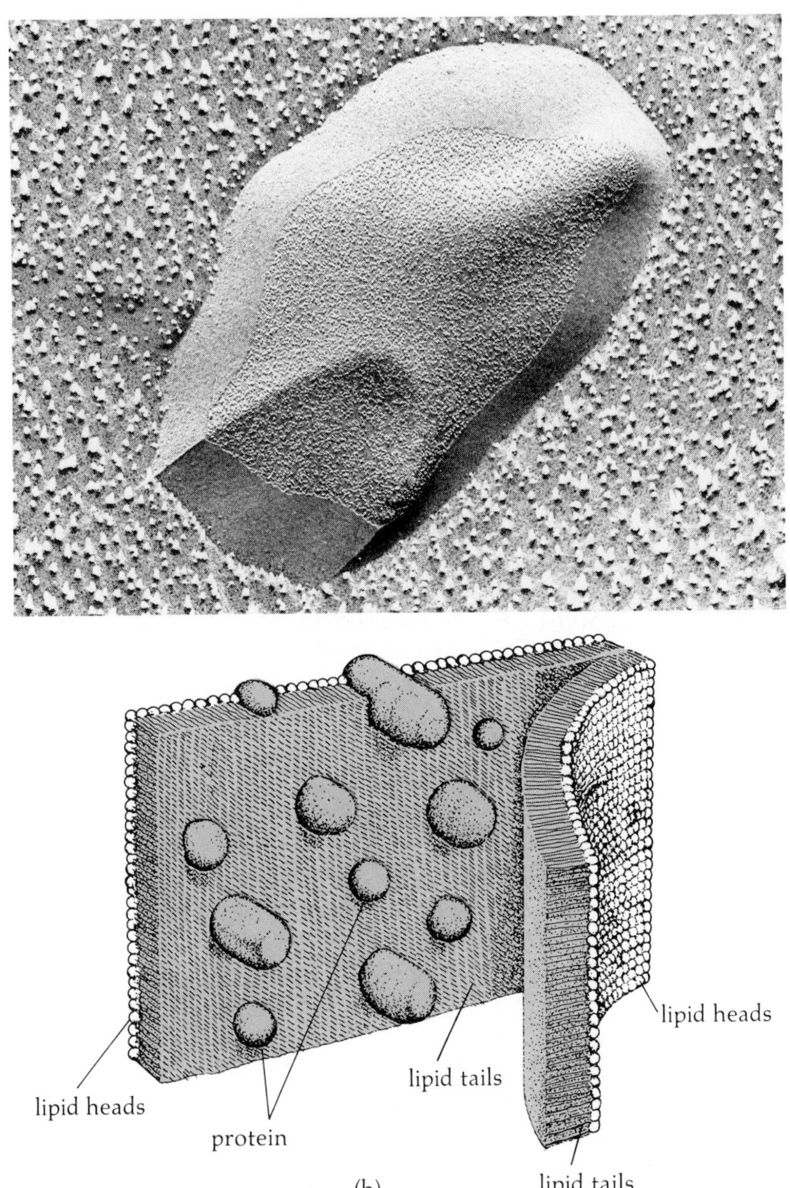

THE RED BLOOD CELL MEMBRANE. (a) The electron micrograph shows the red blood cell with the upper half of the bilayer removed (using the technique of freeze etching). The intact membrane can be seen around the lower edge of the oval-shaped membrane. (b) This drawing shows what has happened in the electron micrograph; the membrane has been pulled apart to expose the protein globs and the "tails" of the phospholipids.

triglyceride of oleic acid, $CH_3(CH_2)_7CH{=}CH(CH_2)_7CO_2H$. You should be able to write out the formula for this triglyceride and also for the corresponding diglycerides and monoglycerides, but it probably is not necessary for you to write them out in detail in order to test your comprehension.

2. Rationalize the experimental fact that fats and oils are not soluble in water.

3. The chemical makeup of natural waxes is similar to the makeup of fats and oils described in this chapter. The differences are that waxes have longer hydrocarbon chains in the fatty acid portion [a typical fatty acid from a wax is cerotic acid, $CH_3(CH_2)_{24}COOH$, an acid with 26 carbon atoms] and the alcohol is typically more complicated than glycerol, although the alcohols often contain only one rather than three —OH groups [example: $CH_3(CH_2)_{26}CH_2OH$, *n*-octacosanol, isolated from the wax on wheat]. Plant waxes also contain long-chain saturated hydrocarbons (paraffins). Waxes are used to protect and enhance finished surfaces of automobiles and furniture. Suggest a reason why it is necessary to rub (or "buff") a wax to bring out the shine. Think about these long hydrocarbon chains.

4. Losing weight or maintaining a trim figure is close to a national obsession in the United States. That concern leads to a careful consideration of the caloric value of foods. We are quite certain that the heat value for complete combustion of food is the same whether done in the body or in a calorimeter (a device that measures the heat evolved in complete combustion using pure oxygen) in the laboratory. Why are we so sure?

5. A sure way to lose weight is to eat less at the same level of physical activity or to increase physical activity while maintaining the same calorie intake. How does that work?

6. Are there chiral carbon atoms in fats? in oils?

Additional exercises for Chapter 20 are given in the Appendix.

A Look Ahead

With this chapter, we conclude our excursion into the study of large molecules with biological significance. The next chapter, 21, discusses other types of large molecules, the synthetic polymers. These molecules can be lumped together under the general term "plastics." It is no exaggeration to say that if humankind first lived in the Stone Age and moved through the Bronze Age to the Iron Age, we are certainly today living in the Plastic Age. Chapter 21 introduces some of the structures, syntheses, and properties of common plastics.

21
SYNTHETIC POLYMERS

Chapter 21

Synthetic polymers (commonly called plastics) are the materials of the modern age. It is more than likely that the clothes you are wearing, the upholstery of the chair you are sitting in, the finishes on your home and furniture, the interior and tires (and many other parts) of the family automobile are synthetic polymers, made of materials that did not even exist before 1930. These synthetic materials have come to replace natural polymers—such as cellulose (wood, cotton), proteins (silk, wool), and natural rubber—which have served for so many years. The natural materials remain unsurpassed for beauty and other desirable qualities but they are expensive, usually require extensive handcrafting in fabrication, and need regular attention to maintain their appearance. Synthetic polymers, in contrast, are uniform, inexpensive, and durable. They are so durable, in fact, that they do not degrade readily from attack by biological organisms, which natural polymers do. Synthetic polymers are an increasing problem for refuse disposal in many countries.

The Plastic Age

This chapter focuses on the chemical preparation and properties of synthetic polymers rather than that particular social issue, however important it may be. There are so many plastics in commercial production that our attention will be restricted by necessity. Most of this chapter will be devoted to two major classes: condensation polymers and addition polymers. These two classes are also referred to more generally as step growth and chain growth polymers.

Condensation Polymers

There are hundreds of condensation polymers, but again we will focus on just two types: the polyamides and the polyesters.

Nylon, a Condensation Polymer

Nylon is an example of a polyamide condensation polymer. The impetus for the development of nylon came from the desire to produce something that would have many of the properties of natural silk (see Chapter 17) yet be much much more inexpensive and be capable of mass production. Nylon has turned out to be a practical answer.

Synthesis of nylon, like that of all condensation polymers, results from the removal of a small molecule (in this case water, H_2O) from two monomer molecules to produce a dimer; the dimer then reacts with another monomer molecule to eliminate another water molecule, thus producing a trimer; and the process can go on until a large number of monomers are linked together, forming the polymer. This sequence is diagrammed below by the formation of one kind of nylon from a particular monomer, 6-amino-caproic acid. The polymer is given the name nylon-6, or polycaprolactam.

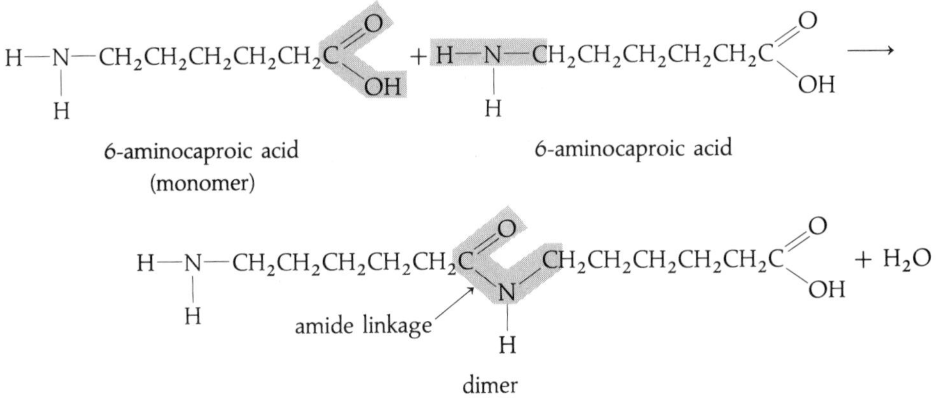

$$\underset{\substack{\text{6-aminocaproic acid}\\ \text{(monomer)}}}{H-\underset{|}{\overset{}{N}}(H)-CH_2CH_2CH_2CH_2CH_2C\begin{matrix}\nearrow O\\ \searrow OH\end{matrix}} + \underset{\text{6-aminocaproic acid}}{H-\underset{|}{N}(H)-CH_2CH_2CH_2CH_2CH_2C\begin{matrix}\nearrow O\\ \searrow OH\end{matrix}} \longrightarrow$$

$$\underset{\text{dimer}}{H-\underset{|}{N}(H)-CH_2CH_2CH_2CH_2CH_2C(=O)-\underset{\text{amide linkage}}{\underset{|}{N}(H)}-CH_2CH_2CH_2CH_2CH_2C\begin{matrix}\nearrow O\\ \searrow OH\end{matrix}} + H_2O$$

The amide linkage is exactly the same as the peptide linkage in proteins. Nylon could also be called a polypeptide, but that term is reserved for polymers of α-amino acids. Further reaction leads to

$$\text{dimer} + \text{monomer} \longrightarrow \text{trimer} + H_2O$$

and

$$\text{trimer} + \text{monomer} \longrightarrow \text{tetramer} + H_2O$$

and so on, until the formula becomes

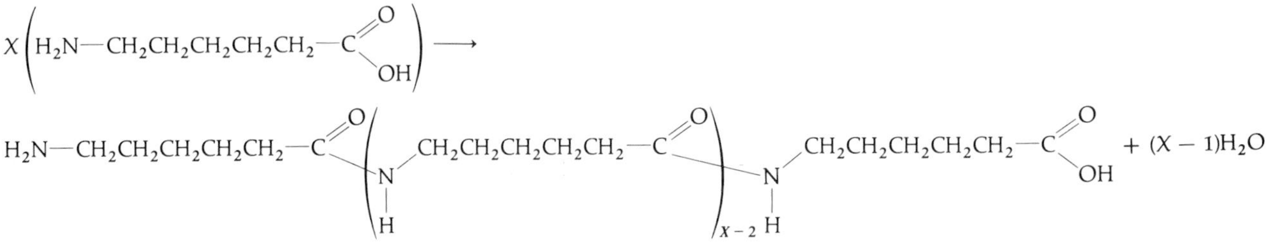

$$X\left(H_2N-CH_2CH_2CH_2CH_2CH_2-C\begin{matrix}\nearrow O\\ \searrow OH\end{matrix}\right) \longrightarrow$$

$$H_2N-CH_2CH_2CH_2CH_2CH_2-C(=O)\left(-\underset{|}{N}(H)-CH_2CH_2CH_2CH_2CH_2-C(=O)\right)_{X-2}-\underset{|}{N}(H)-CH_2CH_2CH_2CH_2CH_2-C\begin{matrix}\nearrow O\\ \searrow OH\end{matrix} + (X-1)H_2O$$

where X is called the "degree of polymerization" and represents the number of monomer molecules incorporated into the polymer molecule. Note that it is possible for the monomer to add to either end of the polymer or for fairly long chains to combine with each other. Degrees of polymerization (X) in the thousands are both possible and desirable.

Another Approach to Nylon

Nylon-type polymers can be prepared in several ways, starting from monomers of the same carbon chain lengths or from monomers of different chain lengths. An example of the first of these is the polymer formed from the reaction of a diamine, 1,6-diamino-hexane (hexamethylene diamine):

$$H_2N-CH_2-CH_2-CH_2-CH_2-CH_2-CH_2-NH_2$$

with adipic acid, a diacid

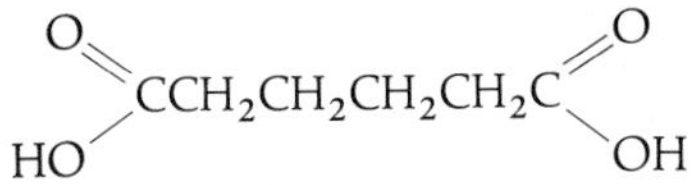

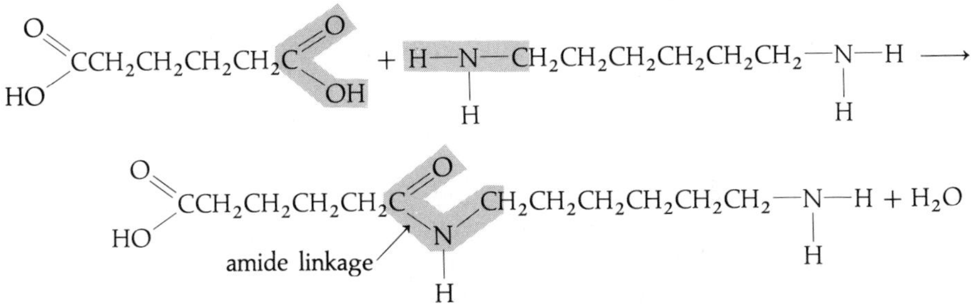

$$\mathrm{HO(O{=})CCH_2CH_2CH_2CH_2C({=}O)OH}$$

$$\mathrm{HO(O{=})CCH_2CH_2CH_2CH_2C({=}O)OH + H{-}NH{-}CH_2CH_2CH_2CH_2CH_2CH_2{-}NH{-}H \longrightarrow}$$

$$\mathrm{HO(O{=})CCH_2CH_2CH_2CH_2C({=}O){-}NH{-}CH_2CH_2CH_2CH_2CH_2CH_2{-}NH{-}H + H_2O}$$

amide linkage

The dimer reacts with the appropriate monomer and the process continues until the polymer is formed. This particular polymer is called nylon-66. (Both monomers have six carbon atoms.) Note that, every time a monomer is added, a water molecule is produced. In the commercial production of nylon, this turns out to be a problem because water is capable of reversing the polymerization reaction: the polymer so carefully made hydrolyzes (reacts with water), which returns it to its monomeric state. This sequence is shown with the dimer as the hydrolyzed molecule:

Hydrolysis of Polyamides

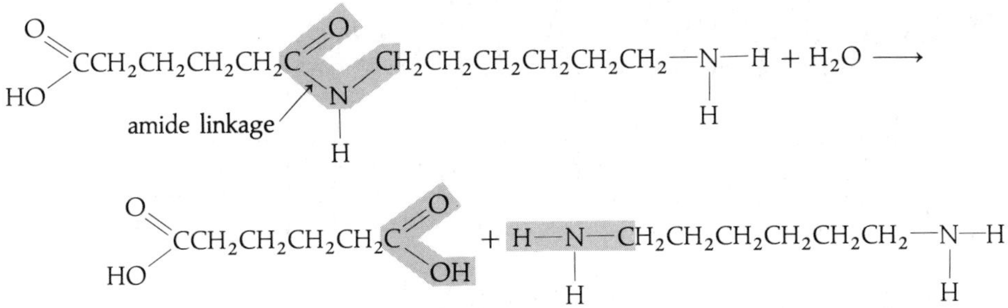

$$\mathrm{HO(O{=})CCH_2CH_2CH_2CH_2C({=}O){-}NH{-}CH_2CH_2CH_2CH_2CH_2CH_2{-}NH{-}H + H_2O \longrightarrow}$$

amide linkage

$$\mathrm{HO(O{=})CCH_2CH_2CH_2CH_2C({=}O)OH + H{-}NH{-}CH_2CH_2CH_2CH_2CH_2CH_2{-}NH{-}H}$$

Nylon manufacturers take steps to remove water molecules as quickly as they are formed to keep the polymer intact.

Gratuitous Information 21-1

SILK

Silk is so expensive because it requires the cooperation of silkworms, food for them to eat, and the work of many people to tend both the worms and the mulberry trees that produce the leaves the silkworms eat and to spin the silk from the cocoons. Silk cultivation remains essentially an Asian industry. There is evidence that the infestation of the gypsy moth in the United States is the result of a misguided attempt to develop the gypsy moth larva as a substitute for the silkworm.

Ring Structures

Another thing you might think about is: Why doesn't a monomer, like 6-aminocaproic acid, just form a ring and forget about polymerization altogether?

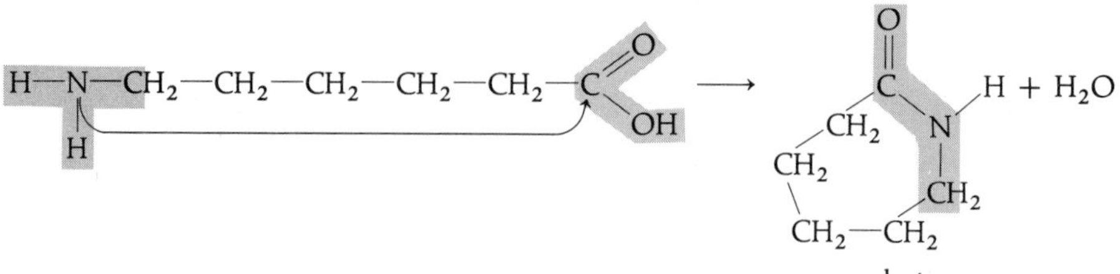

$$\mathrm{H{-}\underset{\substack{|\\H}}{N}{-}CH_2{-}CH_2{-}CH_2{-}CH_2{-}CH_2{-}\overset{\substack{O\\\|}}{C}{-}OH} \longrightarrow \text{caprolactam (ring: } \mathrm{C(=O){-}N(H){-}CH_2{-}CH_2{-}CH_2{-}CH_2{-}CH_2}\text{)} + \mathrm{H_2O}$$

caprolactam

In practice, some ring formation always happens, but the art of making the polymer is in controlling the polymerization conditions so as to limit the number of rings formed and to make it easy for a ring, once formed, to open up again and become part of the polymer chain. In fact, the ring compound above, caprolactam, is the normal starting material for the formation of nylon-6.

Nylon has become pervasive in the economy, replacing silk in nearly all applications (stockings, blouses) except luxury clothing and adding many new applications, such as rope and film, where the strength and durability of nylon outperform those of natural fibers.

Dacron, a Polyester

The other major class of condensation polymers that will be discussed here are the polyesters, and our one example will be dacron, the particular case being poly(ethylene terephthalate). You can see from its "semi-systematic" name why popular or trade names are almost always used in talking about polymers. Dacron is formed when ethylene glycol (1,2-ethanediol) is reacted with terephthalic acid:

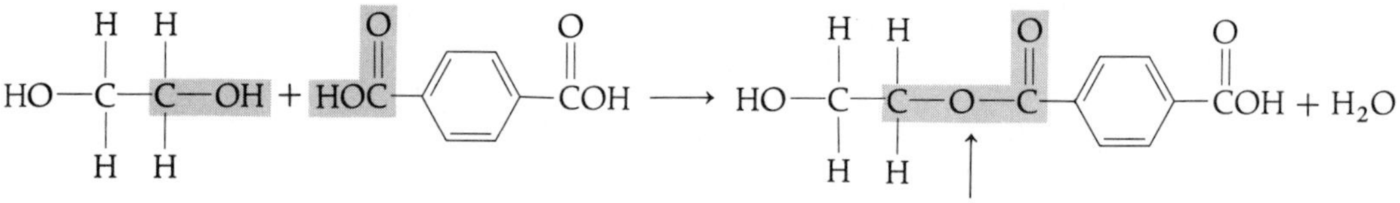

$$\mathrm{HO{-}CH_2{-}CH_2{-}OH} + \mathrm{HO\overset{\substack{O\\\|}}{C}{-}C_6H_4{-}\overset{\substack{O\\\|}}{C}OH} \longrightarrow \mathrm{HO{-}CH_2{-}CH_2{-}O{-}\overset{\substack{O\\\|}}{C}{-}C_6H_4{-}\overset{\substack{O\\\|}}{C}OH} + \mathrm{H_2O}$$

ethylene glycol — terephthalic acid — ester linkage — dimer

The ester linkage and the amide linkage have much in common; the difference is that the

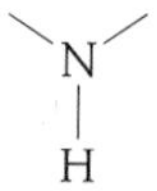

$$\mathrm{\underset{\substack{|\\H}}{N}}$$

group in amides is replaced by the

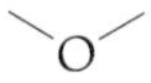

$$\mathrm{{-}O{-}}$$

group in esters. The additional hydrogen atom on the nitrogen atom in amides (or peptides) causes the properties of the polymer chain to be quite different. The interaction of the hydrogen atom on the nitrogen atom with other atoms in polyamides is dealt with in some detail in the chapter on proteins (Chapter 17, pages 424–30).

The overall reaction for the formation of the polymer, once again using the general coefficient X (degree of polymerization), is

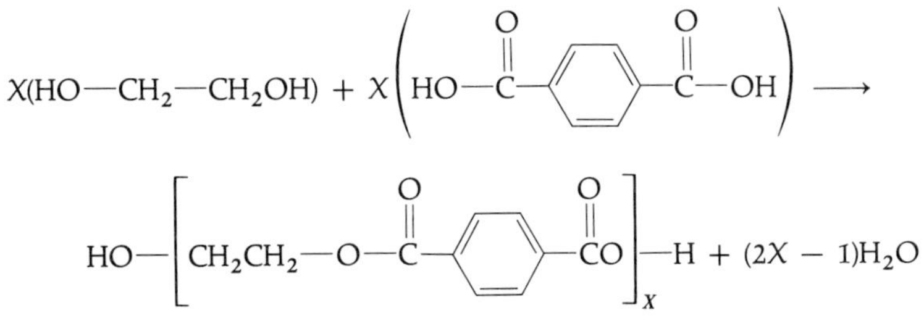

$$X(HO{-}CH_2{-}CH_2OH) + X\left(HO{-}\overset{\displaystyle O}{\overset{\|}{C}}{-}C_6H_4{-}\overset{\displaystyle O}{\overset{\|}{C}}{-}OH\right) \longrightarrow$$

$$HO{-}\left[CH_2CH_2{-}O{-}\overset{\displaystyle O}{\overset{\|}{C}}{-}C_6H_4{-}\overset{\displaystyle O}{\overset{\|}{C}}O\right]_X{-}H + (2X - 1)H_2O$$

This polymer, like the nylons mentioned above, is an example of a linear polymer. The chain proceeds from one end to the other without branching. Such polymers are easily drawn out into filaments (threads) suitable for the manufacture of clothing. Polyester garments were originally thought of as replacements for cottons, but the properties are sufficiently different that polyesters now have an identity of their own.

Polyester–Cotton Blended Fabrics

Cotton is a polymer of glucose (See Chapter 18, pages 446, 448), and the many —OH groups along the polymer chain interact readily with H_2O molecules (such as those produced by our bodies in perspiration). These H_2O molecules spread throughout the cotton fibers and evaporate readily, making cotton fabrics feel "cool" when worn: the fabric is said to "breathe." Polyester chains don't have those —OH groups and consequently do not absorb water. Perspiration tends to remain on the skin, trapped by a layer of fabric that interacts poorly with water molecules. Polyester fabrics don't breathe; on the other hand, they don't wrinkle, either. Most shirts and blouses are now cotton–polyester blends that combine the superior wrinkle resistance and durability of polyesters with the superior "breathing" ability and feel of cotton.

The small molecule removed in the polymerization that forms polyesters is once again the water molecule—by far the most common case—but it is not the only way to carry out a condensation polymerization. Scramble Exercise 2 points out a case where hydrogen chloride (HCl) is the small molecule lost on polymerization.

Addition Polymers

The other class of polymerization reactions to be discussed in this chapter is the case where no molecule is eliminated in the polymerization process, and, from the formula at least, it appears that the monomers just add to each other. This process is typical of those products called vinyl or acrylic polymers. The simplest case is the polymerization of ethene (ethylene):

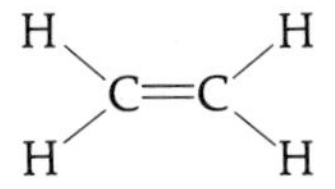

$$\begin{matrix} H & & & & H \\ & \diagdown & & \diagup & \\ & & C{=}C & & \\ & \diagup & & \diagdown & \\ H & & & & H \end{matrix}$$

One way to start the polymerization reaction is by addition of an initiator. One type of initiator is a compound that will readily form a free radical—a compound like benzoyl peroxide.

$$C_6H_5\text{—}\overset{O}{\overset{\|}{C}}\text{—O—O—}\overset{O}{\overset{\|}{C}}\text{—}C_6H_5 \longrightarrow 2\ C_6H_5\text{—}\overset{O}{\overset{\|}{C}}\text{—O}\cdot$$

benzoyl peroxide   the benzoyl radical

Initiation of Addition Polymerization

The bond between the two oxygen atoms in peroxides is unstable and splits readily, even at room temperature. The process is, of course, faster at higher temperatures. The oxygen–oxygen bond in benzoyl peroxide fragments with the two electrons in the bond unpairing, one going with each of the fragments. Compounds having such unpaired electrons (usually called free radicals) are very reactive and are specially good at reacting with compounds containing double bonds. We will call the initiator, whatever it's structure is, simply I· where the dot represents the unpaired electron. The unpaired electron of I· can interact with the *pi* (π) electron pair of the double bond in ethene to produce a new *sigma* (σ)-bond between the initiator and one carbon atom of the ethene molecule:

$$I\cdot + \underset{H\ \ H}{\overset{H\ \ H}{C::C}} \longrightarrow I:\underset{H\ H}{\overset{H\ H}{C:C\cdot}}$$

Polymerization of Vinyl Monomers

The new bond uses only one of the electrons in the π-bond, so the other electron ends up in a sp^3-orbital at the other carbon atom of the ethene molecule. This makes a new initiator and the chain grows as another ethene molecule is added:

$$I\text{—}CH_2\text{—}CH_2\cdot + H_2C{=}CH_2 \longrightarrow I\text{—}CH_2\text{—}CH_2\text{—}CH_2\text{—}CH_2\cdot$$

The process can continue indefinitely as long as enough monomer is supplied and nothing else bonds to the active end of the growing chain:

$$I\cdot + X\left(H_2C{=}CH_2\right) \longrightarrow I\text{—}(CH_2\text{—}CH_2)_{X-1}\text{—}CH_2\text{—}CH_2\cdot$$

Termination of Addition Polymerization

The end of the chain with the unpaired electron is always reactive and the "termination" of the growth of the chain by pairing that unpaired electron can occur in several ways. Only two will be mentioned here.

First, two growing chain ends can meet and react with each other:

$$I\text{—}(CH_2\text{—}CH_2)_{X-1}\text{—}CH_2\text{—}CH_2\cdot + \cdot CH_2\text{—}CH_2\text{—}(CH_2\text{—}CH_2)_{Y-1}\text{—}I \longrightarrow$$
$$I\text{—}(CH_2\text{—}CH_2)_X\text{—}(CH_2\text{—}CH_2)_Y\text{—}I$$

which is the best possible result if the purpose is to produce the longest possible chain (highest molecular weight polymer). The unpaired electrons at the ends of each chain simply combine to produce a new σ-bond, linking the two growing chains and, on the average, roughly doubling the chain length.

Second, the growing chain end may simply react with another initiator I• or any other free radical that might be present to stop the chain growth:

$$I{-}(CH_2{-}CH_2)_{X-1}{-}CH_2{-}CH_2\cdot + \cdot I \longrightarrow I{-}(CH_2{-}CH_2)_X{-}I$$

In typical cases, thousands of monomers are incorporated for each initiation, and the composition of the polymer, except for the initiators stuck on the ends, is essentially $(CH_2{-}CH_2)_X$, X times the composition of the monomer. Hence the term "addition" polymers.

Composition of Common Addition Polymers

There are many addition polymers in commercial use. Polyethylene (polyethene) is well known in packaging materials, with trash bags perhaps the most obvious example. The "vinyl" group is the carbon–carbon double bond with three hydrogen atoms:

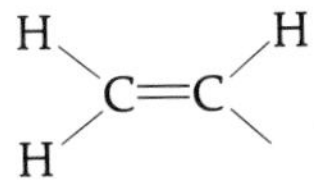

```
H       H
 \     /
  C=C
 /     \
H
```

Poly(vinyl chloride), which is produced by polymerization of vinyl chlorides

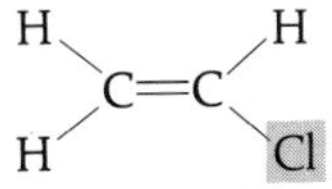

```
H       H
 \     /
  C=C
 /     \
H       Cl
```

is familiar from its use in "vinyl" shower curtains and tablecloths. The monomer vinyl acetate

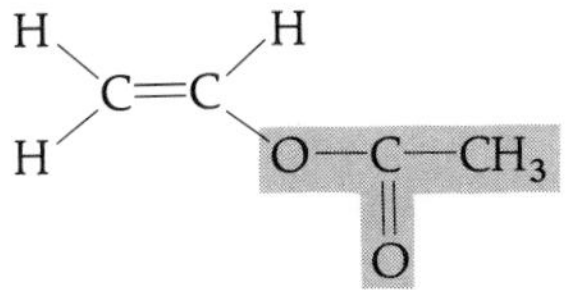

```
H       H
 \     /
  C=C
 /     \
H       O—C—CH3
          ||
          O
```

can be polymerized to produce poly(vinyl acetate). While that polymer has some uses of its own, most poly(vinyl acetate) is hydrolyzed (reacted with water) to give poly(vinyl alcohol):

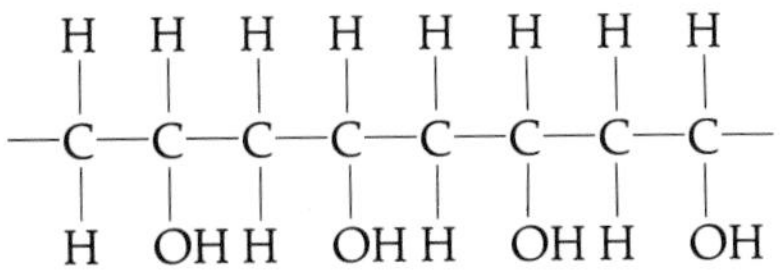

```
  H  H  H  H  H  H  H  H
  |  |  |  |  |  |  |  |
—C—C—C—C—C—C—C—C—
  |  |  |  |  |  |  |  |
  H  OH H  OH H  OH H  OH
```

Vinyl alcohol

```
H       H
 \     /
  C=C
 /     \
H       OH
```

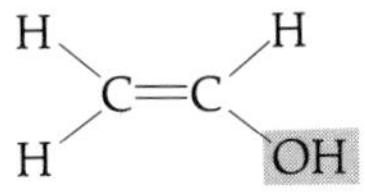

cannot be used as a monomer because it is an unstable compound rearranging to the more stable isomer, acetaldehyde,

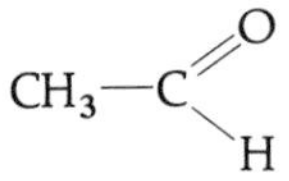

$$CH_3-C(=O)H$$

This makes the two-step route to poly(vinyl alcohol) necessary. Poly(vinyl alcohol) [often mixed with about 10% poly(vinyl acetate)] has many uses as an adhesive and a laminating agent. It is the middle layer in laminated safety glass, for example. Teflon, much used on cookware, is the polymer of tetrafluoroethene:

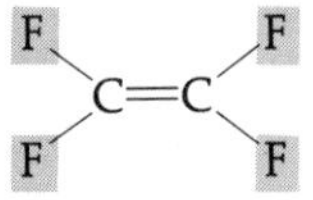

$$F_2C=CF_2$$

Another series of addition polymers is based on the monomer acrylic acid

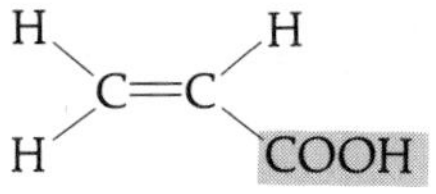

$$H_2C=CH-COOH$$

or its derivatives, such as the esters methyl acrylate

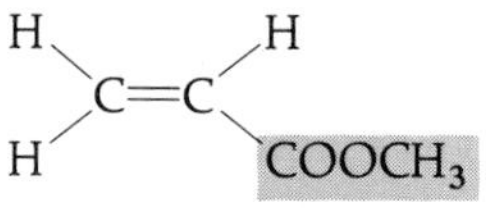

$$H_2C=CH-COOCH_3$$

or methyl methacrylate

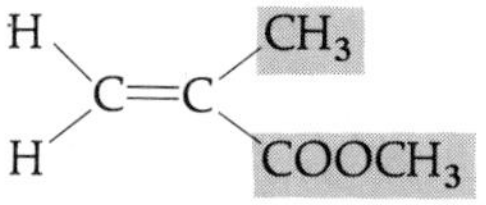

$$H_2C=C(CH_3)-COOCH_3$$

Orlon, often used as a wool substitute, is poly(methyl acrylate), and lucite, which is used in paint and produced to be used in sheets as a glass substitute ("plexiglass"), is poly(methyl methacrylate).

Head to Tail Polymerization

Polystyrene is the result of polymerizing styrene

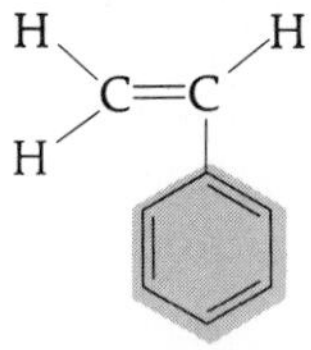

$$H_2C=CH-C_6H_5$$

and the polymer, like nearly all addition polymers, is exclusively "head-to-tail"—the benzene rings "hang off" alternate carbon atoms in the chain—

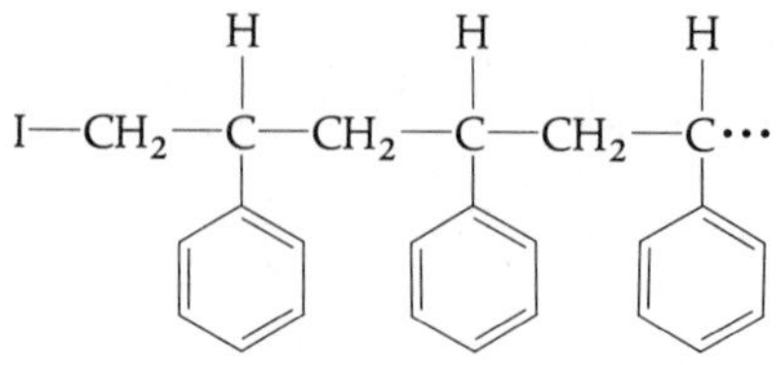

$$I-CH_2-CH(C_6H_5)-CH_2-CH(C_6H_5)-CH_2-CH(C_6H_5)\cdots$$

not any other orientation, such as

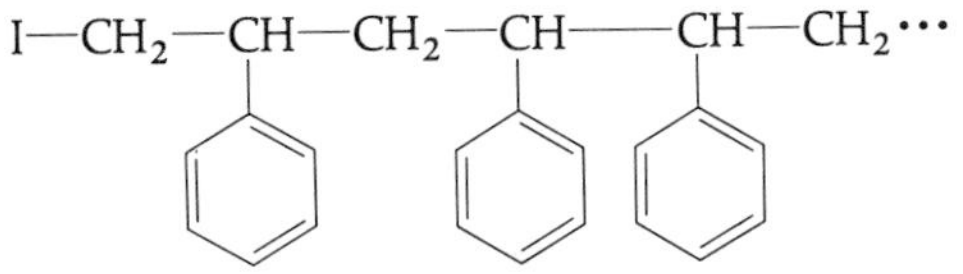

$$I{-}CH_2{-}\underset{\displaystyle C_6H_5}{CH}{-}CH_2{-}\underset{\displaystyle C_6H_5}{CH}{-}\underset{\displaystyle C_6H_5}{CH}{-}CH_2\cdots$$

Polystyrene foam (where air is incorporated into the polymerizing mixture) is very well known from its use in plastic "hot cups" and "foam coolers."

Note that, in all of these addition polymers, it is the chemical reactivity of the π-bond in the vinyl group

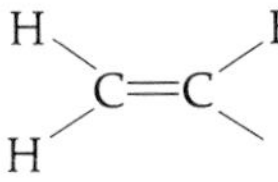

$$H_2C{=}CH{-}$$

that leads to polymerization.

Natural Rubber

Natural rubber is a vinyl addition polymer, the polymer of the monomer 2-methyl-1,3-butadiene, also called isoprene:

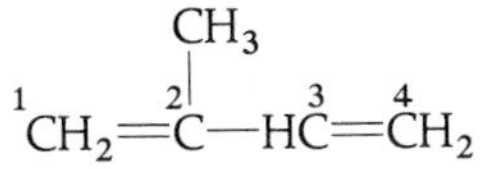

$$\overset{1}{C}H_2{=}\overset{2}{C}(CH_3){-}H\overset{3}{C}{=}\overset{4}{C}H_2$$

Polymerization of Isoprene

Polymerization of this monomer in nature leads to

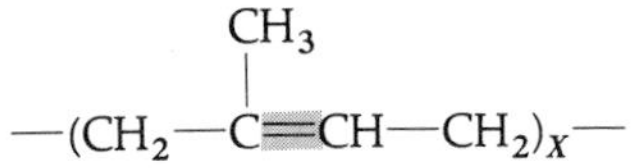

$${-}(CH_2{-}C(CH_3){=}CH{-}CH_2)_x{-}$$

Gratuitous Information 21-2

SYNTHETIC RUBBER

World War II prompted development of synthetic rubber, since most natural rubber came (and still comes) from plantations in Southeast Asia and the western Pacific. Those supplies were cut off by Japanese domination of the seas in those areas. First attempts at making synthetic rubber involved production of a polymer from 1,3-butadiene

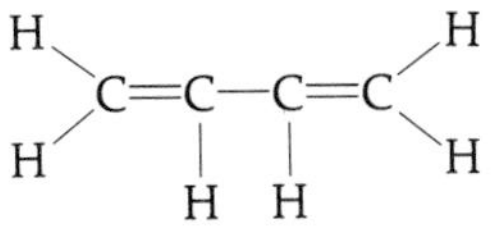

$$H_2C{=}CH{-}CH{=}CH_2$$

also written $CH_2{=}CHCH{=}CH_2$, but the material produced had nowhere near the quality of natural rubber. A better quality synthetic rubber was made by polymerizing chloroprene, a molecule similar to isoprene but with a chlorine atom replacing the methyl group:

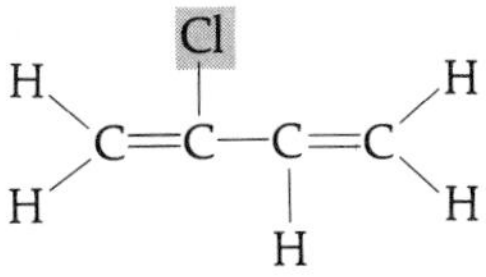

$$H_2C{=}C(Cl){-}CH{=}CH_2$$

chloroprene

$$H_2C{=}C(CH_3){-}CH{=}CH_2$$

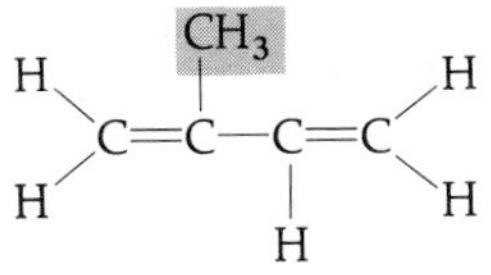

isoprene

The resulting polymeric material is called neoprene and is still used today in the soles of work shoes and in hoses that handle gasoline. The electronegative chlorine atom in neoprene makes this polymer much less soluble in gasoline than natural rubber. Today, it is possible to produce isoprene in quantity and to polymerize it with the double bonds all *cis*, making a synthetic rubber that is essentially identical with the natural material.

Note that the polymer has a double bond in the middle, between carbon atoms 2 and 3, while the monomer has two double bonds at the ends, between carbon atoms 1 and 2 and between carbon atoms 3 and 4. The process that leads to this condition is called 1,4-addition. Using I• again for the initiator, one of the electrons of the π-bond forms a new σ-bond with initiator, just like the initiation of polymerization in ethene:

```
                 H                            H
                 ..                           ..
       H  H:C:H      H              H  H:C:H      H
       ..   ..       ..             ..   ..       ..
I• + C⊙:  C  :  C::C          I:C⊙   :C     :C⊙:C
       ..            .. ..          ..          .. ..
       H             H  H           H           H  H
```

The unpaired electron can now form a new π-bond between carbon atoms 2 and 3, using one of the electrons from the π-bond between carbon atoms 3 and 4,

```
            H
            ..
      H  H:C:H    H
      ..   ..     ..
     I:C  :  C::C⊙·C
      ..        ..  ..
      H         H   H
```

and finally the new unpaired electron between carbon atoms 3 and 4 ends up as an unpaired electron in an sp^3-orbital on carbon atom 4:

```
            H
            ..
      H  H:C:H  H
      ..   ..   ..
     I:C  :  C::C:C•
      ..       .. ..
      H        H  H
```

That unpaired electron on carbon atom 4 can, of course, react with a new isoprene molecule to continue the growth of the chain. Another interesting property is that the orientations at all the double bonds in the natural rubber polymer are *cis*. The all-*trans* isomer also exists in nature and is called gutta percha.

Since the unpaired electron at the end of the growing chain will react with just about any π-bond, it is possible to make polymers that incorporate more than one monomer. These can be incorporated in random order or in blocks of one type of polymer followed by a block of another type. Polymers made from more than one monomer are called copolymers. The symbol A below represents one type of monomer and B represents another type of monomer.

Copolymers

ABAABABBBA	random copolymer
AAAAAAABBBBBBB	block copolymer

An extremely widely used copolymer is the ABS copolymer, a mixture of the monomers acrylonitrile,

```
H        H
 \      /
  C==C
 /      \
H        C≡N
```

butadiene,

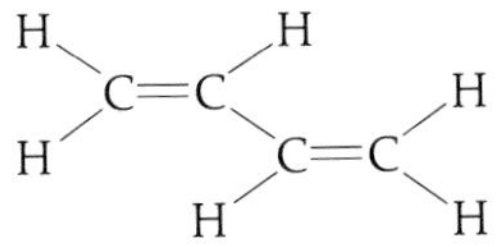

$$\mathrm{H_2C{=}CH{-}CH{=}CH_2}$$

and styrene,

$$\mathrm{H_2C{=}CH{-}C_6H_5}$$

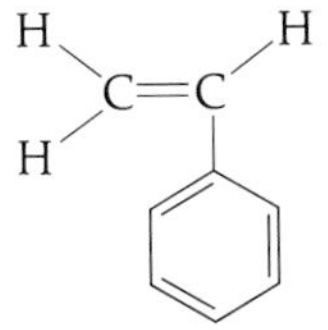

polymerized together. The copolymer has very desirable molding properties and is used extensively for automobile parts like dashboards and for radio and TV cases.

Other Kinds of Initiation

Cationic Initiation

The free radical or unpaired electron initiation of a polymer chain has been discussed above. Two other types of initiation are possible: cationic (positive ion) initiation and anionic (negative ion) initiation. For simplicity, we will once again use ethene as the monomer, although the process is much more effective with other monomers. Reaction of a molecule of ethene with aluminum chloride leads to the formation of a zwitterion

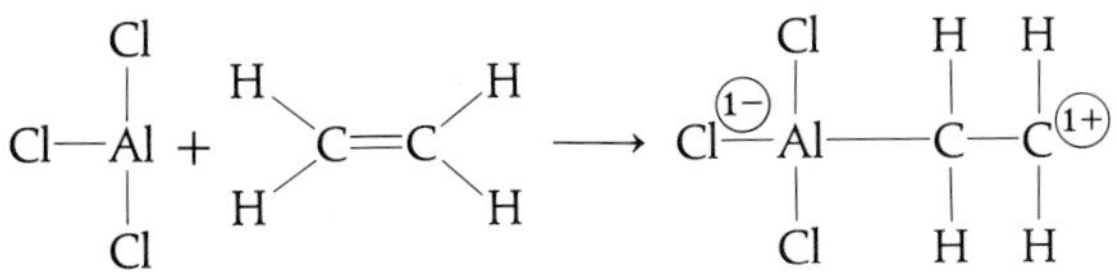

$$\mathrm{AlCl_3} + \mathrm{H_2C{=}CH_2} \longrightarrow \mathrm{Cl{-}\overset{1-}{Al}(Cl)_2{-}CH_2{-}\overset{1+}{C}H_2}$$

or, using the Lewis dot structures,

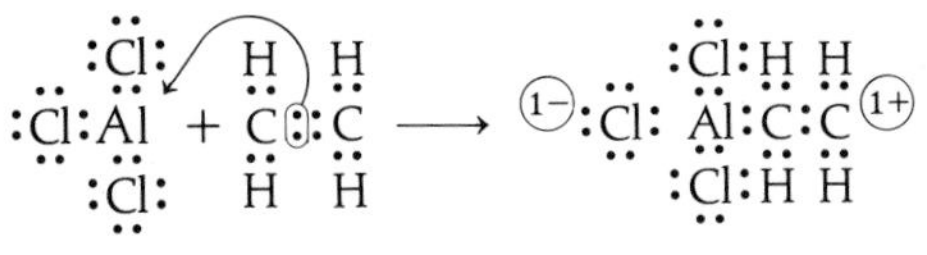

$$\mathrm{:\ddot{C}l:\ddot{A}l:\ddot{C}l:\ (with\ :\ddot{C}l:) + H_2C::CH_2 \longrightarrow {}^{1-}:\ddot{C}l:\ddot{A}l:\ddot{C}:\ddot{C}^{1+}\ (with\ :\ddot{C}l:\ on\ Al,\ H\ on\ each\ C)}$$

where the π electrons of the ethene are shared between one carbon atom and the aluminum atom, leaving only six valence electrons associated with the other carbon atom of the ethene molecule. This end of the ethene molecule is deficient in electrons. It thus becomes a cation that can react with another molecule of ethene to produce a new cation, and the process can continue indefinitely.

$$\text{Cl}^{\ominus}\!\!-\!\text{AlCl}_2\!-\!\text{CH}_2\!-\!\text{CH}_2^{\oplus} + \text{CH}_2\!=\!\text{CH}_2 \longrightarrow \text{Cl}^{\ominus}\!\!-\!\text{AlCl}_2\!-\!\text{CH}_2\!-\!\text{CH}_2\!-\!\text{CH}_2\!-\!\text{CH}_2^{\oplus}$$

There is special interest in this process for polystyrene. As you look at the polystyrene molecule, every other carbon atom in the molecule is chiral (attached to four different groups).

Chiral Carbon Atoms in Polymers—Atactic and Isotactic Polymers

$$\text{I}-\text{CH}_2-\underset{\text{C}_6\text{H}_5}{\overset{\text{H}}{\text{C}^*}}-\underset{\text{H}}{\overset{\text{H}}{\text{C}}}-\underset{\text{C}_6\text{H}_5}{\overset{\text{H}}{\text{C}^*}}-\underset{\text{H}}{\overset{\text{H}}{\text{C}}}-\underset{\text{C}_6\text{H}_5}{\overset{\text{H}}{\text{C}^*}}\cdots$$

(The asterisk indicates chiral carbon atoms.) When the chirality of these atoms is in random order, the polymer is called atactic, and random chirality is the normal result of free radical polymerization. However, appropriate choice of a cationic initiator can lead to all of the chiral carbon atoms in the polymer having the same chirality, in which case the polymer is called isotactic. In cationic initiation, the pair of electrons of the *pi*-bond move together, and in free radical initiation, the pair of π-bond electrons separate and participate in the formation of two different *sigma*-bonds.

Isotactic chains can line up better side by side in the crystal, and thus isotactic polymers are more rigid and have higher melting points than the atactic polymers. Plastic kitchenware that must withstand boiling water is often made of an isotactic polymer.

Crystallization in Polymeric Materials

Crystallization in polymeric materials is an interesting phenomenon and very significant in the determination of their properties. To demonstrate a few of the processes involved, it is convienent to represent a polymer chain very simply as a line. Thus a polyethene chain in the fully extended (zigzag) configuration could be indicated by a straight line: ——————————. Several chains lined up together in a crystal could be represented by several of these lines side by side:

——————————
——————————
——————————
——————————

In the molten state of polyethene (and indeed in most synthetic polymers), the chain is not fully extended but coiled in a random fashion (spaghetti in a bowl is a frequently used analogy), and we could represent several chains in "random coils" as

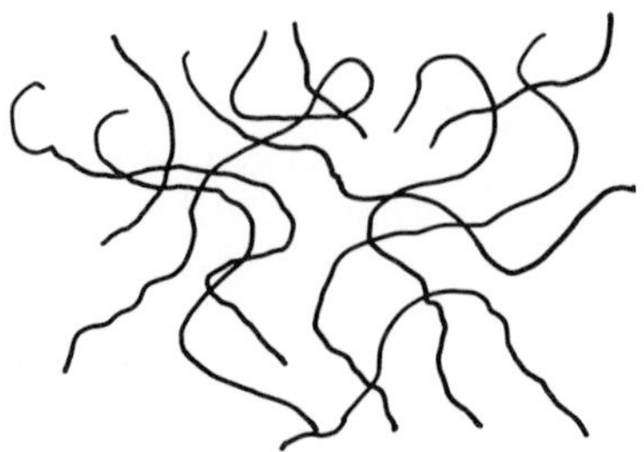

Figure 21-1

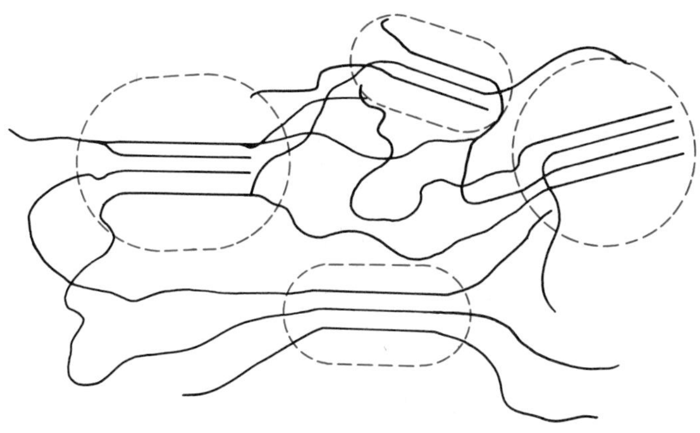

A REPRESENTATION OF PARTIAL CRYSTALLINITY IN A POLYMER. The aligned portions identified within the dotted circles are called crystallites.

As molten polyethene solidifies, portions of the chain align with each other and portions remain in the molten state—a situation represented by Figure 21-1.

The length of any one polymer chain is usually much longer than any of the crystalline regions, so the chain wanders about from crystalline region to molten region (the molten region, where the chains have no particular orientation with respect to each other, is also called the amorphous (without form) region). As the temperature is lowered, the crystalline regions, or crystallites grow at the expense of the amorphous regions; if the temperature is raised, the reverse happens. The presence of many small crystallites interspersed with amorphous regions is the usual cause of the milkiness or opalescence of polyethene films. The more perfect the crystallites and the greater the portions of the chains that are in crystallites, the higher the melting point of the polymer.

Anionic Polymerization

Anionic initiation is also possible using an anion such as $NH_2^{(1-)}$ (the amide ion). Sodium amide is an ionic compound comparable to sodium hydroxide. A pair of electrons of the nitrogen atom is shared with a carbon atom of an ethene molecule, producing a new *sigma*-bond with the carbon atom. That new bond forces the pair of electrons of the π-bond into an unoccupied sp^3-orbital on the other carbon atom and those electrons can then react with another ethene molecule to build up the chain. A simplified equation using ethene as the monomer is

$$\mathrm{H{:}\underset{\displaystyle H}{\underset{\cdot\cdot}{\overset{\cdot\cdot}{N}}}{:}^{(1-)} + CH_2{=}CH_2 \longrightarrow NH_2{-}CH_2{-}CH_2^{(1-)}}$$

or, using the Lewis structures,

$$\mathrm{H{:}\underset{\displaystyle H}{\underset{\cdot\cdot}{\overset{\cdot\cdot}{N}}}{:}^{(1-)} + \underset{\displaystyle H}{\underset{\cdot\cdot}{\overset{\displaystyle H}{\overset{\cdot\cdot}{C}}}}{::}\underset{\displaystyle H}{\underset{\cdot\cdot}{\overset{\displaystyle H}{\overset{\cdot\cdot}{C}}}} \longrightarrow H{:}\underset{\displaystyle H}{\underset{\cdot\cdot}{\overset{\cdot\cdot}{N}}}{:}\underset{\displaystyle H}{\underset{\cdot\cdot}{\overset{\displaystyle H}{\overset{\cdot\cdot}{C}}}}{:}\underset{\displaystyle H}{\underset{\cdot\cdot}{\overset{\displaystyle H}{\overset{\cdot\cdot}{C}}}}{:}^{(1-)}}$$

Branched Chains and Cross-Links

For purposes of simplicity, attention has been restricted so far to linear polymers only. However, polymer chains can have branches or cross-links in them. Using lines to indicate the polymer chains, we can represent some of the possibilities in the simple ways indicated below:

Origin of Chain Branching

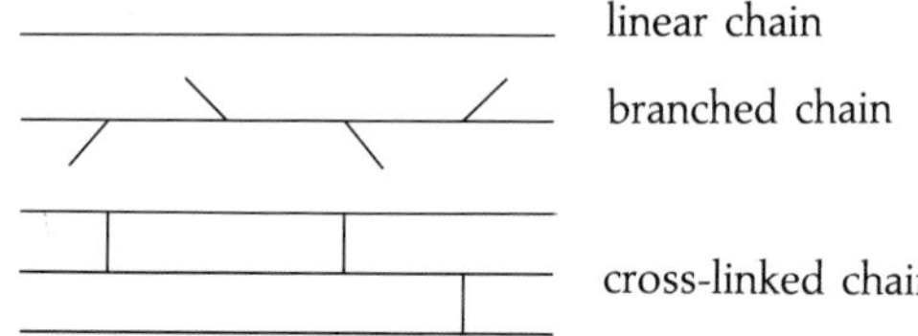

Branched chains can arise in the polymerization of ethene by the following set of circumstances.

The reactive end (a free radical) of a growing polymer chain may react with a hydrogen atom in the middle of another chain (a process called "hydrogen abstraction"). This terminates the first chain and creates a new free radical in the middle of the second chain:

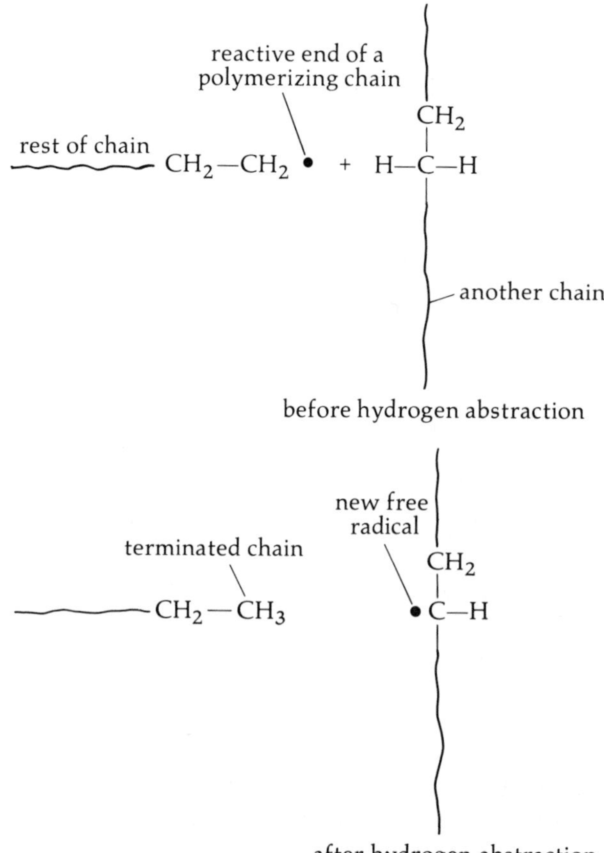

The first reaction of the new free radical with an ethene molecule, $CH_2{=}CH_2$, leads to a growing branch

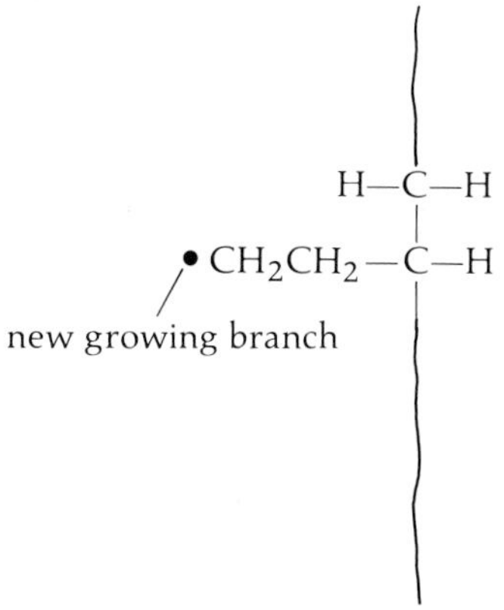

H—C—H

• CH_2CH_2—C—H

new growing branch

Cross-Linking

To make a cross-linked chain, we will use the free radical polymerization polystyrene as an example. A small amount of divinyl benzene is added to the styrene monomer and the mixture is polymerized by the introduction of the initiator free radical I•.

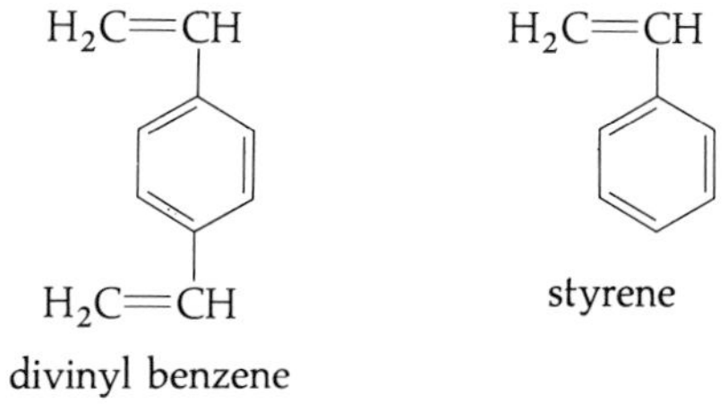

$H_2C{=}CH$ / $H_2C{=}CH$

divinyl benzene

$H_2C{=}CH$

styrene

Since the divinyl benzene can become incorporated into two chains, the process leads to the following:

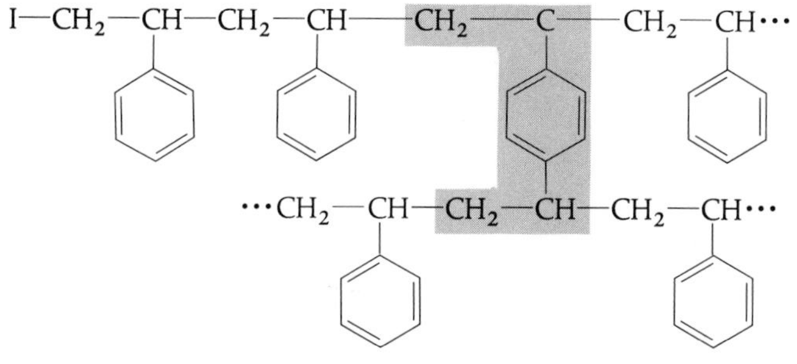

I—CH_2—CH—CH_2—CH—CH_2—C—CH_2—CH···

cross-link by divinyl benzene

···CH_2—CH—CH_2—CH—CH_2—CH···

With enough cross-links present, a three-dimensional network is formed and the materials are hard, rigid, and insoluble in most solvents. An example of highly cross-linked materials formed by condensation polymerization are the phenol–formaldehyde polymers Bakelite and Formica.

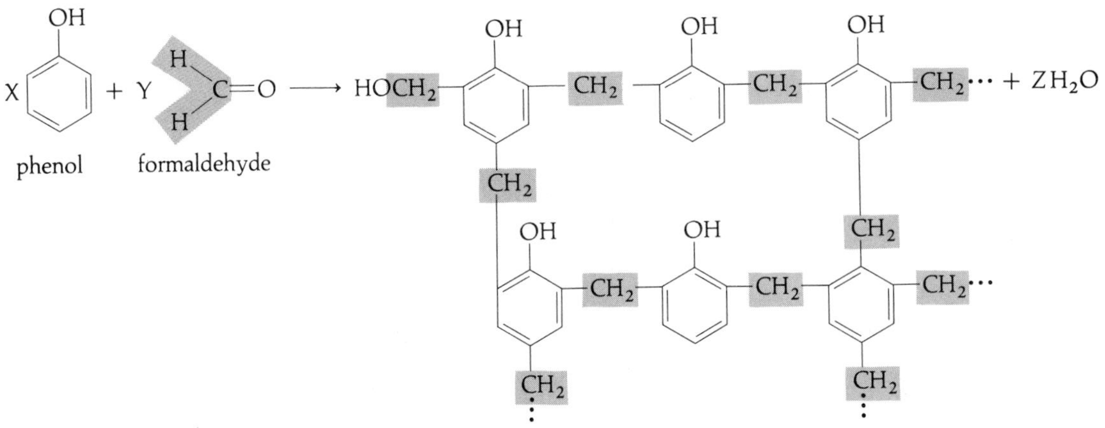

phenol + formaldehyde ⟶ $HOCH_2$–[phenol–CH_2 network]$\cdots + ZH_2O$

Note the distribution of the methylene groups, CH_2, from the formaldehyde throughout the polymer. Bakelite (one of the first commercial polymers) is still used for bottle caps, and Formica, with all the colors added by dye molecules, is used for counter tops and table surfaces.

Silicon–Oxygen Polymers

Polymerization is not limited to carbon compounds, although manufactured polymers nearly always involve carbon atoms in some fundamental way. Many natural minerals (such as the silicate rocks) are based on polymers of silicon and oxygen. Silicon can bond to four oxygen atoms, and each oxygen atom can bond to two silicon atoms. Often, there are fewer bonds than those present in natural minerals, and silicates often have complicated polymeric negative ions. The structures vary from linear chains in asbestos-like minerals

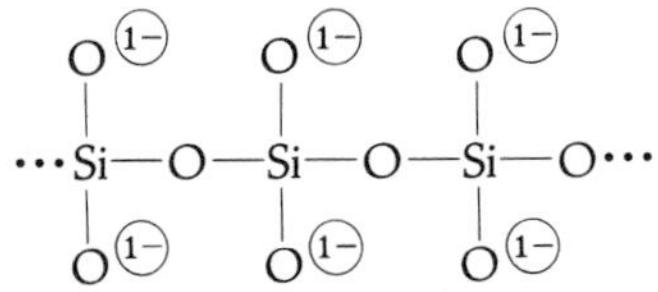

Structures of Silicate Rocks

$$\cdots Si(O^{1-})_2{-}O{-}Si(O^{1-})_2{-}O{-}Si(O^{1-})_2{-}O\cdots$$

to two-dimensional sheets characteristic of talc and mica

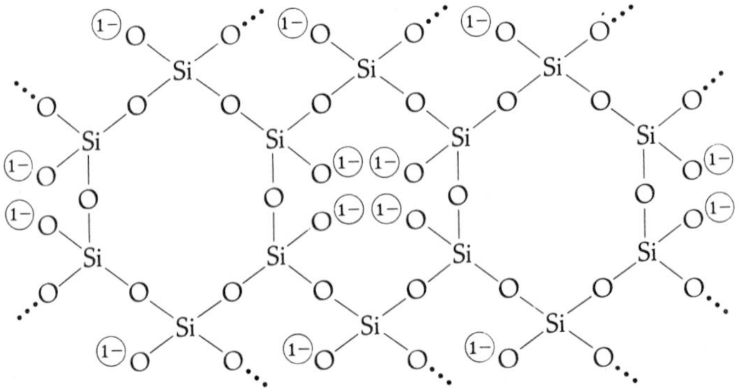

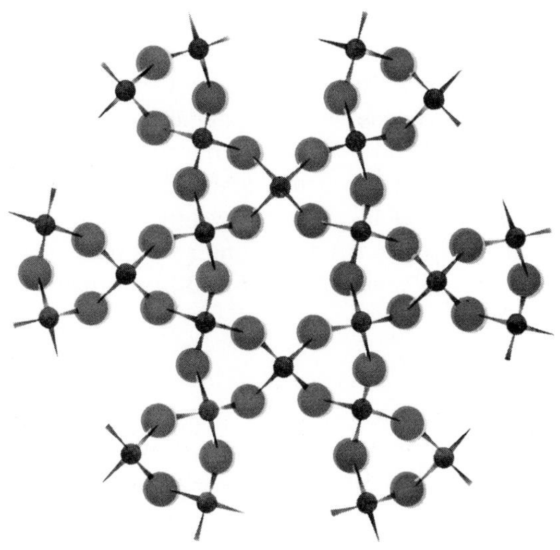

A SMALL SEGMENT OF THE INFINITE THREE-DIMENSIONAL STRUCTURE OF QUARTZ, SiO_2. The small black solid circles represent silicon atoms and the large red circles represent oxygen atoms. The oxygen atoms lie at the corners of tetrahedra around the silicon atoms. The Si—O—Si angles are about 144°, the O—Si—O about 109°. Pure quartz is crystal clear like glass and is not ionic.

Positive ions such as potassium (K^{1+}), aluminum (Al^{3+}), and sodium (Na^{1+}) balance out the negative charges in these minerals, and traces of metal ions such as chromium, manganese, and vanadium lend the attractive colors seen in many minerals.

Infinite three-dimensional networks are characteristic of quartz (Figure 21-2) and feldspar.

In the laboratory, linear silicon–oxygen chains

$$\cdots\underset{CH_3}{\overset{CH_3}{\underset{|}{\overset{|}{Si}}}}-O-\underset{CH_3}{\overset{CH_3}{\underset{|}{\overset{|}{Si}}}}-O-\underset{CH_3}{\overset{CH_3}{\underset{|}{\overset{|}{Si}}}}-O-\underset{CH_3}{\overset{CH_3}{\underset{|}{\overset{|}{Si}}}}-O\cdots$$

have been made by the hydrolysis of compounds like dichlorodimethylsilane

$$Cl-\underset{CH_3}{\overset{CH_3}{\underset{|}{\overset{|}{Si}}}}-Cl$$

Two molecules of HCl are generated for each molecule of H_2O reacting.

These polymers are called silicones and are often superior lubricating oils. The extraordinary temperature range over which they will function as lubricants often justifies their extra expense compared to petroleum-based lubricants.

Silicones

Modern polymer science and technology have permitted the development of a large variety of materials tailored for specific uses. Current developments include the manufacture of synthetic parts suitable for implantation in the body without rejection or other immune response (high durability heart valves, replacement veins and arteries, synthetic hip joints, and many dental materials) and the development of high strength, high temperature resistant, and low weight materials for aircraft and space technology requirements. There is a very strong sense that a lot has happened since the discovery of Bakelite in 1913 and nylon in 1930, and that the best is yet to come.

Scramble Exercises

1. Teflon is a polymer of tetrafluoroethylene, CF_2CF_2. Assuming a free radical mechanism initiated by the thermal dissociation of benzoyl peroxide

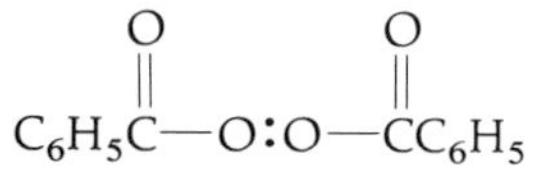

$$C_6H_5\overset{\displaystyle O}{\overset{\|}{C}}-O{:}O-\overset{\displaystyle O}{\overset{\|}{C}}C_6H_5$$

write out a sequence of reactions representing the initiation and continuation of the polymer chain. Indicate what becomes of the pair of electrons in the oxygen–oxygen *sigma*-bond of the peroxide and the pair of electrons in the *pi*-bond of the tetrafluoroethylene.

2. Write out the reaction for the production of nylon from adipyl chloride

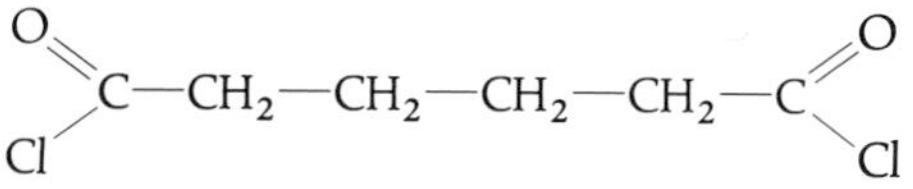

$$\begin{matrix} O \\ \| \\ Cl \end{matrix}\!\!\!C-CH_2-CH_2-CH_2-CH_2-C\!\!\!\begin{matrix} O \\ \| \\ Cl \end{matrix}$$

and 1,6-diaminohexane, $NH_2-CH_2-CH_2-CH_2-CH_2-CH_2-CH_2-NH_2$. The small molecule eliminated is HCl.

3. Suggest a rationalization of the experimental fact that the proportion of crystallites to amorphous material is increased in a given polymer by the physical process of drawing the bulk material into a fiber.

4. Suggest a rationalization of the experimental fact that a high proportion of crystallites increases the tensile strength (resistance to breakage) of fibers. For example, isotactic polypropylene

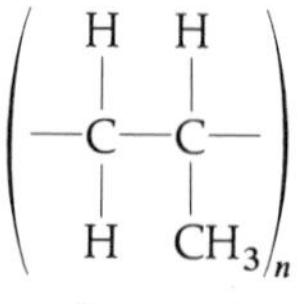

$$\left(\begin{matrix} & H & & H & \\ & | & & | & \\ - & C & - & C & - \\ & | & & | & \\ & H & & CH_3 & \end{matrix}\right)_n$$

fibers make inexpensive ropes that float. The early atactic (randomly oriented methyl groups) polypropylene fibers were much less satisfactory for ropes.

5. Rationalize the experimental fact that plastic articles, such as TV cases, frequently burn with very smoky flames, with very unpleasant odors, and with large globs of a burning liquid dripping to the floor or flowing slowly over the table top. Some of the products of combustion may be toxic.

6. The physical properties of the silicones $[Si(CH_3)_2O]_n$, and quartz, $(SiO_2)_n$, are quite different. There are, however, some similarities in their structures. Discuss the similarities and differences of the bonding involved in these two types of polymers.

7. Carbohydrates, proteins, synthetic polymers, and silicates are extensively used as glues, cements, and the active ingredient on sticky tapes. Speculate on the essential properties of the molecules or ions that can serve these very practical ends.

Additional exercises for Chapter 21 are given in the Appendix.

A Look Ahead

With this chapter, we come to the end of our very brief discussion of molecules with large molecular weights. These molecules, as we have seen, are important in biological systems and in our everyday lives. This chapter has centered on synthetic polymers. A very large percentage of the chemical industry in the United States is devoted to the production of synthetic polymeric materials. Whether the existence and use of these materials has led to a better standard of living for all or not can, of course, be debated. It is, however, essential to realize that our current way of life would be impossible without them.

The final chapter of the text deals with those processes that involve the nucleus of the atom: radioactivity, fission, and fusion. Nuclear fission and fusion reactions have been and are either directly or indirectly the source of all energy in the universe. In recent years, the use of controlled nuclear fission reactors has become an increasingly significant commercial means of generating electric power throughout the world.

22
NUCLEAR REACTIONS

Chapter 22

Nuclear Reactions in the Natural Environment of the Earth

Formation and Decay of Carbon-14

The nuclei bombardment of the most common isotope of nitrogen, $^{14}_{7}N$, by neutrons, $^{1}_{0}n$, leads to the formation of the nuclei of atoms of carbon-14, $^{14}_{6}C$, and protons, $^{1}_{1}H$, the nuclei of the most common isotope of hydrogen:

$$\underset{\text{nitrogen nucleus}}{^{14}_{7}N} + \underset{\text{neutron}}{^{1}_{0}n} \longrightarrow \underset{\text{carbon nucleus}}{^{14}_{6}C} + \underset{\text{proton}}{^{1}_{1}H}$$

Neutrons are a part of cosmic radiation, and this nuclear reaction has proceeded throughout the ages as neutrons from outer space encountered nitrogen atoms—either free atoms of nitrogen or atoms of nitrogen in molecules and ions. The availability of these neutrons from cosmic radiation is greatest in the stratosphere and the upper atmosphere. The number of protons in the nucleus of a nitrogen-14 atom, $^{14}_{7}N$, is reduced from seven to six and the nucleus of the nitrogen atom is thus transmuted into the nucleus of a carbon atom. At the same time, the number of neutrons in the nucleus is increased from seven to eight and the total number of nucleons (protons and neutrons) remains at fourteen. The overall change is replacement of a proton by a neutron in the nucleus of a nitrogen-14 atom, $^{14}_{7}N$.

The nuclei of atoms of the carbon-14 isotope of carbon, $^{14}_{6}C$, are unstable and, in time, are transmuted into the nuclei of atoms of nitrogen-14. $^{14}_{7}N$, by the ejection of *beta* particles, $_{-1}^{0}\beta$:

$$\underset{\text{carbon nucleus}}{^{14}_{6}C} \longrightarrow \underset{\text{nitrogen nucleus}}{^{14}_{7}N} + \underset{\text{electron}}{_{-1}^{0}\beta}$$

Source of beta Particles

The *beta* particles are identical with electrons. *Beta* particles arise in the nuclear reactions and not from the electron atmosphere of the carbon atoms. The *beta* particles emerge from the nuclei with a range of kinetic energies, the maximum energy of which is characteristic of the decay of this particular isotope of carbon. In the process of the ejection of the *beta* particles, neutrons are converted into protons,

$$\underset{\text{neutron}}{^{1}_{0}n} \longrightarrow \underset{\text{proton}}{^{1}_{1}H} + \underset{\text{electron}}{_{-1}^{0}\beta}$$

with the proton remaining in the nucleus and the *beta* particle ejected. In β emission from $^{14}_{6}C$, the number of protons in a nucleus has been increased from six to seven and the nuclei of carbon atoms are transmuted into the nuclei of nitrogen atoms. The number of neutrons in a nucleus is reduced from eight to seven and the number of nucleons remains at fourteen.

Nuclear Equations

The notation for the *beta* particle, $_{-1}^{0}\beta$, the electron (also called a negatron), is designed to be consistent with the notation for the nuclei, $^{14}_{7}N$ and $^{14}_{6}C$. The superscripts designate the number of nucleons. The *beta* particle has no nucleons and the superscript for the *beta* particle is zero. The subscripts indicate the number of protons, also the positive charge of the nucleus. The *beta* particle has no protons but it does have a negative charge. It is this charge that is designated by the subscript -1. In writing equations for nuclear reactions, the subscripts are used to balance charge and the superscripts to balance the number of nucleons.

In the equation

$$\underset{\text{neutron}}{^{1}_{0}n} + {}^{14}_{7}N \longrightarrow {}^{14}_{6}C + \underset{\text{proton}}{^{1}_{1}H}$$

the nucleon balance is 15 and the charge balance is 7. In the equation

$$^{14}_{6}C \longrightarrow {}^{14}_{7}N + {}_{-1}^{0}\beta$$

the nucleon balance is 14 and the charge balance is 6.

In **chemical reactions,** the focus is entirely upon changes in the electron atmospheres of atoms. In **nuclear reactions,** the focus is entirely upon changes involving nuclei of atoms, and we are all quite cavalier about changes involving the electron atmospheres of atoms in the discussion of nuclear reactions. As nuclear reactions occur, changes do occur in the electron atmospheres of atoms: electrons are lost, electrons are picked up, molecular bonds are disrupted, and new molecular bonds may result. The energy difference between nuclear reactions (which involve the nucleus) and chemical reactions (which involve the electron atmospheres) is so large that, with few exceptions, chemical structures and chemical reactions are treated as incidental side effects—so much for the previous 21 chapters of this book.

Carbon-14 in Biological Systems

The two natural nuclear reactions of the production and decay of carbon-14 are much more a part of your life than you may suspect. It is estimated that our bodies contain several hundred moles of carbon atoms distributed in compounds such as fats, carbohydrates, proteins, and nucleic acids. Several hundred nanomoles of that carbon are carbon-14. For every nanomole of carbon-14, approximately 140 thousand atoms of carbon-14 decay each minute by the ejection of 140 thousand *beta* particles, forming 140 thousand nitrogen-14 atoms in our body tissues and body fluids. This is a large number of radioactive events, but it is very small in comparison to the number of atoms of carbon-14, 6.02×10^{14} atoms, in 1 nanomole of carbon-14. One hundred forty thousand events per minute is equivalent to 8.4 million events per hour. Avogadro's number, 6.02×10^{23}, is a very large number.

Half-Life of Unstable Isotopes

The lifetime of radioactive substances is quantitatively expressed in terms of the **half-life** of the unstable isotope. The half-life of carbon-14 is 5730 years. For 1.00 nanomole of carbon-14, half of those moles of carbon-14 (0.50 nanomoles) will still exist, as carbon-14, 5730 years from now. The other half will have become nitrogen-14. Another half-life, another 5730 years, is required to again reduce the number of moles of carbon-14 by one-half (to 0.25 nanomoles carbon-14).

The level of carbon-14 in our bodies is a consequence of a series of processes. Carbon-14 is formed by a natural nuclear reaction in the stratosphere and upper atmosphere, transported to earth in the turbulence of the air, oxidized to carbon dioxide by oxygen of the air, and introduced in the food chain by photosynthesis in plants. We eat plants and we eat animals that eat plants. Over the eons, a **steady state** (a balance) has been attained in which the rate of introduction of carbon-14 into our immediate environment equals the rate of decay of carbon-14 in the environment. There can be perturbations to this balance but in general these perturbations are considered to be small. One perturbation is the dilution of carbon-14 in the atmosphere by the large-scale combustion of fossil fuels in recent years. The ratio of carbon-14 to carbon-12 in fossil fuels is less than the ratio of carbon-14 to carbon-12 at the time the biological material of the fossil fuels ceased to be living materials continuously incorporating carbon-14. In nonliving organic materials, the carbon-14 to carbon-12 ratio continuously decreases as carbon-14 decays, and the measurement of carbon-14 to carbon-12 ratios has become an accepted methodology to determine the age of archeological plant and animal materials. When this methodology was used, the remains of a giant sloth found in Gypsum Cave, Nevada, were determined to be about 10.7 thousand years old (a little short of two half-lives for carbon-14). This giant sloth lived about 9000 B.C.

Carbon-14 Dating

Formation and Decay of Tritium

At the same time that the encounters of neutrons with the nuclei of nitrogen-14 are producing carbon-14, some of the encounters of neutrons with the nuclei of nitrogen-14 produce quite different products. These products are the nuclei of carbon-12, ${}^{12}_{6}C$, and the nuclei of hydrogen-3, ${}^{3}_{1}H$, also called tritium, ${}^{3}_{1}T$:

$$ {}^{14}_{7}N + {}^{1}_{0}n \longrightarrow {}^{12}_{6}C + {}^{3}_{1}H $$

The nucleus of hydrogen-3 contains one proton and two neutrons. Tritium is a radioactive isotope of hydrogen and decays by *beta* emission to give the helium-3 isotope, ${}^{3}_{2}He$:

$$ {}^{3}_{1}H \longrightarrow {}^{3}_{2}He + {}^{0}_{-1}\beta $$

The number of tritium nuclei formed by neutron–nitrogen-14 encounters is about one percent of the number of carbon-14 nuclei formed by neutron–nitrogen-14 encounters. The half-life of the tritium, ${}^{3}_{1}H$, is 12.5 years and the quantity of tritium, from this source, entering the food chain is consequently limited and rapidly decays.

Discovery of Natural Radioactivity

Although nuclear reactions are a part of the natural environment, they were not discovered until 1896. In that year, Henri Becquerel inadvertently discovered natural radioactivity. He was using photographic plates in the investigations of the response of uranium minerals and uranium compounds to exposure by sunlight when he discovered that some of his photographic plates had become exposed while still enclosed in their protective wrappers. This discovery preceded by one year the discovery of the electron by J. J. Thomson. Both were very significant discoveries. They either were or were to lead to the first indications that there are particles smaller than atoms, that some atoms of some elements transmute into atoms of other elements, and that all atoms of an element are not identical. In the research that followed, Pierre

THE STRUCTURE OF NUCLEI

From the standpoint of chemistry, which deals with the distribution of electrons in the electronic atmospheres of atoms, molecules, and ions and with changes in those distributions of electrons in chemical change, it is quite adequate to discuss the structure of nuclei in terms of nucleons, the protons and neutrons, that constitute the nucleus. Such a model is inadequate to rationalize the stability of nuclei and the nature of nuclear reactions. Nuclear physics is concerned with the properties of nuclei and the interactions of one nucleus with another nucleus. Elementary particle physics is concerned with the identification of the elementary particles that are the fundamental building blocks of atoms. If a neutron can be converted to a proton and an electron, then a neutron is not an elementary particle. In fact, neither neutrons nor protons are now considered to be elementary particles. One of the experimental approaches of elementary particle physics is to bombard a particle such as the nucleus of an atom with a high-velocity particle such as the nucleus of a hydrogen atom (the proton) or the nucleus of some other atom and to determine the properties of the products formed as a consequence of the collision. Frequently, the bombarding particles are accelerated to very high velocity using any of a number of instruments known as particle accelerators.

The terminology that has developed for elementary particles is more than a bit whimsical. The current consensus among particle physicists is that matter is built up of two families of particles. One of these is the quarks and the other is the leptons. The quarks come in six varieties, also called flavors: the up quark, the down quark, the strange quark, the charm quark, the top quark, and the bottom quark. The leptons also come in six flavors: the electron, the electron neutrino, the muon, the muon neutrino, the tau, and the tau neutrino. It is not our intent to go into detail about the properties of each of these particles but to indicate the nature and scope of the development in concepts of structure.

The interactions among the particles are approached theoretically in terms of four forces: the strong force, the electromagnetic force, the weak force, and gravity. (In the Schrödinger approach to the relation of electrons to the nucleus, only one type of force is used: the electromagnetic force.) The goal of theorists is to develop models of the role of each of these forces in the interaction of the elementary particles with each other, to develop the combined role of two or more of these forces in the interaction of elementary particles with each other, and ultimately to develop a unified model of the role of the four forces in what is known as a grand unified theory, GUT.

One of the characteristics of the theoretical approach is that each type of force is transferred from place to place by its own carrier particle. These particles are gluons for the strong force; photons for the electromagnetic force; W^+, W^-, and Z° particles for the weak force; and gravitons for gravity. Some of the elementary particles and the carrier particles, called bosons, have been predicted by theorists, and another role of experimentalists is to endeavor to find experimental evidence for the particles and experimentally determine their properties.

The primary attractive force holding neutrons and protons together in nuclei is the strong force. The electromagnetic force (proton–proton charge interaction) is a repulsive force and limits the size of nuclei. In general, the protons and neutrons are considered to be rather uniformly distributed throughout the nucleus, but there is evidence of cluster formation. In the decay of natural radioactive isotopes, neither single neutrons nor single protons escape the nucleus, but *alpha* particles, a cluster of two neutrons and two protons, do escape. This suggests that preformed *alpha* particles exist in the nucleus and that this cluster is less strongly bound to the rest of the nucleus than either a single proton or a single neutron. In fission, the nucleus comes apart into two globs of approximately the same size.

On theoretical grounds, it had been predicted that the nucleus of carbon-12, six protons and six neutrons, could be another preformed group and that there might be a fourth type of radiation exhibited by some natural radioactive isomers. In 600 days of observation of radium-238, 19 events were observed that were interpreted to be carbon-14 nuclei emissions—not the stable carbon-12 nuclei predicted. The rate of emission of the carbon-14 nuclei is very small in comparison to the rate of emission of the *alpha* particle.

Neither the neutron nor the proton is an elementary particle. Each is considered to be a "bag" of three quarks.

and Marie Curie discovered two elements: radium (element 88) and polonium (element 84).

Decay of Uranium-238

Uranium-238, $^{238}_{92}U$, is a naturally occurring isotope of uranium. The nucleus is unstable and undergoes a long series of spontaneous decay reactions. The first eight disintegrations in the uranium-238 series, the half-lives of the isotopes, and the equations for the nuclear reactions are given in Table 22-1.

alpha, beta, and gamma Radiations

In this series, the three types of detected radiations were initally named a, b, and c in Greek before the identities of the three radiations were established, and the use of the terms *alpha, beta,* and *gamma* radiations continues. Alpha particles, $^{4}_{2}\alpha$, are the nuclei of helium-4 atoms. They are made up of two protons and two neutrons, and *alpha* particles are frequently written $^{4}_{2}He$. The *beta* particles, $^{0}_{-1}\beta$, are electrons

Table 22-1

THE URANIUM-238 RADIOACTIVE DISINTEGRATION SERIES

Isotope	Radiation	Name / half-life	Equation
$^{238}_{92}U$		uranium	
↓	α	4.5×10^9 years	$^{238}_{92}U \longrightarrow {}^{4}_{2}\alpha + {}^{234}_{90}Th$
$^{234}_{90}Th$		thorium	
↓	β,γ	24 days	$^{234}_{90}Th \xrightarrow{\gamma} {}^{0}_{-1}\beta + {}^{234}_{91}Pa$
$^{234}_{91}Pa$		protactinium	
↓	β,γ	1 minute	$^{234}_{91}Pa \xrightarrow{\gamma} {}^{0}_{-1}\beta + {}^{234}_{92}U$
$^{234}_{92}U$		uranium	
↓	α	2.5×10^5 years	$^{234}_{92}U \longrightarrow {}^{4}_{2}\alpha + {}^{230}_{90}Th$
$^{230}_{90}Th$		thorium	
↓	α,γ	80 years	$^{230}_{90}Th \xrightarrow{\gamma} {}^{4}_{2}\alpha + {}^{226}_{88}Ra$
$^{226}_{88}Ra$		radium	
↓	α,γ	1.6×10^3 years	$^{226}_{88}Ra \xrightarrow{\gamma} {}^{4}_{2}\alpha + {}^{222}_{86}Rn$
$^{222}_{86}Rn$		radon	
↓	α	4 days	$^{222}_{86}Rn \longrightarrow {}^{4}_{2}\alpha + {}^{218}_{84}Po$
$^{218}_{84}Po$		polonium	
↓	α	3 minutes	$^{218}_{84}Po \longrightarrow {}^{4}_{2}\alpha + {}^{214}_{82}Pb$
$^{214}_{82}Pb$		lead	
↓			
Several more steps			
↓			
$^{206}_{82}Pb$		a stable isotope of lead	

In each case, the maximum kinetic energy of the *alpha* particle or the *beta* particle is a characteristic of the particular isotope from which it is emitted. The *gamma* ray photons are light and have the speed of light. All of the reactions or the products of these reactions are specifically referred to in this chapter.

usually written in chemical equations as e^{1-}. The *gamma* particles, γ, also called *gamma* rays, are high-energy photons with energies in excess of the energy range for X-rays. The primary decay process is either the emission of *alpha* particles or the emission of *beta* particles to give daughter nuclei. For some decays, *gamma* radiation is also emitted. (See Table 22-1.) In the decay of thorium-234 by *beta* emission, the accompanying *gamma* radiation is made up of photons of four different energies (29 thousand, 63 thousand, 70 thousand, and 93 thousand electron volts). The emission of photons of specific energies, such as these, is interpreted in terms of quantized energy states of nuclei and the transition of excited daughter nuclei, protactinium-234 in this case, to lower energy states. If *gamma* rays are included in a nuclear equation, the symbolism is ${}^{0}_{0}\gamma$ or, more simply, just the *gamma* symbol.

The uranium-238 disintegration series, Table 22-1, finally comes to an end with the formation of a stable isotope of lead, ${}^{206}_{92}Pb$. The half-life of the uranium-238 decay to thorium-234 is exceptionally long, 4.5 billion years, and this accounts for the continued presence of uranium-238 and its daughters in the earth's crust. Note that the series of decay processes for uranium-238 included in Table 22-1 involves another isotope of uranium, two isotopes of thorium, and an unstable isotope of lead. This reoccurrence of isotopes of a few elements is characteristic of radioactive disintegration series. *Beta* emission increases the atomic number by one, by increasing the number of protons in the nuclei; *alpha* emission decreases the atomic number by two, by decreasing the number of protons in the nuclei. There are two other natural radioactive disintegration series. One begins with thorium-232, ${}^{232}_{90}Th$ (half-life 14.1 billion years), and ends with another stable isotope of lead, ${}^{208}_{82}Pb$. The other begins with uranium-235, ${}^{235}_{92}U$ (half-life 0.7 billion years), and ends with yet another stable isotope of lead, ${}^{207}_{82}Pb$. Note that the equations given in Table 22-1 are balanced with respect to superscripts (nucleons) and with respect to subscripts (charge).

For lead, atomic number 82, and for each of the elements of higher atomic number, there is at least one known naturally occurring radioactive isotope.

One of the characteristics of *alpha*, *beta*, and *gamma* radiation is that each has the capacity to ionize molecules. The capacity of *alpha* and *beta* particles to produce ionization arises from the momentum with which the particles strike and disrupt molecules and also from the interaction of the electromagnetic field of the moving charge with the electron charge distribution within the molecules. The capacity of *gamma* rays to produce ionization arises from the absorption of these photons (with energies much greater than X-ray photons). Many instruments, such as Geiger counters, used in the detection and investigation of nuclear processes, are designed to detect the ionization produced in a gas by these and other high-energy particles.

The **binding energy of a nucleus** (the energy that holds a nucleus together) can be assessed by comparing the experimental value for the atomic mass of an isotope of an element with the sum of the masses of the protons, the neutrons, and the electrons that constitute the atom of that isotope. On the arbitrary scale of atomic masses where one mole of the carbon-12 has a mass of exactly 12 grams, the masses of the three particles are

electron	0.0005486 grams/mole
proton	1.0072766
neutron	1.0086654

The most abundant isotope of iron, $^{57}_{26}Fe$, has an experimental atomic mass of 56.9354 grams/mole. The sum of the masses of 26 moles of electrons, 26 moles of protons, and 31 moles of neutrons is 57.4721 grams. The experimental mass is 56.9354. One mole of iron-57 has an experimental mass less than the sum of the masses of its parts.

Binding Energy per Mole of Iron-57

summed mass	57.4721 grams/mole
experimental mass	56.9354
difference	0.5367 grams/mole

This difference, of a little more than a half of a gram per mole of iron, is a measure of the nuclear binding energy of Avogadro's number of atoms of iron-57.

Using the Einstein equation (see Chapter 5, page 86)

$$\text{Energy} = \text{mass} \times (\text{velocity of light})^2$$

the energy equivalence of 0.54 grams per mole is evaluated to be 4.9×10^{10} kilojoules per mole. Since joules = kilograms meters^2 seconds^{-2} and the velocity of light is 3.0×10^8 meters/sec, the energy equivalent to 0.54 gram is given by

$$\begin{aligned}\text{Energy} &= 0.54 \times 10^{-3}\ \text{kg}\ (3.0 \times 10^8\ \text{m sec}^{-1})^2\\ &= 4.9 \times 10^{13}\ \text{joules} \quad \text{or} \quad 4.9 \times 10^{10}\ \text{kilojoules per mole}\end{aligned}$$

This is a tremendous amount of energy. The energy changes associated with chemical reactions are of the order of a few hundred kilojoules for the reaction of molar quantities. Mass changes equivalent to a few hundred kilojoules are not experimentally detectable and, as discussed in Chapter 5, page 86, changes in mass associated with chemical reactions are not a practical consideration.

What is the significance of this decrease of 0.54 gram in mass? If 26 moles of electrons, 26 moles of protons, and 31 moles of neutrons would react to give 1 mole of iron-57, 49 billion kilojoules of energy would be released. To convert one mole of iron-57 (less than the amount of iron in the head of a hammer) into electrons, protons, and neutrons would require the investment of 49 billion kilojoules. One mole of iron is much more stable than its constituent parts. The binding energy for Avogadro's number of iron-57 nuclei is 49 billion kilojoules.

Binding Energy per Nucleon for Various Elements

The summed mass minus the experimental mass can be evaluated for any isotope for which the experimental isotopic mass is available. Nuclear binding energies and related quantities are given in Table 22-2 for deuterium (hydrogen-2), helium-4, and uranium-235 as well as iron-57. Use the column for iron-57 to adjust to the table. The top half of the table is the process of arriving at the bottom half of the table.

The mass differences corresponding to the 0.5367 grams/mole for iron-57 are 0.0025 for hydrogen-2, 0.0304 for helium-4, and 1.9152 for uranium-235 (Row a, Table 22-2). These values show that the nuclear binding energies for large nuclei are greater than the binding energies for small nuclei. The more revealing values are the mass differences per nucleon. The value for iron-57 is 1.56×10^{-26} grams/nucleon.

$$0.5367\ \text{grams/mole} \div 6.02 \times 10^{23}\ \text{atoms/mole} = 8.92 \times 10^{-25}\ \text{grams/atom}$$

$$8.92 \times 10^{-25}\ \text{grams/atom} \div 57\ \text{nucleons/atom} = 1.56 \times 10^{-26}\ \text{grams/nucleon}$$

Table 22-2

NUCLEAR BINDING ENERGIES FOR FOUR NUCLEI

	hydrogen-2	helium-4	iron-57	uranium-235
	$^{2}_{1}H$	$^{4}_{2}He$	$^{57}_{26}Fe$	$^{235}_{92}U$
number of protons	1	2	26	92
number of nucleons	2	4	57	235
number of neutrons	1	2	31	143
experimental atomic mass (g/mole)	2.0140	4.0026	56.9354	235.0439
summed masses of particles (g/mole)	2.0165	4.0330	57.4721	236.9591
summed masses minus experimental isotopic mass				
(a) grams/mole	0.0025	0.0304	0.5367	1.9152
(b) grams/atom	4.2×10^{-27}	5.0×10^{-26}	8.92×10^{-25}	3.18×10^{-24}
(c) grams/nucleon	0.21×10^{-26}	1.3×10^{-26}	1.56×10^{-26}	1.35×10^{-26}
(d) binding energy per nucleon (in million electron volts, MeV)	1.2	7.1	8.8	7.6

The corresponding values for the other nuclei are 0.21×10^{-26} for hydrogen-2, 1.3×10^{-26} for helium-4, and 1.35×10^{-26} for uranium-235 (Row c). The mass difference per nucleon, and consequently the binding energy per nucleon, is greatest for iron-57.

Binding energies per nucleon can be expressed in joules or any other energy unit. They are given in the bottom row (Row d) of Table 22.2 in millions of electron volts, MeV. The electron volt is a unit of energy commonly used by physicists and others in the investigation of the structure of atoms and also the structure of ions and molecules. An electron volt is the kinetic energy acquired when an electron falls through a potential difference of one volt. The relation between energy in joules and energy in electron volts is given by the equation for the direct proportion

$$[\text{energy in joules}] = \left[1.60 \times 10^{-19} \frac{\text{J}}{\text{MeV}}\right][\text{energy in million electron volts}]$$

The relation for the direct conversion of mass in grams to energy in million electron volts is given by the equation

$$[\text{energy in million electron volts}] = \left[5.61 \times 10^{26} \frac{\text{MeV}}{\text{g}}\right][\text{mass in grams}]$$

Figure 22-1 presents the relation of binding energy per nucleon, in MeV, to the number of nucleons per nucleus. To orient to this curve, locate on the curve the binding energy values given in the bottom line of Table 22-2. Note that the binding energy per nucleon is greatest for nuclei containing about 60 nucleons. One of these isotopes is iron-57, $^{57}_{26}Fe$. The binding energy of $^{1}_{1}H$ with its single nucleon is zero.

Figure 22-1

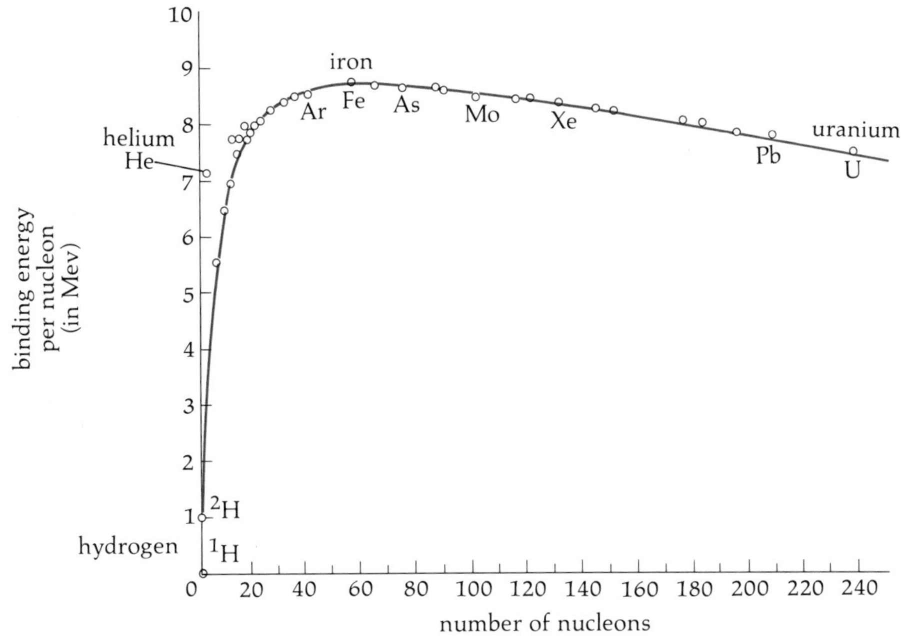

THE RELATION OF BINDING ENERGY PER NUCLEON TO THE NUMBER OF NUCLEONS IN THE NUCLEUS. Isotopes of four elements, hydrogen, helium, iron and uranium, are discussed in detail in the text. Note that all points do not lie on the smooth curve drawn and helium in particular lies well off of the curve.

The presentation of binding energies in Figure 22-1 does not follow the conventions to which we have become accustomed in the presentation of energy states for atoms and molecules. In those diagrams, all energies were presented as negative and the most stable energy state was at the bottom of the diagram. In the presentation of binding energies, all binding energies per nucleon are positive and nuclei having the greatest binding energy (the most stable nuclei) are at the top of the curve. The difference is a difference of convention—not a difference in phenomenon or concept. This difference in convention is an artifact of how the various topics developed. A single set of conventions could have been used.

Spontaneous Nuclear Fission of Uranium-235

A nucleus of uranium-235, ${}^{235}_{92}U$, can undergo **spontaneous fission** into fragments: two smaller nuclei and several neutrons. The nucleus of one fragment is larger than the nucleus of iron-57; the other fragment is smaller than the nucleus of iron-57. One pair of nuclei products is krypton-94, ${}^{94}_{36}Kr$, and barium-139, ${}^{139}_{56}Ba$. The nuclear equation for this spontaneous fission giving this pair of products and two neutrons is

$${}^{235}_{92}U \longrightarrow {}^{94}_{36}Kr + {}^{139}_{56}Ba + 2\,{}^{1}_{0}n$$

We will now use Figure 22-1 to estimate the binding energy for the krypton-94 nucleus and for the barium-139 nucleus. (These estimated values are quite adequate to meet our needs. More precise values are available in handbooks.)

binding energy krypton-94: 8.6 MeV per nucleon

binding energy barium-139: 8.3 MeV per nucleon

We already have the experimental value of binding energy for uranium-235 in Table 22-2:

binding energy uranium-235: 7.6 MeV per nucleon

Expressed in terms of binding energy per atom, the values are

krypton-94 (94 × 8.6):	808 MeV/atom
barium-139:	1154 MeV/atom
uranium-235:	1786 MeV/atom

The change in binding energy for each atom of uranium-235 that undergoes fission to give an atom of krypton-94, an atom of barium-139, and two neutrons is

$$808 + 1154 + 0 + 0 - 1786 = 176 \text{ MeV}$$

The total binding energy of the products is greater than the binding energy of the reactant, and 176 MeV of energy (heat and light) are released for each atom of uranium-235 that undergoes fission. For the fission of one mole of uranium-235, the energy released is

$$176 \text{ MeV} \times 6.02 \times 10^{23} = 1.06 \times 10^{26} \text{ MeV/mole}$$

and the mass decrease corresponding to this energy change is

$$1.06 \times 10^{26} \text{ MeV/mole} \div 5.61 \times 10^{26} \text{ MeV/gram} = 0.19 \text{ grams/mole}$$

According to this mass change estimate, the fission of one mole of uranium-235 is roughly equivalent to the energy produced in the combustion of 600 tons of coal.

The above discussion has been in terms of the formation of the krypton-94 and barium-139 pair of nuclei. In the fission of uranium-235, more than fifty pairs other than krypton-94 and barium-139 are formed with the release of two or more neutrons for each pair and with the release of slightly different quantities of energy for each pair of fragments. On the average, 2.5 neutrons are released and the increase in binding energy per atom of uranium-235 that undergoes fission is about 200 MeV. All of the nuclei formed in these fission reactions are neutron rich—all are radioactive.

The **fission** of uranium-235 is also induced by the capture of neutrons. Again using krypton-94 and barium-139 as the products, the equation for the neutron capture immediately followed by fission is

Induced Fission of Uranium-235

$$^{235}_{92}\text{U} + {}^{1}_{0}\text{n} \longrightarrow {}^{236}_{92}\text{U} \longrightarrow {}^{94}_{36}\text{Kr} + {}^{139}_{56}\text{Ba} + 3\,{}^{1}_{0}\text{n}$$

Since the fisson of one nucleus of uranium-235 yields two or more neutrons, the fission of that one nucleus produces enough neutrons to induce fission in two or more other nuclei of uranium-235 and each of these fissions in turn produces enough neutrons to induce fission in two or more additional nuclei. Thus, the spontaneous fission of one uranium-235 nucleus can initiate a runaway branching chain reaction that proceeds faster and faster unless part of the neutrons fail to be absorbed by other uranium-235 nuclei. However, this buildup of a runaway reaction becomes impossible (1) if the sample is small enough for sufficient numbers of neutrons to escape from the sample without capture or (2) if the concentration of uranium-235 nuclei in the sample is small enough for insufficient numbers of neutrons to collide with uranium-235 nuclei or (3) if a strong neutron absorber, such as cadmium, is present at a sufficiently high concentration.

Biological Effects of Naturally Occurring Nuclear Reactions

Biological effects are the consequence of chemical changes in living tissues brought about by *alpha-*, *beta-*, and *gamma*-ray energy transfers to those tissues. If the source of the radiations is external to the organism, the organism can be protected from the radiation by the introduction of an absorption barrier between the source and the organism. The *alpha* particles (helium nuclei) are the most easily absorbed and ordinary clothing or a sheet or two of paper provides adequate shielding. *Beta* particles (electrons) are more penetrating. Ordinary clothing provides partial shielding and several layers of aluminum foil provide complete shielding. *Gamma* radiation (γ-rays) is by far the most penetrating and several centimeters of lead or several feet of concrete are required for complete shielding. It is the exposure to *gamma* radiation that is the most damaging to biological organisms. Burns very similar to those produced by heat, ultraviolet light, and X-rays are the consequence of intense exposure to *gamma*-ray photons.

If the radioactive materials are inhaled or ingested, they may be deposited upon membrane walls, absorbed into the biological system, and distributed throughout the biological system according to the chemical properties of the material. For example, radium, an *alpha–gamma* emitter, is an alkaline earth element and radium ions tend to concentrate along with calcium ions in the bone structure, placing the source of *alpha* and *gamma* radiation in juxtaposition to the region of formation of red blood cells in the bone marrow. If the dosage is sufficiently high, the consequence can be the destruction of the biological function of the bone marrow leading to death in a short time or to the long-term development of leukemia.

Radium

We have no way of knowing the accumulative effect of the carbon-14 that is so much a part of our bodies. There has never been life without carbon-14. In comparison to the kinetic energies of *beta* particles ejected from some other *beta* emitters, the kinetic energies of the *beta* particles from carbon-14 are relatively small. Another source of ionization radiations from naturally occurring radioactive isotopes is the radioactive element content of native stones, such as granite, that have been and are used as building materials. The portion of the uranium-238 disintegration series given in Table 22-1 indicates some of the radioactive elements that may be present in granite and some of the nuclear reactions that may go on in some natural stone. Note that radon, a member of the inert gas family, is a member of this decay series. It is now known that radioactive radon gas may collect in buildings—particularly in well-

Carbon-14

Radon

insulated buildings. It is also known that radon gas can collect in basements of buildings as radon continuously escapes from soil above uranium ore deposits. Continuous inhalation of this gas leads to the deposit in the lungs of the daughter element, polonium, which is also a radioactive element. It is not known how great the accumulative health effects have been for families living in an atmosphere that contains elevated levels of radon. These are not new risks, but they are risks that can be detected and assessed today using modern instrumentation and methodologies. There never has been life without exposure to *gamma* radiation arising from the natural distribution of naturally occurring radioactive isotopes and to *gamma* radiation that is a part of cosmic radiations. Radiation damage may occur in somatic cells or in germ cells. The first leads to illness and perhaps death in a few days if massive doses are involved, but development of health problems may be delayed for 20 or more years if smaller doses are involved. Genetic damage to germ cells may lead to infertility or to malformed or malfunctioning offspring in either the first or second generation.

Somatic Cells and Germ Cells

Nuclear Reactions and the Sun

The mechanism of generating the high temperatures of the stars, including the sun, and the energy radiated by them are interpreted in terms of nuclear reactions in which protons are converted through a series of reactions into helium-4 nuclei with the liberation of energy equivalent to the mass difference. The overall reaction is believed to be the fusion of four protons to give one helium-4 nucleus:

$$4\ {}^{1}_{1}H \qquad {}^{4}_{2}He$$

Nuclear Fusion

We shall explore this **fusion reaction** in terms of the information given in Table 22-2 and Figure 22-1 and then make comparisons with the energy generated by the fission of uranium-235.

The binding energy of the hydrogen-1 isotope is, of course, zero. There is only one nucleon and consequently no binding among nucleons. The binding energy of helium-4 is 7.1 MeV per nucleon (Table 22-2) or 28.4 MeV per helium-4 nucleus. The mass loss in the formation of the helium-4 isotope is 0.0304 grams per mole of helium-4 (Table 22-2). This corresponds to a mass loss of 0.75% (less than 1%) in the conversion of four moles of protons into a mole of helium-4 nuclei

$$(0.0304 \text{ grams/mole} \div 4.03 \text{ grams/mole}) \times 100 = 0.75$$

and the release of 2.7×10^9 kilojoules of energy for each mole of helium-4 formed. This is about 10 million times the energy changes associated with molar quantities of chemical reactions.

The binding energy of 7.1 MeV per nucleon for helium-4 is exceptionally large for small nuclei. Note the position of the point for He in Figure 22-1 with respect to the binding energy curve.

The variation in binding energy per nucleon with the number of nucleons per nucleus, Figure 22-1, has remarkable significance to life in the solar system. All energy available to us is derived directly or indirectly from nuclear reactions on the sun and nuclear reactions on the earth. A greater understanding of Figure 22-1 can be derived from a comparison of the energy changes related to the fusion of protons to form

Table 22-3

COMPARISON OF ENERGY CHANGES FOR THE FUSION OF PROTONS TO GIVE HELIUM-4 AND THE FISSION OF URANIUM-235 TO GIVE KRYPTON-94 AND BARIUM-139

	fusion of hydrogen-1	fission of uranium-235
decrease in binding energy per atom	28.4 MeV	176 MeV
decrease in mass per mole from Table 22-2	0.0304 grams	0.19 grams
percent change in mass per mole	0.75%	0.08%
energy released per mole	2.7×10^9 kilojoules	17×10^9 kilojoules

helium-4 and the fission of uranium-235 to give krypton-94 and barium-139. (Table 22-3.)

Comparison of Nuclear Fusion and Nuclear Fission

In these transitions, the decrease in binding energy for the formation of one atom of helium-4 is 28.4 MeV, and the estimated decrease in binding energy in the fission of one atom of uranium-235 to give $^{94}_{36}Kr$ and $^{139}_{56}Ba$ is 176 MeV. These values correspond to a decrease in mass of 0.0304 grams per mole of helium-4 formed from protons and 0.19 grams per mole of uranium-235 converted to fission products. However, in terms of the percent change in mass for the reacting materials, the fusion reaction is the more effective source of energy: 0.75% of the mass of the protons in the fusion reaction as compared to 0.08% of the mass of the uranium-235 in the fission reaction.

The atmosphere of the sun was described as gas until the middle of the twentieth century. Today, the atmosphere of the sun is discussed in terms of a plasma, a fourth state of matter—a fourth state in addition to the three states gas, liquid, solid. A plasma is a system of charged particles. The plasma of the sun is primarily protons and electrons with smaller concentrations of nuclei of isotopes of the lighter elements. The pressure is high due to the high gravitational field of the massive sun, and the temperature is high, possibly in excess of 10 000 000 K, as the consequence of nuclear fusion. At this high pressure the protons or other nuclei are so close together, and at this temperature the translational kinetic energies of the nuclei are so high, two nuclei may come sufficiently close together for the strong (short distance) forces to become effective and the two nuclei to fuse. The net reaction to form the helium-4 nucleus from four protons requires a series of nuclear fusions, each involving not more than two nuclei. Two mechanisms are considered to be possible. One is the proposed proton–proton chain and the other is the proposed carbon cycle. See Gratuitous Information 22-2.

The Laboratory Synthesis of Elements

The Rutherford Synthesis of Oxygen-17

Up to this point, we have been discussing naturally occurring nuclear reactions: the formation and decay of carbon-14, the radioactive decay of uranium-238 and other elements, the spontaneous fission of uranium-235, and nuclear fusion reactions in the stars. In 1919, Ernest Rutherford demonstrated for the first time the laboratory synthesis of an element: not the synthesis of gold, the goal of the alchemists, but the laboratory synthesis of an element nonetheless. The element synthesized was oxygen-17, a stable isotope. The process of synthesis was the bombardment of nitrogen-14 with *alpha* particles from a natural radioactive source. The other product was the proton, ${}^{1}_{1}H$:

$$^{14}_{7}N + {}^{4}_{2}\alpha \longrightarrow {}^{17}_{8}O + {}^{1}_{1}H$$

Synthesis of Radioactive Isotopes

In the continuing investigations of the bombardment of elements with *alpha* particles from natural sources, Irène and Frédéric Joliot-Curie (daughter and son-in-law of Pierre and Marie Curie) demonstrated (in 1934) the first laboratory synthesis of a radioactive isotope, ${}^{13}_{7}N$, by the bombardment of boron-10, ${}^{10}_{5}B$, a stable isotope of boron, with *alpha* particles:

$$^{10}_{5}B + {}^{4}_{2}\alpha \longrightarrow {}^{13}_{7}N + {}^{1}_{0}n$$

Nitrogen-13 has a half-life of 10 minutes and emits a positive *beta* particle, ${}_{+1}^{0}\beta$:

22-2 *Gratuitous Information*

MECHANISM OF NUCLEAR FUSION REACTIONS ON THE SUN

Two proposed mechanisms of the fusion of four protons to form a helium nucleus are considered to be feasible. One is a carbon cycle and the other is a proton–proton chain reaction.

The proposed carbon cycle:

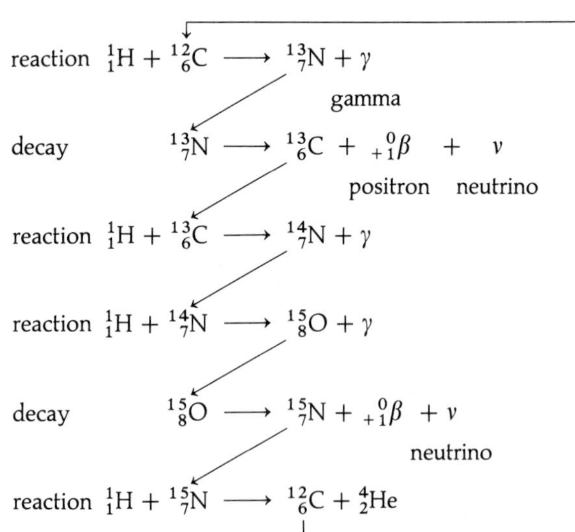

The net result of this proposed carbon cycle mechanism is that a carbon-12 nucleus is regenerated and four protons have been fused to give the helium-4 nucleus.

The proposed proton–proton chain:

fusion $^{1}_{1}H + {}^{1}_{1}H \longrightarrow {}^{2}_{1}H + {}_{+1}^{0}\beta + \nu$ (deuterium, positron, neutrino)

fusion $^{1}_{1}H + {}^{2}_{1}H \longrightarrow {}^{3}_{2}He + \nu$ (neutrino)

fusion $^{3}_{2}He + {}^{3}_{2}He \longrightarrow {}^{4}_{2}He + {}^{1}_{1}H + {}^{1}_{1}H$

The net reaction of this proposed proton–proton chain mechanism is

$$4\,{}^{1}_{1}H \longrightarrow {}^{4}_{2}He + 2\,{}_{+1}^{0}\beta + \text{energy}$$

In both proposed mechanisms, positrons, ${}_{+1}^{0}\beta$, and neutrinos, ν, are formed. Positrons are known to have a very short half-life and to self-destruct into energy. The emission of neutrinos always accompanies the emission of *beta* particles. Neutrinos have no charge and are very difficult to detect. The eventual fate of the neutrinos is still an open question—if indeed they are formed as proposed.

$$^{13}_{7}N \longrightarrow ^{13}_{6}C + ^{0}_{+1}\beta$$

A positive *beta* particle, $^{0}_{+1}\beta$, is a positive electron, also called a positron. The positron has the mass of an electron and positive charge equal in quantity to the negative charge of the electron.

Synthesis of Transuranium Elements

Today, radioactive isotopes of all of the elements known in 1934 have been synthesized in the laboratory, as have the radioactive nuclei of many elements with atomic numbers greater than 92, the largest known atomic number in 1934. In addition to radiation particles from radioactive nuclei, high-energy charged particles produced in instruments called accelerators are used in the synthesis of transuranium nuclei. In some syntheses, sufficiently high kinetic energies are imparted to nuclei such as $^{12}_{6}C$ and $^{14}_{7}N$ to induce nuclear reactions when they collide with targets such as uranium-238, $^{238}_{92}U$, nuclei.

$$^{238}_{92}U + ^{12}_{6}C \longrightarrow ^{246}_{98}Cf + 4\,^{1}_{0}n$$

Californium

$$^{238}_{92}U + ^{14}_{7}N \longrightarrow ^{247}_{99}Es + 5\,^{1}_{0}n$$

Einsteinium

Using these larger particles in place of α-particles produces larger changes in atomic numbers. The quantities of transuranium nuclei produced in this fashion are small—even less than 100 nuclei in some cases.

Much of our empirical knowledge concerning the proton–neutron composition of radioactive nuclei and the composition of stable nuclei is displayed in Figure 22-2. In stable low-atomic-number nuclei, through $^{40}_{20}Ca$, the number of neutrons is equal to or approximately equal to the number of protons, but in stable high-atomic-number nuclei the number of neutrons exceeds the number of protons. In lead, $^{206}_{82}Pb$, a stable isotope, the number of neutrons exceeds the number of protons by more than 50%.

Nuclear Technology

In the preceding 21 chapters dealing with chemical change, the closest we came to a discussion of chemical technology was a section in Chapter 14 dealing with the production of nitrogen fertilizers and a section in Chapter 21 dealing with synthesis of polymers. At no time have we discussed the production of conventional armaments based upon chemical change. Why then do we address nuclear technology in a chapter dealing with nuclear reactions? The answer lies in the exceptionally great magnitude of the energy released by both nuclear fission reactions and nuclear fusion reactions as compared to the energy released by chemical reactions and in the exceptional nature of the potential environmental burden of broad-scale distribution of radioactive isotopes.

Figure 22-2

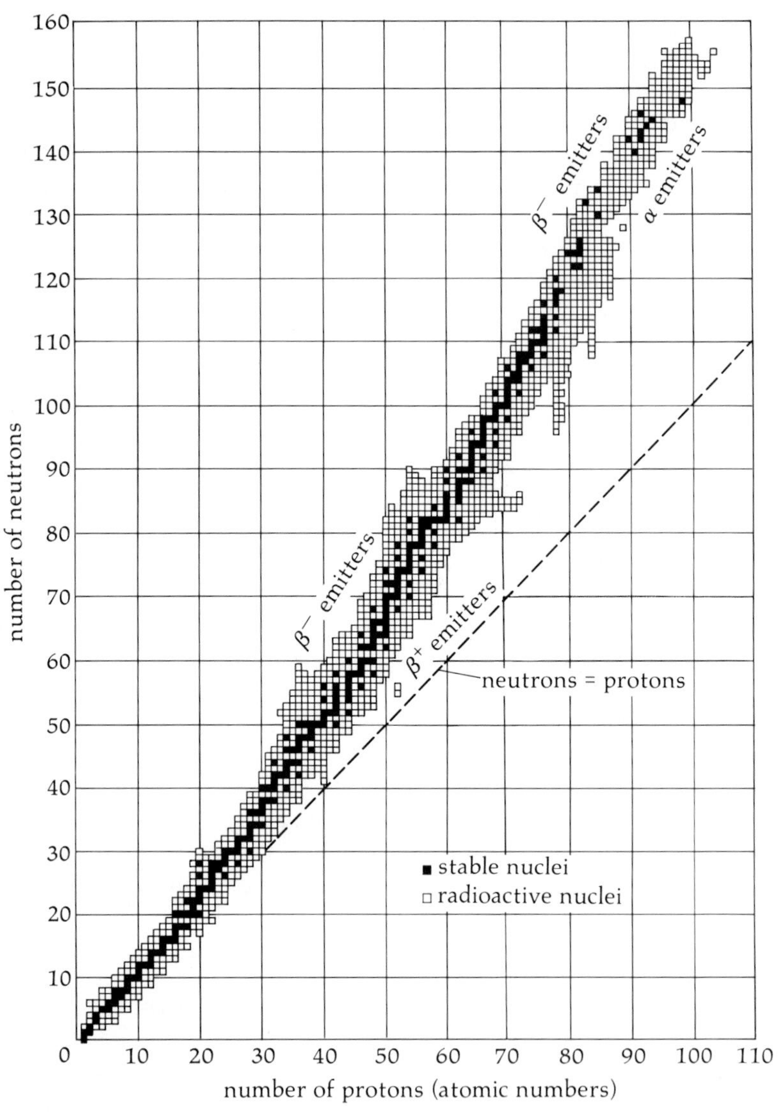

THE NUCLEON COMPOSITION OF STABLE AND RADIOACTIVE NUCLEI.

Nuclear Reactors and Nuclear Armaments

We usually think of nuclear technology in terms of the design, construction, and operation of nuclear reactors to be used for

- the heat to boil water and generate steam,
- the production of a rich neutron source for research,
- the synthesis of isotopes,

and also in terms of the design, construction, assessment, and storage of nuclear armaments to be stockpiled as a deterrent to international aggression. Both nuclear reactors and nuclear armaments depend upon nuclear reactions—in many cases, the same nuclear reactions. But nuclear reactors and nuclear armaments are significantly different in design. Fission reactors are designed to contain essentially all radioactive materials and to make it impossible for nuclear reactors to become nuclear bombs even through misadventure. The explosion of a nuclear fission bomb does produce neutrons, does produce radioactive isotopes, and does generate heat, all of which are dispersed into the environment.

Presumably, nuclear reactors and nuclear armaments could be modeled upon naturally occurring nuclear reactions:

- the neutron-induced fission of high-atomic-number nuclei such as uranium-235, and also
- the fusion of low-atomic-number nuclei such as hydrogen-1 in the sun.

However, at this time the technology and the materials to generate and contain the high-temperature, high-pressure plasma essential to the sustained, controlled production of heat in a nuclear fusion reactor are not available, and nuclear fusion reactors still lie in the future. Nuclear fusion reactions are used in hydrogen bombs, where sustained, controlled reactions are not required and not desired.

Nuclear Fission Reactors

Nuclear fission reactors are installations within which the rate of neutron-induced nuclear fission can be controlled. The primary components are fuel rods containing fissionable isotopes, control rods containing neutron absorbers, a moderator to slow the speed of neutrons, a circulating fluid system around the rods to transfer heat outside the reactor, control mechanisms, and a containment structure. The rods are mounted vertically with the control rods distributed among the fuel rods. The positions of the fuel rods are fixed. The control rods are retractable and are raised (partially removed) at will to permit an increase in the rate of neutron-induced fission or lowered at will to reduce the rate of neutron-induced fission. The reaction in the array of rods, called the core, is said to have "gone critical" when, on the average, one neutron from the fission of one nucleus induces fission in another nucleus. Under these conditions, fission proceeds at a constant rate. When the positions of the control rods are such that, on the average, more than one neutron from the fission of one nucleus induces fission in other nuclei, the reaction has gone supercritical and the rate of fission accelerates.

In order to slow down or to shut down a reactor, the control rods must be lowered to reduce the number of effective neutrons produced below the critical state. The moderator is introduced to increase the efficiency of neutron capture. The probability of capture of the fast neutrons emitted in fission is less than the probability of capture of the much slower thermal neutrons—neutrons that have average kinetic energies approximating the average kinetic energy of gas molecules. A number of substances, such as graphite (a crystalline form of carbon) and water, are effective in slowing the speed of neutrons and are used as moderators. In news reports, diagrams are frequently given of specific nuclear reactors. Even if these designs are complex, you will probably discover that you can identify the components: the fuel rods, the moderator, the control rods, and the circulating fluid system that transfers heat outside of the reactor. A failure of this cooling system without the reactor being shut down (lowering the control rods) could lead to a melting of the fuel and the metal tubes within which the fuel is encased.

The composition of the fuel and also the dimensions and the relative spacing of the fuel rods, the control rods, and the moderator for each reactor must be appropriate to the attainment of the required rate of fission to produce the desired yield of neutrons and energy. Reactors to be used in research and in the synthesis of isotopes are usually operated at lower temperatures than reactors used in the production of electrical power.

Uranium ores also contain the more abundant uranium-238 isotope. Uranium-238 is not fissionable but it does capture neutrons, producing the uranium-239 isotope

$$^{238}_{92}U + ^{1}_{0}n \longrightarrow ^{239}_{92}U$$

which is a *beta* emitter and has a half-life of 24 minutes:

$$^{239}_{92}U \longrightarrow \underset{\text{neptunium}}{^{239}_{93}Np} + ^{0}_{-1}\beta$$

Neptunium-239 also has a short half-life, 2.3 days, and also decays by *beta* emission:

$$^{239}_{93}Np \longrightarrow \underset{\text{plutonium}}{^{239}_{94}Pu} + ^{0}_{-1}\beta$$

Uranium-238 Breeder Reactors

It is this plutonium-239 that is of particular interest. It is radioactive and decays by *alpha* emission with a half-life of 24,000 years. Plutonium-239 also undergoes spontaneous fission and neutron-induced fission. Thus reactors can be designed to "breed" (produce) more fissionable plutonium-239 than the uranium-235 consumed. In all reactors, the fuel rods are periodically replaced and the contents of the used rods can be partially refined and the composition of the material readjusted for use in forming new fuel rods.

Synthesis of Isotopes

The synthesis of isotopes goes on continuously within the fuel rods during the operation of the reactor, but the more common practice in the study of the synthesis of isotopes using neutron sources is to introduce "rabbits" (containers of the substance to undergo neutron radiation) through "rabbit holes" (openings) into cylindrical conduits to the core of a reactor. The synthetic isotopes are extensively used in biological research, physical research, medical diagnostic testing, medical radiation treatments, quality control monitoring of solid objects such as sheet metals and fabricated products.

The use of nuclear reactors to produce electrical power is really the use of nuclear reactors to produce the heat to produce the steam to operate steam turbines to operate electrical generators. See Gratuitous Information 22-3. The use of nuclear fuel has been particularly attractive under circumstances where fossil fuels are not reliably accessible.

The mechanism of achieving explosions is to generate a high concentration of high-temperature gases in a very short period of time. The greater the number of molecules or other small particles, the faster the reaction and the higher the temperature, the bigger the bang, and the greater the lift of a rocket or the greater the destruction of a bomb. Conventional bombs utilize high-speed oxidation reactions that generate gaseous products, and the heat also vaporizes elements and compounds we ordinarily think of as solids. It is the movement of these high kinetic energy molecules and other small particles that is the explosion.

Nuclear Fission Bombs

To achieve the rapid reaction in nuclear fission bombs (called atomic bombs), the fissionable isotope should be of high purity, the escape of neutrons from the fuel should be minimized, and fast neutrons should be moderated. To prevent a premature explosion, the nuclear fuel must be shaped or subdivided into smaller pieces to insure sufficient loss of neutrons from the fuel, preventing spontaneous fission from prematurely initiating the induced fission reaction throughout the fuel. To fire the nuclear

Gratuitous Information 22-3

NUCLEAR POWER PLANTS

Nuclear power plants consist of four units: the nuclear reactor, a high-pressure steam generator, a steam turbine, and an electrical generator:

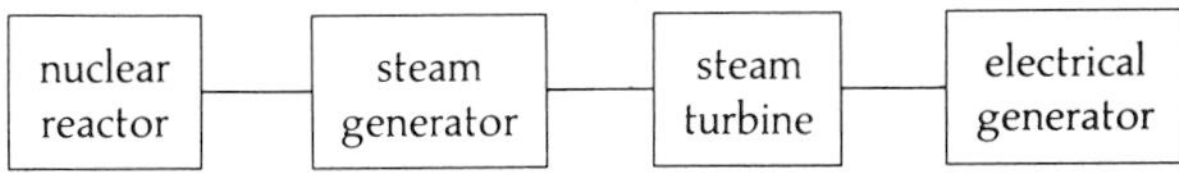

The nuclear reactor produces the heat that generates the steam that turns the turbine that rotates the armature in the electrical generator.

The movement of an electrical conductor in a magnetic field induces a flow of electrons, an electric current, in the electrical conductor. An electrical generator has a stationary electromagnet to produce the magnetic field and an armature made up of coils of insulated wire mounted upon a rotatable shaft. The first three units of a nuclear power station exist for only one reason. That reason is to rotate the armature and thus move the coils of wire through the magnetic field of the electromagnet.

Steam turbines operate on the same principle as those highly colored plastic pinwheel gadgets mounted on the end of a stick and frequently sold in carnival areas to the custodians of small children. Blow on the vanes and the pinwheel rotates. The many vanes of a steam turbine are mounted on a rotatable shaft. Direct high-pressure steam against the vanes and the shaft rotates. In a power plant, the shaft of the armature of the electrical generator is a continuation of the shaft of the steam turbine. The higher the entry pressure of the steam into the turbine, the more efficient the steam turbine.

The linkage between the nuclear reactor and the vessel in which the steam is generated is a heat transfer system. You may be familiar with the heat transfer systems of hot water heating systems. Water is circulated in a closed system of pipes from the furnace through the various rooms in the house, back to the furnace and through the heating coils in the furnace again. If the fluid used in the heat exchange system between the nuclear reactor and the steam generator is water, the closed system must be operated under high pressure to maintain water in the liquid phase at the high temperature of the generator. The heat exchange system operates at temperatures higher than the melting point of sodium and that metal is frequently used as the circulating fluid in the heat exchange system. Heat transfer to the water to be boiled in the steam generator is facilitated by extensive contact of the pipes of the heat transfer system with the water in the steam generator. The nuclear reactor and the steam generator are frequently separated by considerable distance and the heat transfer system must be efficient to transfer heat over that distance. The only difference between a nuclear power plant and the conventional fossil fuel plant is the manner in which water is heated to generate steam.

Nuclear reactors must be engineered to bring about nuclear reactions, either fission or fusion, under conditions such that the rates of the nuclear reactions are always under control and the nuclear reactions can be stopped and started at will.

bomb, a small conventional explosion is used to force all of the nuclear fuel into a single unit having the mass and shape essential for spontaneous fission to initiate induced fission at many sites throughout the nuclear fuel simultaneously.

The devastation inherent in the use of an atomic bomb is much greater than that in the use of a conventional bomb due to the exceptional amount of energy released in the nuclear explosion and also by the ejection of a great variety of radioactive isotopes into the atmosphere. It is the potential biological consequences of these radioactive isotopes that make nuclear fission bombs so uniquely different from conventional bombs. Radioactive isotopes having short half-lives pose the greatest immediate threat, but these isotopes soon decay and it is the radioactive isotopes having long half-lives that add to the continuing long-term environmental burden of naturally occurring nuclear reactions.

Nuclear Fusion Bombs

In hydrogen bombs, there are two nuclear explosions. The first, involving a nuclear fission explosion, produces the high-temperature, high-pressure conditions necessary to fire the nuclear fusion reaction. The nuclear fusion explosion instantaneously follows the nuclear fission explosion. As was pointed out in the discussion of binding energies, nuclear fusion reactions convert a greater percent of the mass of the reactants into energy than nuclear fission reactions. The hydrogen bomb, in terms of the force of the explosion, is more destructive than nuclear fission bombs.

22-1 Editorial Comment

BENEFITS AND BURDENS OF NUCLEAR TECHNOLOGY

The public concern with nuclear technology, as with all technologies, is the balance between benefits and burdens. Very technical questions must be addressed in the assessment of the pros and cons of nuclear technologies, and we have yet to acquire the knowledge to fully resolve many of these questions. Some of the relevant areas of concern:

- the use of nuclear technologies to extend our knowledge of the structure of atoms and the mechanism of chemical reactions;
- the role of medical technology in using nuclear technologies to elucidate biological processes, to diagnose health problems, and to treat health problems;
- the role of nuclear reactors and other sources of energy in making nations energy self-sufficient;
- the role of nuclear reactors and other sources of energy in minimizing climatic changes induced by the greenhouse effect of the carbon dioxide buildup in the atmosphere due to the combustion of fossil fuels;
- the biological consequences of exposure to both low levels and high levels of ionizing radiations;
- the role of nuclear armaments in international relations;
- the capacity and will of engineers and technological institutions to manage risk through engineering processes and safety practices;
- the capacity and will of governments to effectively regulate the procurement, processing, transportation, use, and disposal of radioactive materials.

All of the above and many others involve aspects of the role of nuclear technology in local, national, regional, and world social, economic, and political processes. None of us can escape entirely either the benefits or the risks inherent in the ways nuclear technologies are used. In a democratic society, each of us has both the right and the responsibility to participate in the political processes that will determine how nuclear technologies are used.

There are, however, fewer radioactive isotopes produced by the hydrogen fusion reaction and more of the isotopes produced have short half-lives. Nuclear fusion bombs are consequently said to be much "cleaner" than nuclear fission bombs. These characteristics of nuclear fusion make nuclear fusion reactions attractive for use in nuclear reactors to produce energy—if the technology can be developed.

No technological innovation has posed more fundamental societal issues than nuclear technology, and the control of nuclear technology has been in recent years and is now of concern to more peoples of the world than any other technology. (Editorial Comment 22-1.)

Scramble Exercises

1. You should be able to complete the following incomplete equations for nuclear reactions.

$$^{14}_{7}N + ^{1}_{0}n \longrightarrow \_\_\_\_ + ^{1}_{1}H$$

$$^{226}_{88}Ra \xrightarrow{\gamma} \_\_\_\_ + ^{4}_{2}\alpha$$

$$^{10}_{5}B + ^{4}_{2}\alpha \longrightarrow \_\_\_\_ + ^{1}_{0}n$$

$$^{14}_{7}N + ^{4}_{2}\alpha \longrightarrow \_\_\_\_ + ^{1}_{1}H$$

If you can do these, you understand the balancing of equations for nuclear reactions.

2. The following statement is taken from scramble exercise 2 in Chapter 6. "Radium, Ra, element 88, can also be incorporated with calcium in bone structure. This would probably be of only passing interest if it were not for the fact that naturally occurring isotopes of radium are radioactive and their nuclei undergo spontaneous transformations emitting energy."

   The same statement could be made with respect to barium-139, $^{139}_{56}Ba$, a radioactive isotope of barium. This isotope of barium is a product of the fission of uranium-235, $^{235}_{92}U$. The first step in the decay of barium-139 is the emission of a *beta* particle accompanied by *gamma* radiation. With the background you have now acquired in chemistry and in nuclear reactions, discuss the chemical basis for the deposition of isotopes of radium and barium along with calcium in bone structures and the basis for concern about the deposition of these isotopes in our bone structures.

3. Check the calculations of the numbers given in Table 22-2 for iron-57. Or, if you prefer, check the values given in Table 22-2 for another column of figures.

   It is not important that you acquire the competence to make these calculations on your own, but it is important that you develop an understanding of the concept of binding energies. To follow through the calculations for one isotope of one element is one way to develop this understanding.

4. Be prepared to discuss the production of energy by nuclear fission and by nuclear fusion on the basis of the binding energy curve given in Figure 22-1.

5. Show that 0.0304 gram is equivalent to 2.7×10^9 kilojoules.

6. Trace the derivation of the following sources of energy to nuclear reactions: solar power, wind power, water power, fossil fuels, and geothermal power.

7. Suggest a plausible explanation for the fact that all known uranium-235 ores contain a low concentration of uranium-235.

8. Write the equation for the *alpha* decay of plutonium-239.

Additional exercises are given in the Appendix.

A Look Ahead

The success of your exploration of chemical phenomena and nuclear phenomena can only be measured by your extension of your knowledge and understanding of chemical and nuclear phenomena throughout the remainder of your life—a period of probably more than one-half century—and by your participation in decision making as a citizen in matters related to the use of science, engineering, and technology in the resolution of societal issues. We trust that you will derive satisfaction in what you can and will accomplish.

APPENDIX: SUPPLEMENTARY EXERCISES

Both authors are committed to the proposition that a small number of scramble exercises and the thought processes initiated by them have greater educational value for the general student than a large number of drill problems. Both of us are more than willing to defend the small number of scramble exercises included with each chapter. We seek to encourage students to think—not to encourage students to scurry after numerical answers to repetitive problems.

We recognize that this small number of scramble exercises may not be the choice of all who teach or of all who study. To that end, this appendix expands the number of exercises for each chapter by about 100% and includes some exercises that are clearly drill problems. Here again, many of the exercises address fundamental issues not discussed directly in the text. They provide the opportunity to seek plausible approaches and plausible explanations. Whether the approaches taken and explanations proposed coincide with the currently accepted approaches and explanations of the scientific community is to us quite a secondary matter. After all, it has taken the scientific community years and much effort to reach the current state of understanding of chemical phenomena.

Chapter 1: A Search to Understand

1. Show that the length 9.4×10^{-3} millimeters is shorter than 2.1×10^{-2} millimeters.

2. Determine the exponent of 10 needed to complete the following equalities:
   (a) $0.0032 = 3.2 \times 10^?$
   (b) $9217 = 9.217 \times 10^?$
   (c) $10 = 1.0 \times 10^?$
   (d) $0.1 = 1 \times 10^?$
   (e) $1.0 = 1.0 \times 10^?$

3. An atom of oxygen is reported to have a diameter of 0.132 nanometers. Express this length as 1.32 times 10 with the appropriate exponent for the following units of length:
   (a) nanometers (nm)
   (b) meters (m)
   (c) centimeters (cm)
   (d) millimeters (mm)

4. The diameter of a copper atom is 0.25 nm. How many copper atoms could be lined up side by side in a distance of 1 cm?

5. Few of us realize how big numbers like 1×10^{12} or how small numbers like 1×10^{-12} are. As a brief exercise, assume that a dollar bill is 2.5×10^{-2} millimeters thick (0.025 mm). How tall a stack of one dollar bills is one trillion (1.0×10^{12}) dollars? Express the answer in millimeters, meters, and kilometers. The current federal budget for the United States is more than a trillion dollars.

6. Show that a driving speed of 125 km/hr (kilometers per hour) is approximately 78 mph (miles per hour)—a speed that should get you a speeding ticket in either the United States <u>or</u> Europe.

Chapter 2: One Model of Atoms and Molecules

1. Write out the full structure of the straight-chain hydrocarbon $C_{22}H_{46}$. Hydrocarbons of this size are important components of the paraffin waxes.

2. Write out the structure of an isomer of $C_{22}H_{46}$ that has an eight-carbon-atom-long "branch" in the structure.

3. Long-chain alcohols like isomers of $C_{22}H_{46}O$ are often used in perfumes, particularly to add persistence to the fragrance. Write out three possible alcohol isomers of $C_{22}H_{46}O$.

4. **(a)** Evaluate the expression

$$1.5 \times 10^{14} \times 2.0 \times 10^{-11} \div 6.0 \times 10^{5}$$

   and express the answer in the conventional form (a number with one digit to the left of the decimal point times 10 to an integer power).

   (b) Reevaluate the above expression using a different series of steps.

   (c) Resolve the discrepancy if the values obtained in (a) and (b) are not identical.

5. Four structural isomers of $C_4H_{10}O$ are alcohols. Write out their structural formulas and check for identities by building models.

   How many structural isomers of $C_4H_{10}O$ are ethers? Write their structural formulas and check for identities by building models.

6. By definition of the Celsius temperature scale, the boiling point of water under a pressure of one atmosphere is exactly 100 °C. State the boiling point of water under a pressure of one atmosphere on the kelvin temperature scale.

7. Write out the structural formulas for the *n*-pentane, 2-methyl butane, and 2,2-dimethylpropane structural isomers of C_5H_{12}. Are there other structural isomers of C_5H_{12}? If so, write their structural formulas and check for identities by building models.

8. Write out the structures of three isomers of $C_5H_{12}O$. Identify each one as an alcohol or an ether.

9. The formula $C_5H_{10}O$ is appropriate for a cyclic compound. The oxygen atom can be in the ring or outside the ring. Write out the structural formulas for at least four isomers. Include two isomers that have the oxygen atom in the ring.

10. **(a)** Show that the area of a rectangle that has a length of 1.50 meters and a width of 9.4 millimeters is

    (i) 1.41×10^{2} centimeters^2

    (ii) 1.41×10^{-2} meters^2

    (iii) 1.41×10^{4} millimeters^2

(b) Show that the volume of a metal band 1.50 meters long, 9.4 millimeters wide, and 2.0 millimeters thick is
   (i) 2.82×10^4 millimeters^3
   (ii) 2.82×10 centimeters^3
   (iii) 2.82×10^{-5} meters^3
   (iv) 2.82×10 milliliters
   (v) 2.82×10^{-2} liters

Quite frankly, the evaluation of the piece of metal in some of these units is a bit academic. The volume of small samples of solids is usually expressed in cubic centimeters.

Chapter 3: Extension of the Model of Atoms and Molecules

1. **(a)** Using balls and sticks, endeavor to construct models for cyclobutane, C_4H_8 and cyclopropane, C_3H_6. With springs (bent bonds) for carbon–carbon bonds, again endeavor to construct models for cyclobutane and cyclopropane. Both compounds exist; cyclopropane is a well-known anesthetic.

   (b) Now try to make C_4H_6 cyclobut<u>ene</u> and C_3H_4 cycloprop<u>ene</u>—again, both compounds exist.

   (c) With the springs, you can really let your imagination roam. Try your hand at building "cubane," C_8H_8, with your model set.

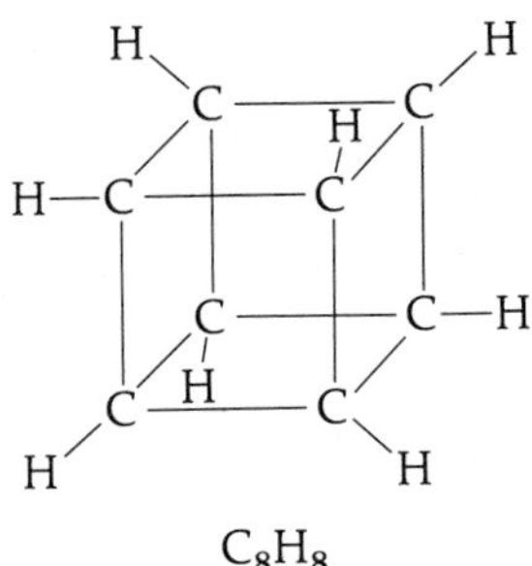

C_8H_8

Originally, this compound was proposed as somewhat of a joke. It has been synthesized and its properties determined.

   (d) The bond angles predicted by the models for these compounds are often quite different from the 109.5° angles found for compounds containing only single bonds or the 120° angles found in compounds containing a double bond. Look at the models and predict the bond angles in C_4H_8, C_3H_6, and C_8H_8.

2. Convert the following volumes to liters:
   (a) 230 mL
   (b) 1284 cm^3
   (c) 2.1 dm^3 (2.1 cubic decimeters)

3. All of the structural formulas given below are isomers of $C_4H_8O_2$. All of the structures given are esters—either the ethyl ester of acetic acid or the methyl ester of propionic acid. That only two compounds are represented by these six structural formulas may not be readily apparent to you. Work with building models until you are convinced that only two compounds are represented by these structures and classify each structure as either the ethyl ester of acetic acid or the methyl ester of propionic acid.

```
                                                           H
                                                           |
                                                        H—C—O
     H  H  O      H           H           H                |   |
     |  |  ||     |           |           |                H   |
(a) H—C—C—C—O—C—H     (b) H—C—O—C—C—H        (c)            C=O
     |  |         |           |      ||   |                H   |
     H  H         H         H—C—H   O   H                 |   |
                              |                         H—C—C—H
                              H                            |   |
                                                           H   H

     H
     |
   H—C—H
     |
     O  H  H                  H  H              H          H  H
     |  |  |                  |  |              |          |  |
(d) O=C—C—C—H       (e) O=C—O—C—C—H     (f) H—C—O—C—C—C—H
        |  |                  |     |  |              |      ||  |  |
        H  H                H—C—H  H  H              H      O  H  H
                              |
                              H
```

4. Show that one of the following compounds is an aldehyde and that the other is a ketone. For each compound, write out the structure in more detail and identify the compound as an aldehyde or as a ketone. Note that the two compounds are isomers.

$$CH_3COCH_3 \qquad CH_3CH_2CHO$$

5. Show that one of the following compounds is a peroxide and the other is an ester. In each case, write out the structure in more detail and identify the compound as either a peroxide or an ester. Note that these compounds are not isomers.

$$CH_3COOCH_3 \qquad CH_3CH_2OOCH_3$$

6. Show that cyclopentane, C_5H_{10}, is an isomer of 1-pentene and 2-pentene. Why is there no 3-pentene?

7. As a part of testing a new race car, the driver cruises the car at a uniform speed several times around a circular track, accelerates for one lap, and again cruises at a uniform speed for several turns around the track. During the initial period of cruising, it takes 50 seconds to complete each turn around the track. During the second period of cruising, it takes 30 seconds to complete each turn around the

track. The acceleration lap required 40 seconds. The distance around the track is 1.61 kilometers.

(a) Show that the speed of the car during the initial period of cruising at a uniform rate is 116 kilometers/hour and that the speed of the car during the second period of cruising at a uniform rate is 193 kilometers/hour.

(b) Show that the average acceleration during the acceleration lap is 6930 kilometers/hour^2.

Note that the actual acceleration must range from zero at the beginning of the lap to some value much higher than 6930 kilometers/hour^2 and back again to zero by the end of the acceleration lap. At the beginning of the lap, the speed of the car is 116 kilometers/hour and the rate of acceleration is zero. At the end of the acceleration lap, the speed is 193 kilometers/hour and the acceleration is again zero. In driving a car, you feel these changes in acceleration. When you press the accelerator down, you feel the change in acceleration, the quick increase in speed; when you ease up on the accelerator to cruise at the higher speed, you feel the drop off in acceleration to zero at the higher cruising speed.

8. List the following lengths in order of increasing magnitude (the longest length at the bottom):

1.34×10^3 meters
9.43×10^{-3} meters
5.89×10^2 meters
6.92×10^{-2} meters
2.13×10^{-1} meters
4.34×10 meters
3.08×10^0 meters

Check your list by writing out these quantities without using 10 raised to an integer power, for example, 1340 meters for 1.34×10^3 meters.

Chapter 4: Fire

1. **(a)** Write the equation for the complete combustion of each of the two-carbon hydrocarbons:

$CHCH$	CH_2CH_2	CH_3CH_3
ethyne (acetylene)	ethene (ethylene)	ethane

The only products of the complete combustion of hydrocarbons are carbon dioxide and water vapor.

Compare the number of molecules of oxygen, O_2, required per molecule of hydrocarbon for the complete combustion for each of these three hydrocarbons.

(b) Each of the two-carbon hydrocarbons

$$\mathrm{HC{\equiv}CH} \qquad \begin{matrix} \mathrm{H} & & \mathrm{H} \\ & \mathrm{C{=}C} & \\ \mathrm{H} & & \mathrm{H} \end{matrix} \qquad \begin{matrix} & \mathrm{H} & \mathrm{H} & \\ & | & | & \\ \mathrm{H-} & \mathrm{C} & \mathrm{-C-} & \mathrm{H} \\ & | & | & \\ & \mathrm{H} & \mathrm{H} & \end{matrix}$$

ethyne (acetylene) ethene (ethylene) ethane

is a gas at room temperature. Once ignited, a small stream of each of these gases flowing through the orifice of a torch into the atmosphere continues to burn but the characteristics of the three flames are quite different. The acetylene flame is yellow and the most untidy, with black soot settling upon the furniture and the experimenter. The ethane flame is the most tidy, with very little, if any, evidence of incomplete combustion. Suggest a plausible explanation for the formation of the clusters of carbon atoms in the combustion of acetylene being so much more pronounced than the formation of clusters of carbon atoms in the combustion of the other two compounds.

Premixed mixtures of each of these hydrocarbons and air flowing from the same torch burn with the pale blue flame characteristic of complete combustion of hydrocarbons.

2. Two of the structural isomers of the five-carbon saturated hydrocarbon pentane, C_5H_{12}, are

$$CH_3CH_2CH_2CH_2CH_3 \qquad \text{and} \qquad \begin{matrix} & CH_3 & \\ & | & \\ H_3C- & C & -CH_3 \\ & | & \\ & CH_3 & \end{matrix}$$

n-pentane 2,2-dimethyl propane

Write the equation for the complete combustion of these isomers and suggest a plausible explanation for the experimental fact that the 2,2-dimethyl propane isomer is more prone to undergo incomplete oxidation than the *n*-pentane isomer.

3. Twelve grams of carbon contain 6.0×10^{23} atoms of carbon.

(a) Show that the mass of one atom of carbon is 2.0×10^{-23} grams.
(b) Show that one gram of carbon contains 5.0×10^{22} atoms of carbon.

4. The statement "A force of one newton gives a mass of one kilogram an acceleration of one meter per second per second" defines the newton as a unit of force. In a similar manner, the statement "A force of one dyne gives a mass of one gram an acceleration of one centimeter per second per second" defines the dyne as a unit of force. Use these two definitions and the force equation, force = mass × acceleration, to show that a force of one newton is equivalent to 1×10^5 dynes.

5. Write equations for the complete combustion of the following compounds. The alcohols and ethers, as well as the saturated hydrocarbons, are saturated. On

the basis of the equations written, can you distinguish among the other three classes of compounds in degree of unsaturation. It may be helpful if you will write out structural formulas for a few of the compounds.

(a) saturated hydrocarbons

C_2H_6	C_5H_{12}
C_3H_8	$C_{15}H_{32}$
C_4H_{10}	

(b) alcohols and ethers

C_2H_5OH	$C_5H_{11}OH$
C_3H_7OH	$C_{15}H_{31}OH$
C_4H_9OH	

(c) unsaturated hydrocarbons

C_2H_4	C_2H_2
C_3H_6	C_3H_4
C_4H_8	C_4H_6
C_5H_{10}	C_5H_8
$C_{15}H_{30}$	$C_{15}H_{28}$

(d) aldehydes and ketones

C_2H_4O	$C_5H_{10}O$
C_3H_6O	$C_{15}H_{30}O$
C_4H_8O	

(e) acids and esters

$C_2H_4O_2$	$C_5H_{10}O_2$
$C_3H_6O_2$	$C_{15}H_{30}O_2$
$C_4H_8O_2$	

Chapter 5: A Model for the Structure of Atoms I: The Nucleus

1. (a) Write out the number of protons, neutrons, and electrons in the following atoms:

$^{9}_{4}Be$	$^{19}_{9}F$	$^{40}_{19}K$
$^{70}_{31}Ga$	$^{93}_{41}Nb$	$^{158}_{64}Gd$
$^{204}_{81}Tl$	$^{244}_{94}Pu$	$^{258}_{101}Md$

   (b) Comment on the relative number of neutrons per proton in the nuclei of these atoms.

2. Find the mass in grams of one mole of acetic acid, CH_3COOH, and determine the number of molecules of acetic acid (a) in 1.0 gram of acetic acid and (b) in 100 grams of acetic acid.

3. Calculate the grams of sulfur dioxide, SO_2, produced by the combustion of 0.100 gram of sulfur. Sulfur dioxide is the only product.

4. Diethyl ether is a very volatile compound: boiling point under one atmosphere pressure, 35 °C. In the early days of its use as an anesthetic, serious explosions occasionally occurred in operating rooms in which the vaporization of ether led to the buildup of an explosive mixture of ether molecules with the oxygen molecules of the air. These mixtures could be ignited by a spark due to the buildup of static charges upon clothing and other materials. Write the equation for the chemical reaction assuming complete combustion of the diethyl ether, $CH_3CH_2OCH_2CH_3$, with oxygen molecules, O_2, to give carbon dioxide and water.

5. The calorie content of a piece of chocolate cake may be 400 Cal (kilocalories) or more. Express this quantity of energy in joules (J) and also in kilojoules (kJ).

6. **(a)** Calculate the sum of the relative masses of the electrons, the protons, and the neutrons that constitute iron-57, ${}^{57}_{26}Fe$.
   (b) Compare this calculated value with the experimental value 56.9354 for the iron-57 isotope. This mass difference is discussed in Chapter 22.

7. An important industrial process is the partial combustion of methanol, CH_3OH, to give formaldehyde, $H_2C{=}O$, and water.
   (a) Complete the equation

$$\_\_\_\_CH_3OH + \_\_\_\_O_2 \longrightarrow \_\_\_\_CH_2O + \_\_\_\_H_2O$$

   (b) Calculate the amount of formaldehyde that can be obtained from 1.0×10^6 grams of methanol. (The figure 1.0×10^6 grams is chosen because it's close to one <u>ton</u> of methanol. Industrial processes are carried out on very large scales.)
   (c) Suppose in the above case the supply of oxygen were limited to 1.0×10^6 grams of oxygen. Would this be enough oxygen to convert <u>all</u> 1.0×10^6 grams of methanol to the products according to the given reaction?

8. Use the equation

$$\text{Energy} = \text{force} \times \text{distance}$$

and the two relations

$$\text{newtons} \times \text{meters} = [\text{energy in joules (J)}]$$

$$\text{dynes} \times \text{centimeters} = [\text{energy in ergs}]$$

to show that one joule is equivalent to 1×10^7 ergs.

Chapter 6: A Model for the Structure of Atoms II: The Electron Atmosphere

1. The family of elements known as the alkali metals includes the elements lithium, ${}_3Li$; sodium, ${}_{11}Na$; potassium, ${}_{19}K$; rubidium, ${}_{37}Rb$; and cesium, ${}_{55}Cs$. All of these elements have similar but slightly different properties. For the ground state atom

of each of these elements, point out the relation of the electron structure of the atom to the electron structure of the ground state atom of the nearest inert gas.

2. The family of elements known as the halogens includes the elements fluorine, $_{9}F$; chlorine, $_{17}Cl$; bromine, $_{35}Br$; and iodine, $_{53}I$. All of these elements have similar but slightly different properties. For the ground state atom of each of these elements, point out the relation of the electron structure of the atom to the electron structure of the ground state atom of the nearest inert gas.

3. To convince yourself that the elements numbered 40 → 48 (zirconium → cadmium) and 72 → 80 (hafnium → mercury) represent elements filling their *d*-orbitals, write out the expected complete ground state electron configuration of as many of these two sets of elements as necessary. From looking at the Periodic Table, you would expect the 4*d*- and 5*d*-orbitals to be involved.

4. To convince yourself that the elements numbered 58 → 65 (cerium → terbium) and 90 → 97 (thorium → berkelium) represent elements filling their *f*-orbitals, write out the expected complete ground state electron configuration of as many of these two sets of elements as necessary. From looking at the Periodic Table, you would expect the 4*f* and 5*f* levels to be involved.

5. To convince yourself that it is the outermost (highest quantum number) energy levels of the ground state atom that determine the position of an element in the Periodic Table, write out the expected ground state electron configurations of as many of the elements across the third and fourth periods as necessary.

6. It is an experimental fact that the differences in chemical properties between elements in the same vertical column (family) of the Periodic Table become smaller as the atomic number increases. For example, the differences between oxygen and sulfur are more spectacular than the differences between sulfur and selenium. Look at the difference in energy levels as quantum numbers increase and see if any correlation suggests itself to you.

7. One of the more bizarre aftermaths of the nuclear accident at Chernobyl, USSR, was the banning of the sale of reindeer meat. Reindeer meat was found to have unacceptably high levels of radioactive cesium. Look at the Periodic Table and suggest why red meat might indeed have a high level of an alkali metal element present. (The connection is through blood and the similarity of Cs to Na.)

8. Evaluate the mass in grams of one atom of iodine and the mass in grams of the diatomic molecule I_2. Should these masses be more or less than 1 gram?

9. Evaluate the number of moles of carbon dioxide formed in the complete combustion of 100 moles of a mixture of octane isomers, molecular formula C_8H_{18}. There is a long way and a short way to get this answer.

Chapter 7: Hydrogen and Other Diatomic Molecules: The Valence Bond Approach

1. Compare the nature of the chemical bonds in the two homonuclear diatomic molecules H_2 and N_2 on the basis of

   (a) the experimental relations

   $$H_2 \longrightarrow H + H \qquad \Delta H = 436 \text{ kilojoules } (25\ ^\circ\text{C})$$

   $$N_2 \longrightarrow N + N \qquad \Delta H = 950 \text{ kilojoules } (25\ ^\circ\text{C})$$

   (b) the valence bond model for the two molecules, and
   (c) the Lewis structures for the two molecules.

2. Using the valence bond approach, discuss the nature of the bond in BrCl using the partially filled $3p$-orbital of Cl and the partially filled $4p$-orbital of Br. Do you expect this molecule to have a dipole moment?

3. Use the relation

   $$HCl(g) \longrightarrow H + Cl \qquad \Delta H = 431 \text{ kilojoules} \quad 25\ ^\circ\text{C}$$

   to estimate the energy of <u>one</u> HCl bond (the energy of the bond in one HCl molecule).

4. Show that the following five statements are consistent with each other.

   (a) Hydrogen gas burns in chlorine gas with the evolution of heat and light.
   (b) $H_2(g) + Cl_2(g) \longrightarrow 2\ HCl(g) + \text{energy}$
   (c) $H_2(g) + Cl_2(g) \longrightarrow 2\ HCl(g) \qquad \Delta H = -184$ kilojoules $\quad 25\ ^\circ$C
   (d) $\text{energy} + 2\ HCl \longrightarrow H_2(g) + Cl_2(g)$
   (e) $2\ HCl(g) \longrightarrow H_2(g) + Cl_2(g) \qquad \Delta H = +184$ kilojoules $\quad 25\ ^\circ$C

5. For each of the following heteronuclear diatomic molecules, predict which end of the molecule is slightly positive and state the basis of your prediction:

   (a) nitrogen oxide, NO;
   (b) carbon monoxide, CO; and
   (c) iodine chloride, ICl.

   Note the relation between the naming of the compounds and the polarity of the compounds.

6. Write out Lewis dot structures for ammonia, NH_3, and borane, BH_3. Ammonia is a well-known compound, whereas BH_3 is unknown. However, the compound $H_3N{-}BH_3$ does exist. Can you rationalize these two statements according to the Lewis rules?

7. If you had success predicting a bond angle for the water molecule H_2O (Scramble Exercise 10), try your hand at the ammonia molecule, NH_3 (or H_3N). What do you expect the angle $\angle$HNH to be? Have the fun of making a prediction now. It does not really matter whether your prediction turns out to be consistent with the experimental fact. Molecules containing three or more atoms are the coming attraction in Chapter 9.

8. Have a go at trying to suggest a rationalization for the fact that compounds containing carbon, nitrogen, and oxygen often have multiple (double or triple) bonds, while carbon compounds of silicon, phosphorus, and sulfur very rarely have multiple bonds.

9. Explain (rationalize) why rotation about a double bond requires "breaking" the π-bond but not the σ-bond.

10. Using the experimental relation

$$H_2(g) + Cl_2(g) \longrightarrow 2\,HCl \qquad \Delta H = -184 \text{ kilojoules} \quad 25\ ^\circ C$$

evaluate how many kilojoules of energy are released during the combustion of 8.24 grams of hydrogen gas, H_2, in chlorine gas, Cl_2.

How many grams of chlorine gas, Cl_2, are required for the combustion of the 8.24 grams of hydrogen gas?

Chapter 8: Vibrational Motion and the Dissociation of Diatomic Molecules

1. In Chapter 4, Fire, the initiation of the reaction of hydrogen gas with oxygen gas

$$2\,H_2(g) + O_2(g) \longrightarrow 2\,H_2O(g)$$

was discussed in some detail. Relate that discussion to the concepts of quantized vibrational energy states.

2. Rewrite the equations given in Chapter 4, Fire, in connection with the exploration of a mechanism for the reaction of H_2 with O_2 to give H_2O, giving the Lewis structure for each reactant and each product. As you know, there is no really satisfactory Lewis structure for diatomic oxygen. Take your choice. Use either :O̤::O̤: or :Ȯ̤:Ȯ̤:.

3. Look at the diagram given in Figure 8-3. Assume the molecule is in the energy state labeled E_4. Does the diagram suggest that it is easier to compress the molecule or to stretch the molecule? Explain (if you can).

4. Does the shape of the energy level diagram (Figure 8-3) suggest a rationalization for the experimental fact that almost all liquids and solids expand when heated? Try out your explanation on a friend.

Chapter 9: Water, Ammonia, Methane, and Other Compounds of Carbon

1. Classify each of the following orbitals as either an atomic orbital or a molecular orbital, state whether the orbital is or is not a hybrid orbital, give the picto-

graph for the orbital, and state whether the orbital has or has not an ∞-fold axis of rotation.

p-orbital	*sp*-orbital
pi-orbital	sp^3-orbital
s-orbital	sp^2-orbital
sigma-orbital	

2. Suggest the hybridization of each carbon atom in the following molecules:

(a) 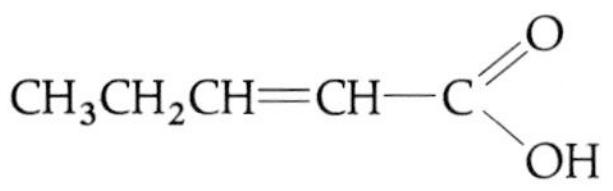 $CH_3CH_2CH{=}CH{-}C(=O)OH$

(b) $CH_3{-}C{\equiv}N$

(c) 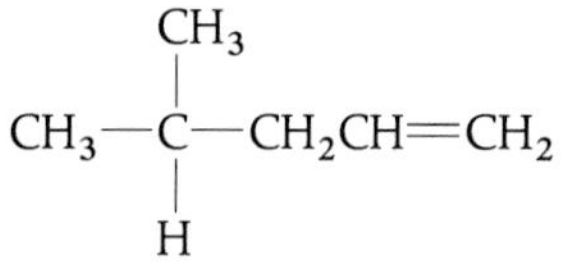 $CH_3{-}CH(CH_3){-}CH_2CH{=}CH_2$ (central C bearing CH_3 above and H below)

(d) 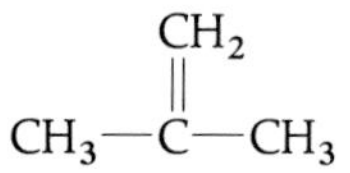 $CH_3{-}C(=CH_2){-}CH_3$

(e) 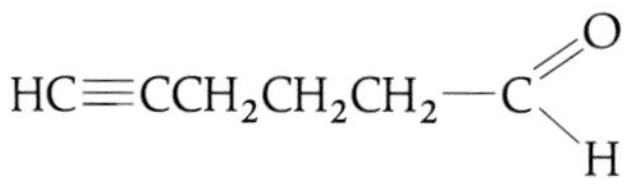 $HC{\equiv}CCH_2CH_2CH_2{-}C(=O)H$

3. For more practice with multiple bonds, write out Lewis structures, build the ball, stick, and spring model, and rationalize according to the valence bond approach the molecule cyanogen, $(CN)_2$. Experimental evidence indicates that the arrangement of the atoms is a straight line:

NCCN

Note the resemblance to hydrogen cyanide, HCN.

4. Discuss the basis of the following statement: "A σ-orbital is to molecular orbitals what an *s*-orbital is to atomic orbitals and a π-orbital is to molecular orbitals what a *p*-orbital is to atomic orbitals."

5. Rationalize the following facts:

(a) The dipole moment of

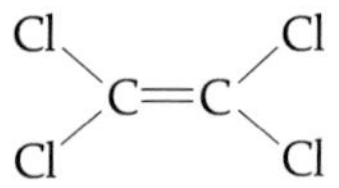

$Cl_2C{=}CCl_2$

is zero.

(b) The dipole moments of CCl_4, CBr_4, and CI_4 are also zero.

6. In terms of the valence bond approach, classify each bond in the following structural formulas as a *sigma*-bond or as a *pi*-bond and specify the probable atomic orbitals involved in its formation.

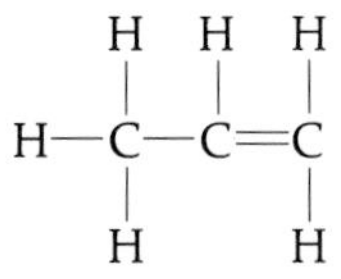

```
   H  H  H
   |  |  |
H—C—C=C          approximate bond angles 109° and 120°
   |     |
   H     H

   H  H
   |  |
H—C—C—C=O        approximate bond angles 109° and 120°
   |  |  |
   H  H  H

   H  H  H
   |  |  |
H—C—C—C—H        approximate bond angles 109°
   |  |  |
   H  H  H

   H  O     H
   |  ‖     |
H—C—C—O—C—H      approximate bond angles 109° and 120°
   |        |
   H        H

   H  H     H
   |  |     |
H—C—C—O—C—H      approximate bond angles 109°
   |  |     |
   H  H     H
```

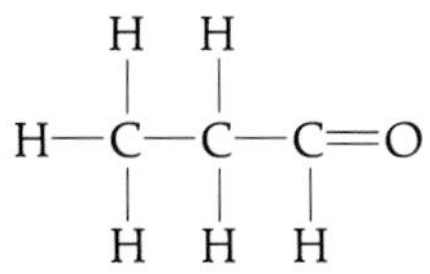

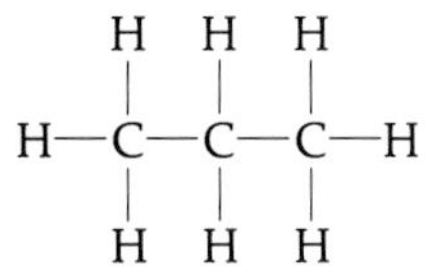

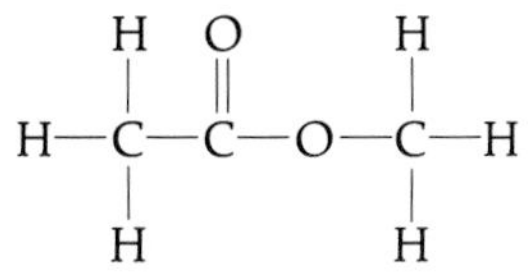

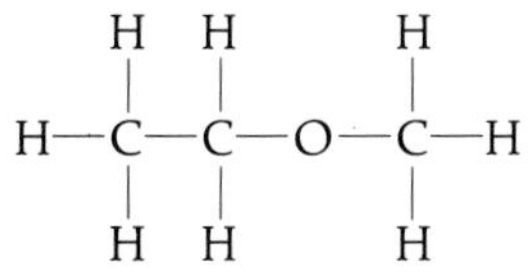

For each of the above compounds, give a Lewis dot formula.

7. Carbon disulfide, CS_2, is a nonpolar molecule. Propose a plausible orientation of the atoms in the molecule and use the valence bond model to discuss the probable nature of the bonds in this molecule.

8. Suggest a valence bond approach to the two isomeric compounds

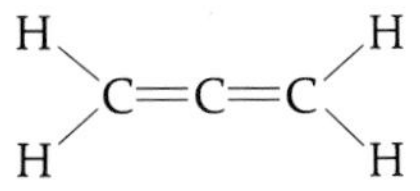

```
H           H
 \         /
  C=C=C
 /         \
H           H
```

1,2-propadiene, or allene

and

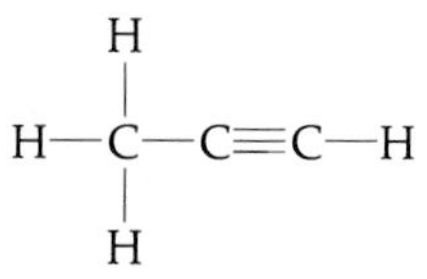

```
   H
   |
H—C—C≡C—H
   |
   H
```

1-propyne

Write out Lewis dot structures and build the ball, stick, and spring models. A particularly interesting feature is the relative orientations of the two planes

$$\begin{matrix} H \\ & \searrow C= \\ H \nearrow \end{matrix}$$

at each end of the allene molecules.

9. The ball-spring-stick model does not predict the carbon monoxide molecule, CO. Try your hand and mind at finding a Lewis electron dot structure and also a possible valence bond rationalization for the bonds in CO.

Chapter 10: Compounds of Sodium, Magnesium, and Related Elements

1. We tend to forget how much of modern science has taken place within the lifetime of one person. For example, astatine (At), element 85, was discovered by a team that included Dale Corson when he was a graduate student. Since that time, Dr. Corson has had a distinguished career as Professor of Physics and a term as President of Cornell University. He remains very active in science and education. Write out the full electronic structure of astatine. Point out the similarities you would expect between astatine and iodine and the other halogens. Astatine has an unstable nucleus; hence the rather recent discovery.

2. Using Table 10-3 for the formulas of transition metal cations and Table 10-6 for the formulas of anions, write the empirical formulas of
   (a) silver(I) chromate
   (b) cobalt(II) carbonate
   (c) copper(II) nitrate
   (d) mercury(II) iodide
   (e) zinc(II) phosphate

3. The melting point of a compound, the solubility of the compound in water, and the electrical conductivity of the aqueous solution are adequate to draw a number of conclusions about an unknown sample. Some of these conclusions:
   - The compound is a low molecular weight covalent compound that dissolves in water but does not react with water to give ions.
   - The compound is either an ionic compound that does not dissolve in water or a very high molecular weight covalent compound that does not react with water to give ions.

   Try your mind at drawing conclusions about the nature of compounds A, B, and C.

   Compound A
   Observation I. Compound A is a colorless liquid. What conclusion can be drawn about Compound A from Observation I?

Observation II. Compound A dissolves in water and the resultant solution does not conduct an electric current. What conclusion can be drawn from Observation II (without considering Observation I)?

What conclusions can be drawn about Compound A from a consideration of both Observation I and Observation II?

Compound B

Observation I. Compound B does not melt below 500 °C. What conclusion can be drawn about Compound B from Observation I?

Observation II. Compound B dissolves in water and the resultant solution conducts an electric current. What conclusion can be drawn about Compound B from Observation II (without considering Observation I)?

What conclusions can be drawn about Compound B from both Observation I and Observation II?

Compound C

Observation I. Compound C melts below 300 °C. What conclusion about Compound C can be drawn from Observation I?

Observation II. Compound C dissolves in water and the resultant solution conducts an electric current. What conclusion can be drawn about Compound C from Observation II (without considering Observation I)?

What conclusions can be drawn about Compound C from both Observation I and Observation II?

4. If an aqueous solution of potassium iodide is added to an aqueous solution of mercury(II) nitrate, a red precipitate (particles of a red solid) forms and gradually settles to the bottom of the container. The red precipitate is mercury(II) iodide, HgI_2. The aqueous solution of potassium iodide contains potassium ions, K^{1+}, and iodide ions, I^{1-}. The aqueous solution of mercury(II) nitrate contains mercury(II) ions, Hg^{2+}, and nitrate ions, NO_3^{1-}. The equation for the precipitation reaction is

$$Hg^{2+}(aq) + 2\,I^{2-}(aq) \longrightarrow HgI_2(s)$$

If 0.0100 mole of HgI_2 solid forms, how many moles of mercury(II) ion have been removed from the solution? How many moles of iodide ion have been removed from solution?

5. The entities S^{2-}, Cl^{1-}, Ar, K^{1+}, and Ca^{2+} form an isoelectronic series (18 electrons each). Predict the relative radii of these entities.

6. Burning magnesium in air produces magnesium oxide, MgO, and appreciable amounts of magnesium nitride, an apparently ionic compound. Suggest a formula for magnesium nitride. Nitrogen as a negative ion (the nitride ion) is fairly uncommon except for compounds with alkali and alkaline earth elements.

7. Describe the nature of the interaction between the ions and the solvent molecules that lead to the ready solubility of $CaCl_2$ (calcium chloride) in water.

8. Suggest reasons why calcium carbonate, $CaCO_3$, might be expected to be much less soluble in water than calcium chloride.

9. In the carbonate ion, CO_3^{2-}, all the atoms lie in a plane. The carbon atom is in the center and the three oxygen atoms lie at the corners of an equilateral triangle.

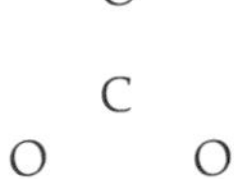

The $\angle$OCO angles are all 120°. The carbon–oxygen bond distances are all 0.132 nm, intermediate between the value 0.143 nm characteristic of alcohols and the value 0.123 nm characteristic of aldehydes and ketones. Try an approach to the bonding between carbon and oxygen in the carbonate ion using the valence bond method. Don't forget that the carbonate ion contains 32 electrons—not 30 electrons. You may find the benzene example helpful.

Chapter 11: Hydrogen and Other Gases

1. Use the ideal gas equation to evaluate the volume in liters of 1×10^{20} molecules of hydrogen gas, H_2, under a pressure of 800 torr and at a temperature of 127 °C. Also evaluate the volume in nanoliters.

2. Use the ideal gas equation to evaluate the volume in liters of 1×10^{20} molecules of methane, CH_4, under a pressure of 800 torr and at a temperature of 127 °C. Also evaluate the volume in nanoliters.

3. Using $PV = nRT$, calculate the volume occupied at 1 atm pressure and 300 K by

$$\left.\begin{array}{r}\text{1 mole}\\ \text{10 moles}\\ \text{100 moles}\\ \text{1000 moles}\end{array}\right\}\text{ of hydrogen gas}$$

For 1 atm pressure and 300K, make a graph of volume (V) vs number of moles (n). Is this a direct proportion?

4. With $PV = nRT$, calculate the volume occupied at 10 atm pressure and 600 K by

$$\left.\begin{array}{r}\text{2 grams}\\ \text{10 grams}\\ \text{50 grams}\\ \text{100 grams}\\ \text{500 grams}\end{array}\right\}\text{ of hydrogen gas}$$

For 1 atm pressure and 600K, plot the volume (V) vs the mass of hydrogen gas. Is this graph a direct proportion?

5. Using $PV = nRT$, calculate the volume of 5 moles of hydrogen gas at 5 atm pressure and 200 K, 300 K, 500 K, 700 K, and 1000 K.

   For 5 moles of hydrogen gas at 5 atm pressure, plot volume versus temperature kelvin.

6. Calculate the volume of 10 grams of liquid water at one atm pressure and 100 °C (density 0.958 g/mL) and the volume of 10 grams of gaseous water at

the same temperature and pressure using $PV = nRT$.

At or near the boiling point, the vapor (gas) would be expected to deviate markedly from ideal behavior. Would the volume calculated for the gaseous phase of water probably be larger or smaller than the experimental value? What is the basis of your answer?

7. Carbon dioxide is an example of a material that sublimes, that is, goes directly from the solid to the gas phase without any appearance of a liquid phase. The temperature at which the solid and gas are in equilibrium at one atmosphere pressure is $-79°$. Calculate the volume of 25 grams of CO_2 at 1 atm pressure and $-79°$ as a solid (density 1.59 g/mL) and as a vapor using $PV = nRT$.

8. Use the ideal gas equation to evaluate the volume of a gas mixture made up of 0.100 mole of hydrogen, H_2; 0.200 mole of helium, He; and 0.100 mole of nitrogen, N_2, when the total pressure of the gas mixture is 1.00 atm and the temperature of the gas is 27 °C.

9. Distinguish among the following easily confused terms: isotopes, structural isomers, isotherms, *cis–trans* isomers.

10. One of the reactions that takes place in the catalytic converter of an automobile is the conversion of carbon monoxide gas to carbon dioxide gas:

$$2\,CO(g) + O_2(g) \longrightarrow 2\,CO_2(g) \qquad \Delta H = -135 \text{ kcal} \quad 25\ °C$$

Show that 17.9 moles of diatomic oxygen are required to convert one kilogram of carbon monoxide to carbon dioxide and that 2.41×10^3 kilocalories are produced in the conversion of the kilogram of carbon monoxide to carbon dioxide.

11. Units of measurement:
Show that

(a) the mass of one mole of carbon dioxide is 44.01 grams;
(b) the mass of one molecule of carbon dioxide is 7.31×10^{-23} grams;
(c) the temperature -33 °C is 240 K;
(d) the pressure of 4.00 torr is 5.26×10^{-3} atmospheres;
(e) the mass of 5.3×10^{-23} grams is 5.3×10^{-26} kilograms;
(f) the energy 1.2×10^{-4} calories is 5.0×10^{-4} joules;

Chapter 12: Oil, Water, and Liquid Solutions

1. In view of the discussion of van der Waals' forces in this chapter, justify the use of hydrogen gas, H_2, in the development of the concept of ideal gases and the experimental evaluation of the ideal gas constant.

2. On the basis of the volume of the molecules and the van der Waals' forces between molecules, which of the following gases do you predict to be the most ideal gas from the standpoint of conforming to the ideal gas equation, $PV = nRT$?

ammonia, NH_3	helium, He
carbon dioxide, CO_2	hydrogen chloride, HCl
formaldehyde, HCHO	krypton, Kr

Give the rationalization for your choice or the rationalization for excluding the other gases from your choice.

3. (a) How many grams of glycerol, $CH_2OHCHOHCH_2OH$ (see scramble exercise 1) are required to prepare 200 mL of a 1.20 molar aqueous solution? Glycerol is added to soap solutions to be used in bubble blowing to improve the quality and lifetime of the soap film. Write out the structural formula of glycerol and, on the basis of its structure, suggest the manner in which it interacts with water.
   (b) If the 200-mL aqueous solution of 1.20 M glycerol has a mass of 205 grams, what is the concentration of this solution expressed in mole fraction glycerol and mole fraction water?
   (c) After the 200 mL of the 1.20 M aqueous solution of glycerol had been prepared, it was decided to add more water to make 500 mL of solution. What is the molar concentration of glycerol in the new solution? Show how your answer is obtained. The definition of molarity (molarity = moles solute/liters of solution) is a very useful relation.

4. Make a graph of the boiling points of the hydrocarbons listed in Table 12-3 versus the molecular weight of the compounds. Is the resulting graph a direct or an indirect proportion or neither? Does any correlation suggest itself to you?

5. Make a similar graph for the boiling points of H_2O, H_2S, H_2Se, and H_2Te (data in Table 12-7) versus the molecular weights of the compounds. Address the same question as in the above exercise.

6. There is a law of gas behavior known as Dalton's law that is frequently stated. "The total pressure of a mixture of gases is equal to the sum of the pressures each gas would have alone in the same volume at the same temperature." Show that this is consistent with $PV = nRT$ applied to a mixture of gases.

7. The concentration of ethyl alcohol in alcoholic beverages is given by the "proof" of the beverage. An approximation is that the "proof" is twice the percent composition. A solution of alcohol and water that is 50% alcohol and 50% water is labelled 100 proof. An estimate of proof of familiar beverages are beer, 12 proof; wine, 24 proof; and whiskey, 80 proof.
   (a) Convert these values to molarity. You need an assumption, and we assume for this problem that 50% water–50% alcohol means 50 grams of alcohol in 100 mL of the solution. (That's not quite right, but it is close enough and it makes the problem easier.)
   (b) Does this information support the assertion often made that a can of beer (12 oz.), a glass of wine (6 oz.), and a shot of whiskey (1.5 oz.) contain about the same amount of alcohol?
   (c) One hundred proof is the lowest concentration of alcohol in water that will burn when the solution is at room temperature. Why does Aunt May heat the brandy in making her famous cherries jubilee (flaming brandied cherries)?

8. This exercise addresses solutions of Substance A, a liquid, and Substance B, another liquid, both at 25 °C.

The vapor pressure of pure Substance A (P^0_A) is 335 torr at 25 °C and the vapor pressure of pure Substance B (P^0_B) is 216 torr at 25 °C. Assume that all solutions are ideal and that, consequently, the Raoult Law relations

$$P_A = X_A P^0_A \qquad \text{and} \qquad P_B = X_B P^0_B$$

can be used. In these relations, P_A is the pressure of the vapor of Substance A and P_B is the pressure of vapor of Substance B in the vapor phase in equilibrium with the liquid solution. X_A and X_B are the corresponding mole fractions of Substance A and Substance B in the liquid solution.

(a) For the equilibrium situation, draw in a line in the graph below to represent the pressure of Substance A in the vapor phase at various mole fractions of Substance A in the solution.

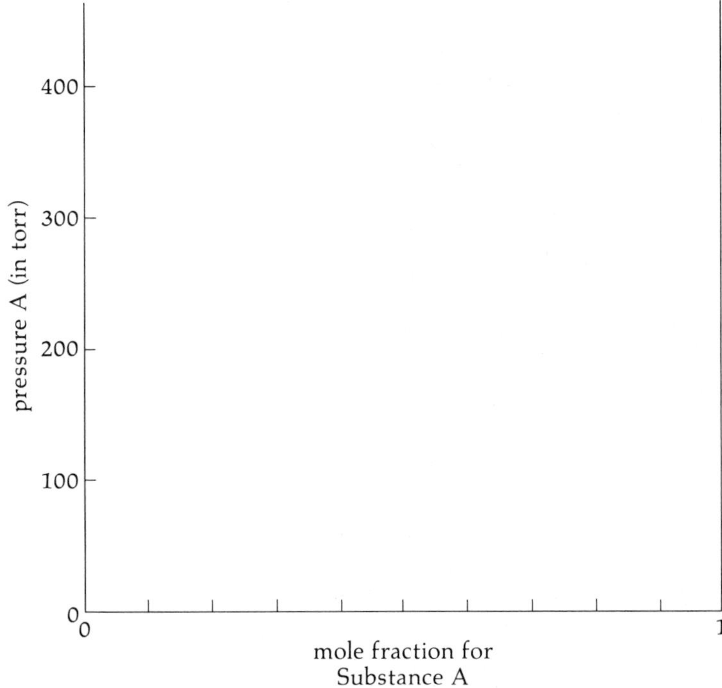

(b) On the above graph, write in a scale for the mole fraction of Substance B. (This scale must be compatible with the mole fraction scale for Substance A.)

(c) Now draw in a second line to represent the pressure of Substance B in the vapor phase at various mole fractions of Substance B in the solution.

(d) If you are not completely lost by now (and it is quite possible you will be lost), have a go at drawing in a third line to represent the total pressure of A and B above the solution for the total range of concentrations. All three lines are straight lines. Two represent direct proportions. The third does not.

Don't give up easily. Just remember that all of the relations represented by by the graph must make sense. For example, if $X_A = 1$, the solution is pure Substance A and $X_B = 0$. If $X_A = 0$, the solution is pure Substance B and $X_B = 1$.

9. As you know, fluoride ion, F^{1-}, is frequently added to municipal water supplies to protect the users, particularly children, from dental cavities. The function of this fluoride ion is to improve the resistance of the tooth enamel to decay.

The concentration specified by regulation in many municipalities is one part per million (ppm): one gram of fluoride ion per one million (1×10^6) grams of water or one pound of fluoride ion per one million pounds of water.

(a) How many fluoride ions are there in a glass of fluoridated water—about 200 mL (approximately 200 g)?

(b) One water molecule weighs 3.00×10^{-23} grams. How many water molecules in a million grams of fluoridated water? How many fluoride ions? How many water molecules for each fluoride ion?

10. If a 50.0-mL sample of an 0.100 M solution of hydrochloric acid is transferred to a flask from its reagent bottle, how many moles of hydrochloric acid have been transferred to the flask?

If water is now added to the above sample in the flask to make a total of 400 mL of a more dilute solution, what is the molarity of the diluted solution?

$$\text{molarity} = \frac{\text{moles solute}}{\text{liters solution}}$$

Chapter 13: Vinegar and Other Sour Solutions: The Brönsted Concept of Acids

1. Oxalic acid

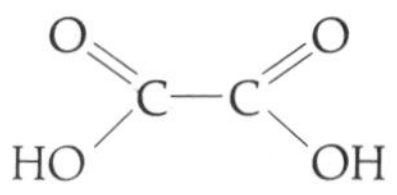

$$HO-C(=O)-C(=O)-OH$$

is a water-soluble white solid.

(a) How many moles of oxalic acid are required to prepare 150.0 mL of an 0.0900 M solution of oxalic acid? How many grams?

(b) If water is added to the above solution to make 400.0 mL of solution, what is the molarity of the new solution?

2. For oxalic acid

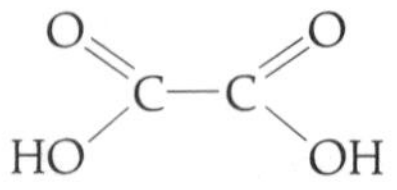

$$HO-C(=O)-C(=O)-OH$$

the pK_{A1} and pK_{A2} values at 25 °C are 1.2 and 4.2.

(a) Rationalize the bonding between all the atoms in this molecule according to the valence bond approach.
(b) Rationalize the fact the pK_{A1} is smaller than pK_{A2}.

3. Ammonium chloride, NH_4Cl, is an ionic compound made up of ammonium ions, NH_4^{1+}, and chloride ions, Cl^{1-}. Dissolved in water, the chloride ion does not react with water. The ammonium ion reacts with water as a Brönsted acid. Only one of the four protons in the ammonium ion is transferred.
   (a) Write the equation for the reaction of ammonium ion with water.
   (b) Identify the conjugate base of the ammonium ion in this reaction.
   (c) Write out the K_A expression for the equilibrium involving the ammonium ion in water.
   (d) The numerical value for this K_A at 25 °C is approximately 5×10^{-10}. Compare the strength of the ammonium ion as a Brönsted acid with the strength of the acetic acid molecule as a Brönsted acid.
   (e) Sketch the percent abundance curves for the ammonium ion and its conjugate base over the pH range 1 to 13.

4. Provide a plausible explanation of the experimental facts that a solution made up by mixing 10 mL of 0.10 M acetic acid and 10 mL of 0.10 M sodium acetate is an effective buffer but that a solution made up by mixing 10 mL of 0.10 M hydrochloric acid and 10 mL of 0.10 M sodium chloride has none of the properties of a buffer solution.

5. Boric acid, H_3BO_3, is a weak acid that used to be used extensively in eyewashes. The three ionization constants have pK_A values of 9.14, 12.74, and 13.80. Sketch the percent abundance vs pH curves for H_3BO_3, $H_2BO_3^{1-}$, HBO_3^{2-}, and BO_3^{3-}. Boric acid is considered a very weak acid. Why? Compare its strength to H_2S and to CH_3COOH.

6. Endeavor to write a Lewis electron dot structure for boric acid, H_3BO_3. The arrangement of the atoms is

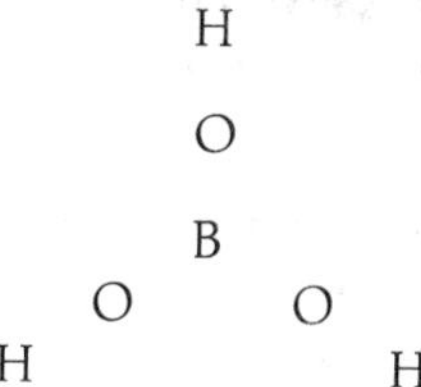

Does the boron atom have a complete octet of electrons? (The chemistry of boron compounds is often unusual, and the Lewis structures are often inadequate to describe boron compounds.)

7. Using pK_A values given in this and the previous chapter, suggest useful buffer systems for maintaining
   (a) a pH near 4.5
   (b) a pH near 6.5
   (c) a pH near 10.3

8. We have discussed in some detail that carbonic acid, H_2CO_3, is a diprotic acid. We have not discussed the facts that carbonic acid is also an unstable compound and dissociates to give carbon dioxide and water:

$$H_2CO_3(aq) \longrightarrow H_2O(l) + CO_2(g)$$

In a sealed container where CO_2 cannot continue to escape, an equilibrium is established:

$$H_2CO_3(aq) \rightleftharpoons H_2O(l) + CO_2(g)$$

Carbonic acid exists only in aqueous solution. Under equilibrium conditions, $H_2CO_3(aq)$ is, of course, also in equilibrium with its ions $HCO_3^{1-}(aq)$ and $CO_3^{2-}(aq)$.

All children, regardless of their age, enjoy the addition of vinegar (dilute aqueous solution of acetic acid) to baking soda ($NaHCO_3$). What is the fizz and how is it formed?

Chapter 14: Ammonia and Related Compounds: The Brönsted Concept of Bases

1. Use the ideal gas equation to determine how many liters of ammonia gas measured at 740 torr and 17.85 °C should be dissolved in water to make 350 mL of 2.00 molar aqueous solution of ammonia. Show how the answer is obtained.

2. According to the equation

$$2\,NH_3(g) + CO_2(g) \longrightarrow (H_2N)_2C{=}O(s) + H_2O(l)$$

urea

how many moles of ammonia gas are required to prepare one kilogram of urea? How many moles of carbon dioxide?

3. Urea

$$(H_2N)_2C{=}O$$

is a Brönsted base in much the same way ammonia, NH_3, is a Brönsted base. Only one of the two nitrogen atoms in the urea molecule accepts a proton from a water molecule.

(a) Write the equation for the reaction of urea with water and set up the expression for K_B. At 25 °C, the numerical value of K_B is 1.5×10^{-14} ($pK_B = 13.82$).

3. Rationalize the fact that the presence of oxygen at very low concentrations can be monitored by a magnetic detection apparatus.

4. Determine the oxidation number of the atoms of the elements other than hydrogen and oxygen in the following molecules and ions.

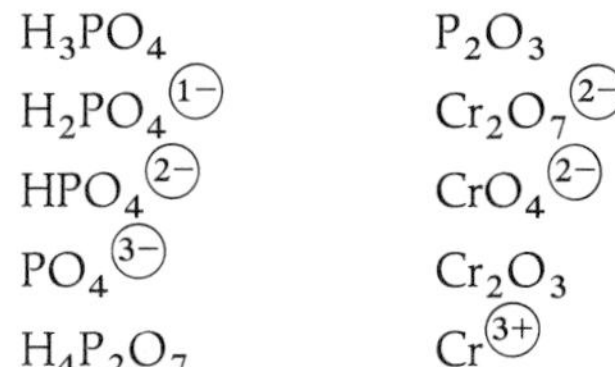

H_3PO_4	P_2O_3
$H_2PO_4^{1-}$	$Cr_2O_7^{2-}$
HPO_4^{2-}	CrO_4^{2-}
PO_4^{3-}	Cr_2O_3
$H_4P_2O_7$	Cr^{3+}

5. For each of the following transitions, determine whether an oxidizing agent, a reducing agent, or neither an oxidizing nor a reducing agent would be required to convert

sulfite ion, SO_3^{2-}, to sulfur dioxide, SO_2;
dihydrogen phosphate ion, $H_2PO_4^{1-}$, to phosphate ion, PO_4^{3-};
sulfate ion, SO_4^{2-}, to barium sulfate, $BaSO_4(s)$;
diatomic bromine, Br_2, to bromide ion, Br^{1-}
sulfide ion, S^{2-}, to sulfur, S.

6. An aqueous solution of potassium permanganate, $KMnO_4$, is frequently used in the laboratory as an oxidizing reagent due to the properties of the permanganate ion, MnO_4^{1-}, as an oxidizing agent in the presence of hydrogen ion (hydronium ion). Like all potassium compounds, potassium permanganate is an ionic compound and is quite soluble in water. The permanganate ion, MnO_4^{1-}, has a magnificent purple color. In acidic solutions, the product of reduction is the colorless manganese(II) ion. Consequently, as the permanganate ion reacts, the color of the solution fades from purple to colorless. The skeleton for the half-reaction of reduction in acidic solution is

$$MnO_4^{1-} + \_\_\_ H^{1+} + \_\_\_ e^{1-} \longrightarrow \_\_\_ Mn^{2+} + \_\_\_ H_2O$$

Balance this expression to give the equation for the half-reaction of reduction and check the number of electrons in the equation against the change in oxidation number of the manganese.

The permanganate ion in an acidic aqueous solution will oxidize a great number of molecules and ions, such as

chloride ion, Cl^{1-}, to diatomic chlorine, Cl_2;
iron(II) ion, Fe^{2+}, to iron(III) ion, Fe^{3+};
sulfur dioxide, SO_2, to sulfate ion, SO_4^{2-};
hydrogen sulfide, H_2S, to sulfate ion, SO_4^{2-}.

The skeletons for these half-reactions of oxidation are

$$Cl^{1-} \longrightarrow \_\_\_ Cl_2 + \_\_\_ e^{1-}$$
$$Fe^{2+} \longrightarrow \_\_\_ Fe^{3+} + \_\_\_ e^{1-}$$
$$SO_2 + \_\_\_ H_2O \longrightarrow \_\_\_ SO_4^{2-} + \_\_\_ H^{1+} + \_\_\_ e^{1-}$$
$$H_2S + \_\_\_ H_2O \longrightarrow \_\_\_ SO_4^{2-} + \_\_\_ H^{1+} + \_\_\_ e^{1-}$$

Complete each of the equations for the half-reactions of oxidation and check the number of electrons used in each case against the change in oxidation number of the element other than hydrogen and oxygen.

For the oxidation of hydrogen sulfide to sulfate ion by an acidic solution of potassium permanganate, combine the equations of the two half-reactions to give the equation for the total redox reaction. Simplify the equation for the total reaction by eliminating or reducing the number of ions and molecules that appear on both sides of the equation.

7. Gold does not dissolve in either hydrochloric acid or nitric acid separately. However, a 50–50 mixture of concentrated aqueous solutions of the two acids, called *aqua regia,* is capable of dissolving gold. The process involves formation of a tetrachlorogold complex ion. The product of oxidation is $AuCl_4^{1-}$. Assume the product of reduction is NO, and write the equation for the dissolution of gold in *aqua regia.* (The translation of *aqua regia* is "royal water" because it dissolves the "noble" metals. One wag referred to *aqua regia* as "the king's drink.") The skeletons without electrons for the two half-reactions are

$$Au(s) + \_\_\_ Cl^{1-}(aq) \longrightarrow \_\_\_ AuCl_4^{1-}(aq)$$
$$NO_3^{1+}(aq) + \_\_\_ H^{1+}(aq) \longrightarrow \_\_\_ NO(g) + \_\_\_ H_2O(l)$$

Complete the two equations for the half-reactions and combine to give the equation for the total reaction.

8. A piece of zinc placed in an aqueous solution of copper(II) nitrate reacts with the copper(II) ion. Metallic copper is deposited upon the metallic zinc. Zinc(II) ions appear in the solution and the concentration of copper(II) ion in the solution decreases.

Write the equation for the half-reaction of reduction.

Write the equation for the half-reaction of oxidation.

Combine the equations for the two half-reactions to give the equation for the total reaction.

Diagram an electrochemical cell that could be used to study this reaction. Indicate the direction of flow of electrons in the outside metal conductor (the wire) and, for each electrode, give the equation for the reaction that takes place at that electrode. The electrical potential for this cell is approximately 1.1 volts if the concentrations of the two ions (Cu^{2+} and Zn^{2+}) are the same and the temperature is 25 °C.

9. The reaction of Cu(II) ions with ammonia molecules to form $Cu(NH_3)_4^{2+}$ ions is discussed in Chapter 15. Does this type of reaction (complex formation) involve oxidation and reduction? How can you tell?

Chapter 17: Proteins

1. In these five chapters dealing with large molecules, it will become apparent that many naturally occurring large molecules and many synthetic large molecules are formed from smaller molecules and that the same type of linkage may occur in substances that seem quite different. One of these, the amide linkage

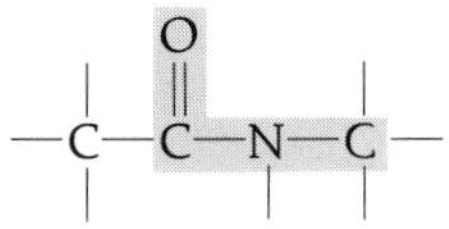

$$-\overset{|}{\underset{|}{C}}-\overset{\overset{\displaystyle O}{\|}}{C}-\underset{|}{N}-\overset{|}{\underset{|}{C}}-$$

is the connecting unit in proteins and nylon. Another, the carboxylic acid ester linkage

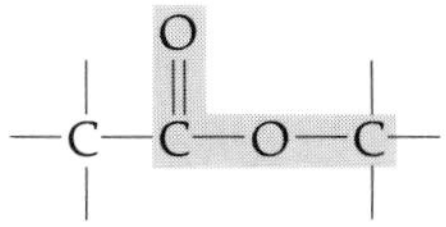

$$-\overset{|}{\underset{|}{C}}-\overset{\overset{\displaystyle O}{\|}}{C}-O-\overset{|}{\underset{|}{C}}-$$

is the connecting unit in fats and polyester filaments.

An amide is the product of the reaction of a carboxylic acid, such as acetic acid, and an amine, such as methyl amine:

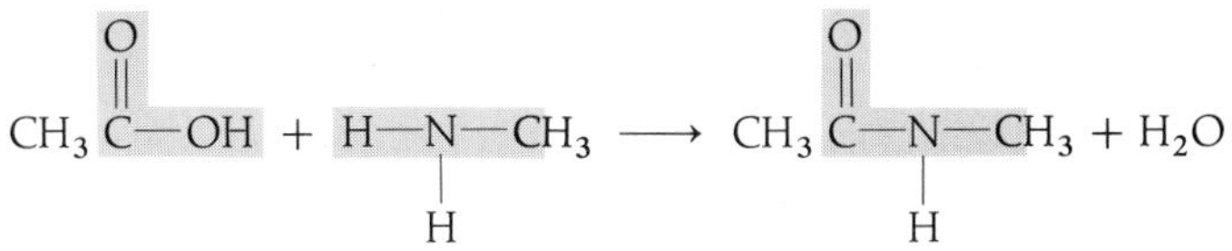

$$CH_3\overset{\overset{\displaystyle O}{\|}}{C}-OH + H-\underset{\underset{\displaystyle H}{|}}{N}-CH_3 \longrightarrow CH_3\overset{\overset{\displaystyle O}{\|}}{C}-\underset{\underset{\displaystyle H}{|}}{N}-CH_3 + H_2O$$

A carboxylic acid ester is the product of a carboxylic acid, such as acetic acid, and an alcohol, such as methanol:

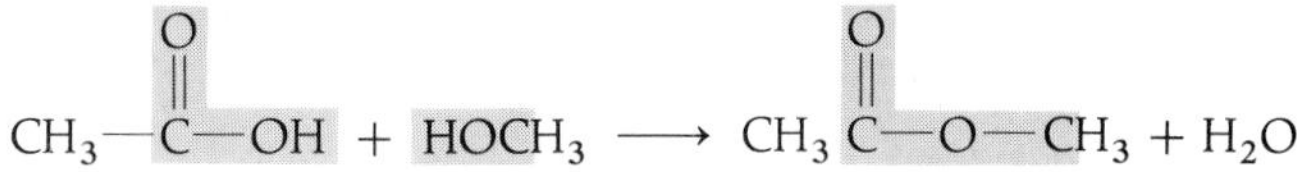

$$CH_3-\overset{\overset{\displaystyle O}{\|}}{C}-OH + HOCH_3 \longrightarrow CH_3\overset{\overset{\displaystyle O}{\|}}{C}-O-CH_3 + H_2O$$

Relate the similarities and the differences of the two groups to the positions of nitrogen and oxygen in the Periodic Table.

2. Schematic representations are given below for three 12-atom chains made up of atoms of carbon or atoms of carbon and nitrogen. In the complete structure, all incomplete bonds are occupied by hydrogen atoms:

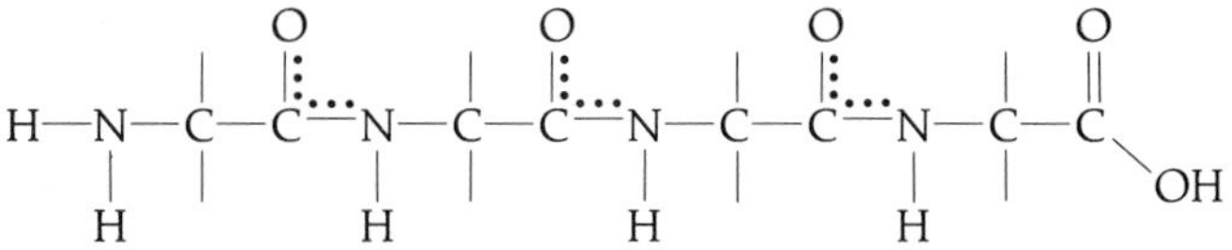

$$H-\underset{\underset{\displaystyle H}{|}}{N}-\overset{|}{\underset{|}{C}}-\overset{\overset{\displaystyle O}{\vdots}}{C}\cdots\underset{\underset{\displaystyle H}{|}}{N}-\overset{|}{\underset{|}{C}}-\overset{\overset{\displaystyle O}{\vdots}}{C}\cdots\underset{\underset{\displaystyle H}{|}}{N}-\overset{|}{\underset{|}{C}}-\overset{\overset{\displaystyle O}{\vdots}}{C}\cdots\underset{\underset{\displaystyle H}{|}}{N}-\overset{|}{\underset{|}{C}}-\overset{\overset{\displaystyle O}{\|}}{C}\diagdown OH$$

a peptide chain showing partial double bonds

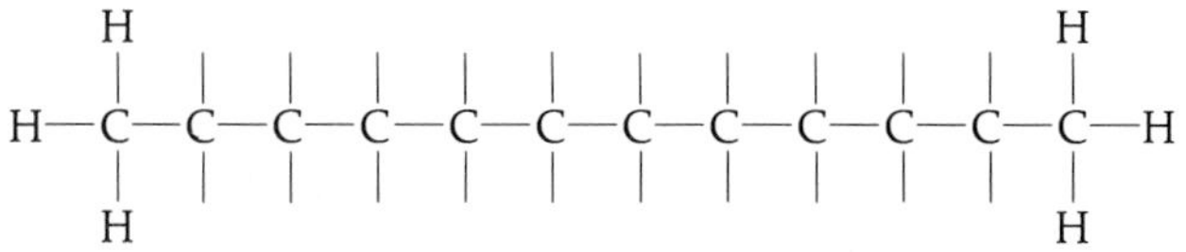

$$H-\overset{\overset{\displaystyle H}{|}}{\underset{\underset{\displaystyle H}{|}}{C}}-\overset{|}{\underset{|}{C}}-\overset{|}{\underset{|}{C}}-\overset{|}{\underset{|}{C}}-\overset{|}{\underset{|}{C}}-\overset{|}{\underset{|}{C}}-\overset{|}{\underset{|}{C}}-\overset{|}{\underset{|}{C}}-\overset{|}{\underset{|}{C}}-\overset{|}{\underset{|}{C}}-\overset{|}{\underset{|}{C}}-\overset{\overset{\displaystyle H}{|}}{\underset{\underset{\displaystyle H}{|}}{C}}-H$$

a saturated hydrocarbon (*n*-duodecane)

$$H_2C{=}CH{-}CH{=}CH{-}CH{=}CH{-}CH{=}CH{-}CH{=}CH{-}CH{=}CH_2$$

an unsaturated hydrocarbon (1,3,5,7,9,11-duodecahexene)

(a) Comment on the comparative flexibility of these chains from the standpoint of freedom of rotation about the bonds connecting the 12 atoms.

(b) Identify groups of atoms in these chains at which either *cis* or *trans* isomerism must occur.

(c) For the complete molecules with hydrogen atoms at all incomplete bonds, identify the carbon atoms that are chiral (asymmetric) and thus make possible two optical isomers.

(d) For the complete molecules, the hybridization of atomic orbitals utilized by most carbon atoms is sp^3. Identify the carbon atoms that utilize sp^3 hybrid atomic orbital and also the carbon atoms that utilize sp^2 hybrid atomic orbitals.

3. Sketch the percent abundance–pH curves for an aqueous solution of alanine at 25 °C for the completely protonated ion

$$^{(1+)}NH_3C(H)(CH_3){-}COOH$$

for the zwitterion

$$^{(1+)}NH_3{-}C(H)(CH_3){-}COO^{(1-)}$$

and for the completely deprotonated ion

$$NH_2C(H)(CH_3){-}COO^{(1-)}$$

$$^{(1+)}NH_3CH(CH_3)COOH(aq) + H_2O(l) \rightleftharpoons {}^{(1+)}NH_3CH(CH_3)COO^{(1-)}(aq) + H_3O^{(1+)}(aq) \qquad K_{A1} = 2.35 \quad 25\ °C$$

$$^{(1+)}NH_3CH(CH_3)COO^{(1-)}(aq) + H_2O(l) \rightleftharpoons NH_2CH(CH_3)COO^{(1-)}(aq) + H_3O^{(1+)}(aq) \qquad K_{A2} = 9.87 \quad 25\ °C$$

Comment on the predominant ion species present in an aqueous solution in the pH range 5 to 9.

4. Amino acids have both an amino group and a carboxylic acid group. Proteins also have many acidic and basic groups in them. The pH at which the number of positive charges on the protein equals the number of negative charges is also called the isoelectric point. Does the protein carry a positive or negative charge at pH's lower than the isoelectric point? Does the protein carry a positive or negative charge at pH's higher than the isoelectric point? (The separation of proteins on the basis of the charge they carry is called electrophoresis and is extremely important in the diagnosis of disease.)

5. The measurement of the rotation of plane-polarized light by a solution of a chiral compound is a summation of the rotation produced by all of the randomly oriented molecules or ions of the chiral compound. The measurement of the rotation of plane-polarized light of a solution of several chiral compounds is a summation of the rotation produced by the randomly oriented molecules or ions of all of the chiral compounds in solution.

   (a) Amino acids (except glycine) are chiral at the α-carbon atom. A solution containing a random mixture of amino acids will usually show the phenomenon of optical rotation. When the same collection of amino acids is organized into a protein chain in solution, the optical rotation is often different. Can you suggest a plausible explanation of this phenomenon?

   (b) The property of proteins discussed in the problem above can be used to follow the configuration of a protein. Explain how the change from a helical structure to a random configuration (a typical type of "denaturation" process) could be followed using the optical rotation of the solution as a monitor.

6. The venoms of wasps and bees contain small polypeptide hormones called kinins. These compounds are the cause of many of the extraordinary biological responses to these venoms. One kinin called Bradykinin has the amino acid sequence

Arg Pro Pro Gly Phe Ser Pro Phe Arg

amino terminal end — carboxyl terminal end

   (a) Write the structure of all the amino acids that would result from complete hydrolysis of Bradykinin.

   (b) Commercial products sold as meat tenderizers contain an enzyme that attacks the peptide bond. Explain why a meat tenderizer is often applied to a bee sting to reduce the biological response. Fresh pineapple also has an enzyme that attacks peptide bonds.

7. Organic compounds called disulfides ($CH_3—S—S—CH_3$, dimethyl disulfide) are similar to the peroxides ($CH_3—O—O—CH_3$, dimethyl peroxide) but are much less reactive. Disulfides are, however, easy to oxidize or reduce with mild oxidizing or reducing agents. For example, dimethyl disulfide reacts readily with iodide ion in acidic solution as follows:

$$2\,H^{(1+)} + 2\,e^{(1-)} + CH_3—S—S—CH_3 \longrightarrow 2\,CH_3SH \qquad \text{reduction}$$

$$2\,I^{(1-)} \longrightarrow I_2 + 2\,e^{(1-)} \qquad \text{oxidation}$$

$$2\,H^{(1+)} + 2\,I^{(1-)} + CH_3—S—S—CH_3 \longrightarrow 2\,CH_3SH + I_2 \qquad \text{total equation}$$

Use this example to help you decide if the change from a disulfide bond between two protein chains

$$\text{—S—S—}$$

protein chains

to two separate sulfhydryl groups

$$\text{—SH} + \text{HS—}$$

is an oxidation or a reduction.

Chapter 18: Sugars, Starch, Glycogen, and Cellulose: The Carbohydrates

1. **(a)** For the ring structures of D-glucose, identify the type of hybrid atomic orbitals used by each carbon atom in forming molecular orbitals.
   (b) For the open-chain D-glucose structure, give the types of atomic orbitals used by each carbon atom in forming molecular orbitals.

2. How many grams of ribose, $C_5H_{10}O_5$, are required to make 300 mL of an aqueous 0.100 molar solution of ribose?

3. For the complete combustion of one mole of sucrose, $C_{12}H_{22}O_{11}$, ordinary table sugar, with oxygen gas, O_2,

$$C_{12}H_{22}O_{11} + 12\ O_2 \longrightarrow 12\ CO_2 + 11\ H_2O \qquad \Delta H = -1350 \text{ kcal}$$

   Evaluate for 1.0 pound* of sucrose.
   (a) how many moles of oxygen gas are required;
   (b) how many moles of carbon dioxide are formed;
   (c) what volume in liters this number of moles of carbon dioxide would occupy at one atmosphere pressure at 27 °C;
   (d) how many kilocalories will be released

   If you eat one pound of sugar and are in good health and don't get fatter, you will exhale this quantity of carbon dioxide and exhale or excrete this quantity of water. Part of the heat you will utilize in chemical processes within your body and the rest you will radiate.

* One kilogram is 2.2 pounds.

4. Simple symbols for complex phenomena enable the experienced chemist to instantaneously visualize large molecules or parts of large molecules with a remarkable degree of detail. In the discussion of the polysaccharides starch and cellulose, the structures of the monomers (molecular formula $C_6H_{12}O_6$) have been represented by the following symbolic structures:

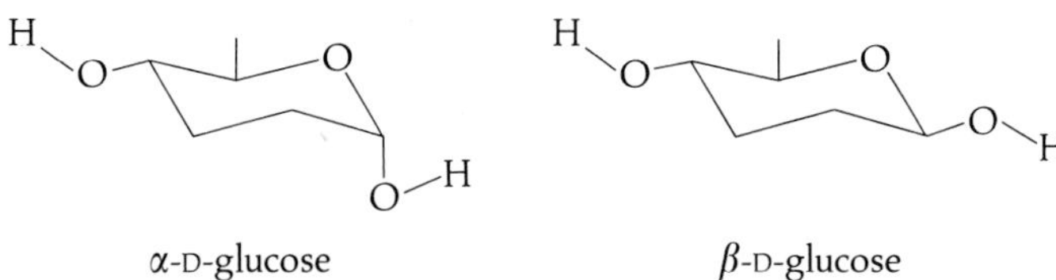

α-D-glucose β-D-glucose

These abbreviated structures in no way specify the relative orientations of the hydrogen atoms, H, and the hydroxyl groups, OH, at carbons 2 and 3. With these exceptions, the quantity of information provided by the abbreviated structures is impressive. Assume the role of an experienced sugar chemist in answering the following.

(a) Translate each of these abbreviated structures into the more conventional full structures.
(b) How many atoms are there in the ring? Is the ring planar or nonplanar?
(c) For the α and the β structures, compare the orientation of the hydroxy group, OH, and the H atom with respect to the ring at carbon-1 and at carbon-4.
(d) What atoms are attached to carbon-5? Is carbon-5 chiral?
(e) Where is carbon-6 and to what atoms is it attached? Is carbon-6 chiral?
(f) Give the full structures for the open-chain molecule of D-glucose. In this open-chain structure, is carbon-1 chiral?

You can of course, check your answers by referring to the text discussion. The goal of this exercise is to think about these questions and work with the text until the answers become very evident.

5. The ether linkage

$$-\overset{|}{\underset{|}{C}}-\overset{|}{\underset{|}{O}}-\overset{|}{\underset{|}{C}}-$$

is an essential part of the cyclic structures of sugars and also is the linkage that joins glucose molecules in forming the starch polymers and the cellulose polymers.

(a) Point out a specific example of the ether linkage in the cyclic structure of a sugar and identify the groups of the open chain sugar that are involved in the formation of the ether linkage.
(b) Using three α-D-glucose units, point out two ether linkages that are frequently formed between pairs of glucose units and identify the groups

of atoms of the cyclic structures that are involved in the formation of each ether linkage.

(c) Identify the chiral carbon atom that is the basis of the *alpha* and *beta* structures of D-glucose.

(d) Identify the chiral carbon atom that is the basis of the difference in structure of the starch and cellulose polymers.

6. Write out the formula for an open-chain ketopentose. If this compound were to form a ring structure, how many atoms would be in the ring?

7. How many isomers are there for the straight-chain ketopentose?

8. The sugar other than ribose involved in the nucleic acids is deoxyribose—a sugar that is the same as ribose except that it lacks the oxygen atom at carbon-2. (H replaces OH at carbon-2.) Write out a plausible open-chain structure and the α and β cyclic isomers of deoxyribose.

9. Write out a plausible structure of the disaccharides ribose–glucose and ribose–fructose using the ring structure. If you encounter a dilemma, point out what that dilemma is.

10. The manufacture of rayon involves the formation of the acetate esters of all the free —OH groups on cellulose (cotton).

(a) Write out a structure of a completely acetylated glucose molecule and enough of a completely acetylated cellulose chain to convince yourself how it works.

(b) Acetate rayon is much more soluble in organic (nonaqueous) solvents than cotton. Suggest reasons for this experimental fact.

(c) Nitrocellulose is the result of esterifying all the free —OH groups in cellulose with nitric acid. Suggest a possible structure for nitrocellulose. (Nitrocellulose is the explosive used in "smokeless powder.")

Chapter 19: DNA and RNA: The Molecules of Inheritance

1. Build models of adenine and cytosine to help convince yourself that all the atoms (or nearly all the atoms) lie in one plane.

2. Two ions are so widely distributed in both plants and animals that the presence of these ions have become synonymous with life itself and evidence of the presence of these ions on other planets is sought as evidence of life beyond this planet. These ions are adenosine 5′-triphosphate and adenosine 5′-diphosphate (known, respectively, as ATP and ADP).

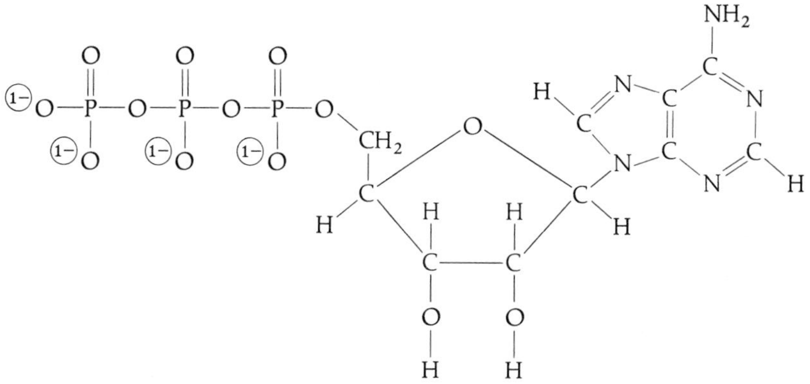

adenosine 5′-triphosphate

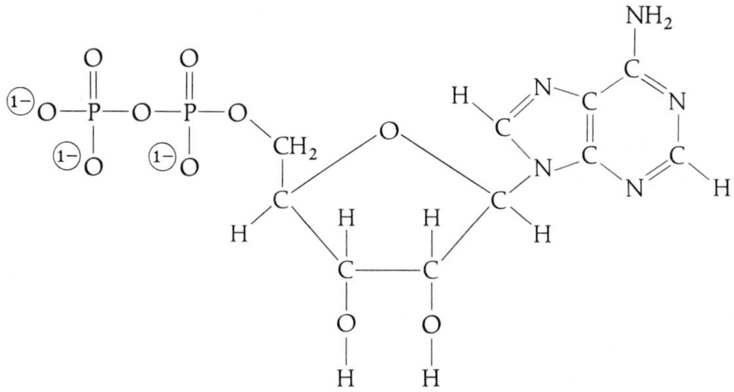

adenosine 5′-diphosphate

Correlate the structure of these ions with the structure of the first phosphate–sugar–base unit given at the top of Figure 19-3 and also with the structures for pyrophosphoric acid given in Chapter 13, page 314.

3. One of the many functions of adenosine 5′-triphosphate ion (see preceding exercise) is to transfer energy to other molecules and ions and to convert other molecules and ions to phosphates, for example, to convert α-D-glucose to α-D-glucose-1-phosphate ion (the mobilized form of glucose discussed in Chapter 18, page 444).

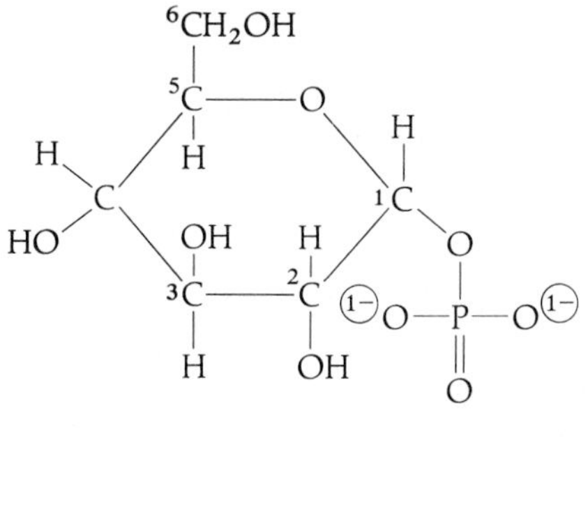

In the process, ATP becomes ADP, adenosine 5′-diphosphate ion. Write out the structures of the α-D-glucose molecule and comment on the bonds broken and the bonds formed in this reaction of ATP with α-D-glucose.

4. The diagram on page 459 is a "ladder schematic" of a short DNA segment. Write out the structures of all the molecules that would result from hydrolysis of all the ester bonds in this fragment. Is the base–sugar bond an ester bond?

5. Give a plausible explanation for the experimental fact that mRNA is more soluble in aqueous solvents than DNA.

6. To test the system of codes involved in synthesis of proteins, assume that it is within your power to program a DNA molecule with a code to synthesize the four-unit peptide lysine–alanine–asparagine–serine.

   (a) Use Figure 19-11 to design the codon code for the messenger RNA including a three-base sequence to initiate the chain and a three-base sequence to terminate the chain.

   (b) Write out the base sequence for this segment of the DNA molecule (in the nucleus) that would have been necessary to generate the desired mRNA codon code.

   Note that more than one sequence of DNA bases would be appropriate for the desired sequence of amino acids since several choices could be made in using Figure 19-11.

   (c) Design the anticodon base sequences for the four transfer RNA molecules required to synthesize the desired four-amino-acid sequence.

7. Gratuitous Information 19-2 gives the structure of some "modified" bases present in tRNA, while Figure 19-2 gives the structure of the unmodified bases.

   (a) Would the substitution of 3-methyl cytidine (m^3C) for cytosine tend to make the substituted tRNA molecule more soluble in water or less soluble? Give your reasoning.

   (b) Repeat (a) for the substitution of N^6-methyladenosine (m^6A) for adenine.

   (c) Repeat (a) for the substitution of wyosine (Y) for guanine.

   Don't be surprised if there is no clear-cut answer to the question. There is a lot involved.

8. Viruses are rather simple biological organisms that seem to consist of little more than a nucleic acid molecule surrounded by a protein coat. A virus reproduces by using the apparatus of a host cell to reproduce its nucleic acid (and its protein). Suggest how the reproduction of the virus' nucleic acid might be accomplished. (See Gratuitous Information 19-1 for at least a hint.)

Chapter 20: Animal Fats, Vegetable Oils, Lipids, and Membranes

1. The complete combustion of 1 gram of sucrose releases about 4 kilocalories. The complete combustion of 1 gram of stearin, $C_{57}H_{110}O_6$ (a saturated fat), releases about 10 kilocalories. Both are compounds of carbon, hydrogen, and

oxygen. Suggest a plausible explanation for so much more heat being released in the conversion of 1 gram of stearin to carbon dioxide and water than in the conversion of 1 gram of sucrose, $C_{12}H_{22}O_{11}$, to carbon dioxide and water.

2. Phosphoric acid and its derivatives are widely distributed in biological systems. Examples:
   - the $H_2PO_4^{1-}$–HPO_4^{2-} conjugate acid–base pair buffer system
   - ADP and ATP (adenosine 5′-diphosphate and adenosine 5′-triphosphate ions) (Both are phosphate esters.)
   - α-D-glucose-1-phosphate ion (a phosphate ester)
   - the nucleic acids (phosphate ester polymers)
   - the phospholipids (phosphate esters)

   (a) Discuss the structure of phosphoric acid as a triprotic acid and identify the predominant ions of phosphoric acids present in an aqueous solution in the 6 to 8 pH range.

   (b) You have been dealing with carboxylic acid esters and phosphate esters. The fact that both are called esters implies that there must be marked similarities. Yet their structural formulas appear to be so different. Point out the way(s) carboxylic acid esters and phosphate esters are similar.

3. A television commercial claims that margarine is just an emulsion of vegetable fat and water and therefore implies that it is of lesser quality than butter. Is there anything fundamentally different about butter?

4. Persons who survive accidents in cold climates when help cannot be provided right away are likely to be those who would be classified as overweight in other circumstances. Can you suggest an explanation for this phenomenon?

5. What are the advantages to a bear of becoming fat and slow-moving in the autumn?

6. Explain the fact that soaps and similar molecules will form monomolecular layers on the surface of water with the polar "head groups" in the water and the non-polar "tails" sticking up.

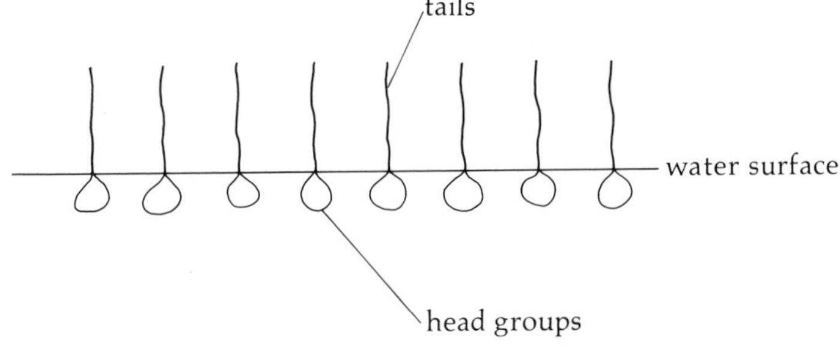

7. In case of an accidental spill of an aqueous solution of sodium hydroxide on the skin, the basic solution should be removed very promptly from the skin with a great excess of cold water. If that solution remains, almost the first effect is that the skin feels "soapy." Rationalize. Remember that skin is rather "oily."

Chapter 21: Synthetic Polymers

1. Extremely durable and tough polymers (used in football helmets and "bullet-proof" glass) can be made by forming polyesters of carbonic acid

$$\mathrm{HO{-}C({=}O){-}OH}$$

and ethylene glycol, $HO{-}CH_2CH_2{-}OH$. Suggest a plausible structure for the material.

2. Compared to biological systems, the chemical industry is a neophyte when it comes to the synthesis of large molecules, including the production of polymers, and chemists have used natural polymers as models in the development of synthetic polymers. Give at least one example of a synthetic polymer that might have been modeled on a natural polymer and explain how you selected that example.

3. Addition polymers are based on the reactivity of the pair of electrons in *pi*-bonds. Contrast the role of a free radical initiator for addition polymerization with the role of either cationic or anionic initiators. Use the polymerization of ethene as an example.

4. Biological polymers are all biodegradable. Synthetic polymers are usually not biodegradable. Suggest a plausible explanation.

   One of the great benefits to society of synthetic polymers is that they are not biodegradable; one of the great burdens to society of synthetic polymers is that they are not biodegradable.

5. We have repeatedly pointed out that the chemist uses whatever model or models best serve the need. Point out one or more examples in which the ball-stick-spring model, the Lewis structures, and the valence bond approach have all been used—almost interchangeably—in the discussion of either the formation or the properties of polymers.

6. Comment on the validity of the statement "Both starch and cellulose are condensation polymers." Also the statement "The phosphate–ribose backbone of nucleic acids is a condensation polymer."

7. Write out the structure (include all the atoms) of a copolymer of styrene and ethene

   (a) as a random copolymer;

   (b) as a block copolymer with five ethene residues followed by five styrene residues. This may be quite a bore but, in actual fact, "blocks" in block copolymers are often much longer—20 to 100 residues.

8. One way to "soften" hard water is to use an ion exchange resin. A cross-linked polymer (natural or synthetic) is prepared that has a lot of carboxylate or sulfate groups attached. Show how such a resin could be prepared by polymerizing

together styrene

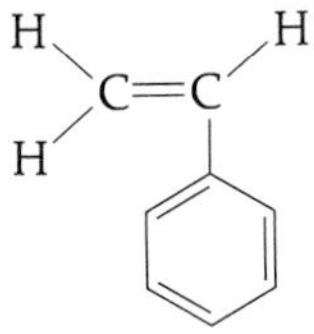

$$\mathrm{H_2C{=}CH{-}C_6H_5}$$

divinyl benzene (a cross-linking agent),

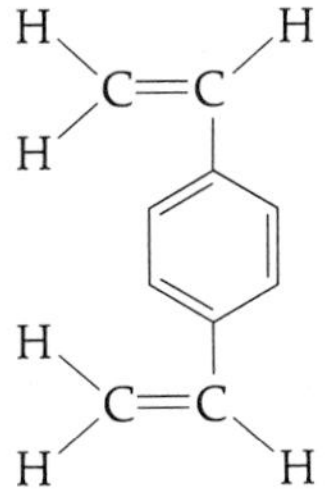

$$\mathrm{H_2C{=}CH{-}C_6H_4{-}CH{=}CH_2}$$

and

$$\mathrm{H_2C{=}CH{-}C_6H_4{-}C({=}O){-}O{-}C_2H_5}$$

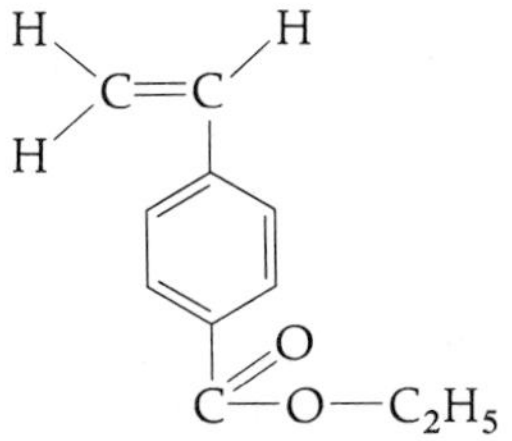

followed by hydrolysis of the ester groups in basic solution.

9. The purpose of an ion exchange resin is to "exchange" the calcium ion (Ca^{2+}) of "hard" water for the sodium ion (Na^{1+}) of "soft" water. A column is prepared containing beads of the resin. The resin is first made in the all-sodium-ion form (for every carboxylate ion found in the resin, there is a sodium ion nearby, $-COO^{1-}\ Na^{1+}$). As hard water containing Ca^{2+} ions is poured over the column, the Ca^{2+} ions tend to replace the Na^{1+} ions on the resin.

(a) Suggest a plausible explanation for how this happens and how the net effect is to "soften" the water.

(b) Suggest a possible way to return the resin to the all-sodium-ion form after most of the sodium ions in the resin have been exchanged for calcium ions.

Chapter 22: Nuclear Reactions

1. One way to experimentally determine Avogadro's number is based upon the *alpha* decay of radium-226 to radon-222:

$$^{226}_{88}\mathrm{Ra} \longrightarrow {}^{222}_{86}\mathrm{Rn} + {}^{4}_{2}\alpha$$

The α particles (helium nuclei) become helium atoms, He, upon picking up the required number of electrons from the equipment. Two types of measurements are required: one set of measurements using a Geiger counter to count the number of *alpha* particles emitted from a specific radium sample and another set of measurements using gas-measuring equipment to determine the moles of monatomic helium gas produced by the same radium sample.The half-life of radium-226 is long and the rate of decay of the specific sample used can be expected to remain essentially constant over a time period such as a week or even a month. Consequently, the Geiger-counting measurements and the gas-measuring experiments do not have to be made at the same time. It is, however, necessary that both the counting measurements and the gas sample collection be made over measured periods of time. Counting periods would be brief due to the large number of *alpha* particles emitted. The gas-collecting periods would be long due to the tiny volume of helium gas formed.

A typical set of data for specific radium sample:

the counting experiment
counting period: 1.00 minute
number of counts: 4.93×10^{14}

the gas-collecting experiment
collecting period: 10.0 hours
volume of gas: 1.20 milliliters
pressure of the gas: 0.0100 atmospheres
temperature of gas: 27 °C

Use these data to calculate Avogadro's number (the number of atoms of helium in one mole of helium).

2. Radioactive isotopes are extensively used in the study of normal physiological systems and in the detection and assessment of abnormalities. For example, iodide ion concentrates in the thyroid glands. If that iodide ion is radioactive, its distribution in the gland can be determined by a *beta*-ray detector positioned beside the throat.

As a diagnostic technique, the patient swallows a very small quantity of sodium iodide-137 in aqueous solution and the take-up of that iodide ion by the thyroid gland is monitored with the detector.

Four radioactive isotopes frequently used in diagnostic tests are given in the table at the top of p. 581. All these radioactive isotopes are *beta*-ray emitters.

(a) For each isotope, identify the daughter nuclei formed on the emission of a *beta* particle and give the nuclear equation for the nuclear transition. In each case, the daughter nucleus is a stable isotope.

(b) Assuming that none of the radioactive isotope is excreted, evaluate for each isotope approximately how many days or hours would elapse after a sample of the radioactive isotope is ingested before the amount of the radioactive isotope in the body would be reduced to one-eighth of the sample of the radioactive isotope swallowed. In actual fact, many of the isotopes are to a large degree excreted.

isotope	half-life	physiological system
iodine-131 $^{131}_{53}I$ (as iodide ion)	8 days	thyroid glands
iron-59 $^{59}_{26}Fe$ (as iron(III) ion)	45 days	red blood cells
phosphorus $^{32}_{15}P$ (as phosphate ion)	14 days	liver and tumors
sodium-24 $^{24}_{11}Na$ (as sodium ion)	15 hours	circulatory system

3. Ions of technetium and thallium concentrate in normal heart tissue and unstable isotopes are used to assess damage to normal heart tissue. The isotopes used, technetium-99 m and thallium-201, do not decay by the emission of either *alpha* or *beta* particles. Instead, they become stable nuclei by quite different processes. In both, *gamma*-ray emission occurs, and it is this γ-ray emission that is monitored with a Geiger scanner.

Technetium-99 m is a high-energy state, called a metastable state, of $^{99}_{43}Tc$ nuclei. A nucleus in this metastable state makes a transition to the lower ground state energy state of the technetium-99 nucleus by the emission of a γ-ray photon. The half-life of technetium-99 m is 6 hours.

A thallium-201 nucleus becomes a stable nucleus by the capture of an orbital electron of the thallium-201 atom accompanied by the release of a γ-ray photon. The half-life of $^{201}_{81}Tl$ is 73 hours.

(a) Decide whether it is appropriate to write an equation for the nuclear transition involved in the decay of technetium-99 m. Give the basis of your decision and, if appropriate, write the nuclear equation.

(b) Repeat (a) for the nuclear transition of thallium-201.

4. Works of art, archeological materials, and criminal evidences are increasingly investigated by a technique known as neutron activation analysis. In this approach, small samples of paint, fiber, or other materials are subjected to neutron bombardment leading to neutron capture by some nuclei to form unstable isotopes that can be identified by their half-lives and also by their *gamma*-ray emission spectra. This approach is particularly useful in distinguishing among elements that have very similar chemical properties, such as adjacent members of the same chemical family. This exercise deals with the use of neutron activation analysis in the analysis for arsenic, As, and antimony, Sb, members of the nitrogen family. (The element antimony was known in the classical period in Rome and its symbol Sb comes from its Latin name *stibnum*.) Arsenic has been used throughout the ages as a poison. Ingested, arsenic compounds are incorporated in hair, and neutron analysis is used to investigate suspected murders by arsenic poison, such as that of Napoleon, long after other biological materials have decomposed.

(a) Activation. When a sample that contains arsenic, ${}^{75}_{33}\mathrm{As}$, is bombarded by neutrons in a nuclear reactor, neutron capture occurs. Complete the equation for the absorption of a neutron:

$$ {}^{75}_{33}\mathrm{As} + {}^{1}_{0}\mathrm{n} \longrightarrow \underline{\qquad} $$

Similarly treated antimony, ${}^{123}_{51}\mathrm{Sb}$, undergoes neutron capture. Write the equation for the neutron absorption by ${}^{123}_{51}\mathrm{Sb}$.

(b) Decay. Arsenic-76 is unstable and decays by β particle emission. Complete the equation

$$ {}^{76}_{33}\mathrm{As} \longrightarrow \underline{\qquad} + {}^{0}_{-1}\beta \qquad \text{half-life 26.5 hr} $$

Similarly, antimony-124 is unstable and decays by β particle emisssion. Write the equation for this nuclear transition. The half-life is 60.9 days.

(c) Detection and Identification by *gamma*-Ray Emission. In both cases, the emission of β particles is accompanied by *gamma*-ray emission and it is this γ-ray emission that is the fingerprint of arsenic-76 and also antimony-124. Their γ-ray emission spectra are given in Figures A-1 and

Figure A-1

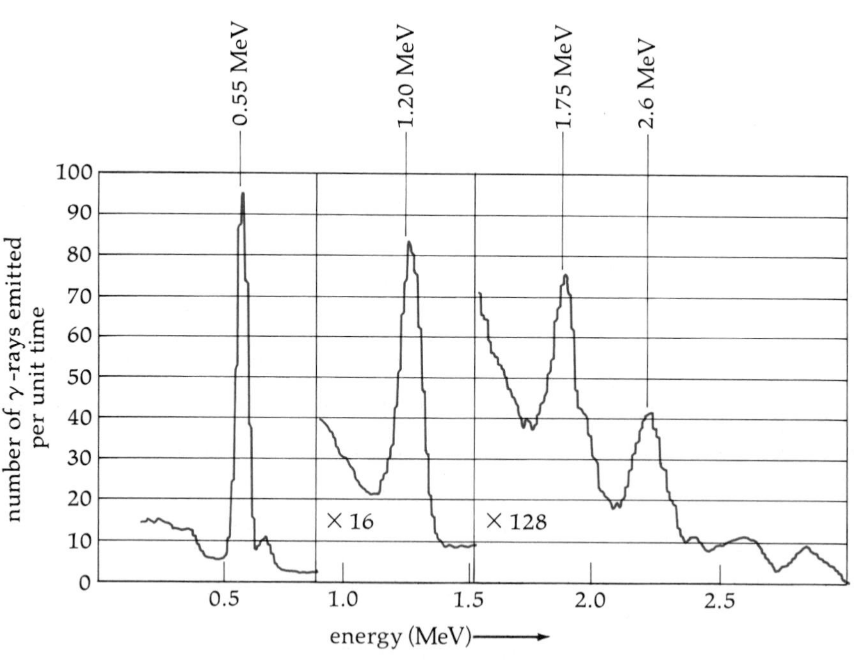

γ-RAY EMISSION SPECTRA OF ${}^{76}_{33}\mathrm{As}$

A-2. The curves are called γ-ray emission spectra. The arsenic spectrum has a major peak at 0.55 MeV and significant smaller peaks at 1.20, 1.75, 2.1, and 2.6 MeV. Note the magnification of the vertical scales between 0.8 and 1.5 MeV and beyond 1.5 MeV. The size of the 1.20-MeV peak has been magnified by a factor of 16 to make it appear comparable in size to the 0.55-MeV peak; the other peaks have been multiplied by a factor of 128. The antimony spectrum has a major peak at 0.60 MeV and smaller peaks at 1.33, 1.69, 2.04, and 2.3 MeV. Again, there is a magnification of the vertical scale between 0.8 and 1.9 MeV and a further magnification beyond 1.9 MeV. The 1.33- and 1.69-MeV peaks have been magnified by a factor of 8 and the 2.04- and 2.30-MeV peaks have been magnified by a factor of 32. See what emission energy peaks (expressed in MeV) you could use to identify that (a) arsenic-75 was originally present in a sample and (b) antimony-123 was originally present in a sample.

(d) Sensitivity of *gamma*-Ray Measurements. The intensity of the γ-rays (number of γ-rays per minute) depends, everything else being equal, on the

Figure A-2

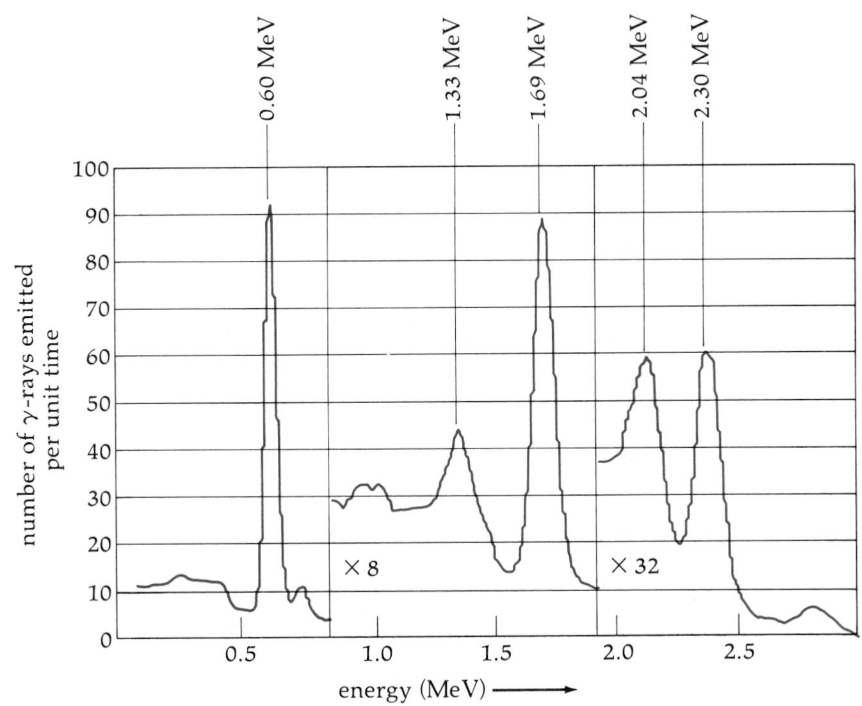

γ-RAY EMISSION SPECTRA OF $^{124}_{51}Sb$

number of arsenic atoms originally present in the sample. The relationship is a simple direct proportion. [Intensity $\propto$ number of arsenic atoms.] If a sample of hair containing 1 milligram (1×10^{-3} g) of arsenic gave a counting rate of 1×10^{8} counts per minute, what would be the expected counting rate for a sample of hair containing one nanogram (1×10^{-9} g) of arsenic?

Counting rates of 100 counts per minute or more are very easily handled with ordinary equipment. A counting rate as low as a few counts per day can be handled with specialized equipment. Counting rates over 1×10^{6} counts per second also require specialized equipment.

(e) Identification Based Upon Half-life Measurements. Suggest another way of distinguishing between the presence of arsenic-75 and antimony-123 in a sample.

5. Be prepared to explain to your friends and enemies the similarities and differences between nuclear reactions and chemical reactions.

Table of Background Information Sections

Table of Gratuitous Information Sections

Table of Editorial Comments

INDEX

The letter *t* following a page number means the information is contained in a table; the letter *f*, in a figure.

A 8
B 9
C 0
D 1
E 2
F 3
G 4
H 5
I 6
J 7

FREQUENTLY USED PROPORTIONALITY CONSTANTS

$$[\text{length in meters (m)}] = \left[1 \times 10^{3} \frac{\text{meters}}{\text{kilometer}}\right][\text{length in kilometers (kg)}]$$

$$[\text{length in centimeters (cm)}] = \left[1 \times 10^{2} \frac{\text{centimeters}}{\text{meter}}\right][\text{length in meters (m)}]$$

$$[\text{length in millimeters (mm)}] = \left[1 \times 10^{3} \frac{\text{millimeters}}{\text{meter}}\right][\text{length in meters (m)}]$$

$$[\text{length in microns } (\mu)] = \left[1 \times 10^{6} \frac{\text{microns}}{\text{meter}}\right][\text{length in meters (m)}]$$

$$[\text{length in nanometers (nm)}] = \left[1 \times 10^{9} \frac{\text{nanometers}}{\text{meter}}\right][\text{length in meters (m)}]$$

$$[\text{volume in milliliters (mL)}] = \left[1 \times 10^{3} \frac{\text{milliliters}}{\text{liter}}\right][\text{volume in liters (L)}]$$

$$[\text{mass in grams (g)}] = \left[1 \times 10^{3} \frac{\text{grams}}{\text{kilogram}}\right][\text{mass in kilograms (kg)}]$$

$$[\text{mass in milligrams (mg)}] = \left[1 \times 10^{3} \frac{\text{milligrams}}{\text{gram}}\right][\text{mass in grams (g)}]$$

$$[\text{mass in micrograms } (\mu\text{g})] = \left[1 \times 10^{6} \frac{\text{micrograms}}{\text{gram}}\right][\text{mass in grams (g)}]$$

$$[\text{mass in nanograms (ng)}] = \left[1 \times 10^{9} \frac{\text{nanograms}}{\text{gram}}\right][\text{mass in grams (g)}]$$

$$[\text{force in dynes}] = \left[1 \times 10^{5} \frac{\text{dynes}}{\text{newton}}\right][\text{force in newtons}]$$

$$[\text{pressure in torr}] = \left[760 \frac{\text{torr}}{\text{atmosphere}}\right][\text{pressure in atmospheres}]$$

$$[\text{energy in ergs}] = \left[1 \times 10^{7} \frac{\text{ergs}}{\text{joule}}\right][\text{energy in joules (J)}]$$

$$[\text{energy in joules (J)}] = \left[4.18 \frac{\text{joules}}{\text{calorie}}\right][\text{energy in calories (cal)}]$$

$$[\text{energy in ergs}] = \left[1.60 \times 10^{-12} \frac{\text{ergs}}{\text{electron volt}}\right][\text{energy in electron volts (eV)}]$$

$$[\text{energy in joules (J)}] = \left[1.60 \times 10^{-19} \frac{\text{joules}}{\text{electron volt}}\right][\text{energy in electron volts (eV)}]$$

$$[\text{energy in million electron volts (MeV)}] = \left[5.61 \times 10^{26} \frac{\text{million electron volts}}{\text{gram}}\right][\text{change of mass in grams (g)}]$$

AVOGADRO'S NUMBER

$$[\text{number of atoms}] = \left[6.023 \times 10^{23} \frac{\text{atoms}}{\text{mole}}\right][\text{number of moles of atoms}]$$

$$[\text{number of molecules } (N)] = \left[6.023 \times 10^{23} \frac{\text{molecules}}{\text{mole}}\right][\text{number of moles } (n) \text{ of molecules}]$$

UNIVERSAL GAS CONSTANT R IN THE EQUATION $PV = nRT$

For volume in liters and pressure in atmospheres: $R = 0.08205$ liter atmospheres mole^{-1} kelvin^{-1}

For volume in milliliters and pressure in torr: $R = 6.240 \times 10^{4}$ milliliters torr mole^{-1} kelvin^{-1}